Lecture Notes in Electrical Engineering

Volume 827

The book series *Lecture Notes in Electrical Engineering* (LNEE) publishes the latest developments in Electrical Engineering - quickly, informally and in high quality. While original research reported in proceedings and monographs has traditionally formed the core of LNEE, we also encourage authors to submit books devoted to supporting student education and professional training in the various fields and applications areas of electrical engineering. The series cover classical and emerging topics concerning:

- Communication Engineering, Information Theory and Networks
- Electronics Engineering and Microelectronics
- Signal, Image and Speech Processing
- Wireless and Mobile Communication
- Circuits and Systems
- Energy Systems, Power Electronics and Electrical Machines
- Electro-optical Engineering
- Instrumentation Engineering
- Avionics Engineering
- Control Systems
- Internet-of-Things and Cybersecurity
- Biomedical Devices, MEMS and NEMS

For general information about this book series, comments or suggestions, please contact leontina. dicecco@springer.com.

To submit a proposal or request further information, please contact the Publishing Editor in your country:

China

Jasmine Dou, Editor (jasmine.dou@springer.com)

India, Japan, Rest of Asia

Swati Meherishi, Editorial Director (Swati.Meherishi@springer.com)

Southeast Asia, Australia, New Zealand

Ramesh Nath Premnath, Editor (ramesh.premnath@springernature.com)

USA, Canada:

Michael Luby, Senior Editor (michael.luby@springer.com)

All other Countries:

Leontina Di Cecco, Senior Editor (leontina.dicecco@springer.com)

** This series is indexed by EI Compendex and Scopus databases. **

More information about this series at https://link.springer.com/bookseries/7818

Jason C. Hung · Neil Y. Yen · Jia-Wei Chang

Editors

Frontier Computing

Proceedings of FC 2021

Set 2

Springer

Editors
Jason C. Hung
Department of Computer Science
and Information Engineering
National Taichung University of Science
and Technology
Taichung, Taiwan

Neil Y. Yen
School of Computer Science
and Engineering
The University of Aizu
Aizuwakamatsu, Japan

Jia-Wei Chang
Department of Computer Science
and Information Engineering
National Taichung University of Science
and Technology
Taichung, Taiwan

ISSN 1876-1100   ISSN 1876-1119   (electronic)
Lecture Notes in Electrical Engineering
ISBN 978-981-16-8051-9   ISBN 978-981-16-8052-6   (eBook)
https://doi.org/10.1007/978-981-16-8052-6

This Springer imprint is published by the registered company Springer Nature Singapore Pte Ltd.
The registered company address is: 152 Beach Road, #21-01/04 Gateway East, Singapore 189721,
Singapore

Preface

This LNEE volume contains the papers presented at the International Conference on Frontier Computing (FC 2021) virtually held online on July 14, 2021. This event is the 11th event of the series, in which fruitful results can be found in the digital library or conference proceedings of FC 2010 (Taichung, Taiwan), FC 2012 (Xining, China), FC 2013 (Gwangju, Korea), FC 2015 (Bangkok, Thailand), FC 2016 (Tokyo, Japan), FC 2017 (Osaka, Japan), FC 2018 (Kuala Lumpur, Malaysia), FCABH 2019 (Taichung, Taiwan), FC 2019 (Kitakyushu, Japan), FC 2020 (virtual event), and FC 2021 (virtual event). This conference is expected to bring together researchers and practitioners from both academia and industry to meet and share cutting-edge development in the field. One colocated event, namely International Conference on Machine Learning on FinTech, Security and Privacy (MLFSP 2021), is jointly held with high appreciations from the participants.

The papers accepted for inclusion in the conference proceedings primarily cover the topics: database and data mining, networking and communications, Web and Internet of things, embedded system, soft computing, social network analysis, security and privacy, optics communication, and ubiquitous and pervasive computing. Many papers have shown their academic potential and value and indicate promising directions of research in the focused realm of this conference. We believe that the presentations of these accepted papers will be more exciting than the papers themselves and lead to creative and innovative applications. We hope that the attendees and readers will find these results useful and inspiring to your field of specialization and future research.

On behalf of the organizing committee, we would like to thank the members of the organizing and the program committees, the authors, and the speakers for their dedication and contributions that make this conference possible especially to FC conference group and Korean Institute of Information Technology, Korea Institute of Information Technology and Innovation (KIITI), and SIEC Korea Chapter. We appreciate the contributions from these experts and scholars to enrich our event. We would take the chance to thank and welcome all participants, and hope that all of them enjoy the technical discussions within the conference period, build a strong friendship, and establish ties for future collaborations.

Jason C. Hung
Neil Y. Yen
Jia-Wei Chang

Contents

Contents

Analysis and Prediction of Higher Vocational Students' Employment Based on Decision Tree Classification Algorithm

Fang Fang[(✉)]

Zhejiang Tongji Vocational College of Science and Technology,
Hangzhou 311231, China

Abstract. According to the actual situation and characteristics of employment in higher vocational colleges, this paper preprocesses the data related to students' employment and analyzes the employment decision-making. It uses ID3 classification algorithm to generate the decision tree model of employment prediction, extract the prediction rules, and evaluate the accuracy. It mines and analyzes the students' employment data to obtain the potential information of regularity, Finally, the prediction model of students' employment and employment unit type is generated. According to the results of entrance examination, comprehensive quality evaluation, the region and nature of employment units, the employment situation of students is classified and predicted, and the employment prediction rules are obtained, which lays the foundation for further decision-making analysis and application. The experimental results show that the prediction model of student employment and employment unit type established by decision tree classification I3 algorithm has high accuracy and efficiency.

Keywords: Vocational college students · Employment analysis · Prediction classification · Decision tree · Employment model

1 Introduction

Recently, with China's higher education entering a new stage of popularization, the employment situation of college graduates is becoming more and more severe, and more and more higher vocational college students are facing the crisis of unemployment as soon as they graduate.

The reason is influenced by the enrollment expansion of colleges and universities and the changes in the demand of China's employment market. How to improve the employment rate and quality of higher vocational college students is imperative. If we can mine the data of the training process of higher vocational students, we can get the factors that affect students' employment and the prediction model of employment, The employment guidance department of higher vocational colleges can carry out targeted employment guidance work according to the mined model [1]. At the same time,

J. C. Hung et al. (Eds.): FC 2021, LNEE 827, pp. 973–977, 2022.
https://doi.org/10.1007/978-981-16-8052-6_126

teachers should design or adjust the training program and teaching plan according to the four factors affecting employment, so as to improve students' employment quality and employment rate.

2 Basic Knowledge of Data Classification Mining

2.1 Classification Based on Neural Network

The development of neural network originates from the exploration of neurobiologists and psychologists for the research and testing of simulated neural computation, aiming at the development and testing of neural computational simulation. The neural network is composed of a group of related input and output units, each of which has a corresponding weight. By changing the weight of each connection in the network, the information can be stored and processed. In the neural network, each neural unit represents not only an information storage unit, but also an information processing unit. The processing and storage of information are integrated. Under the joint action of each neural unit, the network composed of the Shenjing unit can be divided into two parts, In the process of recognition and memory of input pattern, with the input of pattern, the connection weights between neurons are constantly adjusted, and the statistical rate of environment is reflected and maintained in the neural network structure to achieve the memory of input pattern [2]. After learning the input pattern, the neural network extracts the features of the input pattern and generates memory.

2.2 Decision Tree Classification Algorithm

Decision tree classification algorithm came into being at the end of 1970s. Professor J Ross Quinlan proposed the famous algorithm. The main goal of the algorithm is to reduce the depth of the classification tree, but it ignores the study of the number of leaves. C4.5 algorithm is improved on the basis of ID3 algorithm, especially in the missing value processing of prediction variables and pruning technology. It not only meets the classification problem, but also adapts to the regression problem. The core content of decision tree algorithm is to construct a decision tree with high precision and small scale. The expected information is:

$$I(s_1, s_2, \ldots s_m) = -\sum_{i=1}^{m} P_i \log(P_i) \tag{1}$$

The gain or expected information of attribute variable a splitting into subsets is:

$$E(A) = \sum_{j=1}^{v} \frac{s_{1j} + \ldots s_{mj}}{S} I(s_{1j} + \ldots s_{mj}) \tag{2}$$

3 Establishment of Higher Vocational Students' Employment Prediction Model Based on Decision Tree Classification Algorithm

3.1 An Overview of Student Employment Prediction Model

The employment rate is directly affected by the utilization of employment and the quality of employment. It will play a positive role in the construction of employment and employment unit type prediction model based on data mining sub algorithm. The main work of data preparation stage is to summarize and sort the required data, select the appropriate data records and attributes, preprocess them, and provide them to the data mining model. In this paper, the basic attributes of students and employment units that affect the employment of higher vocational students are screened and summarized [3]. The information of students' college entrance examination categories, entrance scores, students' regions, previous years, age, gender, comprehensive quality evaluation, successful graduation, enterprise address, unit nature and so on are cleaned up, converted and integrated to generate the model, This paper uses fmeasur's evaluation method to evaluate the accuracy of the samples after classification, and verifies it with the prediction model of students' employment and employment unit type, so as to obtain the quantitative basis comparison, so as to ensure the effectiveness of the prediction model.

3.2 Data Preprocessing

In order to improve the quasi effectiveness and scalability of classification, we preprocess the basic situation of higher vocational education data before classification, mainly through data cleaning, correlation analysis and data transformation. In this step, we fill in the vacancy value and correct the noise data which are inconsistent between the employment unit address and the actual employment address, and whose comprehensive quality evaluation results are not up to the standard, and improve the accuracy of training set learning in the data. There are 22 attributes in the student employment information section, which may not be related to the classification task or have little correlation, The results show that the decision tree will have a greater impact on the learning rate of the decision tree, and the attribute correlation weights are calculated. The two classification attributes of employment and the type of employment unit have relatively large weights, such as the total score of enrollment, the quality evaluation results of previous years, the nature of the unit, and the address of the unit. The purpose of correlation analysis is irrelevant or redundant.

4 Experimental Analysis

4.1 Purpose of the Experiment

Based on the above theory, this experiment constructs the employment and employment unit type prediction model based on decision tree classification algorithm, and tests the model to verify the accuracy of the model. This experiment is divided into the following steps: first, by inputting the training set, the prediction model of student employment and employment unit type is generated. Then, by comparing the results of model prediction and classification with the sample data, the change trend of model prediction accuracy and prediction efficiency with the amount of data is analyzed [4]. This experiment also tests the efficiency of the algorithm to verify the adaptability of I3 algorithm for the prediction model of employment and employment unit type of higher vocational students.

4.2   Experimental Steps

As the data comes from the three departments of enrollment, educational administration and employment, their respective data management systems are not the same. Therefore, before the experiment, we must summarize the required data, and then select the original student training data for preprocessing, which is provided to the data mining model. First, according to the decision tree classification algorithm, we select the students' entrance scores, students' gender, comprehensive quality evaluation score correlation and enterprise nature as the classification attributes, and then clean and preprocess the data set, The students' employment data is divided into training test set (see table 35 for the range of specific attributes). There are 306 samples in this experimental training set, and the employment prediction attributes of positive example (successful employment y) are 288 samples, and counter example (unemployed, n) are 288 samples. The prediction attributes of employment unit types: Class A (state-owned enterprises and institutions) has 61 samples, class B (individual and private companies) has 173 samples, and class A (entrepreneurship or free employment) has 61 samples. The corresponding decision tree model is shown in Fig. 1.

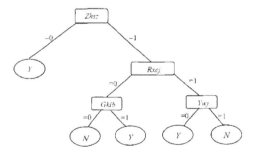

Fig. 1. Decision tree model of employment measurement

5   Summary

In recent years, the first employment rate announced by most higher vocational colleges inside and outside the province has exceeded 90%. On the surface, the employment rate is indeed very high, but it is not. From the employment quality of higher vocational students, the proportion of industry counterparts and the effectiveness of employment, it is far from the problem that the employment rate can solve. This text tries to consider the factors that affect students' employment as much as possible, and reduce the adverse effects brought by the limitations of factors. It carefully studies the ideas and theories put forward by domestic experts and papers. Combining the employment system of higher vocational students with decision tree classification is not only the practical application of decision tree classification theory, but also the guidance of theory to practice.

References

1. An analysis of the causes and Countermeasures for the employment difficulties of shipping graduates. J. Nantong Ship. Vocat. Tech. College, (4), 91–94 (2010)
2. Huang, X.: Problems and countermeasures of employment guidance service in higher vocational colleges. J. Shazhou Polytechnic **1**, 52–56 (2008)
3. Dai, Q.: Establishment of teaching quality monitoring and evaluation system in Higher Vocational Education. Vocat. Tech. Educ. **12**, 66–67 (2007)
4. Wu, Y., Jiang, H.: Application of data mining in college student management. Netw. Secur. Technol. Appl. **10**, 81–82 (2008)

Computer Genetic Algorithm in Oil and Water Layer Identification

Qiuxin Chen[(✉)]

Geological Brigade in No. 7 Oil Production Plant of Daqing Oilfield
Company Ltd., Daqing 163458, China
chenqiuxin@petrochina.com.cn

Abstract. Based on the physical and electrical properties of the reservoir, the physical properties of the strata, combined with the logging data, and on the basis of previous interpretation, the reservoir is comprehensively analyzed and understood from more detailed, profound and influential factors. A set of theoretical methods for oil and water layer identification is proposed, which has been verified and applied in the oilfield development and production practice.

Keywords: Oil and water layer identification · Reservoir characteristics · Electrical properties · Rock properties · Fluid properties · Logging data · Oilfield development

1 Analysis of the Influence of Lithology on Interpretation of Oil-Water Property of Reservoir

The lithology of an oil field or an oil reservoir has a direct influence on the discrimination of oil-water property of the reservoir, such as the hydrophilic bedrock, the type and content of the cement, the property of the surrounding rock, etc. In the identification of oil and water in reservoir, we should master these factors comprehensively and analyze the main influencing factors comprehensively, so as to make a correct judgment on the oil-water property of reservoir.

1.1 Petrophilic and Hydrophilic Properties of Bedrock

The oil-bearing and hydrophilic properties of reservoir minerals and their pore structure have great influence on the electrical characterization of oil-bearing reservoirs, and the direct characterization factor of the formation. The saturation particle diameter of the rock skeleton. If the particle diameter of the local layer is very small (generally less than 100 um) or the clay mineral mainly composed of hydromica and montmorillonite is very large, the formation has the characteristics of. The irreducible reflects the hydrophilic and lipophilic characteristics of the reservoir. electrical logging data, the oil and water bearing capacity of reservoir is determined by its water saturation, which is a comprehensive characterization of irreducible water and flowing water. For the hydrophilic reservoir, the irreducible water saturation is higher than that of the oleophilic reservoir, and the irreducible water saturation is lower [1]. Formation water in

J. C. Hung et al. (Eds.): FC 2021, LNEE 827, pp. 978–982, 2022.
https://doi.org/10.1007/978-981-16-8052-6_127

reservoir includes irreducible water and free water, while electric measurement of water saturation is a comprehensive reaction of irreducible water and free water. Therefore, the formation with high water saturation but low oil saturation can not be simply judged as oil or water layer because of high water saturation, but should be judged by referring to. If the high of the formation is caused by the high, it can be judged that if the water saturation of the formation is high and the irreducible water saturation is low.

1.2 Effect of Occurrence and Content of Argillaceous Cements on Reservoir Electrical Properties

It is proved that high layer often misjudges oil layer and oil-water layer as water layer in some special reservoir identification. This is due to the formation of low resistivity in the following three types of formations: (1) in the siltstone formation with little clay mineral content and very small particle size; (2) in the argillaceous sandstone formation with dispersed clay minerals mainly composed of hydromica (illite) and montmorillonite; (3) thin interbedding of sand and mudstone. In the Jurassic and Triassic reservoirs in the Shaan Gan Ning basin, the diagenetic epigenetics is very significant. Although the total shale content of the reservoir has little effect on the reservoir resistivity, if the main component of the argillaceous cement is water-bearing mica (illite), it will cause low reservoir resistivity [2]. Argillaceous cement reduces the resistivity of reservoir. The resistivity of this kind of low resistivity reservoir is generally lower than that of surrounding rock, and has little difference with the resistivity and water saturation of water layer. It is difficult to distinguish them from water layer in electrical property, which often leads to misinterpretation. In the actual development, we should consider the shale adsorption property of reservoir and comprehensively identify the oil-water content of reservoir.

2 Genetic Algorithm

2.1 Introduction of Genetic Algorithm

Genetic an efficient random search and, heredity and evolution. It simulates the evolution process of our real biological population from low to high. Under the constraints of living environment, each generation of biological population continuously selects, mates and mutates according to the biological evolution rule of "survival of the fittest", and finally obtains the optimal population in a certain environment. Compared with traditional algorithms, it has the characteristics of simplicity, versatile, robustness, simultaneous interpreting and group optimization. On the other hand, based on multiple unrelated and measurable anycast QoS routing problems, after analyzing its characteristics, it is found that genetic algorithm is an effective optimization algorithm to solve such NP hard problems. This paper introduces several new genetic algorithms to solve the delay constrained anycast Q σ s routing problem, and analyzes and compares them from the aspects of coding, initial population and fitness function [3].

2.2   Main Calculation Methods of Genetic Algorithm

In the adaptive genetic algorithm, P_c and P_m are adaptively adjusted according to the following formula:

$$P_c = \begin{cases} \frac{k_1 (f_{\max} - f')}{f_{\max} + f_{avg}} \\ k_2 \end{cases} \tag{1}$$

$$P_m = \begin{cases} \frac{k_3 (f_{\max} - f')}{f_{\max} - f_{avg}} \\ k_4 \end{cases} \tag{2}$$

3   Influence of Reservoir Fluid Properties on Interpretation of Reservoir Oil-Water Properties

In oilfield development, the salinity of formation water varies greatly in different regions and geological ages. The salinity of formation water directly affects the test results of formation resistivity. Because the formation water in the reservoir contacts with rocks and oil, and is affected by sedimentary conditions, all kinds of metal salts (such as sodium, potassium, calcium, magnesium, etc.) are dissolved in varying degrees. These salts dissociate into positive and negative ions in water, so they have conductivity. High salinity formation water has good conductivity and low resistivity effect in reservoir electrical logging. For the reservoirs with fine lithology, most of them are fine sandstones. Even for pure oil layers, the reservoirs generally contain high irreducible water. Combined with the influence of high salinity, the reservoirs with good physical properties have low resistivity. Therefore, their low resistivity can not be used as the basis for oil layer identification. The oil-bearing and water bearing properties of the reservoirs should be determined by integrating other physical and electrical properties.

4   Influence of Drilling Mud Invasion on Judgment of Oil-Water Properties of Reservoir

For permeable formation, mud invasion is an unavoidable phenomenon. The invasion degree is related to the mud property and formation property. The result is that the invasion zone with complex electrical property is formed near the well zone, which causes different degrees of reservoir pollution and affects the test of reservoir electrical property. In the well zone with serious mud loss, the logging data can be completely distorted.

4.1   Effect of Mud Salinity on Reservoir Electrical Properties

Drilling mud mainly includes fresh water mud (salinity <2000 mg/L) and salt water mud (>2000 mg/L). When mud intrudes into the formation, the mud and formation water mix together to affect the formation conductivity. When the salinity of mud is lower than that of formation water, the formation resistivity will be higher, otherwise, the formation resistivity will be lower. The latter is easy to cause low formation resistance and low interpretation degree of oil layer, so as to misjudge oil layer.

4.2 Influence of Mud Loss on Reservoir Electrical Properties

Mud loss in drilling is related to reservoir physical properties and formation pressure. For high-permeability and low-pressure reservoirs, mud loss is relatively more, and there is obvious shrinkage phenomenon on the radius curve. For low-permeability and high-pressure reservoirs, mud loss is relatively less, and the shrinkage phenomenon on the radius curve and latitude is not obvious. Due to the difference between the salinity of drilling mud and formation water, the reservoir resistance is distorted in different degrees. The degree of distortion is directly related to the leakage and salinity [4]. The greater the leakage, the greater the difference between the salinity of drilling mud and formation water, the higher the degree of distortion.

5 Application Example of Oil and Water Layer Identification

Low resistivity reservoir development in Anwu oilfield. The reservoir electrical characteristics of Anwu oilfield generally show low permeability and low resistivity, and the electrical logging interpretation is low, which brings many inconveniences to the formulation of rolling production plan, and even misleads the decision-making. According to the analysis of high-pressure mercury data, reservoir physical property, water property and electrical property, it is found that the main reasons for low resistivity of oil layer in Anwu area are small pore size in reservoir pores, high irreducible water saturation and higher formation salinity than other areas. These factors synthetically form the characteristics of Reservoir Electrical property different from other areas.

The laboratory calculation results show that the irreducible water saturation of Anwu oilfield is high. The irreducible water saturation of Chang 2 reservoir in Jing'an oilfield is 355%. Well A21 is a well with high production test in Chang 2 reservoir of Anwu oilfield. Its daily pure oil production is 1828t. Its resistivity is low (560m). Double induction (i.e. eight lateral) is a typical negative difference. The water saturation of core analysis is low, with an average of 378%. However, the irreducible water saturation calculated by multiple regression of mercury injection data, porosity and permeability is 394%, which is equivalent to the analysis of water saturation. This fully shows that there is no free water in this layer and it is a reservoir with high irreducible water saturation.

References

1. Sima, L.: Logging Geological Application Technology. Petroleum Industry Press, Beijing (2002)
2. Wu, T.: Logging Data Analysis Manual. Petroleum Industry Press, Beijing (2003)
3. Li, D.: Development of Low Permeability Sandstone Oilfield. Petroleum Industry Press, Beijing (1997)
4. Wang, D.: Development Technology of Low Permeability Oil and Gas Fields in Ordos Basin. Petroleum Industry Press, Beijing (2003)

Computer Simulation System in ASP Flooding Design

Lijun Dong[✉]

No.4 Oil Production Plant of Daqing Oilfield Company Ltd., Daqing 163000, China
donglijun@petrochina.com.cn

Abstract. In the design, according to the characteristics of ASP flooding produced fluid and the actual development of the oilfield, the preliminary ASP experiment results were applied and verified. A number of new processes, new technologies, new equipment and new materials were adopted, and the layout of oil and water was reasonable. The construction investment was saved, the energy consumption and operation cost were reduced, and obvious benefits were achieved, It can accumulate practical engineering experience for ASP flooding produced fluid treatment, preparation and injection.

Keywords: ASP flooding · Productivity · Technology optimization · Environmental protection

1 Research and Development of Mechanical Automatic Descaling Heating Device to Solve the Problem of Medium Scaling and Furnace Efficiency Reduction

In the past, the conventional heating furnace heating ASP flooding medium has the problem of maladjustment. The scaling and thermal efficiency reduction of ASP flooding medium are serious. In order to prolong the service life of the heating furnace, reduce the maintenance cost of the heating furnace and reduce the production cost, the mechanical automatic descaling heating device is developed. The steam enters into the heat exchange tube (tube side) of the heat exchanger through the connecting pipe, and the ASP flooding medium enters into the heat exchange tube (shell side) of the lambda heat exchanger through the connecting pipe. In the heat exchanger, the steam fully exchanges heat with the medium to be heated. The cooled steam flows into the phase change heating furnace through its own weight and is reheated to raise the temperature [1]. After the medium to be heated absorbs heat and raises the temperature, it is pumped out from the shell side of the heat exchanger. The descaling system is composed of pin gear drive mechanism, transmission shaft and descaling device. The pin gear drive mechanism transmits power to the driving pin gear fixed on the transmission shaft. The pin gear is meshed with the teeth to drive the descaling device connected with the rack. The descaling device intermittently moves back and forth on the heat exchange tube bundle to remove the dirt on the heat exchange tube bundle, The heavy scale is deposited on the lower part of the heat exchanger body by self weight and discharged

J. C. Hung et al. (Eds.): FC 2021, LNEE 827, pp. 983–987, 2022.
https://doi.org/10.1007/978-981-16-8052-6_128

from the sewage outlet. By adopting this mechanism, the problem of unadaptability of ASP flooding medium heated by heating furnace will be fundamentally solved, the service life of heating furnace will be extended, the maintenance cost of heating furnace will be reduced, and the production cost will be reduced.

2 Introduction of Computer Simulation System

The computer age has come. Now many centers are processing files. Document turnover management stipulates that reliable information should be reasonably processed by high-level experts where necessary, and be made into a clear form, and processed in the information equipment network by optical and magnetic systems. This paper mainly discusses the intelligent management of document turnover. At present, the file turnover intelligent management system can process a large amount of information at the speed of 100000 bits/second. The 560000 bit/s high speed processing of telegraph system will also become a reality. If it took a few hours to transmit a large amount of data in 1980, it will only take a few minutes now. With the help of V.34 modem, a standard small picture can be transmitted in 20 s. During 1979 and 1990, the range of information recording increased 199 times. At the same time, it can create a wide range of conditions for storing information in many places, which can simplify the access of files. This paper explains the possibility of developing file turnover intelligent management system, such as adopting broadband file processing system, reducing file transmission time, improving business volume, compressing information, reducing file identification time, making extensive use of servers, monitoring information processing process and using navigation system, and describes the future file turnover intelligent management system.

$$F_1 = GL \cos x / d_1 \tag{1}$$

Considering the uncertainty of soil parameters, the elastic modulus E of foundation soil is modeled as a random field which obeys lognormal distribution. The uncertainty of Poisson's ratio μ is not considered temporarily, and is taken as the determined value 0.3. The statistical results assume that the relevant indexes of elastic modulus random field $E\ (x,\ y)$ have been obtained statistically, and the mean value $\mu\ \varepsilon = 30$ MPa, the standard deviation $\sigma\ \varepsilon = 9$ Mpa, and the correlation distance $\theta\ \varepsilon = 0.9$ m (for the convenience of analysis, It is assumed that the foundation soil is isotropic, that is, the correlation distance in horizontal and vertical directions is equal), and the autocorrelation function is:

$$\beta(\tau_1, \tau_2) = \exp(-\frac{2}{\theta_E} \sqrt{\tau_1 + \tau_2}) \tag{2}$$

With the increase of the correlation distance, the correlation of the discrete values of the adjacent units in the field increases, and the variation characteristics of the random field slows down, which is reflected in the graph, that is, the color of the adjacent units in the field is closer and closer, and the change between colors is more and more gentle, and the area of the same color of the adjacent units in the field is gradually increasing, When the correlation distance tends to infinity, it means that any two points in the field are completely correlated, that is, the sample function of each sampling is a constant, which is essentially a random variable. Through the Monte Carlo stochastic finite element analysis of an engineering example, it is found that the statistical variability of the settlement calculation results will be reduced if the foundation soil is layered artificially, and the more the layers are, the more significant the reduction will be; the closer the upper soil layer is, the greater the influence of the uncertainty of soil parameters on the probability response characteristics of the whole foundation settlement; in addition, the higher the probability response characteristics of the whole foundation settlement will be, The engineering example also shows the application process of random field simulation in geotechnical reliability analysis.

3   Application and Practice of Computer Simulation Technology

According to the requirements of national classified protection technology, this paper considers the factors of identity authentication, access control, security audit, system security detection, border security protection and so on when carrying out the simulation technology practice, so as to basically meet the simulation requirements of enterprise information security production environment, It mainly involves the following aspects: (1) research on security domain partition: according to the requirements of hierarchical domain protection, seven security domains are divided in the simulation environment, and the security domain partition and switch ACL access control list configuration technology are studied to limit the mutual communication between security domains. (2) Border access control technology research: according to the requirements of security domain border isolation, deploy firewalls at the border of security domain, study the configuration of firewall access control policy, and realize the isolation of security domain by firewall. (3) research on terminal security protection technology: install and deploy the same security products as office network, including host audit system, anti-virus system, domain management, etc. [2]. To achieve the same protection level of the office network, at the same time, to study the configuration of various security policies, to further improve the security of the terminal, and test and verify the compatibility of various security products and security policies, as well as the impact on system performance. First, research on domain control technology: Research on the erection technology of domain control and sub domain control and the setting of domain policy, To improve the rationality and security of domain policy settings.

4 Develop Water Quality Stabilizer and Sulfide Removal Agent to Improve Crude Oil Dehydration and Sewage Treatment Level

In view of the difficulty in removing suspended solids from ASP flooding produced water, a water quality stabilizer suitable for ASP flooding produced water was developed for the first time. In view of the frequent wellbore acid pickling or reservoir acidizing operations in ASP flooding production wells due to scaling in the wellbore and near wellbore, and the carbonate and ferrous sulfide particles are easily produced when the produced liquid is mixed with the surface water in the initial stage of acidizing and acid pickling operations, which leads to the serious deterioration of the treatment effect of ASP flooding produced liquid and produced water, a sulfide remover is developed for the first time [3]. Two sets of demulsifier dosing devices, one set of antiscaling agent/water quality stabilizer dosing device and one set of sulfide remover dosing device are designed and constructed in the new Sanyuan oil transfer and drainage station in the block. The operation results show that the crude oil dehydration and produced water treatment effect are significantly improved.

5 Using Renewable Packing Three-Phase Separator to Improve Oil and Gas Treatment Capacity

5.1 The Advantages of Using this Method

Due to the large degree of oil-water emulsification of ASP flooding produced fluid, the emulsification strength of produced fluid is greater under the action of multi-component chemical agents, and the interface property is more complex. Moreover, the high viscosity fluid carries a large amount of mud, sand and rock debris from the reservoir, and the mechanical impurities and scaling substances in the produced fluid increase, which is easy to cause free water coalescence, and the filler is easy to block. In view of the problems existing in the process of free water removal, a three-phase separator suitable for ASP flooding is developed. The use of coalescence packing with renewable function can fully reduce the water content of the oil after removal, avoid the impact of high water content emulsion on the electric dehydrator, realize in-situ regeneration of coalescence packing, and solve the problem of replacing packing due to siltation, The cost of production and operation is reduced. The dehydration efficiency is 20% higher than that of conventional three-phase separator.

5.2 Combined Electrode and Frequency Conversion Pulse Electric Dehydrator are Used to Improve Crude Oil Dehydration Efficiency

Due to the complex composition of ASP flooding produced liquid and the unstable operation of dehydration electric field, the frequency conversion pulse high-speed crude oil dehydration device is used for ASP flooding produced liquid dehydration. In the process of pulse electric dehydration, the water droplets in crude oil are not only subject to the dipole coalescence of DC electric field, but also subject to the oscillating demulsification coalescence of AC electric field, which is conducive to the water phase coalescence of water in oil emulsion and improves the dehydration efficiency [4]. At the same time, the short circuit between electrodes caused by liquid long conductive chain is avoided, and the electric field is stable. Compared with AC/DC power supply dehydration, the annual power saving of a single unit can reach 93×104 kwh, Two sets of equipment in a project can save 11.16 million yuan in annual operation cost. Total investment + present value of ten-year operation cost saved 63.06 million yuan.

6 Concluding Remarks

The advanced nature of the design of Daqing oil displacement drive is analyzed, which is of great reference and guidance for design.

Acknowledgements. National key S&T Special Projects- Enhanced oil recovery demonstration project in extra high water cut of Changyuan Daqing oilfield.

References

1. He, H., Cheng, J.: Research progress of polymer flooding produced fluid treatment technology in tertiary oil recovery. Petrol. Refin. Chem. Ind. **33**(9), 29–32 (2002)
2. Jiang, L.: Problems and solutions of heating furnace in use in Daqing oilfield. Surf. Eng. Oil Gas Fields **30**(10), 51–52 (2011)
3. Chen, J.: Practical application of information security technology in advanced fields in enterprises. Netw. Secur. Technol. Appl. (09), 101+103 (2014)
4. Zhang, X., Wei, Z., Yang, X.: Optimization of investment casting process for cast steel bracket. Cast. Technol. **39**(5), 1039–1041 (2018)

Network Node Coverage Optimization Based on Improved Ant Colony Algorithm

Xueli Feng[(⊠)]

Shandong Technician College of Water Conservancy, Shandong 255130, China

Abstract. This paper studies the optimization problem of wireless network node cover. Due to the redundancy of network nodes and the limitation of power supply energy, it is difficult to solve the hot zone problem of the current wireless network due to the influence of the life of the network. An ant colony algorithm is proposed to solve the problem. Firstly, the nodes in the network are divided into non-uniform parts, and the pheromones are placed on the nodes of the transmitter network. Each node is given two pheromones to indicate two different information quantities of the node. Effective heuristic information is set in the algorithm to guide the search behavior of ants. The simulation results show that compared with the current classic network node coverage algorithm, the algorithm improves the network coverage, reduces the network energy consumption, and has a good effect on optimizing the network node coverage.

Keywords: Wireless sensor network · Frontal region optimization mosquito swarm algorithm

1 Introduction

In recent years, many researchers extend the life cycle of network by increasing the density of network nodes. However, with the continuous in-depth research, it is found that with the increase of network nodes, the growth rate of network lifetime becomes slower, but at the same time, it brings the problem of node redundancy. Scholars have also conducted in-depth research on the problem of node redundancy, and adopted the method of maintaining network coverage node scheduling. After that, some scholars have made some improvements on the basis of it. This method is mainly based on the classic leach [1]. Because some of the current algorithms are based on the random part of nodes, these methods make the network energy waste higher.

2 Node Coverage in Wireless Sensor Networks

2.1 The Principle of Network Node Overlay

Generally speaking, sensor nodes in WNS have a fixed sensing radius. Any point in the monitoring area is in the range of sensing radius, so the monitoring area is divided into two parts: one is coverage area, the other is blind area. In this paper, when building this wireless sensor network, the coverage of the network is the most basic problem in the

J. C. Hung et al. (Eds.): FC 2021, LNEE 827, pp. 988–992, 2022.
https://doi.org/10.1007/978-981-16-8052-6_129

process of its implementation, that is to say, under the condition of keeping the network unblocked, using as few nodes as possible to get large coverage. The traditional WNS node coverage graph is shown in Fig. 1.

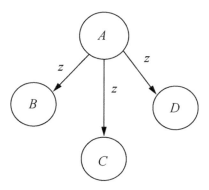

Fig. 1. Node coverage of wireless sensor network

In Fig. 1, all the information in the sensing area with a node s as the center and R as the radius can be sensed [2]. In the communication coverage mode, any two nodes R_A and R_B must satisfy the divergence ability of the distance within R_C to ensure that the information sensed by the node can be accurately transmitted, And connected coverage is that any node s has exactly adjacent nodes, which can be perceived only by arranging them in a certain way.

2.2 Mathematical Model of Coverage Optimization

Let e^1 be the energy consumption of a node sending 1bi data, and e^2 be the energy consumption that has been accepted. In WNS, the number of nodes on L1 is n. therefore, according to the above assumptions and network model, the energy consumption en of the outermost ring L in a certain period of time is expressed as:

$$E_R = N_R M(e_1 + e_2) \tag{1}$$

However, other nodes in the circle not only need to send certain data generated by themselves, but also need to forward data from other circles. Therefore, the energy consumption e of the whole network node can be expressed as:

$$E_i = M(\sum_{k=i+1}^{R} N_k(e_1 + e_2) + N_i e_1) \tag{2}$$

The goal of WNS node coverage optimization is to maximize the coverage of the network and reduce the utilization of sensor nodes as much as possible. The sub objective function can be transformed into the overall objective function to define the fitness function, which is expressed as:

$$f_2 = |C'|/N = \sum_{i=1}^{N} a_i/N \tag{3}$$

3 Research on Improved Ant Colony Optimization Algorithm

In practice, most of the optimization problems are multi-objective problems with constraints. The traditional ant colony algorithm always leads to low accuracy and instability. The effective exploration and development of feasible region also depends on the search ability of the algorithm, which is ultimately realized by the search operator. This paper proposes an improved ant colony algorithm for solving multi-objective optimization problems. When selecting the definition of evolutionary algorithm, a certain number of individual information sources in the population are used as the diffusion center, and there is a certain distance between multiple center points; other individuals in the population belong to one of the pheromone diffusion sources according to the principle of the closest distance from the original individual; according to the pheromone diffusion algorithm, each individual in the pheromone diffusion source obtains the pheromone from the center point; The central point of each generation population is reserved to the next generation population, which ensures the convergence and maintains the diversity of the population [3]. To improve the search ability of the algorithm, and used in the optimization of wireless sensor network coverage model.

3.1 Information Source Selection and Distance Determination

A certain number of sources are selected in each generation, and the population size and fitness function determine the number of sources. The number of individuals from the experimental data source is usually about 1/5 of the population size. After determining the number of sources, the distance between sources should also be determined. This paper designs an adaptive distance formula:

$$D_{\min}(t) = D_0/T^\beta \tag{4}$$

In the formula, $D_{\min}(t)$ is the minimum distance between the center points of the T generation source individuals, D_0 is the minimum distance between the center points of the first generation source individuals, and B is to adjust the distance between the center points to change with algebra. It is concluded from the experiment that the value of β between 0.15–1 is more appropriate.

3.2   The Diffusion of Pheromone

Pheromone diffusion is the key for ant colony to find the shortest path:
When R > S:

$$H_i(Y) = H_i^{\min} + H_i(Y_0) * ((R - S)/R) \tag{5}$$

When R $\leq$ S:

$$H_i(Y) = H_i^{\min} \tag{6}$$

Where R is the diffusion radius of the source and S is the distance between the center points x and X of the source. H (y) is the smallest pheromone in the first generation chromosome. When calculating the pheromone of the source diffusion of a point x, we first find the individual X of the source central point nearest to the point, and then the pheromone of X is obtained by the source central point x diffusion.

4   Ant Marching Selection Strategy

For the discrete space optimization problem, the ant moves directly from one node in the solution space to another node, which is not suitable for the optimization problem in continuous space. In this paper, we define a range R for ants, that is, ants can only move in the range of radius R. When the distance between the ant and the target point exceeds its range R, the ant can only move a distance of R to the target point. However, after the ant moves, it can not reach the target point directly. It is also affected by a random disturbance α. The value of disturbance factor α affects the convergence of the algorithm [4]. A larger α value is conducive to increasing the search range, and a smaller α value is conducive to local optimization. Therefore, at the beginning of the algorithm, this paper gives a relatively large value range of α, which is used to increase the search range of the algorithm; with the increase of the number of iterations, the value range of α can be reduced linearly to enhance the local convergence ability of the algorithm.

5   Analysis of Simulation Results

All the experiments in this paper are done on the computer of pcp4t23101.86g, 2gram182865g graphics card, the experimental environment is mat0, for the effectiveness of the improved algorithm, the root line sensor network coverage method, the area to be monitored is 80 m, the sensing radius is 12 m. According to the area and sensor parameters of the area to be monitored, 40 sensor nodes are located in the monitoring area, and then 61 sensors are randomly planted in the monitoring area.

In this paper, several common network node coverage optimization algorithms are compared and analyzed. The improved ant colony algorithm, basic genetic algorithm, basic particle swarm algorithm and basic ant colony algorithm are used to optimize the coverage of wireless sensor networks in the monitoring area. The parameters are set first, and the number of nodes activated is reduced on the basis of considering the largest coverage area, The fitness function is (5), L1 = 0.9, W2 = 0.1.

The comparison results show that the coverage rate of the proposed algorithm is at least 2% higher than that of the other three algorithms. See the effectiveness of the algorithm in this paper, which greatly improves the f-network coverage performance.

6 Conclusion

The sensor nodes in linear sensor networks have a fixed sensing radius. If the sensing radius of any sensor node in the monitoring area is within the sensing radius, the main technology of its coverage has great application prospects in various fields. Network coverage is one of the key problems of sensor networks, In this paper, based on the in-depth study of the research progress in the field of wireless sensor coverage at home and abroad, and making full use of the advantages of ant colony optimization algorithm, a node coverage scheduling optimization algorithm based on improved ant colony algorithm is proposed. Through the comparison of several common optimization algorithms, the improved ant colony algorithm optimization mechanism proposed in this paper can jump out of the local optimum more effectively and obtain more accurate results, so as to better and more effectively realize the global optimization of wireless sensor network layout. It can maintain a high network coverage in a long time, so as to extend the network life cycle, better balance the energy consumption of nodes, and improve the network coverage and service quality.

References

1. Mao, Y., et al.: Research on wireless sensor network coverage control technology. Comput. Sci. **3**, 20–22 (2007)
2. Wang, X., et al.: Research on grid based coverage in wireless sensor networks. Comput. Sci. **33**(11), 38–39,78 (2006)
3. Qu, B., Hu, D.: Research on energy efficient routing protocol for wireless sensor networks. Comput. Simul. **25**(5), 113–116 (2008)
4. Zhou, Y., Li, P., Mao, Z.: A new hybrid method and its application in constrained optimization. Comput. Eng. Appl. 48–50 (2006)

Design of Non Significant Region Adaptive Enhancement System for 3D Image of Martial Arts Movement

Qingkai Guo[✉]

Tianjin University Renai College, Tianjin 301636, China

Abstract. Aiming at the problems of low efficiency and poor effect in traditional enhancement system, an adaptive enhancement system based on reti η ex enhancement model optimized by center surround method is proposed. In the image space domain, the mean filtering method is used to denoise the three-dimensional image of martial arts movement. The gradient operator is used to calculate the gradient vector which can reflect the three-dimensional image of martial arts movement. In the image space domain, the differential operator is used to sharpen the three-dimensional image of martial arts movement. On this basis, the SSR algorithm is used to solve and weight the non significant areas in the three-dimensional image of martial arts movement, the center surround method is introduced to estimate the three-dimensional image of martial arts movement, and the Retinex enhancement model is introduced to adaptively enhance the non significant areas in the three-dimensional image of martial arts movement.

Keywords: Dynamic dimension image of martial arts · Non significant regional stress · Enhancement system · Mean Hongbo

1 Introduction

The three-dimensional images of martial arts movements have developed into a very active research field in the field of image research. Each kind of three-dimensional images of martial arts movements has its fixed characteristics. They are located at a certain point in the three-dimensional images of martial arts movements, but due to the interference of many uncertain factors, It makes the three-dimensional image of martial arts movement have non significant area increases the difficulty of the research on the three-dimensional image of martial arts movement, and the adaptive enhancement of the non significant area is the most effective method to solve the problem of the research on the non significant area of the three-dimensional image of martial arts movement, which has become the focus of many scholars.

J. C. Hung et al. (Eds.): FC 2021, LNEE 827, pp. 993–997, 2022.
https://doi.org/10.1007/978-981-16-8052-6_130

2   Main Requirements of System Design

An adaptive enhancement method based on wavelet transform for non significant region of 3D image of martial arts movement is proposed. The method transfers the image to HSV space, and uses discrete wavelet transform to analyze the subband of the image. Secondly, the bilateral filter is used to estimate and remove the image quickly to reduce the non saliency of the image and enhance the non saliency area of the image automatically. However, this method has the problem of poor enhancement effect., an image non significant adaptive enhancement method based on the mixture of artificial fish swarm and particle swarm algorithm is proposed. This method optimizes the nonlinear enhancement parameters of the image by the mixture of artificial fish swarm and particle swarm algorithm, so as to avoid the problem of incomplete regional enhancement. This method has high adaptability, but it is easy to fall into local optimization and slow convergence speed. In this paper, an image adaptive enhancement method based on shuffled frog leaping optimization is proposed [1]. This method uses the optimization mechanism of partial information exchange and all information exchange in shuffled frog leaping algorithm to automatically search for the best gray transformation parameters and obtain an optimal gray transformation curve to realize the adaptive enhancement of non significant regions of the image. However, this method ignores the influence of the image background on the target, an adaptive enhancement system for non significant regions of Wushu movement 3D images based on the Retinex enhancement model optimized by center surround method.

3   Three Dimensional Image Processing of Wushu Movement

3.1   Denoising Processing

In this paper, we design the adaptive enhancement system of the non significant region of the three-dimensional image of martial arts movement, and focus on the design of its software. Before the software design, it is necessary to denoise and sharpen the three-dimensional image of martial arts movement. In real life, when collecting the three-dimensional image of martial arts movement [2].

Two dimensional windowed Fourier transform is used to extract the spectral information of each part of the infrared image:

$$F(i,j,u,v) = \int_{-\infty}^{+\infty} f(x,y)W(x-i,y-i)e^{-j(ux,uy)}dxdy \tag{1}$$

The mean filtering method is used to filter the three-dimensional image of martial arts movement:

$$g(x,y) = \frac{1}{M}\sum_{(i,j)\in s} f(i,j) \tag{2}$$

3.2 Three Dimensional Image Sharpening of Martial Arts Movement

In the double Gaussian filter is a kind of nonlinear filter, which is mainly used to smooth the image. Because the residual noise of the image while retaining the image edge information:

$$R_s = \frac{1}{k(s)} \sum_{p \in \Omega} f(p - s) g(I_p - I_s) I_p \tag{3}$$

To the three-dimensional image of martial arts movement by using the mean filtering method, the three-dimensional image of martial arts movement is sharpened by using the operator in the spatial.

4 3D Digital Image Correlation Method

The 3D digital image correlation method is an optical non-contact high-precision measurement method, which uses two high-speed cameras to synchronously collect the surface image of the model to be measured, combined with binocular stereo vision for deformation measurement. The basic idea is to obtain the three-dimensional displacement and strain distribution on the surface of the measured model by taking and processing the two-dimensional images before and after the deformation of the sub region of the measured model image [3]. The calculates the three-dimensional displacement field by matching the different image sub regions of the two speckle images before and after the deformation of the measured model, and obtains the strain field by numerical differential calculation.

4.1 Binocular Vision Imaging Model and Camera Calibration

The main principle of binocular vision technology is based on parallax. Two cameras with known spatial position and attitude relationship are used to collect images of measured features in different directions at the same time. Image processing and homonymous point matching are used to obtain homonymous image point pairs corresponding to measured features. The imaging ray equation is established by using camera imaging model, and the ray triangulation intersection constraint is constructed, The mathematical model of binocular vision measurement is established to calculate the three-dimensional coordinates of the measured feature space.

4.2 Stereo Matching

The right image is the before deformation, and the right image is the speckle image of the object after deformation. The rectangular reference image sub area with the pixel size of $(2m + 1) \times (2m + 1)$ centered on a point P $(x0, Y0)$ is selected in the speckle image, and the same size samples are selected in the image after deformation. According to the calculation results of the standardized covariance cross-correlation coefficient, the cross-correlation coefficient is calculated, In order to find the sub region point P (x0, Y0) corresponding to the maximum absolute value of covariance correlation coefficient, the displacement components in X and Y directions are determined [4]. When using the digital image for practical calculation, the displacement of each virtual network is calculated in the form of virtual grid to obtain the three-dimensional map of the whole field.

5   Analysis of Experimental Results

In this method in the non significant region enhancement of Wushu movement three-dimensional image, a comparative experiment is designed for analysis. Figure 1 shows the MSSM.This system is compared with the hybrid frog leaping optimization method and the wavelet transform enhancement system. In each group of experiments, in order to ensure the image enhancement and further increase the time efficiency of the method, the three-dimensional image of each martial arts movement is selected as the research object.

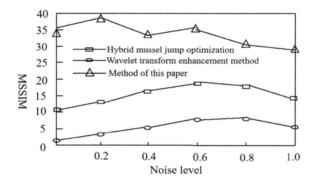

Fig. 1. MSSM index of image enhancement processing under different noise levels

It can be seen that the processing effect of this method is better than that of the traditional enhancement system in terms of adaptive enhancement. In the case of image distortion after enhancement by the traditional method, the image enhanced by this method. Experiment 2 robustness contrast experiment in the experiment, the spatial sequence and standard sequence are selected, the unified mask of 3 × 3 is used for enhancement processing, and each image is twice sampled. Because the three-dimensional image of martial arts movement has certain noise, the experiment is based on the noise level. Under different noise levels, the hybrid frog leaping optimization method, wavelet transform enhancement method and the proposed enhancement method are compared, and PSNR and MSSIM are used as indexes for experimental analysis.

Compared with shuffled frog leaping optimization method and wavelet transform enhancement method, the proposed method proposed method has good enhancement performance and high robustness.

6   Conclusion

In order to solve the problems of low efficiency and poor effect of traditional rein-forcement system, this paper proposes an adaptive reinforcement system based on the Retinex reinforcement model optimized by the center surround method for the non significant area of three-dimensional image of martial arts movement, and makes an experimental comparative analysis. The experimental results show that the enhance-ment efficiency and effect of the proposed method are better than those of the tradi-tional method, and it has certain advantages.

References

1. Ren, Y., Dong, X.: Image salient region extraction algorithm based on adaptive manifold similarity. J. Shandong Univ. (Eng. Ed.) **47**(3), 56–62 (2017)
2. Xie, H., Tang, T., Xiang, D., et al.: Scale adaptive saliency detection method for SAR images. Comput. Eng. Appl. **51**(20), 145–152 (2015)
3. Wang, R., Chen, J., Jiao, L., et al.: Block adaptive compressed sensing algorithm based on visual saliency. J. Huazhong Univ. Sci. Technol. (Nat. Sci. Ed.) **43**(1), 127–132 (2015)
4. Ke, H., Shao, W., Liang, C.: A moving target tracking method based on saliency region. Innov. Appl. Sci. Technol. **34**(9) (2017)

Design and Implementation of Intelligent Building Programmable Control System Based on Linux Platform

Chuanke Li[✉]

Hainan College of Software Technology, Qionghai 571400, China

Abstract. In this paper, Linx operating system is used to build a control unit of intelligent building programmable control platform to facilitate the control and management of other units in the intelligent building. The control unit is programmable, secondary programming users can control the script code through simple python, upload it to the control unit through the web page, and control the units of different protocols and obtain unit information through a single control platform. It can also obtain the network setting information and abnormal log information of the control unit of the control platform by using CGI technology through the web page. By testing the external interface and web page, according to the test results, the intelligent building programmable control platform has achieved the goal of the research content and function.

Keywords: Linux · Intelligent building control platform · Programmable · Python · Web development

1 Introduction

With the development of China's social economy and science and technology, people's living standards continue to improve, generally hope to improve the current quality of life, improve life and living conditions, yearn for intelligent buildings. The basic connotation of intelligent building is: Based on generic cabling system, computer network system as a bridge, each functional subsystem in the integrated building realizes the management of communication office automation system, all kinds of equipment in the building (air conditioning, heating, water supply and drainage, power transformation and distribution, lighting, elevator, fire safety), etc. According to the actual experience of the project, through the implementation of high-efficiency energy-saving equipment and automatic control system, the building energy consumption can be reduced by 30%–50%, but without relevant energy consumption monitoring control and maintenance management department, the building energy consumption and energy saving system will not be able to carry out its own real-time control, resulting in the building energy saving effect can not be continuously and stably regulated and managed.

2   Key Technology

Building programmable control UX system development, through the current popular Apache server technology w, through the browser can access the intelligent building programmable control platform, the platform network equipment programming control and the platform hardware information settings. The key technology is mainly lnux web programming technology, python programming.

2.1   Linux

Linux is a popular operating system in the field of engineering and information technology, and it is more and more popular in the business world. Linux has the following characteristics.

Free and open source. Stable and efficient performance; supporting file system.; Support network communication.; Rich equipment management; Support process management; Size can be cropped. The calculation expression of cache hit ratio (HR) is shown in Eq. (1):

$$HR = \frac{\sum_{i=1} req_i}{N} \tag{1}$$

Creating set match contains function pairs that are exactly matched; creating set contains function pairs that are not exactly matched; creating set match contains function pairs that are not exactly matched; creating file internal call graph file f call relation:

$$G_f = \{Node_1, Node_2, \ldots Node_n\} \tag{2}$$

2.2   Web Programming

HTML, called hypertext markup language, is a markup language used to make web pages. It does not need to be compiled and is executed directly by the browser. HTML pages are all suffixed with Hm1 or HM. When the browser finds the correct suffix, it will parse the file [1]. An HTML file contains not only text content, but also some tags, which are called "tags" in Chinese. HTML tags are keywords surrounded by angle brackets, such as < HTM >. HTML tags usually appear in pairs, such as start tag < b > and end tag < b >. Each label represents a format, and each label is attached with all the attributes of this format. The same label with different attributes can show different display effects. In web page typesetting, these tags are combined to make the page look more beautiful and neat.

3   Requirement Analysis

3.1   Functional Requirements Analysis

AIn order to save energy, manpower and material resources, intelligent management building, developers need a set of intelligent building programmable control platform solutions. The specific requirements are as follows.

(1) It needs to be able to manage different equipment in the building remotely through simple programming.
(2) Security is required.
(3) The product is stable and the price is low.
(4) Web interface is provided to view and modify the information of control platform remotely.
(5) The programming script of the control unit can be updated at any time.

In order to achieve the above functions, the function of the intelligent building programmable control platform is divided into three parts, namely, the function of daemons, the function of providing external interface, and the function of web page display. The external interface function includes providing communication interface and auxiliary function interface [2]. Web page display function includes fault record display, file upload function, network setting function, clock synchronization function and local restart function. The level of detail function diagram is shown in Fig. 1.

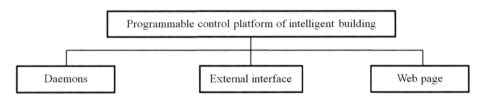

Fig. 1. Hierarchy diagram of intelligent building programmable control platform

3.2   Feasibility Analysis

In order to facilitate the operation and management of secondary programming users, web pages must be friendly and easy to learn. Python CGI programming makes the development of Web sites simple and convenient, and provides convenient system and Apache service for the multi-function of the page, which can bring the stability and security of the network server. Python, CGI, HTML and CSHE are all mature technologies, and the invention technology is feasible. With the increasingly fierce competition of enterprises, good management can bring better benefits [3]. Therefore, in order to effectively save energy and material resources, ensure the good development of enterprises, the investment and development of control system is essential. The construction and expansion of the system only need common computer equipment, conventional personnel management and simple development system investment. Therefore, the programmable control platform is feasible economically.

3.3   Non Functional Requirement Analysis

The programmable control platform has the following functions:

(1) High reliability, friendly practicality, high efficiency and portability. The specific requirements are as follows: (1) if the power is cut off abnormally, the programmable control platform can recover. (2) Good interaction with users, good user experience.
(2) Compared with the current control platform, the response speed is faster and the amount of information access is larger.
(3) Use mature and free third-party middleware or Python functional modules as much as possible.
(4) Considering that it is easy to add and change functions in the future, we need to design highly modular and try to use class methods to realize functions.
(5) The hardware may be changed in the future, and the possibility of hardware configuration change should be considered in the design.
(6) The portability of operating system changes should be considered in design.

4   Programming

When the programmable control platform daemons start, the action process is: initial system information, keep the information, start the door dog thread, TP clock synchronization, secondary programming, script uploaded by users, start Apache service, monitor CGI message. The cycle of NTP synchronization and WD cycle is 1 s, but the execution process will cause cycle blocking. The blocking time is determined by the network and N server response and other reasons, and it needs to start a thread separately instead of the maximum delay time of the watchdog. For security protection, file permissions are used to allow authorization orders, files, programs, and system resources. Because CGI has no right to write important files in the programmable control platform, it can add, delete and modify important files and give them to the daemons. Ntpdate is also operated by the daemons [3, 4]. The CGI messages monitored by the daemons include: delete the local log file, set NTP synchronization time, execute NTP synchronization, start NTP automatic synchronization and stop clock synchronization [4].

5   Summary and Prospect

5.1   Summary

The functions of the daemons include: initial control unit information, log record acquisition and deletion, starting dog watching thread, dog feeding operation, starting NTP clock synchronization thread to process NTP clock synchronization command, starting secondary programming, user programming script to provide external interfaces, including serial communication interface destruction, sending, receiving, network creation, client destruction, client sending Client section receiving, creating service destroying server side), interprocess communication (creating channel, destroying channel, sending data, receiving data, locking channel and unlocking channel). Web page EB page functions include: fault record screen display, delete log file upload screen to realize Python script document page upload function of programming user programming; network setting screen to realize the setting of two network ports (network information (p address, p address prefix, network s server) by IPv4 and IPv6, and display and modify the name of the machine; The NTP clock synchronization setting screen can execute and stop NTP clock synchronization, and set NTP server information.

5.2   Expectation

Due to the relatively tight design time, there are still many defects to a certain extent, such as the beautiful interface, higher security of the website and so on, which need to be further improved and improved. Next, we need to provide communication verification interface to ensure the accuracy of data transmission. And support remote online system upgrade to facilitate future system updates.

References

1. Ding, H., Pan, L., Ni, Y., Liu, H.: A multifunctional burner with touch screen. Electron. World, (24), 202–203 (2020)
2. Wu, L., Li, X.: Content design of Guangxi digital TV library. Radio TV Netw. **27**(12), 76–78 (2020)
3. Pang, C., He, L., Huang, Q., Wang, J., Liao, J., Fu, J.: An optimization method of operating system to improve the channeling ability of 428XL. Petrol. Pipe Instrum. **6**(06), 78–82 (2020)
4. Wang, W., Huang, S., Huang, T., Ma, J.: Simulation and solution of clock jump fault in Linux time server. Power Inf. Commun. Technol. **18**(12), 66–70 (2020)

Alternative Routing Configuration of Key Services in Power Communication Network Based on Genetic Algorithm

Lei Li, Jun Sun, Yi He, Ge Tang[✉], and Lu Ye

Information and Communication Branch of State Grid, Hubei Electric Power Company, Zhejiang 430077, China

Abstract. For a large number of key services in the power grid which need to be properly planned and configured, considering the service performance requirements and network wind factors, this paper proposes an alternative routing mechanism based on relay algorithm. Firstly, the channel pressure index is modeled for the key service routing configuration problem, and the routing elixir setting mathematical model with minimum pain is constructed. Then, combined with the characteristics of the model, the genetic algorithm is used to solve the problem, and the simulation experiment is carried out based on the existing network. The experimental results show that for the large-scale network scenario, compared with the traditional system, it can obtain the alternative routing scheme with lower global channel pressure.

Keywords: Power communication network · Alternative routing configuration · Transmission algorithm

1 Introduction

With the development of smart grid, power communication network has attracted extensive research. In 2006, the European sustainable, competitive and safe power strategy emphasizes that smart grid technology is a key technology and development direction to ensure the power quality of the European grid. In December 2012, the IEEE802 standard development project team was launched to support the worldwide power industry in the construction of smart grid data communication infrastructure. In order to promote the transformation of China's power grid to smart grid, the "12th Five Year" special plan of major science and technology industrialization project of smart grid puts forward the goal of building a smart grid characterized by informatization, automation and interaction. Although the research of electric power communication network has achieved remarkable results, there are still some problems, among which the network's ability to deal with risks is insufficient, which leads to accidents all over the world. In 2003, the North American blackout caused a direct economic loss of 30 billion US dollars, affecting 50 million people, and the cause of the accident is only due to human factors leading to some key point failures in the network. In 2008, the snow disaster in southern China led to power grid paralysis, resulting in economic losses of 150 billion yuan, highlighting the vulnerability of China's power network to cope with

© The Author(s), under exclusive license to Springer Nature Singapore Pte Ltd. 2022
J. C. Hung et al. (Eds.): FC 2021, LNEE 827, pp. 1003–1006, 2022.
https://doi.org/10.1007/978-981-16-8052-6_132

natural disasters. Network vulnerability refers to any vulnerability in the network that can be used as the target of attack, so as to degrade the network performance. Network vulnerability exists in all aspects of the network system, such as network topology, hardware, software and so on. Because of its physical structure, business data characteristics, distribution range and other factors, the vulnerability of power communication network is inevitable, which threatens the safe and stable operation of smart grid [1].

2   Key Service Routing Configuration Modeling

2.1   Network Topology Model

When the service paths of all services have been determined, the service path corresponding to the service can be obtained: $P_n(v_n^{start}, V_n^{end})$, where n sub as follows:

$$T_n = \frac{L_n \gamma}{c} + \sum_{v \in P_n} t(v) \tag{1}$$

2.2   Definition of Channel PressureA

It can be used to effectively distinguish the high voltage link from the low voltage link:

$$\Pr(e_{ij}) = \delta_{eij} \sum_{s_i} n_{ij} \times d(s_i) \tag{2}$$

3   Simulation Experiment

In order to verify the effectiveness of the proposed mechanism and the performance compared with other optimization algorithms, the simulation is carried out under the topology close to the existing network. In the experimental simulation, according to reference, 100 generations of cyclic algebra are designed, the population number is 100 groups of genes corresponding to each business, the coding method is real number coding, the crossover probability is 60%, and the mutation probability is 0.5%.

In the experiment, the number of network nodes is set to 10 and 20 in turn, and the node connection is set according to the actual situation of the current network, that is, the connection ratio between nodes is 30%. Then, in each kind of network with different number of nodes, different number of services to be configured are set for multiple simulation experiments. The starting node and the ending node of the services to be configured are the same, which are high important key services such as line protection service or security and stability service [2].

At the same time, the network randomly deploys the initial services according to the existing network rules, and generates the initial channel pressure value and link delay. The delay of each link is generated randomly from 0 to 1 ms, and the delay threshold is set to 10 ms. 75% of the link channel pressure values are randomly generated from 0 to 90, and 25% of the link channel pressure values are randomly generated from 90 to 250. After the algorithm is executed, the increment of global channel pressure is recorded [3].

When the number of services to be configured is low, the optimization performance of the two algorithms is similar [3, 4]. However, when the number of services to be configured and the number of network nodes gradually increase, the optimization effect of Dijkstra algorithm and genetic algorithm deteriorates rapidly, and the performance is worse than that of genetic algorithm, which indicates that this algorithm has better optimization performance than Dijkstra algorithm in the environment of multi node and multi service. The poor performance of genetic algorithm as an intelligent optimization algorithm is due to the weakening of global search ability by genome coding mode, resulting in the decline of optimization performance after the expansion of network scale. At the same time, when the number of network nodes is low and the network size is small, the performance of the two algorithms is similar. When the network scale expands, the performance of genetic algorithm is better than Dijkstra algorithm.

There are many traditional network vulnerability analysis technologies, which vary with specific problems, and there is no unified standard. Among them, the representative analysis methods are graph theory based analysis method, attack graph based analysis method, network survivability analysis method, probability and risk theory based analysis method. Although these methods have corresponding application scenarios and advantages, they also have some disadvantages. For example, the graph theory based analysis method is suitable for simple network structure scenarios, which can measure the nature of network vulnerability from a macro perspective, but it ignores the vulnerability in detail, and does not get an intuitive quantitative result. With the development of smart grid, the scale of smart power communication network is also expanding, the relationship between networks is more complex, and gradually evolved into a large-scale complex network. The traditional network vulnerability analysis technology is more effective and feasible in the analysis of simple network, but it is not suitable for the vulnerability analysis of complex network [4].

The key nodes and links of Gangluo refer to the key nodes and links which will cause more loss to the network performance than other nodes and links when they fail due to disasters. In other words, the importance of network nodes and links to maintain network performance is defined according to the impact of their failure on network performance. The greater the impact of nodes and links, that is, the more vulnerable the network, the more critical the corresponding. This kind of network vulnerability research 4546 * 448 is to identify the key nodes and key links of the network under the scenario of single node and single link failure caused by random failure. There are two identification methods. One is to investigate the importance of network nodes and links according to the local vulnerability index liu9 in the statistical characteristics of complex networks. Although this method is simple and feasible, it can only evaluate the importance of nodes and links relative to other nodes and links in the network topology. Secondly, the attack vulnerability method is adopted, that is, some components (nodes, links) are selectively broken down, and the importance of these components for maintaining network performance is quantified by comparing the network performance loss caused by them. The precondition of using this method is to have a global vulnerability index that can accurately quantify network performance [3, 4].

4   Concluding Remarks

At present, the research of network vulnerability is mainly divided into three aspects: first, the global vulnerability of the network is studied. This research focuses on the network model abstracted from the real network (ER model, SM model, etc.) to study the network topology properties. This leads to the lack of pertinence in the related research work, and the research results can only reflect the topological properties of power communication network as a general network. Second, the network vulnerability research based on the identification of key nodes and key links, most of which use attack vulnerability method to identify key nodes and links. Once these nodes and links fail, the network performance will be greatly reduced. But the premise of attack vulnerability method is to have a global vulnerability index which can accurately quantify the network performance. Thirdly, network vulnerability research based on key area identification, which fully considers the geographic location information of the network, identifies the key area of the network with the help of the minimum coverage circle theory and attack vulnerability method, but the key area found by using the minimum coverage circle theory only considers the role of nodes. In addition, this is a relatively new field of network vulnerability research, Many factors are not considered, such as the uncertainty of the network. This paper also uses the attack vulnerability method to study the network vulnerability based on the identification of key areas. On the basis of the research on the power communication network system, it points out that the business is the essential feature which is different from the traditional communication network. Considering the business of power system, this paper proposes the vulnerability research method of key area identification network based on joint node and link and the vulnerability research method of key area identification network based on uncertain network, and makes theoretical analysis and simulation verification of these two methods.

References

1. Sun, Y.: Male, doctor, Institute of electric power science, Yunnan Power Grid Co., Ltd. his main research interests are power system communication, distribution network communication and automation (1987)
2. Hu, J.: Male, master, senior engineer, Yunnan power dispatching and control center. His main research direction is power system communication and power communication operation management (1966)
3. Zhao, P., Duan, H., Zheng, S., Zhang, Q., Ji J., Wu, C.: Multi integration coordination planning of distribution network based on improved genetic algorithm. Zhejiang Electr. Power **39**, (12), 44–49 (2020)
4. Dong, Y., et al.: Research on intelligent operation and maintenance system of power communication network based on Internet of things. Mach. Electron. **38**(12), 51–54+59 (2020)

Modeling and Control Algorithm of Automobile Electronic Hydraulic Braking System

Hailiang Liu[✉]

Weifang Engineering Vocational College, Qingzhou 262500, Shandong, China

Abstract. Research on Modeling and control algorithm of automobile electric hydraulic braking system nowadays, people's life is inseparable from the automobile, which brings a lot of convenience at the same time, traffic safety problems also follow. Since the car entered people's lives, in order to prevent traffic accidents, the research on automobile safety technology has never stopped. With the increase of vehicle speed, the research on automobile active safety is more important. The vehicle braking system is the main actuator of active safety control system, such as anti lock braking system, stability control system and drive anti-skid system. Therefore, the vehicle braking system has been paid attention to by people, and has become a hot spot of vehicle active safety research in recent years.

Keywords: HB system · Vehicle stability · Wheel cylinder pressure · Bang bang fuzzy PI control · Predictive control

1 Introduction

With the development of automotive electronic technology and automotive control technology, X-by-wire, which was once used in aerospace field, is introduced into automotive driving. X-by-wie changes the traditional hydraulic or pneumatic driving mode to the electric driving mode, which improves the controllability and response speed, and is easy to realize the integrated control of chassis, which makes it the trend of future brake system development.

According to the different ways of providing wheel brake pressure, the brake by wire system is divided into electronic hydraulic brake (EHB) and electronic mechanical brake.

The two types of EMB (electronic mechanical brake) systems no longer use brake fluid and hydraulic components. The brake signal is transmitted by wires, and the braking torque is realized by the motor braking installed on four wheels respectively. Therefore, the 12 V voltage of existing vehicles can not meet the requirements, and the 42 V voltage system needs to be established. EHB system is a transitional product between traditional braking system and EMB system. It retains the original hydraulic braking part [1]. At the same time, the 12 V on-board power supply can also meet the demand. Therefore, the existing vehicle braking system has less changes and lower cost, so it has become a research hotspot of major automobile manufacturers and

© The Author(s), under exclusive license to Springer Nature Singapore Pte Ltd. 2022
J. C. Hung et al. (Eds.): FC 2021, LNEE 827, pp. 1007–1011, 2022.
https://doi.org/10.1007/978-981-16-8052-6_133

research institutions. However, due to the monopoly of EHB technology in foreign countries, the research results of EHB technology in China are relatively less, so the research on EHB system is of great significance to the development of China's automobile industry.

2 Establishment of EHB Hydraulic System Model

2.1 Structure and Working Principle of Automobile EHB System

EHB system, as a transitional product between traditional braking system and EMB system, cancels the vacuum assist unit and related hydraulic components in structure, and the brake master cylinder and brake wheel cylinder are no longer directly connected. The establishment of brake pressure is after the driver's braking intention is obtained through the electronic pedal, The electronic control unit (ECU) analyzes, calculates and sends the control signal to the control unit. This not only makes the system more compact and more modular, but also makes the braking response time smaller, has greater freedom in the four-wheel braking pressure distribution, and improves the braking efficiency of the vehicle. The structure and working principle of EHB system are introduced below.

The pressure control unit of automobile EHB system is generally placed at the front end of the engine he EHB system is mainly composed of brake operating unit (bou), hydraulic control unit (HCU) and sensor detection unit (SDU).

2.2 Mathematical Model of Hydraulic Components in EHB System

For the structure diagram of high-speed solenoid valve, by providing a certain driving voltage for the high-speed solenoid valve, when the solenoid valve coil is powered on, the corresponding electromagnetic suction will be generated, which will drive the armature to overcome the spring force, brake fluid resistance and friction resistance to move upward, so as to open the valve; when the solenoid valve coil is powered off, it will drive the armature to move downward under the action of the return spring [2]. Through the analysis of the working principle of high-speed solenoid valve in EHB system, it is divided into three subsystems as shown in Fig. 1: circuit system, magnetic circuit system and mechanical system.

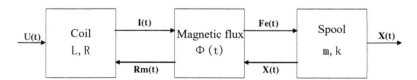

Fig. 1. Coupling relationship of high speed solenoid valve

(1) The circuit system can be obtained from Kirchhoff's law of voltage:

$$U = iR + \frac{d\varphi}{dt} = iR + \frac{d(N\phi)}{dt} \tag{1}$$

Where, u is the driving voltage of the solenoid valve; I is the current in the coil; R is the circuit resistance of the solenoid valve; Φ is the flux and flux in the magnetic circuit; n is the coil turns of the solenoid valve.

(2) The magnetic circuit system can be obtained from Kirchhoff's law of magnetic circuit.

$$N_i = \phi R_m \tag{2}$$

The total magnetic resistance R of magnetic circuit consists of the magnetic resistance of ferromagnetic material, the magnetic resistance of working air gap and non working air gap of solenoid valve. Because the magnetic field intensity of solenoid electromagnet is almost concentrated in the inside of solenoid, it is considered that the magnetic field intensity of the whole magnetic circuit is uniform in order to conveniently calculate and ignore the magnetic resistance of non working air gap and yoke.

3 Research on Simulation Experiment of Vehicle Stability Control

3.1 Co Simulation Environment of CarSim and Simulink

Cardin provides users with a parameterized interface, which defines the type of simulation by setting the vehicle parameters, road parameters and simulation scene. The vehicle can be controlled by open-loop or closed-loop through the driver model. Simulink is a software package of MATLAB, which is generally used for modeling, simulation and analysis of dynamic system. In order to make full use of the advantages of carin and Simulink software in all aspects, this paper selects the car IM and Simulink co simulation to build the vehicle stability control system, in which the car IM is used to build the vehicle dynamics model, and the Simulink is used to build the control system model.

The schematic diagram of the interface between CarSim and Simulink, the vehicle model of CarSim software will generate a DLL file, and Simulink will regard the DLL file as an s function, select the co simulation with Simulink model in CarSim interface, and select the input and output parameters of car SIM vehicle dynamics model [3]. In Simulink, the S-function model representing cars im vehicle dynamics model is connected with the built ideal two degree of freedom vehicle model, stability analysis module, stability control module and EHB system module to build the whole stability control system.

3.2 Building Diagram of Simulation Model

It is a simulation model of vehicle stability control system, including cars im vehicle dynamics model, ideal 2-DOF vehicle simulation model, stability analysis module, stability control module and EHB system module. In this paper, cars IM software is used to build the vehicle model. The stability analysis module can use the steering wheel angular rate, vehicle speed and road adhesion coefficient of the vehicle to calculate the stability threshold of the combination of vehicle yaw rate and mass center sideslip angle error by fuzzy control method, and then compare with the actual error combination of vehicle yaw rate and mass center sideslip angle, and output the stability signal to the stability control module. If the actual error combination is less than the stability threshold, the vehicle is in a stable state, and the stability analysis module outputs the stability signal 0; if the actual error combination is greater than the stability threshold, the stability analysis module outputs the stability signal 1. In this chapter, under the simulation environment of Simulink and CarSim, the vehicle is simulated under three working conditions: sharp turn, emergency avoidance and closed-loop control of pedestrian vehicle road, and each working condition is simulated with rain road and dry road [4].Through the analysis and comparison of the experimental data, it can be seen that the vehicle with stability control can improve the stability of the vehicle, and the vehicle stability control system based on predictive control proposed in this paper has better control effect on the vehicle stability.

4 Concluding Remarks

(1) The construction of EHB hydraulic system model introduces the structure and working principle of EHB system, discusses the theoretical model of high-speed switch solenoid valve, high-pressure accumulator, brake wheel cylinder and brake hydraulic pipeline in EHB hydraulic system, and establishes its mathematical model. On this basis, the mathematical model of pressure increasing and pressure reducing process of BB hydraulic system is established, The simulation model of EHB hydraulic system is established by matla B/Simulink software.

(2) Research on the control algorithm of wheel cylinder braking pressure firstly, the bang bang control and fuzzy PI control are introduced respectively, and the corresponding controllers are designed. Because the bang bang control and fuzzy PI control have their own advantages and disadvantages, this paper combines their advantages and designs the bang bang fuzzy PI controller to control the wheel cylinder braking pressure. From the simulation results, it can be seen that the Bangbang FUZZY P control has not only the advantages of fast response of bang bang control, but also the advantages of high precision and good stability of fuzzy P control. Its control effect is obviously better than that of a single controller, which provides a strong guarantee for the realization of brake pressure tracking control.

5   Prospect and Prospect

Due to the limitation of time, energy and experimental conditions, there are still some deficiencies in the research of modeling and control algorithm of electro-hydraulic braking system.

(1)   As BHb system is a complex nonlinear system, this paper simplifies the EHB system model and ignores the influence of hydraulic pump and brake pipe on the system and the instantaneous impact of brake fluid when the solenoid valve is switched on and off. In order to make the model closer to the real system, we need to further improve the EHB system model.

(2)   In this paper, fuzzy control is used in the control algorithm, and fuzzy control is based on operating experience. Due to the inaccuracy of experience, there will be some shortcomings in the process of making fuzzy control rules. In order to make the control effect better, we need to accumulate the corresponding experience knowledge in long-term practice, and then improve the fuzzy control rules.

References

1. Zhang, Z.: Research on automobile ESP control strategy. Changan University, Xi'an (2011)
2. Wang, Y.: Simulation and experimental study on dynamic characteristics of automotive EHB hydraulic system. Nanjing University of Aeronautics and Astronautics, Nanjing (2010)
3. Zong, C., Liu, K.: Development of automotive drive by wire technology [Ding]. Autom. Technol. **3**, 1–4 (2006)
4. Chen, Z.: Current situation and development trend of automotive chassis control technology. Autom. Eng. (2), 105–113 (2006)

Partner Selection of Collaborative Product Design Chain Based on Ant Algorithm

YanPing Liu[1(✉)] and Xie Jing[2]

[1] Jiujiang Gongqing College, Jiujiang 332020, Jiangxi, China
[2] College of Communist Youth, Nanchang University, Jiujiang, Jiangxi, China

Abstract. The optimized design chain can shorten the time to market and gain competitive advantage. Partner selection is an important link in the construction of design chain. The solution of partner selection in design chain is a combinatorial optimization problem. The combination of ant colony algorithm and genetic algorithm is used to optimize partner selection. The fast global search ability of genetic algorithm is used to generate pheromone distribution, and ant colony algorithm is used to seek precise solution and complement each other. The satisfactory results are obtained.

Keywords: Partner selection · Ant algorithm · Genetic algorithm · Collaborative product design chain

1 Introduction

With competitiveness and cost of products on the market [1]. A mature and reliable supply chain can significantly reduce the cost of products, but an optimized design chain can shorten the time of products on the market, so as to obtain the competitive advantage of the market and customers.

Collaborative organizations (manufacturers, suppliers and customers) cooperate provide services to meet the needs of users. The construction of collaborative product design chain is the basis of design chain management, and the establishment and selection of design partnership is an important part of design chain. The selection of appropriate design partners is the key to the success of design chain construction.

2 An Overview of Partner Selection in Collaborative Product Design Chain

2.1 Evaluation Index of Collaborative Product Design Chain Partners

Collaborative equal and mutually beneficial and information sharing network organization based on innovation and defining their respective tasks in the form of contract. The selection of partners generally.

The evaluation index of partner selection in collaborative product design chain includes the time required to complete the design task, the cost to complete the design task, the quality of product design and the comprehensive ability of candidate partners

[2]. The specific comprehensive capabilities include technology (technical require-ments, innovation history and intellectual property rights), R & D capability, process compatibility (similarity of R & D process, automation of R & D process and compatibility of design tools), organizational structure adjustability.

2.2 Partner Selection Method of Collaborative Product Design Chain

The partner selection problem has been deeply discussed, but most of them focus on the, artificial neural network (ANN), genetic (GA), etc. But the former methods are subjective, which generates the, but the genetic algorithm can not effectively use the feedback information. When the solution reaches a certain stage, it often performs.

Ant some defects, such as difficulty in determining pheromone at the initial stage and speed of solution.161 in order to overcome these problems, this paper adopts the method of combining genetic algorithm and ant algorithm to select design partners, that is, genetic algorithm is used to generate pheromone distribution, ant algorithm is used to obtain precise solution, Compared with the traditional leech mosquito algorithm, the computational complexity is relatively small.

3 Partner Selection of Design Chain Based on Combination of Genetic Algorithm and Ant Algorithm

3.1 Problem Description

As for the selection of design partners in the design chain, there is a group of design tasks with certain time-series dependence. partner enterprises the best (global optimal), The design task is completed in the best state. Specifically, in the development of new products, R & D tasks are divided into n subtasks, $Task = \{r_1, r_2, \ldots \ldots . r_n\}$. Each subtask is represented by a node in the design chain network. The design of subtasks is outsourced to suppliers [3]. There are m candidate manufacturers in each node, and the appropriate partners are selected from m candidate manufacturers, Finally, the best combination of partners among n nodes is selected by integrating various factors.

Let the set of all candidate partners be E, then:

$$E = \{e_{ij}|j \in [1, m_j]\}, i \in [1, n] \tag{1}$$

Namely:

$$\min B = W_1 T + W_2 C + W_3 Q + W_4 S \tag{2}$$

The above problem is a multi-objective optimization problem, in which W_k is the weight and $\sum_{k=1}^{n} w_k = 1$ is taken. If the sub tasks in the design chain are executed in sequence, then

$$T = \sum_{i=1}^{n} \sum_{j=1}^{m_j} t(r_i, r_{ij}) u_{ij} \tag{3}$$

3.2   Partner Selection of Design Chain Based on Combination of Genetic Algorithm and Ant Algorithm

Each candidate partner is regarded as a path for the ant to climb, so problem is how to select a group of paths according to the order of tasks, so that the shortest group of paths for the ant to climb.

(1) Genetic algorithm rules solving the optimal chromosome coding problem. Each bit in the coding string represents the status of a candidate partner. $E_{ij} = 1$ indicates that the first candidate partner of the task is selected, and $E = 0$ indicates that the candidate partner is not selected.

(2) Rules of ant mosquito algorithm
    Every time an ant climbs a path, it means that it selects a partner and leaves a local pheromone on the path. When an ant climbs the whole path, it means that it selects a group of partners and leaves a global pheromone on the whole path. All mant ants complete the whole process. After this cycle, the path with the highest pheromone concentration is the optimal partner combination. and leave a higher concentration pheromone on the path.

4   Application Examples

Matlab 6.5 is used for simulation. Due to the limitation of space, the detailed process is brief. After calculation, the optimized design chain partner combination is (E_{12}, E_{21}, E_{34}, E_{42}). In this example, different values of mant are compared and analyzed. It is found that the value of man has an impact on the accuracy of the calculation results, but the impact is smaller than that of ant algorithm alone [4]. The results may be local optimization results, and the larger the man value is conducive to get the global optimal solution, but the amount of calculation will also increase; using genetic algorithm to generate pheromone initial value can improve the efficiency of ant algorithm, at the same time, because the method of randomly generating population is used in the genetic algorithm stage, it avoids trapping λ local optimal solution and improves the solution accuracy.

In this example, the different values of mant are compared and analyzed. It is found that the value of mant has an impact on the accuracy of the calculation results, but the impact is smaller than that of ant algorithm alone. When ant algorithm is used alone, if the mant value is small, the result may be a local optimization result. When mant value is large, it is conducive to obtain the global optimal solution, but the amount of calculation will also increase; Using genetic algorithm to generate pheromone initial value can improve the efficiency of ant algorithm. At the same time, using the method of randomly generating population in the genetic algorithm stage can avoid falling into the local optimal solution and improve the solution accuracy.

5 Concluding Remarks

Regional vulnerability is studied in order to find the key areas in the network. As the network vulnerability research based on key area identification is a relatively new research field, on the basis of the research content of this paper, there are still some areas worthy of in-depth study. This section summarizes them as a reference for the follow-up research. The vulnerability analysis of critical area identification network under dynamic uncertain fault model is studied. In this project, the fault area is modeled as a statically determined circle, and the nodes and links covered by the fault area fail with probability 1. However, in the actual scenario, the failure probability of nodes and links is related to the degree of disaster, so it is more valuable to establish a dynamic uncertain failure model to determine the failure probability according to the degree of disaster. Network vulnerability analysis considering fault propagation is studied. The loss of network performance caused by disaster can be divided into two kinds: direct loss and later loss caused by fault propagation. As the power communication network is a complex network composed of multi-level and multi system, regional faults will propagate in the network according to the dynamic behavior of the network, resulting in the expansion of network performance loss. Considering the fault propagation can more truly and accurately evaluate the vulnerability of the network.

Acknowledgements. A Study on the Strategy of Construction of Yongqing University City and Creative Industry Park, A Study on the Operating Mechanism of "Five-in-one" Creative Industry Park—Taking Yongqing University City as an Example.

References

1. Li, Q., Xu, X.: An adaptive genetic algorithm for partner selection in dynamic alliance. High Tech Commun. (10), 66–69 (2001)
2. Li, M., Xu, B., Kou, J.: Combination of genetic algorithm and neural network. Theory Pract. Syst. Eng. **19**(2), 65–69 (1999)
3. Wang, X., Cao, L.: Genetic Algorithm Theory, Application and Software Implementation. Xi'an Jiaotong University Press, Xi'an (2002)
4. Zhou, M., Sun, S.: Principle and Application of Genetic Algorithm. National Defense Industry Press, Beijing (1999)

Design and Implementation of Virtual Experiment Environment for Communication System Based on Internet

Jing Xie[1(✉)] and YanPing Liu[2]

[1] Gongqing College of Nanchang University, Jiujiang 332020, Jiangxi, China
[2] Gongqing College of Nanchang University, Jiujiang, Jiangxi, China

Abstract. A virtual experiment environment of communication system based on Internet is designed and implemented by using the development model of virtual experiment system based on CORBA technology. In the process of realizing the virtual experiment loop, the client uses the technology of embedding Java applet in the browser, and the experimental equipment is developed and called in the form of Java beans components. The server uses matrix laboratory (matlab) as the computing background, The system also uses extensible markup language (ML) to save the configuration information, C++ to call the distributed processing of MATLAB computing engine CORBA, and the design of general parts. Compared with the existing virtual experiment system, the experimental environment not only has a great improvement in versatility, autonomy and reusability, but also has good interactivity. At the same time, the development speed of components in the experimental environment is also greatly improved. The practical application results show that the virtual experiment environment can provide a good interactive experiment platform for users, and users can use the virtual experiment equipment to construct various experiments.

Keywords: Virtual experiment environment · Common object request broker architecture · Matrix lab · JavaBeans · Etc

1 Introduction

Virtual experimental environment is the key factor to improve the quality of modern distance teaching. Virtual experiment environment provides a new way for students to carry out experiments. Students can simulate all kinds of experiments on the computer through the network. At present, many researchers at home and abroad have carried out the research and construction of virtual experiment system. The research results show that for most of the virtual experiment environment, they can only transfer information between the same object in the system implementation, and can not realize the communication between different objects, which causes the system structure is only suitable for the development of a certain kind of virtual experiment environment, and the development workload is large, and can not make full use of the existing third-party software [1]. Based on the Common Object Request Broker Architecture (CORBA) technology, the development architecture of online virtual experiment system can

J. C. Hung et al. (Eds.): FC 2021, LNEE 827, pp. 1016–1020, 2022.
https://doi.org/10.1007/978-981-16-8052-6_135

effectively integrate the third-party software written in Java, C++ and other languages, and has good scalability, platform independence and software reusability. Therefore, the author establishes the virtual experiment environment of communication system based on Internet.

2 System Architecture of Virtual Experiment Environment

The virtual experiment environment adopts the development model of online virtual experiment system based on CORBA technology. The model takes Java apple as the client and CORBA as the intermediate communication bridge. It can integrate Java beans components, Matrix Laboratory (matlab) and COM/DCOM as the computing background, greatly improving the development efficiency of virtual experiment environment. The system structure includes client and server, as shown in Fig. 1. The client uses Java applet embedded in the browser, which makes the client of virtual experiment environment have the characteristics of platform independence and security of Java language [2]. The server mainly includes CORBA service components (servants) written by MATLAB computing engine, dynamic link library generated by compiling matlab source files, comdcom and Java beans components provided by the third party.

2.1 Client Software Design

The design of client software includes the realization of Java applet and virtual device. The main function of the applet is to realize the user operation interface. The virtual device is realized in the form of lava beans. Each java bean completes the service object acquisition and method call. The client interface mainly includes: device bar, device attribute edit bar and experimental operation window. The device bar displays all the actual devices provided by the virtual experimental environment for users; the device attribute edit bar is used to set the parameters of the selected devices: in the experimental operation window, the user can set the selected experimental devices, and connect each device according to the experimental requirements to form an experimental flow chart. The client interface is shown in Fig. 1.

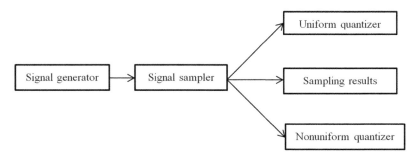

Fig. 1. Client interface of virtual experiment environment

2.2   Design of Server Software

The software design of server mainly includes the main program entrance, the realization of object method and the call of MATLAB. The service main program is responsible for initializing the object request broker (ORB), creating a portable object adapter (POA) with appropriate policies to manage service objects, and then generating service objects according to the definition in the CORBA interface definition language DL file, common principle ID, and waiting for receiving calls. The implementation of object methods is based on inheriting the base class generated when compiling the common principle. ID file, Write C++ code to implement the defined method. In order to consider the long-term fairness of users, Eq. (1) is modified to combine the current request rate with the previous average transmission rate, so as to consider the long-term fairness of users:

$$P_i(t) = \frac{r_i(t)r_i(t)}{t_i(t)} \tag{1}$$

$$r_i(t) = (1 - \beta)r_i(t - 1) + \beta r_i(t - 1) \tag{2}$$

3   Running Mode of Virtual Experiment Environment Components

The virtual experiment environment provides abundant algorithm components related to communication system, It includes signal generator, random signal generator, signal sampler, uniform quantizer, non-uniform quantizer, PCM encoder, PCM decoder, DM encoder, DM code evaluator, digital modulator, digital baseband signal code converter, spectrum analyzer, analog oscilloscope, digital oscilloscope, etc., The general process of analog-to-digital (AD) conversion of analog signal with information is sampling, quantization and coding. Here, we will take one of the key steps as an example to introduce the operation mode of components, assuming that the uniform quantization method is adopted. In the experiment process of this communication system, the virtual experimental equipment used for quantization is the component uniform quantizer provided in the simulation experimental environment. The following are introduced from the client and server respectively.

3.1   Client

Java beans, which represents uniform quantizer, is defined as a uniform quant class in virtual experiment environment [3]. The input and output of iform quantify class are defined as double precision array format. The data that need to be converted by ad is less than the sampler, quantizer and encoder in turn. Assuming that the sampled sequence is x, the sampled sequence x is saved in an array, and the sampled value x is sent to the quantizer for quantization through the interface between the sampler and the quantizer. Similarly, the processed quantized sequence is saved in an array y, and Y is sent to the encoder for coding through the output of the quantizer.

3.2   Server Side

Using mquant iflmp class to implement the method u defined in the above interface, the server first obtains the parameters from the client, including the sampling sequence x, the maximum signal amplitude dat am and the quantization level number quantify, and converts the data stored in the array or single variable into the data format recognized in Ma Π lab.

4   Key Technology

4.1   Using XML to Save Configuration Information

The platform of virtual experiment environment provides many algorithm components. The related information of these components needs to be read and written frequently, such as the name, label, attribute and description information of the components, which requires finding an appropriate method to save the information. Because the structure and label of XML file are self defined and semantic, it is convenient to write, read and maintain the file. The structure of the XML file given below illustrates the information saving method of the component digital oscilloscope.

4.2   Calling Matlab Calculation Engine with C++

Matlab provides abundant functions in the toolbox of signal processing and communication computer. In other environments, if you can call the files or functions in Matlab toolbox, the realization of some algorithms will be greatly accelerated, and its reliability will also be improved. In this virtual experimental environment, C++ calls matlab to calculate the Citation Method to realize the direct call of MATLAB functions or M files. Matlab engine adopts client and server computing mode. In the implementation of the system, the C++ language program in VC is used as a home computer. It transmits commands and data information to the Matlab engine, receives data information from the Matlab engine, and interacts with the client through the following functions: eng open, eng get array, eng putarray, eng EVA string, eng output buffer, enclose. The method of using C++ to call the matlab calculation engine is simple and efficient, The accuracy can also meet the requirements of the system, so the matlab application program interface is realized in this way.

4.3   Using CORBA Technology to Realize Distributed Processing System

CORBA defines an open distributed object bus (ORB) standard, which allows interoperation between distributed object applications, No matter what language these applications are written in or where they reside [4]. The inherent cross language and cross platform characteristics of CORBA make it an appropriate choice to build a simulation experimental environment system. CORBA is used to realize the communication among JAA, MATLAB and C++.

5 Conclusion

By using the object-oriented programming method and component technology, a powerful and rich virtual experiment environment based on Internet is realized. The client of the virtual experiment environment is implemented in Java language, which has the characteristics of platform independent, safe and robust. The devices in the experiment environment are developed in the form of components, which improves the development efficiency, The software reuse server is realized, and MATLAB is used as the computing background to speed up the implementation of component algorithm and improve the reliability of the system. The virtual experimental environment is used in distance teaching, and has a good application prospect in digital signal processing, basic principles of communication and basic technology learning and research.

Acknowledgements. A Study on the Strategy of Construction of Yongqing University City and Creative Industry Park, A Study on the Operating Mechanism of "Five-in-one" Creative Industry Park—Taking Yongqing University City as an Example.

References

1. Wang, J., Pei, H., Chen, S.: Architecture design of virtual laboratory platform based on Internet. J. Central South Univ. Technol. Nat. Sci. Ed. **33**(5), 530533 (2003)
2. Jiao, B., Yin, C.: Research and implementation of web oriented CAD interoperability technology based on CORBA/IIOP. Small Microcomput. Syst. **22**(2), 219, 221 (2001)
3. Qi, Y., Ma, L., Qi, X., et al.: Research on domain framework based on CORBA business component. Minicomput. Syst. **22**(10), 1213, 1215 (2001)
4. Zhang, Y., Luo, X., Yu, B.: Distributed discrete event simulation environment based on component. J. Syst. Simul. **14**(8), 10191021 (2002)

Application of 3D Geological Modeling in Geological Exploration

Fengzhi Zang[✉]

Daqing Oilfield Limited Company No.7 Oil Production Company,
Daqing 163000, China

Abstract. Geological mapping is very complex and difficult, and there are usually three-dimensional entities. Due to the continuous deepening of scientific research, three-dimensional geological simulation has attracted great attention of the global scientific community. At present, a variety of three-dimensional geological simulation software is being developed, which has been widely used. It is also applied in various fields, such as geology, mineral resources, hydrology, environmental science and so on. This paper systematically analyzes the three-dimensional characteristics of Geology from the aspects of geological structure, beauty and distribution, and summarizes the development status of three-dimensional geological simulation software in the fields of geological structure, geological engineering, mining exploration and physics at home and abroad.

Keywords: Three dimensional geological modeling technology · Geological mapping · Application

1   Theoretical Basis of 3D Geological Modeling Technology

The shape of geological structure has two basic forms: "number" (structural elements, such as occurrence and scale) and "shape" (spatial form). Complex geological structure can always be modeled by point, latitude and surface. 3D geological modeling technology can collect and analyze many elements, especially 3D shape interpretation and analysis in spatial coordinate system. As a method of managing three-dimensional geological phenomena, the main geological phenomena should be considered as strata, defects and ore bodies. Strata are strata or rocks in a specific geological age. The interconnected strata are located between interfaces, and can be located in front of flat stairs with trend, slope and depth data. However, the structural interface is not a real stable surface, but an incomplete surface due to the trend and trend change. In order to obtain the gradient, well data, well measurement data and vibration data are usually used to determine the surface area of wave defects [1]. Defects are similar to strata, which divide rock mass into upper and lower walls, but usually have a specific surface, a specific degree and a specific angle. The detection and description methods are consistent with the earth's surface. In order to determine the range of minerals, it must be determined by surface survey, underground excavation and geological sounding.

It is usually calculated by the inclinometer data of the borehole, and the borehole is placed according to certain rules (according to the vertical and horizontal sections),

J. C. Hung et al. (Eds.): FC 2021, LNEE 827, pp. 1021–1025, 2022.
https://doi.org/10.1007/978-981-16-8052-6_136

passing through the three-dimensional coordinates (XY, b) of the borehole and the top and bottom of the ore body. The digital surface model can be described by DM model. In fact, some existing GIS software also use this method. The data observed in the geological survey, including strata and ores, are very different in spatial distribution, with only a few different points in many cases. Due to the complexity of the situation, tardive motion distortion, damage and other factors, if there are discontinuities such as defects, the continuity and integrity of the data will be damaged, and the original distribution of the data will sometimes change. However, you can't carry out continuous exploration in the whole area, because it starts the whole body. For example, the development of defected structures is regular, such as joints and defects in plane and profile. The development of defected structures usually has square, band, conformal and zoning rules. Therefore, when approaching the surface in the direction of space, we must follow the geological rules with due consideration of the role of geological movement. For example, in the case of a defect, the effect of the defect on the lipid of the relevant layer.

2   Research Results of 3D Geological Modeling

2.1   Domestic Research Status

In China, the research on 3D geoscience imaging has begun, but many useful explorations have been carried out. Recently, the natural science foundation of China strongly supports the research on 3D geological modeling, including supporting "geological institutions such as 3D modeling and mapping research continue to raise complex funds" and oil storage geophysics, including 3D seismic mapping and image method, geospatial space-time information, dynamic and image modeling and their applications. In 1996, the Institute of Geology and Geophysics (Institute of Geology and Geophysics, Chinese Academy of Sciences) and Shengli Oilfield began to track the main project "Complex Geological Institutions" of the National Natural Science comprehensive plan. Eview geographic information system developed by China University of Geosciences can promote the management, processing, calculation and analysis of geoscience information, and support three-dimensional decision-making [2]. China Petroleum Exploration and Development Research Institute has developed a deep-sea resources development system. This system can be recovered through the development of 3D structure. During the Eighth Five Year Plan period, China developed 3D geological model software, although it is quite different from similar software abroad.

2.2   Key Technologies of the Algorithm

The latitude position of the spherical mesh on the model (not the true latitude) is the function of the spherical triangle coding, and the latitude position can be expressed as the row data. It should be noted that the edges of the sgog mesh are geodesic arcs, which do not coincide with the latitude arcs. The latitude band here only serves as the direction of filling sequence, not the grid scope in the strict sense. Filling is realized by coding, so the edge of sgog grid does not affect the implementation of the algorithm. For a given level of sgog subdivision, four spherical triangles (belonging to an upper parent triangle) are divided into two latitudes, which are respectively recorded as the 0'and 1' latitudes. Then, when the code of the sub subdivision unit of this level (the modified direction coding method is adopted here), the latitude band mark value of the current level is returned, index = 1; when the code of the sub subdivision unit of this level is 0, 2, 3, the latitude band mark value of the current level is returned, index = 0, that is:

$$index = \begin{cases} 0, code[i] = "0", "2", "3" \\ 1, code[i] = "1" \end{cases} \tag{1}$$

The latitude zone here is divided in ascending order from low latitude to high latitude according to the level of spherical subdivision. The code of low latitude zone is 10', and the code of high dimension zone is 1 (in the case of northern hemisphere). When the spherical triangle corresponding to the upper level is the upper triangle (that is, the upper level code is 1), the sub triangle coded as 1 is in the high latitude position (as shown in Fig. 1).

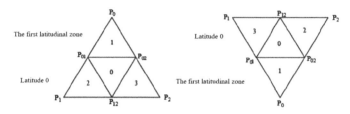

Fig. 1. Division of latitudinal zones of upper and lower triangles

By calculating the latitude band mark value of each level of sgog grid coding, a latitude band mark value sequence index with the same length (number of digits) as the grid coding can be obtained. For each digit of the sequence, the latitude band value of the sgog spherical coding can be obtained according to the following formula, namely:

$$N_{lat} = \sum_{i=0} index[i] \times 2^{n-i-1} \tag{2}$$

3   Problems Found Temporarily

3.1   Hardware Issues

The hardware of 3D geological modeling is not mature yet, but the geological software can not build complex ground based on the surface model of specific observation points or geological laws, and may produce other unpredictable maps. If the capture time is too early, it is impossible to capture large-scale geological and deep-sea data. Modeling quality depends on geological modeling experts, which not only affects the effectiveness of software, but also limits the improvement of modeling automation. Nowadays, in South Korea, where geological research is relatively advanced, scholars have not yet developed a relatively low-cost 3D Earth simulation software, because the software combines data management, information introduction, interactive process and geological analysis, which makes the cost of 3D imaging hardware expensive. There are hundreds of geological exploration and research departments in China [3]. If they do not develop GIS 3D geological software with independent copyright, it is unrealistic to rely too much on foreign software. We need our staff to carry out cooperative research, develop our own 3D imaging software of geological science, and carry out international research at the same time, so as to integrate into the world as soon as possible.

3.2   Disadvantages of 3D Geological Simulation

This kind of complex structure and spatial relationship is different because of many points, lack of continuity, and complex geological structure. In the field of Earth Science, there are many defects, such as geology, water wells, ores, tunnels and so on. Different experts can draw different conclusions according to different geological theories. Geological data are different in structure and non system, mainly including gravity data, geomagnetic data, atmospheric electric data, geothermal data, oil well data (Geological and geophysical information), seismic data and satellite movie. The coordinated development of crustal data, regional geological data (Geological control point data, collected data, rock thickness data, phantom data, etc.) and digital map makes the data classification, interpretation and processing more difficult [4].

4   Application of 3D Geological Modeling Technology

4.1   3D Spatial Data Model

In the in-depth study of global three-dimensional g | s data structure, the main three-dimensional g | s data structure is composed of three-dimensional grid (aray) and solid structure geometry (CSG). Network structure (ten) represents a three-dimensional vector data structure based on volume technique such as network (ten), which can accurately represent spatial entities and fully describe spatial topological relations. However, ten itself is difficult to construct data structure and more complex. Octree is a data model based on size description, which is an extension of three-dimensional solid model. In this model, protons are used as small units to describe space objects, which is a simple and easy to analyze structural model. The accuracy of this characteristic table is low, but higher accuracy will lead to a substantial increase in the number of data and a decrease in processing speed. The model shows the obvious defects between the site and the spatial entity statement. In the field of Geosciences, space objects are very complex, with irregular defects in three-dimensional description and display. Geology, borehole, ore and tunnel are very complex, and the vector structure is difficult to solve the internal inhomogeneity of objects.

4.2   Technology and Method of 3D Geological Simulation

At present, three-dimensional measurement simulation technology is summarized as follows: section modeling, surface modeling, celestial mass modeling, frame modeling, solid modeling and interface vision technology. Section method is a three-dimensional problem, but it is incomplete. The surface method is used in DTM (digital terrain model), but its disadvantage is that the information about the characteristics can not be expressed in the ground quality. The 3D model balances accuracy and information better. The line is a combination of surface and block method. Its advantage is to describe minerals in any way. Its disadvantage is to readjust and divide the surface when the control point code changes the geological interface. Radiation imaging technology is developed according to the scientific assumption of computing. A typical example of the application of external imaging technology in 3D ground simulation is GCAD software developed by Nacy University in France and 3D geological simulation in the United States. Including the application of 3D geological survey simulation technology in China, most scholars put forward the challenges of 3D geological simulation technology and physical imaging, and studied the key technologies of biological imaging method, algorithm display and optical model.

5   Conclusion

The wide application of 3D geological modeling technology in various fields has effectively improved the work efficiency and engineering quality. In recent years, China's 3D geological modeling software has developed rapidly and its functions are becoming more and more perfect. However, there are still many disadvantages in the application and development of the software, which requires software developers to make better use of 3D geological modeling technology and continuously improve the quality of products.

References

1. Deng, H.: Geological mapping modeling and analysis method based on 3S technology. Resour. Inf. Eng. **6**, 110–111 (2016)
2. Wang, Z.: Engineering Geology of Large Deposits in Deep Valley in Southwest China. Chengdu University of technology, Chengdu (2015)
3. Xu, Q.: Application of "3S" Technology in Glacier Mapping of Qinghai Tibet Plateau. Chengdu University of technology, Chengdu (2008)
4. Zhang, D., Wang, Z., Mao, Z.: Research progress of polylysine. Amino Acids Biol. Resour. **27**(2), 48–51 (2005)

Computer Management of University Library Based on Deep Learning

Cong Lin[✉]

Zaozhuang University, Zaozhuang 277160, China

Abstract. With the development of science and technology, the information technology is advancing. The preservation of library documents is not the only standard to evaluate the quality of library work. The continuous progress of science and technology and the development of education promote the library to effectively change the traditional working mode. Computer management has become an important guarantee for the automatic management of the library, It has a great impact on the traditional library management. This paper analyzes the problems existing in the implementation of computer management in the library, and puts forward some countermeasures for the optimization of computer management, so as to lay a good foundation for improving the effectiveness of library automation management.

Keywords: Library · Computer management · Problems · Countermeasures

1 Introduction

The use of computer in library management is not only conducive to improving the efficiency of library work, but also conducive to improving the effectiveness of library automation management. The work of literature statistics, reader statistics, literature retrieval, workload statistics and book borrowing and inquiry in the library need to spend a lot of time and energy, which leads to a huge waste of human and material resources [1]. The use of computers can simplify the working environment and improve the effectiveness of library management. For example, in the library book borrowing work, the traditional library management needs to spend a lot of time and energy on the classification of borrowing and returning readers, that is, it needs to be classified according to the Department, level and specialty, and then the library card and borrowing card are classified, but the workflow is very complex, which provides a great challenge to the staff. Using the computer to manage the library can save the manual steps, and use the bar code scanning method to automatically classify the book bar code and the reader bar code. Therefore, when the reader returns the book, as long as the reader gently scans the book bar code, it can do a good job of recording, and the actual management method is more convenient. Considering the efficiency of computer management, most libraries use computers for management, which is conducive to improving the service level and management level of the library. However, some problems gradually appear in the actual operation process. If they are not solved in time, the library will face development difficulties.

J. C. Hung et al. (Eds.): FC 2021, LNEE 827, pp. 1026–1030, 2022.
https://doi.org/10.1007/978-981-16-8052-6_137

2   Problems Existing in the Implementation of Computer Management in Libraries

2.1   There are Some Problems in Library Computer Technology

At present, the computer ability of library staff is relatively low, and they are prone to failure in the process of operation. In addition to the technical staff of the library, the staff of other departments are often lack of computer ability, and there are often operational errors in the process of computer operation, which is easy to make the library's management system either fail to run or crash. In the face of computer failure, the staff can not directly eliminate it due to their own ability limitations, It directly affects the library work. Due to the problems of the library's book management system itself and virus invasion, the book management data is easy to be lost, or the instability of computer power supply voltage or poor network contact directly affect the data integrity.

2.2   There are Some Problems in the Circulation of Library Computer Management

There are some problems in the circulation of library computer management.

The circulation department of the library is more responsible for the management of book borrowing. In the process of computer management, on the one hand, book acquisition and editing, book number printing, and book information data input all need to build a literature database. At the same time, it also needs to do a good job in reader information data input and barcode distribution, and build a reader information database. In the process of book borrowing, readers only need to scan the book bar code and the reader's reading card bar code, and improve the effectiveness of book borrowing management through the corresponding relationship between the two. In the process of readers' borrowing books, one of the more common problems is the problem of readers' impersonation. Due to the influence of personal friendship or colleague relationship, some readers borrow other people's library card to borrow books, and may not even inform them in time, which makes readers deny it when the library urges them to return books. There are also some readers who have lost their borrowing cards and have been replaced by others, but they have not reported the loss in time, which makes the phenomenon of borrowing books very serious. As an important mark of a book, a bar code corresponds to a book. However, the computer can only recognize the bar code, and some readers with relatively low comprehensive quality change the bar code of a book in order to occupy the book, that is, change the bar code of a book to other books. When returning the book, the management staff receives more readers, This makes it often pay attention to the book barcode scanning work, but did not pay attention to the problem of book loss or book replacement [2].

2.3   System Design

System design is the core of the whole software development process, and it is also the largest part of the workload, which is directly related to the effect of software design. Clear design of the actual needs of the software application, detailed design of each functional module in the system, complete design of a set of management information system.

2.4   Overall Design

First of all, the overall design goal of the library management information system should be made clear: it aims to realize an efficient automatic management information system which can improve the old management and office state of the library with the help of school LAN and computing equipment. It can not only help librarians in their work, but also improve the efficiency of serving readers, and play a positive role in the later management of the library [3].

Borrowing and returning management mainly includes four function modules, namely borrowing, returning, renewing and booking management. The structure of borrowing and returning management function module is shown in Fig. 1.

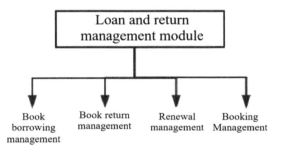

Fig. 1. Structure of loan and return management function module

2.5   Database Design of Management Information System

In the University Library Management Information Department to be developed in this paper, the main work content of background data of the system revolves around all kinds of information data related to books, such as information data of books, information data of borrowing amount, information data of readers, information data of overdue fine, etc. In view of the limited amount of data in university library, the centralized storage mode can be adopted [4].

The useful features of the image are extracted. The output formula of pooling characteristics of pooling layer can be expressed as follows:

$$x_j^l = f(\max(x_j^{l-1}) * w_j^l + b_j^l) \tag{1}$$

By arranging the two-dimensional feature map data output by pooling layer into one-dimensional feature vector for expression, and classifying the data after the output of full connection layer, the full connection output formula can be expressed as follows:

$$x^l = f(w^l x^{l-1} + b^l) \tag{2}$$

When the data training set is large, the convolutional neural network may have higher training accuracy, but when the test set is input to the trained neural network for testing, it will not achieve very good recognition accuracy, that is, the network model over fits the training set and cannot be well generalized to the test set.

2.6   Overdue Fines

In computer management, the problem of overdue fine often appears. According to the regulations of university libraries, overdue books will be fined or stopped. Usually, according to the settings, the reader can pay the fine in time after expiration. The staff can click "OK" after collection, or click "exit" when the reader does not have the fine, so as to "hang up" for the reader and pay the fine before the next borrowing. However, when the overdue time is long and the amount of fine is large, a few readers often raise objection and refuse to pay the fine.

There are many reasons why readers refuse to pay fines. First, readers' borrowing date is not recorded on their library card or the books they borrow. If readers do not deliberately remember their borrowing date and fail to check their returning date from the Internet in time, the phenomenon of overdue is inevitable. Second, with the extension of China's legal holidays, when the overdue records contain holidays, some readers will raise objections and ask for less or no fines. Third, limited by the funds for purchasing books, the number of copies of books purchased each time is limited, and some readers are afraid that it will be difficult to borrow the reference books borrowed for preparing for the examination or for curriculum design, graduation design and other reasons, so they prefer to choose the treatment of a fine for losing books when they have to. Fourth, readers can't return books on time because of business trip, off campus internship, tree planting, military training and other reasons, which often become reasons for readers to refuse to pay fines.

We think that the following measures should be taken to solve the above problems: in the last blank page of each book, paste the reminder table of book return date, and ask the readers to immediately cover the date of book return in the table. In addition, eye-catching signs can be set up on the cashier desk to remind readers of the return date of the borrowed books on that day. Second, in addition to the winter and summer holidays deducted from the period of borrowing books, we should also consider the legal holidays deducted from the period of borrowing books. Thirdly, it is suggested that when a powerful software company is researching and making library integrated management system software, can it refer to the form of "supermarket shopping voucher" and consider that when readers borrow books, they can print out the record form of borrowing and returning secretary at the same time, and design the reader's name or library card number, borrowing date, especially the date of returning books and the amount of fines, so as to reduce many people's unhappiness.

3   Conclusion

With the rapid development of science and technology information technology, modern technology is widely used in library management, which makes the library enter the stage of modern management. The use of computer management can improve the efficiency of library work, and the use of automatic processing can effectively improve the efficiency of library management. The computer management of the library involves many factors, so we need to pay attention to the library management, pay attention to the professional ethics education of the library staff, enhance the sense of responsibility of the library staff, formulate effective management system, and then strictly operate the computer operation process of the library, so as to avoid the loss caused by human reasons, At the same time, we also need to increase the investment in computer hardware and replace hardware equipment in time to provide guarantee for the normal operation of the library computer system.

References

1. Zaituna: Problems and countermeasures of jimusaer county library after implementing computer management. Western Region Library Forum (1) (2010)
2. Liu, X.: On the problems and countermeasures in the implementation of computer management in library cataloging work. J. Dezhou University, (2) (2001)
3. Xi, H.: Problems and countermeasures of circulation department after the implementation of computer management in library. Intelligence (12) (2009)
4. Zou, J.: Analysis of the problems and countermeasures in the computer circulation management system of University library. High. Archit. Educ. (1) (2002)

Parallel Route Design and Performance Analysis Based on Beidou Satellite Based Augmentation System

Huanchang Xiao[✉]

Number46, Section 4, Nanchang Road, Guanghan, Deyang, Sichuan, China

Abstract. GNSS plays an indispensable role in today's world's economic and military fields. The world's powerful countries have built their own GNSS systems one after another. According to the needs of civil aviation, the International Civil Aviation Organization (ICAO) puts forward the required navigation performance (RNP) of navigation system, which is reflected in four aspects: accuracy, integrity, continuity and availability. In order to meet the performance requirements of civil aviation, countries all over the world have developed their own satellite based augmentation system (SBAS) in order to provide navigation services with higher accuracy and better integrity for wide area users. In view of the increasingly mature Beidou satellite navigation system in China, it is necessary to research and develop our own satellite based augmentation system to provide higher quality navigation services for civil aviation users in China and its surrounding areas.

Keywords: Bidou · SBAS: Grid ionosphere · Integrity

1 Introduction

Since the 1950s, countries around the world headed by the former Soviet Union have been scrambling to develop and launch their own man-made earth satellites. Since then, a new era of human development in space science and technology has opened. The study of navigation and positioning by using satellites is of great strategic significance for countries around the world to improve their national defense and economic strength. Satellite navigation system can provide continuous and real-time navigation, positioning and monitoring services for all kinds of land, sea of navigation and positioning, accurate and continuous time service. Such advantages and characteristics make its applications benefit all fields of national economic development and social life, and it has become one of the most potential and fastest growing information industries in the world [1].

Many different types of satellite navigation systems are, which can send users time, space benchmark and real-time information related to positioning. GNSS provides important technical support for promoting national economic development and safeguarding national security, and is the embodiment of the comprehensive national strength of a modern country. As GNSS applications are related to many industries and

J. C. Hung et al. (Eds.): FC 2021, LNEE 827, pp. 1031–1035, 2022.
https://doi.org/10.1007/978-981-16-8052-6_138

closely related to the background of the communication industry, its derivatives and applications are widely used in many fields of the national economy and closely related to people's daily life, which can better provide services for people's daily life and improve people's quality of life.

2 Interpretation of RINEX Format Files and Calculation of Satellite Coordinates

2.1 RINEX Format

RINEX also known as independent exchange format, is widely used in GNSS measurement applications. It is an internationally recognized standard data format. This format stores the data in the form of text file, and the data formats of different types of receivers made by different manufacturers are the same, so it brings great convenience for GNSS data processing and sharing [2].

The origin of this standard is that the University of Berne in Switzerland proposed the concept of RINEX format when comprehensively dealing with a large-scale GPS joint measurement project. The project involves 60 different GPS receivers provided by four manufacturers. If the data format can be unified, it can provide a lot of convenience for the project. After development and continuous improvement, riiex format has been widely used in GNSS measurement applications. Different types of receivers made by different manufacturers can convert their own data format files into RINEX format files, and the data content can be directly read and processed by most software. Such normative documents bring great convenience for us to read and process various observation documents and navigation information.

2.2 Calculation of Satellite Space Position

In the process of Beidou navigation and positioning, users need to know the position of the satellite first, then they can use the satellite as a known point to determine the user's position. Therefore, solving the position of the satellite is a very critical and premise problem. According to the parameters needed, the position of the satellite in the orbital coordinate system can be accurately calculated.

The average angular velocity n of satellite motion is calculated:

$$n_0 = \sqrt{\mu/A^3} \tag{1}$$

According to the perturbation correction Δ n provided by the ephemeris, the average angular velocity n of the satellite motion after correction is calculated as:

$$n = n_0 + \Delta n \tag{2}$$

It is necessary to extrapolate the position of the satellite at time t according to the orbit parameters of the satellite at time t. Therefore, it is necessary to reduce the observation time to:

$$t_k = t - t_{oe} \tag{3}$$

Firstly, the paper explains the file format of RINEX, including the origin of file format, different updated versions and file types. This paper focuses on two formats: observation file and Beidou navigation message file. Finally, how to use the ephemeris parameters in the Beidou navigation message in the orbit is explained in detail, so as to prepare for the subsequent solution of the real.

3 Principle and Error Analysis of Beidou Satellite Based Augmentation System

The Beidou satellite four parts: space part, ground end, user end and data link. The space part refers to the space constellation of Beidou system, including GEO, MEO and IGSO. All satellites can send basic navigation information to users, while GEO satellite can also send wide area differential information, grid ionosphere information and integrity information of Beidou system to users. The ground part includes monitoring station, central processing station and upload injection station. The reference station uses a dual frequency receiver to receive the code and carrier phase of the frequency. At the same time, it also collects meteorological parameters, which are sent to the central processing station after preprocessing. The central processing station calculates the collected data to obtain the error correction information and integrity information, and sends them to the geostationary communication satellite through the upload and injection station. The main task of the client is to receive the navigation information of satellite and the differential and integrity information of geo broadcast at the same time, and use these information for positioning and integrity analysis. The data link includes the communication link between the reference station and the central processing station and the communication link between the satellite and the user. Among them, the reference station and the central processing station are connected by public data transmission network, telephone network or special communication mode; satellite communication mode is adopted between satellite and users, and the broadcast information needs to be edited according to a certain format[3].

3.1 Basic Principle of BDS

(1) All visible Beidou satellites are monitored by the reference stations distributed in all regions of the country, and the monitoring data is sent to the central processing station through the ground communication network;

(2) The central processing station uses the collected data to calculate the error correction information and integrity information. The error correction information includes satellite clock error, satellite ephemeris, vertical delay of ionospheric grid

point, and the integrity information includes user differential distance error UDRE, grid ionosphere vertical correction error gift and regional user distance accuracy rura;

(3) The information from the central processing station is encoded and sent to the GEO satellite through the upload and injection station, and the GEO satellite broadcasts to the user through the satellite communication link;

(4) According to the received integrity information, users can get the integrity status of Beidou satellite and satellite based augmentation system, and get accurate positioning and navigation parameters according to the received error correction information and Beidou satellite observation data.

3.2   Characteristics of Beidou Satellite Based Augmentation System

Different from other navigation augmentation systems, Beidou satellite based augmentation system has the following characteristics[4]:

(1) It covers a large area and can provide navigation services for civil aviation users in various flight stages such as route, terminal and approach;

(2) The satellite clock error, ephemeris error, ionospheric delay and tropospheric delay of Beidou satellite are calculated separately;

(3) The grid model is used to correct the ionospheric delay, and the user can obtain more accurate ionospheric delay and integrity parameters by using the correction value on the grid;

(4) The error correction information is broadcasted by GEO satellite, and the satellite integrity information is also transmitted in addition to the error correction information;

(5) GEO Satellite not only broadcast error and integrity information to users, but also broadcast basic navigation information, so that the satellite can also be used for ranging, which increases the number of satellites in the navigation constellation and improves the continuity and availability of the system.

The idea of grid ionospheric model is to imagine the ionosphere covering the earth as a uniform shell, and divide the ionospheric reference surface into grids according to certain rules. Different grid points correspond to different ionospheric delays. The user's ionospheric delay can be calculated from the ionospheric delays of these grid points according to a certain algorithm. The average height of the ionosphere from the ground is 375 km.

4   Conclusion

With satellite will become like the GPS of the United States and GLONASS of Russia. Compared with other navigation systems, global navigation satellite system (GNSS) has many advantages, such as global coverage, all-weather, high precision and so on, but it can not be used alone in civil aviation and other fields. The reason is that the accuracy, integrity, continuity and not fully meet the requirements of all aviation flight stages. Therefore, the enhancement system to improve GNSS performance arises at the historic moment.

At present, and are two augmentation systems based on differential technology to improve navigation performance. Although the precise approach for users, it is impossible to provide continuous navigation augmentation service for users in wide area because of its small scope of action and considering the economic cost and construction difficulty. Satellite based augmentation system can overcome this defect, it can provide civil aviation users with a wide range of positioning accuracy and integrity improvement with less hardware equipment and economic investment. Therefore, the satellite based augmentation system is sought after by aviation powers, and the development of China's independent satellite based augmentation system has become a development trend.

References

1. Zhao, L., Ding, J., Ma, X.: Principles and Applications of Satellite Navigation. Northwestern Polytechnic University Press, pp. 1–10 (2011)
2. Gang, X.: GPS Principle and Receiver Design, pp. 80–85. Electronic Industry Press, Beijing (2009)
3. Fei, N.: Research on the Theory and Method of GNS Integrity Enhancement. PLA University of Information Engineering, Zhengzhou (2008)
4. Cao, H.: Research on the theory and method of GNSS integrity monitoring. Chang'an University (2013)

Analysis and Comparison of Automatic Image Focusing Algorithms in Digital Image Processing

RuiWu Jia[1(✉)] and Jing Tang[2]

[1] School of Mathematics and Statistics, Zhaotong University,
Zhaotong 657000, China
ztjrw2003@tom.com
[2] Yongfeng Middle School, Zhaoyang District, Zhaotong 657000, China
ztxystxyjrw@tom.com

Abstract. This paper first briefly summarizes the imaging principle and focusing principle of digital image processing automatic image focusing, then expounds the digital image processing automatic image focusing evaluation function from the aspects of image information entropy function, gray gradient function, frequency domain evaluation function, other evaluation functions. Finally, the digital image processing window area selection and focus search algorithm are described from the aspects of depth of field and focal depth, algorithm selection, algorithm improvement direction and so on.

Keywords: Digital image processing · Auto-image focusing algorithm · Evaluation function

1 Introduction

Automatic image focusing technology is very important in digital image processing. If the level of automatic image focusing technology is low, the function of digital image processing system will also be greatly reduced. Only by improving the sensitivity of automatic image focusing technology, reducing its complexity and dispersion, can it really play the role of digital image processing system.

2 Principle of Automatic Image Focusing for Digital Image Processing

(i) Principle of imaging
   Although automatic image focusing technology is advanced, its imaging principle is basically the same as that of convex lens imaging. The formula of convex lens imaging principle is as follows:

$$\frac{1}{u} + \frac{1}{v} = \frac{1}{f} \tag{1}$$

J. C. Hung et al. (Eds.): FC 2021, LNEE 827, pp. 1036–1040, 2022.
https://doi.org/10.1007/978-981-16-8052-6_139

The meaning of each index in formula (1) is as follows: the meaning of u is the distance between convex lens and object; the meaning of v is the distance between convex lens and imaging plane; the meaning of f is the focal length of convex lens. According to the principle of convex lens imaging, the convex lens imaging model can be obtained as shown in Fig. 1:

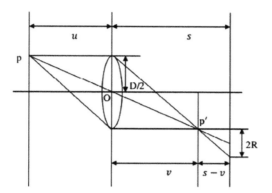

Fig. 1. Image model of convex lens

Figure 1 u, v, f, D, p, R refer to object distance, image distance, focus, convex lens diameter, object position, imaging radius, respectively. When the digital image processing system defocuss, the distance between the imaging and the convex lens will gradually decrease from s to v, object imaging will leave a fuzzy image on the image detector. The distance between the focus plane and the convex lens is s-v.. The distance between the focus plane and the convex lens is If the value of the s-v continues to increase, the image on the image detector will be more blurred. According to the similar triangles in Fig. 1, the imaging scaling factor formula can be obtained as follows:

$$q = \frac{2R}{D} = \frac{s-v}{v} = (\frac{1}{v} - \frac{1}{s}) \tag{2}$$

Formula (2) q refers to the imaging scaling factor. The following formula can be obtained from the convex lens imaging formula and the imaging scaling factor formula:

$$R = q\frac{D}{2} = S\frac{D}{2}(\frac{1}{f} - \frac{1}{u} - \frac{1}{s}) \tag{3}$$

The formula (3) shows that when the q > is 0 and s > v, the imaging surface is Rs > v 0, and when the Rs > vq > is 0 and Rs > vq > q > v, the imaging surface is 0, and the imaging surface is in front of the positive focus position. Therefore, digital image processing can realize auto-focusing according to formula (3) principle [1].

(ii) Principle of focusing

The development process of digital image processing is divided into two stages. The first stage mainly adopts the traditional image automatic focusing system, and the second stage mainly adopts the automatic image focusing system. The traditional image automatic focusing principle first adjusts the lens to include the target and then enters the PC machine or embedded system by the CCD/CMOS camera. The embedded system determines whether the lens is readjusted through the motor control module according to the image definition. The auto-image focusing system is divided into two situations: focusing depth and defocusing depth. In focusing depth, the search algorithm is used to focus, then the image processing module is used to determine whether the image is clear or not. Finally, the defocusing depth is calculated by collecting defocusing image parameter information or defocusing image degradation model and fuzzy graphics. Finally, the image definition can be adjusted to the best [2].

3 Evaluation Function of Automatic Image Focusing for Digital Image Processing

(i) Information entropy function for images

The formula of image information entropy function is as follows:

$$F = -\sum p_i log_b(pi) \tag{4}$$

The meaning of each index in formula (4) is as follows: the meaning of pi is the probability of characterizing information; the value of b is 2. In digital image processing, the gray level of auto-focusing image is independent, so the probability of representation information of each gray value is different. Based on this, the probability of gray value in gray histogram can be calculated [3].

(ii) Grayscale gradient function

The change of gray scale fluctuation and absolute change of gray scale have a certain function relation with the gray value of a certain point in the image and the pixel of image scale, while the gradient vector mode square function is also related to the change of gray scale fluctuation and absolute change of gray scale. Therefore, the gray gradient vector mode function can be obtained according to gray fluctuation and absolute change:

$$F = \sum_{x}^{M} \sum_{y}^{N} \left\{ [g(x+1,y) - g(x,y)]^2 + [g(x,y+1) - g(x,y)]^2 \right\}^{1/2} \tag{5}$$

The meaning of each index in formula (5) is as follows: the meaning of M*N is image scale pixel;(x,y) is a point in image; the meaning of g (x,y) is the gray value of a point in image.

(iii) Frequency domain evaluation function

Based on Fourier transform, the frequency domain evaluation function can be obtained as follows:

$$F = \sum_X^M \sum_Y^N \left(\sum_X^M \sum_Y^N g(x,y) W_{MN}^{xyXY} \right) - \varphi \qquad (6)$$

After comparing the sensitivity, precision, deviation, complexity, signal-to-noise ratio, time and other parameters of each function, the following conclusions can be obtained: the image information entropy function has long focus time, poor focus position, short focus time, high focus dispersion, and strong focus sensitivity of frequency domain evaluation function. Therefore, the most suitable function for automatic focusing of digital image processing is the frequency domain evaluation function, but the function obtained from Fourier transform is not good in terms of the complexity of the function. Frequency domain evaluation functions need to be further optimized or explored for other functions, such as wavelet analysis, which are also obtained by Fourier transform.

4 Area Selection and Focus Search Algorithm for Digital Image Processing Window

The deeper the depth of field and focal depth of the digital image processing window, the more blurred the image is, the larger the depth of field of the camera window is, the smaller the aperture is, the distance, focusing parameters and imaging clarity will be affected [4].

1. Blind Mountain Climbing Algorithm

   The principle of blind mountain climbing algorithm is to judge the position of mountain peak during mountain climbing, which can determine the best focus position of image definition. The algorithm can optimize the automatic image focusing evaluation function of digital image processing, improve the image focusing speed and reduce the deviation of focusing imaging.

2. curve fitting algorithm

   The principle of curve fitting algorithm is to synthesize the original complex curve function into the simplest clustering evaluation function by simple function, and then the extreme point of the original curve function can be obtained by the extreme point of the near fitting function. The algorithm can improve the accuracy of the image, but it has certain requirements for the maximum value of the image data [5].

3. Fibonacci Search Algorithm

   Fibonacci search algorithm is a search algorithm, which can use the hypothesis principle to analyze the most suitable points in the process of auto-focusing, and then determine the best auto-focusing interval by theoretical calculation. Although the algorithm can improve the focusing speed, it is easy to appear larger focusing deviation in the process of moving direction change.

5 Conclusion

Accuracy is one of the criteria for evaluating whether the automatic image focusing evaluation function conforms to the digital image processing. If the imaging image is fuzzy, the minimum gradient value can be adjusted according to the numerical change of gradient value. This can reduce the impact of minimum gradient image definition and improve the accuracy of evaluation function.

Signal-to-noise ratio (SNR) represents the anti-noise interference ability of digital image processing auto-focusing algorithm, and the increase of SNR can reduce the probability of auto-focusing algorithm. Therefore, the image processing can directly take out most of the gradient values, and then use a simple algorithm to bring them into the gradient matrix to calculate, so that the SNR can be improved.

Conclusion: To sum up, this paper mainly analyzes and compares the image information entropy function, gray gradient function, frequency domain evaluation function, other evaluation functions and other digital image processing automatic image focusing algorithms. The image information entropy function has the disadvantages of inaccurate focus position and long focus time. Therefore, the image information entropy function and gray gradient function are not suitable for the automatic image focusing algorithm of digital image processing. Although the frequency domain evaluation function has some advantages, there are still some shortcomings in the automatic focusing time. These digital image processing automatic image focusing evaluation functions can be further improved.

References

1. Mao, X., Zhang, S., Tong, Y.: Development of special CAD/CAE system for large parts packing case. J. Jiangsu Univ. Sci. Technol. **20**(01), 69–72 (2006)
2. Chen, Z., Sun, C., Huang, L.: Research on the design system of transport wooden box structure based on C++ and open GL U. Packag. Eng. **28**(9), 52–54 (2007)
3. Peng, G.: Transport packaging m Printing Industry Press, 19952-73
4. Li, T.: On the selection of China's logistics pallet standards
5. Han, Y.: Packaging Management Standards and Regulations, p. 53. Chemical Industry Press, Beijing (2003)

Library Automatic Borrowing System Based on STM32 and Java

Cong Lin[(✉)]

Zaozhuang University, Zaozhuang 277160, China

Abstract. The purpose of this paper is to promote the effective use of information resources, strengthen the deep processing of information resources, improve the performance and ease of use of the retrieval system, and realize the intelligence of information retrieval and the automation of information management.

Keywords: Artificial intelligence technology · Digital library · Personalization · Information service system

1 Introduction

In recent years, with the deepening of the research on artificial intelligence technology, the application of artificial intelligence technology is more and more widely. The main application fields of artificial intelligence technology include: language learning and processing, knowledge representation, intelligent search, reasoning, planning, machine learning, knowledge acquisition, neural network, complex system, genetic algorithm and human thinking mode. The personalized information service system of Digital Library Based on artificial intelligence belongs to intelligent search, reasoning, planning and so on [1].

In the current personalized information service system, users can fully express their needs and customize library learning resources according to their actual situation. But now the library information resources are digital network resources covering the whole world, so it is a very hard work for users to find and screen the information resources that meet their needs in the mass of information. Artificial intelligence technology can solve this problem well. The neural network algorithm in artificial intelligence technology has strong learning ability. It can infer the user's intention according to the actual situation and operation behavior of the user, so as to search and select the resources that meet the user's needs intelligently. This not only improves the efficiency of personalized information service system, but also improves the intelligent degree of information service system, so that the user's personalized information needs are further met.

2 Distribution of Information Resources

The current personalized information service system requires users to manually select the required information categories level by level, and then submit these choices to the system. The system displays the corresponding information according to the categories

J. C. Hung et al. (Eds.): FC 2021, LNEE 827, pp. 1041–1045, 2022.
https://doi.org/10.1007/978-981-16-8052-6_140

submitted by users. This kind of service mode is not only inefficient, but also increases the load of the system. In the network era, there may be thousands of people online at the same time, and people choose and submit demands at the same time, so the information service system has the risk of collapse. Artificial intelligence technology can obtain and analyze the user's demand information intelligently, and then automatically display the information resources on the page. Some users are not clear about the information they want after entering the library, just to find resources to solve some practical problems. For this situation, artificial intelligence technology can play a great advantage. The artificial intelligence system gradually understands the real needs of users by learning the user's operation behavior, and actively pushes the information resources that meet the user's needs according to these needs [2].

ROI can be expressed as:

$$ROI = \frac{r(\alpha_0^*)}{B(e^*)} \tag{1}$$

$$ROI = \frac{\beta^2 \alpha_1 (p^*)^2}{x} \tag{2}$$

There are two ways of intelligent acquisition of personalized information: one is static acquisition, that is, the reader's personal basic information stored in the library service system.

3   Key Problems to be Solved in the Application of Artificial Intelligence Technology in Personalized Information Service of Digital Library

3.1   Intelligent Acquisition of Personalized Information

It includes the basic information that readers fill in when they register, such as education background, major and hobbies. When readers first use the library personalized information service system, they need to register an account to fill in their personal basic information, which can be modified later. According to these basic information, the system can make a preliminary speculation on the personalized information needs of readers. Another way to obtain personalized information is dynamic acquisition, which is to track the daily operation of users. Through the user's operation behavior, the information system can further infer the user's personalized information needs, and then establish or update the user's personalized information needs database [3].

3.2   Intelligent Screening of Personalized Information

The intelligent screening service system of personalized information can intelligently screen the digital network information in the information service system according to the user information. The system makes a preliminary screening through the user's basic information and establishes the user's interest knowledge base, and then further filters the personalized information and updates the user's interest knowledge base according to the user's operation behavior and other dynamic information. The reason why this kind of layer by layer screening, step by step update interest knowledge base, mainly because in the context of contemporary network, digital library information resources are massive, changeable and repetitive. These three characteristics of information resources determine that it is not easy for users to filter personalized information. It is not easy to find the information users need from the massive information resources, and then actively recommend these information to users. And some of these massive information are repeated, and some are updated at any time, which adds a lot of difficulty to the information screening. In the artificial intelligence technology, the artificial neural network algorithm is used to screen these information intelligently, and the practice has proved that the effect is good [4].

Interaction designer Gillian C. Smith also proposed four dimensions of interaction design language, namely interaction design language model (as shown in Fig. 2–2). Kevin silver later extended this model with the fifth dimension. He thought that interaction design (IxD) includes word (1D), visual representations (2D), physical objects or space (3D), time (4D) and behavior (5d). Designers can use five dimensions to consider the interaction between users and products in a holistic way. As shown in Fig. 1.

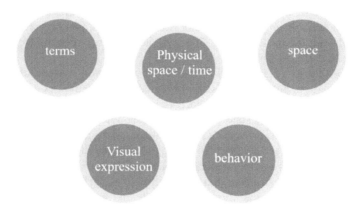

Fig. 1. Interaction design language model

At present, most of the digital library service systems are passive services for users, and most of them are not standardized for the establishment of user models, and the acquisition or analysis of users' personalized information needs is not deep enough. For various reasons, the service functions of the current digital library personalized information service system are not perfect. Based on the personalized information service system of artificial intelligence technology, this paper proposes to establish a "user information needs knowledge base" model, which uses artificial intelligence technology to intelligently obtain and analyze user personalized information needs, and intelligently filter these information to establish the model. At the same time, the self-learning ability of the information service system is also gradually updated to improve the knowledge base, so as to improve the intelligent service ability of the service system.

4   Integration of Information Resources

The use of search engine, that is, users in the establishment of the engine on the web page to enter the key words of the information they need, or use the list of directory links on the page, through the engine server in its internal database to find the relevant information, and according to certain rules after sorting through the network to the user's local machine is an online service mode. In recent years, the development of Internet search engine is very fast. According to statistics, there are more than ten search engines in the whole network, and hundreds of other search engines according to professional fields.

The integration of information resources is a system strategy composed of computer information network and related technologies. It means that according to the needs of users, it analyzes the retrieval conditions, chooses to enter the local area network information database or Internet information database, locates the location of the day mark information (the physical address of the server where the information resources are stored), starts the indexing system, translation system, information navigation system, search engine, etc., After sorting and eliminating redundancy, the retrieval results are fed back to users. The integration of information resources should include three interrelated functional subsystems.

The establishment of information intelligent navigation system, that is, as an embedded function module, automatically or under the stimulation of users, carries out on-site intelligent guidance service in each stage of the process of user retrieval. Its functions are as follows: (1) state analysis: the user information retrieval process is represented by a standardized state. (2) Knowledge analysis; determine the required knowledge of each state and its problem solving; determine the structure and source of these knowledge; determine the relationship between the corresponding knowledge of each state; determine the structure and transfer mode of each knowledge module and the whole knowledge base; the knowledge module and knowledge base can be automatically updated. (3) State and problem matching: connecting the state with the specific link or step of the retrieval process, analyzing and determining the problems in the specific link or step; analyzing or necessary interactive analysis for the problems; accurately matching the problems and knowledge; transferring the relevant knowledge to the user, and matching the new state and problem according to the user behavior.

5   Conclusion

Intelligent information search and automatic information extraction are the development needs of the information age. The integration of information resources in LAN and Internet can promote the effective utilization of information resources, strengthen the deep processing of information resources, improve the performance and ease of use of the retrieval system, and realize the intelligent information retrieval and automatic information management, To better realize the co construction and sharing of online information resources: automatic document indexing and multilingual automatic identification have the function of analyzing retrieval conditions and information retrieval automation in the integration of information resources. Users complete the retrieval task in the same interface, and make the information fed back to users not miss, mistakenly check and recheck as much as possible; In the integration of information resources, information intelligent navigation technology has the functions of intelligent selection of information source, automatic positioning of target information position, without manual pointing to LAN or Internet address, reasonable allocation of network resources, etc., which can improve the speed and efficiency of retrieval.

References

1. Hui, Z.: Library information resources construction under network environment. Library **2**, 30–34 (2000)
2. Wang, Y., Gu, X.: Automatic indexing of Chinese document subject. Acta Informatica Sinica **17**(3), 219–225 (1998)
3. Su, X., et al.: Research on automatic indexing of document information. Digit. Libr. Technol. **1**, 2326 (2000)
4. Feng, Z.: Machine translation. Chinese Transl. (4,5) (1999)

Performance Analysis of Beidou Precision Landing Technology for Runway Airport

Huanchang Xiao[✉]

Number46, Section 4, Nanchang Road, Guanghan, Deyang, Sichuan, China

Abstract. The key technology for CAAC to implement PBN is compass G2 and GBAS. This paper describes the composition and work flow of GBAS, constructs the GBAS simulation system based on compass, and gives the design scheme and development program diagram. Through static test, airport runway sports car test and airport area sports car test, the collected data are analyzed to verify the effect of signal enhancement. Based on the improvement of signal accuracy by GBAS, a performance evaluation system of Beidou landing system is constructed to guide aircraft precision approach (GIS) in terminal area. The ground and air inspection methods for evaluating the guidance approach performance of Beidou system are given.

Keywords: PBN · Compass · GBAS · Experimental system · Integrity

1 Introduction

The implementation roadmap of performance based navigation (PBN) of civil aviation of China (CAAC) issued in October 2009 proposes to realize the key application of PBN before 2012 and the comprehensive application of PBN from 2013 to 2016. It is expected that GNSS (Global Navigation Satellite System) and its augmentation system will have precision approach capability, GNSS will become the main navigation facility for PBN operation, and CAAC will use GNS on the basis of multilateral cooperation, including Beidou and its augmentation system [1].

The civil aspect mainly focuses on the application optimization of LAAS system based on GPS constellation. The military aspect, such as dod-h4e JPALS joint precision approach and landing system, is the latest high precision approach and landing system developed by Raytheon based on GBAS. Relevant foreign research institutions take technical blockade measures for its key technologies and core algorithms. Domestic research institutions are mainly Beijing University of Aeronautics and Astronautics and 20 Research Institute of China Electronics Technology Group, but the work focuses on satellite navigation constellation layout planning simulation and hardware construction. For civil aviation navigation using Beidou system, the analysis and evaluation of signal accuracy, integrity, continuity and availability are rarely involved. In this paper, a GBAS simulation system based on compass is constructed, and the method to evaluate the precision approach performance of the system is given.

J. C. Hung et al. (Eds.): FC 2021, LNEE 827, pp. 1046–1050, 2022.
https://doi.org/10.1007/978-981-16-8052-6_141

2 Sorting Out the Deformation Monitoring Data of Comparison Bucket

Because the XYZ coordinate data in WGS84 coordinate system can't directly observe the deformation variables of monitoring points, it needs to be converted into the enum in station center coordinate system to reflect the three-dimensional deformation variables of each monitoring point. The process involves the use of SQL statements to query and process the data in the database. The SQL statements used in the process are helpful to the compilation of the later software.

2.1 Using Beidou Data to Interpolate InSAR Data in Time Domain

Using the point data of fitting curve with the preprocessed Beidou monitoring data, the inverse distance from PS point to Beidou monitoring point is taken as the weight, and the proportion of settlement to the total settlement as the weight object, the InSAR settlement data is interpolated and encrypted in the time domain, which can be expressed by Formula 1 and formula 2:

$$Z_p = \frac{\sum_{i=1}^{n} \left(d_i^{-u} \times R_i\right)}{\sum_{i=1}^{n} d_i^{-u}} \times Z \tag{1}$$

$$R_i = \frac{b_i^p}{b_i} \tag{2}$$

When the difference between the fitting results of more than 50% Beidou monitoring data and the corresponding PS monitoring results is greater than or equal to $\pm$ 5mm, according to formula 3, the InSAR settlement data is translated by a fixed displacement Δ, and then interpolated.

$$\Delta = \frac{\sum_{i=1}^{n} \left(h_i^B - h_i^I\right)}{n} \tag{3}$$

After interpolating the spatial continuous InSAR monitoring data with the time continuous Beidou monitoring data, the continuous settlement results in time and space are obtained. The first purpose is to more comprehensively and accurately reflect the historical settlement situation in the field area, and the second is to more accurately predict the future settlement trend, which is also the fundamental purpose of the "interpolation" method. So we need a reasonable extrapolation method based on the existing data results. The extrapolation method used in this paper is based on the practical creep algorithm, combined with genetic algorithm and least square method, which can maximize the use of existing data and predict the future through reasonable extrapolation [2].

3 Precision Navigation of RNP for Arrival and Departure

RNP is a relatively new aircraft take-off and landing guidance technology, which mainly uses global positioning system and aircraft airborne navigation equipment. At present, countries are still strengthening the research and practice in this field, and it has been determined that it will be the main development trend of civil aviation navigation in the future. In the past ten years, RNP technology has been applied to the aircraft arrival and departure, and has played a good navigation efficiency. If the Beidou satellite navigation system can be applied to civil aviation navigation, through the flight computer and the corresponding procedures, the aircraft can enter the terminal tunnel very accurately. Compared with the traditional civil aviation navigation technology, this kind of navigation can make civil aviation aircraft get rid of the limitation of ground navigation facilities, and get in and out of the field accurately and safely. Even under the influence of weather conditions, the visibility and visibility of the aircraft entering and leaving the field are poor, which will not have any impact on it. This is of great significance for airports in areas with changeable climate and complex terrain, such as the Western Plateau Airport in China [3].

In view of the current actual situation, China's civil aircraft approach and landing mainly relies on the instrument landing system. Under the condition of low working pressure, the performance of the system is still relatively good. However, the development of China's civil aviation market is growing, and the number of civil aircraft take-off and landing flights is increasing, which exposes various shortcomings of the instrument landing system, The application of Beidou satellite navigation system in civil aviation navigation can help solve this problem. Precision navigation and positioning, flight guidance and landing are the three main functional modules of the current precision approach and landing system. With the support of Beidou satellite navigation system, the precision approach and landing system can meet the requirements of a class of precision approach, and provide horizontal and vertical deviation guidance. For the ILS, one instrument runway only corresponds to one ILS approach procedure of the same level, while for the precision approach and landing system, one instrument runway can correspond to multiple ILS approach procedures of the same level, Precision approach and landing system can bring obvious approach efficiency and better navigation service.

The precision approach and landing system of ground-based augmentation system has typical advantages compared with the traditional instrument landing system. For example, the approach is more precise, the approach data block can flexibly define the approach track, and a set of precision approach and landing system can meet the requirements of all instrument runways and different types of aircraft independent approach and departure. In addition, it has higher flexibility and adaptability. In airports where the ILS is difficult to operate, the precision approach and landing system is more adaptable. The precision approach and landing system supports curve approach, and the data block can customize the approach procedure according to the temporary situation, so as to improve the flexibility of flight and control.

4 Finite Element Back Analysis Deformation Prediction Method of UH Model Considering Time Effect

It can predict the deformation of the whole airport area, but because the practical creep algorithm is essentially a curve fitting algorithm, it can only predict the settlement of the monitoring points, but can not predict the settlement of the unknown points, and the whole interpolation process does not produce new monitoring points in space, so there is a problem of insufficient attention to the key areas and local areas. Therefore, the finite element back analysis method based on the constitutive model is selected to supplement the analysis and prediction of local deformation. This method not only considers the characteristics of soil, but also takes into account the fitting of the measured data. It can obtain the three-dimensional prediction data of any point in the finite element model, and the prediction results are more comprehensive and have more physical significance, It can analyze the local deformation of the key area of the airport more accurately. In view of the complex geological conditions of Beijing Daxing International Airport, unclear bedrock location, large thickness and uneven distribution of underground compressible layer, there are also problems in the process of using this method, which are difficult to accurately establish the foundation model. This chapter will also provide corresponding methods to solve them [4].

In order to accurately predict the deformation of Yichang Road, the finite element model of Yichang road must be established as accurate as possible. However, because the airport area is located in the middle and lower reaches of the river, the foundation soil is very thick alluvial proluvial soil, the distribution is discontinuous and irregular, and the geological conditions are very complex; Even the thickness of compressible soil layer is not clear, so for the airport as a whole, it is impossible to directly establish a model that can accurately reflect the overall geological conditions of the site. But for the establishment of local finite element model, we can refer to the relevant geological data to establish a more accurate stratum model. In order to simplify the modeling process, facilitate the convergence of the finite element calculation of the viscoelastic plastic model, and determine the model parameters according to the settlement monitoring data from 1998 to 2011, the local finite element model of the airport is established based on the following considerations.

5 Conclusion

The applicability of various monitoring methods in airport deformation monitoring is fully investigated, and the Beidou monitoring and InSAR monitoring technologies with good applicability are selected. Based on the complementary advantages of the two technologies, the airport runway deformation monitoring method using Beidou and InSAR joint monitoring is proposed. The relevant standards and specifications of civil aviation airfield are fully investigated, and the complete Beidou deformation monitoring scheme of Beijing Daxing International Airport is designed and implemented, including the layout of monitoring points, the design of monitoring equipment and its construction and installation. The engineering experience of applying Beidou monitoring method to airport deformation monitoring is accumulated, It has reference significance and value for other airport Beidou monitoring scheme design.

References

1. CAAC: CAAC performance based navigation implementation roadmap. CAAC, Beijing (2009)
2. Han, S.: Introduction to New Navigation System. Nanjing University of Aeronautics and Astronautics Press, Nanjing (2007)
3. Tian, F., He, Y.: Positioning performance verification of Beidou satellite navigation system. Railw. Surv. (03), 11–13 (2017)
4. Li, Y., Qian, X., Li, S.: Development status and Prospect of Beidou satellite navigation system. Commun. World **14**, 217–218 (2016)

Stability Analysis of Electronic Circuit Based on Complex Neural Network Theory

Xiaolin Chen[⊠]

Wuhan Railway Vocational College of Technology Wuhan, WuHan 430000, Hubei, China

Abstract. With the rapid development of microelectronics technology, the research of analog circuit diagnosis is facing challenges. The purpose of this paper is to combine neural network and analog circuit fault diagnosis to explore the method of analog circuit fault diagnosis. Taking an analog circuit as an example, MATLAB software is used to train the sample. Through learning, the network threshold is adjusted, and the sample characteristic parameters to be identified are input to the network, The output value of the network is calculated. The simulation results show that the fault diagnosis method of analog circuit based on neural network theory is feasible and reliable.

Keywords: Analog circuit · Neural network · Fault diagnosis

1 Introduction

In the early 1960s, people began to study the automatic analog circuits have tolerance and nonlinearity slow. Neural network science based on nonlinear mathematical theory provides a new idea. Neural network is a kind of non-linear dynamic operation model, which imitates the work of human brain. It has outstanding ability of distributed storage and parallel collaborative processing of information, which makes it widely used in complex environment, unknown background and irregular problems. The application of neural network in fault diagnosis is the best example. Neural network has high self-learning and self-organization ability, This makes the operation model global, which becomes an effective means of fault diagnosis. The method is based on the process parameters obtained from the test, and the neural network model is used to establish the correlation between the measurement space and the fault space, so as to make the circuit fault diagnosis [1].

2 Neural Network

It is the most model at present. It can learn by itself without knowing the mapping relationship between input and output in advance. BP feed-forward network composed of an input layer, multiple hidden layers and an output layer. As shown in Fig. 1, the input node of the neural network is x, which is the zero layer of the neural network and has no calculation function. The node n of each layer represents the calculation unit with calculation function. The figure shows that the network from the front to the back

J. C. Hung et al. (Eds.): FC 2021, LNEE 827, pp. 1051–1055, 2022.
https://doi.org/10.1007/978-981-16-8052-6_142

is the first to the N layers, forming the m-layer forward network, and the network output is, The first node layer and the output node layer are the visible layer, and the middle layer is the hidden layer. As shown in the figure above, neural networks usually have one or more hidden layers [2].

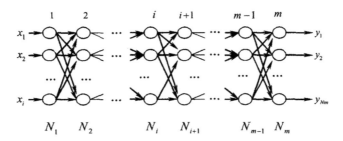

Fig. 1. BP neural network

The learning process of BP neural network is divided into two processes: forward and backward. The sample value of the signal is input from the input layer and then reaches the output after being processed by the multi hidden layer. This process is forward transmission. When the output is inconsistent with the expected value, the signal turns to back propagation. At this time, the output error is transmitted forward layer by layer through the multi hidden layer. In this process, the output error is allocated to each layer unit, The forward transmission of signal and the reverse transmission of error together constitute the learning process of neural network. This process is continuously circulating, and the weights are gradually corrected, so that the error reaches an acceptable value. It is precisely because of this learning process that the neural network has a high degree of self-adjusting ability.

Through the above analysis, it is concluded that the error of neural network is the function of the weights of each layer and the input samples, such as function 1. The error function E is a complex surface in multi-dimensional space. The flat area of the surface indicates that the error decreases slowly and is not sensitive to the change of weights. There are concave and convex points in the surface, that is, the minimum point of the function, where the error gradient is 0, When the model is trained, it often falls into these minima, and it is difficult to converge to the given value. The standard BP algorithm is a simple fast descent optimization algorithm, which does not take into account the previous accumulated experience, showing slow convergence speed, local extremum and other phenomena [3].

$$E = F(X^P, W, \theta, t^P) \tag{1}$$

Through the above analysis, it is concluded that the error of neural network is a function of the weights of each layer and the input samples, such as function formula 1. The error function E is a complex surface in multidimensional space. The flat area of the surface indicates that the error decreases slowly and is insensitive to the change of weights. The number of iterations increases and the adjustment time is long. There are concave and convex points in the surface. The concave point is the minimum point of the function, where the error gradient is 0, When the model training process often falls into these minima, it is difficult to converge to a given value. The standard BP algorithm is a simple fast descent optimization algorithm, which does not take into account the previous accumulated experience, showing slow convergence speed, local extremum and other phenomena.

2.1   Adjust Step Size

In the standard BP algorithm, η step (also known as learning rate) is a fixed value. In the process of analog circuit fault diagnosis, it is difficult to find a suitable value to adapt to the error adjustment of the whole circuit network. Combined with the previous error surface analysis, for flat areas, the step size is too small to increase the number of training, and then we hope that η value is larger; in concave and convex areas, the step size is too large to cross narrow concave, There is shock training, increasing the number of iterations, different regions have different requirements for step size, the learning rate is best to have the ability of adaptive, accelerate convergence. For example, it can be adjusted according to the total error of the network: after a weight adjustment of the network, if e increases, the adjustment is invalid. If e decreases, it is effective.

2.2   Introducing Steepness Factor

In the flat area of the error function surface, because of the saturation characteristic of the transfer function s, the process of adjusting the weight is slow, so the steepness factor is introduced to prevent the saturation. The specific method is that when the training reaches the flat area, the output is far away from the saturation area of the transfer function s by compressing the net input:

$$f(x) = \frac{1}{1 + e^{\frac{-net}{\lambda}}} \tag{2}$$

When e $\rightarrow$ 0 and is still large, enter the flat area, at this time, make x > 1; when exit the flat area, make the transfer function return to its original state.

3   Fault Diagnosis Method Based on Quantum Neural Network

At present, artificial neural network is widely used in fault diagnosis of power electronic circuit. However, when there is the problem of cross data pattern recognition between fault patterns, neural network shows its shortcomings. Both neural network and wavelet neural network mentioned above have the defect of high fault false diagnosis rate. These shortcomings urge many researchers to explore new neural network structure and theory. And there is a combination of neural network and other theories. Among them, the combination of neural network and quantum computing has a very attractive prospect, which is a frontier topic in the development of artificial neural network theory. Next, we will describe how quantum theory and neural network are combined to realize fault diagnosis of power electronic circuits.

3.1 The Basic Theory of Quantum Computing

In the early 1980s, beno6 and Feynman proposed the concept of quantum computing, then shor proposed the large number prime factorization algorithm, Grover proposed the quantum search algorithm of unordered database. Since then, quantum computing has attracted wide attention with its unique computing performance, and quickly become a research hotspot. Quantum computing takes advantage of the superposition, entanglement and interference of quantum states in quantum theory. Through quantum parallel computing, it is possible to reduce the complexity of some problems with high computational complexity on classical computers (such as NP hard problem). At present, the research on quantum computing and its models is still in its infancy in the world, and there have been some explorations in this direction, For example, in 1995, Kak first proposed the concept of quantum neural computing, in the same year, Narayanan et al. Proposed the quantum derived neural network model, and in 1998, Ventura et al. Proposed the quantum associative memory model. All these attempts will help to broaden the application scope of quantum computing and improve the performance of traditional neural computing [4].

3.2 Quantum Neural Network

Quantum neural network (QNN) appeared in the 1990s. It introduces the idea of quantum mechanics into the research of neural network and overcomes the shortcomings of traditional neural network. Quantum neural network is an extension of classical neural network. It makes use of some advantages of quantum computing, especially the parallel computing characteristics of quantum computing. Quantum neural network has stronger parallel processing ability than classical neural network, and can process larger data sets. It has unprecedented potential advantages in data processing. Compared with the classical neural network, it has the following advantages: 1) exponential memory capacity and recall speed; 2) fast learning and high-speed information processing ability; 3) the ability to eliminate catastrophic amnesia because there is no interference between modes; 4) high stability and reliability.

4   Conclusion

According analog circuit, the adaptability of neural network is improved by adjusting the step size, adding momentum term and introducing steepness factor. The improved neural network method is introduced into diagnosis of analog circuit is verified by testing the experimental circuit in Matlab environment.

References

1. Zou, R.: Principle and Method of Analog Circuit Fault Diagnosis. China University of Technology Press, Beijing (1989)
2. Guo, X., Zhu, Y.: Evolutionary neural network based on genetic algorithm. J. Tsinghua Univ. **40**(10) (2000)
3. Yu, C.: Application of improved neural network in analog circuit fault diagnosis. Comput. Simul. **28**(6), 239–242 (2011)
4. Xu, X., Fu, X.: Research on analog circuit fault diagnosis based on wavelet decomposition and BP network **34**(19), 171 (2011)

Internal Crack Detection Algorithm of High Speed Multi Layer Asphalt Pavement Engineering

Huimin Gao[1(✉)] and Xien Yin[2]

[1] Henan Boda Engineering Management Consulting Co., Ltd.,
Xuchang 461000, Henan, China
[2] Henan Junfei Construction Engineering Co., Ltd.,
Xuchang 461000, Henan, China

Abstract. Asphalt highway pavement crack automatic detection technology is of great significance for the entire highway pavement maintenance management system. The traditional manual detection method can not meet the needs of modern detection. With the development of computer technology, road detection technology has been combined with computer technology and entered the automatic detection stage. In this paper, the author participated in the Key Laboratory of the Ministry of Education - image detection, recognition and automatic evaluation method research project of asphalt pavement cracks, and studied the image detection system of asphalt highway pavement cracks.

Keywords: Pavement crack · Median filter · Edge detection

1 Introduction

With the development of China's economy and the increase of the number of cars, highway transportation has become more and more important in the national economy and people's life. Highway transportation is the most extensive and basic mode of transportation. It is an important infrastructure for the development of national economy. It has a broad development prospect in the national comprehensive transportation system. In recent years, China's transportation industry continues to develop rapidly, especially the development of high-grade highway, which accounts for an increasing proportion in the total mileage of highway.

Expressway is a new type of traffic facility with safety, high speed and large carrying capacity, which began to appear in the western developed countries in the 1930s. It can generally adapt to the speed of 120 kmh or higher. It uses asphalt concrete or cement concrete high-grade pavement, has the width of 4–6 lanes, and has a dividing belt in the middle. The construction of expressway can greatly shorten the travel time, reduce the traffic accident rate, speed up the transportation turnover speed between regions, and at the same time make the original residents evacuate to both sides of the expressway, which can effectively alleviate the problem of urban population density [1].

© The Author(s), under exclusive license to Springer Nature Singapore Pte Ltd. 2022
J. C. Hung et al. (Eds.): FC 2021, LNEE 827, pp. 1056–1060, 2022.
https://doi.org/10.1007/978-981-16-8052-6_143

Although the total mileage of China's highway traffic has been in the forefront of the world, the technologies and means of road maintenance, road condition monitoring, data collection and performance evaluation are far behind the developed countries in the world, which obviously can not meet the needs of highway maintenance. At the national highway maintenance and management conference which ended in 2001, it was clearly put forward that the concept of "construction is development, maintenance and management is also development" should be firmly established in highway traffic work. At the same time, it also puts forward that we should strengthen the research of road maintenance technology, actively research and develop advanced and practical new technology, new equipment and new technology of road maintenance, and apply modern science and technology, so as to comprehensively improve the technical level and work efficiency of road maintenance. At present, road detection technology is developing from manual detection to Automatic Non-destructive detection technology, which makes road quality monitoring, evaluation and damage analysis faster, and road maintenance more reasonable and economic.

2 Asphalt Pavement Damage Detection Method and Evaluation Index

2.1 Pavement Damage Classification

When all kinds of road surface damage types are in different development stages, they have different degrees of influence on the pavement performance. For example, when cracks appear, the cracks are small and the materials at the edges are complete, so it has little impact on the driving comfort, and there is still a high load transfer capacity between cracks; in the later stage, the cracks become very wide, the edges are seriously fragmented, the driving is bumpy, and there is almost no load transfer capacity between cracks. Therefore, in order to distinguish the different effects of the same damage on the pavement performance, all kinds of damage are divided into several grades (generally 2–3 grades) according to the severity of the impact. For fracture or crack damage, the degree of impact on the structural integrity is mainly considered in the classification, which can be characterized by the gap width, edge fragmentation degree, crack development and other indicators, Considering the influence on the driving comfort, the flatness index can be used for grading; for the surface damage, there is no need for grading. The specific indicators and grading standards are determined according to the characteristics of each region and other factors [2].

2.2 Pavement Damage Measurement Method and Calculation

In order to quantitatively compare the damage condition or severity of each road section, a comprehensive evaluation index should be used. Generally, pavement condition index (PCI) is used to evaluate pavement damage. Deduction method is a common method to determine the pavement condition index PCI, such as Formula 1. It is to stipulate different deduction values for different damage types, severity and scope. After accumulating the deduction values according to the damage status of the road section, the remaining values are used to represent the integrity of the road surface and evaluate the quality of the road surface.

$$PCI = 100 - \alpha_0 DR^{\alpha_1} \tag{1}$$

$$DR = 100 \frac{\sum_{i=0}^{21} W_i A_i}{A} \tag{2}$$

Dr – damage rate of asphalt concrete pavement, which is the ratio of the sum of the equivalent areas of various damages and the investigated pavement area.

3   Image Enhancement Algorithm of Road Image

Digital image enhancement is a processing method that highlights some information in an image according to specific needs and weakens or removes some unnecessary information at the same time. Its main purpose is to make the processed image more suitable for a specific application than the original image. In the process of image generation and transmission, we often receive the interference and influence of various kinds of noise (such as the reflection of sandy materials on the road, the internal noise of sensitive components, the particle noise of sensitive materials, the jitter noise caused by electrical machinery movement, etc.). These noises worsen the image quality, blur the image, even submerge the features. If they are not removed seriously, it will bring great difficulties to the subsequent crack extraction. In order to suppress the noise and improve the image quality, the image must be enhanced. When processing the noise, the edges and various details of the image should not be damaged as much as possible, so that the processed image is more suitable for human visual characteristics or machine recognition system [3].

3.1   Improved Median Filtering Algorithm

Sometimes the effect of median or mean smoothing is not very good. Although it removes some noise, it also makes the edge of the image blurred, which is mainly related to the selected window. When median filtering is applied to the image, if the window is symmetrical about the center point and contains small and medium points, the median filtering can keep the jumping edge in any direction. For this reason, we propose a smoothing template which can not only keep the edge clear but also eliminate the noise, as shown in Fig. 1.

3.2   Edge Detection of Gray Image

The most basic feature of an image is edge. The so-called edge refers to the set of pixels whose gray level changes step by step or roof like. It exists between target and background, target and target, region and region, and primitive and primitive. One of the main methods to determine the edge of an object in an image is to detect the state of

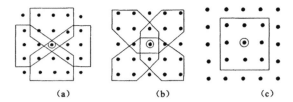

Fig. 1. Improved image smoothing template

each pixel and its immediate neighborhood to determine whether the pixel is on the edge of an object. Pixels with the desired characteristics are labeled as edge points. When the gray level of each pixel in an image is used to reflect the degree that each pixel meets the requirements of edge pixels, this kind of image is called edge image or edge image [4]. It can also be represented by a binary image which only represents the position of the edge point and has no intensity. In the process of crack disease detection, the quality of edge detection algorithm will greatly affect the detection effect and accuracy.

Edge detection is usually performed by spatial differentiation operator, which is completed by convolution of template and image. There are always gray edges between two adjacent regions with different gray values. Gray edge is the result of discontinuity (or mutation) of gray value, which can be easily detected by derivative.

3.3 Summary of this Chapter

Firstly, this chapter mainly introduces two classical image enhancement methods of road image features, compares and proposes an improved median filtering algorithm. The improved median filter smoothes the road background, removes most of the noise points, and provides high quality digital image for edge detection. Then it focuses on the edge detection of road image. Firstly, the classic Sobel operator algorithm is selected to detect the image. However, because Sobel has only horizontal and vertical templates, the algorithm has directionality and can not effectively identify and detect the irregular cracks in road diseases. Therefore, 8-direction Sobel edge detection algorithm is adopted, which has no directivity and can detect transverse cracks and longitudinal cracks better, but it is easy to produce false edges. Canny criterion is an important criterion of edge detection, which is used to detect the image, but it is greatly affected by the noise. Finally, an improved Canny algorithm combined with threshold selection method is applied to detect the edge of pavement crack image. The experimental results are satisfactory, which lays a good foundation for the subsequent recognition of pavement image types.

4   Conclusion

The research content of this paper is mainly aimed at the research project "image detection, recognition and automatic evaluation method of asphalt pavement cracks" of the Key Laboratory of the Ministry of education. According to the characteristics of asphalt pavement crack image, this paper makes a deep research and Discussion on image filtering, edge extraction and other aspects, puts forward an improved median filtering algorithm, studies the improved cany algorithm combined with threshold selection method, and tests it according to the actual situation of the project. The experimental results show that the algorithm can detect and process the edge of pavement crack image better.

References

1. Huang, X., et al.: Highway Engineering Inspection Manual. People's Communications Press, Beijing (2004)
2. Zhang, Q., Jing, G., Yang, F.: Research on 3D integrated detection technology for pavement damage and subgrade disease of expressway. Appl. Technol. Ed. Highw. Traff. Sci. Technol. **93**, 3741 (2006)
3. Chu, X.: Research on image recognition method of asphalt pavement damage. Doctoral dissertation. Jilin University (2003)
4. Wang, R., et al.: Research progress of pavement damage image recognition. J. Jilin Univ. (Eng. Ed.), **32**(491–97) (2002)

The Application of Computer Technology in Zen Brand Packaging Design

Xiaojie Min[✉]

Gongqing College of Nanchang University, Jiujiang 332020, China

Abstract. Computer technology in packaging decoration design constantly explore new technology, so that packaging decoration design from manual into the era of computing. Starting from the superiority of computer technology in packaging design, this paper explores the superiority of computer in packaging decoration in three aspects: rapidity, accuracy and stronger integrity.

Keywords: Packaging decoration · Computer application · Future prospects

1 Introduction

Personalization is the need to add unique and own characteristics [1]. The radius of the key region found under three global vulnerability indexes is smaller than that of the traditional key region recognition method. Therefore, it can be proved that the two key area recognition algorithms proposed in this paper can identify the key areas of the network more accurately. Finally, by comparing the proposed algorithm with the traditional key area recognition algorithm, the simulation results under a certain global vulnerability index show that, whether it is the key area recognition algorithm whose center is limited by the network node, or the key area recognition algorithm whose center is not limited, with the increase of the fault impact rate, The difference between the radius of the key area identified by the proposed algorithm and the traditional algorithm is getting smaller and smaller.

2 The Superiority of Computer Technology in Packaging Decoration Design

2.1 Quick and Changeable Advantages

The emergence of computer makes the world enter a new era. Computer operation as a representative to promote the construction of the information highway, and make people's life into a fast-paced era. The biggest advantage of computer is its fast and accurate. First of all, what computing brings to people is the speed of calculation, and in the packaging and decoration design, the entry of computer also makes it enter the era of speed [2]. Packaging decoration design from the past manual painting design into the computer age, so that the past manual work into the era of software. Packaging decoration design with drawing software not only changes the previous design mode, but also changes the previous design concept, so that packaging decoration design has entered a new field, the design content is more complex and rich, but the operation is more and more simple, and in the operation can be changed at any time, no waste, direct change, this is manual design can not do. This is the greatest advantage of computer technology in packaging decoration design.

2.2   Improvement of Accuracy

Another advantage of computer technology in packaging decoration design is its accuracy. The main feature of computer technology is accuracy. In the packaging decoration design, computer technology has played its strong advantages. In the past manual design, the length of lines and the size of patterns are calculated or measured by hand, and the accuracy of computer design is far greater than that of manual design. This shows the superiority of computer technology.

3   Application of Computer Technology in Packaging Decoration Design

3.1   The Application in Color Selection

The application computer technology in color selection in packaging decoration design brings great convenience to design. In the past, the selection of color in manual design depends on the combination of the designer's experience and the overall design idea, and the selection of color depends on the comparison of the design samples one after another. With the colors can be changed at will. All colors can be selected in the computer software. There is no need for sample design and color deployment, but the colors are richer and more accurate, and the color spraying is more accurate and uniform, which makes the package decoration design more delicate and improves the overall aesthetic sense. In color, it can achieve the effect of "color" first [3].

3.2   The Application of Color in Pattern Design

The outstanding. The random exchange of various graphics and graphics attached to computer software makes the design of decoration pattern more convenient. Both geometric figures and irregular figures can be obtained directly by drawing software, and the proportion of each part can be changed arbitrarily according to the needs. The transformation, superposition and deformation of the pattern can be easily achieved, and the pattern that the designer wants to use outside the software can be drawn by himself, which does not hinder the designer's personality design. The application of computer technology has brought great convenience and speed to packaging decoration design, and also added unlimited aesthetic elements.

The CCD recording device will record the same number of element images as the lens array, and the image recorded on the CCD is called element image array. The acquisition process satisfies the principle of Gaussian imaging, which can be expressed by Formula 1:

$$\frac{1}{f_1} = \frac{1}{g_1} + \frac{1}{l_1} \tag{1}$$

When a meta image beyond adjacent lens, the redundant image will be observed, that is, the flipped image. Therefore, the number of meta segmentation is limited, and the integrated 3D image cannot be seen by the observer outside the viewing angle. Viewing angle a can be expressed as:

$$a = 2\arctan(\frac{p}{2g}) \tag{2}$$

Where p is the lens spacing (horizontal and vertical spacing are equal), and G is the distance [4].

The range of the reconstructed object is represented by the depth of the edge image. The edge as:

$$\Delta z_m = 2l\frac{p_l}{p} \tag{3}$$

The formula is used when $g \neq f$ and $g = f$:

$$\Delta z_m = 2g\frac{p}{p_x} \tag{4}$$

Where l is the distance from the lens array to the central depth plane.

3.3   The Application of Computer in Character Design

Packaging decoration is to beautify the packaging, to a large extent is to make a beautiful dowry clothes for a product. The use of words is an important part of decoration. In the past, characters were designed by hand, and the design of characters also depended on the designer's conception and originality. The application of computer technology fundamentally changed the previous concept. Computer software stored a characters, and it could also carry out various deformation designs of characters. There were ready-made samples for three-dimensional, shadow, circle and personification, The existing samples can also be changed according to the needs of the design. In this way, the composition design of fine arts characters is greatly enriched, and the inspiration of designers can be inspired to make the design more prominent personality and highlight the aesthetic effect.

4   The Future of Computer Technology in Packaging Decoration Design

The application will make the packaging decoration design in the future have more exquisite design operation software, and more powerful special computers to meet the needs of all kinds of design; there will be more complex design content and convenience, and the capacity of professional computer will store more design patterns for designers to use, making the design rich, exquisite and convenient; 3D printing will replace the traditional design and production. In the future packaging decoration design, it will integrate the overall packaging and design as a whole. 3D printing technology will be widely used to make packaging enter a new field. At the same time, packaging decoration design will also enter a new field.

One of a wider viewing angle can also be obtained. As the operating frequency increases, the complexity of using mechanical motion based technologies increases. A moderate frequency of 2030 Hz images.

Now the convenience of product design, manufacturing and assembly is due to the continuous modernization of the development of processing and manufacturing industry, which leads to higher and higher standardization of parts design. The idea of parametric design and drawing is used in the CAD/CAE system of wood packaging. Users can complete the modeling design of wood packaging by modifying the relevant parameters of wood packaging modeling. The wood packaging required by designers is the standard wood packaging model or general model selected from the wood packaging library. The selected standard wood packaging model or general model fully meets the design requirements, Then parametric design is carried out to obtain the required wood packaging structure model. This process can greatly reduce the number of repeated labor, improve the design efficiency and shorten the research cycle of new products.

5   Conclusion

With the every field will enter the computer era, packaging and decoration will also enter a new field in the continuous development of computer technology, especially the emerging 3D printing technology, which has a great trend to change the world. The future packaging field will be occupied by 3D printing technology, integrating design and production, which not only reduces the design cost, but also reduces the production cost, and can change at any time, adapt to the fast pace of life in the future, adapt to the needs of people's pursuit of personality, but also meet people's aesthetic needs.

Acknowledgement. This paper is based on the subject YG2020195 of Jiangxi Province culture, Art and Science Planning Project in 2020, "Application research of Chinese Zen Taoist Culture in Packaging Design".

References

1. Li, H.: Application of computer graphics software in modern packaging and decoration design. J. Zhengzhou Inst. Aeronaut. Ind. Manag. (Soc. Sci. Ed.), 05 (2004)
2. Wang, C.: Application problems and Countermeasures of founder epack packaging structure design software. Printing Technology, (14) (2008)
3. Wang, Q.: 3D Display Technology and Devices. Science Press, Beijing (2011)
4. Kim, C., Chang, M., Lee, M., et al.: Depth plane adaptive integral imaging using avarifocal liquid lens array. Appl. Opt. (2015)

Software Reliability Model Considering Different Software Failure Process Deviations

Enliang Wang[(⊠)], Ming Cheng, and Defeng Tu

School of Electronic Engineering, Anhui Xinhua University, Hefei 230088, China

Abstract. Software reliability analysis is to predict and evaluate software reliability through reasonable modeling based on software failure data and other information, The modeling of software failure data is essentially to make it become a sample trajectory of a random process. In this paper, a software reliability model considering different software failure process deviations is established, and NHPP process is used to represent the change trend of mean function of failure process, ARMA process represents the deviation sequence of actual failure process to mean process. Experiments on two groups of published real data sets show that the new model has better fitting ability and applicability than some widely used NHPP Software reliability models, and maintains better prediction ability.

Keywords: Software reliability model · Sample trajectory · Non-homogeneous Poisson process model · Stochastic process · Autoregressive moving average process

1 Introduction

Personalization is the need to add unique and own characteristics on the basis of popularization. It emphasizes the service and demand with individual interest characteristics. Personalized recommendation system can provide users with personalized resource list according to their historical behavior data. With the rapid popularization and rapid development of the Internet, people have entered an era of information explosion. In this era, while enjoying the convenience of information, they are also trapped in the mire of information overload. Personalized recommendation system is proposed to solve these problems and has been developed rapidly. A complete recommendation system consists of user behavior recording module, user preference analysis module and recommendation algorithm module [1].

With the continuous expansion of software scale and the increasing complexity of software structure and function, people pay more and more attention to software quality. Software reliability is one of the most important indexes to measure software quality, which refers to that under the specified conditions and within the specified time, Probabilistic software reliability model without software failure is a very active field in software reliability research. From Hudson's work to the publication of jelinski moranda (J-M) model in 1971, more than 200 models have been published. They are not only powerful tools for prediction, distribution, analysis and evaluation of software

J. C. Hung et al. (Eds.): FC 2021, LNEE 827, pp. 1066–1070, 2022.
https://doi.org/10.1007/978-981-16-8052-6_145

reliability, The purpose of software reliability model research is to give software reliability estimation or prediction value by statistical method according to software reliability data.

2 Research on Software Reliability

According to the definition of avizienis, failure refers to the failure of a system to provide correct service according to the specified function; error refers to the state that may lead to system failure, which can be generally understood as the difference between the calculated, observed or measured value and the real, specified or theoretically correct value; Fault is usually considered as the root cause of errors, including software and hardware design defects, software code vulnerabilities, external environment interference and other factors. The failure of a system to provide services correctly may be manifested in different ways, which are usually called failure modes [2].

Reliability refers to the ability of the system to continuously provide the right service. It can be defined as a probabilistic measure, that is, the probability of correctly completing its function. Its value can be expressed as 1 minus the probability of failure. Reliability is an important measure of software quality, which is widely valued by programmers, engineers and researchers. These researches can be divided into two categories: software reliability growth model and software reliability analysis/prediction model.

In the process of software debugging, software reliability growth model takes the whole software as a black box model, obtains software failure data through testing, and then predicts mean time to failure (MTTF). The black box model only focuses on the interaction between the software interface and the external environment, and does not care about the internal structure of the software. Therefore, the software reliability growth model cannot obtain other data except the failure data. Once the software is slightly modified or upgraded, the whole testing process needs to be carried out again.

Mean square of fitting error (MSE) and predicted error (PE) were used to measure the fitting ability of the model:

$$MSE = \frac{\sum_{i=1}^{n} (y_i - \hat{y}_i)^2}{n} \tag{1}$$

$$PE_i = Actual_i - predicted_i \tag{2}$$

Where n is the number of failure samples in the failure data set.
Relative error (RE) is used to evaluate the prediction ability of the model:

$$RE = \frac{m(t_q) - q}{q} \tag{3}$$

Because it is impossible to know the behavior of the program for each possible input in advance and to predict the future input of the program accurately, the software failure process is indeed random. The traditional software reliability model based on random process always assumes that the software failure process should be described by different decay curves, For example, exponential decay type, S-type, Weibull type, inverse linear type, geometric decay type, etc., However, computer software products have their own characteristics, and their failure behavior is much more complicated than that of physical products.

The main reasons for the low accuracy and poor adaptability of traditional stochastic process models are as follows [3]: (1) researchers can never get all the information about the program in advance, so the statistical components in reliability problems can not be described by a single statistical distribution function, and with the change of software projects, Many prior assumptions made by stochastic process model on the attributes of software failure and software failure process are not always accurate; (2) generally speaking, software reliability failure process is not reproducible, even for the same software system, there will be different failure processes under different test conditions and use environments, In fact, the failure time or failure interval data used in reliability modeling is only a sample trajectory of software failure process. The existence of these problems makes the stochastic process reliability model produce some deviations in describing the failure process of different programs, while the traditional stochastic process reliability model just ignores these deviations.

3   Reliability Model Considering System Call

In reality, most of the software modules are called more or less. Although in the macro view, the software is only the execution of a module, but from the micro view, there are different stages of the execution of the module code and system call alternately. For the module with system call, its reliability is not only related to its own code, but also closely related to its dependent system call.

Intuitively, the reliability of the module with system call seems to be obtained by multiplying the reliability of its own code with the reliability of system call. However, due to different users or different tasks, the number of system calls in the module may be random, and simply multiplying the reliability of the two can not reflect the real situation; at the same time, hiding the failure of system calls in the module, it is difficult to find the real reason for the vulnerability of the module. In addition, the code of system call does not belong to the module itself, the code style is very different, it may not be open source, and the test method is completely different, so it should be considered separately. Therefore, the system call is extracted from the module and processed as call and return style. In other words, each system call is regarded as an independent special module in the software, that is, the module that determines the control flow (returns to the calling module) after execution.

3.1 Failure Mode and Error Propagation Modeling

Based on the reliability model of system call, this paper attempts to model various failure modes and error propagation in software. For convenience, assume that the software extracts system call modules and forms N modules in the state diagram. Set self represents all the code modules extracted from the system call, and set call represents all the system call modules.

Because the common operating system only uses two privilege levels, corresponding to the kernel state and user state of the operating system, two different failures can be caused: user state failure and kernel state failure, numbered 1 and 2 respectively. At the same time, for the convenience of description and the unity of expression method, no failure is regarded as a special failure mode, labeled as 0. In any module, the three failure modes may be converted to each other.

Based on the classical software reliability analysis and prediction model, the influence of running environment on reliability is considered. The system call in software is extracted from the software modules as a special module, and a variety of failure modes are considered. A more practical software reliability analysis model is proposed. Although the state space of the model is increased compared with the traditional model in the application process, the state transition matrix is a sparse matrix, which only needs to calculate specific non-zero elements. At the same time, the matrix can be calculated by using mathematical software such as MATLAB. In addition, due to the different reliability of kernel code in different operating systems, the reliability of software operation is also affected. This model can effectively reflect the impact of different operating systems on software reliability [4]. At the same time, the model is suitable for common system architecture and common operating systems such as windows, Linux, macosun "X", and has certain application value when selecting the running environment for specific software.

The applicability of the new model is verified on more data sets by using stochastic process models other than NHPP, and how to more accurately describe the deviation between the traditional reliability model and the actual software failure process is studied, so as to further improve the prediction ability of the model.

4   Conclusion

In this paper, experiments show that the fitting ability of the new model is better than that of other models, which is particularly obvious in data sets, and the new model has good applicability to two groups of failure data with different characteristics. Theoretically, the new model can improve the adaptability of software reliability model because it does not make any statistical assumption on the change of software failure strength, Compared with other stochastic process models, many NHPP models have advantages in fitting effect, structure and application. Therefore, we can conclude that the model selected in this paper is representative, and the new model can still achieve similar results compared with other stochastic process models, The method proposed in this paper is not to fundamentally solve the problem of low accuracy and poor adaptability of traditional stochastic process model, but to make up for the deviation between traditional reliability model and actual software failure process by data pre-processing, Two complementary stochastic processes are used to model the software failure behavior. Firstly, the expected value of failure is obtained by fitting. Then, the autoregressive moving average process is used to control the error. The autoregressive moving average process is used to improve the fitting ability and prediction ability of the model.

Acknowledgements. This paper is supported by Anhui provincial fund for outstanding young talents in 2019 (gxyq2019133), 2016 Anhui quality engineering project "science and technology dream" maker Laboratory for college students (2016ckjh095).

References

1. Liu, H., Yang, X., Qu, F.: A software reliability growth model based on bell shaped fault detection rate function. Acta Comput. Sinica **28**(5), 908–913 (2005)
2. Zhao, J., Zhang, R., Gu, G.: Research on software reliability growth model considering fault correlation. Acta Computa Sinica, **30**(10), 1713–1720 (2007)
3. Zou, F., Li, C.: Software reliability chaos model. Acta Computa Sinica **24**(3), 281–291 (1999)
4. Zhang, Y., Sun S.: Software reliability modeling based on unascertained theory. Acta Softw. Sinica, **17**(8), 1681–1687 (2006)

Cyclic Study of Graph Search Algorithm in Lower Bound of Ramsey Number

Jiali Xu[✉]

School of Mathematics, Shanghai University of Finance and Economics,
Shanghai 200433, China

Abstract. In this paper, we study the algorithm of searching the lower bound of the equal Ramsey number by constructing the circular ingenious graph. We present an efficient algorithm, which has been programmed and implemented. From this, we obtain a circular ingenious graph with 46 points (4, 7), and prove that R (4.7) $\geq$ 47.

Keywords: Ramsey number · Cyclic ingenious graph · Algorithm · Lower bound

1 Introduction

AGV (automatic guided vehicle), also known as AGV, is equipped with electromagnetic or optical automatic guidance device, which can drive along the specified guidance path, AGV is an automatic transport car with car programming and parking selection device, safety protection and various transfer functions. The transport system composed of AGV is called AGV. The system plays an increasingly important role in the modern computer integrated manufacturing system (CIMS). At the same time, the system can be used in some special places, such as film warehouse, cold storage, vault, ammunition warehouse, etc., With the continuous progress of science and technology, AGVs has broad application prospects. The research of AGVs has become the focus and difficulty of scholars at home and abroad in recent years. The loop deadlock problem of the system is one of the important problems in the research of AGVs [1].

2 Regular Cyclic Graphs

A subgraph $G_j(S_i)$ is called regular if the parameter set (1) satisfies the regularity condition [2].

$$\begin{cases} 0 = a_0 < a_1 < a \cdots < a_{t-1} \leq 1 \\ a_i = \min\{a_j - a_{j-1} : j \in [1, t-1]\} < m/(t-1) \end{cases} \tag{1}$$

The set of all such n-PART partitions is denoted W. two n-PART partitions are specified.

J. C. Hung et al. (Eds.): FC 2021, LNEE 827, pp. 1071–1075, 2022.
https://doi.org/10.1007/978-981-16-8052-6_146

$$a = \{B_1, B_2, \cdots B_n\}, |B_t| = t_1, i \in [1, n] \qquad (2)$$

Now when people seek the lower bound of the two-color Ramey number, they have noticed that the circulant graph surface of prime order is not the circulant graph of general order, and have demonstrated the advantages of doing so. However, this paper has not explored more properties of the circulant graph of prime order. The four lower bounds R (5, 7) $\geq$ 80, R (5, 9) $\geq$ 114, R (4, 12) $\geq$ 98 and R (4, 15) $\geq$ 128 are obtained, Only the first one is still the best at present, the second one has been surpassed by R (5, 9) $\geq$ 1161, the third one and the fourth one, we get R (4, 12) $\geq$ 1287. As for finding the lower bound of multi-color Ramsey's teachings by using circulant graphs of prime order, we have not got the results of R (3, 3, 4) $\geq$ 30 and R (3, 33, 4) They are all obtained by using the method of General Order Circulant Graphs.

3   A Brief Introduction to the Research at Home and Abroad

3.1   Explanation of the Definition

In the definition of 1agvs, in the time interval [a, b], the resources occupied by AGV have not changed. We call AGVs in the state time t.

Definition 2 the so-called task conflict between two AGVs means that when an AGV performs one of the tasks on the same resource, other AGVs cannot perform another task [3]. For example, two conflicting tasks conflict with each other.

Definition 3 the so-called task blocking of two AGVs means that when one AGV executes one of the tasks, the other AGV can execute another task, but when one of the tasks cannot be executed normally, the other AGV cannot execute either.

3.2   A Survey of AGVs Deadlock

As for the deadlock problem, it first refers to how to solve the problem of blocking or cyclic waiting between processes caused by the unreasonable scheduling of the internal operating system of the computer. Because AGVs also has the problem of blocking and cyclic waiting between Ag V to avoid collision caused by the unreasonable scheduling of tasks, AGVs stops running, Therefore, AGVs also has deadlock problem. How to effectively solve the problem of AGVs cyclic deadlock is the focus and difficulty of AGVs research. According to the situation of AGVs deadlock research, Nai VIII proposed resource oriented colored Petri net (cropn) to solve AGVs cyclic deadlock problem. This method takes AGV as the object to decompose cropn, At the same time, according to the control rules developed by cron to control the occurrence of cyclic deadlocks, simulation experiments show that new cyclic deadlocks will be generated when they are in use. There is a kind of cyclic deadlocks that can not be searched by the domestic cyclic deadlock search algorithms, As shown in Fig. 1, Liu Bin proposed a method to realize AGVs deadlock free operation by controlling the number of AGVs in buffer stations. This method can control cyclic deadlocks, but it may cause new cyclic deadlocks. The fundamental reason is that this method can not observe the relationship between cyclic deadlocks. In many AGVs, two kinds of cyclic deadlocks exist in AGVS.

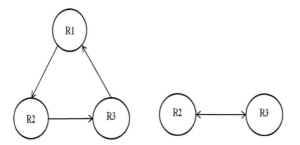

Fig. 1. Schematic diagram of circular deadlock in T-R diagram

From the current research situation, there is no effective algorithm for deadlock search, and the current search algorithm has some limitations. In addition, although the current deadlock search algorithm can find circular deadlocks, the relationship between them can not be effectively expressed, The simulation results show that in some ACVs, using the control rules proposed by Nai et al., new cyclic deadlocks can be triggered [4].

4   Discussion on Improved Algorithm of Cyclic Deadlock Search

4.1   Algorithm for Judging Whether There is a Cyclic Deadlock

Example: an AGVs, for the T-graph at the state time, g = <R, t> asks whether there is a cyclic deadlock in the AGVs at this state time.

G = <R, t> less than a directed graph. If there is a cycle deadlock in AGVs, there must be a cycle (loop) in the graph G. to judge whether there is a cycle in a graph G, we can do depth first traversal to g. once we find a backward path, there is a loop in the graph. Check whether the tasks in the loop can be executed normally. If there are abnormal tasks, the resource capacity in the loop will reach the maximum, There must be a cyclic deadlock.

From the above algorithm, it can be seen that only one vertex can be accessed twice at most, and the other vertices can be accessed once at most. Therefore, the time complexity of the algorithm is $O(N_{Tn})$, where N_T is the number of all state moments of the system:

① Do depth first traversal to tr graph G under state time t;
② If there is a backward direction, turn to ③, otherwise turn to ⑤;
③ Check whether the tasks in the loop can be performed normally. If so, turn to ⑤, otherwise turn to ④;

④ Check whether the loop with abnormal task execution causes the resource capacity in the loop to reach the maximum. If so, record the state time, otherwise, switch to ⑤;

⑤ Check whether all the state moments in the system are completed. If it is 6, otherwise it is 1;

⑥ Confirm the existence of cycle deadlock in AGVs, and output at which state the cycle deadlock exists.

4.2 Algorithm for Searching All Cyclic Deadlocks in T-R Graph

Example: an AGVs, for the T-graph of state time, g = <R, t>.

Question: how many loops with cyclic deadlock feature exist in the graph:

① For a given graph G, depth first search is performed, and each vertex is numbered according to the sequence of recursive calls;

② Change the direction of each edge in G to construct a new digraph G^T;

③ According to the vertex number determined in (1), the depth first traversal of G^T is started from the vertex with the largest number. If all the vertices in G^T are not visited in the traversal process, the vertex with the largest number is selected from the never visited vertices, and the depth first traversal is continued from this vertex;

④ In the final depth first generated forest of G, the vertices of each tree just form a strong connected branch of G^T;

⑤ For the strong connected branches that have been found, the transition enabling and triggering of each branch and the change of resource capacity are detected to determine whether it constitutes a circular deadlock;

⑥ Repeat steps (1) and (5) until all the states are checked.

4.3 Case Analysis

The T-R diagram of a certain AGVs at a certain state time t is shown in Fig. 2.

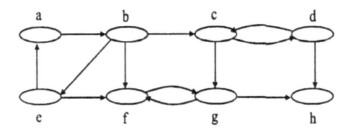

Fig. 2. T-R diagram of a certain state time

In CD GF EB, there is a task that can not be executed normally, and there may be a circular deadlock. Call the judgment algorithm to judge, and find that the backward loop contains tasks that can not be executed normally, and find that the resource capacity of the backward loop reaches the maximum. Therefore, there is a circular deadlock. Call the search algorithm to search all the circular deadlocks.

5   Conclusion

For the search algorithm of cyclic deadlock, whether it has been proposed in China or abroad, there are many unnecessary operations. When the state of the system is normal, all the reported search algorithms of cyclic deadlock need to be operated. In the improved algorithm, when the state of the system is normal, there is no need for all operations, At the same time, because the state time is used as the basis for finding the cyclic deadlock, it overcomes the shortcomings of the existing cyclic deadlock search algorithms at home and abroad. The cyclic deadlock search algorithm based on the state time can clearly express the relationship between the cyclic deadlocks of each AGV, Therefore, the control rules developed on this basis can completely avoid the generation of new cyclic deadlocks. In addition, with a slight improvement of this algorithm, all deadlocks in AGVS can be searched.

References

1. Wu, N.-G., Meng, Z.: Resource-oriented petri nets indeadlock avoidance of AGV systems. In: Proceedings of IEEE International Conference on robotics and Automation, pp. 64–69 (2001)
2. Wu, N.-A., Zhou, M.: Modeling and deadlock control of automated guided vehicle systems. IEEE/ASME Trans. Mechatron. 9(1), 50–57 (2004)
3. Zhu, F., Tan, D.: Multi robot collision avoidance and deadlock prevention based on elementary motion. Acta Computa Sinica 24(12), 1250–1255 (2001)
4. Liu, B., Wu, N., Cao, Y.: Deadlock free operation of AGV system based on colored Petri net model. Robotics, 23(7), 746–752 (2001)

Design and Implementation of Management Information System Based on Network Environment

Xiaohong Zhang[(✉)]

Jiangxi University of Engineering, Xinyu 338000, Jiangxi, China

Abstract. The main purpose of developing the student management information system based on the school factor network environment is to realize the automatic processing of university student management information. This paper introduces and describes the student management information system of Shandong Institute of architecture and engineering from the aspects of development background, design date standard, information sharing, function design, technical characteristics and operation environment. While focusing on the organization, storage and extraction of data, the system also takes a variety of measures, such as the setting of multi-level permissions to ensure data security; the use of triggers and stored procedures to ensure data consistency; the automatic generation of combination conditions and the setting of various code bases to ensure system flexibility.

Keywords: Network examination system · B/S three-tier system · ASP.NET

1 Introduction

With the computer Internet, especially the application and popularization of modern education technology, computer distance education and computer network examination have been paid more and more attention. Efficient and scientific examination students' learning, understanding students' mastery of knowledge, evaluating teachers' teaching effect, standardizing and guiding teachers' teaching behavior, and further optimizing teaching mode. The research on online examination at home and abroad has gone through a long period of time, and at this stage, People have perceived the advantages and convenience brought by this new form of examination [1].

The network examination system overcomes the shortcomings of the traditional examination, such as time-consuming, laborious and space-consuming. At the same time, it also avoids the influence of teachers' subjective factors in the marking process. The development of test paper generation technology and item bank management technology makes the network examination more scientific and reasonable; paperless examination reduces the consumption of manpower, time and resources for printing and distributing test papers; the application of modern multimedia in the network examination system makes invigilation easier; the marking process is complete or mostly realized by computer, which reduces the influence of subjective factors of

J. C. Hung et al. (Eds.): FC 2021, LNEE 827, pp. 1076–1080, 2022.
https://doi.org/10.1007/978-981-16-8052-6_147

manual marking; the score statistics are more accurate, Student information management is more perfect. Therefore, network examination has become technology.

2 System Architecture

The network examination system adopts the current mainstream B/S (Browser/server) architecture. This system structure divides the system into three layers: presentation layer and data layer.

B/S structure is to add a "middle layer" between client and database, also known as "component layer". It is an of C/S (client/server). Figure 1 shows the B/S three-layer architecture.

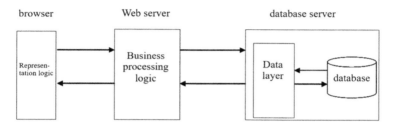

Fig. 1. B/S architecture

B/S mode is three levels:

The browser, which is the. It mainly completes the client and the background, displays and receives the data input by the user, outputs the final query results, and provides an interactive operation interface for the user. The client makes a service request to the designated web server, and the wweb server transmits the required information to the user according to the HTTP protocol, which is displayed on the client's WWW browser.

The is the, which, providing business logic, transaction scheduling and database connection, serving as the bridge between the client and the database. The web server accepts the client's request and connects with the background database for business processing. The background database returns the processing result to the web server and then transmits it to the client.

The third layer is the database server, which is the data access layer. It mainly realizes the function of managing the background database, completing the request of the web server, operating the tables and other files in the database, and finally returning the results to the web server [2].

Suppose there is an NN pixel image g (x, y) determined by Eq. (1):

$$g(x,y) = \frac{1}{M} \sum_{m,n \in s} f(x,y) \tag{1}$$

According to vector analysis, the direction of gradient is defined as:

$$g(x, y) = \arctan(G_x/G_y) \tag{2}$$

If the magnitude of the gradient is independent of the direction of the edge, such an operator is called an isotropic operator.

3   System Analysis and Design

3.1   Analysis of System Function Structure

The administrator module includes login management, user management, authority management and test paper management. The teacher module includes paper generation management, marking management and student management. The student module includes login examination, online examination and paper handing management. The system includes all kinds of questions under the traditional written examination. At the same time, it can score the students' management information system simulation operation, And comprehensively test the students' mastery of theoretical knowledge and practical application ability. Different from the traditional examination, the information-based examination process under the system is more automated, and students' objective questions and simulated operation questions are given scores immediately after submitting the examination. The reviewers only need to evaluate the subjective questions, which greatly reduces the manpower and time of evaluation, and shortens the cycle time of score release.

3.2   System Business Process Analysis

The business process of the system is divided into teacher operation, examinee operation and marking. Teachers can not only manage the examination information, student information and score information, but also customize the examination content and assign the reviewer after the examination; support students to simulate the operation and scoring of the management information system; be able to query and manage the score according to the course, admission number and reviewer; support the backup and recovery of the examination data [3].

4   Requirement Analysis of Management Information System

4.1   Summary of Demand Analysis

Requirement analysis is the first step of system development. It is mainly to collect and document users' real and actual requirements for MIS. The end user of MS is the user. Whether the system can fully meet the user's expectation of the system function is success development. Therefore, requirement whole development process of MS. In system development, the user's requirements for the system may be very specific, or very vague; it may be the current urgent need, or the expectation for the future.

Generally speaking, the initial user requirements are not completely determined, not systematic, multifaceted and messy, and may change with the development of demand analysis. All these increase the complexity and difficulty of requirement analysis. In the stage of requirement analysis, developers must fully investigate and systematically understand the real requirements of users for the system. In the demand analysis of scientific research management information system of a certain unit.

4.2   Business Process Analysis

When analyzing system, and optimize the business process on the basis of the new system [4].

Drawing business flow chart is a way to describe business process as little as possible and as simple as possible. Because of its simple symbols, it is very easy to read and understand the business process. However, its disadvantage is that it lacks sufficient performance means for some professional business processing details. It is more suitable for reflecting the business process of processing type. At present, the main business management processes of scientific research management are: horizontal project management, vertical project management, fund project management, project fund management, achievement/patent management.

Vertical projects mainly refer to the research tasks assigned by the relevant departments at a higher level and the national high-tech and major safety basic projects. For this kind of topic, a certain unit takes a more rigorous attitude, and implements it in strict accordance with the plans and relevant documents issued by the superior. First, it conveys information to each department, then it summarizes the application materials and conducts a comprehensive review. After the project is basically determined, it will report in time, prepare for defense and other related matters, and then it will manage the projects approved by the superior and allocate funds, And be responsible for regular inspection, comment, summary and acceptance.

5   Conclusion

With the rise and popularity of the Internet, as well as the development of network information, network examination system as a typical web system also develops rapidly and goes deep into the school teaching. The rapid development of computer and network technology will bring more convenience to the network examination. The network examination system will also be promoted and gradually replace the traditional examination. ASP.NET As an important part of. Net framework, it has many advantages in network development. We can make full use of net framework class library to build a new generation of network examination system.

Acknowledgements. Humanities and social sciences research project of Jiangxi universities in 2019 (No. JY19113), Research project on the teaching reform of colleges and universities in Jiangxi Province in 2019 (JXJG-19-28-4).

References

1. Yu, J.: Exploration of examination methods in new curriculum reform. Contemp. Educ. Sci. **12**, 28–29 (2006)
2. Ouyang, W.: Design and implementation of network examination system based on. Net. Central South University, Changsha (2009)
3. Jiayi, R.M.: Fundamentals of Educational Technology. Educational Science Press, Beijing (1992)
4. Chen, X.: Research and implementation of online examination system. Xi'an University of petroleum, Xi'an (2010)

Design and Implementation of Intrusion Detection System Based on Data Mining Technology

Wangping Zou[✉]

ChiZhou Vocational and Technical College, Chizhou 247100, China

Abstract. This paper proposes a method of building intrusion detection system based on data mining technology, and discusses the key technologies and solutions in the implementation of the system, including: data mining algorithm technology, feature selection technology, intrusion detection model construction technology and data preprocessing technology, In an intrusion detection experiment based on data mining for network tcpdump data, the effectiveness of this method is evaluated, and the future research direction is summarized.

Keywords: Data mining · Feature selection · Intrusion detection · Data preprocessing

1 Introduction

In recent years, with the rapid development of tenet, computer networks play an increasingly important role in modern society, and they have become the targets of many malicious attacks. All kinds of network intrusion events promote the development of network security technology. The main network security technologies include firewall technology, data encryption, authorization and authentication, intrusion detection and vulnerability detection technology. Intrusion detection technology is a very important method to detect intrusion [1].

Intrusion detection system collects information from the internal system and network, analyzes whether the computer has security problems from the information, and takes corresponding measures. Intrusion detection analysis technology is divided into misuse detection and anomaly detection. Misuse detection uses known intrusion methods to match and identify attacks. Anomaly detection refers to those behaviors that deviate from the normal way of using the system.

The development of computer network requires that a good intrusion detection system should be accurate, easy to expand and have good adaptability. Since all kinds of user behaviors, including attacks, are recorded in different audit data of the system, an intrusion detection system should be able to combine the information collected from various parts of the network system, such as the host and subnet. At present, most of the security related maillist www network stations publish new system vulnerabilities and intrusion methods, so the intrusion detection system should be able to integrate the information collected from various parts of the network system, An intrusion detection

J. C. Hung et al. (Eds.): FC 2021, LNEE 827, pp. 1081–1085, 2022.
https://doi.org/10.1007/978-981-16-8052-6_148

system must be updated in time. As an intrusion detection system needs to deal with a large amount of information, it also needs to be constantly updated, so it is a complex and huge project to construct an effective intrusion detection system. The people who build the system largely rely on their experience to choose the statistical means for anomaly detection. Recently, some intrusion detection systems and commercial products have begun to use built-in mechanism to customize and expand the system, such as NFR, which can complete data collection, analysis and storage on a platform. But it still needs security experts to analyze and classify attack means and system weaknesses in advance, and then manually write corresponding rules and patterns to detect misuse. Due to manual analysis and coding, the scalability and adaptability of many intrusion detection systems including NFR are limited.

2 Research on Data Mining Technology

2.1 Overview of Data Mining

Due to the progress of network technology, database technology and hardware implementation technology, a large number of information can be placed in the computer for efficient retrieval and query. The emergence of web technology promotes the use of hypertext format to integrate text, image and other information, enriches information resources, and enhances the ability of information generation and data collection. Therefore, thousands of databases are used in business management, administrative office, scientific research and engineering development, and the massive historical data stored in the system has caused new problems. For example, Wal Mart, an American retailer, has to process 20 million transactions a day, and the earth observation system launched by NASA in 1999 has to generate 60 GB of image data per hour, Mobil Oil Company is developing a data warehouse system that can store 100TB of data related to oil exploration. These huge databases and massive data are extremely rich information sources. However, the traditional data retrieval mechanism and statistical analysis methods can not meet the needs of effective information extraction. In order to make the data information truly become an effective resource, we need to improve the efficiency of data retrieval, Only make full use of it for business decision-making and strategic development services, otherwise it can only become a burden. Therefore, data mining, the core technology of knowledge discovery in database, emerges as the times require [2].

2.2 Classification Analysis

The nonlinear SWM transforms the vector in the input space into the vector in the feature space by nonlinear transformation, and constructs the optimal hyperplane in the feature space:

$$\langle W, \phi(X) \rangle + b = 0 \tag{1}$$

To separate positive and negative sample data. Here, W is the vector in the feature space, b ∈ R. In other words, the hyperplane constructed should satisfy the following requirements:

$$y_i\left[\left\langle W,\phi(\vec{X_i})\right\rangle + b\right] + \xi_i - 1 \geq 0 \quad \xi_i \geq 0 \quad i = 1, 2, \cdots, d \tag{2}$$

To solve this problem, Lagrange optimization method is used to transform the problem into its dual problem:

$$\sum_{i=1}^{d} y_i\alpha_i = 0 \quad 0 \leq \alpha_i \leq C \quad i = 1, 2, \cdots, d \tag{3}$$

This paper discusses the basic concept, architecture and basic algorithm of data mining technology; at the same time, it analyzes the possibility and relationship of data mining applied in intrusion detection: it briefly introduces the association rule analysis method and DRC-BK algorithm.

3   Intrusion Detection System Framework Based on Data Mining

3.1   Research on the Architecture of Intrusion Detection System

The intrusion detection system of centralized architecture is composed of a central station (client) and several host monitoring agents (servers), which adopts the client server mode [3]. Host monitoring agent is responsible for collecting local audit data and converting it into a format independent of the operating system, which can support heterogeneous network environment. Which audit data to collect can be determined in advance, which can be determined by the central station according to the requirements of data analysis. The audit data is sent to the central station for remote analysis and decision-making by data analysis system (usually expert system). In this structure, the relationship between the central station and the monitoring agent is that of controlling and being controlled. The central station provides security management for the monitored hosts, controls their audit function, obtains their new audit data through polling, and returns the analysis decision of the analysis system.

In the centralized architecture, the function of data analysis is in the central station with high performance. Because the data is analyzed centrally, it is easier to carry out global detection and decision-making. In addition, the central station has higher and more stringent security measures than the remote monitored host, which can ensure the more reliable implementation of intrusion detection function. In addition, the centralized policy management significantly simplifies the security management work, so as to ensure that the effectiveness and consistency of security policies can be improved, so the centralized architecture has the advantages of security.

3.2   Hybrid Hierarchical Architecture

The logic of intrusion detection system based on hierarchical architecture is tree structure: leaf node is the collection, analysis and detection entity of local (host) basic events (host based or network-based); intermediate node or root node is the collection, analysis and detection entity of comprehensive events in monitoring domain (root node is usually also the management console). Each detection entity is composed of a group of basic detection units called autonomous agents with independent detection functions. The leaf node only detects the intrusion based on the basic event information collected from the local host. The leaf node in the same monitoring domain further integrates, extracts and reduces the local basic events, and then transmits them to the same parent node in the middle layer responsible for the local monitoring domain. According to these comprehensive event information, the parent node can detect intrusion behaviors such as multi host scanning and cooperative scanning. Such a number of parent nodes can form the monitoring domain of higher intermediate nodes, so that more holistic and global detection and decision-making can be made according to more abstract comprehensive information. And so on, up to the root node responsible for the highest level monitoring domain. The root node is usually used as the management console to evaluate the attack status and respond. The system administrator can understand the system status and send commands. By reducing information layer by layer and control layer by layer, the hierarchical structure can achieve effective communication effect [4].

3.3   Detection System Client

Data preprocessing module. The data in intrusion detection comes from a large number of data packets in the network collected by network sensors and h-logs generated by the local host system collected by host sensors. The amount of calculation is very large, so data preprocessing must be carried out first. Data preprocessing includes data cleaning, data integration, data reduction, data transformation, correlation analysis, feature selection, discretization and concept stratification. Its main function is to transform the data into a form suitable for data mining through the standardized analysis of a large number of data in the network, the processing of data omission, the elimination of dirty data, the correction of data inconsistency, the storage and consolidation of data, and the processing of subset selection. The accuracy of data analysis not only depends on the algorithm used, but also depends on the quality of the data processed to a great extent, so the module plays an important role in the whole system.

Database module. The database adopts Oracle database system, which mainly accepts and stores two parts: one is the data processed by data preprocessing module, the other is the generated rules and models. The database can choose real-time, or choose to send the newly generated data to the data mining module for analysis every certain period of time or after a certain amount of data, and accept the features and detection models mined by the data mining module. If new features and detection models are found, they are sent to the detector module to update in real time. At the same time, these new features and detection models are sent to the central control platform to update the global feature database and detection model publishing module.

In addition, the database also sends the uncertain edge data suspicious data in the data mining module to the central control platform for secondary mining to discover cooperative attacks and distributed intrusion behaviors.

4   Conclusion

Data mining algorithm is used to process a large number of computer audit data, and the intrusion behavior and normal behavior patterns of computer system are obtained. In this way, the administrator does not need to analyze and write the intrusion mode manually, and does not need to guess the characteristic items by experience when establishing the normal use mode. And it can process data from different data sources, such as network monitoring data, host monitoring data or monitoring data of some new attack, and combine them into a new detection mode. Therefore, it has a good scalability.

Acknowledgement. Quality Engineering Project in Anhui Province in 2019 (Subject No: 2019cxtd045), Subject Title: Computer application technology teachers teaching innovation team.

References

1. Jiang, J., Feng, D.: Principle and technology of network security intrusion detection. National Defense Industry Press, Beijing (2001)
2. Lu, Y.: Intrusion detection system framework based on data mining. J. Wuhan Univ. (Sci. Ed.) **2**, 63–66 (2002)
3. Zhang, B.: Computer Communication and Protection Technology. Machinery Industry Press, Beijing (2003)
4. Tang, T.: Introduction to Military Intrusion Detection Technology. China Machine Press, Beijing (2004)

Computer Software Java Programming Optimization Design

Yinchao Li[✉]

Yunnan Technology and Business University, Kunming 445000, China

Abstract. With the rapid development of science and technology and the network world, people have a higher and higher demand for the functions of computer software, and the updating of computer software is also dazzling. In the development process of computer software, Java language is more and more widely used, and has a very broad space and application prospects, This paper starts with the Java language features of computer software design, explores the characteristics of Java programming and related technologies, hoping to provide effective insights for the development of computer software.

Keywords: Construction system · Multi-objective optimization · Ant colony algorithm

1 Introduction

In recent years, computer technology is more and more widely used, network technology and computer software has also obtained a lot of breakthrough progress, people have got a lot of convenience in life and work. The reason why the computer can play such a big role is to rely on relevant software. With people's growing material and cultural needs, computer software must be updated in time. Among many computer software, the prospect of Java VI a language application is the most extensive. Therefore, relevant designers should actively develop java programming to promote the development of computer software. With the continuous development of Internet technology, the popularity of computer applications for all sectors of society provides great convenience, and in the process of practical application of computer, whether the software function is perfect directly affects the use effect of the audience [1]. Therefore, for software developers, we need to use stable and safe programming language to continuously optimize the functions of computer software.

2 Introduction of Java Programming

2.1 Overview of Java Programming Language

Java language is by far the most influential program in the computer field in the world, and it is also one of the most popular computer programming languages. First of all, it was launched by Sun company, which has won the favor of many computer software

J. C. Hung et al. (Eds.): FC 2021, LNEE 827, pp. 1086–1092, 2022.
https://doi.org/10.1007/978-981-16-8052-6_149

developers. In the aspect of computer software programming, Java language has incomparable advantages over other software. Different from the past v F and v B, Java VI a language can not only fully support a variety of operating systems, but also synchronously implement the work of software writing under the condition of network. Compared with the complexity and inflexibility of v b and V f languages, Java VI a language is more complex and inflexible, JA VI a language has become the best popular programming language in the field of computer software development.

2.2   Overview of AVA Programming of 1 Computer Software

In the field of computer software research and development, the main programming language used to write software programs is JA VI A. this programming language has been widely praised and recognized by researchers in the field of software research and development with its own advantages. From the effect of practical use, Java programming language has the advantages, performance and characteristics that other programming languages lack [2]. Taking VF and B as the contrast object, the advantages of Java programming language are: on the one hand, it can adapt to the new requirements of different operating systems for software functions; on the other hand, it can realize the writing of software based on network. In this process, the disadvantage of VF and B programming language is too complex, and based on JAA programming language, the whole software programming work is not only simple and convenient, but also relatively high stability and security, which meets the new requirements of current users for software functions.

3   Characteristics of Java Programming Language

3.1   Simplicity

Compared with the previous C++ language, Java programming language no longer supports multi-level inheritance, automatic coercion, operator overloading and other functions, because these functions are less used in the whole application process, and there is often confusion. Java programming language has neglected these programs, which greatly reduces the complexity of programming. In addition, JA VI a language also adds some new functions, such as garbage collection in memory space, which greatly increases the practicability of programming language. On the one hand, it reduces the complexity, on the other hand, it increases the practicability, so JA VI a language is more reliable and simpler in the whole computer software development work. The class file filter designed in this paper is mainly used to filter encrypted class files. Firstly, through the designed algorithm, some binary strings with fixed length are selected from the system, and then compared with the key string by special bit operation, a candidate string with great possibility is obtained. According to the special pair conditions that the candidate string meets, the class file is encrypted, otherwise it will be encrypted.

3.2 Platform Independence

The biggest advantage of JA VI a language is platform independence. Java VI a language is the "Java virtual machine" as its own ideological guidance, first compiled into the intermediate code, and then expand the verification, loading, and finally implemented in the machine code of the cutting stone computer. Therefore, it can completely shield the specific characteristics of the platform environment. In the process of running, as long as it is a virtual machine that can support Java VI a, it can run Java program 23 object-oriented. With the development and innovation of all kinds of computer software, object-oriented has formed a kind of programming thinking and become a widely used programming concept. All kinds of object-oriented programming technologies have the characteristics of polymorphism, encapsulation and inheritance [3]. By classifying some objective elements, encapsulating parameters and using members' variables, the properties and states of elements are described, and some methods are used to function the software function behavior. Java language not only inherits the above advantages, but also fully meets the dynamic binding characteristics, so that the whole advantage of object-oriented technology is more widely used.

4 The Specific Advantages of Java Programming in Computer Software

4.1 Simplicity and Independence

First of all, simplicity means that JA VI a programming language is easy to operate in the practical application process, which is based on C and C programming language, but to some extent, it is the upgrade and optimization of these two programming languages. Even beginners can quickly master this programming language. The simplification of the programming process makes the corresponding programming operation simple and convenient, At the same time, the pointer is replaced by reference, which further reduces the difficulty of programming and improves the security and reliability of programming. In addition, Java programming language has many new functions, which strengthen its practicability. Secondly, independence means that the corresponding platform based on Java programming language has the characteristics of independence. In the actual application process, it can run after a compilation, and does not need to be changed, there is no requirement for hardware devices. Therefore, it can avoid the requirements of the platform environment, as long as the system can support the JA VI a virtual machine, it can run. However, in some cases, based on the differences of the platform itself, in order to solve the differences in the code conversion process, it needs to be fine tuned in the compilation process.

4.2 Safe, Reliable and Flexible

Easy to expand first, safe and reliable. This is one of the biggest advantages of Java programming language. The programming language can ensure the security and reliability of the software through the application of encryption technology. Once the corresponding display changes, it can use this security technology to disconnect the program from accessing the data, so as to ensure the security and integrity of the data. Especially in today's increasingly complex computer network environment, the invasion of network viruses and Trojan horse programs will directly threaten the security of computer software programs, and the use of Java programming language can provide technical basis for solving this problem. Second, it is flexible and expandable. Java programming language can be continuously innovated and optimized in the application process, which will not interfere with the operation of the original program [4]. Therefore, it can effectively expand the software program with the help of flexible programming.

5 Analysis of the Key Technologies Involved in Java Programming of Computer Software

5.1 Java. D. C Technology

JA VA ad.c technology provides the foundation for the mutual access of associated data, and lays the foundation for the connection between databases. In the actual programming, JA VI a programming language requires the corresponding programmers to debug according to the actual situation after compiling the program, so as to realize the connection between the corresponding databases. Based on this, through the application of Java. D.C technology, it can provide technical support for the effective management of the background database. At the same time, through the construction of the corresponding data connection tools, it can ensure that the software program has background data support services in the actual operation process.

Java class file B calculation:

$$K_B = Y_A^{X_B} \bmod p_0 \tag{1}$$

Get session key:

$$K_A = Y_A^{X_A} \bmod p_0 \tag{2}$$

Encrypting bilateral computation:

$$Y_A = a^{X_A} \bmod p, 0 < Y_A < P \tag{3}$$

In the filtering function, the main algorithms involved are abstract algorithm and bit operation. The abstract algorithm is completed by RSA algorithm, and a 128 bit length of the base string is generated by the information. In the base, there are 128 bit keys, which can be used for bit operation of binary strings, and a candidate string of 28 bit length is generated, which is convenient for the following calculation. The ratio of candidate strings generated is 1. It is found that if the length of this type is larger than the encryption degree, it is found that the encryption operation is carried out for the files of this type.

As a necessary embedded system for computer installation in China, it has strong practicability and wide application. This special computer system requires high volume and power consumption. Main principles of device embedded as shown in Fig. 1. Around the center of application, computer technology is the foundation of it, regardless of software and hardware can be tailored. This seriously reduces the service life and quality of the system, but the embedded system equipment has the advantages of high efficiency and fast running speed, and it will determine the completion of the work quickly after receiving the signal. The application of Java programming technology will help the embedded system to receive the signal again to a great extent. It can complete the tasks of various performance and indicators efficiently, quickly and purposefully, ensure the orderly completion of work steps, and improve the efficiency of system operation and work.

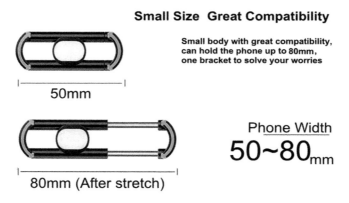

Small Size Great Compatibility

Small body with great compatibility, can hold the phone up to 80mm, one bracket to solve your worries

50mm

Phone Width

50~80mm

80mm (After stretch)

Fig. 1. Main principles of device embedded

Nowadays, mobile devices have become indispensable items in people's lives, so the industry pays more and more attention to the wireless application of computer Java programming technology, and considering the continuous promotion of Java and mobile, people's recognition and use of wireless are accelerating at this stage. Therefore, in the field of programming, we also improve the status of Java language in programming. In order to maximize the benefits, mobile manufacturers will invest a lot of energy and capital to support Java technology on mobile devices [5]. This kind of application technology will also be sought after, and the market demand for this kind of application will also be growing. The understanding of the relationship between wireless Java and mobile phone is equivalent to software and computer, and the application without VA improves mobile devices. It can be predicted that in the future mobile devices, Java applications will be closely combined with relevant operating systems to promote the use of devices and gradually create a new situation in the mobile device market environment. The different requirements and goals of all walks of life promote the development of Java programming technology, and fill in the loopholes of this technology bit by bit. In the process of standardizing Java programming, the corresponding mobile manufacturers will also make the basic requirements and specifications of various details applied in Java programming technology, such as application speed and load balancing, It effectively solves the problems of Java programming.

5.2   Java Annotation and Java Remote Invocation Technology

The application of Java annotation technology can associate the parameters and variables in the corresponding program language, and build the corresponding system, which has high stability and security; the application of Java Remote it vocation technology is the technical foundation to realize the distributed collation of the program, which lays the foundation for ensuring the safe and reliable operation of the whole program by calling objects and categories. From the perspective of software developers, the biggest advantage of using java remote vocation technology is to achieve a high degree of integration of information resources, which lays the foundation for ensuring the reliable operation of the corresponding programs using java programming language and the integrity of the program itself. In general, if the overall performance of the software is good, and its random encryption is part of the system class files, then you can protect the class files by software code. If the decompiled intruder wants to get the global information of the system, he must get all the code of the system, and then steal the file information according to the code. In view of this situation, we can encrypt some class files in the system at any time to ensure the security of the code and improve the efficiency of decryption.

6   Concluding Remarks

With the gradual development of computer technology, programming language has higher and higher standards. In this trend, Java VI a language is prominent in all computer programming languages, which has been widely promoted and applied. It also shows that Java is the most appropriate and efficient programming technology in the work of computer software development. I believe that it can provide more and more reliable functional support in the future development process. In addition, we should continue to deepen the discussion and analysis of programming technology, and explore the most suitable software program to meet the needs of society, so as to improve the overall level of computer software development. Thus, with the continuous development of China's computer level, Java programming language plays an important role in the process of scientific research and development. Therefore, we must pay more attention to it and fully realize that it is very important and urgent to develop java programming technology of computer software. There is no doubt that the development of computer software based on Java programming language will have a bright future.

References

1. Cao, H.: Characteristics of Java programming in computer software and thinking of its technology. Inf. Comput. (Theoret. Ed.) **11**, 41–42 (2018)
2. Liu, B.: Characteristics of Java programming in computer software and thinking of its technology. Bus. Story **10**, 16 (2018)
3. Huang, B., Wang, G.: Characteristics and technology of Java programming in computer software. Computer Fan, (04), 116 (2018)
4. Du, D.: Characteristics and technical analysis of Java programming in computer software. Comput. Knowl. Technol. **13**(36), 215–216 (2017)
5. Yu, B.: Computer software Java programming characteristics and technology thinking. Southern Agric. Mach. **48**(23), 123+127 (2017)

Application of Improved Quantum Evolutionary Algorithm in Computer Network Routing

Yinchao Li[✉]

Yunnan Technology and Business University, Kunming 445000, China

Abstract. With the continuous development of science and technology, computer network is widely used in various social fields. Accordingly, in the specific planning and expansion of the Internet, efficient routing of Internet communication network links has become an important issue. At the same time, the application of related improved quantum evolutionary algorithm plays a very important role in solving this problem. Therefore, the author analyzes the topic of application analysis of improved quantum evolutionary algorithm in computer network routing.

Keywords: Computer network · Routing · Quantum evolutionary algorithm

1 Introduction

With the development of economy and science and technology in our country, not only the technical level of computer network has been improved, but also its application scope has been expanded. Accordingly, the problem of routing has become increasingly prominent. In order to make the development of computer network conform to the requirements of social development and the law of development of corresponding things, it is necessary to put the selection of corresponding routes in an important position. We need to improve the quantum evolutionary algorithm to solve the problem of routing optimization. So that the computer network can be optimized and better applied in people's production and life. Internet computer application in various fields of society, at present, in the specific planning and expansion of the Internet, the important problem will be faced with is, in the case of the communication needs of each node of the Internet, How to choose the efficient routing of Internet communication network link is a very messy nonlinear programming problem, which is constrained by many constraints. It belongs to NP class problem in combinatorial optimization [1]. At present, there is no way to solve this kind of problem in the previous mathematical theory. Through the specific algorithm, on the basis of Internet routing, this paper compares the traditional quantum algorithm with the improved quantum algorithm, and concludes that the selection of Internet computer routing should be carried out from these aspects, which are: improving the convergence speed and the optimization ability.

J. C. Hung et al. (Eds.): FC 2021, LNEE 827, pp. 1093–1099, 2022.
https://doi.org/10.1007/978-981-16-8052-6_150

2   Computer Network Router

2.1   Overview of Computer Network Router

As we all know, in the computer network, the router occupies an important position, is one of the necessary equipment to realize the interconnection between networks. Router mainly refers to the computer network, in the corresponding OS or RM network layer above the relevant work. On this basis, it forwards, divides and stores the relevant data between different networks. At the same time, it makes corresponding decisions on the router orientation when transmitting data between networks. Network routing can not only realize the connection between networks, but also can be used to transfer information between different networks. In general, routers can only receive information transmitted by other routes. At the same time, it can connect two or more | P subnet logical ports, and it also needs a corresponding physical port. It mainly includes input and output ports, switching network, routing processor and so on. Different parts play their respective roles to make the network in an orderly operation. The router will help each relevant data frame to find the best transmission path, and on this basis, use the path as a bridge to transmit the corresponding data information to the destination node. How to choose the best path is the key to the problem. Routing algorithm is the most important part.

2.2   Two Word Evolutionary Algorithm

Quantum evolutionary algorithm (QEA) is a new evolutionary algorithm based on probabilistic evolutionary algorithm. It uses qubit chromosome coding, evolves chromosomes through quantum gate mutation, then observes the state of quantum chromosome to generate binary solution, and finally realizes evolutionary operation through the effect of superposition state of quantum gate. Traveling salesman problem (TSP) is one of the well-known combinatorial optimization problems, which has a wide range of practical applications, such as urban pipeline laying optimization, vehicle scheduling optimization in logistics industry, cutting path optimization in manufacturing industry and power distribution network reconfiguration. In this paper, an improved QEA algorithm is proposed to solve TSP, and a new TSP coding is proposed. The simulation results show the superiority of the algorithm.

3   Improved Quantum Evolutionary Algorithm in Internet Computer Routing

In the aspect of quantum evolutionary algorithm, it mainly uses this principle to complete the corresponding quantum planning algorithm by using the corresponding quantum revolving gate. Use search method to force the current solution to the optimal solution. The results can be preserved in the form of corresponding probability increase. Accordingly, we can use the method of probability reduction to delete the useless results and make the results in the best state.

3.1   Traditional Rotation Angle

3.1.1   Traditional Rotation Angle

In the improved quantum evolutionary algorithm, the traditional algorithm generally uses the way of looking up the table, the rotation angle is discontinuous, and the space search with jumping is not comprehensive and precise enough [2]. The strategy of dynamic adjustment will be used for optimization. The expression of is:

$$\Delta\theta_i = 0.01\pi \cdot 50\frac{f_h - f_x}{f_x} \qquad (1)$$

The fitness of B is expressed by F, and the fitness of X is expressed by F. According to the analysis in (1), a represents an interval value and individual fitness. If the value of B is larger, it means that the distance between the individual and the optimal individual is far, and then it means that the Internet search ability is greater, so it is required to improve the speed of network search correspondingly; if the value of B is smaller, it means that the distance between the individual and the optimal individual is closer, and then it means that the Internet search ability is smaller.

$$|\varphi \geq a|0 > + \beta|1 > \qquad (2)$$

$$U(\theta) = \begin{bmatrix} \cos\theta & -\sin\theta \\ \sin\theta & \cos\theta \end{bmatrix} \qquad (3)$$

3.1.2   Use Function to Adjust

Based on this point, the definition of qubit is meaning 1: meet the normalization condition to realize the real number pair of (a, b), and then corresponding to the corresponding probability of qubit. 2: the qubit of definition 1 is corresponding to the two-dimensional space, and the specific phase angle is arctan (b/a), which can be represented by a (x $\neq$ 0, me (−m2, M/2). According to the above definition, a strategy is designed, which includes the rotation of qubits in the specific two-dimensional space quadrant and phase angle. As shown in Fig. 1. Compared with the standard genetic algorithm, the algorithm uses a very small population size, the four initial populations can get satisfactory results in the experiment, and the number of iterations is also low. In 2000 iterations, the results can basically reach the optimal solution, and in large urban scale problems, the optimal value can often be achieved. In order to achieve the same performance, the classical genetic algorithm needs a large population size and a high number of iterations, compared with the conventional QEA. As shown in Fig. 1.

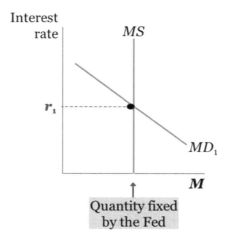

Fig. 1. Use function to adjust

3.2   Major Improvements

In order to accelerate the convergence of qubits, the reward function f (x) is introduced to improve the quantum revolving gate in the process of evolution. According to certain rules, the bits that are beneficial to evolution in prior knowledge are rewarded appropriately, and the bits that are not beneficial to evolution are punished. In order to replace the genetic evolutionary algorithm (GEA) which uses the current optimal individuals to update the square of the quantum gate, the optimal individuals are retained by the optimal retention mechanism as the rotation angle of the updated quantum gate [3]. based on the strategy of the genetic algorithm, the first n optimal populations are crossed, rearranged and migrated to generate x new populations, The (n + x) populations are observed and the top m (m $\leq$ n) optimal populations are selected according to the fitness function to expand the optimal population and expand the search range of the population. In the remaining population, select (n − m) populations into the optimal population with a certain probability to improve the population diversity. According to the above improvement strategy, the specific process is as follows.

4   Improvement of Quantum Evolutionary Algorithm

4.1   Ways to Improve

In the computer network, the powerful computing power of quantum computing has become one of the hot topics. In the computer network routing, the main problem in the traditional quantum evolutionary algorithm is that they mostly search the corresponding table to find a suitable solution. In this way, the correlation between the corresponding rotation angles is not close. At the same time, in the aspect of problem search, it has certain jumping, which is not conducive to the normal operation of the computer network. In order to solve the routing problems better, the quantum evolutionary algorithm needs to be improved. One is to adjust and optimize its rotation angle to make the final value more conducive to routing. The expression of the improved rotation angle is as follows:

$$\min \sum_{i=1}^{n-1} d(c\pi_i, c\pi_{(i+1)}) + d(c\pi_n, c\pi_1) \tag{4}$$

4.2 Improvement Effect

According to the improved expression, we can know that when the rotation angle is different, we will get different results. In other words, different values of rotation angle represent different meanings. For example, the smaller the rotation angle is, the closer the distance between the optimal individuals is and the smaller the search network is. In this case, the optimal solution can be found by using fine search; when the value of correlation rotation angle is larger, the distance between the corresponding individual and the optimal individual is increasing. In this case, we need to improve the search speed in order to make the routing network run better. The second is to optimize and adjust the corresponding function. The adjustment can adopt the combination optimization method to make the function in the optimal state and provide the corresponding conditions for obtaining the optimal solution. By using the method of combinatorial optimization, it can be concluded that there is no strong correlation between individual genes. Therefore, in the selection of computing network routing, we can adjust and optimize the function of quantum evolutionary algorithm. It now realizes the corresponding real property pairs under the condition that the normalization condition is satisfied, and corresponds them to the probability amplitude of a qubit [4]. The corresponding qubits can correspond to the two-dimensional space, and are represented by related symbols. On this basis, the corresponding quantum evolutionary algorithm can be simulated and tested to see whether the improved quantum evolutionary algorithm has more advantages than before.

5 Design Results and Analysis of Wide Angle EUV Multilayer Based on QEA

The optimal solution of wide angle MoSi multilayer membrane design based on QEA and QEA is compared. The results show that the structures of wide angle Mo/s multilayers based on the two quantum algorithms are completely different, which provides another better structure for the research and development of wide angle MOS multilayers and lays the foundation for further experiments.

5.1   Comparative Analysis of Design Results of Wide Angle EUV Multilayers Based on QEA and QEA

Based on QEA and the optimal wide angle MOS multilayers designed by QEA, the reflectance spectra are obtained by inversion. The reflection spectrum platform based on QEA is smoother, less volatile, and can achieve higher reflectivity, which shows that the wide angle MOS multilayer design method based on QEA has certain advantages. The design efficiency of wide angle m σ Si multilayers based on QEA is obviously better than that of α EA, which has the advantages of fast convergence speed and high accuracy. This shows that the algorithm uses gradient information of evaluation function to obtain adaptive heuristic algorithm, which speeds up the efficiency of the algorithm. In IQEA, individual gene coding also adopts quantum coding. A single chromosome can represent multiple Superposition States, which increases the diversity of the population and makes it search and optimize under the condition that the population size is less than the number of optimization parameters. The above analysis shows that the theoretical design method of wide angle EV multilayer based on EA is an efficient and feasible method.

5.2   Wide angle QEA Based on Different Population Sizes

However, when the search information is sufficient, increasing the number of population has little effect on the efficiency of solution, and too large number of population will lead to too long calculation time. The optimal solution of wide angle MoSi layer design based on QEA under different population size can reflect the reflection spectrum. The reflection spectra based on the optimal solution of different population sizes can be used to design wide angle Mo/Si multilayers with incident angle of $0°$ to $18°$ and reflectivity of about 50%. It is proved that QEA is feasible to design wide angle EU multilayers [5]. At the same time, the reflection spectrum platform based on the algorithm with population of 20 and 30 is smoother, less volatile, and can achieve higher reflectivity, which indicates that the algorithm has higher accuracy when the population size is larger. Therefore, in the design process of wide angle EU multilayers based on I · EA, selecting the appropriate population can make the algorithm approach the optimal solution faster, improve the efficiency and accuracy of the algorithm, and reduce the optimization time.

6   Conclusion

TSP is a combinatorial optimal 1 with a wide range of references, which can improve the solution of this kind of problem and has guiding significance in engineering. Quantum evolutionary algorithm itself has some good performance, which makes it more suitable for solving this kind of group. In this paper, we improve the coding method of TSP for the information it has. On the basis of quantum evolutionary algorithm, we improve the design of quantum revolving gate through prior information, and guide the evolution process through the global optimal solution. The experimental results show that this improvement can significantly improve the efficiency of solving

TSP, and has faster convergence speed and better global optimization ability. In the future, we need to expand the following work: combining with the evolutionary perspective of quantum revolving gate, we study the setting of penalty function and its influence on different problems; applying the improved quantum evolutionary algorithm to some more practical engineering models, Such as large-scale TSP, multi traveling salesman problem, multi traveling salesman problem with repeatable paths, etc.

References

1. Wang, Y.: Application of improved quantum evolutionary algorithm in computer network routing. Comput. CD Softw. Appl. **18**(03), 307–308 (2015)
2. Zhao, R.: Application of improved quantum evolutionary algorithm in computer network routing. Sci. Technol. Commun. **6**(24), 148+152 (2014)
3. Song, M., Yu, H., Chen, H.: Application of improved quantum evolutionary algorithm in computer network routing. Sci. Technol. Bull. **30**(01), 170–173 (2014)
4. Zhang, J.: Research on production scheduling problem based on improved quantum evolutionary algorithm. East China University of Science and Technology (2013)
5. Gao, H., Zhang, R.: Improved real coded quantum evolutionary algorithm and its application in parameter estimation. Control Decis. **26**(03), 418–422 (2011)

Analysis and Research on the Construction of Book Retrieval Platform Based on Disconnected Apriori Algorithm

Zhu Zhang[✉]

Guangzhou Vocational and Technical College of Foreign Economics,
Guangzhou, Guangdong, China

Abstract. This paper introduces an improved Apriori algorithm. The algorithm can directly get the final frequent itemset by finding the maximum frequent itemset which is greater than the minimum holding count. The improved algorithm is applied to the library bibliography recommendation service. The analysis and Research on the construction of the book retrieval platform under the disconnected Apriori algorithm, The performance of the improved algorithm and Apriori algorithm is analyzed and the running time of the experimental data is compared. The experimental results show that the improved algorithm has a significant improvement in running speed and mining performance compared with the classical Apriori algorithm.

Keywords: Data mining · Apriori algorithm · Book retrieval platform

1 Introduction

The amount of books in modern library is increasing. If there is no good recommendation system, a large number of useful books in the library may be idle. Therefore, many libraries use data mining technology to achieve bibliographic recommendation service. Generally speaking, data mining (DM) is a process of discovering potential, valuable and interesting knowledge discovery in database (KD) from a large number of incomplete and noisy data stored in a large database or data warehouse. Association rule is one of the main methods of data mining, and the most classic algorithm of association rule mining is Apriori algorithm proposed by Agrawal in 1993.

2 Apriori Algorithm

Apriori algorithm is a recursive method based on frequent set theory, which is also an iterative method called layer by layer search. As a classic frequent itemset generation algorithm, Apriori algorithm plays a milestone role in data mining research, but it has two defects: ① the algorithm must spend a lot of time processing a huge number of candidate itemsets; ② the algorithm must frequently scan the transaction database to calculate the support of candidate itemsets. These two defects are the hot and difficult points of current association rule mining, and also the bottleneck of constraint system

J. C. Hung et al. (Eds.): FC 2021, LNEE 827, pp. 1100–1105, 2022.
https://doi.org/10.1007/978-981-16-8052-6_151

performance. Many foreign scholars have also proposed algorithms based on hash, partition, sampling, dynamic itemset counting, matrix algorithm of association rules, etc. these algorithms have optimized Apriori algorithm of association rules in different aspects. Many domestic researchers have also improved the Apriori algorithm, mainly including: ① reducing the number of database scans; ② reducing the number of candidates that cannot become frequent itemsets; ③ reducing the number of frequent itemsets connected. Although these methods improve the classical Apriori algorithm to a certain extent, they do not really solve the bottleneck problem of Apriori algorithm. Confidence level: determine the frequency of Y appearing in the included transaction.

$$c(X \rightarrow Y) = \sigma(X \cup Y)/N \tag{1}$$

Probability description: confidence of item set X to item set y.

$$(X == > Y) = P(X/Y) \tag{2}$$

3   System Outline Design and Detailed Design

3.1   Database Conceptual Design

Because this system is a small library retrieval system developed for colleges and universities, considering the problems of funds and use, this system uses Microsoft SQL Server database. Microsoft SQL server is the most popular open source database in the world. It is a fully networked cross platform relational database [1].

SQL, which means structured query language. The main function of SQL language is to establish and communicate with various databases. According to ANSI, SQL is regarded as the standard language of RDBMS. SQL statements can be used to perform a variety of operations, such as modifying and deleting data in the database. Nowadays, most of the relational database management systems in the market adopt SQL language standard. For example: Oracle database, Sybase database, aces database and so on. Even though some databases have redeveloped and extended SQL statements, most of the standard SQL commands including select, insert, update, delete, create and so on can still be used to complete almost all database operations.

3.2   Interface Design

The interface design principle of the system is that the function is more intuitive and the user is clear, so that the ordinary staff can see all the functions clearly, and the system can be used conveniently without a lot of training. The interface design takes the user as the core, let the user control the workflow and work response, rather than the developer to design the operation process and impose it on the user.

The information display of the system should not let the data surround the user, but should use the information display mode that the user can easily absorb. The operation of the system should display meaningful information, and do not let the user think it is English garbled. Different information should make users feel separated. In order to show the user interface in the most efficient way, the consistency of the interface should be kept in the design. Consistency includes not only the use of standard controls, but also the use of corresponding information representation methods, such as ensuring consistency in font, label style, color, terminology, display error information, etc. [2].

3.3   Database Security

Database security is divided into two parts: security and confidentiality. Security refers to the reliability and stability of the database. Confidentiality refers to data encryption and data access control. The system's data will be regularly backed up, data files and log files will be regularly backed up to different physical devices, so as to ensure the reliability of data. If there is a data problem, you can recover the data [3]. The system's database hierarchical set permissions, so that different managers have different access rights, for external personnel, the database will be rejected. This also ensures that the data access rights will not be used illegally. Database security is shown in Fig. 1.

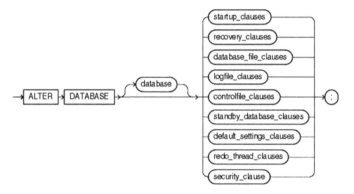

Fig. 1. Database security

4   Book Retrieval Module

Enter one or more keywords in the query box. And, or and other conjunctions can be used to connect multiple search terms. If you want to retrieve a phrase, enclose it in double quotation marks. Select the search field. This inquiry provides book title, book abstract, book author, book keyword, etc. Note that the search field is the scope of the search area in the database. Select sort type: if you want the search results to be displayed in a certain order, two choices are provided in the drop-down menu. Relevance and publication date are in reverse order by default. Select the number of records to be displayed on each page. By default, 10 records will be displayed on each page. Click the search button or press enter directly. The retrieval process of simple retrieval function is to convert the data entered by users into Lucene index format, and then through Lucene component, high-speed retrieval is carried out in the index file. The retrieval result of the file is converted to the data source format and bound to the table control of the presentation layer.

There are multiple text boxes in advanced retrieval that enter query conditions, and one or more keywords are entered in the first query box. If you want to retrieve a phrase, enclose it in double quotation marks. The right side of the query box is the list of search fields. Click the drop-down list to select the search fields including book title, book digest, book author, book keyword, etc. Note that the search field is the scope of the search area in the database. If necessary, you can continue to enter the query keywords in the second and third query boxes. Select the search fields corresponding to the second and third query boxes. Select the cloth operator (and, or, not) to combine the relationships between the characters retrieved by each input box. Select the sort type correlation or release date. Click the search button or press enter directly. Add search conditions to browse and improve retrieval [4].

Book classification is to set the classification details of books. And complete the increase of all books, book classification is the basis of book management system. The basic operation of book classification includes adding, modifying and deleting. The newly purchased books should be added and updated in the book classification in time. In order to maintain the consistency of library management. The basic operations include "add same level classification", "add lower level classification", "Edit classi-fication", "delete classification", etc. In the classification management, there will be a tree type classification on the left, and the information modification area on the right. The tree type classification on the left should clearly show the hierarchical structure, classification and sub classification. When a classification is selected, the information modification area will list the basic information, classification name, classification number, classification description of the classification. Modify the information directly on the left side. The name, number and number cannot be empty. Serial number can only be numeric. When the name and number are empty, the system will give a prompt. The name and number cannot be empty. When it is correct, the system will directly save and update the data on the left. You can also add sub items. When you click Add sub item, the data will be cleared, and the upper level data will display the currently selected classification. After entering the relevant data, click Save to directly add the sub classification.

5   Function Test of the System

Software testing technology can be divided into white box testing and black box testing. Software testing is the process of executing a program in order to find errors in the program. Software testing in the process of software development across two stages: usually after the completion of the development of a module, it is necessary to make the corresponding test, such a test is called unit test. General module developers are also responsible for the unit testing of this module. Generally speaking, the goal of software testing is to find out all defects and errors systematically with the least manpower and time. If the goal of testing is to prove the correctness of the software, then people will choose some correct data to test. If the purpose of testing is to prove that the software is wrong, the tester will choose some error prone data as test cases. The so-called good and bad or success and failure test also has some psychological problems.

The general idea of testing is to find out all the errors and defects in the software, but it is impossible, even the simplest program is not good. Software testing also has a certain risk. If you can't test all the situations, you have chosen the risk. The principle that software testers should master is how to reduce the endless possibilities to the controllable range. And make a reasonable test program. Developers and testers are different jobs. Development is the act of creating or building, and testing is to prove that the system is abnormal. There is a contradiction between the two. Developers' participation in testing activities should be limited. Developers are only suitable for the lowest level of testing, namely unit testing. If the software testing is undertaken by an independent department, there are many advantages. The so-called independent department means that the development department and the testing department are relatively independent in economy and management. Independent departments can prevent software developers from participating in testing. If test and development are in the same department, there is a lot of work pressure from the same management department, which interferes with the test process. It is of great significance to improve the overall quality of software by using independent testing methods, both in technology and management.

6   Conclusion

In this paper, the association mining is used in the book lending database, which can find a lot of useful association information, which can be used to guide librarians to choose books. Using the potential information, it can recommend related books for readers, and optimize the display position of related books. In the daily business of the library, there are still a lot of data worthy of mining, and the library service is becoming more and more personalized. It also needs to draw personalized information from the daily information, so it is worth further exploring and studying the data mining technology. Since Apriori algorithm needs to scan the database once every time it looks for frequent itemsets, the efficiency of the system is greatly affected with the increase of the database. Therefore, in the future work, take other methods to improve the aprior algorithm, in order to improve the efficiency of the system. The author's innovative point of view: This paper applies the association rule mining method in data mining to the analysis of book circulation data, and uses the classical algorithm Apriori algorithm of association rules to mine and discover the implicit rules in readers' borrowing behavior from library book borrowing records, so as to guide administrators to choose books, Recommend relevant books for readers to optimize the placement of books.

Acknowledgements. Project source: one of the research results of the research project "Innovation and Development of Information Literacy Education Mode of University Library under the Background of Mass Innovation and Innovation", which is planned by the Third Council of Guangdong Vocational and Technical Education Association for the year of 2019–2020 (Project No.: 201907Y61).

References

1. Goofy. Research on association rules mining algorithm. Xi'an University of Electronic Science and technology (2001)
2. Peng, Y., Xiong, Y.: The application of association mining in the analysis of historical data of literature borrowing. Information Technology. two thousand and five
3. Chen, Y., Cao, D.: And an improved apoi algorithm for mining frequent itemsets. Microcomputer information 20108 - three: one hundred and thirty-eight - one hundred and forty
4. Li, C., Ling, Y.: Data mining and its application in library. Small Information Technology, twenty thousand and twenty-six: thirty-three – 35

Application of Deep Learning in Supply Chain Management of Transportation Enterprises

Zhengteng Hao[✉]

School of Economics and Management, Qinghai Nationalities University,
Xining 810007, Qinghai, China

Abstract. In the increasingly fierce international competition, the rapid development of high and new technology, the rapid change of customer demand and the complex market environment, how to adapt to the new competitive environment of enterprise management has become the focus of theoretical and business circles. The emergence of supply chain management provides a strong theoretical basis for enterprises to solve the above problems. At present, most of the supply chain innovation comes from product manufacturing and retail enterprises. The next wave of change will sweep the service industry (such as banking, entertainment, medicine and health care, etc.). Through creative supply chain management, service enterprises can quickly gain competitive advantage. Transportation enterprise is not only a service enterprise, but also an important member of the supply chain. If it has high performance and fast response supply chain, it can help customer enterprises to improve the performance of supply chain management.

Keywords: Information fusion · Agile supply chain · Transportation enterprise · Management mode

1 Introduction

Since the 1990s, technological progress and demand diversification have accelerated the upgrading. Superior supply chain management capabilities enable enterprises to capture fleeting market opportunities and make rapid response. Therefore, today's supply chain management is crucial for enterprises, and it has become a strategic issue that needs to attract the attention of senior managers [1].

There are a large number of Companies in the forefront of supply chain management in the world. Through effective supply chain management, they have won the recognition of the market and gained obvious returns in terms of revenue. For example, Benetton (rapid response), Wal Mart (extremely low supply chain cost), Microsoft (virtual manufacturing and logistics), whirlpool (direct door-to-door delivery) and Peapod (online shopping) are examples of efficient companies that utilize and create the best supply chain management practices in the world.

2   Agile Supply Chain Management

2.1   Overview of Supply Chain Management

With the advent of economic globalization and information age, in the increasingly fierce international competition, the rapid development of high and new technology, the rapid change of customer demand and the complex market environment, how to adapt to the new competitive environment of enterprise management has become the focus of theoretical and business circles. The emergence of supply chain management provides a strong theoretical basis for enterprises to solve the above problems, and guides the management practice of enterprises [2].

2.2   The Development trend of Supply Chain Management

With special needs of customers, the research and application of supply chain management pay more and more attention to the direction of globalization, agility, networking and green, so that the supply chain member enterprises can stand in a dominant position in the market competition.

The concept of virtual enterprise was first put forward in the report "strategy of manufacturing enterprises in the 21st century" of Caspian University in 1991. Professor R. dove, one of the proponents of virtual enterprise, once pointed out: "agile can also be connected with virtual enterprise, which means smooth supply chain and various ways of connection.". In other words, the original intention of global virtual enterprise also implies the premise of supply chain. Coupled with the complexity of the relationship between supply, production and sales, the process will involve more and more manufacturers in different regions, and eventually become global. It is based on this theoretical origin and practical demand that the global supply chain emerges. The formation of global supply chain management will make logistics, to take the initiative and have a say in the competition by virtue of the integration advantage of the huge supply chain.

3   Agile Supply Chain Management Mode of Transportation Enterprises

3.1   Overall Framework of Agile Supply Chain in Transportation Enterprises

With the trend of market globalization, China's transportation enterprises are facing great challenges and opportunities. Enterprises are in urgent need of modern management methods and technologies to help them change their operation mechanism, so that enterprises can change from large and comprehensive to specialized and refined, from closed to open, from repetitive to high-level optimal combination, so as to strengthen the competitiveness of enterprises through cooperation. Agile supply chain management is emerging to adapt to market globalization and customer demand diversification. It emphasizes the integration of entities and their activities in the supply chain, so as to coordinate the relationship between entities in the supply chain, effectively and quickly control, so as to achieve a flexible and stable relationship between supply and demand.

3.2 Fusion Decision Method of Transportation Decision Center

When the transportation decision-making center makes decisions, it generates a series of decision-making schemes through information collection, preliminary calibration and preliminary fusion judgment. These schemes form a set, and then it is ultimately to select an optimal decision-making scheme from the set. Among many information fusion methods, DS method is the evidence theory put forward by Dempster and Shafer in the 1970s. It mainly makes decision by combining multiple information and makes reasonable information theory explanation for reasoning. It is a decision theory [3]. It can not only deal with the uncertainty caused by inaccuracy, but also deal with the uncertainty caused by ignorance. Therefore, the main application of transportation decision center is the fusion decision.

The brief calculation process of DS algorithm is as follows:

$$M(\Phi) = 0 \quad \sum_{A \subset \Omega} M(A) = 1 \tag{1}$$

M (A) is called the basic probability assignment of a, which indicates the precise degree of trust in A.

$$Bel(A) = \sum_{B \subset A} M(B) \quad (\forall A \subset \Omega) \tag{2}$$

The fusion decision method based on evidence theory is as follows:

$$M(A_1) = \max\{M(A_i), A \subset U\} \tag{3}$$

$$M(A_2) = \max\{M(A_i), A \subset U \text{ and } A_j \neq A_1\} \tag{4}$$

At present, the main idea of agile study on the basis of limited links (the supply relationship and behavior between two entities), that is, to determine the corresponding supply chain structure, management and operation mode based on the limited entities participating in the supply chain and their specific supply relationship in the actual environment, which is called serial management mode. Serial management mode and its related planning and control process are suitable for the static enterprise structure and relatively stable external environment, but in the competitive, cooperative and dynamic environment, serial planning and control process often lags behind the changes of the environment, which is not suitable. So, next, we discuss the agile supply chain management mode of integrated transportation enterprise based on transportation decision center.

3.3   Strategic Dynamic Alliance of Transportation Enterprises

Therefore, in the process of implementing agile supply chain management, transportation enterprises should establish some horizontal strategic dynamic alliances according to customer needs to strengthen their service capabilities.

The risk of strategic dynamic alliance refers to the possibility that the member enterprises of strategic alliance may lose because of the uncertainty and complexity of the internal and external environment of strategic alliance system. There are two types of risks in strategic dynamic alliance, one is relationship risk, the other is operation risk. Operational risk refers to those factors that threaten the realization of strategic objectives. Operational risk exists in any organization's operation. Relationship risk refers to the failure of alliance enterprises to fulfill their commitments in the expected way. It includes those alliance relationship problems that prevent the realization of alliance goals, such as the opportunistic behavior of alliance members, the loss of core competence or competitive advantage, the possibility of merger and alliance failure, etc. The partner selection of strategic dynamic alliance should ensure the minimum risk of the whole alliance [4].

Transportation enterprise is a service enterprise which provides transportation, storage, packaging, handling, information and so on. The service quality provided by its strategic dynamic alliance partners is directly related to the competitiveness of the alliance, so a standard of partner selection is to enable the alliance to provide high-quality services. Generally, service quality is positively correlated with cost, and the improvement of service quality may be at the cost of cost. In the case of low service quality, increasing the cost may make the quality grow faster, but when the service quality reaches a certain level, the continuous increase of cost may only make the quality increase in a small proportion. At this time, it is also a good way to provide high-quality service through the establishment of strategic dynamic alliance.

Determine the selection range of alliance partners. Different strategic dynamic alliances of transportation enterprises have different selection ranges, which should be determined according to the development strategy of transportation enterprises. For example, for an alliance within a region, its partners can consider the cooperation of enterprises within the region; for a cross regional or national alliance, it is necessary to select suitable enterprises between regions or even across the country. In short, the larger the scope of the alliance, the more complex the selection process.

4   Conclusion

Agile supply chain (ASC) is an important research topic in the field of business and theory. Agility is an important characteristic of ASC. At present, most of the supply chain innovation comes from product manufacturing and retail enterprises. The next wave of change will sweep the service industry (such as banking, catering, medicine and health care, etc.), and this wave will closely follow the manufacturing enterprises to provide excellent supply chain service capabilities to a generation of consumers who require more diversification. Through creative supply chain management, service enterprises can quickly win competitive advantage.

References

1. Stevenson, W.J.: Operation Management, pp. 47–69. China Machine Press, Beijing (2000)
2. Lan, B., Zheng, X., Xu, X.: Supply chain management in e-commerce era. China Manag. Sci. (3), 1–7 (2008)
3. Anbang, D., Zhiying, L.: Review of supply chain management. Ind. Eng. **5**(5), 16–20 (2002)
4. Zhixiang, C., Shihua, M., Yifan, W.: Customer satisfaction evaluation model and empirical analysis. Syst. Eng. **3**, 43–70 (1999)

Machine Learning Algorithm Based on Relevance Vector Machine

Xuan Luo[✉]

Wuchang Shouyi University, Wuhan 430064, Hubei, China

Abstract. Correlation vector machine (RM) is a sparse probability model based on Bayesian theory. It uses the idea of conditional distribution and maximum likelihood estimation to transform the nonlinear problem in low dimensional space into linear problem in high dimensional space through kernel function. It has the advantages of good learning ability, strong generalization ability, flexible kernel function selection and simple parameter setting. Because of its excellent learning performance, it has become a research hotspot in the field of machine learning. This paper introduces the classical relevance vector machine algorithm and its improved model, focuses on the ideas, methods and effects of using relevance vector machine algorithm to solve the classification and prediction problems in fault detection, pattern recognition, Cyberspace Security and other fields, summarizes the existing problems, and prospects the future research direction.

Keywords: Correlation vector machine · Machine learning · Bayesian theory · Autocorrelation decision theory · Sparse probability

1 Introduction

Relevance vector machine is a supervised machine learning algorithm based on Bayesian statistical learning theory proposed by tipping in 2001, which can be used to solve classification and prediction problems. After the theory of relevance vector machine was put forward, it soon became the research hotspot of new statistical learning theory, and achieved rapid development and wide application, and gradually became an independent research direction in machine learning algorithm. Compared with support vector machine (SVM) 2, RVM algorithm can give probabilistic output, and has the advantages of stronger generalization ability, better sparsity, more flexible kernel function selection and simpler parameter setting. Based on Sparse Bayesian framework, relevance vector machine (RVM) uses conditional distribution and maximum likelihood estimation in mathematical statistics to transform low dimensional nonlinear problems into high dimensional linear problems through kernel function, and uses automatic correlation decision (ARD) to constrain the model, so as to obtain a more sparse model than support vector machine (SVM). Different from support vector machine, relevance vector machine does not need to satisfy the Mercer theorem that the kernel function is positive semidefinite, and can give a probabilistic output. In 2003, a fast sequence sparse Bayesian learning algorithm was designed, which significantly improved the training speed of the model. Thayananthan constructs multi class

J. C. Hung et al. (Eds.): FC 2021, LNEE 827, pp. 1111–1116, 2022.
https://doi.org/10.1007/978-981-16-8052-6_153

classifiers by combining two class classifiers, which is used to solve the training problems of multiple output regression and multi class classification [1].

2 Relevance Vector Machine Model

The correlation vector machine model is as follows:

$$t_n = y(x_n; w) + \xi_n \tag{1}$$

To represent the relationship between the input x and the target t, and satisfy the following Gaussian distribution:

$$\xi_n \sim N(0, \sigma^2) \tag{2}$$

Using graph theory knowledge, the relationship between the above parameters can be expressed as the following directed acyclic graph As shown in Fig. 1.

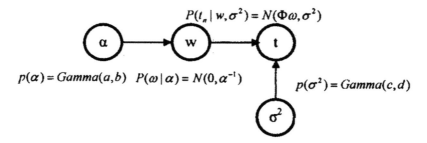

Fig. 1. Graphical model of relevance vector machine

The classical relevance vector machine algorithm reduces the computation of kernel function, and has the characteristics of good sparsity and high learning efficiency. However, it is still sensitive to outliers far away from the decision boundary and disturbed by noise data. In order to enhance the robustness of relevance vector machine model and further expand its generalization ability, scholars have improved the traditional relevance vector machine algorithm model [2].

A generalized correlation vector machine (grvm) model for classification and regression is proposed. It inherits the advantages of generalized linear model (CLM), such as the unified model structure, the same training algorithm, convenient task specific model design and so on, as well as the characteristics of relevance vector machine, such as probability output, good sparsity, super parameter self occupation and so on. Grvom extends RVM to a wider range of learning tasks besides classification and ordinary regression by assuming that the conditional output belongs to the exponential family distribution (EFD), and solves the difficult problem of reasoning in Bayesian analysis by using Laplace approximation method, which has good performance.

3 Research and Application of Relevance Vector Machine in Classification and Prediction

The algorithm model of relevance vector machine can be used to solve the problems of classification and regression prediction. The main difference between the two models is the different types of target variables. A framework for solving classification and prediction problems using relevance vector machine algorithm. Classification problems mainly include fault diagnosis of motor bearing and transformer, hyperspectral image classification, data mining and text recognition, speech recognition, face recognition, network traffic classification and network security intrusion detection. For the classification problem, the relevance vector machine gives a new pattern, infers the corresponding category according to the training set, and the classification target variable is generally discrete. Prediction and regression problems mainly include time series prediction, battery remaining life prediction, water quality prediction, equipment foamability evaluation and prediction, network traffic prediction and network security situation prediction. For this kind of regression prediction problem, the output of relevance vector machine model can be any real number.

3.1 Research and Application of Classification Model Based on Correlation Vector Machine

Fault detection refers to the use of various inspection and testing methods to find out whether there are faults in the system and equipment, so as to further determine the approximate location of the fault. Aiming at the problem of fault detection, reference [9] uses a heuristic algorithm of quantum particle swarm optimization (QPSO) to optimize the parameters of relevance vector machine, and applies it to the fault detection of rolling bearing. In this paper, the bearing vibration signal after preprocessing is used as the input vector of qpsa-rvm to find the global optimal solution of RVM parameters and build the fault diagnosis model. Through the simulation experiment of wine and other data sets, the accuracy of classification can reach 98.8 ‰ with qpso-rvm model, but the time cost of RVM optimized by QPSO is high in the training stage [3].

A rolling bearing fault diagnosis method based on eemd-pe and multi classification relevance vector machine (mcvm) is proposed. The vibration signal of rolling bearing is decomposed by integrated empirical mode decomposition (EMD). Under different fault modes, the eigenvectors of vibration samples of rolling bearing are extracted by displacement entropy method, and the mcvm model is constructed. The experimental results show that compared with eemd-sesvm and other four methods, this method can improve the fault recognition accuracy of rolling bearing, but the fault modes selected in this paper are relatively less and less representative.

Combined with wavelet packet energy entropy method to extract fault features of motor bearing, the improved particle swarm optimization algorithm is applied to RVM, which overcomes the shortcomings of RVM easily falling into local optimal solution, and improves the accuracy by 2%–4% compared with the traditional pso-rvm algorithm. A multi classification relevance vector machine is proposed, and particle swarm optimization algorithm is introduced to optimize the parameters of kernel function, which effectively improves the efficiency of fault classification.

3.2 Pattern Recognition

Pattern recognition is to study the automatic processing and interpretation of patterns by means of mathematical techniques. A noisy face recognition algorithm based on RVM is designed, which broadens the application field of relevance vector machine. The ORL face database is used as training and testing samples. The features of samples are extracted by two-dimensional wavelet transform and k-transform, and three kinds of noise adding methods are designed: random noise, salt and pepper noise and mixed noise. Experiments show that the RVM theory applied to face image recognition system has the advantages of high accuracy and insensitive to image noise [4].

This paper proposes a model of text summary recognition based on relevance vector machine. In this method, eight text features, such as sentence length, position, subject words and digital data, are selected, and the relevance vector machine of Gaussian kernel function is used as the classifier. The experimental results show that RVM algorithm improves the accuracy by 2% and has better sparsity than SVM algorithm, but the number of text documents used as experimental data is relatively small, and the sample space of test set is small, It is not enough to fully reflect the classification effect of RVM model.

4 Prediction in Network Security

Aiming at the problem of network traffic prediction, a local prediction method based on nonlinear time series and relevance vector machine model is proposed to predict small-scale network traffic data. This paper proposes an improved simulated annealing algorithm to optimize the super parameters of relevance vector machine, and applies it to network traffic prediction. By reconstructing the phase space and dividing the training and test sample set, the psa-rvm network traffic prediction model is constructed. To a certain extent, this method solves the problems of nonlinearity and high complexity in network traffic prediction, and the prediction accuracy and stability are further improved compared with particle swarm optimization algorithm combined with relevance vector machine algorithm.

In order to further improve the prediction accuracy of network traffic, the hybrid kernel method is used to replace the traditional single kernel function in RVM, and the cuckoo optimization algorithm is used to optimize the parameters to build cs-rvm model. Simulation results show that this method has smaller error and stronger generalization ability than model.

Network security is time-varying, nonlinear and difficult to predict and evaluate. RVM based on harmony search algorithm (HS) is introduced into network security situation prediction. This method reconstructs the phase space of situation samples, optimizes the hyper parameters of relevance vector machine by HS algorithm, and constructs hs-rvm model with high prediction accuracy and high speed, which can accurately predict the change law of network security situation. In the network security situation prediction, the traditional single kernel function vector machine prediction accuracy is low. The RVM model, which combines polynomial kernel function and Gaussian kernel function, is proposed for network security situation prediction, which overcomes the problem that simple single kernel function can not adapt to different sample data. This method can obtain more accurate network security situation prediction description, which is conducive to network maintenance personnel to take timely measures to solve the potential network security trends and network security incidents, but lack of relevant theoretical derivation.

In addition to battery life prediction, fault prediction and network security prediction, relevance vector machine algorithm has made great achievements in the fields of biology and online water quality prediction. In bioinformatics, protein-protein interaction (PPI) is the key of protein function and regulation in cell cycle and DNA replication. A new method for PPI prediction based on sequence is proposed. This method extracts evolutionary features from protein position specific scoring matrix, and inputs these features into R v m classifier to distinguish interacting and non interacting protein pairs. The accuracy of this method is 94.56%, which is a tool for proteomics research.

5 Conclusion

With the deepening of domestic and foreign research, relevance vector machine has not only achieved fruitful results in theoretical research, but also has excellent performance in specific engineering practice. Compared with other traditional machine learning algorithms, RVM algorithm has the advantages of better generalization performance, more sparse solution space and less computation. Although the algorithm is constantly improving and updating, the relevance vector machine still has some limitations.

There is no perfect theoretical guidance on how to construct an appropriate kernel function, so the current methods mainly rely on experience and artificial selection. The kernel functions used in most literatures are Gaussian kernel function, polynomial kernel function, linear kernel function and sigmoid kernel function. These kernel functions have their own advantages and disadvantages, such as Gaussian kernel function has the advantages of good locality, small number of parameters, small amount of calculation, but it is easy to over fit and weak generalization ability; linear kernel function is relatively simple, strong interpretability, but it can only solve linear problems; polynomial kernel function belongs to global kernel function, which has strong generalization ability, but it needs to select more parameters and weak learning ability. Some literatures also use the method of mixed kernel function, which combines two or more kernel functions according to a certain weight to "learn from each other", but the combination of kernel function increases the number of parameters, increases the computational complexity and the training time of the model, and reduces the learning efficiency.

References

1. Dai, K., Yu, H., Ma, X., et al.: Summary of feature selection algorithm based on support vector machine. J. Univ. Inf. Eng. **15**(1), 85–91 (2014)
2. Peng, L., Chen, J., Wu, Y.: Face recognition algorithm based on best discriminant feature and relevance vector machine. J. Jilin Univ. (Sci. Ed.)
3. Lu, W., Wang, H., Shu, J.: Bearing fault diagnosis based on quantum particle swarm optimization and relevance vector machine. Comput. Appl. Softw. **36**(1), 611 (2019)
4. Ma, W., Hu, P.: Probability based parallel particle swarm intrusion detection. Appl. Electron. Technol. **42**(11), 119–121125 (2016)

Research on Power Big Data Analysis Technology and Application Based on Cloud Computing

Dajiang Ren[✉], Xiaoxiao Hao, Yajie Zhao, and Mengen Wang

Zhangjiakou Power Supply Company of Wangjibei Electric Power Co., Ltd., Zhangjiakou, China

Abstract. In order to solve the problem that the current electricity service strategy can not provide accurate marketing services for electricity customers, this paper carries out exploration and research in three aspects. One is to study the technical principles of energy substitution, build the potential model for electricity substitution, and evaluate the potential of electricity substitution, The third is to use big data mining technology, combined with customers' electricity consumption and electricity cost, to analyze their potential to participate in electricity market transaction. By analyzing the characteristics of customers' electricity consumption behavior contained in the data, we can mine the characteristics of different types of customers' electricity consumption behavior in an all-round, multi angle and multi-level way.

Keywords: Electricity cost optimization · Electricity operation and maintenance optimization · Electricity substitution · Demand side response · Market transaction

1 Introduction

1.1 Current Situation of Electric Power Development

At present, in the energy consumption structure of the whole society, the proportion of coal is too high and the pollution emission is serious, so it is urgent to realize the clean, low-carbon and efficient utilization of energy. How to quickly and accurately grasp the characteristics of customers' electricity consumption behavior, find potential target groups of energy substitution customers, and guide customers to carry out energy substitution work is a difficult problem that hinders the effective development of comprehensive energy service. At the same time, due to the support of the state for new energy to replace traditional energy in recent years, the overall situation of the power market is oversupply. Therefore, under the new market situation, power supply companies must break the traditional mode of power sales, actively adapt to the new requirements of energy supply side reform and power system reform, adjust power sales strategy, actively participate in market competition, and constantly explore new power market, so as to obtain new profit model.

J. C. Hung et al. (Eds.): FC 2021, LNEE 827, pp. 1117–1123, 2022.
https://doi.org/10.1007/978-981-16-8052-6_154

1.2   Research and Development Strategy

Therefore, Shaanxi Electric Power Company of State Grid actively carries out the research on comprehensive energy service strategy. Based on the data of users' electricity consumption behavior, it makes full use of the results of big data analysis and artificial intelligence research to carry out power marketing scientifically and effectively, provide intelligent and efficient methods and means for power marketing management, and improve the accuracy of power marketing service, Promote the digital level of power marketing [1]. In this paper, the State Grid Shaanxi electric power company's comprehensive energy service strategy is taken as an example to introduce three research directions of energy alternative technology research and alternative potential evaluation in comprehensive energy service, customer power optimization service and target customer group analysis of power marketization.

2   Research on Energy Substitution Technology and Its Potential Evaluation

2.1   Research Objective

This paper studies the electric energy substitution technology in different fields, analyzes and calculates the substitution potential, and provides strong support for the State Grid Shaanxi Electric Power Company to vigorously implement electric energy substitution, promote the application of new technology and new equipment, and increase supply and sales. The research idea is to carry out research on electric energy substitution technology in various fields, including clean heating, industrial (agricultural) production and manufacturing, transportation, power supply and consumption, and household electrification. According to different alternative technologies, the comprehensive benefit analysis model of electric energy substitution is constructed, and the practical value of the comprehensive benefit analysis model is verified through case analysis.

2.2   Model Building

Taking the technology of electric vehicle replacing fuel vehicle in transportation field as an example, this paper constructs the potential calculation model of electric energy replacement.

Alternative energy calculation:

$$E = E_{unit} \times l \times 0.01 \tag{1}$$

Energy saving cost calculation:

$$C = O_{unit} \times l \times 0.01 \times q_0 - E \times q_e \tag{2}$$

Calculation of emission reduction benefits:

$$M_{co} = \eta_{co} \times l \tag{3}$$

The formula is carbon monoxide emission coefficient is hydroxide emission coefficient, C is hydrocarbon emission coefficient, NPM is PM emission coefficient, and pollutant emission coefficient refers to national five standards.

Case analysis: Taking BYD E6 electric vehicle driving 100000 km instead of Dihao GS gasoline vehicle as an example, the 100 km power consumption of BYD E6 electric vehicle is 215kwh, and the price of household charging point is 04983 yuan kwh, while the 100 km power consumption of Dihao GS gasoline vehicle is 85L, and the gasoline price is 6.88 yuan per hour [2]. According to the relevant formula calculation, the alternative benefits are shown in Table 3. It can be seen from the table that replacing natural gas vehicles with electricity requires 21500 kwh of alternative energy, and the energy consumption cost is 10498 yuan, which not only reduces the energy consumption cost, but also realizes the zero emission of carbon monoxide, hydrocarbons and other pollutants. The substitution effect is good and can be vigorously promoted.

3   Customer Power Optimization Service

3.1   Research Objective

The first is to optimize the electricity cost of customers and the operation and maintenance of special line customers, comprehensively analyze the differences of customers' electricity consumption behavior characteristics, find potential service optimization target customers, and explore the relevant reasons. Second, during the peak period of power consumption, data analysis is carried out on the load change trend of each region and the power consumption situation of relevant customers, and the demand side response needs to be mined. Research ideas: according to the composition of electricity charge, the optimization of electricity cost for high-voltage customers is analyzed according to the optimization of basic electricity charge and electricity charge.

3.2   Horizontal Construction

High voltage customer electricity cost optimization) to build the basic electricity cost optimization analysis model. For the high-voltage consumers who charge the basic electricity charge according to the capacity, the actual maximum monthly demand is analyzed, and the basic electricity charge generated according to the actual maximum demand is calculated according to the calculation formula of the basic electricity charge. If the total value of the difference for 12 consecutive months is positive, it indicates that the user is the actual maximum demand user of the potential capacity reform, and the monthly difference is the estimated electricity saving after the actual maximum demand of the capacity reform. It is suggested that the capacity after the capacity reduction is 80% of the maximum demand of the user in one year, and the proposed capacity after the capacity reduction is compared with the current capacity, If the proposed capacity after capacity reduction is less than the current capacity, it is potential capacity reduction [3]. The optimization analysis model of electricity consumption and tariff is constructed. This paper analyzes the catalog tariff of high-voltage users in their tariff, analyzes the average catalog tariff of users for 12 consecutive months, and compares it with the average catalog tariff of users of the same electricity type and voltage level, so as to find out the users whose average catalog tariff is higher than the average tariff of the same category.

3.3   Operation and Maintenance Optimization of Special Line Customers

Statistical analysis of potential optimization distribution transformer area. Secondly, through the statistical analysis of the heavy load and overload situation of the dedicated line user's station area, find out the station area where the overload phenomenon occurs, that is, the potential optimization distribution transformer area (2) build the comprehensive evaluation model of the heavy load and overload of the station area. The selection of data mining model has a direct impact on the effect and accuracy of distribution transformer heavy overload prediction $\{\theta\}$. In this paper, the distribution transformer heavy overload situation rating model is constructed by combining the statistical value division and the τ OPS evaluation algorithm θ. The heavy load level of the area is divided into four levels: A, B, C and D. Grade A is the best with slight heavy overload, grade D is the worst with serious heavy overload. See Table 1 for details.

Table 1.   Classification Standard of heavy load level in station area

Second level	Rule	Remarks
D	R average daily weight overload point > 85	All p1-p96pr are overloaded or overloaded at g time point
C	60 < R long day weight overload point < 85	P1-p96pr are overloaded or overloaded at 60 to 85 time points
B	Heavy overload < 0.8	In addition to tc9d grade 1 unit area, consider heavy load and overload 1 point (, times ((continuous duration)
A	Heavy overload evaluation a > 0.8	In addition to tc9d grade 1 unit area, consider heavy load and overload 1 point (times (continuous duration))

4   The Necessity of Big Data for Power Equipment Condition Monitoring

The improvement of power equipment condition monitoring by big data is mainly reflected in the key technology of big data, big data thinking method and big data algorithm. The key technologies include distributed computing technology, inner computing technology, stream processing technology, batch processing technology, distributed storage technology and non relational database technology, which can be directly used for the calculation and storage of massive condition monitoring data.

4.1   Key Technologies of Big Data

The key technologies of big data include data integration management technology, data processing technology, data analysis technology, data presentation technology, etc. most of these technologies come from the Internet and information technology field where big data is emerging [4]. With the development in recent years, they have high maturity and portability, Therefore, the application of these key technologies in the field of power equipment condition monitoring will have strong vitality. Big data processing technology includes distributed computing technology, memory computing technology, stream processing technology, etc. These technologies can complete the processing of PB and ZB massive data in a relatively short time, and can be directly applied to the processing of massive raw data of power equipment condition monitoring.

4.2   Big Data Thinking Method

The ideas and methods formed in the process of big data development have reference value for power equipment condition monitoring, which is mainly reflected in four aspects: global, distributed, parallel and low-cost. 1.2.1 the analysis of global big data from the perspective of the whole rather than the individual is one of the key factors with strong vitality. Therefore, big data is comprehensive data, and the core idea is to explore big value from large samples, so as to solve the problem of "data explosion and lack of knowledge". Literature 22 illustrates the global idea of big data with the fable of blind people feeling the elephant, which vividly shows that the small sample analysis method, like blind people, can only recognize the local characteristics of the huge system and diagnose the wrong conclusion.

5   Distributed

Distributed file management system and distributed database system are widely used in big data to store raw data and parameters to meet the challenge of massive data. Typical cases include Hadoop of Apache foundation and GFS of Google. GF adopts a distributed file system and is deployed all over the world.

5.1   Parallelization

Parallelized MapReduce programming mode is widely used in big data processing. Its implementation principle is to decompose tasks into small modules through map, send them to more nodes for parallel computing, and then summarize the results through reduce to get the final results. Its core idea is "divide and rule, mobile logic, shielding the bottom layer, processing customization". The originator of parallelization can be traced back to Cao Chong's story of calling an elephant. The typical MapReduce idea is to disperse the weighing elephant into measurable stones, process the results in parallel with small tools, and then summarize the results to get the results of big data.

5.2   Power Equipment Status

The cost of monitoring system is generally too high in condition monitoring of power equipment. The cost of some condition monitoring systems of power equipment is 10% or even higher than that of main power equipment, which is one of the important reasons why condition monitoring of power equipment is not popular, Therefore, how to reduce the cost of sensors, local monitoring unit, data server, software and system maintenance, and make the condition monitoring system "civilian" is an urgent problem in the promotion of condition monitoring. For example, it is an effective way to reduce the monitoring cost to replace the expensive industrial control computer with cheap single chip microcomputer and application specific integrated circuit [5]. Overall architecture of power equipment condition monitoring based on big data the overall architecture of power equipment condition monitoring system based on big data mainly includes power equipment online monitoring system, big data source, big data storage, big data processing, big data analysis, advanced application functions and big data display modules. The main idea is to take the relevant data of power equipment condition monitoring as the data source, use big data storage technology, big data processing technology and big data analysis technology to realize the advanced application of big data in the field of power equipment condition monitoring, and visualize the results through big data display technology.

6   Concluding Remarks

This paper aims to explore integrated energy services and service strategy formulation, which is better than the actual needs of integrated energy services industry. Using the relevant data of power grid operation and consumption collection system, combined with big data mining technology, this paper carries out energy substitution related technology research and its substitution potential evaluation, customer electricity optimization service analysis, power market agent trading target customer group analysis, And establish the relevant model. The research results of this paper have been used in national and provincial electric power companies, and preliminary good results have been achieved. This method is conducive to improving the company's big data analysis and application ability and professional lean management level, and its practical experience has certain reference significance.

References

1. Hu, C.: Management mode reform of power enterprises in the era of big data. Hum. Resour. (22), 28–29 (2020)
2. Lan, X.: Power big data is promising. Securities Daily (B03), 23 November 2020

3. Cao, M., Bai, Z., Ju, J.: Comprehensive energy service analysis and service strategy formulation based on power big data analysis. Power Big Data **23**(11), 72–78 (2020)
4. Wang, L., Luo, S., Jiang, Y., Li, J.: Research and application of digital analysis of market monitoring based on real-time power calculation. Big Data Electr. Power **23**(11), 79–85 (2020)
5. Li, T., Wang, B., Tu, K.: Operation and maintenance strategy of power communication network under big data. Electron. Technol. Softw. Eng. (22), 181–182 (2020)

Application of Cloud Computing and Internet of Things Technology in Power System

Dajiang Ren[✉], Yiqing Huo, Zhilu Bai, and Jiawei Chen

Zhangjiakou Power Supply Company of Wangjibei Electric Power Co., Ltd., Zhangjiakou, China

Abstract. The development trend of equipment maintenance in power system is from symptomatic maintenance to condition based maintenance. The results show that condition based maintenance can overcome the shortcomings of regular maintenance, find out the potential safety hazards that can not be found in time during the symptomatic inspection, avoid the equipment damage caused by excessive maintenance, reduce the waste of manpower and material resources caused by blind maintenance, find out the faults in time, and actively take countermeasures. In this paper, RfD technology and wireless ad hoc network technology are applied to condition based maintenance. This paper designs a perception layer based on wireless sensor and RFD tag on the device; a multi hop self-organizing network composed of wireless sensor nodes, which combines the power system private network and wireless public network to realize the seamless transmission of information; the role of application layer is to analyze, sort out, summarize and evaluate the status of the device according to the collected information.

Keywords: Internet of things · Power equipment · Condition monitoring · Condition assessment · Condition based maintenance

1 Internet of Things and Smart Grid

1.1 The Foundation of Internet of Things

The early definition of the Internet of things refers to the network formed by the combination of radio frequency identification (RFD) technology and equipment, according to the agreed communication protocol and Internet technology, so as to realize the intelligent identification and management of goods information and realize the interconnection of goods information. After development, the generalized Internet of things can realize the perception and recognition of objects, and achieve the purpose of intelligent processing and high intelligent decision-making through network interconnection. Internet of things technology is a new network based on the development of Internet technology [1]. It is a network formed between any object to realize the communication and exchange of information. The main solution is the interconnection of thing to thing (T2T), human to thing (h2t) and human to human (H2H), which develops into a dimension.

J. C. Hung et al. (Eds.): FC 2021, LNEE 827, pp. 1124–1130, 2022.
https://doi.org/10.1007/978-981-16-8052-6_155

1.2   Key Technologies of Internet of Things

Radio frequency identification (RFID) is a kind of non-contact automatic identification technology, which can identify the target object through radio frequency signal and obtain the relevant data information of the target. The identification process is independent, does not need manual intervention, and can withstand harsh environment. A typical RFID application system consists of tag, reader and information processing system. When an article with an electronic tag passes through an information reader, the tag is activated and then transmitted to the reader through radio waves to complete the information collection. The information processing system processes and controls the corresponding information according to the required role [2]. According to different frequencies, RfD can be divided into low frequency (LF) and high frequency (HF), FF), ultra high frequency (UHF) and microwave (MW). The specific parameters are shown in Table 1.

Table 1. RFID classification and specific parameters

Parameter	Low frequency (LF)	High frequency (HF)	Ultra high frequency (UHF)	Microwave (MW)
Frequency	125–134 kHz	13.56 MHz	433 MHz, 860–960 MHz	2.45 GHz, 5.8 GHz
Operating distance	<10 cm	1–20 cm	3–8 m	<10 m
Main applications	Access control and anti theft system	Smart card, electronic ticketing, etc.	Automatic control, warehousing management, logistics tracking	Road pricing

Cloud computing is to realize computing through a large number of distributed computers, rather than through a single local or remote server. Users can access the computing storage system according to their needs. "Cloud" is the computer center, each computer center includes hundreds of thousands or even millions of computers. Through the organization, distribution and use of resources, it is conducive to improving the utilization of computer resources and equipment.

2   Smart Grid Technology

2.1   Definition of Smart Grid

There is no unified definition of smart grid i38.31 among different countries and organizations. The American Academy of Electric Power Sciences proposes that smart grid is a power system composed of many automatic transmission and distribution systems, which realizes all grid operations in a coordinated, effective and reliable way, has self-healing function and intelligent communication architecture, and provides reliable power services for users.

The development of smart grid in developed countries in Europe and the United States focuses on the technical transformation of power distribution and consumption. Due to the rapid economic development in China, the power infrastructure is not perfect, the energy distribution and economic distribution are not balanced in China, the western region is rich in coal resources and water resources, and the load center is in the central and eastern regions. Therefore, the key point of our country is to establish a strong grid structure and intelligent organic combination.

2.2 The Role of Internet of Things and Smart Grid

Through the perception, identification, analysis and control of objects and natural environment, the Internet of things can further realize the networking and intelligence of people's production, life and resource management, so as to better deal with the constraints of natural environment and resources [3]. Applied to the smart grid, through the monitoring of the state of the power grid operation equipment and the environment, the information of all aspects of the power grid can be interconnected. Through the support of various data, it can play a full role in dealing with faults, controlling reasonable economic operation and controlling energy consumption.

Among the three levels of the Internet of things, the perception layer realizes the information of equipment and operating environment by arranging a large number of sensing devices such as RFID devices and intelligent sensors; the network layer mainly focuses on the private network of power system, supplemented by wireless public network and wireless ad hoc network, so as to realize the accurate and timely transmission of the information of the perception layer and the processing of the three application layers.

3 Development and Research Status of Condition Monitoring Technology for Power Transmission and Transformation Equipment

3.1 Research Status

At present, it is in the transition stage to condition based maintenance, which requires on-line monitoring technology as the basis, combined with preparation analysis, in which equipment information comprehensive classification, equipment failure rate and a large number of factors such as comprehensive analysis. It is one of the important characteristics of online monitoring line diagnosis of equipment, and equipment condition diagnosis is to judge the type, location, severity and cause of equipment fault based on the results. In order to have the ability to meet the various characteristics of the application, home system and intelligent diagnosis, transmission and transformation equipment needs to carry out systematic and comprehensive comparison and analysis of equipment cumulative condition monitoring. Only when there is a unified standard can it enter the design stage, So this process takes a long time. How much time will it take for the process of "intelligent degree of condition monitoring system, measuring result degree, and long process level required to meet the requirement".

3.2 Smart Grid Requirements

Smart grid requires the ability to obtain the status information of all equipment in the whole network through safe and reliable information channels, so as to achieve real-time acquisition, analysis, integration and sharing of the information of the whole power system. In this way, through the analysis and diagnosis of the information of the whole network, we can provide more comprehensive and accurate operation status for the operators of the power grid, and give the optimal solution, so as to ensure the efficient, safe and reliable operation of the power grid. With the development of smart grid, it has become very important to establish a perfect monitoring system for its equipment [4]. The State Grid Corporation of China has put forward the corresponding technical specifications and formulated the corresponding technical guidelines, which has laid the foundation for the scale and standardization of condition monitoring technology. The equipment condition monitoring technology provides technical support for the realization of comprehensive condition monitoring.

4 Condition Monitoring Based on Internet of Things

The premise of condition based maintenance of equipment is to get the status information of the equipment, and it is necessary to find out the information that can reflect the characteristic quantity of the equipment status among the large amount of information taken in the perception layer. By tracking these information, the health status of the equipment can be judged, and the plan of condition based maintenance can be made. The main content of this chapter is, under the support of the Internet of things technology, to monitor the condition of power system equipment, analyze the methods of monitoring these equipment and the technical means of condition monitoring.

4.1 Condition Monitoring of Power Equipment

The working conditions of power equipment, especially primary equipment, are high voltage and high current, so in the working process, these equipment will always be affected by electrical, thermal, mechanical and other factors. These conditions will cause aging and wear of power equipment under the condition of long-term work. If it is allowed to continue to develop, it will gradually develop into a failure. These changes are not sudden changes, but the performance degradation, loss increase, failure rate increase and so on caused by long-term bad operating environment. Therefore, it is necessary to monitor these power equipment. If there are abnormal signs in the equipment status, attention should be paid to monitoring. If it continues to deteriorate and leads to the possibility of equipment failure, the system should issue a warning and arrange maintenance as soon as possible.

4.2 Condition Monitoring of Primary Equipment

The primary equipment of the system refers to the equipment directly involved in the production, transformation and distribution of electric energy. The secondary equipment mainly includes generators, transformers, circuit breakers, disconnectors, fuses, buses, mutual inductors, motors, power lines, etc. The common feature of the primary equipment is that it is directly related to the high current. The insulation and safety of the detection equipment should not affect the safe and reliable operation of the primary equipment or cause other potential safety hazards. Although the role of the generator and motor in the power system is completely different, one is that the production of electric energy is the consumption of electric energy, However, the two devices have the same operation characteristics, both of them work through rotating parts, and the operation is accompanied by high voltage and large current. Therefore, the same means can be used in the monitoring. The structure of generator and motor is relatively closed, so the conventional method will encounter problems, and the internal structure of the equipment can not be observed. Using the Internet of things technology, through the use of infrared imaging technology, wave sensor and other devices, to detect the internal structure of the generator or motor, to achieve the internal visualization of the equipment.

5   Research on Cable Condition Assessment Method

To carry out the condition based maintenance of cables, we must first grasp the cable state. Through the condition monitoring system, the temperature state of the cable joint can be mastered. For the judgment of cable state, many aspects of information are considered. A few days ago, the condition evaluation of cable is based on the comparison between the test data and the data set in the specification. The cable state can not be accurately evaluated. This chapter establishes a new evaluation method to evaluate the cable.

5.1   Scoring Principle of Preventive Test Information

The cable is tested according to the preventive test procedure of the cable. The result of each test has a scoring factor A. when scoring α, the standard is the standard value in the preventive test procedure. A score of 0 indicates that the cable needs to be replaced, and a score of 100 indicates that the cable is in normal condition. The other states are in the middle of 0–100, which has no influence. For two cables with the same pre-test information, the degradation degree is different, and the future state of cables is very different. By introducing the dynamic evaluation score, the cable condition can be evaluated more accurately. First, the static evaluation time-sharing α is made according to the measured value of the pre-test. According to the analysis of the historical test data, the deterioration factor, that is, the dynamic evaluation score, is obtained.

According to the influence of dynamic evaluation score and static evaluation score on cable condition, different weight coefficients are given to them. Through the product of the score and the weight, the score of a certain influence information of the cable is obtained:

$$a_i = a_i w_i + a_i^n w_i^m \tag{1}$$

The above is the scoring method of a certain pre experiment information, and the cable condition evaluation is the common evaluation of multiple factors. According to the different importance and proportion of each factor, the common score of pre-test information is calculated:

$$A = \frac{\sum_{i=1}^{n} a_i w_i}{\sum_{i=1}^{n} w_i} \tag{2}$$

5.2   Scoring Principle of Family Defects

Family defects refer to the same kind of defects in the same manufacturer, different models and different series. Two factors should be considered when scoring the family defect, one is the probability of a family defect in design or manufacturing, the other is the fault severity e caused by the defect [3]. The scoring range of C is also 0–100. A score of 0 indicates that the defect will definitely lead to cable failure, and a score of 100 indicates that the defect will hardly affect the operation of the cable.

In general, the scoring formula of family defect is as follows:

$$r_i = \frac{\sum_{j=1}^{k} c_j w_j}{\sum_{j=1}^{k} w_j} \tag{3}$$

6   Summarize the Work Done in this Paper

This paper mainly studies the process of condition based maintenance, determines the advantages of condition based maintenance relative to regular maintenance, and the future development direction is condition based maintenance. Compared with regular maintenance, condition based maintenance has the advantages of high reliability of power supply, high interest rate of equipment, long service life of equipment, maintenance cost, high safety of operation and maintenance personnel. At present, the problem of condition based maintenance is that the existing technical conditions can not fully realize the condition based maintenance of power equipment. RFD technology, wireless ad hoc network technology, M2M technology of Internet of things technology are applied to the power grid to realize the condition based maintenance of power equipment. According to the function of the equipment in the power system, there are no secondary equipment and secondary equipment.

References

1. Zhang, J.: Research on Application of Internet of Things Technology in Distribution Line Monitoring System. Jiangsu University (2016)
2. Xu, L.: Research on Smart Grid Equipment Management System Based on RFID Internet of Things Technology. North China Electric Power University (2016)
3. Wang, C.: Unattended Substation SF_6 Research on Condition Monitoring of Circuit Breaker. Northeast Agricultural University (2014)
4. He, Y.: Research on Key Technology of Power Transmission and Transformation Equipment Condition Monitoring Based on Internet of Things. Hunan University (2014)
5. Li, W.: Application of Internet of Things Technology in Condition Based Maintenance of Electrical Equipment. Zhengzhou University (2014)

Design of Risk Prevention and Early Warning System for EPC Mode of University Infrastructure Project

Fangyan Yu[(⊠)], Xiong Na, Liang Chen, and Yanping Xu

Nanchang Institute of Technology, Nanchang 330044, China

Abstract. The research results of this paper are as follows: firstly, it provides a reference for the in-depth study of EPC mode. This model is a new type of compound contract organization, which can give full play to the subjective initiative of all parties and overcome the state of separation of management and self governance in the traditional model. It has great adaptability and superiority in the construction of modern colleges and universities. Secondly, it constructs the risk evaluation index system for university infrastructure projects under PPP + EPC mode. Through the investigation and screening of risk factors, 7 first level risk indicators and 25 second level risk indicators are finally determined, which can provide reference for relevant risk management research. Finally, the risk evaluation model under EPC mode is constructed and a case study is carried out. The construction of the model includes risk assessment and risk quantification. The overall risk level and index risk level of the case are determined by risk assessment, and the corresponding fault tree and Bayesian network are established according to the assessment results, and the risk quantification results are obtained. The results show that the model is effective.

Keywords: EPC model · University infrastructure projects · Risk assessment · Fuzzy comprehensive evaluation · Bayesian network analysis

1 Introduction

University Infrastructure refers to the facilities that provide public services for social production and residents' life, including transportation, posts and telecommunications, water and power supply, commercial services, scientific research and technical services, landscaping, environmental protection, culture and education, health and other public facilities. The infrastructure of colleges and universities is the foundation of the development of the national economy, which is very important to the development of China's economy [1]. However, since 2018, China's GDP growth has continued to slow down, and the growth rate of university infrastructure investment has also slowed down to 3.7%. In the face of the downward pressure of the economy, the general office of the State Council requests to strengthen the infrastructure construction of colleges and universities in the nine fields of poverty alleviation, railway, highway, water transportation, airport, water conservancy, energy, agriculture and rural areas, ecological and environmental protection, so as to further improve the infrastructure construction of colleges and universities and play a huge regulatory role of infrastructure

J. C. Hung et al. (Eds.): FC 2021, LNEE 827, pp. 1131–1137, 2022.
https://doi.org/10.1007/978-981-16-8052-6_156

investment in Colleges and universities in China's economy. In recent years, China's economic growth rate has declined and entered a downward cycle. Infrastructure investment in Colleges and universities is still the key method to solve the problem of economic cycle under the background of the new normal. In recent years, the cumulative investment in university infrastructure and the year-on-year comparison are shown in Fig. 1.

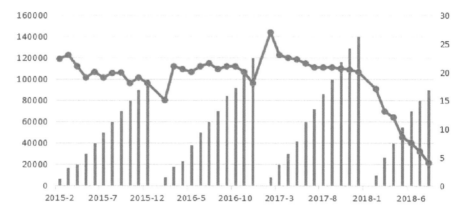

Fig. 1. Accumulated investment in infrastructure construction of colleges and universities from 2015 to 2018

Under the background of the "new normal" of China's economy and the high debt ratio of the government, due to the long cycle and large funding gap of university infrastructure projects, PPP (public private partnership) mode, a long-term cooperative relationship between the government and enterprises on the basis of concession contract, provides a more effective way for local government financing. From January to September in 2018, the total investment in infrastructure was 5503308 billion yuan, and the financing of PPP mode has reached 21%, which not only reduced the financial burden and investment and financing risks of the government, but also laid a foundation for mutual benefit and win-win for government departments, investors and consumers. In recent years, the national level has increasingly supported the application of EPC mode (general contracting). The general office of the State Council [2] and the Ministry of housing and urban rural development [3] clearly proposed to strengthen the promotion of EPC mode, and suggested that government investment projects and state-owned capital holding projects should give priority to EPC mode. Under the current common pricing convention of total price contract, it can effectively control and avoid the "three exceeding" problem of traditional design and construction separation mode, and improve the efficiency of project construction. As one of the most important general contracting modes in China, EPC mode has attracted the attention of university infrastructure experts. In practice, EPC is also combined with PPP mode to form a compound with PPP mode - EPC mode.

In China, EPC mode has not only been vigorously promoted in practice, but also gained support in policy.

Hold. In 2017, the finance office's No. 92 document, the state owned assets supervision and Administration Commission's No. 192 document and the new asset management policy (Draft) were issued one after another, which not only provided impetus for investors to participate in the construction of university infrastructure projects through the mode, but also eased certain financial pressure for the government. In the university infrastructure projects, the adoption of EPC mode is not only to comply with the trend of the times, but also to meet the needs of the public. However, practice always precedes theory. Although the advantages of EPC mode are obvious, EPC mode is not a simple addition of PPP mode and EPC mode, and its risks cannot be simply added. Therefore, how to realize the integration of the two modes in the university infrastructure project, what risks will be produced by the integration of the two modes, and how to scientifically evaluate these risks are very important for the smooth implementation of the university infrastructure project under the EPC mode. The purpose of this paper is to explore the connotation of EPC mode from the perspective of project company, analyze the risk characteristics of university infrastructure projects under EPC mode, identify the risk, evaluate the overall risk level of the project, quantify the high-risk indicators, and finally make targeted suggestions on the risk quantification results to build a complete risk evaluation system.

2 Related Theories of Risk Assessment

"China project management knowledge system and international project management professional qualification certification standard" defines risk management.

It is a systematic process to identify, analyze and respond positively to project risks. As an important part of risk management, risk assessment is to decide whether to take appropriate measures on the basis of risk identification and assessment, considering the probability of risk occurrence. For example, Jiang Xin and Li Qi (2014) conducted risk assessment on engineering projects and quantified the estimation interval in risk assessment study; pan bin and Huang Jing (2015) conducted risk assessment on university infrastructure projects and ranked them; Li Qiang and Han Juntao (2017) conducted risk identification in railway projects, The main risk factors are determined by the model. But the content of risk assessment is generally composed of identification, evaluation, analysis, quantification and other links. In order to study risk more comprehensively, based on the collation and reference of relevant literature, this paper divides the content of risk assessment into risk identification, risk assessment and risk quantification. Its logic shows that risk identification is the basis of risk assessment, and risk assessment is the basis of risk quantification. The three links complement each other and constitute the content of risk assessment. Therefore, this chapter describes the risk identification, risk assessment, risk quantification theory and the method selection.

Risk assessment refers to the process of determining the project risk level and ranking according to the impact of risk on the project objectives on the basis of risk identification of the project. The degree of project impact describes the degree of risk impact on the project objectives if the risk occurs. The degree of influence is generally described as very high, high, low, general, low and very low. The risk assessment results should determine the overall risk level of the project, and evaluate the overall risk degree of the project by comparing its risk level [4]. The risk of the project will have different effects on different participants, so different activity subjects have different emphasis on risk management and different bearing capacity for risk assessment results.

For risk assessment, risk assessment is not static, but a scientific, dynamic and circular process. With the continuous change of the assessment subject, the environment is also changing, and the risk factors are also changing, so the risk content of each stage is different. This change includes not only the change of risk factors, but also the change of risk nature and consequence.

3   Construction of University Infrastructure Project Risk Evaluation Model Under EPC Mode

3.1   Model Building

Under EPC mode, the construction of university infrastructure risk evaluation model is based on risk identification, which is composed of risk assessment and risk quantification. The risk evaluation model is based on the risk evaluation index system, and it is also an important basis for risk response.

In the first step, based on the risk evaluation index system, the hierarchical structure model is established, and the judgment matrix is constructed to calculate the risk evaluation index and determine the index weight;

The second step is to determine the risk assessment measures to define the value of high risk;

The third step is to establish the grey evaluation matrix and carry out the fuzzy comprehensive evaluation to obtain the overall risk level and the risk level of each index;

The fourth step is to evaluate the index risk level. If the result of index evaluation is low risk and within the controllable range, it will be transferred to Risk database. If the result of index evaluation is high-risk, it can not be ignored, that is to say, it will be transferred to risk quantification;

In the fifth step, the high risk index is taken as the top event, and the basic events causing the risk are analyzed layer by layer to construct the fault tree;

The sixth step is to construct Bayesian network with fault tree, calculate the posterior probability, get the occurrence probability of basic events, trace the risk source, and quantify the risk factors in the form of probability.

This model has the following characteristics:

First, the model is based on the risk evaluation index system. AHP is used to determine the weight of each index, and then fuzzy comprehensive evaluation method is used to evaluate the risk of the project to get the overall risk level of the project and the risk level of each index.

Second, risk quantification is based on the results of risk assessment. In this paper, the reliability theory is introduced into risk quantification, and the project is regarded as a risk system. The Bayesian network analysis method is used to quantify the risk assessment results, further analyze the reliability of each subsystem in the system, and calculate the posterior probability of risk events.

Third, combining qualitative and quantitative methods to build the model. Fully considering the logic and relevance of risk factors, this paper uses fault tree to carry out qualitative analysis of risk, further studies the causes of risk factors, and uses Bayesian network diagram to more intuitively describe the relationship between risk factors, through qualitative analysis of causal relationship, quantitative calculation of risk probability.

3.2 Model Building

Based on the in-depth analysis of university infrastructure projects under EPC mode, according to the risk evaluation index system established in Sect. 3, the risk of university infrastructure projects under EPC mode is set as the top level, namely the target level; the first level risk index is set as the middle level, namely the criterion level; the second level risk index is set as the bottom level, namely the factor level. The related factors are decomposed into several levels from top to bottom according to different attributes. The factors of the same level are subordinate to the factors of the upper level, or have a certain degree of influence on the factors of the upper level. At the same time, they dominate the factors of the next level or are affected by the factors of the lower level.

The weights of the two levels will be compared. In order to reduce subjectivity, relative scale is used to minimize the difficulty of comparison between indicators due to their different nature. In a certain criterion layer, each index is compared in pairs, and the evaluation grade is the result of importance comparison according to its importance degree. The nine importance levels and their assignments are listed in Table 1.

Table 1. Matrix element scaling method

Index I is better than index J	Quantized value
Equally important	1
A little more important	3
Strong and important	5
Very important	7
Extremely important	9
The middle value of two adjacent judgments	2, 4, 6, 8

The matrix formed by pairwise comparison is called judgment matrix. The judgment matrix has the following properties

$$a_{ij} = \frac{1}{a_{ij}} \tag{1}$$

The scaling method of judgment matrix elements is as follows:

$$\begin{pmatrix} b_{11} & b_{12} & \cdots & b_{1m} \\ b_{21} & b_{22} & \cdots & b_{2m} \\ \vdots & \vdots & \ddots & \vdots \\ b_{m1} & b_{m2} & \cdots & b_{mm} \end{pmatrix} \tag{2}$$

4 Simulation Analysis

In this paper, based on the gray evaluation matrix of whitening weight function, whitening weight function refers to a triple line or S-shaped curve in rectangular coordinates. Quantitative description of the degree to which an evaluation object belongs to a grey class (commonly called weight function), that is, the relationship between the evaluation object and the size of the evaluated index or sample point value. The typical whitening weight function is a continuous function of left ascending and right descending with the starting point and end point determined. In practice, three kinds of special whitening weight functions derived from typical whitening weight functions are mainly used, i.e. lower bound measure whitening weight function, moderate measure whitening weight function and upper bound measure whitening weight function. Several kinds of function graphs and expressions are as follows. As shown in Fig. 2:

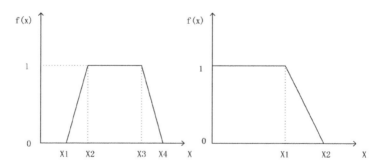

Fig.2. Whitening weight function diagram

5 Conclusion

Firstly, it provides an objective reference for the further study of EPC mode. EPC mode is a new type of compound contract organization. This special organizational structure makes the project company, project sponsor, government, general contractor, creditors and other stakeholders more closely united and become an interest whole. In this mode, all parties give full play to their subjective initiative and actively participate in the project management, which overcomes the traditional mode of management separation. This mode can not only guarantee the quality of the project, reduce the cost and speed up the construction progress, but also improve the economic benefit and maximize the profit. In the development of modernization, it has great adaptability and superiority to apply EPC mode to university infrastructure projects. Secondly, it constructs a risk evaluation index system for university infrastructure projects under EPC mode. In this paper, through reading the relevant literature to determine the scope of risk types, and to some city related staff issued 150 questionnaires, through two rounds of question-naire results to screen and determine 7 The first level risk indicators are: bidding management risk, design management risk, procurement management risk, operation risk, construction management risk, financial risk and political environment risk. Under the seven first level risk indicators, there are 25 second level risk indicators. The first level risk index and the second level risk index constitute the university infrastructure project risk under the EPC mode, which provides a reference for the related risk management research.

Acknowledgements. Science and Technology Research Project of Jiangxi Education Department in 2019, Project name: innovative Application Research of BOT financing Model in University Infrastructure Project, Project number GJJ191018.

References

1. Qin, R.: Analysis on contract management and risk prevention of construction projects. Low Carbon World **11**(03), 254–255 (2021)
2. Hao, Q., Wang, X., Li, R., Qin, R.: Risk management in engineering projects. Smart City **6** (23), 77–78 (2020)
3. Zhong, Z., Wang, C., Qiu, W., Song, T.: Research on bidding management of university infrastructure projects. Sichuan Build. Mater. **47**(01), 248–249 (2021)
4. Pan, J., Wang, Z.: On budget management of university infrastructure projects. Mod. Econ. Inf. **21**, 216 (2014)

Judgment of Computer Software Infringement and Possible Civil Liability

Yangchuan Bao[✉]

Sichuan Judicial and Police Officers Professional College, Deyang 61800, Sichuan, China

Abstract. In the information age, there are many kinds of computer software. Nowadays, the number of software is extremely large, and there are many kinds of software. Some of the software is favored by people, and has become a common tool in people's life and work, which makes the developers or operators of such software get a lot of benefits. However, this phenomenon may involve civil liability, so we must combat this phenomenon, but it needs to make clear the judgment conditions of tort and the civil liability that may be undertaken. In this regard, this paper will carry out relevant research, discusses the two forms of computer software infringement, and analyzes the judgment conditions of the two forms, and finally puts forward the possible civil liability.

Keywords: Computer software · Software infringement · Civil liability

1 Introduction

Under the protection of copyright, any act of stealing others' achievements without permission or publicizing oneself as the copyright owner will cause infringement on the image, reputation, economy, spirit and other aspects of the real developers of computer software. This phenomenon is the standard form of software copyright infringement [1]. For example, a computer software has been loved by users since it was developed, which has brought a lot of profits to the enterprises and related personnel behind it. However, this has made other personnel "envious" and developed similar software highly similar to the computer software. This action has not been allowed by the enterprises or related personnel behind the software, so this behavior infringes the copyright of the computer software enterprises and related personnel Right to work [2].

Most of the computer software in the market belongs to application software, which mainly provides services for users. If users want to use the software, they must download the software to the user computer. At this time, the software enters the user computer. Some of the software may infringe the rights and interests of users for some purposes, and the most representative one infringes the rights and interests of users The most common phenomenon is the infringement of users' privacy. In theory, the user's right to privacy is mainly to protect the privacy information of computer users, such as personal photos and personal information. Any act of checking and obtaining the user's privacy information without the user's initiative or permission can be regarded as infringing the user's right to privacy, which is also a form of computer software infringement. This form of infringement is common in modern social environment

J. C. Hung et al. (Eds.): FC 2021, LNEE 827, pp. 1138–1142, 2022.
https://doi.org/10.1007/978-981-16-8052-6_157

Fresh [3], for example, the recent incident of "Tencent QQ software scanning user's browser history" is a typical violation of user's privacy. QQ software will scan user's browser history without the user's knowledge. The official explanation for this move is "to prevent malicious login", but this explanation can not deny that QQ software does not infringe user's privacy The behavior of the software is still a standard violation of user privacy.

2 Judgment Conditions of Computer Software Infringement

2.1 Judgment Conditions of Infringing Software Copyright

In the judgment of whether computer software infringes the copyright, the main concept to be distinguished is "similar" and "plagiarism", that is, there are similarities between some computer software and other software, but it can not be judged that the later infringes the copyright of the former just because there are arbitrary similarities between the two software. Only when there are too many similarities between the two can the later infringe the copyright of the former The latter infringes the copyright of the former, so these two concepts are the key to judge such cases. This can be judged according to five conditions, as shown in Table 1.

Table 1. Judgment conditions and conditions of infringing software copyright

Judging conditions	Summary
Comparison of floppy disk content, directory and file name	Compare the floppy disk content, directory and file name of the software (if possible, it is better to compare them all). If the two softwares in the comparison project are completely consistent or highly similar, it is judged that the latter infringes the copyright of the former
Comparison of installation process	By comparing the software installation process, if the relevant information of the two software is consistent or highly similar in the process, it is judged that the latter infringes the copyright of the former
Comparison of directory and files after installation	The purpose of the software after installation and related documents are compared. If the meaning of the relevant information of the two software is highly similar, it is judged that the latter infringes the copyright of the former
Comparison of using process after installation	If the process is consistent or highly similar, it is judged that the latter infringes the copyright of the former
Code comparison	If the codes of the two softwares are consistent or highly similar, it is judged that the latter infringes the copyright of the former

In addition, it is mentioned in Table 1 that if software is highly similar, it can also be judged as infringing copyright. In order to define this point accurately, there must be a clear numerical standard. According to the current laws and regulations of our country, if the degree of similarity between the two is more than 20%, it can be generally judged as infringing copyright, but the standard has little effect, and the specific value remains to be discussed [4, 5].

2.2 Judgment Conditions of Infringement of User Privacy

User privacy is a traditional law and regulation in China, which has been relatively sound in the long-term development. Its judgment conditions are systematic, which can be divided into four main points. See Table 2 for details.

Table 2. Summary of judgment conditions and conditions of infringement of user privacy

Key points of judging conditions	Summary
Subjective existence fault	It needs the computer software or related actors who have made the behavior of illegal access to other people's privacy subjectively. Whether the behavior is intentional or negligent, it can be judged that the software or actors have violated other people's privacy
Behavior violates social obligation	The social obligation in the right of privacy refers to: any unspecified person shall not infringe upon the privacy of others. If the act violates the obligation, it means that the act infringes upon the consciousness of others. Generally, it can be convicted and punished, unless there is a legitimate defense, such as the infringed privacy has insulted the actor, the actor shall not obtain evidence to prove his innocence by violating the privacy of the former

3 The Possible Civil Liability in the Infringement of Computer Software

3.1 Civil Liability for Infringing Software Copyright

According to Article 30 of the regulations on the protection of computer software, there are six kinds of behaviors that will infringe the software copyright, which can be judged according to the judgment conditions. The six kinds of behaviors and the possible civil liabilities are shown in Table 3.

Table 3. Infringement of software copyright and possible civil liability

Infringement of software copyright	Possible civil liability
Without the authorization or consent of the software copyright owner, it can be judged as infringing the software copyright to publish his software works without permission, to publish others' software as his own software works, and to publish the software developed in cooperation with others as his personal works	According to paragraph 1 of Article 32 of the regulations on the protection of computer software, any one of the six acts should bear the civil liability, that is, the infringer must destroy all the infringing achievements, stop all the infringing activities, and help the infringed to make up for the losses. If the losses cannot be made up, economic compensation should be made. The amount of compensation will be released by the legal arbitration institution, and the infringer has the right to pay compensation Responsibility and obligation to perform
Signing in the software developed by others or modifying the signature on the software developed by others can be judged as infringing the software copyright	
Without the authorization or consent of the software copyright owner, it can be judged as infringing the software copyright to modify, translate or annotate the software works without permission	

3.2   Civil Liability for Infringing User's Privacy

According to the general provisions of the civil law of China, infringement of users' privacy rights will bear three kinds of civil liabilities, as shown in Table 4.

Table 4. Civil Liability for infringement of user privacy

Civil liability	Summary
Stop all direct infringement and indirect infringement	Stop all direct infringement: stop publishing others' privacy information in public, delete others' privacy published in the past, stop continuing to obtain others' privacy; stop all indirect infringement: appeal to the public not to continue to spread others' privacy, ask the third party to inform them to continue to use or spread others' privacy
Make an apology	In accordance with the Supreme People's court "on the trial of the right of reputation cases to answer a number of issues" provisions of the apology
Compensation for loss	The contents of compensation include mental damage compensation and privacy infringement compensation

3.3   Precautions

In the two forms of computer software infringement, there are two major issues that need attention: (1) in the civil liability of any form of infringement, if there is a request for the infringer to make economic compensation for the infringed, the specific amount of compensation must be set according to the general provisions of the civil law. At the same time, the arbitration institution should negotiate with the infringed to understand the actual situation of the infringed, and then make compensation (2) when judging whether any act infringes copyright or privacy, the authenticity of the evidence information must be made clear to avoid the phenomenon that the infringed exaggerates the facts or the infringer conceals the facts, and the infringer will be punished by other laws.

4   Conclusion

To sum up, this paper analyzes the judgment of Computer Software Infringement and the possible civil liability, expounds the two forms of computer software infringement, discusses the judgment conditions of the two forms, and finally puts forward the possible civil liability for the two forms. Through the analysis, it can be seen that computer software may infringe the copyright of other software and the privacy of users. This kind of phenomenon violates the law. Although it has not risen to the level of criminal law, the relevant personnel must also bear their own civil liability. According to the discussion in this paper, the possible civil liability is clarified. Combined with the judgment conditions of infringement, it can play an accurate role in judging whether the infringement and how to deal with it Secondly, the role of tort.

References

1. Luo, J., Chang, Q.: Research on data preprocessing method of soft sensing technology. Control. Eng. **13**(4), 298–300 (2006)
2. Yu, J., Zhou, C.: Soft sensing technology in process control. Control Theory Appl. **23** (1996)
3. Huang, F.: Soft sensing thought and technology. J. Metrol. **31** (2004)
4. Zhu, X.: Soft sensing technology and its application. J. South China Univ. Technol. (Nat. Sci. Ed.) **12** (2002)
5. Pan, F., Jin, X.: Research on soft sensing of citric acid evaporation process based on neural network. Measur. Control Technol. **23**(8), 17–20 (2004)

Parameter Algorithm of Sociological Intractable Problem Based on D-Approval Rule

Hui Bao[✉]

University of Qingdao, Qingdao 266000, China

Abstract. Voting problem in sociology has been studied for a long time. Now it has been widely used in the field of computational theory, and plays an important role in artificial intelligence, bioinformatics and graph editing. The theory of parameter calculation is a new method to solve NP hard problems accurately, which has been widely concerned. This paper mainly studies the application of parameter calculation theory in sociological voting problem, and discusses the influence of control and bribery on voting problem under d-approval rule.

Keywords: Sociology · NP complete · D-approval · FPT

1 Introduction

Traditional complexity theory shows that for NP hard problems, unless P = NP, there is no accurate algorithm in polynomial time, which frustrates people's confidence in solving difficult problems [1]. However, the problems in practice are almost completely NP hard problems, so people have been looking for solutions to NP hard problems. In this context, heuristic strategy, approximate strategy and stochastic strategy emerge as the times require. However, although these methods solve the NP hard problem to a certain extent, they have their own shortcomings. The mining of heuristic rules in heuristic algorithm relies too much on engineering experience, the accuracy or effectiveness of approximation algorithm can not meet the requirements, and it is difficult to establish the probability distribution model in stochastic algorithm. The key point is that we can't find the exact solution of the problem. Based on the deficiency of heuristic strategy, approximate strategy and random strategy, people put forward the theory of parameter calculation. Parameter calculation theory is a new method to solve NP hard problems, which can provide accurate solutions. The theory is based on the fact that many NP hard problems are actually related to the changing parameters of a cell. According to the selected parameters, people construct the corresponding parametric form of the optimization problem. Parametric problems often contain good properties, which is helpful to solve NP hard problems. Parametric problems are divided into fixed parameter solvable (FPT) and fixed parameter unsolvable. The following gives the relevant definitions and important terms in the theory of parameter calculation.

J. C. Hung et al. (Eds.): FC 2021, LNEE 827, pp. 1143–1147, 2022.
https://doi.org/10.1007/978-981-16-8052-6_158

2 An Overview of Sociology

Voting system has been studied for a long time in sociology. As early as 500 BC in ancient Greece, voting was introduced into government elections. In this way, citizens have more rights to choose their ideal politicians and prevent dictatorship and autocracy. With the development of society and the popularity of voting, people pay great attention to the voting system. A variety of voting rules emerge as the times require because of the diversity of system and culture, and have been widely used. With the development of sociology, the application of voting system is not limited to election. For example, dwork (2001) proposed to design meta search engine, regarding search engine as voting, web page as candidate, and applying the idea of voting to the design of search engine, so as to provide tools and basis for better design and optimization. Therefore, voting system has been widely used in the field of computing theory, and will have a profound impact on the research and development of computing theory [2].

2.1 Kemeny Voting System

The importance of Kemeny problem in sociology has prompted people to pay great attention and research. People analyze the complexity of the problem from the perspective of different parameters. Based on KT distance, the definitions of maximum K-T distance and average K-T distance are given below.

The maximum K-T distance of selection (V, C) is defined as:

$$d_{\max} = \max_{u,v \in V, u \neq v} dist(u, v) \tag{1}$$

The average K-T distance of selection (V, C) is defined as:

$$d_a = \frac{1}{n(n-1)} \sum_{u,v \in V, u \neq v} dist(u, v) \tag{2}$$

2.2 Control, Manipulation and Bribery

There is such a phenomenon in the social election. For personal purposes, people often change the election process through certain means, so as to change the election results, and let the given candidate become the final winner. It's not fair, so people have been looking for the "perfect" voting system. Bartholdi et al. Introduced the following concept for the first time. If any form of attack can not affect the voting result, then the voting system is immune, otherwise it is vulnerable. For example, the voting system under the plurality rule is vulnerable [3]. However, the voting systems used by people are vulnerable, so it is of great value to study the influence of different manipulation methods on the voting system. If there is a voting system immune to all manipulation methods, then the voting system is immune. The common manipulation methods in social election are control, manipulation and bribery. Control is to add, delete and divide votes or candidates so that the given candidate C is the winner in the new voting system. Manipulation is to select a subset from the unregistered voting set, change the priority list of voting in the subset and add it to the original voting system, so that C is the winner. Bribery mainly bribes voters to support the given candidate C, so as to change the election result. The parameter complexity of these three manipulation methods under different voting rules is different, and people have paid great attention and research. There are two control strategies in this kind of problems, namely constructive control and destructive control. Constructive control is to let candidate C become the winner, while destructive control is to prevent candidate C from becoming the winner. The following is a case study of candidate control.

3 Parameter Complexity of CC-DV-D-Approval Problem and Its Dual Problem

3.1 The Definition and Research Status of the Problem

This chapter mainly discusses the influence of deleting voting on the election. Delete voting is one of the classic forms of control problem in voting. In order to achieve personal purpose, remove the voters who are unfavorable to themselves, and make the given candidate become the final winner.

In view of the validity of the problem structure, a bipartite graph G = (A ∪ B, E) is constructed. As shown in Fig. 1.

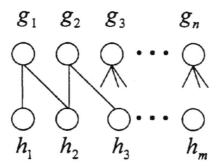

Fig. 1. Structure diagram of G

The voting systems studied in this paper are all based on d-approval rule. In the bipartite graph G, the degree of each vertex in a is D. the change of D value will affect the complexity of the problem. When d = 1, the problem is obviously polynomial time solvable. When d = 2, the candidate is regarded as the vertex, and the voting is regarded as the edge between the corresponding vertices of two candidates. It can be proved that the problem is also polynomial time solvable by transforming it into simple b-edge problem. When D⩾3, the problem is NP complete. However, the research on the parameter complexity of this problem is still an open problem. Next, we will mainly discuss its parameter complexity.

3.2   Research on Parameter Complexity

This section mainly studies the parameter complexity of c-dv-d-approval problem when d⩾3. Firstly, different from the proof in literature 16, by reducing the hitting se problem, it is proved that the problem is NP complete when d⩾3, which is more concise than the proof in literature. Then the voting and candidates are classified, unnecessary structures are deleted, and the core of the problem is obtained. Finally, a fixed parameter algorithm based on dynamic programming technology is proposed [4].

In the simplified example, the score of each candidate is not less than s (c). In order to make a given candidate C a winner, the score of each candidate is reduced by at least 1, that is to say, the voting related to each candidate must intersect with the solution set v. The maximum number of supports V can give is, which means that the size of candidate set in the simplified instance is not more than DK.

In the multi-stage decision-making problem, the solution of each stage has an impact on the decision of the next stage. In order to solve the multi-stage decision-making problem effectively, people put forward the idea of dynamic programming. Different from the divide and conquer method, there are too many subproblems in the divide and conquer method, and some subproblems are repeatedly calculated for many times. Dynamic programming can effectively reduce the solution space of the problem. The dynamic programming technique is used to get the fixed parameter algorithm of the problem, which has better time complexity than enumeration.

This section mainly studies the dual problem of cc-dv-d-approval problem. The original problem takes the number of deleted votes as the parameter, and the dual problem takes the number of reserved votes as the parameter. Although the parameters of the dual problem may be a large number, it is still of practical significance to study its parameter complexity, which is of great help to better understand the impact of deletion voting on the election. In this section, we first prove that the problem is NP complete when d = 3. Then, the structure of the problem is analyzed, the simplified rules are given, and the relationship between the simplified problem and set packing is found.

4   Conclusion

The theory of parameter calculation is a new method to solve NP hard problems accurately, which has been widely concerned. This paper briefly introduces the application of parameter calculation theory in sociological voting problems from three aspects, namely, winner determination, possible winner determination and voting control. On this basis, this paper focuses on the influence of control and bribery on voting. The structure of the voting problem is deeply analyzed, its properties are mined, and the core rules are given to simplify the example of the problem. A fixed parameter algorithm based on dynamic programming and local greedy is proposed. The problems studied in this paper are all based on d-approv rule.

Through the core technology, the voting problem is simplified, and there is a certain relationship between the simplified problem and the set packing problem. The classical set packing problem is to find a set in which the subsets do not intersect each other, that is, each element can only appear once in the solution at most. This paper only considers the case of unique winner, that is, the goal is to make a given candidate become the only winner, but if the goal is changed to MWLI winner, that is, there can be multiple winners in the voting system, then what is the parameter complexity of these problems? It is worth studying.

References

1. Kemeny, J.G.: Mathematics without numbers. Daedalus **88**(4), 577–591 (1959)
2. Mahajan, M., Raman, V.: Parameterizing above guaranteed values: MaxSat and MaxCut. J. Algorithms **31**(2), 335–354 (1999)
3. Young, H.P.: Extending condorcet's rule. J. Econ. Theory **16**(2), 335–353 (1977)
4. Dodgson, C.: A method of taking votes on more than two issues. Pamphletprinted by the Clarendon Press (1876)

Visual Representation of Scene Pattern Based on Computer Aided Design

Jia Chao[(✉)]

Leshan Normal University, Leshan 614000, Sichuan, China

Abstract. Aiming at the problem that the original art aided design does not accurately extract the known visual scene image information, resulting in the output image is not clear, an art aided design method based on scene visual understanding algorithm is studied. The known visual scene image is color segmented according to the set threshold, and the image is segmented by morphological processing, so as to reduce the impact of noise and fracture on the obtained connected area, The extraction rules are established to screen candidate regions. More SFT target features in candidate regions are extracted in the form of dense sampling, and feature points are matched. The coordinate system of known image information is integrated into a unified coordinate system, and the design image is extracted to complete art aided design. The simulation results show that the art aided design method based on the scene visual understanding algorithm is more accurate than the original method in extracting the information of the known image, which helps to solve the problem of clear image and improve the overall design effect of the image.

Keywords: Auxiliary vision · Scene extraction · Scene classification

1 Introduction

Software design and hand-painted design are usually used in art design. With the rapid development of multimedia technology, art design is more and more inclined to combine with computer technology to produce a variety of auxiliary design tools and design software with the function of sample combination. There are more and more design software with more and more functions, which can repeatedly polish and open design works and reduce the cost of manual design. At the same time, through two-dimensional drawing software and three-dimensional stem design software to make design samples, it can enrich the creativity of visual effect, better present the designer's design concept and innovative ideas, change the original form of art design, improve the working mode of art design, and bring great changes to traditional art design.

In the computer technology which integrates with art design, scene visual understanding is one of the technologies widely used in art design. Scene vision understanding is to use computer to simulate the visual function of human clothing, and use computer to replace human eyes and brain to perceive, recognize and understand the three-dimensional scene and objects in the objective world [1]. It is used to analyze the complex distribution of objects in field images. In the process of artistic creation, it can effectively help designers solve the problem that the output image is not clear due to the

J. C. Hung et al. (Eds.): FC 2021, LNEE 827, pp. 1148–1153, 2022.
https://doi.org/10.1007/978-981-16-8052-6_159

inaccuracy of the extracted scene image information data, A suitable algorithm for scene vision understanding is proposed.

2 Art Aided Design Method Based on Scene Visual Understanding Algorithm

In complex scenes, the target information or objects are blocked by other objects, the color difference changes, and the shape size is different, which is not conducive to the target information feature extraction. Therefore, in the art aided design method based on scene visual understanding algorithm, firstly, the scene image is segmented according to the set threshold value; secondly, the candidate region is obtained by morphological processing of the effective description region; then, the target candidates in the candidate region are extracted by SFT features and input into the design system for combination design, and the qualified results are screened to realize art aided design.

2.1 Rough Segmentation to Obtain Candidate Region of Scene Image

To segment the known scene image according to the set threshold of one or more color spaces, R is usually used. The specific scene image segmentation algorithm formula is as follows:

$$(x, y) = f_1(x, y) \& f_2(x, y) \tag{1}$$

Where XY is the abscissa and ordinate of the plane of the image. There are different values of F_1 (XY) under different thresholds. When the R channel of the input image is less than the optimal threshold in RGB color space, F_1 (XY) $= 255$, otherwise F_1 (x, y) $= 0$. Through F_1 (XY), the interference of the blue area in the image can be effectively eliminated, especially in the outdoor scene image, the interference effect of the sky on the blue area can be obviously controlled. However, F_1 (x, y) has no effect on dark areas such as black, gray and brown [2]. Therefore, the introduction of F_2 (x, y) into the formula to solve $61F_2$ (x, y) can eliminate the interference of colors such as green, red, orange, which are very different from blu.

In the HSV color space, when the H channel of the input image is in the upper and lower limits of the optimal threshold, the value of F_2 (x, y) is 255. But usually the image in the input software is RGB color space, so the conversion between the two color spaces is needed. The conversion formula of H is as follows:

$$H = 60 \times \frac{G - B}{\max - \min} \tag{2}$$

$$H = 60 \times \frac{G - B}{\max - \min} + 120 \tag{3}$$

After the color segmentation of the scene image, the segmented scene image is easily affected by external factors, resulting in multiple noise and fracture. Therefore, morphological processing is carried out on the areas effectively describing the shape in the image, such as shell and frame, to reduce the influence of noise and obtain connected areas.

2.2   Extracting Internal Targets from Candidate Regions

Feature scene image candidate region target feature description uses a computer vision algorithm SFT feature, which has the advantages of position information, scale information and rotation invariance of image local features. 0. It detects and describes the candidate region local features, reduces the amount of computer processing data, and retains the key visual information to extract the scene internal target features, Generally, the gradient and direction of each pixel in the 16×16 window are calculated with the SFT feature point as the center in the candidate region. In order to make the sampling more sufficient and enhance the image clarity of the design output, this paper adopts the form of dense sampling to obtain more SFT feature points, as shown in Fig. 1.

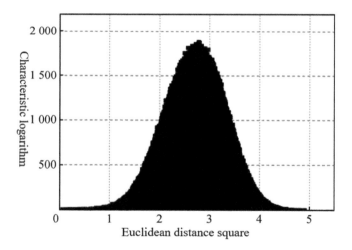

Fig. 1. Schematic diagram of SIFT feature point extraction

2.3   Generating Output Image by Matching Feature Points

After extracting the feature vectors of the image, the coordinates and vectors of the feature points are input into the matching algorithm, and the closest matching vector pair is obtained according to the Euclidean distance. First search algorithm is used to find the two nearest feature points, and then the ratio of Euclidean distance between the feature point and the two feature points is calculated separately. If the ratio is less than the specified value, the matching is successful, otherwise it fails. The two points that match successfully are a group of matching points in the image pair. The matching points in the adjacent images are in their respective image coordinate systems. According to the relationship of the coordinate systems, they are integrated into the same coordinate system [3]. After the transformation from the original image to the target image, an image is generated after the matching. So far, the research on art aided design method based on scene visual understanding algorithm is completed.

3 Algorithm Description

Firstly, the image is decomposed by three times wavelet transform to get the detail component information of the image at each scale, and the local features of each component image are extracted. Then, the salient region of each component image at each scale is detected, and the local features are filtered and weighted by the detection results.

3.1 Frequency Domain Transformation

According to the characteristics of human vision, the content of the scene image can be obtained only through the general picture, without detailed description of its details, which can simplify the image recognition and calculation. For example, wavelet transform can be used to obtain low-frequency profile information of images in multiple scales and directions. The low frequency part of the image after three times of wavelet transform can be seen many times of wavelet transform, and the low frequency part can still judge the general content information of the scene image. Then, the same component information at different scales is fused, The multi-scale fusion windowed weighted SFT feature (wssift) is obtained and sent to SVM for training. Because the image sizes in each dataset are different, and only the detailed texture information of the image is used, the image needs to be scaled to the same size and converted to gray image before the algorithm in this paper.

3.2 Visual Perception Characteristics

According to the characteristics of human visual perception, when we observe the surrounding scene, we often only focus on those areas and objects that are significantly different from the surrounding things, and ignore those contents that have no obvious color or texture changes. This is the basic principle of saliency detection. Those areas that attract people's visual attention and interest are saliency areas, The process of extracting the region is significance detection. For example, when we see the cars on the highway in the outdoor scene, the cabinets in the kitchen in the indoor scene and so on, these objects can not only attract people's attention, but also fully represent the content period of the scene. The saliency detection algorithm mainly extracts the contour information of the saliency object, and does not describe the saliency area and the target area, Or only consider the accurate segmentation of the target region. On the one hand, some detection algorithms ignore the regional concern of the visual perception characteristics, and people observe the things themselves.

3.3   Feature Extraction

Image features play an important role in image description. The accuracy and comprehensiveness of feature extraction affect the effect of image post-processing. Early scene classification methods mainly focus on the expression of global spatial features. Although global features are easy to implement and low-cost, they have poor robustness and generalization ability, and lack of description of local spatial details. Especially for the scenes to be classified, the changes between classes are small, while the differences within classes are large, the local detail texture features are often very important, Therefore, on the basis of fully considering the overall visual salient region in the previous visual perception characteristics, the local features are used to optimize and enhance the description. On the one hand, the salient information is used to retain the overall information of the image and the primary and secondary relationship between the surrounding things. On the other hand, the local information is used to describe the regional details in detail, In this way, the global and local information are considered to improve the accuracy of image classification.

4   Analysis of Experimental Results

In order to verify the effectiveness of our algorithm, we test it on three common standard datasets, including 8 types of motion scene datasets 121, 15 types of natural scene datasets 813.141 and 67 types of indoor scene datasets 15. In order to compare with the results of other literatures, we train and test according to the division ratio of training and test data commonly used in similar literatures, In order to verify the effectiveness of the algorithm, the final result is the average of the five test results. In the test process, for each training image, four groups of features are obtained by using the above method, which are respectively sent to four trained SM classifiers for testing, and four classification results are obtained. The voting dominant category is the final discrimination result. If two of the results are the same, the CA feature discrimination result shall prevail, Because the low-frequency features have more detail information, there are 8 kinds of motion scene data sets (SE). There are 8 kinds of motion events and 1579 color images in the data set. These include badminton, rock climbing, polo, bowling, rowing, gateball, skateboarding, and sailing. Each class uses 70 images for training, and 60 images are used as test = 5 class scene data set (LS). The data set includes 15 kinds of scene images, including 4485 images, such as suburbs, forests, kitchens, offices and so on. Each class uses 100 images for training, and the remaining images are used as test images.

5   Epilogue

The art aided design method based on scene visual understanding algorithm designed in this paper can effectively extract the known visual scene image information accurately, help to solve the problem that the output image is not clear, at the same time, it can improve the output quality of the image, and improve the visual effect of art aided design. Aiming at the problems existing in scene classification, such as using image features for modeling, insufficient feature extraction and ignoring the correlation between things in the scene, an effective scene classification method is proposed [4]. This method fully considers the characteristics of human visual perception, and fuses the saliency detection algorithm on the basis of retaining the advantages of local low-level feature patterns, The overall sensitivity information of the image is used, which not only considers the relationship between the whole scene things, but also enhances the description of the local information.

References

1. Liu, L.: Computer aided Putonghua test system design in complex environment. Mod. Electron. Technol. **44**(01), 149–152 (2021)
2. Jiang, L.: Construction and implementation of the integration of course competition and Teaching – Taking the course of "computer aided design of clothing effect drawing" as an example. Textile Clothing Educ. **35**(06), 525–530 (2020)
3. Liu, F., Yu, W.: Application of computer aided technology in agricultural machinery design. Southern Agric. Mach. **51**(24), 42+50–51 (2020)
4. Wang, T., Liu, J., Sun, H., Sun, L., Dong, J.: Experimental course construction of computer aided drug design for postgraduates. Lab. Sci. **23**(06), 114–117 (2020)

Artistic Image Style Based on Deep Learning

Jun Chu[(⊠)]

Shandong Management University, Jinan 250300, Shandong, China

Abstract. With the popularity of smart phones and people's pursuit of the beauty of art, the application of image art is very popular. In addition to the realization of image art in hand application, it also has a large number of applications in game rendering, animation production, advertising design, film production and other fields. Because of the strong visual appeal and cultural connotation of the non photorealistic image after the artistic image, it has the significance of research and development. The art algorithm of image style based on deep learning enables the computer to simulate the form of human creation. This intelligent algorithm can quickly generate images of various art styles. Through the implementation of the existing image style art algorithm based on deep learning, we find that there are some problems.

Keywords: Construction system · Multi-objective optimization · Ant colony algorithm

1 Introduction

In today's life, more and more people use the photo software of style painting to produce images similar to the style effect of famous paintings, such as B612, prism, Meitu, etc. In the pursuit of artistic beauty, image stylization is more and more worthy of our in-depth study. Movies with novel oil painting style themes are also beginning to be favored by the public, so in the research of image stylization, the research of dynamic image stylization is indispensable. In recent years, due to the rapid progress of graphics processor, the deep learning based on convolutional neural network has ushered in the second spring [1]. Compared with the traditional image art stylization algorithm using rendering and filtering, the image stylization algorithm based on deep learning is more suitable for different styles of image processing. In this paper, we have done a lot of experiments on image stylization based on deep learning, and put forward solutions to different problems. At the same time, we have done corresponding research and Implementation on video stylization. The solution to this problem is of great significance to the research of neural art. The research background and significance of image stylization, and the research status at home and abroad will be the focus of this chapter.

J. C. Hung et al. (Eds.): FC 2021, LNEE 827, pp. 1154–1158, 2022.
https://doi.org/10.1007/978-981-16-8052-6_160

2   Still Image Stylization

Based on deep learning, the principle of image stylization originated from the neural algorithm paper on artistic stylization by gays, which is also the beginning of realizing the artistic image style by neural network. At the beginning of this chapter, we will mainly introduce the convolutional neural network and the basic principle of image stylization. Then, according to the problem, we propose some style transmission methods, such as color preserving style transmission, multi style transmission and local style transmission.

2.1   Image Stylization Principle Based on Deep Learning

Convolutional neural network is the core of deep learning. As time goes on, the layers of convolutional neural network become deeper and deeper, and more and more features can be learned, which can be used in image recognition, speech analysis and other aspects. The name and principle of convolutional neural network are derived from the transmission mode of biological neurotransmitter. On this basis, with the weight sharing and the improvement of computer computing ability, the layers of convolutional neural network are getting deeper and deeper. The earliest convolutional neural network started from lenet, but at that time, the graphics processing ability of computer was not matched, so convolutional neural network could only process a small number of images. Subsequently, the development of neural network is faster and faster, and it is the further improvement of alexnet at the beginning. It has five convolution layers, each convolution layer is followed by a maximum pooling layer, which reduces the size of the feature graph by half, and three fully connected layers [2]. The first eight belong to the learning layer, and the last one is the softmax layer of 1000 classes. Figure 1 shows the structure of alexnet:

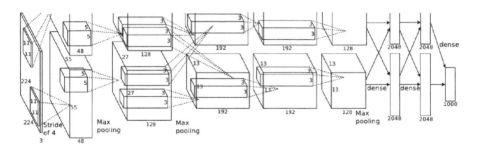

Fig. 1. Alexnet network structure

In the formula (1) of the content loss function, X is the image to be generated with white noise, P is the content image, 1 is each convolution layer, and F and P are the matrix composed of n m-Size response results of 1 layer. F is the image to be generated:

$$L_{content}(\vec{P}, \vec{x}, l) = \frac{1}{2} \sum_{i,j} \left(F_{i,j}^l - p_{i,j}^l \right) \tag{1}$$

When two vectors are multiplied, the more relevant they are, the greater the multiplication result value is, and the smaller the uncorrelated value is. When two vertical vectors are multiplied, it is 0. Style can be expressed as the correlation between pixels, then the style texture can be measured by the gramme matrix:

$$G_{i,j}^l = \sum_k F_{ik}^l F_{jk}^l \tag{2}$$

2.2 Optimization Method of Deep Learning

When the final loss function is obtained, it is necessary to iteratively optimize the loss function so that the final loss tends to a global minimum. In the literature, the stochastic gradient descent method (SGD), constrained quasi Newton algorithm (lbfgs), conjugate gradient method (CG) and so on are introduced, and the most basic method is the stochastic gradient descent method. The iterative optimization method of loss function often determines the degree of convergence, even the length of training time. Therefore, it is very important to select an appropriate optimization method in the training process of convolutional neural network. We know that the direction along the gradient is the direction in which the multivariate function changes the most and rises the fastest in any direction, so the bottom can be reached by calculating the opposite direction of the gradient at each step, which is the core principle of the gradient descent method.

3 Dynamic Image Stylization

Dynamic image stylization is actually a more complex image stylization, because dynamic images, such as GIF or video, can be divided into one frame of image, and the stylization of each frame of image based on deep learning can get the result of dynamic image stylization. But the result of this method is not only very time-consuming, but also jitter and flicker, which makes the final effect beyond recognition, so the following will introduce the related dynamic image stylization processing methods.

3.1 Optical Flow Method

Due to the displacement between the frames in the video, local flicker and jump will appear after each frame is connected. In order to get a good result of dynamic image stylization, optical flow method is needed to predict and constrain the motion.

Usually, when we describe the motion of an object in three-dimensional space, it means that after a period of time, a point in three-dimensional space reaches a new position in three-dimensional space, and the displacement can be described by the physical concept of motion field. In the image processing of computer, if we don't add the depth information supplemented by the camera, we mostly receive the two-dimensional image information. The optical flow field needs to express the three-dimensional motion field into the two-dimensional plane to describe the motion information of the object. Generally speaking, the optical flow method is the movement of the object that the eye can feel, so the optical flow method can sense the movement information of the target object.

3.2 Deep Low Optical Flow Method

Deepflow is a descriptor matching algorithm inspired by the optical flow algorithm of brox and Malik. The matching algorithm uses a six layer structure, interleaving convolution layer and maximum pooling layer, which is similar to deep convolution network. Brox and Mali found that if we add the descriptor matching term to different optical flow algorithms, we can deal with better large displacement optical flow estimation. The specific idea is to estimate the optical flow through the consistency of sparse descriptors [3]. Of course, someone proposed to extend sparse matching to dense matching under local affine reduction. It is found that dense matching of descriptors can improve the final optical flow estimation accuracy.

Depth matching algorithm is used in deep flow, which combines descriptor matching algorithm with large displacement optical flow method to estimate optical flow. This depth matching algorithm uses dense sampling descriptor matching algorithm, which uses a series of feasible non rigid corresponding points, and can calculate the effective comparison between non rigid descriptors at the same time of dense matching. Therefore, the algorithm is robust for computing large displacement optical flow. The specific description matching includes two steps in image matching. Firstly, the local descriptors within the frame are extracted, and then the non rigid dense matching is used between the whole frame images.

3.3 Dynamic Image Stylization Based on Deep Low Optical Flow Method

In the realization of dynamic image stylization, dynamic image stylization based on deep flow optical flow method is adopted. Its basic principle is still based on the still image stylization of gays, still using the vgg-19 model. As the stylized image of each frame is calculated directly, there will be a lot of jitters and jumps in the final result, which will affect the final dynamic image stylization result [4]. Therefore, on the basis of the original static image stylization, the optical flow method is added to predict the motion correlation between frames, so as to reduce jitters, improve the calculation speed, and get better dynamic stylization results.

4   Concluding Remarks

The image stylization algorithm developed by gays, the ancestor of neural art, only realizes the common image stylization. The basic principle is to learn the generated image in the optimization iteration by establishing the loss function of the distance between the generated image and the style image and between the generated image and the content image, so that the final generated image is similar to both the style image and the content image. This paper follows the theory of the stylization, but finds that there are different problems in the stylization, and solves the problems respectively, and finally obtains the comprehensive artistic image style. The problems solved in this paper are as follows.

References

1. Sun, Z., Xue, L., Xu, Y., et al.: Review of deep learning research. Comput. Appl. Res. **29**(8), 2806–2810 (2012)
2. Liu, W., Liang, X., Qu, H.: Study on learning performance of convolutional neural networks with different pooling models. Chin. J. Image Graph. **21**(9), 1178–1190 (2016)
3. Qian, X.: Overview of non photorealistic art rendering technology based on image. J. Graph. **31**(1), 12–18 (2010)
4. Lu, S., Zhang, S.: A new algorithm for stylized rendering of image oil painting. In: The 4th National Conference on Geometric Design and Computer Science (2009)

Analysis and Research of 4G and WLAN Convergence Network Access Authentication Protocol

Juanjuan Gong[✉]

Leshan Normal University, Leshan 614000, Sichuan, China
gjj0223@sohu.com

Abstract. EAP-AKA is a multi network convergence authentication protocol defined by 3GPP, which is adopted in wlan-4g convergence network. Among them, the wireless broadband access technology represented by WLAN has the characteristics of high bandwidth, low cost, free frequency band, flexible networking, easy expansion and so on. It has the natural integration advantages with LTE. Based on the heterogeneity of 4G and WLAN, 3GPP uses EAP-AKA as the security authentication mechanism for ue to switch between the two networks. But most of the existing authentication mechanisms are still very complex and vulnerable to network attacks, so it is difficult to meet the security and performance requirements at the same time. Based on the analysis and research of authentication process of multi network convergence protocol, this paper analyzes its security and defects, establishes 4g-wlan access authentication model, and studies and analyzes the improved EAP-AKA protocol to ensure the authentication security of wlan-4g convergence network.

Keywords: WLAN 4G convergence · EAP-AKA · Security authentication

1 Introduction

With the rapid development of wireless communication technology, the requirements of terminal equipment for mobility are higher and higher. At the same time, the requirements of users for network performance and quality of service are also higher and higher, which brings great challenges to the development of mobile communication technology. With the extensive construction of 4G network in China, the coverage and users of 4G network are growing day by day. More and more people enjoy the fast and convenient service brought by 4G network. With the advantages of low cost, high speed and rapid deployment, WLAN, which belongs to the same broadband wireless access technology as 4G, has not only become an important technology for telecom operators to integrate networking, but also occupies a large share in the field of government and enterprise informatization, as well as in the home and personal market. The land of the mat. Therefore, the integration of WLAN and 4G network will play a complementary role in common development. The third generation partner Technology (3GPP) organization has put forward the corresponding standard scheme for 4G and WLAN network convergence access. In order to realize the security of convergence

J. C. Hung et al. (Eds.): FC 2021, LNEE 827, pp. 1159–1165, 2022.
https://doi.org/10.1007/978-981-16-8052-6_161

access, the extensible authentication protocol authentication and key agreement (EAP-AKA) is adopted. EAP-AKA protocol is the authentication and key distribution protocol of 4G and WLAN convergence interconnection, which is the basis to ensure its security. In addition, EAP-AKA protocol is also widely used in wlan-3g and other multi network convergence authentication, so EAP-AKA protocol involves the security of multi network convergence authentication.

This paper first introduces the wlan-4g fusion authentication protocol EAP-AKA, then analyzes the security of EAP-AKA protocol, and finally analyzes the improvement of EAP-AKA protocol [1].

2   The Research Significance of WLAN

The rise of unlicensed spectrum technology, represented by WLAN, provides a choice for operators to expand their capacity. It can be used as a solution to reduce the load of cellular infrastructure and provide users with reasonable quality of service (QoS). In the early days, WLAN was only used as an alternative to cellular data, which was independent of the research, development and standardization of cellular network. Although WLAN is unsatisfactory in quality and security, and can not provide seamless connection and roaming, it has the characteristics of high bandwidth, low cost, free frequency band, flexible networking, easy expansion and so on. It has natural integration advantages with cellular network. Therefore, in recent years, the idea of operators began to change, and realized that WLAN can supplement cellular data well. So operators and WLAN standardization organizations began to work closely to make better use of WLAN for cellular devices. Mobidia and maravedis rethink have made statistics on the distribution of mobile data traffic in relevant reports, as shown in Fig. 1.

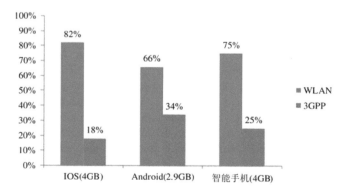

Fig. 1. Mobile data traffic distribution statistics

As can be seen from Fig. 1, no matter which application system is used, the data traffic used by users through accessing WLAN is far more than the cellular data traffic. Cisco also predicted in its VNI that by 2018, more than half of mobile data traffic will be diverted to other alternative technologies, such as WLAN, WiMAX, etc. Therefore, the typical network in the future should be the combination of various wireless network technologies, and WLAN and small base station will be used to unload traffic and improve coverage. 3GPP started the research of WLAN and cellular convergence technology as early as 2002, from release 6 to the latest release 14, and put forward a variety of solutions according to different needs. After WLAN is connected to the core network, UE can access the related services of the two networks by accessing WLAN, which can not only reduce the load of cellular network, but also supplement the authentication, billing and other functions of WLAN, making the access of users more secure and reliable.

3 WLAN-4G Convergence Authentication Protocol

The optimal integration of wlan-4g is to combine the two networks into one. Users do not need to perceive what kind of network they are, but have what kind of business experience. Therefore, in 3GPP R8, 3GPP organization proposed the untrusted access scheme of WLAN converged access 4G network; in 3GPP r1i, it proposed the trusted access scheme of WLAN converged access 4G network. Compared with trusted access, the biggest difference of untrusted access is the establishment of IPSec tunnel between terminal and network [2].

In order to ensure effective authentication between different networks during wlan-4g network fusion, 3GPP organization stipulates that EAP-AKA protocol is adopted as authentication protocol for wlan-4g network convergence. The realization of EAP aka protocol is completed by UE (terminal of receiving WLAN network), wlan-an (visit network of WLAN network), 3GPP AAA server and HSS. In order to more effectively explain the protocol flow of EAP aka, Algorithm 1 is set here to generate message authentication code, Algorithm 2 is used to calculate the expected response value in message authentication, Algorithm 3 is used to generate encryption key, Algorithm 4 is used to generate integrity key, and Algorithm 5 is used to generate anonymous key. During the interaction, EAP aka protocol has the following messages: UE + WLAN an: NaI (network receiver flag, including UE temporary flag); WLAN an → 3GPP aaa: nai; 3GPP AAA → WLAN an: rand (random number), auth UE temporary flag and message identification code; wlan-an → ue:rand, auth.Ue temporary flag and message identification code; WLAN UE + WLAN_ An: res, message identification code; WLAN an → 3GPP AAA:RES, message identification code; 3GPP aaa- → WLAN an: UE authentication result, wlan-an and UE shared key; wlan-an → UE: UE authentication result.

The following describes the protocol in detail. The schematic diagram of the protocol flow is shown in Fig. 2, which can be seen in the figure:

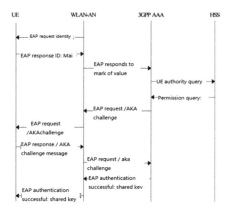

Fig. 2. Flow diagram of EAP-AKA protocol

Step 1: wlan-an sends an EAP request message to UE.

Step 2: after receiving the message, UE sends EAP response identity message to wlan-an, which carries Nai identity.

Step 3: wlan-an sends the received EAP response identity message to 3GPP AAA server.

Step 4: after receiving the UE's identity mark, 3gppaaa first queries the HSS whether the UE has permission to use the WLAN network, and then obtains the authentication vector AV related to the UE from HSS and obtains the new temporary mark corresponding to the UE's IMSI, among:

$$AV = RAND \parallel XRES \parallel CK \parallel IK \parallel AUTH \tag{1}$$

$$XRES = Algorithm2(RAND) \tag{2}$$

$$CK = Algorithm3(RAND) \tag{3}$$

$$IK = Algorithm4(RAND) \tag{4}$$

$$AUTN = SQN \oplus AK \parallel AMF \parallel MAC \tag{5}$$

$$AK = Algorithm5(RAND) \tag{6}$$

SQN it's the serial number,AMF is the authentication management domain.

3GPP AAA generates shared key from IK and CK again and constructs EAP request/aka challenge message, which includes Rand, auth and temporary flag, and calculates message authentication code. 3GPP AAA finally sends EAP request/aka challenge message to WLAN - AN.

Step 5: wlan-an sends the received EAP request/aka challenge message to UE.

Step 6: UE first verifies auth and confirms whether the received Sqn is in the valid range. If it is correct, it realizes the authentication of 4G network. A shared key is generated from IK and CK to verify that the message identifier is correct and save the received temporary flag. Re = Al-algorithm 2 (rand) is calculated and combined with res to construct EAP response/aka challenge message. At the same time, message authentication code is calculated. Finally, EAP response/aka challenge message is sent to wlan-an [3].

Step 7: WLAN an sends the EAP response/aka challenge message received to 3GPP AAA server.

Step 8: 3GPP AAA first verifies the message identification code, then calculates Xres and compares it with the res received. If correct, UE authentication is passed and message of EAP authentication success is sent to wlan-an. At the same time, the shared key for confidentiality and consistency protection in WLAN communication is sent.

Step 9: wlan-an saves the shared key, which will be used for confidentiality and one-to-nature protection when communicating with UE, and send the message of EAP authentication success to UE.

WLAN and 4G network realize bidirectional authentication between the terminal ue of WLAN network and 4G network through EAP aka protocol, and share session key between the terminal UE and WLAN network to realize the encryption transmission and the verification of the authenticity between the two.

4   Security Analysis of EAP-AKA Protocol

Although EAP aka' uses pre shared key K to realize the mutual authentication between UE and 3GPP AAA, there are still some hidden security vulnerabilities:

(1) UIM clone attack: there is no shared key update mechanism in eap-aka', which may lead to UIM clone attack once K is leaked.

(2) Middleman attack: an attacker may pretend to fake AP to launch an intermediate attack. The attacker first uses the relevant attack means to attack AP, then disguises himself as AP, obtains the session key sent by UE to AP, which leads to the UE communication no longer secure. In addition, the attacker can send unrecognized identity to 3GPP AAA, which leads to 3GPP AAA constantly requesting to generate authentication vector AV from HSS, which causes the overload of HSS and crashes. In this process, fake AP may obtain the message of the session between 3GPP AAA and HSS server by eavesdropping, obtain the HSS pass to 3GPP AAA Authentication vector AV, and get ck' and ik'. After that, fake users access the network, and can communicate and access network resources normally.

(3) User identity information disclosure: during the whole authentication process of EAP aka', UE uses temporary anonymous ID instead of IMSI, but the target network has not defined the flag in advance, UE still needs to send IMSI in clear text [4].

In addition, if LTE network can not recognize the temporary symbol of UE, UE also needs to send IMSI in clear text. This undoubtedly gives an attacker a chance to steal the user IMSI after monitoring, so that the location and bill information of the user will be disclosed, and the confidentiality of the user identity will be threatened.

According to the above analysis, this paper designs the access authentication mechanism of MEAP aka on the basis of EAP aka', aiming to achieve the following objectives:

(1) The unreasonable trust hypothesis is eliminated, and the mutual authentication between UE, Waaa and 3GPP AAA is realized, which effectively prevents various attacks in the authentication process.
(2) To protect the identity information of users from being disclosed, LSK keys are generated locally in UE, Waaa and 3GPP AAA for encryption or decryption of transmission messages. A dynamic ID update mechanism is designed to ensure the freshness and anonymity of ID at each authentication time, and avoid IMSI sending in clear text.
(3) In the process of re authentication, considering the network type of switching, the fast re authentication protocol is proposed, which is suitable for inter domain switching and intra domain switching respectively, which can reduce the authentication delay, signaling cost, communication cost and energy consumption.
(4) The security of MEAP aka is verified by using the formal verification tool Avispa to ensure the security of the improved mechanism.

5   Security Improvement of EAP-AKA Protocol

In view of the problems in the above scheme, the EAP aka protocol is improved by public key mechanism, and the following changes are made in the EAP aka protocol interaction process.

(1) In the second step of the original process, UE extracts the address of 4G network from Nai, and then calculates the ciphertext EK (randulnai) by using session key K and sends it to wlan-an.
(2) In the third step of the original process, wlan-an generates random message man and random number randan, then generates ciphertext EK (randanman), and finally sends it to 4G network with EK (randuelinai). 4G network decrypts the two ciphertext after receiving these two ciphertext, so as to authenticate wlan-an and obtain Nai information of UE so as to query UE permission information at HSS.
(3) If you need to update the session key, 4G networks use random to calculate IK and CK, then k' with IK and CK', and finally k' to calculate ek' (random) and send it to UE. UE decrypts and obtains random after receiving ek' (random) and compares it with the rangue sent before it. If the same, it indicates that the session key is successfully updated. UE calculates H (k') by hash function and sends it to wlan-an. After WLAN an receives H (k'), it performs anti hash operation to get k' and compares it with k' obtained from 4G network. If the same, it indicates that session key at UE is more successful.

The EAP aka protocol is improved by using public key mechanism. By adding shared key between.Wlan and 4G network, mutual authentication between them is realized, and the counterfeit attack of wlan-an is prevented. Through the encryption transmission Nai, the encryption protection of IMSI is realized, and the leakage of mobile user identity information is prevented. Through the mechanism of introducing key update, the shared key between UE and 4G network is updated safely.

6 Conclusion

In the future, the integration of 4G network and WLAN network is one of the main trends in the development of access network. The security of the integration largely depends on the EAP-AKA fusion authentication protocol. This paper first analyzes and introduces the authentication process of EAP-AKA, the wlan-4g network convergence authentication protocol, then analyzes the security and defects of EAP-AKA in detail, and finally analyzes the improvement of EAP-AKA protocol. This paper has a certain reference for the construction of wlan-4g convergence network.

References

1. Zhu, H., Li, B.: Research on the integration of ltefdd and TDD-LTE. Mob. Commun. **12** (2008)
2. Han, Y., Cheng, G., Peifei: The research on the integration and development of LTE and Internet of things. Mob. Commun. **19** (2012)
3. Damnjanovic, A., Montojo, J., Wei, Y., et al.: A survey on 3GPP heterogeneous networks. IEEE Wirel. Commun. **3** (2011)
4. Zhang, S., Xu, A., Hu, Z.: Analysis and improvement of EAP-AKA protocol. Comput. Appl. Res. **07** (2005)

Application of Neural Network Data Mining Method in Network Marketing

Xin Guo and Han Wang[✉]

Xing Zhi College, Hubei University, Wuhan 430070, China

Abstract. Neural network is recognized as a high-precision classifier at present. This paper introduces a data mining classification method based on feedforward neural network, and gives an application example in network marketing. The limitation of the method is analyzed and the way of improvement is given.

Keywords: Neural network · Data mining · Classification · BP algorithm · Network marketing

1 Introduction

Classification is an important problem in data mining. Its purpose is to find out the hidden classification rules through the analysis of a large number of data. In network marketing, classification is widely used in customer rating, in order to carry out targeted marketing. It can also be used to predict who will respond to the promotion methods such as mailing advertisements, product catalogues and coupons. The classification methods mainly include decision tree method, statistics method, nearest neighbor method and neural network method [1].

Artificial neural network (ANN) technology imitates the function of human brain, and finds patterns for prediction and classification by repeatedly learning training data sets. Metanetwork is especially good at solving complex problems.

This paper first introduces the related concepts of neural network and BP algorithm, and then introduces the application of neural network in customer grading and classification through an example. Experiments show that the method is feasible and effective. Finally, the author puts forward some ideas to improve the algorithm.

2 Overview of Artificial Neural Network

Artificial neural network refers to a network system that simulates the structure and function of biological neural system, uses a large number of processing components, and is established by artificial means. It is established on the basis of the study of brain nervous system. Human brain is the prototype of artificial neural network, and artificial neural network is the simulation of brain nervous system.

J. C. Hung et al. (Eds.): FC 2021, LNEE 827, pp. 1166–1170, 2022.
https://doi.org/10.1007/978-981-16-8052-6_162

2.1   Connections Between Biological Neurons

In the nervous system, a biological neuron first receives signals from other neurons through its dendrites; then all input signals are simply processed (such as summation) by the neuron cell body; finally, the processing results (i.e. output signals) of neurons are transmitted to other biological neurons through axons and synapses [2].

2.2   Artificial Neuron

Just as biological neuron is the basic processing unit of biological neural network, artificial neuron (hereinafter referred to as "neuron") is the basic processing unit of artificial neural network as the simulation of biological neuron.

Based on the in-depth study of the structure and characteristics of biological neurons, psychologist W. McCulloch and mathematical logician W. Pitts first proposed a simplified neuron model, M-P model, in 1943. The model puts forward the following assumptions

1) Each neuron is a multi input and single output information processing unit;
2) There are two types of synapses: excitatory type and inhibitory type;
3) Neurons have threshold characteristics;
4) The synaptic strength of neurons is constant.

The mathematical model of artificial neuron is produced by mathematicalizing the neuron with the above characteristics. The input and output mathematical expressions of the model are as follows:

$$y = f(u) = f(\sum_{i=1}^{n} w_i x_i - \theta) \tag{1}$$

The output of neurons is expressed as:

$$y = f(X^T \cdot W - \theta) \tag{2}$$

The excitation function of linear neuron is as follows:

$$f(x) = x \tag{3}$$

2.3   Learning Mechanism of Artificial Neural Network

Learning is an important feature of neural networks. Only the trained neural network can solve all kinds of problems in practical application.

1. Biological background of artificial neural network learning. According to biologists' observation, in the learning process of the brain, if it is repeatedly stimulated by a certain external information, the neurons receiving the information will have some growth process or metabolic changes, which will lead to changes in the strength of synaptic connections, making the influence of external information on

themselves change. Therefore, scientists believe that the essence of neural network learning lies in the adaptive change of synaptic connection strength of nerve cells. Similarly, for an artificial neural network with fixed topology, its learning process is mainly an adaptive process to achieve a desired purpose by adjusting the connection weights.

2. Classification of learning algorithms neural networks can be divided into two categories: learning with teachers and learning without teachers. When a teacher is learning, in order to solve the problem in the practical application of artificial neural network, he must first select some expected output known sample data from the application environment (this kind of sample is usually called "learning sample"), and then adjust the connection weight of neural network according to the input-output relationship contained in these samples. Because this kind of learning process needs to have learning samples as teachers to teach and supervise the network learning, it is called teacher learning.

3   Application Examples

Our data is directly from the business data of an online bookstore from 2002 to 2003. Now we need to use the customer's registered identity information (including age, education, income, etc.) and the customer's actual annual purchase amount calculated by the website to rate the customers, so as to determine which customers' needs need to be developed. For example, if a customer rated as having medium purchasing power only purchases goods equivalent to the purchase quantity of low purchasing power customer group, then its demand has development potential. So the key of the problem is to determine an appropriate classification algorithm, according to the customer's registered identity information and the actual annual purchase amount [3].

The learning module of classification program completes the learning work. According to the 010100 value of the input mode string, the program first calls the threshold function to calculate the output of the input layer, and passes the output to the hidden layer. Similarly, the output of the hidden layer is passed to the output layer, and the output layer calls the threshold function to generate the output. According to the output mode, each output node gets a value close to 0, and the error is allowed within 0.1. Similarly, when the ninth input mode is input, the first output node must output a value close to 1, and the second output node must output a value close to 0. However, the actual output can not meet the requirements immediately. So we must modify the network to make the output meet the requirements. According to the formula, the weight correction is calculated, and the network is changed until the output error meets the requirements. Then enter the next mode. Until all modes are entered. At this time, it is still necessary to input and calculate the output value from the first mode. If it is found that there are still outputs that do not meet the requirements (because the latter mode is likely to modify the network generated by the previous mode, it is difficult to meet the error requirements in one learning), it is necessary to learn from the current state of the network until the next mode that meets all the inputs is found. Until all modes are entered. At this time, it is still necessary to input and calculate the output value from the first mode. If it is found that there are still outputs that do not meet the requirements (because the latter mode is likely to modify the network generated by the previous mode, it is difficult to meet the error requirements in one learning), it is necessary to learn from the current state of the network until a mode that meets all the input modes is found, The network required by the output mode (i.e. weight set and threshold set). The end of the study. At this time, it can be considered that such a rule about human classification has been recognized by the computer [4].

The network generated by the learning process contains rules, and saving the network parameters also preserves the rules. Once the rules are extracted and classified, it can start. If you enter 000101, you will get the output of 01. 01 identifies the category of annual book purchase amount of 100–500 yuan. If the unlearning value is input, the classification can be predicted easily, which is the advantage of neural network classification algorithm. According to the results, the corresponding annual purchase amount of 100–500 yuan is compared with the actual annual purchase amount (set as a, whose value comes from the business data of the website). If a <100, it means that the customer's purchase potential needs to be tapped.

4 Improvement and Conclusion

In the running of the program, we find that the running time is long, and the three-layer network structure is completed after 105 times of cyclic learning and the total number of learning is more than 7000 times. After expanding to 5-layer network structure, the situation has not been improved. This shows that BP network is an empirical network, the increase of network layers can not improve the performance, only the network structure is related to specific problems can improve the learning speed. We should improve it from the following aspects.

First of all, in the process of algorithm learning, the accuracy value is constant, which leads to some patterns can not be recognized in the accuracy range, and even deviate from the accuracy range in the next cycle. Therefore, it can be considered to keep its precision value in the process of pattern learning, and when the cumulative sum of a cycle control variable reaches, the learning will end with the minimum precision value, so as to shorten the learning time. This may bring some deviation to classification prediction, but it has little effect on classification.

Second, for complex classification problems, we can preprocess the data first, and then train BP network, so as to reduce the learning time. Firstly, rough set theory is used to optimize the quantified attributes to delete redundant attributes and attributes without valid information, and then BP network is trained to achieve better efficiency and accuracy. Attribute selection will become an effective means of pattern classification.

Second, for complex classification problems, we can preprocess the data first, and then train BP network, so as to reduce the learning time. In paper [2], the rough set theory is used to optimize the quantified attributes to delete the redundant attributes and the attributes without valid information, and then the BP network is trained, which achieves better efficiency and accuracy. Attribute selection will become an effective means of pattern classification.

Thirdly, in large-scale data mining, it is unrealistic to realize neural network classification with single machine. In addition, BP algorithm is simple, so distributed parallel processing can be used to improve the speed. We can divide the large-scale problem into subproblems. For example, dividing the problem with 10 features into three subproblems is equivalent to defining a hidden layer to communicate in three subproblems, so the scale of the problem with 10 features is reduced to four. But considering that BP network is a kind of feedback network, the nodes of each layer can not be calculated in parallel simply. We can find a way to use another threshold function to estimate the feedback quantity of the network in advance, so as to realize the pipeline calculation between each layer.

References

1. Li, B., Song, H.: Application of data mining in customer relationship management. Comput. Appl. Res. **10** (2002)
2. Chen, Z.: Rough set neural network intelligent system and its application pattern recognition and artificial intelligence (1999)
3. Jiao, L.: Application and Realization of neural network. Xi'an University of Electronic Science and Technology Press (1996)
4. Haykin, S.: Synthesis Basis of Neural Network. Tsinghua University Press (2001)

A Decision Tree Classification Algorithm in University Archives Management

Zhilin Han[(✉)]

Shandong Institute of Management, Yantai 250357, China

Abstract. Using computer technology to realize the digitization of archives management is a key point of archives management at present. Through the research on the decision tree classification algorithm of data mining technology, combined with the example of archives management in Colleges and universities, this paper shows that using D3 algorithm to realize the scientific classification of digital archives management in Colleges and universities.

Keywords: Archives digitization · Data mining · Decision tree

1 Introduction

With modern information technology emergence of electronic documents, the research and construction of digital archives has become a hot topic in the field of archives in China. The so-called digital and information archives refer to the written materials formed by people in social activities, which are carried by chemical and magnetic materials such as computer disks, disks and optical disks. It mainly includes electronic documents, e-mail, electronic reports, electronic drawings, video, audio and so on. From August 11 to 13, 1998, the Institute of archival science and technology of the State Archives Administration held an expert seminar on the management of electronic documents. The central topic of the meeting is to revise, supplement and improve the draft of "Introduction to electronic file archiving and electronic file management" and "office automation electronic file archiving and electronic file management method" and "CAD electronic file CD storage, archiving and file management requirements", Electronic documents and digital, information-based archives are widely used in some enterprises and institutions, especially [1]. How to effectively manage, classify and apply digital and information-based archives is the focus of archives management department. technology, this paper applies decision number algorithm to classify different archives, which effectively realizes the classification management of electronic documents of different archives. At the same time, it can also find the relationship between different types of archives.

J. C. Hung et al. (Eds.): FC 2021, LNEE 827, pp. 1171–1175, 2022.
https://doi.org/10.1007/978-981-16-8052-6_163

2 Classification Algorithm Based on Decision Tree

2.1 The Meaning of Data Mining

With the development of education informatization and lifelong learning, open, online and distance education has become the field of rapid development of modern education. Its potential influence in the field of education transmission can be reflected by ICT. Especially in the education activities carried out on the platform of Internet, ICT has become an internal component of open, online and distance education. It technology promotes and strengthens the trend of two aspects. One is that more and more learners have the opportunity to participate in various forms of lifelong learning activities; Second, more and more educational resource providers can use ICT technology to carry out teaching activities. ICT has changed the traditional learning mode, learning method and education transmission mechanism, and promoted the realization of lifelong learning (Regan, 1998). From cramming teaching to discovery teaching, from linear learning to multimedia learning, from teacher centered to student-centered, from learners' passive acceptance of learning to active learning planning and learning to learn, from school education to lifelong learning, the enabling of technology is mainly reflected in: building basic teaching platform and transmitting teaching materials with the help of ICT infrastructure (network, man, LAN, etc.); Display, manage, track and transmit online learning content with the help of learning content system; With the help of learning management system, we can carry out qualification certification, performance management, individual career development planning and record learning activities; With the help of learning support tools (timely communication tools, forums, expert led discussions, online meetings, etc.) to carry out real-time teacher-student communication and interaction [2].

2.2 ID3 Decision Tree Classification Algorithm

Decision most useful method classification data mining. This needs to build process. Once the tree is built, it can be applied to tuples in the database and the classification results can be obtained. In the method of decision number, there are two. The classification method divides the search space into some rectangular regions, and then classifies the tuples according to the region in which they fall. The following definition gives the definition of decision number classification method.

   Definition 1 gives a database $D = \{t_1, \ldots, t_n\}$ the attributes $\{A_1, A_2, \ldots, A_n\}$. At the same time, given the class set $C = \{C_1, C_2, \ldots, C_m\}$. For refers to the tree with the following properties.

(1) Each internal node is marked with an attribute a.
(2) Each marked, which can be applied to the properties of the corresponding parent node. (3) each marked of C. using decision tree to solve the classification problem includes two steps.
(3) Stop criteria. When the training data is classified correctly, the tree generation process should stop. In order to prevent the production of too large trees, it is sometimes desirable to stop ahead of time. In addition, early stop can prevent over

fitting. There is a trade-off between classification accuracy and performance. It is even conceivable that if there are known data distributions that cannot be expressed in the training data, more levels can be established in the tree.

(1) Decision tree induction, using training data to build a decision tree.

(2) For each tuple t ∈ D, the decision tree is applied to determine the category of tuples. In the database, its attributes will be used to mark the nodes in the tree, and the split attributes are called split attributes [3]. The predicates used to mark arcs in a tree are called split predicates. The performance of decision tree construction algorithm mainly depends on the size of training set and how to choose the best splitting attribute. Most decision tree algorithms need to face the following problems.

(3) Pruning. After a tree is built, it needs stage, too many comparisons may be deleted or some subtrees may be deleted for better performance.

Definition 2 given probability $P_1, P_2, \ldots P_s$, where $\sum_{i=1}^{s} P_i = 1$, then entropy is defined as:

$$H(P_1, P_2, \ldots P_s) = \sum_{i=1}^{s} (p_i \log(1/p_i)) \tag{1}$$

Given represents a measure $S = \{D_1, D_2, \cdots D_S\}$ The entropy of these states can be recalculated. Each step in DD3 selects the splitting state that maximizes the order. If all tuples in a database belong to the same class, the database state is completely ordered. D3 selects splitting and the information needed for correct classification after splitting. Clearly, fragmentation should minimize the information needed for correct classification. By determining the weight of the sub data set after the original data is the proportion of the subset in the whole data set. D3 algorithm calculates the information gain of a specific split according to the following formula.

$$Gain(D, S) = H(D) - \sum_{i=1}^{s} P(D_i)H(D_i) \tag{2}$$

3 Application of ID3 Classification Algorithm in Student Status File Management

For example, the classification of graduate students' year-end examination files in a university, the following table is the data of graduate students' year-end examination results, and the application of ID3 algorithm is explained by the following table. As shown in Table 1.

Table 1. Registration form of year end assessment results of Postgraduates

Full name	Gender	Types of candidates	Year end results	Output 1	Output 2
Lory	Male	This year	75	Good	Excellent
Torn	Male	This year	99	Good	Excellent
Job	Female	This year	80	Good	Good
Wory	Female	This year	85	Excellent	Excellent
Kitty	Female	On the job	45	Excellent	Excellent
Amy	Male	On the job	60	Excellent	Good
susan	Male	On the job	65	Qualified	Good

In the student file, the classification method of output 1 is relatively simple, and the simple segmentation method is used for in-service graduate students and fresh graduate students.

$85 \leq$ excellent results
$75 \leq$ grade ≤ 84 good
$60 \leq$ score ≤ 74 qualified
Grade ≤ 59 poor.

The result of output 2 uses a more complex segmentation method, which refers to the two attributes of examinee category and score.

For the training data in the above table, the initial state is: excellent (2/10), good (4/10), qualified (210), poor (210). So the entropy of its initial set is:

$$2/10 \log(10/2) + 4/ \log(10/4) + 2/10 \log(10/2) = 0.587 \qquad (3)$$

The subsets of current members are as follows:

$$1/6 \log(6/1) + 3/6 \log(6/3) + 2/10 \log(10/2) = 0.583 \qquad (4)$$

Therefore, this attribute has a large information gain. Therefore, the performance attribute is chosen as the first split attribute. From the above analysis, it can be seen that for the management of university student status records, taking the score attribute as the first split attribute, and then classifying it according to the type of candidates will improve the efficiency of file management [4]. Of course, this is only a simple explanation of university student status file management, University Archives Management includes student status files, scientific research management, teacher files, teaching security and so on. The informatization and digitization of archives management is imperative. The application of data mining technology will greatly improve the scientificity of electronic archives management. First, the safety awareness of managers is not high enough, There is no special protection for the privacy of archives. Under the temptation of others, it is easy to leak the non personal information to others, causing some security risks. Second, the monitoring system for managers is not perfect. At present, the management level of human resource archives in most schools is still at the primary stage. A staff member has to deal with several aspects of work, and there is no standardized process of division of labor and cooperation, The system of individual responsibility cannot be implemented. Once there is a problem, it is difficult to find the main person in charge. The so-called management monitoring is mostly a mere formality, only through some monitoring equipment to implement, lack of systematic and perfect management and supervision system. In the information age, the security of human resource archives management in Colleges and universities is facing more technical threats. Managers should be fully aware of this and be vigilant.

4   Concluding Remarks

When ID3 algorithm constructs decision tree, according to the principle of maximum information gain, the first attribute can not provide more information. There are many improved algorithms about D3 algorithm, including binary tree decision algorithm, estimation method based on gain ratio, estimation method based on classification information and estimation method based on partition distance. Here is just a description of how to use computer data mining technology to scientifically manage digital and information-based university archives. There is still a lot of work to do in the practical application of data processing technology to build university archives management system.

References

1. Feng, H.: Archives Management. China Renmin University Press, Beijing (1996)
2. Dunham, M.H.: Data Mining Course. Tsinghua University Press, Beijing (2005)
3. Li, X., et al.: Data Mining and Knowledge Discovery. Higher Education Press, Beijing (2003)
4. Tong, J., Zhou, E.: Analysis of human resources management in Internet plus era: comment on the Internet and new media era: employment management of enterprises - human resource legal management. Forest Prod. Ind. **6**, 122–129 (2019)

Bioengineering Algorithm of Dynamic Fuzzy Neural Network

Xue Hu[✉]

Changchun Institute of Architecture, Changchun 130607, Jilin, China

Abstract. Dynamic fuzzy neural network engineering is a biochemical reaction process with high nonlinearity, time-varying and hysteresis, and its internal mechanism is very complex. The traditional bioengineering algorithm is difficult to measure some key variables in the process online, which makes it very difficult to optimize the control of the whole fermentation process. Soft sensing technology is an effective way to solve this problem, Three important variables in fermentation process were predicted. The simulation results show that the soft sensor model can accurately estimate the output value, and the model has strong stability in the case of disturbance.

Keywords: Dynamic fuzzy neural network · Bioengineering · Algorithm research

1 Introduction

In but dynamic, that is, there is no fuzzy rule before learning, and its fuzzy rules are gradually formed in the learning process [1]. Compared with the common fuzzy neural network method, the fuzzy rules obtained by this method do the increase of input variables; especially, this method can automatically model the system and extract fuzzy rules without domain expert knowledge. Because the structure of the is small, the phenomenon of over fitting is avoided, so it brings great convenience to users.

2 Dynamic Fuzzy Neural Network

2.1 Dynamic Fuzzy Neural Network is Proposed

The existing research on is still a time-consuming problem, which is still focused on the expression of fuzzy neural network parameters. An unreasonable structure will lead to over fitting and over training. So far, there is no unified guiding theory to determine a fuzzy neural network with good generalization ability. In practical application, a trial and error method is usually used to determine a system (such as the determination process of fuzzy neural network shown in Fig. 1), which has great uncertainty [2]. Therefore, a person who does not master certain fuzzy system and neural network theory and has no deep understanding of practical application system can not design an effective system. In addition, the current fall into local minima.

J. C. Hung et al. (Eds.): FC 2021, LNEE 827, pp. 1176–1179, 2022.
https://doi.org/10.1007/978-981-16-8052-6_164

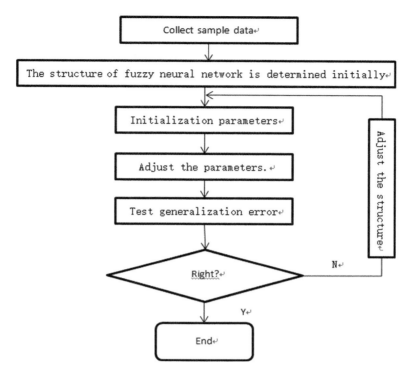

Fig. 1. Determination process of fuzzy neural network

2.2 Characteristics of Neural Network

Therefore, network structure is more optimized, the phenomenon of over fitting is avoided, and the generalization ability of the network is improved [3]. Dynamic fuzzy neural network can automatically model the system and extract fuzzy rules without the knowledge of domain experts. At the same time, it has the characteristics of online learning method, simple and effective learning algorithm and fast learning. Because of this, dynamic fuzzy neural network can deal with nonlinear and time-varying systems online and in real time, which is in sharp contrast with fuzzy ability [4].

3 Structure Model of Dynamic Fuzzy Neural Network

The network studied in this in Fig. 2.

The following is a detailed of each layer of the network:

(1) The layer: input represents an input variable.
(2) The layer: fuzzy layer, each most widely used membership functions are trigonometric function and Gaussian function:

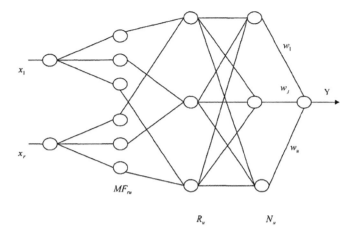

Fig. 2. Structure of dynamic fuzzy neural network DFN

$$\mu_{ij}(x_i) = exp\left[\frac{(x_i - c_{ij})}{\sigma_j^2}\right] \tag{1}$$

(3) The third layer: also reflects the total of the system. The output of rule j is:

$$\varphi_j = exp\left[\frac{\sum_{i=1}^{m}(x_i - c_{ij})}{\sigma_j^2}\right] = exp\left[-\frac{\|x - c_j\|^2}{\sigma_j^2}\right] \qquad j = 1, 2, \ldots, u \tag{2}$$

It can be seen from the above formula that each node of this layer is equivalent to a RBF neuron, where $X = (x_1, x_2, \ldots, x_n)$, $C_j = (c_1, c_2, \ldots, c_j)$ is the center, and the number of neurons.

4   Structure Model of Dynamic Fuzzy Neural Network

For dynamic, if the number rules is too small, it can not completely contain an input-output state space, and the performance of the trained DFNN will be much worse. complexity of the system, which will lead to the poor generalization ability of DFNN. It can be seen that the system output error will be an important parameter to determine whether to increase error of the system can be described as follows: assume that the ith sample data of the system is (x_i, t_i), where x, vector of the neural network, t, represents the expected output, and the total output y_i of DFNN is calculated according to the above formula. Therefore, the output error of the system can be given by the following formula:

$$\|e_i\| = \|t_i - y_i\| \tag{3}$$

When the output error e_i satisfies the following conditions, a rule is added.

$$\|e_i\| > k_t \tag{4}$$

In the formula, k_t According to the expected accuracy of DFNN system, it is preset.

Gaussian function has good local characteristics, when the distance between the output and the center is larger, the output is smaller; on the contrary, if the distance between the output and the center is smaller, the output is larger. When the membership function of DFNN system adopts Gaussian function, it is equivalent to dividing the input space of the system with many Gaussian functions. When a new sample is input into the system, if it can be covered by the existing Gaussian function, it means that the sample is in the accommodation boundary. At this time, the new input rules or new RBF cells.

5   Conclusion

In algorithm of dynamic, as well as the improved dynamic fuzzy neural network and are fully studied. The main is that its structure is not preset, but changes dynamically in the learning process, that is, fuzzy rules are gradually formed in the learning. In addition, the method can model the system and extract fuzzy rules without domain expert knowledge, and the extracted fuzzy rules are easy to understand and use. Generally speaking, it has the following characteristics:

(1) Dynamic learning, parameter estimation in the learning process, dynamic adjustment rules, according to the rule generation criteria, dynamic generation rules, through the construction strategy put forward unimportant rules.
(2) The hierarchical learning strategy and self-organizing learning method are adopted, which improves the effectiveness and speed of learning.
(3) In order to avoid over training and over fitting and improve the generalization ability of compact and the unimportant rules are put forward.

References

1. Shi, P.: The prevention countermeasures of catastrophe from the causes of ice and snow disaster in South China. Disaster Reduction China **02**, 12–15 (2008)
2. Liu, X.: Research on Network Vulnerability Analysis and Simulation Verification Technology of Complex Information System. Beijing University of Posts and Telecommunications (2013)
3. Wu, J.: Network Vulnerability Analysis. Beijing University of Posts and Telecommunications (2006)
4. Li, S.: Reliability Evaluation of Power Communication Network Based on Business Importance. North China Electric Power University (Hebei) (2010)

Application of BP Neural Network in Civil Engineering Cost Estimation Method

Hongyong Kan[✉]

Shandong Management University, Jinan 250357, Shandong, China
kanhongyong@sdmu.edu.cn

Abstract. The investment estimation in the stage of project establishment is the highest limit of quota design, and its accuracy is of great significance to the establishment and approval of construction projects. The traditional project investment estimation method, although has certain accuracy, but the workload is large. It shows the disadvantages of low efficiency under special circumstances. In order to meet the needs of the actual project and its accuracy and efficiency, the BP neural network theory is used to give the BP neural network model for the construction project cost estimation with examples. The results are obtained by substituting the data into the back frame, which is convenient to grasp the overall project budget, and can also be used as a reference for controlling the production cost, which is of great significance to the civil engineering project.

Keywords: Civil engineering · BP neural network · Fuzzy mathematics · Project cost estimation

1 Introduction

Construction project cost estimation is an important technical and economic activity in project management. have different characteristics. At present, there are two main types of engineering estimation models. One is to use computer simulation technology to establish simulation models. The basis of the simulation model is that many factors affecting the project cost are uncertain, so we should not pursue a certain value, but estimate the probability of the actual cost falling within a certain range; the second type is to establish an expert system for project cost estimation by using artificial intelligence and knowledge base technology, but this model mainly relies on the knowledge of experts to estimate the project cost, The estimation results are easily affected by the subjective consciousness of the estimators and the limitations of experience and knowledge, and are prone to personal bias and one sidedness [1].

Artificial information processing method developed in the 1980s inspired by the research of biological neural system. It simple neurons. It uses engineering technology to simulate. It is an effective means to deal with complex nonlinear problems. Therefore, it is of practical significance to establish the project cost estimation model by using neural network method.

There models network, CPN network, art network and so on.

J. C. Hung et al. (Eds.): FC 2021, LNEE 827, pp. 1180–1184, 2022.
https://doi.org/10.1007/978-981-16-8052-6_165

2 The Basic Principle of Artificial Neural Network

Artificial subject developed rapidly. It is a new type of information processing system. From 1943, W. Pitts and W. s. moculloch studied and proposed M-P neuron model to today, artificial neural network has developed rapidly. Its application has penetrated into biology, electronics, mathematics, physics, construction engineering and other disciplines, and has a broad application prospect.

At present, neural network has been widely used in signal processing, pattern recognition, data compression, intelligent control and other fields. BP neural network is a multi-layer neuron structure, which is connected between layers, but not connected in layers. The network learning algorithm is based on gradient descent error back propagation algorithm. (BP algorithm) that the three-layer BP network with nonlinear transformation function can approximate any nonlinear function with any precision. However, in complex applications, BP to fall into "over fitting", resulting in the calculation results deviate from the reality, thus affecting the reliability of the model. Therefore, measures must be taken to avoid it.

The structure of BP neural network is composed of DE.Rumelhart In 1985, PDP (parallel distributed processing) group proposed a neural network model, which eliminates errors by error back propagation, and has three levels: as shown in Fig. 1. In the training, the error expected value is calculated to solve the generalized error of the output layer unit In this case, the weight and threshold will not change. Thus, the unknown information of similar projects can be detected.

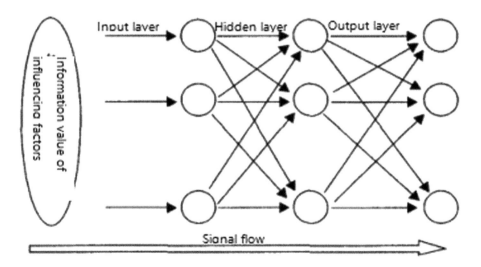

Fig. 1. Topological structure of BP neural network

3   Establishment of BP Neural Network Model

(1) The structure parameters of BP neural network model are determined, that is, the number.

(2) Original data normalization and hidden unit input module. Since the neural network can only deal with the data between -1 and 1, it is necessary to normalize the original data when it is input into the network:

$$\text{net}_j = \sum_{i=1}^{M} w_{ij} \cdot x_i \tag{1}$$

Formula: w_{ij} is from the i node to; x_i is the input information of the input unit; M is the number of input nodes connected to the j neurons of the hidden layer.

(3) Hidden unit output module.

According to the operation model of neural network, the *net* signal is processed by an excitation function f to value of the hidden unit.

$$O_j = f\left(net_j\right) \tag{2}$$

Formula: O_j is layer node.

(4) Output node input module.

For the output cell node, the input information of each hidden layer cell in the previous layer is weighted sum:

$$\text{net}_k = \sum_{j=1}^{S} v_{ij} \cdot o_j \tag{3}$$

Formula: v_{ij} is cells connected to the k-th node.

(5) The final output of neural network.

Similar to the third step, the output value Y of neural network is net_k. The excitation function g of the generation output layer is obtained:

$$Y = g(\text{net}_k) \tag{4}$$

Formula: Y is the final output value of the neural network.

(6) Adjustment of weight matrix (error propagation analysis).

Artificial neural network can approximate the mathematical mapping of any space. If in the n-dimensional Euclidean space R,

$$F: A \subset R^m \rightarrow R^n \quad Y = F(X) \tag{5}$$

For the training sample set A = {X, Y}, we can find an optimized approximation mapping G to approximate F by learning.

4 Cost Estimation Model of Subway Civil Engineering Based on BP Neural Network

In recent years, a work has been done on the model of project cost estimation at home and abroad, and some mature cost estimation methods have been produced, but each method has its own applicable conditions and certain limitations. Artificial neural network with its strong nonlinear mapping ability, large-scale parallel processing ability, self-learning ability, adaptive and fault-tolerant ability, can overcome some shortcomings of other estimation methods, and put forward a new method for engineering cost estimation in feasibility study stage. In this paper, fuzzy mathematics and artificial neural network theory are introduced into the subway project cost estimation, and the subway civil engineering cost estimation [2].

4.1 Index System of Engineering Cost Estimation

Due to the different functions and use requirements of construction projects, the factors affecting the project cost are also quite different. Therefore, the index system of estimation objects should be established according to the classification of construction projects. This residential building, according to the size of the influencing factors of the residential project cost, selects 10 main influencing factors, including structure type, foundation type, construction year, construction period, interior and exterior decoration [3], doors and windows, to establish the residential project index system. The index system is composed of qualitative index and quantitative index, and the qualitative index is quantified, in Table 1.

Table 1. Quantitative table of qualitative indicators

Index	Quantized value			
	1	3	5	7
Structure type	Brick concrete structure	Frame structure	Frame shear wall structure	
Foundation type	Concrete strip foundation	Cast in place pile foundation	Raft foundation	Box foundation
Internal finish	General painting	Spraying white mortar		
Doors and windows		Aluminum alloy windows	Steel doors and windows	Aluminum alloy windows
Construction site	According to the bustling degree of the project site, it can be divided into four levels: 1.3.5.7			

4.2 Neural Network Method for Engineering Cost Estimation

The problem of construction project cost estimation can be regarded as a nonlinear mapping from input (index system of project cost estimation) to output (unit cost of the project). Because BP neural mapping relation with any precision, this paper adopts three-layer BP neural network. The the indicators that affect the project cost, a total of 10 neurons; the hidden layer is a layer, a total of 20 neurons; the output layer has only one neuron, which represents the unit cost of the project. The as the excitation function. Experiments show that the output error of these two functions as the excitation function is the smallest. The neural network needs a certain number of known samples as the training set to estimate the project cost, and then it can estimate other projects [4–6]. The sample set of neural network is usually obtained from the data of completed projects. For the project to be estimated, the unit cost of the project can be given in the output layer as long as the value of each index in the index system is input and estimated by neural network.

5 Conclusion

The investment estimation of the civil engineering project is the highest limit of the quota design of the construction project, and its accuracy is of great significance to the construction project. Traditional investment estimation methods are mostly based on relatively simple linear mathematical model, which can not reflect the complex non-linear relationship between project investment estimation and influencing factors. Therefore, the traditional engineering investment estimation method has the disadvantages of low accuracy and heavy workload. In practical engineering and accuracy requirements under special circumstances, based on BP neural network model, this paper discusses an engineering investment estimation method with highly nonlinear relationship with the main characteristic factors of the project.

BP adaptability and strong fault tolerance. This paper attempts to apply BP neural network model to construction project cost estimation, which successfully overcomes the limitations of traditional project estimation. Through the example, it can be seen that the artificial neural network method has the characteristics of simple, accurate and fast calculation. The neural network estimation needs certain samples to train the network. Therefore, the selection of Engineering eigenvectors and training samples needs to be further improved. However, the application prospect of neural network model in the field of modern economic nonlinear is very broad.

References

1. Plateau: Uncertainty Graph and Uncertainty Network. Tsinghua University (2013)
2. Izaddoost, A., Heydari, S.S.: Enhancing network service survivability inlarge-scale failure scenarios. J. Commun. Netw. **16**(5), 534–547 (2014)
3. Transmission OO E. Grid 2030: A national vision for electricity's second 100years (2010)
4. Lu, Z., Jiang, J.: Interpretation of the "grid 2030" power grid vision of the United States. China Power Enterprise Manag. (5), 38–41 (2004)
5. Zhang, J.: Fuzzy analytic hierarchy process. Fuzzy Syst. Math. **14**(2), 80–88 (2000)
6. Shiji: Power communication and its application in smart grid. Digit. Technol. Appl. **06**, 50–51 (2012)

Identity Authentication Mechanism Based on Hash Algorithm in E-commerce

Ding Li[✉]

Yunnan Communications Vocational and Technical College, Kunming, China

Abstract. With the rapid development of Internet of things, the number of Internet of things devices and sensing data increase exponentially. In this situation, people are very concerned about how to protect the security of these data effectively. In the Internet of things, due to the intelligent and open characteristics of intelligent sensors, the security problems such as data theft, forgery and deception easily occur in the process of data transmission, storage and identity authentication. Once the storage server is attacked by malicious attackers, the security of the Internet of things data will be greatly threatened, especially in the centralized storage server. In this state, people have an urgent need for the security of the management, transmission and storage of massive sensor data in the Internet of things.

Keywords: Internet of things · Data transmission · Distributed storage · Collaborative identity authentication · Sensing

1 Research Background and Significance

With technology, the Internet has affected and changed all aspects of human society. of things is the information interaction between people, people and things, or between things [1]. The basic feature of the Internet of things is to collect data through a large number of sensors, and then reliably transmit and store the data to the server for calculation and intelligent processing.

2 Security Problems and Countermeasures in E-commerce

2.1 The Concept and Technical Characteristics of E-Commerce

Electronic are two main sub commerce modes, Bob and BOC mode. The latter mode is the most widely used and the most complex technology, and it is also the key to the success of e-commerce. Its application range is e-store, online shopping, e-payment, online banking, etc. To ensure the smooth implementation of e-commerce, we need to solve the following problems: authentication: before the electronic transaction, the buyer and the seller carry out identity authentication, which is guaranteed by a third party. Security: build a secure channel between businesses and consumers, encrypt sensitive information sent by both parties, and ensure that it will not be stolen by a third party.

J. C. Hung et al. (Eds.): FC 2021, LNEE 827, pp. 1185–1189, 2022.
https://doi.org/10.1007/978-981-16-8052-6_166

2.2   Network Security Issues

Information leakage: information is leaked or disclosed to an unauthorized entity. This kind of threat mainly comes from the destruction of complex information detection attacks: the consistency of data is caused by unauthorized creation, modification or destruction of information or other resources; the attacker attacks the system through the amount of load that can't succeed at all, which leads to the system's resource matching, and the user can't be caused by the attack: the attacker visits the system illegally and the root can't be used. It can also be used by an unauthorized person or in an unauthorized way [2]. This kind of threat intrudes into a computer system, the attacker will use this system as the starting point of stealing business or as the starting point of invading other systems.

3   Authentication of E-commerce

3.1   Certification is the Key Link

The Internet enables both sides of communication to communicate at any place and at any time, no matter how far apart they are. They can even deal with strangers, discuss a topic or cooperate with each other. However, the following problems arise: how do people know that the other party is the person or entity they claim, not his name? Even if the information they get really comes from the claiming party, how can they ensure that it has not been tampered with, that is, to ensure its integrity and authenticity? Without solving these problems, there will be no authenticity on the Internet, and the obtained network will not be able to develop. In e-commerce, both the provider and the consumer must confirm the identity of the other party and the origin of the data, otherwise who dares to use this service? Therefore, security authentication is the first line of defense of e-commerce security protection, and also the core issue of its development.

$$(BIIC_k| \Rightarrow P(U)) \Rightarrow BIMM \tag{1}$$

$$(BIMM \leftarrow (BIIC_k \updownarrow P(U) = BIMM \tag{2}$$

$$BIMM \otimes \updownarrow N = P^*Q \tag{3}$$

Data encryption and decryption, data the security storage mechanism of intelligent sensing image, and makes a detailed analysis and mathematical proof of the described security function.

3.2   Password Based Authentication Mechanism

Based on the principle of cryptanalysis, the verifier's convincing claim is claimed by the claimant because the claimant knows a secret key. Both symmetric cryptography and public key cryptography can be used. The simplest method based on symmetric cryptography is to share a symmetric key between the claimant and the verifier. The key is used to encrypt or encapsulate a message. If the verifier can successfully decrypt the message or verify that the encapsulation is true, the verifier believes that the message is from the claimant. The content of encrypted or encapsulated message usually includes a non duplicate value to resist replay attack. The method of using public key technology is [3]. The declarant signs a message with his secret key, and the verifier checks the signature with the declarant's public key. If the signature can be checked correctly, the verifier believes that the claimant is himself. In addition, messages can contain duplicate values to resist replay attacks.

3.3   Traditional Encryption and Public Key Encryption

Encryption is the process of transforming data and information (called plaintext) into unrecognizable form (called ciphertext), so that people who should not understand the data and information can not know and identify. If you want to know the content of the ciphertext, and then change it into plaintext, this is the decryption process. These two processes often rely on a certain number of keys, which is called the key. Currently, there are two popular encryption methods: one is the traditional encryption method represented by idea (international data encryption algorithm), and the other is the public key encryption system represented by RSA algorithm. As shown in Fig. 1. In the traditional encryption method, both encryption, which brings difficulties to the key distribution and management. The emergence of public key encryption technology represented by RSA algorithm is a major breakthrough in computer cryptography [4].

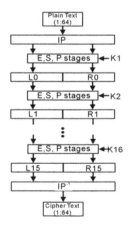

Fig. 1. Traditional encryption and public key encryption

4   Secure Storage of User's Intelligent Sensing Image

4.1   Secure Storage of User's Intelligent Sensing Image

In this section, the model of distributed secure storage machine is constructed according to the user's intelligent sensing image storage procedure. Secondly, on the basis of safety:

In this mechanism, the objects of data service are users (user sensing image end, image sensor server). Data service "and image sensing application image, block partition of user sensing image, data transmission of user sensing image block to cloud server, and secure storage of user sensing image. In the traditional data transmission process, many security cases are cheated by the theft and malicious attackers to obtain cloud service authentication and attack the security problems in the data transmission process such as non disclosure of user data stored in the server. The existing mechanism will encrypt the data that users need to upload first, so as to ensure the security of user data in the transmission process. With of things cloud storage, people need more and more cloud services. How to transfer and store massive user data safely and effectively has become one of the important problems in the development of cloud storage.

In order to solve the security problem of centralized cloud server storage, this chapter proposes a secure storage mechanism of user intelligent sensing image. The mechanism is based on the principle of distributed storage. Firstly, the user's image information is intelligently sensed by the back-end image sensor, and then transmitted to the back-end image manager locally. Secondly, the back-end image manager divides the sensing data into intelligent data blocks. Then, the smart signature algorithm is used to sign different blocks of data and transmit them to multiple cloud servers. The mechanism sets multiple cycles, and the private key of each cycle is different. In different periods, the private key will be updated by intelligent private key update algorithm.

4.2   Security Goal of Secure Storage

The system model needs to assume that the cloud server is secure, that is, (1) the cloud server will not actively destroy and disclose the user sensing image data that has been safely stored in the cloud server; (2) the cloud server will not maliciously tamper with and intercept the signature information of the sensing image data block; (3) for each correct signature, the information of the sensing image data block will be used, The goal of this system model is to overcome the security problems of the traditional centralized storage mechanism of cloud server, that is, the user's data is easy to be completely destroyed, stolen or other illegal operations in the process of transmission and storage, so as to improve the security of user data storage. Before transmitting user sensing image information, the system will divide user sensing image information into intelligent blocks, and use different private keys to sign and encrypt all user sensing image data blocks. Its security goal is that even if one user sensing image data block is stolen or damaged, it will not affect the safe transmission and storage of other user sensing image data blocks.

4.3   Detailed Algorithm of Secure Storage

Firstly, the secure storage mechanism of user intelligent sensing image is described in detail. In this machine.

E-commerce is a new business model operating in an open network environment, and it is also a new development stage of organization information construction in a broad sense. At present, it is widely concerned by people. But its output value is only a tiny part of the global GDP. According to statistics, the output value of us e-commerce reached 95 billion US dollars in 1999, accounting for only 1.10% of US GDP. Even the estimated output value of e-commerce in 2001 is 299 billion US dollars, which is less than 3.4% of US GDP in 1999 [11, 31]. People may attribute this fact to the low average level of global networking, but this is only part of the reason. In the United States, where the level of network is quite high, in fact, most companies with high level of network do not use e-commerce as a new way of business. Many investigations show that one of the main obstacles to the wider development of e-commerce is its security. Therefore, in the process of changing from traditional paper-based trade to electronic trade, how to keep electronic trade as safe and reliable as traditional trade is the focus of people's attention, and it is also one of the key issues in the comprehensive application of e-commerce.

References

1. Lin, J., Zhang, Z., Yuan, Z.: Research on authentication and encryption mechanism based on IBC in Internet of Things. Inf. Secur. Commun. Confidentiality (08), 95–101 (2020)
2. Liu, L., Shen, Y.: Intelligent device authentication scheme based on blockchain technology. Comput. Digit. Eng. **48**(07), 1722–1726 (2020)
3. Li, H.: Research and Implementation of Home Internet of Things Service Architecture and Security Mechanism. Harbin Institute of Technology (2020)
4. Tu, Y.: Secure Storage and Authentication Mechanism for Smart Sensing Image in Internet of Things. Hangzhou University of Electronic Science and Technology (2020)

The Application of Support Vector Machine in the Calculation of Pavement Structural Modulus

Hao Li[(✉)]

CCCC Second Highway Survey and Design Institute Co., Ltd.,
Xi'an 430056, China
ccshcc@ccshcc.cn

Abstract. In this paper, support vector machine (SVM) is introduced into back analysis, and the inversion method of pavement modulus based on SVM is discussed. The main work can be summarized. In the process of modulus optimization inversion, the forward analysis process needs to be called repeatedly, which results in a large amount of inversion calculation. In this paper, support vector machine is introduced into the back analysis, and a new idea of pavement modulus inversion based on support vector machine is proposed, that is, firstly, the support vector machine model is established through the learning of samples, Then, the trained support vector machine model is used to calculate the pavement deflection corresponding to the modulus parameters instead of the numerical model, so as to reduce the amount of back analysis calculation. Taking advantage of the strong global search ability and high search efficiency of particle swarm optimization algorithm, this paper combines it with support vector machine, and successively applies it to the optimization selection of support vector machine model parameters and intelligent search of pavement modulus parameters. The analysis results show that with the help of particle swarm optimization algorithm, the support vector machine model with the best prediction performance and the optimal parameter inversion results can be effectively obtained.

Keywords: Road · Support vector machine · Particle swarm optimization · FWD

1 Introduction

With the continuous progress of China's economic construction and the gradual increase of the scale of national investment in infrastructure, the highway transportation industry has entered a period of rapid development, and the scale of road network is expanding. By the end of 2008, China's highway mileage had reached 373.02 million kilometers, including 6.03 million kilometers of expressways; in 2008 alone, China's fixed asset investment in transportation reached 83354.2 billion yuan. In the face of such a huge scale of investment and road network, it can be predicted that the highway detection and evaluation work in China will face increasingly heavy tasks. However, due to the relatively short construction time of China's high-grade highway, the relevant detection and evaluation methods are still relatively backward, which can not meet the needs of the rapid development of China's expressway, seriously restricts the level of road maintenance, and brings a greater waste of resources [1]. Therefore, it is of great significance to improve the pavement detection and evaluation methods as soon as possible for controlling the road construction quality, deeply understanding the long-term performance of the pavement, improving the pavement design, optimizing the road reconstruction scheme and improving the maintenance level of the road network.

2   The Basic Principle of Support Vector Machine

2.1   Machine Learning Model

The purpose of machine learning is to use the given training samples to estimate the internal relationship between the input and output of a system, so that it can make the prediction of the unknown output as accurate as possible. The basic model is shown in Fig. 1, where the system is the object of study. It obtains the output y under the given input x, and uses the input and output to form a sample pair (x, y) to train the learning machine [2].

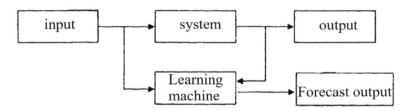

Fig. 1. Basic model of machine learning

In order to obtain the expected risk, we need to know the joint probability distribution F(x, y) between input and output. However, in practical problems, the joint probability distribution F(x, y) is often unknown, so the expected risk cannot be calculated directly. In the traditional learning method, people usually use the arithmetic average of training samples to replace the mathematical expectation in the formula, so the machine learning risk can be written as follows:

$$R_{emp}(w) = \frac{1}{l}\sum_{i=1}^{l} L(y_i, f(x_i, w)) \qquad (1)$$

In fact, the empirical risk minimization criterion has not been fully demonstrated in theory, it is just a reasonable and intuitive way to take it for granted. At present, people usually use the large number theorem to explain the relationship between the empirical risk minimization and the expected risk minimization:

$$\lim_{l \to \infty} R_{emp}(w) \to R(w) \tag{2}$$

The large number theorem requires that the number of training samples tend to infinity, which can not be satisfied in practical problems. Even if a large number of samples can be obtained, it is still far from infinity. Therefore, it is difficult to guarantee that the results obtained by using empirical risk minimization criterion under limited samples can approach the real risk.

2.2 Support Vector Machine Classification

Support vector machine (SVM) is a concrete realization of structural risk minimization in statistical learning theory. It was originally proposed from the optimal separating hyperplane of binary classification problem in the case of linear separability. In the data classification problem, the usual neural network method randomly generates a hyperplane and moves it until the points belonging to different classes in the data set are located on different sides of the hyperplane. If the final classification surface can be located in the center of the two categories, the classification ability of the learning machine will be maximized. The support vector machine method has achieved this goal cleverly. Its principle can be described as follows: to find a classification hyperplane which meets the classification requirements, so that the hyperplane can maximize the blank area on both sides of the hyperplane while ensuring the classification accuracy, so as to achieve the optimal classification of known separable data.

3 Inversion of Pavement Modulus Based on SVM

With the development of pavement detection technology, falling weight deflectometer (FWD) is widely used in Highway Engineering in China. How to use a large number of data detected by FWD to calculate the modulus of pavement structure layer, and then make a reasonable evaluation of road bearing capacity has become the focus of road engineering research. At present, the research of pavement back analysis in China has achieved some theoretical results, and many theories and methods of modulus back calculation have been put forward. The basic idea is to use FWD deflection detection data, based on the pavement structural mechanics analysis model, and with the help of certain optimization technology, to find a group of pavement modulus parameters that can minimize the measured deflection and calculated deflection error [3].

3.1 Basic Idea of Modulus Inversion Based on Support Vector Machine

Under the condition of deflection detection data and given the pavement mechanics model, the key to carry out the inversion of pavement modulus is to adopt effective parameter adjustment algorithm (i.e. optimization technology). At present, the conventional iterative method or intelligent optimization algorithm is often used in modulus inverse calculation. However, the inversion results of traditional iterative methods are greatly affected by the initial value and the convergence speed is slow. Although the intelligent optimization method can overcome the dependence on the initial value, it is necessary to call the positive analysis process repeatedly in the random search process, which brings a lot of calculation work [4].

In order to reduce the amount of calculation, the neural network method is introduced into the inversion of pavement modulus, forming a new inversion idea. Firstly, a small number of samples are obtained through the numerical model, and the neural network model is established by learning the samples to replace the numerical model to calculate the deflection basin, and then the optimization algorithm is used to search the optimal parameters, which not only overcomes the control effect of the numerical analysis process on the inversion efficiency, Moreover, it can make full use of the latest research results of the optimization algorithm to reduce the amount of inversion calculation and improve the efficiency. But the problems of neural network are: the network training is complex, mostly based on gradient descent algorithm, prone to local extremum problems; at the same time, because of the slow convergence speed of gradient descent algorithm, the network training time is too long; the training process is greatly affected by the initial value, prone to divergence and oscillation; the selection of the number of hidden layer neurons in the network is still lack of theoretical guidance.

3.2 Basic Steps of Modulus Inversion Based on Support Vector Machine

In order to determine the parameters to be inversed and construct the learning samples for training the SVM model, a certain number of numerical experiments are needed, usually using orthogonal or uniform experimental design. The basic characteristics of orthogonal and uniform experiments are briefly introduced below. Orthogonal experimental design is a method used in multi factor experiment, which selects some representative points from the comprehensive experiment. The requirements of orthogonal experimental design are as follows: (1) the same number of experiments should be carried out at all levels of any factor; (2) the same number of experiments should be carried out at any combination of two factors. Orthogonal table is the basic tool for the optimization of orthogonal experimental design. Orthogonal array has two characteristics of "balanced dispersion and comprehensive comparability". Balanced dispersion means that the test points are evenly distributed in the scope of the test, and each pilot has a certain representativeness. Comprehensive comparability is uniform comparability. Because of the orthogonality of orthogonal array, the occurrence times of each level in any column are equal, and the occurrence times of all possible combinations between any two columns are equal, so that the test conditions of all levels of each factor are the same, and the influence of different levels of each factor on the test indexes can be comprehensively compared, So we can analyze the influence of each factor and its interaction on the index and the change law.

3.3 Training Support Vector Machine Model

When the learning samples are given, how to use the limited samples to establish the support vector machine model which can fully approximate the complex mapping relationship between modulus parameters and pavement deflection, so as to ensure the accuracy and reliability of the inversion results, is the primary problem that the back analysis method based on support vector machine must solve. The reasonable setting of model parameters is one of the key factors to be considered. The support vector machine model contains two types of parameters, one is the basic parameters of the model, and the other is the parameters related to the kernel function, such as the kernel width a in the radial basis function, the constant C and the exponent D in the polynomial kernel function. Whether these parameters are selected properly or not has a great influence on the generalization performance of the model.

In recent years, researchers pay more and more attention to a new heuristic algorithm called global optimization algorithm to improve the performance of SVM. Chen (2004) and Zheng (2004) proposed two SVM parameter selection methods based on GA by using different generalization ability estimation as the fitness function of genetic algorithm. The results show that GA not only reduces the calculation time of parameter selection, but also reduces the dependence on the initial value selection; Du Jingyi (2006) realized the automatic selection of SVR parameter by using GA; Du Jingyi (2006) realized the automatic selection of SVR parameter by using GA; Chen Guo (2007) used GA to realize the parameter adaptive optimization of SVM classifier model. Particle swarm optimization (PSO), a new heuristic global optimization algorithm, has attracted more and more attention. Particle swarm optimization (PSO) is a new branch of swarm intelligence. It was first proposed by American scholars Eberhart and Kennedy in 1995. 4. The algorithm is based on the study of birds' predation behavior. Different from GA based on evolutionary theory, PSO finds the optimal solution through the cooperation among individuals, which is not only suitable for all occasions where GA can be applied, but also has the advantages of simpler concept, less need to adjust parameters, higher efficiency and easier algorithm implementation. Therefore, it has attracted the attention of many scholars at home and abroad, and has been applied to the optimization of support vector machine model parameters, and achieved good results, Therefore, in this paper, particle swarm optimization algorithm is considered to search the optimal parameters of support vector machine.

4 Conclusion

In this paper, support vector machine (SVM), a new machine learning method, is introduced into the back analysis of pavement modulus, aiming at the problems existing in the back analysis of pavement modulus. The introduction of SVM in the back analysis provides a new idea for the back analysis of pavement modulus, which can not only reduce the calculation amount of back analysis, but also reduce the calculation cost, Moreover, it can give full play to its advantages such as based on the structural risk minimization criterion, suitable for dealing with small sample problems, and strong generalization ability, which is helpful to improve the accuracy and reliability of parameter inversion. The results show that the combination of particle swarm optimization and support vector machine can give full play to the strong global search ability and high efficiency of particle swarm optimization, and obtain the best prediction performance of support vector machine model and the optimal modulus parameter inversion results.

References

1. Wang, F., Liu, W.: Report on nondestructive testing and CAE technology of high-grade highway, national key science and technology project of the eighth five year plan, theory and method of back analysis of pavement structure. Zhengzhou University of Technology (1996)
2. Xi, Y., Chai, Y., Yun, W.: Review of genetic algorithm. Control Theory Appl. **13**(6), 697–708 (1996)
3. Chen, J., Guo, D., Xu, N., et al.: Review of genetic algorithm theory. J. Xi'an Univ. Electron. Sci. Technol. **25**(3), 363–368 (1998)
4. Li, Y., Li, B., Fang, T.: Research on nonlinear combination forecasting method based on wavelet support vector machine. Inf. Control **33**(3), 303–306 (2004)

Application of Machine Learning in the Design of Ecological Environment Detection System

Ling Li[(✉)]

Wuhan Donghu University, Wuhan 430212, China

Abstract. In the new media era, there are more ways of media communication. Both the content and the way have changed greatly, which brings challenges to the traditional news work and puts forward higher requirements to the ecological environment news propaganda work. The article thinks that we should optimize the traditional way of eco-environmental news communication to ensure the role of new media. This paper analyzes the problems existing in the work of eco-environmental news and publicity, and puts forward some targeted suggestions combined with the actual situation under the development background of the new era.

Keywords: New media · Ecological environment · News propaganda

1 Introduction

The arrival of the new media era has brought people a more convenient way to receive information. News communication has the characteristics of timeliness, interaction and community. The era is developing, the level of science and technology is constantly improving, and the Internet is widely welcomed by the public. The arrival of the Internet makes people's life services and information access more convenient [1].

Compared with the traditional media, the new media is more extensive in the scope of communication, involving a wider audience. Because of the fast update speed and complete information content of new media, the traditional way of broadcasting has been changed, which can be directly broadcast. The continuous upgrading and transformation of the mode of communication can break through the limitation of time and space, and carry out all-round news reporting. Even if the public is not at the scene, it can receive effective information at the first time. At the same time, the dissemination of new media also contains a huge amount of information capacity, storage capacity is very large, and the database is more complete.

Compared with TV, radio and other media, new media can not only realize real-time monitoring and information exchange with the audience during the news broadcast, but also ensure the real-time update and improvement of news propaganda information based on the use of modern advanced means. Moreover, in the process of communication with the public, we can timely analyze and understand the psychological state of the public, and give correct public opinion guidance. In the interaction and exchange, it can also share and solve the existing contradictions in time.

J. C. Hung et al. (Eds.): FC 2021, LNEE 827, pp. 1196–1200, 2022.
https://doi.org/10.1007/978-981-16-8052-6_168

These characteristics not only provide greater convenience for the implementation and development of news publicity work, but also make the publicity work more direct and effective [2].

The input vector X is compressed to the feature vector a by the encoder, and its mathematical expression can be recorded as follows:

$$A = f(W_1X + b_1) \tag{1}$$

Similarly, the process of decoder restoration can be recorded as follows:

$$\widehat{X} = g(W_2A + b_2) \tag{2}$$

In order to minimize the reconstruction error, the objective function of self encoder can be defined as the two norm of input and output:

$$\min\left\|X - \widehat{X}\right\|_F^2 \tag{3}$$

There is only one fully connected layer from input to eigenvector, which is just a simple linear transformation. In order to increase the ability of model expression and feature extraction, we can add more layers of full connection layer to make the model have more complex nonlinear representation ability. Such a depth self encoder is called stacked autoencoder (SAE). The training process of stackable self encoder is usually not completed at one time, but through layer by layer fixed parameters training, so as to obtain good restoration ability.

2   The Deficiency of Ecological Environment News Propaganda

In the new media era, great changes have taken place in the way of news communication, which, to a certain extent, provides technical convenience and broad space for news propaganda. In the new media era of ecological environment news propaganda work, the network new media has become the main information dissemination platform, at the same time, the traditional media are also developing in the direction of integration with the new media. In such a process of media integration and the transformation of information transmission mode, the development of ecological environment news and publicity work also faces some practical problems.

2.1   Lack of Innovative Change

In the actual work of ecological environment publicity, there is often a lack of innovation and change. Therefore, only by making full use of new media resources, can we ensure the improvement of the level of ecological environment news and publicity work, and provide broader development space for ecological environment news and publicity work. Based on the analysis of the communication concept, the publicity work should be based on the development needs of the public, with the user as the main body; at the same time, we should also make clear its important nodes to enhance the publicity power of news by means of special topics; we should also recognize the key points, and increase the application of new media on the basis of providing high-quality services for the public. Therefore, for some influential websites, we can guide them to strengthen the forwarding of information, so as to promote the acquisition of media credibility.

2.2 Insufficient Sustainability

At present, the eco-environmental news publicity can only be carried out through various environmental protection days, such as "5.22 International Biodiversity Day, 6.5 World Environment Day" or the central eco-environmental protection inspector. During the theme day, a large number of theme reports, series reports and in-depth reports will be planned [3]. However, after this stage, the eco-environmental news and publicity went to a low ebb, with few reports and obviously insufficient sustainability.

2.3 Lack of Technical Talents

Even with the continuous progress and development of new media, there is still a shortage of professionals who are familiar with the operation of new media platform, far from keeping up with the pace of the times. The development of new media has a long way to go. Nowadays, new media has become a kind of media generally accepted by the majority of the audience, which also means that the operation of this new type of media puts forward more stringent requirements for news propaganda personnel. The application of new media not only requires propagandists to have a solid foundation in writing, photography, political thought and theoretical knowledge, but also requires eco-environmental journalists to have professional eco-environmental knowledge. But in the current situation, this kind of compound talents is very few. Therefore, we should strengthen the team construction of ecological environment news work talents, promote the innovative development of propaganda methods, and improve the level of eco-logical environment news propaganda work.

2.4 Pressure on Public Opinion Guidance

With the continuous improvement of modern science and technology, some digital products and electronic products have achieved intelligent development. Through the application of network platform, we can actively obtain news information. The advent of China's Internet age has not only changed the development of new media, but also changed people's way of life, the most obvious of which is the application of network new media platform. The widespread development of network information dissemi-nation is based on its universality and timeliness. Through the use of the network, the public can not only actively receive information, but also release information without the limitation of time and space. Although this way gives the audience a more com-prehensive right of information dissemination, and the audience's opinions and feelings can also be exchanged and expressed in real time, because of this, the network media platform is easy to increase the bad guidance of public opinion, which has a negative impact on the society.

3 Measures of Ecological Environment News Propaganda in New Media Era

In the context of the development of the new media era, combined with the problems existing in the ecological environment news propaganda work, we need to improve the new media, strengthen the renewal of ideas, and in practice, make full use of the new media combined with the actual situation, so as to promote the optimal implementation of the news reporting work [4].

We should change the traditional concept of eco-environmental news appropriately, deal with the existing eco-environmental problems actively, and maintain the sensitivity of news appropriately. According to the analysis of the ecological environment news propaganda work, the ecological environment news is facing the problems of small coverage, slow response and low efficiency of work implementation, which can not promote the innovation and development of the ecological environment news propaganda work. In view of these problems, first of all, we need to train the eco-environmental journalists to ensure their understanding of the basic eco-environment related content, improve their practical skills in the eco-environmental news work, and promote their formation of keen thinking.

Secondly, in the process of eco-environmental news gathering and writing, we should have a keen perspective of eco-environmental news, and conduct in-depth mining for a large number of eco-environmental materials and related news materials, so as to fully grasp and understand the news information and ensure to obtain more accurate, true and complete information. In addition, we should establish a new type of ecological environment news thinking, and strengthen the staff's ability to explore the ecological environment news. News has timeliness, and the news about the ecological environment is updated and changed every day, so it is more necessary for the staff to broadcast the ecological environment news in real time. In the ordinary ecological environment events, how to stimulate people's interest in watching and avoid boring news is a problem that every journalist needs to focus on. We should think from multiple dimensions and angles. For example, in daily work, by holding activities such as "the most beautiful environmentalists" and mining the most beautiful grassroots environmentalists, we can record the work of grassroots environmental protection workers for a day, and use short videos and animations to arouse the public's interest in reading ecological environment news.

4 Conclusion

China has entered the era of new media. In this era of network society, mobile Internet and big data, the characteristics and problems of human social relations have changed. In January 29, 2019, Xi Jinping delivered an important speech on the integration and development of the media in the twelfth collective learning of the Political Bureau of the CPC Central Committee, which reflected the Central Committee's high regard for the impact of the development of new media on social life. The development of social relations is an important part of man's all-round development. The richness and harmony of social relations restrict the degree of man's all-round development. The research on the development of social relations has become an important reference for the analysis of man, society and historical development. Therefore, it is of great significance to study the development of people's social relations in the new media era for people's all-round development.

References

1. Fu, Y.: How to play the role of news propaganda in grassroots ideological and political work. News Res. Guide **8**(19), 291 (2017)
2. Zhang, Y.: Thoughts on strengthening the application of new media and doing a good job in grassroots news propaganda of state owned enterprises. Corp. Cult. (Punishment in Chinese) (10), 1–3 (2017)
3. Hao, J.: Brief analysis of new media application and strategy in news propaganda of electric power enterprises. Enterp. Cult. (second issue) (2), 246 (2019)
4. Zhan, H.: My opinion on the application of new media in grassroots news propaganda. Petrol. Polit. Work Res. (3), 66 (2016)

A Method and System of Rural Tourism Recommendation Based on Season and Location Characteristics

Hongyan Li[✉]

Shandong Institute of Commerce and Technology, Jinan 250103, China

Abstract. A rural tourism recommendation method and system based on season feature and location feature, the method includes: extracting user evaluation time based on historical user evaluation of each scenic spot, and extracting season feature based on the user evaluation time; obtaining longitude and latitude information of each scenic spot according to the name and location information of each scenic spot, and mapping the longitude and latitude information to the corresponding tag block, Constructing the geographical spatial location feature of each scenic spot; forming the feature vector according to the user evaluation time, season feature and geographical spatial location feature, and taking the user score record in the historical user evaluation as the label; based on the feature vector and the label, using the pre established factor decomposition machine algorithm model, estimating the user's prediction score of each scenic spot, And a scenic spot recommendation strategy is generated based on the prediction score.

Keywords: Seasonal characteristics · Location characteristics · Rural tourism

1 Introduction

Productive landscape, originated from human life and production labor, is the most natural and common landscape type, which has both production and landscape functions. The garden of China and the garden of Eden in the West are considered to be the origin of productive landscape, which first organically combines agricultural production with landscape leisure. The rapid development of modern cities has led to the diversification of urban landscape and the rise of rural tourism, which has aroused people's high attention to agricultural productive landscape. At present, many rural areas with characteristic agricultural productive landscape resources are speeding up tourism development, setting up scenic spots one after another, or combining with the development of beautiful countryside, in pursuit of efficient use of agricultural productive landscape, but their landscape construction is still at a simple primary level [1]. According to the characteristics of productive landscape resources with strong seasonality, how to examine the construction of productive landscape from the perspective of comprehensive development of scenic spots is worthy of our consideration.

J. C. Hung et al. (Eds.): FC 2021, LNEE 827, pp. 1201–1207, 2022.
https://doi.org/10.1007/978-981-16-8052-6_169

2 Productive Landscape

Productive landscape comes from life and productive labor, integrates productive labor and labor achievements, including man's production and transformation of nature and reprocessing of natural resources. It is a kind of productive landscape with life, culture, long-term inheritance and obvious material output, and relies on agricultural production.

2.1 Main Production Landscape

Agricultural productive landscape is based on production. There will be one or two main crops in a certain area, which will become the main body of productive landscape in this area. Generally, it has the characteristics of large scale, high purity and strong seasonality. For example, Mailang, paddy field, Huatian and hehuadang, which are common in our country, are easy to catch people's eyes and form a certain shock because of their large scale and high color purity, Become the main resource of tourism. The main productive landscape is a living body, which will present different scenes with the alternation of seasons, showing people the beauty of season and nature. According to its production content, it can be divided into food crops, flowers and seedlings, tea, vegetables and other landscape types. The common agricultural productive landscape elements include orchards, vegetable fields, farmland, fish ponds, agricultural facilities and so on. From the perspective of landscape design, agricultural productive landscape can be divided into three parts: main production landscape, production facilities landscape and production activities landscape.

2.2 Landscape of Production Facilities

In the process of agricultural production, the working people rely on their wisdom, in order to improve the production efficiency, and constantly improve the production mode and production conditions, there are auxiliary production related facilities, such as ditches, trunk canals, dams, wells, fish ponds, fences, water trucks, scaffolding tools and so on. These facilities are the appendages of agricultural production, but they can also arouse the interest of tourists, form landscape independently or together with the main production content, and have the characteristics of strong experience and prominent regional characteristics. The facility agriculture supported by modern science and technology has brought new production modes, such as hydroponic vegetables, soilless vegetables and fruits, etc. the main body of production is highly integrated with the facilities, which is also a new facility landscape=) in the process of agricultural production, there are still some non-material elements, such as lotus picking, lotus root selecting, fishing, shrimp catching, tea picking and so on, It also includes people's festival activities to celebrate the harvest and customs formed in the long-term production process. This kind of non-material elements rooted in a certain environment, with a strong local flavor, can also be transformed into visual landscape through appropriate ways to enrich the connotation of productive landscape.

3 Analysis on the Measurement Index of Tourism Seasonal Fluctuation

3.1 Main Definitions

Tourists' perceived value is a comprehensive judgment of products based on the balance of payment and benefit in the process of traveling (Graf et al. 2008). Quality and cost perception are the basis and premise of tourism perceived value (Sheth et al. 191). Lee et al. (2007) studied the destination of War sites and summed up three dimensions: function, emotion and integrated value. Oliver (1999) based on the expectation theory, pointed out that tourists' perception is affected by both cognitive factors such as expectation and expectation inconsistency, and positive or negative emotional factors. The positive factors include efficiency, quality, aesthetics and social negative emotions, including money payment, risk perception and time consumption [2]. Using SEM model, based on the case of Shanghai World Expo, this paper makes an empirical study on the relationship between tourists' perception and behavior intention, and holds that the perceived value of tourism grudge comes from utility, pleasure, service and convenience in turn. In addition, principal component analysis (PCA), analytic hierarchy process (AHP), topss and other methods are also widely used. Qian Ye et al. (2011) used fuzzy PA analysis method to quantitatively measure and evaluate the tourists' perception of Shanghai World Expo, believing that cultural perception and service perception have more influence weight than the perception dimensions of creativity, infrastructure and technology.

3.2 Main Research Methods

First of all, as an important part of cellar perception, climate comfort plays a key role in travel decision-making. Tourism is more dependent on climate than other industries (BODE et al. 2003). Many traditional holiday resorts are famous for their pleasant climate (Kwun et al. 2004). Mieczkowski (1985) put forward the concept of tourism climate index (TC). The key role of climate change in the process of travel activities and destination selection has always been the focus of academic attention (Hamilton et al. 2005; amelung et al. 2006). Comfortable climate and sufficient sunshine will increase the attraction of the destination (Shoemaker 1994). Domestic research on tourism climate environment mainly focuses on the analysis of climate resources and comfort. Xi Jianchao et al. (2011) evaluated the significant impact of climate change on passenger flow in five southern provinces. Liu Qingchun et al. (2007) drew the deviation map of climate comfort index of 44 cities in China. In addition to climate comfort, resource taste and abundance (Wu Jing et al. 2012), safety and service also have an important impact on tourists' perceived value. Wang Lan et al. (2010) took Jiuzhaigou Scenic Area as an example to verify the impact of destination accessibility on tourists' perception.

4 Weight Determination of Evaluation Index System

4.1 Weight Determination, Consistency and Dimensionless Processing

The index system of tourism seasonality (a) includes four modules: θ 1 tourism climate comfort, B2 tourism resource perception, θ 3 tourism demand perception, B4 destination image and brand cognition. In the time dimension evaluation, based on the series calibration scale of AHP, the entropy weight correction path is used to obtain the relevant weight coefficient, and the judgment matrix of B and C layer (two and three levels) system is constructed based on expert opinions. The maximum eigenvalue mmax of judgment matrix A is obtained by using maab, and the corresponding eigenvector is the relative weight of each index, so we get the weight coefficient allocation structure. Based on the second level weight, the third level weight can be obtained immediately. (In Fig. 1) At this time, the weight of the third level index (layer C) in the first level index measure (a) can be obtained. The relevant climate indicators are processed by interval transformation and reciprocal deformation.

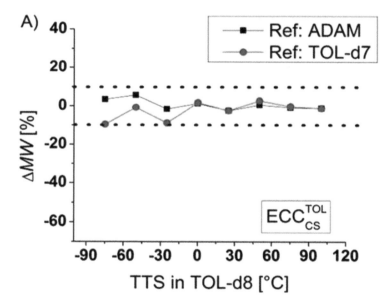

Fig. 1. Weight determination, consistency and dimensionless processing

4.2 Evaluation and Consistency Test of Tourism Seasonal Volatility

Based on the annual time interval, we calculate the off-season tourism market potential evaluation index. We use the weighted average calculation path $e = \sum qp1$, e is the tourism seasonal market potential, q is the weight coefficient, P is the index measure value, and N is the number of variables. In addition, in view of the complexity of the evaluation object and the subjectivity of the judgment matrix may lead to logical inconsistency, it is necessary to complete the consistency test. Based on the previous calculation results, the consistency index of AB layer C1 = (mmax − n)/(n − 1) = 0.009116 (here mmax refers to a)_ B is the special root of the judgment matrix (maximum), n is the order of the matrix, n = 4), and the mean value of consistency R1 = 0.90. Therefore, it can be known that the random consistency ratio CR = C1/R. = 0.0101 < 0.1, so the consistency result is acceptable. From the same calculation process, the consistency test of B-C layer is also consistent.

Steps of entropy weight correction.

Based on the calculation path of entropy weight correction, the weight coefficient correction of the above time series is realized to reduce the weight information loss of monthly data in the sample space. Select m (=12) samples, 26 provinces, autonomous regions and cities monthly sample space, n = 15 indicators, construct matrix (1) the index coefficient proportion of the first city:

4.3   The First Index Coefficient Proportion of the First City

The first index coefficient proportion of the first city:

$$p_{ij} = r_{ij} \sum_{i=1}^{m} r_{ij} \tag{1}$$

Entropy weight WJ of the third index: L

$$w_j = (1 - e_j) / \sum_{j=1}^{n} (1 - e_j) \tag{2}$$

The comprehensive weight β of measurement index:

$$\beta_j = \frac{a_i w_i}{\sum_{i=1}^{m} a_i w_i} \tag{3}$$

Based on the change of the influencing factors of tourists' perception in the space-time dimension, combined with the results of the questionnaire survey on tourists' travel motivation, four dimensions of climate comfort, resource endowment, demand perception and image perception are screened out, and the tourism seasonality measurement index system is constructed, and then the entropy weight modified AHP method is used in the empirical analysis of 26 provinces and cities. The results show that the tourism market development of the underdeveloped areas in the central and western regions is under developed in many months of the year, and the potential space is large, while the eastern region is characterized by the coexistence of under development and over development. It can be seen that the development of tourism is not entirely dependent on the level of economic development, and the impact of non economic factors such as resource endowment and climate environment can not be ignored.

5 Empirical Analysis on the Seasonality of Destination

5.1 Case Study

After the compound addition of the fluctuation effects of the above four tourism seasonal influencing factors modules B1-B4 (climate, resources, demand and image perception), the seasonal comprehensive effect of passenger flow in 26 provinces, autonomous regions and cities is formed. It can be seen from the seasonal intensity value in the table that in the tourism seasonal cycle from January to December, there are great fluctuations in different regions in different months, and the temporal and spatial differences provide a huge space for the seasonal development and connotative development of tourism industry, We match the seasonal intensity index obtained from the above analysis with the actual tourist flow scale in different months of each region, and take the difference between the two as the potential space for tourism development. The value obtained is distributed around zero value [3]. A positive value indicates that the month in this region is the peak season, while a negative value indicates that the passenger flow in this region is insufficient.

5.2 Main Performance

This paper takes February and March as examples, where the spatial value of seasonal market development potential ranks first. For example, in February, Zhejiang, Guangxi, Guangdong, Gansu, Guizhou and other provinces have relatively large seasonal development potential, which is the highest in the figure; while Sichuan, Yunnan and Hainan are in the peak tourism season, In other words, in February, the potential for further development of these areas is small. In March, Xinjiang, Tianjin, Gansu and Guangxi ranked first, with great potential of tourism seasonal market; however, in this period of time, the tourism market in Zhejiang Province has been in a state of reception saturation, so it is necessary to control and divert the tourism flow. Due to the limited space, the regional pattern of seasonal potential in other months is no longer described.

6 Epilogue

In the process of analysis, 26 provinces, autonomous regions and cities are taken as the research object, and their monthly data are taken as the sample space, and the entropy weight modified AHP is used for quantitative measurement. AHP process has the characteristics of clear logic and robust results. The calculation path increases the degree of fitting with the actual through the weight information to improve the credibility. The empirical results are basically consistent with the seasonal characteristics of tourist reception in 26 provinces, autonomous regions and cities, which verifies the effectiveness and reliability of the index system. It is a measurement framework with promotional value. Limited by the current situation of data collection and research, there are still some problems in this paper, such as the lack of data information in small-scale questionnaire survey and the unreasonable setting of indicators, which need to be adjusted and corrected in the future application research [4]. It should be pointed out that the analysis process focuses on the short-term change trend and law of each influencing variable, and does not identify the medium and long-term trajectory, although the medium and long-term seasonal trend also exists objectively. Based on the medium and long-term trend, the seasonal intensity is not constant.

References

1. Yu, X., Zhang, Y., Zhu, G., Li, D., Wang, J.: Tourism seasonality measurement framework and its empirical research. Econ. Geogr. **39**(11), 225–234 (2019)
2. Sun, X., Ni, R., Feng, X.: Research on the seasonal characteristics of urban inbound tourism and tourist source market – an empirical analysis based on Shanghai. J. Tourism **34**(08), 25–39 (2019)
3. Wang, G.: Research on the impact of tourism seasonal intensity on ticket prices of scenic spots – based on the data of 5A scenic spots in China. Stat. Inner Mongolia (03), 30–33 (2019)
4. Wu, J.: Analysis and prediction of tourism seasonality in Wuhan. Economics **2**(3) (2019)

Design of AHP-Delphi Emergency Capability Evaluation Index System Model in Management Stage

Pingfen Li[✉]

Guangxi Vocational Normal University, Guilin 30007, China

Abstract. On the basis of studying the design principle and composition of the evaluation index system model of urban disaster emergency response capability, aiming at the shortcomings of the existing evaluation methods, the evaluation index system of urban disaster emergency response capability is constructed by using AHP Delphi integration method, the factor set of the evaluation index system of urban disaster emergency response capability is determined by Delphi method, and the hierarchical diagram of the system structure is constructed by using AHP method. This paper decomposes the factors of the complex evaluation index system to directly reflect the subordinate relationship among the internal factors of the system, introduces the average random consistency index value to calculate the weight of each index factor, and tests the consistency of the judgment matrix to achieve satisfactory results. It shows that the AHP Delphi integrated method is applicable and scientific to build the evaluation index system model of urban disaster emergency response capability.

Keywords: Construction system · Multi-objective optimization · Ant colony algorithm

1 Introduction

Under the background of globalization, strengthening urban disaster emergency response capacity has become one of the important symbols to measure the government's governance capacity. Experts in disaster science all over the world believe that to achieve safe and sustainable development of a country, the first thing to solve is to improve the city's disaster prevention and disaster resistance capacity. It will certainly involve the development and implementation of the emergency capability evaluation index system. At present, experts at home and abroad have begun to study the evaluation criteria, evaluation procedures, evaluation methods and related operating systems of emergency capability evaluation. This paper attempts to use the AHP delplh integration method to study the model design of urban disaster emergency capability evaluation index system [1]. Network courses not only provide completely.

J. C. Hung et al. (Eds.): FC 2021, LNEE 827, pp. 1208–1214, 2022.
https://doi.org/10.1007/978-981-16-8052-6_170

2 Design Principle and Composition of Evaluation Index System Model

However, after the investigation, it is found that there are many problems in the situation design, activity design, content organization design and layout design of the current online courses. The author thinks that educational technology researchers need to deeply study the guiding role of relevant theories in the design of online courses. This paper will discuss the application of various theories in network course design.

2.1 Model Design Principles of Urban Disaster Emergency Response

The emergency management of urban disasters should first guarantee the safety of people's lives. The life safety here refers to the life safety of every citizen. It should embody equality, justice and benevolence, pay attention to the assistance of vulnerable groups, especially the elderly and children, and never give up or abandon the life in danger. This is not only the basic moral requirement of humanitarianism, but also the fundamental embodiment of the people-oriented value of socialism. To ensure the safety of life, great attention should be paid to the safety and smooth flow of water supply, power supply, transportation, communication and other lifeline systems, to ensure the normal operation of public safety facilities, and to maintain the basic social order, so as to ensure that more people are free from the threat of disasters, secondary disasters and related events. 2. The fairness and justice of property protection requires that the property should have personal property, personal property, personal property, personal property, and so on It can be divided into collective property and state property. In property protection, except for the key public facilities (because this part of property is the guarantee of the safety of the whole group), personal property, collective property and state property should be treated equally. Equal treatment is also reflected in how to minimize property losses, here, more reflects the priority of fairness, pay attention to efficiency, that is, on the basis of fairness and justice, pay attention to efficiency, try to save more property. Fairness and justice reflect the requirements of socialist morality. Making a correct value orientation for the public is conducive to uniting the people and maintaining social stability and harmony.

2.2 Reasonable Construction of Urban Disaster Emergency Capability Evaluation Based on Index System

The evaluation n disaster takes urban government t system as the evaluation object. According to a certain disaster system is comprehensively and comprehensively reflected and described. disaster capacity consists of three parts. (1) evaluation index set evaluation index set is mainly composed of a series of internally related and representative indexes, (2) basic data set is the basis of quantitative analysis and evaluation index, which includes the original statistical data needed to analyze the evaluation index factors. (3) evaluation model evaluation model is a certain quantitative analysis model, which uses mathematical model to describe the relationship between various factors of the system. The analysis object of the evaluation model is the set of evaluation index factors.

3   Model Design Method of Urban Disaster Emergency Response Capability Index Evaluation System

3.1   Design Overview

As urban disaster emergency management is an interdisciplinary subject of geography, environment, safety, meteorology, economics and sociology, the evaluation of urban disaster emergency capacity should not only study the physical entities such as space, resource allocation and disaster sources, but also study their interaction, influencing factors and information flow, as well as human factors. The whole system has such phenomena as uncertainty, randomness, mutation and fuzziness. For this complex project, its evaluation needs the guidance of correct methodology [2]. AHP method fully embodies the idea of system analysis and system synthesis. 34.6.6 decomposes the solution method of a complex system into the solution of much simpler subsystems. Then the evaluation factor set is established by deph, and the weight of each index factor is calculated by AHP.

3.2   AHP is Used to Determine the Weight of Evaluation Index

1)   The judgment matrix is constructed, the scale method of 1–9 is used to compare the importance of any two factors, the ratio of relative importance of factors is judged, the quantitative AHP method is given, and the ratio scale of 1–9 is designed (see Table 1). The ability to distinguish things qualitatively can be expressed by five attributes of importance degree. When higher accuracy is needed, two adjacent values can be adopted, so nine values can be obtained. That is to say, the problem that should be paid attention to in 9-scale is that at most 9 factors can be compared each time.

Table 1. Factor comparison scale

The importance of B_i compared with B_j	Scale
Equally important	1
A little more important	3
Obviously important	5
Strongly important	7
Absolutely important	9

Suppose several factors (ns9), compare two of them with 1–9 scale method, and use B to express the ratio of relative importance of factors and j to get the expression of judgment matrix:

$$C = [C_{ij}]_{m \times n} = \begin{bmatrix} C_{11} & \cdots & C_{1n} \\ \cdots & & \cdots \\ C_{n1} & \cdots & C_{nn} \end{bmatrix} \tag{1}$$

2) The square root method is used to solve the eigenvector and the maximum eigenvalue of the judgment matrix C. first, the nth root of the product of the elements in each row of the judgment matrix C is calculated:

$$W_i = \sqrt[n]{\prod_{i=1}^{n} C_{i,j}} \, i = 1, 2, \ldots, n \tag{2}$$

3) Calculating the maximum eigenvalue of judgment matrix:

$$\lambda_{\max} = \sum_{i=1n}^{n} \frac{(CW)_i}{nW_i} \tag{3}$$

The main advantage of AHP is to quantify the decision-maker's qualitative thinking process, but the consistency of judgment thinking must be maintained in the process of modeling. On the judgment matrix C, each element C should meet the following requirements: for any 1sksn. There is C = ckck, which is the precondition that the eigenvector corresponding to the maximum eigenvalue of the judgment matrix is the factor weight [3]. In order to measure whether the judgment matrix of different orders has satisfactory consistency, the average random consistency value is introduced, as shown in Table 2.

Table 2. Average random consistency index

Stage n	3	4	5	6	7	8	9
RI	0.58	0.9	1.12	1.24	1.32	1.41	1.45

3.3   Analytic Hierarchy Process Model Method

The goal of AHP model is to decompose a complex system, determine the relative weight of all factors in the system, and test. First of all, have a clear and profound understanding of the problem to be solved and the system to be studied, including the scope of the system, the factors involved, and the relationship and subordination between the factors. Then, we can start to construct a hierarchical structure diagram, and use a hierarchical tree structure diagram to represent the factors included in the system and the subordination between the factors. When decomposing the AHP model, we should pay attention to the following problems: (1) each factor is associated with no more than 9 factors at the next level; (2) the factors at the same level should be of the same kind and meet the requirements of the upper level.

4   Experiment

The setting of urban disaster emergency capability evaluation index must cover the background of urban emergency and various natural and social factors in space, which is the realistic basis and objective basis for determining the content of emergency capability index. In terms of factor composition, it must cover many factors, such as natural factors to social factors, system design to public behavior, organization efficiency to engineering ability, etc.

5   AHP is Used to Establish the Hierarchical Structure and Calculate the Initial Weight

5.1   The Basic Principle of AHP

AHP method is a comprehensive evaluation method of system analysis and decision-making, which was put forward by t lsaaty in 1970s. It is a flexible and practical multi criteria decision-making method, which can reasonably deal with qualitative problems quantitatively. By establishing a hierarchical structure, AHP transforms people's judgment into the comparison of the importance of several factors. The principle is to classify all kinds of selection indexes and schemes according to their properties, and divide them into several levels, so that the problem can be transformed into a ranking problem of relative merits of each index scheme. By constructing a judgment matrix, the single ranking structure of a certain level factor relative to each factor of the previous level and the total ranking weight relative to the previous level can be calculated.

5.2   The Hierarchical Structure of the Evaluation System of Network Course

According to the principle of AHP, on the basis of systematic analysis of network course, it is decomposed into various parts composed of elements, and these elements are divided into several groups, forming some factors of the next level, and at the same time, they are upper level. This dominating relationship from top to bottom forms a hierarchical level. The network curriculum quality can be divided into three levels.

In the "three-level model", the top W is the quality characteristics of some aspects of online courses [4]. For each identified quality characteristic, there must be a set of quality sub characteristics U1, U2 Un. Each quality sub feature is composed of several quality problems, which are the software quality metrics U11 to u1r 1 to unr. This three-level tree framework provides a top-down structure and a bottom-up measurement mechanism for software quality measurement. The judgment matrix, weight and

Cr of the secondary indicators under the secondary indicators will be given in the table respectively. The algorithm of the judgment matrix, weight and Cr value of the secondary indicators is the same as that of the primary indicators. The course content, interface design, technical secondary index judgment matrix and weight are calculated in turn, and the consistency test is carried out (the chart is omitted). Finally, the weight of each level relative to the total goal is synthesized.

6 Evaluation Index Set of Emergency Response Capability in Management Stage

6.1 Using Delph Method to Establish Urban Disaster Emergency Response Capacity Evaluation Index Set

The factors covered are listed as the basis for the design of the consultation form. Because there are too many factors for the evaluation of urban disaster emergency response capability, they are not listed one by one in this paper. 2) the application of delph method. The composition of the expert group requires a high degree of authority and a wide range of experts. Generally, 40–50 people are suitable. 3) design the consultation form to design all the factors covered by the evaluation index capability to the members of the expert panel for careful filling.

6.2 Evaluation Index System Model of Urban Disaster Emergency Response Capability

The evaluation of urban disaster emergency response capability is a cyclic process. The urban pre disaster early warning ability includes the ability of monitoring existing disasters, the ability of monitoring and forecasting possible disasters, the ability of disaster early warning technology, and the ability of disaster mitigation measures; Emergency response capability in disaster includes disaster identification capability, disaster emergency rescue capability, emergency response capability and disaster behavior response capability of urban residents; post disaster recovery capability includes social security system, disaster loss assessment capability, urban post disaster recovery system and post disaster reconstruction capability [5]. It can be divided into four layers: general target layer, quasi target layer, criterion layer and index layer.

7 Conclusion and Prospect

This study uses AHP to determine the layered model of network curriculum quality factors, and calculates the initial. The rationality of the weight design of each index is further analyzed and verified by de ψ pH method. Finally, four evaluation dimensions are formed: teaching design, curriculum content, interface design and technicality. Each dimension is divided into several secondary indicators, including 11 indicators, such as motivation motivation of course positioning; 8 indicators including scientificity and content block; interface design includes seven indicators such as style system screen layout; technical dimension includes six indicators, such as operation environment description, installation and unloading. The whole system has 32 indicators. The rationality of weight setting is proved and tested, and the validity and reliability of the index system are improved.

Acknowledgements. Guangxi Philosophy and Social Science Planning Research Project "Strategy Research on Improving the Emergency Response Ability of Party Member Leading Cadres in Border Areas under the Background of Risk Society", 2020 (Project No.: 20BGL013).

References

1. Wang, Y.: Research on Supplier Evaluation and Optimization of Company A. Yunnan University of Finance and Economics (2020)
2. Zheng, M., Chen, J., Lin, J., Liu, B.: The concept of the emergency capacity building system for the prevention and control of ship pollution in waters of Shantou Special Economic Zone [a]. Inland waterway Maritime Professional Committee of China Maritime society. Outstanding papers of 2020 maritime management academic annual meeting. Inland waterway Maritime Professional Committee of China Maritime society, vol. 3 (2020)
3. Gao, K., et al.: Safety evaluation of well control and Countermeasures for improving blowout emergency capability of high sulfur gas wells. Progress Chem. Ind. **39**(S2), 83–88 (2020)
4. Zhong, Y., Yang, L.: Establishment of an evaluation system for nurses' ability to cope with public health emergencies in traditional Chinese medicine hospitals. Henan Tradit. Chin. Med. **40**(11), 1629–1632 (2020)
5. Yun, T., Nie, W., Wang, L.: Research progress of nurses' emergency response ability and intervention strategies for public health emergencies. Mod. Clin. Nurs. **19**(11), 68–74 (2020)

Construction of Credit System Prediction Model Based on L-M Optimization Algorithm

Qing Li[✉]

Baoshan College, Baoshan 678000, Yunnan, China

Abstract. Using the relationship between the students' historical score data and the curriculum, this paper constructs a credit score point prediction model based on BP neural network, which has certain theoretical and practical value. In the course, the neural network can be used to predict the complex relationship between credit and grade point. The prediction results of network trained by different algorithms are compared. It is found that the L-M optimization algorithm has the best preview performance. Finally, the function is used to simulate, and then the simulation results are compared with the sample data. It is verified that the L-M optimization algorithm has high accuracy and can be used to predict grade points.

Keywords: Grade point · Prediction model · Neural network · LM optimization algorithm

1 Introduction

With the in-depth application of computer big data storage technology in Colleges and universities, students' scores have been completely saved in electronic form, but a large number of performance data have not been well utilized. For each student, they can only query their current grades, but can't know their current learning status and the trend of their future academic achievements. For teaching managers, only through personal work experience to guide students, there is no data to support, and can not provide effective and reasonable learning suggestions. Therefore, academic early warning has become a hot spot in the management of college students. However, the difficulty of academic early warning is to find scientific, effective and accurate credit score prediction methods. In this paper, we use the relationship between the students' historical score data stored in the computer and the curriculum, and construct the character grade point prediction model based on neural network. Through the data comparison test, it is found that the prediction model results using LM optimization algorithm have the highest consistency with the actual situation, and can be applied to students' academic early warning.

© The Author(s), under exclusive license to Springer Nature Singapore Pte Ltd. 2022
J. C. Hung et al. (Eds.): FC 2021, LNEE 827, pp. 1215–1219, 2022.
https://doi.org/10.1007/978-981-16-8052-6_171

2 Curriculum and GPA

2.1 Grade Point

According to the teaching management regulations of our university, one of the important criteria for undergraduates to award bachelor's degree is that the average GPA should reach 1.8 when students complete their studies.

2.2 Relationship Between Courses

When the school formulates the teaching and training plan, it not only stipulates the type and nature of the curriculum, but also sets the starting sequence of the course. Therefore, students need to understand the internal relationship between the courses and the lead-by relationship between the courses when learning to choose courses. The general courses and professional basic courses of civil engineering are the main and core courses of the major, which need to be completed in the first four semesters. These courses are connected with each other and are the leading courses of other courses [1]. For example, the relationship between advanced mathematics A1 and advanced mathematics A2 is the leading and follow-up relationship. The learning results of advanced mathematics A1 directly affect advanced mathematics A2, and at the same time, they also affect the study of college physics B. College Physics B is related to the study of theoretical mechanics, material mechanics a and structural mechanics. Therefore, there are obvious hierarchical and network relationships between the courses.

2.3 Relationship Between Curriculum and GPA

Students' final academic GPA is the accumulated credits of all courses in their major. All the courses are closely related to each other. When a course is the leading course of other courses, if the score of the course is not ideal, the scores of subsequent courses will be affected, and the final GPA is relatively low.

3 Credit Grade Point Prediction Method

The relationship between courses and between courses and between courses and GPAs is very close. Therefore, by constructing the relationship model between courses and between courses and between courses and GPAs, and determining the correlation coefficient between courses, we can predict the follow-up course scores and GPAs according to the scores and grade points of leading courses, so as to predict and warn students in the process of learning. The specific methods are as follows.

Suppose to predict a student's score and grade point of a course B_n. The leading courses are: $A_1, A_2, \ldots A_n$, B_n on $A_1, A_2, \ldots A_n$. The correlation coefficient vector of an is:

$$\{\beta = \beta_1, \beta_2, \ldots, \beta_n\}^{\mathrm{T}} \tag{1}$$

The scores of $A_1, A_2, \ldots A_n$, an of the students to be tested are $S_1, S_2, \ldots S_n$; Then the student's score of B_n in a course is:

$$S_B = \sum_{i=1}^{n} \beta_i S_i \tag{2}$$

The GPA is available.

In fact, there are many problems in the actual prediction. On the one hand, there must be no missing items in the scores of S1, S2 and Sn of the leading courses. On the other hand, it is necessary to determine the curriculum relevance, clarify the leading and subsequent relationship among all courses, and construct the relationship chart between all courses [2]. Secondly, it is necessary to accurately set the value of the relationship coefficient between courses. However, these are very difficult. The model of using curriculum correlation coefficient to predict GPA has limitations, so we can use machine learning tools to simulate GPA prediction model.

4   Determination of Credit Grade Point Prediction Model

4.1   Construction of Credit Grade Point Prediction Model

In the process of learning, if the average score of the students is predicted in advance, the neural network will be used to predict the students' academic achievement points in the process of learning, There is a correlation between courses, which is very suitable for the prediction of GPA.

4.2   Determination of Model

(1) The specific design of the input and output GPA prediction model is S1S2 SN is used as the input, and the final GPA GN result is used as the output. After model learning and training, the relationship between the course score and GPA is obtained. For the students to be predicted, the final average GPA 3 can be obtained by taking the course scores of the students who have obtained credits as the input, and the trained GPA prediction model.

(2) In BP neural network, it is difficult and important to determine the number of hidden layer nodes. If the number of hidden layer nodes is too small, the network performance will be poor or even unable to train. If the number of hidden layer nodes is too large, the system error of neural network can be reduced, but the network training time will be increased, resulting in local minima, and the fault tolerance and generalization ability of the network will be weakened. Therefore, it is very important to select the exact number of hidden layer nodes [3]. The number of nodes in the two layers is relatively fixed and can not be changed at will.

Therefore, the number of hidden layers and the number of nodes should be determined in the whole network structure.

4.3   Data Preprocessing

(1) The results of each course in the input data processing teaching plan truly reflect the students' learning situation of the course, so the model uses the course score as the input. The input data of model training needs to be normalized. This paper uses the following formula:

$$P = \frac{x - x_{min}}{x_{max} - x_{min}} \tag{3}$$

Where x is the course score, x_{min} is the minimum value of the course score, generally 0 x_{max} is the maximum value of the course score, and generally 100 P is the value between (0,1) after the course score is normalized.

(2) The model of the output data office uses the average grade point as the model output, and analyzes the database in the teaching management system of our university. The score database contains the credits, grades and grade points of each student's selected courses.

$$T = \frac{y - y_{min}}{y_{max} - y_{min}} \tag{4}$$

According to these data and the GPA calculation formula, the average GPA of students can be calculated. In this paper, we use the following formula to normalize the output data of the model between [0, 1], where y is the average grade point, y_{min} is the minimum average grade point, generally yma is the maximum course score, and generally 5T is the normalized value between (0, 1).

5   Performance Test of GPA Prediction Model

In order to test the basic performance of GPA prediction model, SIM function is used in matlabi to test the accuracy of the training network. The course score of the sample is input to the input layer of the model, and the output result is obtained in the output layer through the operation of the model. The actual output is compared with the expected output. If the error between the two meets the requirements, the accuracy of the model is high [4].

6   Conclusion

In this paper, a prediction model of average grade point based on BP neural network is constructed. The results obtained by students in each semester are used as the input of the model, and the average grade point value of students is used as the output of the model. When the number of hidden layer nodes is 9, the model meets the error requirements between the actual output and the expected output of the network, that is, the model can map the functional relationship between the course grade and the average grade point. Finally, the experimental results show that LM optimization algorithm is superior to other algorithm in the construction of GPA prediction model, which can be applied to intelligent academic early warning, and has certain theoretical and practical application value.

References

1. Practice and exploration of GPA system in seven universities affiliated to Ministry of Wuhan. J. Henan Inst. Educ. (Philos. Soc. Sci.) 131–132 (2012)
2. Zhai, X., Yin, J., Lin, L.: Construction of satisfaction model of flipped classroom in China from the perspective of structural equation. Explor. High. Educ. (05), 65–72 (2015)
3. Lu, Y., Zhao, X., An, L.: Application of fast BP algorithm in annual runoff prediction. J. Water Resour. Water Eng. 96–97 (2012)
4. Ma, Z., Wang, T., Zhou, J.: Prediction of spatiotemporal variation of eutrophication in Taihu Lake based on BP neural network. J. Changzhou Univ. (Nat. Sci. Ed.) 63–64 (2013)

Analysis of Risk Asset Model Based on Artificial Intelligence Algorithm

Zhuojie Liang$^{(\boxtimes)}$ and Yixian Liu

Nanfang College of Sun Yat-sen University, Guangzhou 510970, China

Abstract. In the framework of mean variance model, this paper establishes the WCVaR model with worst case var (WCVaR) instead of variance as the risk measurement index. At the same time, the logarithmic membership function is introduced into the model to maximize the expected return rate and minimize the WCVaR of the portfolio, and the logarithmic satisfaction degree of the fuzzy decision-making portfolio selection model is established, which better reflects the value intention of investors to the target value. According to the actual data of Shanghai Securities, the genetic algorithm is used to simulate the calculation, and the effectiveness of the model is verified.

Keywords: Portfolio optimization · Worst case value at risk · Fuzzy · Genetic algorithm

1 Introduction

In 1952, Markowitz put forward the "mean variance" theory of portfolio, which provides a feasible quantitative means for the trade-off between risk and return. In this model, the variance is used as the risk function to find the portfolio with the minimum variance under a certain income level. But there are some problems in measuring risk with variance. Therefore, many scholars have improved the "mean variance" model. One of the main ways is to use new risk index instead of variance to establish a new mean risk model. For example, Konno's "mean absolute deviation" portfolio model, which takes the expected absolute deviation as the risk function, Mao and swalm's "mean lower semivariance" portfolio model, which takes the expected value of the square of the negative deviation relative to the mean as the risk function, is developed into "mean lower semiabsolute deviation" portfolio model by Speranza; Young and Cai take the minimum order statistics and Yao Jing take "VaR based financial asset allocation model". These are portfolio selection models based on probability theory, and the uncertainties considered are all random uncertainties, which indicates that the research on portfolio selection based on random uncertainty has developed to a considerable extent [1, 2].

However, the uncertainty in financial market is more fuzzy. At present, some scholars have used the fuzzy set theory to study the portfolio problem, which is still in its infancy. In this paper, we introduce a new risk measurement function, worst case VaR, into the portfolio model, and use it to describe the risk, which overcomes the assumption that the distribution of return is normal distribution. At the same time, in

J. C. Hung et al. (Eds.): FC 2021, LNEE 827, pp. 1220–1225, 2022.
https://doi.org/10.1007/978-981-16-8052-6_172

order to more vividly describe the satisfaction degree of investors to investment income and investment risk, this paper uses the fuzzy number with logarithmic membership function to describe the target level of investors to investment income and investment risk, and puts forward the mean WCVaR fuzzy portfolio selection model [3]. In the model, a practical method is used to transform the complex model with nonlinear constraints into a simple and easy to calculate programming problem. Finally, we use genetic algorithm to make an empirical study on investor selection and verify the effectiveness of the model.

2 Related Work

At the same time, we should establish a commercial sustainable operation mechanism to continuously assist the development of seed industry and clarify the support focus. In the regional aspect, we should focus on the key links such as breeding technology research and development, pay attention to the progress and financial service demand of seed industry enterprises in the whole gene selection breeding, transgenic technology, mulberry editing and other biological breeding technology innovation, and take the initiative to connect with key breeding projects, major biological breeding scientific research projects, and other research projects Modern seed industry upgrading project and other national major projects. In terms of customers, in response to the trend of increasing concentration of seed industry, focusing on the integration of breeding, breeding and promotion enterprises, high credit evaluation enterprises and seed enterprises with technology accumulation and sales channel advantages in regional characteristic crop varieties, we continue to adjust the credit structure of the industry to ensure the healthy and sustainable development of credit business.

Strengthen product and mode innovation [4]. Do a good job Based on the design of basic products, combined with the seed production cycle of crop varieties, we will launch short-term and flexible liquidity loan products or credit service schemes to improve the accuracy of credit support. The financial services of the industrial chain should be based on seed production orders and purchase and sale contracts, with seed enterprises as the core and extending to the upstream and downstream. The financial products such as loans and factoring should be comprehensively used to provide the overall services of the industrial chain, especially for the agricultural production and operation entities such as farmers and farmers' professional cooperatives. Focus on the financing needs brought by the improvement of industry concentration, actively follow up the M & A projects of large seed enterprise customers, promote the combination of strong and strong through M & A loans, fund operation and other forms, promote the development and growth of leading seed enterprises, and help improve the quality and efficiency of industry development. We should explore new ways of guarantee. With the continuous strengthening of intellectual property protection in the industry, the importance of new plant variety rights as the core assets of seed enterprises has become increasingly prominent [5]. Commercial banks should strengthen the market research of seed industry, gradually explore to set up mortgage loan of new plant variety right through local pilot, cooperate with institutions or market entities with data accumulation advantages in value evaluation of new plant variety right, continuously improve the internal value evaluation and review ability, and activate the mortgage circulation performance of core assets of seed enterprises, Solve the problem of mortgage guarantee for seed enterprises. We will strengthen cooperation with agricultural credit guarantee companies, insurance companies and other third parties, promote special guarantee funds for seed industry in advantageous regions, and constantly improve the credit enhancement mechanism of the industry. As shown in Fig. 1:

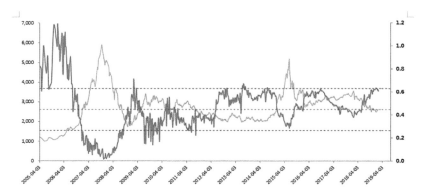

Fig. 1. Model analysis table

3   Portfolio Model

Assumes that are averse. He assumes a very ideal securities market. In this market, there is no transaction cost and tax, and there is no limit on interest rate. The number of shares of securities is infinitely divisible. We call this kind of market. There are n kinds of tradable risk assets in the frictionless securities market. The time series of return rate of these n kinds of risk assets in the past t period is as follows:

$$r_{i1}, r_{i2}, ..., r_{iT}(i = 1, 2, ..., n) \tag{1}$$

Then the expected return rate μ I of the i-th risk asset is

$$\mu_i = \frac{1}{T} \sum_{j=1}^{T} r_{ij}(i = 1, 2, ..., n) \tag{2}$$

The variance of the i-th risk asset () and the covariance of the IK risk asset () are as follows:

$$\sigma_{ii} = \frac{1}{T} \sum_{j=1}^{T} (r_{ij} - \mu_i)^2$$
$$\sigma_{ik} = \frac{1}{T} \sum_{j=1}^{T} (r_{ij} - \mu_i)(r_{kj} - \mu_k) \tag{3}$$

Assuming that Xi is the investor's investment share in the i-th risk asset, the time series of risk asset in the past t period is as follows:

$$\sum_{i=1}^{n} x_i r_{i1}, \sum_{i=1}^{n} x_i r_{i2}, ..., \sum_{i=1}^{n} x_i r_{iT} \tag{4}$$

The expected return rate of venture capital investment in t period is estimated as follows:

$$R(x) = \frac{1}{T}\sum_{j=1}^{T}\left(\sum_{i=1}^{n} x_i r_{ij}\right) = \sum_{i=1}^{n} x_i\left(\frac{1}{T}\sum_{j=1}^{T} r_{ij}\right) = \sum_{i=1}^{n} x_i \mu_i = X^T \mu \tag{5}$$

The variance is as follows

$$V(x) = \frac{1}{T}\sum_{j=1}^{T}\left(\sum_{i=1}^{n} x_i r_{ij} - \sum_{i=1}^{n} x_i \mu_i\right)^2 = x^T \Gamma_x$$

$$\text{Namely: } V(x) = \sum_{i=1}^{n}\sum_{j=1}^{n} x_i x_j \hat{\sigma}_{ij} = x^T \Gamma_x \tag{6}$$

In general, the expected return of investors is the largest and the risk is the smallest, which mathematically.

$$\begin{cases} \max & R(x) = \sum_{x=i}^{n} x_i \mu_i \\ \min & V(x) = \sum_{i=1}^{n}\sum_{j=1}^{n} x_i x_j \hat{\sigma}_{ij} \\ s.t. & \sum_{i=1}^{n} x_i = 1 \quad x_i \geq 0 (i = 1, ..., n) \end{cases} \tag{7}$$

Given different returns r according to formula (7), the investor is assumed to be risk averse, so there is a maximized, the feasible solution obtained by the solution. Although the expected rate of return of investors is satisfied, the risk of this model does not necessarily reach the expected level of investors.

Therefore, this paper proposes a dual objective portfolio model with worst-case VaR as risk index.

Because investors always want to control the and risk in a certain range in actual investment, that is to say, the objective function shows a certain range of fuzzy value in a certain range.

Therefore, the fuzzy portfolio problem of introducing fuzzy concept into the objective function of the securities problem

$$f(x) = \frac{1}{1 + \exp(-\alpha x)} \tag{8}$$

$$\mu_R(R(X)) = \frac{1}{1 + \exp(-\alpha_R(R(X) - R_M))} \tag{9}$$

The smaller the target risk is, the better

$$\mu_V(WCV_aR(X)) = \frac{1}{1 + \exp(-\alpha_V(WCV_aR(X) - WCV_aR_M))} \tag{10}$$

4 Experimental Verification

Matlab7.0 genetic algorithm toolbox is used for simulation. The results show that the probability of crossing is 0.8, the probability of variation is 0.2, the population size is 100, the evolution algebra t = 100, and the error of investment ratio is less than 5%. In order to illustrate the effectiveness of mean WCVaR fuzzy portfolio selection model (10), we compare the calculation results of mean variance portfolio selection model (7). In the aspect of valuation, institutions or market entities with the advantage of data accumulation should cooperate to continuously improve the ability of internal value evaluation and review, activate the mortgage circulation performance of core assets of seed enterprises, and solve the mortgage guarantee problems of seed enterprises. Strengthen cooperation with agricultural credit guarantee companies, insurance companies and other third parties, promote the establishment of special guarantee funds for seed industry in advantageous regions, and continuously complete the credit enhancement mechanism of seed industry. As shown in Table 1 and Table 2.

Table 1. Weight distribution of mean variance model

A	B	C	D	E	F	G
0.0039	0.0011	0.01054	0.0037	0.00062	0.002411	0.003772

Table 2. The return and risk of portfolio in mean variance model

	Portfolio income	Portfolio risk
Mean variance model portfolio	0.0045	0.00052123

5 Conclusion

This paper describes the mean WCVaR model using WCVaR as risk index, which is an extension of mean variance model. At the same time, considering the influence of information, mentality, environment and other factors, investors' expectations of expected return and risk often change, with some fuzzy characteristics. A fuzzy Bi objective combination optimization model with expected return and WCVaR as objective functions is established.

1) Choosing WCVaR as the risk index is not affected by the type of return distribution, and this risk can more directly reflect the investors' risk tolerance. From the perspective of risk control, the WCVaR values are all lower than the risk limit indicators under 95% guarantee degree.

2) The parameters α R and α V reflect the fuzzy scale of investors' expected. The larger the value is, the smaller the fuzzy scale is, which better reflects the investors' intention to value the target value. Under five fuzzy scales, the investment proportion of eight stocks in the portfolio is obtained. The results show that the portfolio investment is effective.

3) Using nonlinear investment risk can more vividly describe investors' satisfaction with investment return and investment risk. It can be seen that the obtained solution λ $(0 < \lambda < 1)$ is the solution on the effective boundary. Because of the use of logarithmic membership function, the solution of logarithmic satisfaction model is quite convenient and practical. Investors can adjust the parameters to get their own investment strategy to the greatest extent.

References

1. Lin, A.: An analysis of Wang Lu's ways to cultivate college students' entrepreneurial ability. J. Ankang Univ. **23**, 223–228 (2011)
2. Niku, S.B.: Introduction to Robotics. Electronic Industry Press, Beijing (2004)
3. Cai, Z.: Robot Guidance. Tsinghua University Press, Beijing (2009)
4. Bi, S.: Development status of industrial robot at home and abroad. Mechatron. Door **2**, 6–9 (2006)
5. Yuan, K.: development status and trend of industrial robots. Mech. Eng. **7**, 5–7 (2008)

Traffic Parameter Detection Based on Computer Vision and Image Processing

Yongxing Lin[1]([⊠]) and Quan Wen[2]

[1] Keyi College of Zhejiang Sci-Tech University,
Hangzhou 310018, People's Republic of China
[2] Zhejiang Institute of Economics and Trade,
Hangzhou 310018, People's Republic of China

Abstract. In this paper, a method of traffic flow detection based on computer vision and image processing is proposed. By analyzing the vehicle and scene information in the image acquired by CCD TV camera, the traffic parameters can be detected effectively. This method can complete the vehicle counting and vehicle speed detection of two-way four lane Expressway and urban road. The experimental results show that the accuracy of vehicle counting is 94%. The accuracy of vehicle speed detection is 92%, and it has satisfactory real-time characteristics.

Keywords: Vehicle count · Vehicle speed detection · Computer vision · Image processing · Label analysis

1 Introduction

Traffic flow parameters mainly include traffic flow, traffic density, vehicle speed and queue length parameter. There are many methods to detect traffic flow parameters, such as ultrasonic detection, infrared detection, annular induction ring detection and computer vision detection. The accuracy of ultrasonic detection is not high, which is easily affected by vehicle occlusion and pedestrians, and the detection distance is short (generally not more than 1 m); the infrared detection is affected by the heat source of the vehicle itself, so the anti noise ability is not strong, and the detection accuracy is not high; The ring sensor has high detection accuracy, but it is required to be set in the civil structure of the road, which has damage to the road, inconvenient construction and installation, and a large number of installation. In recent years, continuous, image processing, artificial intelligence, pattern recognition and other technologies, computer vision detection has been more and more widely used in traffic flow detection [1].

Compared with other traffic flow detection technologies, computer vision detection has the following characteristics: 1) it can extract high-quality vehicle and traffic scene information from video digital image; 2) video sensor can detect larger traffic scene area, reduce the number of sensors, with less investment and low cost; 3) video sensor is easy to install and debug, and will not damage the road and civil facilities.

J. C. Hung et al. (Eds.): FC 2021, LNEE 827, pp. 1226–1231, 2022.
https://doi.org/10.1007/978-981-16-8052-6_173

2   System Structure and Processing Flow

2.1   System Structure

Six white sign lines are set at equal intervals along the road direction (the interval is 10 m, with a total processing of 50 m) perpendicular to the road direction. The sign lines run through four lanes. The width of the sign line should be set to ensure that it takes up three pixels in the image. The sign lines are marked as 1, 2 Each CCD TV camera processes three sign lines. Six virtual lines parallel to the sign line are set in the image, and the six virtual lines correspond to six sign lines, The width of the virtual line is 1 pixel and is close to the corresponding sign line (for example, in lane a of the image [2], if the sign line K, K $\pm$ 1, then the virtual line K + 3, that is, there is a difference of two pixels between the sign line and the virtual line). The detection system is as shown in Fig. 1.

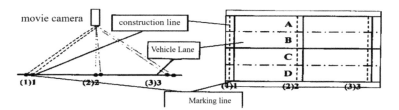

Fig. 1. System structure diagram

2.2   System Processing Flow

In the vision system, the CCD TV camera acquires the traffic scene and vehicle images. At the same time, the image acquisition card collects the images. The images are stored in the computer or processed directly. The image acquisition card collects 50 frames per second. The image frame difference is the most critical technology in image processing. The image frame difference is the gray difference of two images collected by the CCD TV camera at different times, It can detect the appearance and movement of vehicles, and the virtual line state detection can further reduce the error of detecting the appearance and movement of vehicles by image frame difference. On the basis of image frame difference and virtual line state detection, the vehicle count, speed detection, queue length detection and queue condition detection can be further obtained by symbol analysis.

3   Traffic Parameter Detection Method

If the input image contains no vehicle, it is the same as the reference image, and the frame difference is 0; on the contrary, if the input image contains vehicle, it is different from the reference image, In this case, the image frame difference is not a q-threshold, and its function is to reduce the influence of noise and light changes. In this paper, the method of dynamically updating the reference frame is used to adapt to the changes of environmental conditions [3]. In order to reduce the amount of calculation, the image frame difference is only carried out on 10 sign lines. In most cases, the image frame difference can detect the movement and appearance of vehicles, But because of the humidity of the road, the shadow of the vehicle and the color of the car body are close to the sign line, the detection results are sometimes incorrect. Further analysis of the virtual line state can solve this problem.

4   Vehicle Speed Detection

There are two methods for vehicle speed detection: the first method is to detect on the same sign line and virtual line; the second method is to detect on two adjacent sign lines and virtual line.

4.1   Vehicle Speed Detection on the Same Sign Line and Virtual Line

The vehicle length is taken as the equivalent length based on statistics. For example, the length of medium-sized bus is taken as the average vehicle length L (usually 5N for expressways and urban roads). In this way, on the premise that the vehicle length is known, the calculation of vehicle speed (actually the average speed of vehicle) can be directly operated on a single sign line and virtual line to detect the vehicles that appear and pass, As long as the passing time of the vehicle is obtained:

$$V(k) = T^{END}(K) - LT^{START}(K) \tag{1}$$

Here K is the secondary serial number of the vehicle.

In many traffic places, it is usually only necessary to calculate the average speed of vehicles in a certain period of time (such as hourly average speed, daily average speed, etc.):

$$AVG - V = \frac{1}{N_2 - N_1} = \sum_{k=N_1}^{N_2} V(K) \tag{2}$$

4.2   Vehicle Speed Detection on Two Adjacent Sign Lines and Virtual Lines

Since the distance between two adjacent sign lines and virtual lines is known (set as 10 m in this paper), the by calculating the time difference between the two adjacent sign lines and virtual lines. Assuming that the time distance between two vehicles and the head exceeds 10 m, the vehicle passes the sign line and virtual line first, and then passes the sign line and virtual line [4].

Model can not accurately describe the geometric relationship of camera imaging, such as in the case of close range and wide angle. Therefore, it is necessary to consider the linear or nonlinear distortion compensation in order to be more reasonable as the pinhole model imaging process, and use the calibrated model for three-dimensional reconstruction in order to obtain higher accuracy, which constitute the main problems of traditional camera calibration research. As shown in Fig. 2. The basic method is on the undetermined the sh calibration reference object, the image is processed, and a series of mathematical transformation and calculation methods are used to obtain the internal and external parameters of the camera model. There are two basic methods based on single frame image and stereo vision method based on multi frame known correspondence. Another important application background is that in many cases, due to the requirement of frequent camera adjustment and the unreality of setting known calibration reference, a so-called camera self calibration method independent of calibration reference is needed.

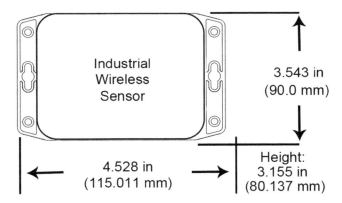

Fig. 2. Model of linear velocity detection

5   Traffic Flow Prediction Based on Neural Network

5.1   The Basic Principle of BP Neural Network

This paper uses BP neural network algorithm to predict traffic flow. The main reason why BP algorithm is adopted is that it is widely used, and the results are quite remarkable, and the structure is simple, easy to realize, and the function is strong. BP neural network is proposed by rumelbart, Hinton and williams. It is a multi-layer feedforward network, which solves the learning problem of hidden unit connection weight in multi-layer network. BP neural network is trained by BP (back propagation) algorithm. The network has an input layer, an output layer and at least one hidden (middle) layer. The results show that increasing the number of hidden layers does not necessarily improve the accuracy and expression ability of the network. Generally, one hidden layer is enough.

5.2   Traffic Flow Prediction Based on BP Neural Network

There is a certain relationship between the traffic flow of the current period on a certain section and the traffic flow of the previous period on this section. At the same time, the traffic flow of a certain section is also affected by the traffic flow of the front and back sections. Therefore, the traffic flow on the road section has an internal relationship with the traffic flow in the first few periods of the connected road section. In this way, we can use the traffic flow of the first few periods on the road to predict the traffic flow of the last period, and we can also use the traffic flow of the first few periods on the front and back road to predict the traffic flow of the future period.

In order to predict the traffic flow based on BP neural network, it is necessary to collect a fixed scale sample set. Then BP algorithm is used to train and determine the network weight to make the network converge to the predetermined accuracy. Finally, for the input part of the data to be measured, the forward propagation algorithm of BP neural network is used to calculate the network output. After processing, the predicted traffic flow is obtained.

According to the above experiments, it can be seen that it is feasible to use BP neural network to forecast traffic flow. The error of prediction is about 8%, and it can also reach 5%. Of course, the prediction accuracy of traffic flow may be higher in the actual traffic monitoring. However, the traffic flow prediction based on BP neural network is easy to realize, and the low cost can provide a good basis for traffic control, and it has a good application prospect, and it can also improve the prediction accuracy by improving BP training algorithm.

6   Conclusion

A large number of experiments are carried out on the counting and speed detection method of vehicles in this paper. A total of 50 experiments are carried out on the actual urban road. 10 groups of images are collected continuously by CCD camera, and the acquisition time of each group is 2 min. The image acquisition card collects 50 original images per second, At the same time, the experiment is carried out under different vehicle flow speeds. The results of many experiments show that: when the vehicle flow speed is within 60kmn/h, the accuracy of vehicle counting is 94%, and the accuracy of vehicle speed detection is about 92%; when the vehicle flow speed is 6080 km/h, the accuracy of vehicle counting and vehicle speed detection is reduced to 93% and 90%, respectively; when the vehicle speed is 80–100 kmnh, the accuracy of vehicle counting and vehicle speed detection is about 92%, The accuracy rates of vehicle counting and vehicle speed detection are 92% and 88% respectively. When the vehicle flow speed exceeds 100 km/h (this experiment is carried out on the urban expressway), the accuracy rates of vehicle counting and vehicle speed detection will be greatly reduced, which are less than 89% and 85% respectively. In fact, the vehicle flow speed on the urban road will not exceed 80 km/h, Therefore, the accuracy of this method for vehicle counting and vehicle speed detection can meet the requirements.

References

1. Zhu, Y.: Technology and development of road traffic accident treatment. Highw. Traffic Sci. Technol. (3), 66–68 (1999)
2. Ding, Y., Zhao, Y.: Research on moving target detection technology. Opt. Technol. (7), 311–312 (2002)
3. Yang, H., Yao, Q.: A method of object segmentation in natural scenes. Chin. J. Image Graph. (8), 646–650 (1998)
4. Yang, P.: Image processing for intelligent transportation system. Comput. Eng. Appl. (9), 4–7 (2001)

Research on EHB System Algorithm and Controller Implementation of Energy Vehicle

Yan Liu$^{(\boxtimes)}$ and Bingxian Li

Weifang Engineer Vocational College, Weifang 262500, China

Abstract. This paper first analyzes the basic composition and working principle of EHB system, and on this basis, establishes the mathematical model of EHB system hydraulic components and the Simulink hydraulic model. Based on the theory of fluid physics, this paper establishes the mathematical model including high-speed solenoid valve, accumulator, brake wheel cylinder and brake pipeline to describe the dynamic performance of the main hydraulic components of the system. On this basis, the mathematical model of the pressurization and decompression process of EHB hydraulic system is established, and the simulation model of single wheel hydraulic braking system is established in Simulink.

Keywords: EHB system · Vehicle stability · Cylinder pressure · Bangbang fuzzy PI control · Predictive control

1 Introduction

With the rapid development of new energy vehicles and Internet industry, automobile enterprises are gradually transforming from traditional manufacturing industry to intelligent service industry, The research and development of new energy vehicles rely on the whole set, analysis and utilization of vehicle data and user data. Understanding the operation status of new energy vehicle industry and user use law has become an important way to enhance the competitiveness of automobile enterprises. Based on a number of data of pure electric passenger vehicles in China FAW new energy vehicle charging big data visualization analysis system, this paper makes a comprehensive and systematic analysis and carding of charging behavior law, user habits and charging fault statistics. Through the quantitative data, it is found that the charging behavior preferences of different user types are quite different. Before designing the charging system, we should develop and design the product function and performance according to the actual usage habits of the target users [1]. From the perspective of users, we should think about the optimization design of functions in specific scenarios, improve the charging safety performance, and play a supporting and borrowing role in formulating product structure and charging function development for new energy industry.

J. C. Hung et al. (Eds.): FC 2021, LNEE 827, pp. 1232–1238, 2022.
https://doi.org/10.1007/978-981-16-8052-6_174

2 Big Data Visualization Analysis System for New Energy Vehicle Charging

The premise of the combination of big data technology and new energy vehicles is to establish a big data platform to efficiently collect massive data resources. In order to solve the safety problems of new energy vehicles in China, improve the supervision of new energy vehicle industry and promote the development of new energy vehicle industry, the Ministry of industry and information technology established a national monitoring and management platform for new energy vehicles in Beijing in 2016.

2.1 Data Resources

Based on the big data visualization analysis system of China FAW new energy vehicle charging, this paper collects data from 14.48 million samples of pure electric vehicles from January 2017 to November 2019, including 161 models and 43.67 million charging times. In order to get a more effective data analysis system, this paper considers the geographical distribution, usage distribution, activity distribution and charging mode when selecting vehicle samples, and the system can filter and query from the time dimension, environment temperature dimension, battery information and vehicle use, driving range and charging mode dimensions. At present, there are 192 analysis indicators (Fig. 1). This paper mainly analyzes 11 indicators of charging behavior data, deeply mines user behavior and charging scenarios, and uses data resources to optimize the design of charging system.

Fig. 1. Analysis section of charging big data system

2.2 Step Introduction

Firstly, the system filters, cleans and processes the original data of the platform, and provides data fragments including alarm (fault) fragments, charging fragments and driving fragments. Then, according to the specific analysis and statistics of fault distribution, charging data visualization and big data distribution, the system develops the data architecture and data processing flow, and designs the visualization scheme of charts and reports. The specific process is divided into three parts: data screening, data processing and visualization display. The login interface of big data visualization analysis system for new energy vehicle charging and multi analysis results display.

3 User Charging Behavior Data Analysis Report

The charging big data in the system is accurately identified, and the charging rules and influencing factors are analyzed from the dimensions of user group, charging mode, vehicle type, region and temperature. Through the multi perspective analysis of user charging behavior data and vehicle strategy performance data, we found the optimizable items of charging related function definition, control strategy and product performance design, so as to realize the accurate research and development of charging system and reduce cost and increase efficiency.

3.1 Charging Mode

According to the analysis of charging modes for different vehicle uses and activities, about 82% of the charging times are AC charging, of which about 80% for taxis are DC charging; about 60% for high activity users who drive more than 100 km per day are DC charging, and about 90% for low activity users who drive less than 30 km per day are AC charging, It can be seen from the data that there is a great difference between the charging mode of transport users and private users, and the development of charging system for private passenger cars and operating cars should be designed according to user preferences.

3.2 Charging Start SOC

State of charge (SOC) at the beginning of charging vehicles in different regions and driving range Charge.soC) According to the analysis, the SOC distribution of charging start-up of users in different regions is similar, mainly distributed in the SOC of 20%–80%, of which the SOC of 35%–45% accounts for the largest proportion. The charging start-up SOC of users with long driving and large electric fuel (500 km, 70–90 kw) is mainly above 20%, It can be seen that users of long-term driving vehicles still have mileage anxiety. The SOC value of AC charging is higher than that of DC charging. It is speculated that the AC charging of users is mainly home charging, charging at any time. The SOC of charging start-up is basically the same at different temperatures, and charging start-up at higher SOC is not due to low temperature. The SOC of high activity users is lower than that of low activity users, which indicates that operating vehicles will start charging at a lower SOC than private users.

4   Analysis Conclusion and Enlightenment

The charging behavior preferences of different user types are quite different. Before designing the charging system, the product performance life design, function definition and strategy development should be carried out according to the actual usage habits of users in the target market [2]. Through the quantitative data of user's charging behavior, the targeted product is designed to avoid the user's complaint caused by insufficient design and waste caused by over counting.

4.1   Strategy Development of Charging System

Through the analysis of charging mode and "battery temperature data from charging start", different charging strategies should be designed for different users from the aspect of strategy development. For example, private users use less DC charging, and the battery system can properly improve the DC charging request current to shorten the time of fast charging; for operation users, the daily use of DC charging requires that the battery system properly reduce the charging current. To ensure the service life of the power battery; according to the statistical results of the battery temperature at the beginning of charging, the charging heating control strategy can be optimized, for example, when the battery temperature is 0 °C, the function of heating while charging can be started, so that the total time of heating and charging above 0 °C is the shortest, so as to ensure the use experience of most users. Because of the large difference between private car and operating vehicle users, the vehicle charger can be cancelled and portable DC charging equipment can be configured for the operating vehicles. In addition, the configuration of battery heater (PTC) can be determined based on the vehicle use area and user type. For example, due to the low DC charging utilization rate and high battery temperature, PTC can not be configured for private users in South China.

4.2   Design of Product Performance Life of Charging System

Through the analysis of "monthly charging times", "monthly charging time length" and "single charge amount data, the performance and life design of the product can be provided. For example, the product life and reliability of onboard charger (OBC) can be determined according to the AC charging time; based on the charging frequency data of different users, the plug-in life and reliability design of charging base can be determined, Ensure the safety performance of charging seat temperature rise; through the data statistics of DC charging times, 5000 plug and plug life tests with load can be conducted on the basis of standard life requirements of charging base, and the safety risk of relevant assemblies can be early warning according to big data statistics.

5 Mathematical Model of Hydraulic Components in EHB System

5.1 High Speed Solenoid Valve

By providing a certain driving voltage for the high-speed solenoid valve, when the solenoid valve coil is powered on, the corresponding electromagnetic suction will be generated, driving the armature to overcome the spring force, brake fluid resistance and friction resistance to move upward, thus opening the valve; when the solenoid valve coil is powered off, the armature will be driven to move downward under the action of the reset spring, thus closing the valve [3]. By analyzing the working principle of high-speed solenoid valve in EHB system, it is divided into three subsystems: circuit system, magnetic circuit system and mechanical system.

The magnetic circuit system can be obtained from Kirchhoff's law of magnetic circuit:

$$N_i = \Phi R \tag{1}$$

At the working air gap. Therefore, the formula for calculating the total magnetoresistance is as follows:

$$R_m = R_\delta + R_{mf} = \frac{\delta - x}{\mu_0 S_0} + \frac{l_m}{\mu_0 S} \tag{2}$$

It can be obtained from Maxwell's electromagnetic suction formula:

$$F_m = 2 \frac{\Phi^2}{2\mu_0 S_0} = \frac{\Phi^2}{\mu_0 S_0} \tag{3}$$

Combined with the above formula, the mathematical model of high speed switching solenoid valve can be obtained.

5.2 High Pressure Accumulator

In this paper, when analyzing the EHB hydraulic system, we first simplify it, and think that the EHB hydraulic system is composed of the basic hydraulic components analyzed in the above section through the connection of the hydraulic pipeline. Firstly, we establish the model of the single wheel hydraulic braking system. Because of the complexity of the hydraulic braking system, we first simplify it. (1) ignore the influence of the brake pipe on the brake fluid pressure and flow, All the pressure loss of the pipeline is accumulated into the brake pipeline model. Q) the influence of the instantaneous impact of the brake fluid on the system when the high-speed solenoid valve is switched is not considered. 3) the influence of the hydraulic pump on the system is ignored, and it is considered that one brake process can be satisfied with one hydraulic charge. 4) the elastic deformation of the brake wheel cylinder is ignored.

6 Research on Simulation Experiment of Vehicle Stability Control

6.1 Co Simulation Environment of CarSim and Simulink

CarSim, the co simulation environment of CarSim and Simulink, provides users with a parameterized interface. The type of simulation can be defined by setting the parameters of the whole vehicle, road surface parameters and simulation scene. The driver model can be used for open-loop or closed-loop control of the vehicle. Simulink is a software package of MATLAB, which is generally used for modeling, simulation and analysis of dynamic system. In order to make full use of the advantages of carin and Simulink software in various aspects, this paper selects carin and Simulink co simulation to build the vehicle stability control system, in which CarSim is used to build the vehicle dynamics model and Simulink is used to build the control system model. The diagram is the interface diagram of CarSim and Simulink, and the vehicle model of CarSim software will generate a DLL file, Simulink will regard DLL file as an s function, and choose to co simulate with Simulink model in CarSim interface, and select input and output parameters of CASM vehicle dynamics model, In Simulink, the S-function model representing CarSim vehicle dynamics model is connected with the built ideal two degree of freedom vehicle model, stability analysis module, stability control module and EHB system module to build the whole stability control system [4].

6.2 Simulation Experiment of Vehicle Stability Control Based on Predictive Control

The simulation research of vehicle stability control can not only shorten the development time of stability control system, but also reduce its development cost. Therefore, stability simulation experiment is an important part of the development of stability control system, and is the basic work of the development of stability control system. This experiment is carried out under the conditions of sharp turn, emergency avoidance and double lane change on the rainy road and dry road. Through the extraction and analysis of the experimental results, the rationality and effectiveness of predictive control for vehicle stability control are verified.

7 Concluding Remarks

Based on the national platform data, a high-quality sample database is established. Through the accurate identification and data mining of big data, the data mining and influencing factors analysis of charging rules are carried out from time, space, human and vehicle dimensions. The charging safety design is driven by data analysis. Combined with data resources and big data analysis technology, the rich value behind the data is mined, In order to improve the current safety problems of xinnengyou automobile and promote the development of xinnengyou automobile industry. This paper is the embodiment of the data information flow and the acceleration of the development of electrification, networking, temporary energy and sharing in the R & D field of new energy travel vehicles. It complements the offline data and provides strong data support for the optimization design of charging system and charging safety design.

References

1. Liu, Y.: New energy vehicles have entered the society as adults and have to pass the consumer experience. Sci. Technol. Daily **23**(002), 467–475 (2021)
2. Zhang, J.: Developing intelligent vehicle operating system to overcome "neck jam" technology. Gov. Procurement Inf. **44**(007), 231–243 (2021)
3. Zhang, F.: Market value of new energy vehicles is unique. Chin. Brands **56**(03), 54–55 (2021)
4. Zhang, F.: Is the market value of new energy vehicles a flash in the pan. Chin. Brands **65**(03), 56–57 (2021)

Application of Machine Learning in Blood Pressure Prediction and Arrhythmia Classification

Jingsong Luo[✉], Tingting Deng, Haojie Zhang, and Sliu Bozhang

Chengdu University of Traditional Chinese Medicine, Chengdu, Sichuan, China

Abstract. Blood pressure and ECG are two important indicators of human health in clinical detection. Wrist sphygmomanometer Measurement and ECG observation are the conventional diagnostic methods. The process of diagnosis depends on professional medical staff. The controllability of weight measurement time is poor, which easily aggravates the shortage of medical resources. Therefore, this paper uses machine learning method to classify blood pressure prediction and arrhythmia abnormality, and proposes a blood pressure prediction model based on support vector machine and an adaptive improvement algorithm based classification model of arrhythmia.

Keywords: Machine learning · Blood pressure prediction · Arrhythmia classification · Support vector machine regression

1 Introduction

The traditional shallow neural network can approach any nonlinear function theoretically only when the number of hidden layer elements is very large. Increasing the number of hidden layer elements will bring a series of calculation problems. Their improved versions solve the problem of network structure adjustment and optimization through kernel method and intelligent algorithm respectively [1], and make the shallow neural network to the extreme, and achieve good results for specific patients or ECG arrhythmia recognition with small sample size. Although the traditional neural network and support vector machine methods have made many achievements, due to the rapid onset of heart disease, ECG interference, individual differences and other characteristics, ECG signal automatic recognition still has the following problems: (1) the effect of feature extraction directly affects the classification results of the classifier; (2) shallow neural network classifier for big data trainingThe efficiency is not high, nonlinear fitting ability is limited, accuracy is limited; (3) many methods in the standard data set (such as MIT data set) verification effect is good, clinical data set effect drops a lot, lack of clinical data verification. (4) At present, ECG remote monitoring equipment mainly analyzes ECG simply, extracts heart rate, St interval and other information according to the traditional feature extraction method, and gives rough suggestions. It does not realize the clinical application of automatic diagnosis of a variety of arrhythmias. Deep learning integrates feature extraction and classification, which avoids part of the

J. C. Hung et al. (Eds.): FC 2021, LNEE 827, pp. 1239–1246, 2022.
https://doi.org/10.1007/978-981-16-8052-6_175

deviation in the process of feature extraction. In particular, there are lead shedding and common noise interference in clinical ECG. Establishing the mapping relationship between ECG morphological characteristics and diseases means fitting a highly complex nonlinear decision function. With the increase of other fields, but also been proposed for ECG disease diagnosis. Compared with the traditional shallow neural network, convolution neural network, cyclic neural network and deep confidence network have stronger nonlinear fitting ability, and have better recognition effect on single lead, partial data sets and few types of ECG heart beat classification. Although the deepening of the network is conducive to improving the recognition rate of the network, with the stacking of network layers, the convergence and optimization of deep network is still a thorny problem. Deep neural network model suitable for large-scale ECG data training. In this chapter [2, 3], a real-time ECG diagnosis system suitable for remote out of hospital monitoring is proposed, and the core algorithms of ECG signal filtering, processing and intelligent diagnosis are studied and verified by experiments. The scheme combines wavelet adaptive filtering and designs deep residual network algorithm for ECG arrhythmia diagnosis. Firstly, the ECG signal is filtered by real-time wavelet adaptive filtering to remove baseline drift, EMG noise and other interference signals, and then the heart beat is segmented automatically. Then, a deep residual neural network model suitable for large-scale ECG data training and testing is constructed by using residual block, and a deep residual neural network model suitable for large-scale ECG data training and testing is constructed by using residual block local depth neural network structure unit, which alleviates, and adds batch standardization in network designThe algorithm ensures the smooth convergence of the network. Finally, large-scale ECG data were used to verify. The algorithm is suitable for wearable intelligent monitoring and diagnosis scenarios, as shown in Fig. 1.

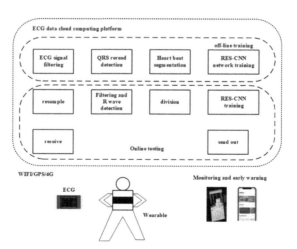

Fig. 1. ECG remote diagnosis scheme based on deep residual network

2 Deep Convolution Neural Network

Deep learning is a kind of deep network structure simulating human brain. It achieves the cognition of information by selecting and sampling the received information layer by layer. Convolutional computer network is a special deep feedforward computer network, which is inspired by the concept of "receptive field" in the field of biological computer science. It provides an end-to-end training model. The trained network can learn the features in the image or signal and complete the feature extraction and classification. Convolution neural network makes the extracted features more distinguishable by convolution transform of data layer by layer. The structure adopts local connection and weight sharing, which has good translation, scaling and distortion invariance. The research content of this chapter is to classify and identify ECG signals. The input is multi lead ECG heart beat signal, which can be regarded as the image data of one channel input [4].

2.1 The Structure of Convolutional Neural Network

The traditional convolutional neural network structure is composed of input layer, multiple alternating convolution and pooling layers, full connection layer and output layer. The simple diagram is shown in Fig. 2.

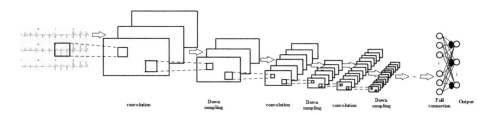

convolution   Down sampling   convolution   Down sampling   convolution   Down sampling   Full connection   Output

Fig. 2. Convolution neural network structure

Sent to the next layer of neurons connected with it as input. Assuming that the input (net activation) of channel J of the convolution layer is, and the output is represented by, then.

$$\begin{cases} o_j^l = f(z_j^l) \\ z_j^l = \sum_{i \in M_j} x_i^{j-1} * k_{ij}^l + b_j^i \end{cases} \tag{1}$$

$$\begin{cases} o_j^l = f(z_j^{l+1}) \\ z_j^l = a_j^{l+1} pool(x_j^l) + b_j^{i+1} \end{cases} \tag{2}$$

After several alternating convolution of the m-th neuron:

$$\begin{cases} o_j^m = f(z_j^m) \\ z_j^m = w_j^m x_j^{m-1} + b_j^m \end{cases} \qquad (3)$$

2.2   Training Algorithm of Convolutional Neural Network

The goal is to estimate network parameters according to training samples and expected output, including convolution kernel parameters, down sampling network weight, full connection layer network weight W and bias of each layer. The idea of the algorithm is to calculate the effective error of each layer according to the output error function (loss function) of samples and the chain derivative Faber ij, and then update the network parameters such as kernel parameters to make the network output more close to the expected output (minimum loss function), that is, to enhance the nonlinear fitting ability of the network through layer by layer learning and parameter adjustment. Because the deep learning network generally uses the batch gradient descent method to train the network, the overall square error loss function of each sample can be defined as:

$$J(k, \alpha, w, b) = 1/2N \sum_{i=1}^{N} \|t_i - y_i\|^2 \qquad (4)$$

By using the chain rule of partial derivative, the error signal is transmitted forward layer by layer, and the sensitivity of each layer is obtained:

$$\delta^l = \frac{\partial J}{\partial z^l} \qquad (5)$$

Assuming that the convolution layer is the first layer, then the +1 layer is the pooling layer. Since the sensitivity of the first layer is related to the sensitivity and weight of the +1 layer and the derivative of the activation function of the current layer, the sensitivity of the first layer is expressed by the sensitivity of the +1 layer, and then the partial derivative of the error J to the convolution layer parameters (kernel parameters and bias terms) is calculated. Because the sensitivity of each pixel in the +1 layer is related to the n * N pixel block in the layer, in order to facilitate the calculation, it is necessary to upsample the feature map of the +1 layer to make it consistent with the size of the feature map of the layer, and then calculate the sensitivity. The formula is shown in (6), where "." Is the multiplication of elements

$$\delta_j^l = \alpha_j^{l+1} (f(z_j^l) \circ up(\delta_j^{l+1})) \qquad (6)$$

The operation example of upsample is shown in Fig. 3. With the sensitivity of each pixel, the partial derivative of error to convolution layer parameters can be obtained according to the chain derivative rule and formula (1). For the bias term, because the bias term contributes to each pixel, according to the polynomial derivation rule, the derivative of the bias term of each feature graph in this layer is the sum of the sensitivities of all pixels of the feature graph, and the formula is shown in (7).

$$\frac{\partial J}{\partial b_j} = \sum_{u,v} (\delta_j^l)_{u,v} \tag{7}$$

For the partial derivative of the error to the kernel parameter, according to formula (1), the derivative can be obtained in chain form

$$\frac{\partial J}{\partial k_{ij}^l} = \sum_{u,v} (\delta_j^l)_{u,v} (P_i^{l-1})_{u,v} \tag{8}$$

Among them, it is the element multiplied by element by element when calculating the output characteristic graph.

3   Deep Residual Network

Neural network can increase the depth of the network to improve the nonlinear fitting ability. However, with the increase of the number of layers and neurons, simple stacking network layers will cause gradient dissipation, network convergence and optimization. When constructing a deep network, the accuracy of the model will decrease with the deepening of the depth, which indicates that the accuracy of the network model has reached saturation. The phenomenon that the accuracy decreases with the deepening of the network is called network degradation, which indicates that it is difficult to solve the parameters of an approximate identity map by using the deeper nonlinear network layer. If the added layer can be constructed by identity mapping, the training error of a deepening model will be smaller than that of the original shallow network model. In fact, the training of neural network is to adjust a series of parameters to make the network more accurately fit a nonlinear Abstract mapping relationship. However, due to the complex structure of deep network and the existence of a large number of parameters, it is difficult to adjust the parameters to the optimal, that is, the implicit nonlinear.

Accuracy of the network model has reached saturation. The phenomenon that the accuracy decreases with the deepening of the network is called network degradation, which indicates that it is difficult to solve the parameters of an approximate identity map by using the deeper nonlinear network layer. If the added layer can be constructed by identity mapping, the training error of a deepening model will be smaller than that of the original shallow network model. In fact, the training of neural network is to adjust a series of parameters to make the network more accurately fit a nonlinear Abstract mapping relationship. However, due to the complex structure of deep network and the existence of a large number of parameters, it is difficult to adjust the parameters to the optimal, that is, the implicit nonlinear mapping relationship is difficult to optimize. Deep residual network is to seek another solution, as shown in Fig. 3, (x) is the implicit mapping relationship that you want to learn. Deep residual learning is to let multiple continuous stacked nonlinear computing layers (such as two-layer convolution) to fit the residual (x) = H (x) − X between the input data and the mapped output data (assuming that the input and output are of the same dimension), and this residual is more and more compellingNear 0 means that the closer the feature extracted from the network is to the original input. If you want to fit the identity mapping, just reset the weight to zero; if you want to fit the approximate identity mapping, through the parameters [5]. As shown in Fig. 3.

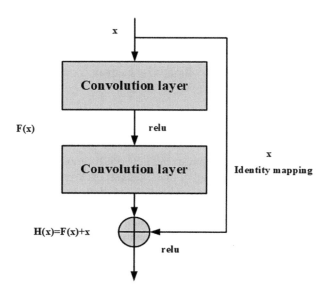

Fig. 3. Residual learning: stacking model

4   Recognition of Arrhythmia Based on Deep Residual Network

4.1   Experimental Data Analysis and Evaluation Criteria

The experimental data in this section are from MIT-BIH arrhythmia database, which has 48 records (from 47 patients, of which 201 and 202 records are from the same person), each record is 30 min, and the sampling rate is 360 Hz. Each record consists of two leads. The lead of each record is not exactly the same, only 40 records including the same II lead and VI lead, the actual application should consider the lead consistency, so this section of the experiment uses all the heart beats of the 40 records. In order to be more scientific and convenient for comparison, 15 types of arrhythmias were divided into five categories in strict accordance with the standards of the American Association for the promotion of medical devices (AAMI), namely normal (n, including normal heart beat, left and right bundle branch block, etc.), supraventricular extrasystole (s, including atrial premature beat, boundary premature beat, etc.), ventricular extrasystole (V, including ventricular premature contraction and ventricular escape), and fusion chamberThe detailed classification. The number of 2-lead heart beats in five categories was 94091. In addition, this experiment tests the recognition effect of the proposed method through the detection of ventricular premature beat (veb) and supraventricular premature beat (sveb). The performance of the classification model is measured by sensitivity (SE), overall accuracy (ACC) and specificity (SP), where accuracy is the proportion of heart beats correctly classified.

4.2   Data Preprocessing

The original data of MIT-BIH has been processed by removing the power frequency signal notch and 0.1–100 Hz band-pass filtering, and the sampling frequency is 360 Hz. The energy of the useful signal is mainly concentrated in 1–40 Hz. The adaptive wavelet filtering algorithm is adopted, which includes three steps: wavelet decomposition, adaptive denoising and reconstruction. Since the Symlet wavelet function family is similar to the shape of the egg signal, sym6 is selected as the wavelet function.

Deep learning can mine internal features from the signal and extract them automatically, so the signal preprocessing stage only needs simple filtering. On the one hand, it can enhance the generalization ability of the network and reduce the distortion of the signal as much as possible. Therefore, firstly, the 10 level decomposition of the egg signal is carried out, and the Mallat algorithm is used to calculate the detail coefficient and approximate coefficient. Baseline drift disturbance is caused by human breathing, movement and other factors. When the subject is still, the baseline drift is usually lower than 1 Hz. The approximate coefficients of the 9th and 10th layers are set to 0 to remove the baseline drift. For high-frequency noise, the high-frequency detail coefficients of the two highest frequency layers are set to 0 to remove the high-frequency components. Finally, the signal is reconstructed by inverse wavelet transform. After simple filtering, ECG data is segmented. As the MIT-BIH arrhythmia database provides the R-wave position of ECG signal recording, 120 and 229 points were intercepted from left to right, that is, 350 points were intercepted from each heart beat. In this experiment, 2-lead data were used, and each sample size was 2 * 350.

5   Conclusion

Based on the shallow neural network such as Gemini support vector machine, the classification of EGR arrhythmia with standard database and few martial arts athletes has achieved good results. Due to the strong interference of clinical big data, lead shedding and other problems, in order to overcome the limited nonlinear fitting ability of shallow neural network and the influence of feature extraction on the recognition effect of classifier, a recognition algorithm of ECG arrhythmia based on wavelet adaptive filtering and deep residual network for clinical big data is proposed. Wavelet adaptive de-noising algorithm can effectively remove baseline drift and other interference, and keep the shape of egg undistorted. A deep residual network model based on multiple residual blocks is designed. Based on the deep residuals network, the multi classification of ECG arrhythmia was verified by using the 100 000 order of magnitude ECG data in mie-bih database and clinical 2-lead ECG data. In the same experimental environment, several popular deep neural network algorithms are compared. The experimental results show that the proposed method can effectively improve the overall recognition accuracy, sensitivity and specificity of ECG arrhythmia signal. The results show that the classification rate of arrhythmia is 87.3885%, which has obvious reference and application value. A scheme of real-time diagnosis system for clinical big data arrhythmia is proposed. The scheme can collect the ECG signal through the wearable terminal and send it to the cloud computing platform for remote real-time transmission. After data transformation, wavelet adaptive de-noising, QRS complex recognition, segmentation and other preprocessing, the real-time diagnosis of arrhythmia is carried out through the deep residual network and returned to the client. This kind of arrhythmia automatic diagnosis method based on deep learning can be combined with wearable devices, Internet of things and wireless communication technology to further promote the development of new smart medicine, extend the prevention, monitoring and diagnosis of heart disease to the family, nursing home and other out of hospital scenes, make up for the weakness of medicine, provide efficient services for patients, and greatly save medical capitalSource.

Acknowledgements. Sichuan Provincial Department of Science and Technology's key project X-nurse - Nursing butler around disabled elderly people No. 2021YFS0152.

References

1. Ren, H.: Application of machine learning in human blood pressure prediction and classification of arrhythmia. Yanshan University (2019)
2. Liu, S.: Study on prediction model of blood pressure in patients with hypertension based on deep learning. Qingdao University of Science and Technology (2020)
3. Wei, Z.: Research on blood pressure prediction and analysis method based on human physiological data. Yanshan University (2019)
4. Zhao, J., He, Y., Li, X., Ren, R., Ren, J.: Prediction method of human blood pressure based on support vector regression. J. Yanshan Univ. **41**(05), 438–443 (2017)
5. Wang, S.: Research on classification algorithm of arrhythmia with different data domain distribution. Tianjin University (2018)

Automobile Emission Fault Diagnosis Method and Device Based on Fuzzy Reasoning and Self Learning

Cong Luo[⊠]

Bortala Polytechnic, Bole 833400, Xinjiang, China

Abstract. Heating furnace is a typical complex industrial controlled object, which has the characteristics of multivariable, time-varying, nonlinear, strong coupling, large inertia and pure lag. Moreover, it is difficult to accurately model and control the furnace temperature distribution because of the difficulty in measuring and many external disturbance factors. With the increasing process requirements, the traditional control methods can not meet the requirements of energy conservation and environmental protection. In this paper, combined with the characteristics of heating furnace, taking furnace temperature control as the research object, a self-tuning control method of P parameter control based on fuzzy logic is proposed.

Keywords: Heating furnace · Fuzzy control · Self-learning · Fuzzy PI controller

1 Introduction

In the process of rapid development, China has gradually improved the strength of the society and the quality of life of the people. Cars have gradually entered all levels of society and thousands of households. However, the pollution brought by cars is more and more serious. Now, environmental protection has become a very big problem we have to face. Among them, automobile exhaust pollution is a very important aspect in all kinds of pollution, so it is very necessary to pay attention to automobile exhaust emission at all levels of society. In all urban pollution, automobile exhaust emission accounts for more than 70%. Therefore, the control and governance ability of automobile exhaust emission needs to be solved immediately at all levels of today's society [1]. Moreover, China's automobile oil consumption accounts for more than 13% of the national oil consumption. At the same time, the number of automobiles is continuously increasing, and the total amount of automobile pollutant emissions is also increasing. The air pollution caused by automobile emissions has seriously affected people's life and physical and mental health. Therefore, we need to seek a new balance between the development of automobile industry and environmental protection, and we need to seek new policies to solve the problem of automobile exhaust pollution. China has issued a series of documents and adopted certain measures to improve the current situation according to the existing vehicle quantity, current road conditions and emission problems.

J. C. Hung et al. (Eds.): FC 2021, LNEE 827, pp. 1247–1252, 2022.
https://doi.org/10.1007/978-981-16-8052-6_176

2 Self Learning Method

Self learning, also known as adaptation, is a feature that organisms can change their habits to adapt to the new environment. Adaptive control is to modify its own characteristics to adapt to the changes of the dynamic characteristics of the object and disturbance. Model parameter self-learning is divided into short-term self-learning and long-term self-learning. The short-term self-learning is used to modify the parameters from plate to plate in the same batch number. The learned parameters automatically replace the original parameters and are used for the next piece of the same kind of rolled piece. Long term self-learning is used to modify the long-term parameters of the same rolled piece with different batch numbers. The learned parameters can selectively replace the original parameters.

2.1 Short Term Self-learning

In short-term self-learning, the self-learning coefficients of the nearest m steel plates are selected and weighted according to their heating time sequence. The average value of the weighted m steel plates is taken as the short-term self-learning coefficient of the current steel plate. The main process is as follows:

(1) Check the correctness of measured data;
(2) Select self-learning coefficient and carry out limit check;
(3) The parameter self-learning calculation is carried out according to the following formula:

$$A_{n+1} = \frac{a[m] \cdot A_n + a[m-1] \cdot A_{n-1} + \cdots + a[1] \cdot A_{n-m+1}}{\sum_{i=1}^{m} a[i]} \tag{1}$$

The specific steps of model parameter self-learning are as follows:

(1) Reliability test of measured value. That is, the upper and lower limits and consistency of the measured values required by self-learning are tested to ensure the reliability and availability of the measured values;
(2) Treatment of measured value. That is to say, some available indirect measured values can be calculated based on the measured values with some relations;
(3) Choose the gain of self-learning. In order to ensure the stability and convergence rate of self-learning parameters, the appropriate gain should be selected for different self-learning parameters and different heating and rolling conditions.

2.2 Data Analysis Based on Finite Difference Method

In general, the ratio of the increment of a continuous function f (x) to the increment of an independent variable is defined as a finite difference quotient. Obviously, when the increment of the independent variable approaches zero, the limit of the finite difference quotient is the derivative of the function [2]. In general, the finite difference quotient can be used as an approximation of the derivative, As shown in Fig. 1. that is, the finite difference quotient can be used instead of the derivative.

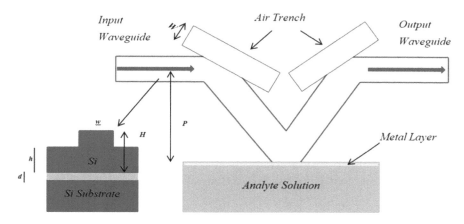

Fig. 1. Data analysis model of finite difference method

$$\frac{df}{dx} = \frac{f(x + \Delta x) - f(x)}{\Delta x} \tag{2}$$

When the finite difference method is used to solve the partial differential equation, the whole region of the object should be meshed first. Domain discretization only needs to get the approximate value of discrete points, not the effective solution in a region. Region discretization is to mesh the region of the research object to form nodes and control units. How to accurately describe the set shape of objects is the key to ensure the accuracy of calculation. In the finite difference method, the standard difference scheme is usually divided into orthogonal grids: one-dimensional problem is equivalent to line segment, two-dimensional problem is equivalent to rectangle, and one-dimensional problem is equivalent to cuboid. In this case, the elements are line segment, rectangle and cuboid. If the temperature gradient of each point in the region is very different, the grid should be more dense in the place where the temperature changes sharply, and sparse in the place where the temperature changes slowly. The value of the step size depends on the specific problems, such as the calculation accuracy, the stability of the difference equation and the calculation workload.

3 Parameter Auto Tuning Method of PI Controller Based on Fuzzy Reasoning

3.1 Overview of Fuzzy Control

The basic idea of its application is to first summarize a set of complete control rules according to the operator's manual experience, and then calculate the control quantity according to the current motion state of the system through fuzzy reasoning, fuzzy decision and other operations, so as to realize the control of the controlled object.

The idea of fuzzy control or fuzzy reasoning is often combined with other relatively mature control theories or methods to give full play to their respective advantages, so as to obtain the ideal control effect [3]. Because fuzzy rules and language are easy to be widely accepted, and fuzzy technology can be easily implemented in microprocessor and computer, this combination shows strong vitality and good effect. The improvement methods of fuzzy control can be roughly divided into fuzzy compound control, adaptive and self-learning fuzzy control.

3.2 Adaptive and Self Learning Fuzzy Control

Self tuning fuzzy controller: a self-tuning fuzzy controller that modifies the control rules. From the evaluation of response performance index, it uses the translation of fuzzy set or the change of membership function parameters to achieve partial or comprehensive correction of control rules, and it can also be adjusted by modifying the rule table or membership function itself, It includes the linguistic method of system model identification based on fuzzy set theory, the method of system fuzzy relation model identification based on reference fuzzy set, and the fuzzy rule model based on Io data, which is the basis of self-tuning controller design. Parameter self-adjusting fuzzy control: the fuzzy control of self-adjusting scale factor introduces the function of performance measurement and scale factor adjustment, changes the parameters of fuzzy controller online, greatly enhances the adaptability to environmental changes; PD self-adjusting control based on fuzzy reasoning, such as parameter self-adjusting fuzzy PI control, and similar P and PID control. Model reference adaptive fuzzy controller: the output of the fuzzy controller is modified by the deviation between the output of the reference model and the output of the system under control, including the scale factor, the strategy of defuzzification, the fuzzy control rules, etc. Fuzzy control with self-learning function: including a variety of fuzzy control methods with self-learning function and self optimizing fuzzy controller, which affect the performance of external disturbances or repetitive tasks. The key lies in the design of learning and optimization algorithm, especially to improve its speed and efficiency. Self organizing fuzzy controller: fuzzy model reference learning control which combines reference model with self-organizing control, adaptive hierarchical fuzzy control and other advanced self-organizing forms have great development potential [4].

3.3   The Combination of Fuzzy Control and Other Intelligent Control Methods

Although there are still many controversies on the concept and theory of fuzzy control, since the 1990s, due to the participation of many famous scholars in the world and the success in a large number of engineering applications, especially for complex systems that can not establish accurate mathematical models with classical and modern control theories, it has made remarkable achievements, which has led to more extensive and in-depth research, In fact, fuzzy control has been determined as an important branch of intelligent control.

Neural network can realize partial or total fuzzy logic control functions. The former uses neural network to realize fuzzy control rules or fuzzy reasoning, while the latter usually requires more than three layers of network. Adaptive neural network fuzzy control uses neural network learning function as model identification or controller directly; The methods of obtaining membership functions and reasoning rules based on fuzzy neural network and fuzzy neural network with fuzzy connection strength are all applied in control; the controller design method combining fuzzy system with genetic algorithm provides a more novel idea. In addition, the research of fuzzy predictive control, As shown in Fig. 2. fuzzy variable structure method, fuzzy system modeling and parameter identification, fuzzy pattern recognition and so on, also belong to the more advanced research direction.

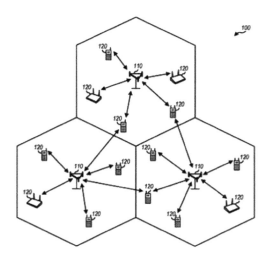

Fig. 2. Combination model of fuzzy control and other intelligent control methods

4   Conclusion

For the portable NOx detector equipment unified management, and real-time receiving detection instrument equipment upload detection data information for analysis and storage: can unified management of integrated handheld terminal equipment, authority control audit, and push the detector detection data analysis to the handheld terminal for display, data management platform according to the national standard city area code for regional division, It can realize the cooperation between different regions. Finally, the data platform realizes data collection and distribution, equipment comprehensive control and authority division. The data management system can combine the big data analysis thinking correlation principle to establish a mathematical model for data collection, find the law in the big data for subsequent data application, can give the evaluation results of environmental pollution, can give the prediction of vehicle growth rate, can give the prediction of future traffic conditions and so on.

References

1. Tan, G.: Application of portable time of flight mass spectrometer in on line detection of automobile exhaust. Acta Mass Spectrom. **37**(03), 193–2000 (2016)
2. Luo, H., Luo, P.: Plug in diesel electric hybrid vehicle rule control and processor in the loop test. China Mech. Eng. (2019)
3. Zhang, Z., Chen, H., Qu, J.: Multi component automobile exhaust quantitative detection system. Electron. Devices **40**(3), 773–779 (2017)
4. Xiao, Q.: Design of fermentation tail gas detection system based on Arduino. Jiangsu Agric. Sci. **44**(3), 452–454 (2016)

Application Analysis of Computer Network Security Technology Based on Network Security Maintenance

Xiang Ma[✉]

Xijing University, Xi'an 710123, China

Abstract. With the rapid development of information technology, computer network technology has a very obvious development, and gradually goes deep into each leading city of social development, creating a good condition for people's life. It can be said that computer network technology makes people's life more colorful, but the network security problems also interfere with netizens' friends, such as viruses and Trojans. Some lawless elements invade the computer through the network to search for personal information or even personal property. Starting from the computer network security, this paper discusses the hidden dangers of computer network security, and the application strategy of computer network security technology in security maintenance.

Keywords: Information technology · Network security technology · Computer

1 Introduction

With the development of computer network technology, people's life style has been changed to a great extent. It breaks the limitation of traditional communication in time and space, and makes information transmission and data acquisition more smooth. However, due to the network system itself has some problems, coupled with the existence of illegal elements, the information security of network users can not be effectively guaranteed. In this context, we should analyze many factors that affect the security of computer network, grasp the crux of threatening computer network security, and use computer network security maintenance technology to minimize all kinds of security risks in the use of computer, and create a green network technology security application environment.

Network security is a very concerned problem in the world. How to develop computer network security technology in the future has become a key research topic in all walks of life. Therefore, we should make clear the importance of network security maintenance and attach importance to the application and research of computer network security technology.

J. C. Hung et al. (Eds.): FC 2021, LNEE 827, pp. 1253–1258, 2022.
https://doi.org/10.1007/978-981-16-8052-6_177

2 Overview of Network Security Technology

The further analysis of the content of computer network security shows that it is mainly a man-made protection system, which can protect the information security in a certain network environment, so as to make users more reliable, secure and complete. We should pay attention to the security of computer network from three aspects: security, confidentiality and integrity. The first is security, which is mainly caused by hardware and software, such as software vulnerabilities, operating system vulnerabilities, and so on; security problems caused by human factors, such as artificial secrets, the slow use of viruses. Generally speaking, for the integrity of information, it means to ensure the integrity of information. As for information confidentiality, it is aimed at the user's private information to ensure that the information will not be stolen by others, so that users can use all kinds of information software more safely. In addition, for information security, it is mainly combined with the software and hardware of the computer to protect the user's information, so as to ensure that it will not be broken [1].

Continuously and effectively carried out. This system is to connect the communication line with the equipment, and then connect multiple sets of computers with multiple independent functions, which exist in different geographical locations. Through information exchange, network operating system, network communication protocol and other multi-functional network software, the information in the network can be resource efficient. The computer network mainly includes resource subsystem Network and communication subnet.

3 Common Factors Affecting Network Security

The current operating system, whether windows or Linux or vista system, has strong expansibility, which will provide opportunities for hacker attacks and the spread of computer viruses, and threaten network security. It is precisely because of such technical defects that the whole computer network system has problems. At present, the TCP / IP protocol is still commonly used on the Internet, this protocol does not fully consider the security of the network, it is precisely because people are particularly familiar with this protocol, so they can know the vulnerability of this protocol, so as to attack and invade the computer network. The following figure is the cloud diagram of network security intrusion. As shown in Fig. 1.

3.1 Openness of Computer Network

Because the computer network itself has the characteristics of wide area and openness, it is difficult to achieve the confidentiality of information. In addition to the quality of communication and network wiring and other factors, it has led to many security problems. As an international network, the Internet has increased the difficulty of network information protection [2].

For the computer, it must be in an open network environment to give full play to its functions. Therefore, in the process of the application of all kinds of software, outside criminals or software with virus may enter the system from the system's own

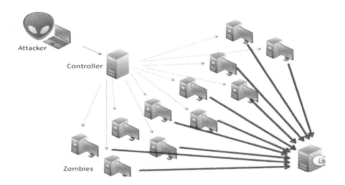

Fig. 1. Cloud picture of network security intrusion

vulnerability, which poses a threat to the information security of network users to a large extent. Moreover, because the overall speed of system development at this stage is relatively slow, it is not able to accurately identify some modern viruses, so that network users often let information be obtained by the outside world without knowing it.

3.2 Hacker Attack

With the continuous development of computer network system, hacker attack is one of the important reasons for the security problems of computer network operation, such as the leakage of computer storage information, the destruction of internal files, and the interception of transmission information. According to the destructive situation of hacker attacks, computer network security.

Hackers are also coming out. Hackers in order to maximize their own interests. Often with the help of information technology. Artificial invasion of some computer network users, to a large extent, the security and stability of the network system caused damage. Moreover, for this kind of personnel, it will not only cause great losses to network users, but also seriously hinder the long-term development and progress of network technology.

3.3 The Invasion of Viruses

Computer virus is highly infectious and destructive, usually developed by some lawless elements for profit. For the virus, it will invade the user's computer from the loopholes of the computer system, and has a long latency. After getting the instructions of the criminals, it will break the user's computer system, such as Trojan horse virus and script virus. Trojan horse virus makes users install certain programs by luring them to make their accounts threatened and steal their account passwords, which brings huge security risks to users. If it can not be processed in time, it will seriously affect the security and reliability of user information. Figure 2 shows the virus attack process [3].

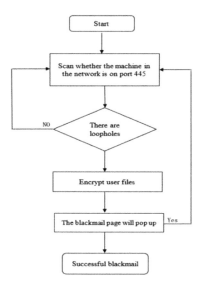

Fig. 2. Virus attack process

4   Computer Network Security Application Technology

In the face of the current computer network security problems, in order to better avoid, in the future development process, we must strengthen the application of computer network security technology [4].

(1) Firewall technology

At present, the most extensive computer network security protection technology is firewall, which mainly includes two kinds: packet filtering firewall and application level firewall. Packet filtering firewall scans and filters the data uploaded to the host by the router to effectively block the data. Firewall is not only a kind of software but also a kind of hardware, but also a kind of security component, Firewall is a perfect integration of software and hardware, which can isolate the internal network from the external network. When the corresponding network data is transmitted, the firewall can filter the data and information according to its own access configuration, so as to protect the network security and isolate the external intrusion So that the internal input is useful and valuable information.

(2) Data encryption technology

Data mining encryption technology and access rights technology. Data encryption technology is an important technical means to protect data. It uses secret key control to transform data. The encrypted data can only be viewed by the corresponding secret key decryption. It is a very important means of data protection for the government and enterprises. Its advantages lie in low cost, good effect, and various encryption methods. It is a very effective technology for data protection. Access right technology uses the authorization operation of network access right to ensure security, such as controlling the access right to a file according to one's own needs.

(3) Anti virus software prevention technology

The use of antivirus software can remove the virus and Trojan horse in the computer system to the maximum extent, and ensure the safe and stable operation of the computer. At present, there are many computer antivirus software for users to choose from, and users can choose and use them flexibly according to their own needs. It should be noted that although antivirus software can identify and remove potential viruses and Trojan horse programs in computer system, it can not completely remove some viruses with strong concealment and destructive. Therefore, users should not only pay attention to the selection of antivirus software with strong security function, but also regularly upgrade the function of antivirus software, so that the killing function of antivirus software can be continuously updated and improved. At the same time, users should develop the habit of using anti-virus software to detect computer system regularly to ensure the frequency and effect of anti-virus software.

(4) Intrusion Prevention Technology

First, the detection system of the system itself. For this system, it is mainly with the help of detection technology, effective logarithm detector and the user's local system machine to detect and improve the effect of information security. In the detection period, the accuracy is high, but there are some defects, which will lead to the phenomenon of missing inspection.

Second, network intrusion detection system. For this kind of system, first of all, it needs to collect the detected object machine, and then on the basis of comprehensive analysis of logarithm machine, it effectively studies all kinds of data about to enter the user's system. If there is a problem with the data, it needs to warn the user in time and let it process quickly. This technology has strong defense ability, and can achieve the purpose of real-time detection. At the same time. In the computer system. Virtual use of this technology. In addition to self-detection network outlier, it can also scientifically collect some data, and continuously optimize the detection behavior according to the collected data. Figure 3 shows the application layer of computer network security technology.

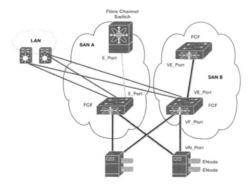

Fig. 3. Application layer of network security technology

5   Conclusion

Computer network is changing people's life, but network security has always affected the development and experience of network technology. Generally speaking, due to the rapid development of computer network technology, the network environment faced by users in the process of network activities is gradually complex, which makes the user's resource information face the risk of being destroyed and disclosed. Therefore, in order to effectively avoid such problems, in the future development process, we must strengthen the promotion of user's network security awareness. Effectively detect each configuration, scientifically use modern network security technology, in order to reduce unnecessary losses, relevant technical personnel should strengthen the maintenance awareness of network security, and do a good job in network security maintenance, Combined with anti-virus technology, firewall technology, encryption and access control technology and other technical means to protect the computer network security.

References

1. Xiao, C., Zhang, Y., Li, W.: Application of data encryption technology in computer network communication security. Sci. Technol. Wind (36), 71 (2019)
2. Wang, S.: Application of virtual private network technology in computer network security. Sci. Technol. Wind (35): 94+102 (2019)
3. Yue, S.: Application of computer network security technology in network security maintenance. Satell. TV Broadband Multimed. **17**(21), 137-1 (2019)
4. Xu, W.: Computer network security technology application analysis group based on network security maintenance. Comput. Program. Skills Maintenance **331**, 174–176 (2020)

Reservoir Data Modeling Technology Based on Well Seismic Information Fusion

GuoPing Pei[✉]

Daqing Oilfield Limited Company, No. 10 Oil Production Company,
Daqing 163000, China

Abstract. The research process of reservoir modeling is a multi-disciplinary comprehensive research process, which involves many data types and data, and needs the assistance of spatial data mining technology. Based on the process of reservoir modeling, this paper analyzes the application opportunity of teaching data mining in reservoir modeling, which is mainly used in the two stages of fine reservoir geological research, reservoir model optimization and subsequent use. According to the characteristics of commonly used spatial data mining methods, it is pointed out that statistical analysis method, spatial analysis method, three-dimensional visualization method and nonlinear method are more effective data mining methods in reservoir modeling. The spatial analysis method is applied in the reservoir fine geological research of Wenxi No. 1 block, and satisfactory results are obtained.

Keywords: Reservoir modeling · Data mining · Spatial analysis

1 Introduction

The process of reservoir modeling is a multi-disciplinary comprehensive research process, which involves many types of data, large amount of data, and complex data flow direction. If it only depends on the thinking of researchers, without the help of other auxiliary tools, it is not conducive to improve the efficiency of reservoir modeling, but also to deepen the reservoir geological knowledge obtained in the process of reservoir modeling. Therefore, it is very necessary to use λ spatial data mining in the process of reservoir modeling to provide auxiliary tools for each subject research. Spatial data mining, also known as data mining based on spatial database and a new branch of knowledge data mining, is to extract the general relationship between spatial patterns and non spatial data that users are interested in and other general data features hidden in the database. In short, spatial data mining refers to the process of extracting implicit spatial and non spatial patterns, general features, rules and knowledge from spatial numbers [1]. In short, the task of data mining in reservoir modeling is to discover knowledge from spatial data warehouse and provide geological knowledge constraints for reservoir modeling.

J. C. Hung et al. (Eds.): FC 2021, LNEE 827, pp. 1259–1265, 2022.
https://doi.org/10.1007/978-981-16-8052-6_178

2   Stage Characteristics of Data Mining Application in Reservoir Modeling

2.1   Main Analysis Contents of Reservoir Modeling

The reservoir modeling work starts with the analysis of the monoparallel sedimentary microfacies, then carries out the analysis of the profile microfacies, and then carries out the plane microfacies combination according to the characteristics of the profile microfacies and the sedimentary model, and so on, until a satisfactory reservoir conceptual model is established.

The data involved in single well sedimentary microfacies analysis include qualitative data, such as mudstone color, sedimentary structure characteristics, etc., as well as definite display data. Such as the physical parameters of the reservoir. In this stage, the possible knowledge obtained by data mining method is the correlation knowledge between sedimentary microfacies and parafacies, and the sedimentary knowledge of various rock types. After the relationship model between sedimentary microfacies and logging facies is established, the facies model of the established section is established in turn. In reservoir modeling, it is mainly the division of Danbi county. According to the sedimentary facies, the force ratio in Siju formation is calculated to ensure the accuracy of sand group and single sand layer calibrated by network curve. The research method of measuring and sedimentary facies is used for small layer correlation to ensure the accuracy of plane tracking in Danbi county.

2.2   Analysis Process

After obtaining the reservoir geological knowledge base, the reservoir geological model can be established. At this stage, it only needs the support of reservoir geological knowledge base, and does not need the help of data mining methods. Since then, because of the establishment of multiple equal probability reservoir models, model selection needs the support of data mining methods, in order to find out the candidate models that meet the reservoir geological knowledge and provide them for reservoir numerical simulation. At this time, the model can also be used to guide the development and production of oil and gas fields, such as the distribution range of high permeability zone, and so on, so data mining method is indispensable.

The establishment process of reservoir model should be said to be the quantitative process of reservoir geological research results, that is, the synthesis and generalization of geological knowledge [2]. This process is inseparable from the data mining method, but it does not always act on the whole process of reservoir modeling, but on the whole process of reservoir modeling.

Therefore, it is not necessary to establish a data mining subsystem in the development of reservoir modeling software. Instead, it is integrated into the process of reservoir modeling as a function (Fig. 1).

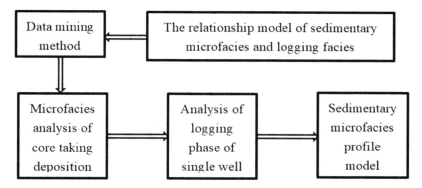

Fig. 1. Application opportunity of data mining method in reservoir modeling

3 Basic Knowledge of Information Fusion

3.1 Introduction

With the rapid development of computer technology, network technology, signal processing technology, control technology and so on, a large number of multi-sensor systems for various complex and changeable application background also appear, Because there is a huge amount of information, a variety of information representation, information diversification, the complex relationship between information and the more stringent requirements on the real-time, accuracy, integrity and reliability of information processing, With the help of computer technology, the information fusion technology of intelligent processing, estimation and decision-making of multi-source information under certain constraints has been developed rapidly [3]. More accurately, information fusion technology is emerging with the development of automatic command system and information processing. Information fusion involves computer technology, artificial intelligence, signal processing technology, communication technology, military technology, mathematical automatic control theory and other interdisciplinary.

3.2 Information Fusion Technology

The idea of information fusion is how to use multi-source information together to achieve a more accurate and objective understanding of things or goals. In reality, because the collection of original information is often incomplete, lost, disordered, and even wrong, only by fusing a large number of diverse information to make up for each other, can we get correct, relevant, and useful information. Animals in the biosphere, including human beings, are a relatively advanced and complete information fusion system. Ears, eyes, nose, hands and feet are multi-sensor systems, and the brain is the fusion center to coordinate these sensors to perceive all aspects of information of things, and gather these information into the brain to analyze, synthesize and judge by using the existing knowledge in the brain, In order to get a correct and comprehensive understanding of things.

The most important part of information fusion system is the information fusion algorithm, which is a mathematical method and a mathematical model to comprehensively analyze and process the information from multiple information sources in order to get more accurate results. At present, a large number of information fusion algorithms have emerged. This paper briefly introduces several commonly used information fusion algorithms.

$$A_i \cap A_j = i \tag{1}$$

For any event B, P (b) > 0, there is:

$$P(A_i|B) = \frac{P(A_iB)}{P(B)} = \frac{P(B_i|A)P(A_i)}{\sum\limits_{j=1}^{m} P(B|A_j)P(A_j)} \tag{2}$$

DS evidence theory is a generalization of Bayes theory, which was first proposed by Dempster in 1967 and promoted by Shafer in 1976. In the process of multi-source information fusion, the algorithm obtains data from each information source, and uses the obtained data to generate the probability of some characteristics of the environment. These probability values become the evidence of the method. Compared with Bayes theory, DS evidence theory can solve the uncertainty caused by "inaccuracy" and "ignorance", and its adaptability is stronger than Bayes theory.

Definition: let u be an identification frame, which is the basic probability assignment on u, and define the function n as:

$$\begin{cases} m(\varnothing)=0 \\ \sum\limits_{A \subset U} m(A)=1 \end{cases} \tag{3}$$

4 Research on Reservoir Data Modeling Based on Multi Well Data Interpolation

With the rapid development of oil exploration and exploitation technology, the increasing demand for oil, the gradual reduction of underground oil reserves, more and more complex reservoirs, the technical difficulty of oil development is also increasing. Researchers in the field of petroleum pay more and more attention to the accurate description of reservoir by using the existing data. In recent decades, Kriging has been widely used in the field of oil exploration, and has gradually become an important part of the field of oil exploration.

4.1 Kriging Interpolation

Using instruments and equipment to carry out field measurement, the data obtained after measurement needs to establish a digital map. When generating the digital map, the data need to be generated into DEM form of uniform grid (grid refers to some squares on the plane or cubes in three-dimensional space) terrain elevation (height) file, However, in three-dimensional space, the data obtained from these field measurements are usually irregular and scattered, so it is necessary to apply some relevant interpolation algorithms to obtain the height of each grid point based on the data obtained from field measurements. The traditional method of DEM Spatial interpolation is distance inverse proportion weighting, which is relatively simple and convenient to use in practice, but it ignores the correlation between data points in space, so the final calculation result is not accurate enough, which is not in line with the actual situation. Kriging interpolation algorithm does not have this shortcoming [4]. In DEM Spatial interpolation, height can be regarded as a regional variable parameter. Starting from the spatial variation law and variation characteristics of height, Kriging interpolation algorithm is used to solve the elevation value of each grid point.

4.2   Implementation and Result Analysis

The basic principle of least square method is to realize the approximation of the minimum error value by using the feature that the first partial derivative of the error surface at the extreme point is zero. This method is correct when the error surface has only one global minimum. However, when it comes to complex problems, there will be many local extremum points on the error surface. At this time, the least square method can not get the correct results. Because the least square method can not distinguish the local and global extremum, the least square method has the defect of falling into the local minimum. RPSO can make up for the defect of least square method, which has strong global optimization ability, can avoid local optimal solution in the process of calculation, and then make the formula of variation function designed have better fitting effect.

5   Research on Reservoir Data Modeling Based on Well Seismic Information Fusion

Aiming at the problem of reservoir data modeling, a well seismic information fusion model based on DS evidence theory is established. In order to solve the problem of wavelet extraction, a high-order cumulant seismic wavelet extraction method based on stochastic particle swarm optimization is proposed. The key technologies of seismic inversion, such as time depth conversion, reflection coefficient calculation and seismic record synthesis, are studied.

5.1   Introduction to Seismic and Logging

Artificial earthquake refers to the artificial use of instruments on the ground to send out earthquakes. The seismic wave generated belongs to elastic wave. What we use is the propagation principle of elastic wave in the stratum. When the elastic wave propagates underground, it will be reflected back to the ground when it meets the reflecting surface. The geophone on the ground receives the information reflected by the reflecting layer. The information received is related to the density, lithology, porosity and other properties of the formation. The information that can reflect the formation properties can be obtained by processing the received data information such as dynamic correction and static correction. This information is my own What we call seismic records for seismic inversion is also called seismic data or seismic data.

Seismic records have the characteristics of band limitation in frequency domain, strong nonlinear relationship with stratum properties, reflecting transverse and longitudinal information of stratum, close relationship with wave impedance of stratum and reflecting information in time domain.

Well logging is a way to detect the formation properties around the wellbore by means of instruments and equipment in the well head. There are many logging methods, including density logging, image information logging, acoustic logging and so on. Therefore, logging only reflects the formation property information near the well wall, but it can reflect a series of property information such as lithology, sound velocity, density and so on.

Logging records have the characteristics of wide frequency domain, strong non-linear relationship with formation properties, good reflection of formation longitudinal information, good reflection of wave impedance interface, diversity and complexity, and reflecting information in spatial domain.

5.2   Brief Introduction of Well Seismic Joint Inversion

Well seismic joint inversion has been widely used in the field of petroleum exploration. Seismic data information can not only describe the horizontal information of the formation, but also describe the vertical information, but the resolution of seismic data information is very low, especially in the vertical, but in the horizontal, seismic data information has a very strong tracking ability [5]. The lateral resolution of well logging data is very high upward, so it is put forward that the combined inversion of well logging and seismic data not only uses the characteristics of high vertical resolution of well logging, but also makes full use of the characteristics of strong lateral tracking ability of seismic data, and fully excavates the signal of well logging and seismic data.

6   Conclusion

Parallel CORBA technology is a successful example of the combination of distributed computing technology and parallel computing technology. It realizes the combination of the two technologies and gives full play to their respective advantages. Parallel CORBA is not the standard feature of CORBA, but OMG has set up a high-performance CORBA group, and it is believed that there will be a formal standard on the market soon. Parallel CORBA itself is still in continuous development, there are still many places to be improved.

References

1. Deng, L.: Application Research on logging intelligent interpretation and 3D geological modeling of coalbed methane reservoir. Hefei University of Technology (2020)
2. Xing, K.: Research on geostatistical inversion method based on information entropy and cuckoo algorithm. University of Electronic Science and Technology (2020)
3. Chen, B.: Application of multi-core learning in reservoir modeling. Xi'an University of Petroleum (2019)
4. Li, Z.: 3D geological modeling and visualization of superimposed gas bearing system. China University of Mining and Technology (2019)
5. Zhang, Z.: Improvement of sand body superposition pattern recognition method and its application in seismic driving modeling. Southwest Petroleum University (2019)

Research on Equity Financing Valuation Model of Growth Enterprises Under Data Mining Technology

Xueqiong Tan[(⊠)] and Xiaojuan Chen

Yunnan University of Business Management, Kunming, Yunnan, China

Abstract. Enterprise valuation is the core content of equity financing. Reasonable valuation helps to alleviate the conflict between investment and financing, and promote the enterprise to achieve leapfrog development. Based on the traditional valuation method of price to book ratio, this paper constructs a valuation model for growth enterprises to take equity financing for market development. This paper discusses the optimal valuation of an enterprise and its influencing factors under the game between investors and financiers. The valuation model in this paper establishes the game theory basis for the price to book ratio valuation method, and points out that the "most undervalued value" of an enterprise is not the "optimal valuation" of investors. It is found that there is a win-win cooperation space to ease the contradictions and conflicts between investment and financing parties. The optimal price to book ratio can form an incentive for enterprise managers with high growth, light assets or high industry price to book ratio, boost the rapid growth of enterprises and improve the return on investment of investors.

Keywords: Equity financing · Enterprise valuation · Game theory · Growing enterprises

1 Introduction

When there is a serious deviation between enterprise value evaluation and value creation, the contradiction between investment and financing parties is inevitable. Once the founder loses control due to the underestimation of the enterprise, the differences between the two sides in the enterprise development strategy will seriously hinder the growth of the enterprise. In 2009, it paid 80% of its equity to introduce Ping An's 80 million yuan equity investment (with an enterprise valuation of 100 million yuan), and its turnover achieved explosive growth. Since then, Wal Mart has taken a stake in No. 1 store for many times and completely controlled it. Because it is contrary to Wal Mart's business philosophy, the founder of No. 1 store chose to leave and pursue a new development direction. The No. 1 store, which lost its founder, was finally acquired by Jingdong (with an valuation of about 9.5 billion yuan). When Ping An's equity investment was introduced, store 1 was extremely undervalued. The lack of funds made Ping An reduce the valuation of store 1 to the greatest extent, forcing the founder to lose the control of the enterprise and restricting the development of the enterprise from

J. C. Hung et al. (Eds.): FC 2021, LNEE 827, pp. 1266–1270, 2022.
https://doi.org/10.1007/978-981-16-8052-6_179

the strategic level. Although Ping An and Wal Mart have made achievements in the equity investment of store 1. However, the practice of underpricing makes store 1 lose the opportunity to compete with Jingdong, and Ping An and Wal Mart lose more than they gain. Therefore, when the enterprise is facing development opportunities but in financial difficulties, investors should not take the opportunity to suppress the enterprise valuation, but should examine the value added brought by the enterprise growth from a long-term perspective, and give a reasonable valuation level to boost the rapid growth of the enterprise and achieve win-win cooperation [1].

2 Problem Description and Benchmark Model

In this paper, we consider a two-stage supply chain, in which the retailer's fixed assets is a, its own capital is η, facing the market demand d = A − B, the order quantity is Q, and the sales price of the product is; the wholesale price of the supplier is u, and the unit cost is C. before financing, the retailer's own capital can just meet the normal operation of the supply chain, that is, η = WQ, Retailers can make efforts to develop the market by level E, but due to the lack of their own funds, they can make equity financing from equity investors (VC/PE), and the amount of financing is B [2].

Before financing, the retailer and the supplier enter into a wholesale price contract. The supplier sets the wholesale price as the leader and the retailer sets the sales price as the follower to maximize their respective profits:

$$\pi R(q) = pD - wq = -\frac{1}{b}q^2 + (\frac{a}{b} - w)q \tag{1}$$

The supplier's profit function is as follows:

$$\pi R(w) = (w - c)q \tag{2}$$

The optimal wholesale price of the supplier is determined by Stackelberg game.

3 Valuation Model of Equity Financing

When retailers are faced with better market opportunities (for example, there are obvious unmet real needs in a large-scale market, or the whole industry is faced with good development prospects and market space), they can carry out market development by paying effort level E. market development expands the market scale, and the new market demand D (e) can be subdivided into basic demand Q* and demand growth brought by market development efforts. This paper focuses on the impact of enterprise value creation on enterprise valuation, and does not consider the response of consumers to the increase of products and the competitive behavior among multiple retailers

$$D(e) = q^* + \beta e \tag{3}$$

Because the retailer's own capital can only meet the operation demand before the market development, after the market development, the retailer chooses equity financing, and the financing amount is B(e). According to the capital needed for supply chain operation and the value it creates, we can make a reasonable decision. Too much financing will lead to some idle funds, which will reduce the shareholding ratio of retailers and their assets [3].

$$B(e) = w^* \beta e + \frac{1}{2} se^2 \tag{4}$$

If the retailer valuates according to the P/B method in the process of financing, the P/b value of the enterprise before financing.

In equity financing, the income obtained by Shuangfang financing does not cover the profit dividends generated by the enterprise through operation (in fact, in the process of rapid development of the enterprise, the profits of the enterprise are more used for the enterprise expansion in the next cycle, and the profit dividends are relatively less), but more through holding the equity of the enterprise, However, retailers are usually in a weak position in the process of equity financing, and the valuation of enterprises is dominated by investors (very few enterprises may be favored by many investment institutions, so they occupy a dominant position in the financing process. This paper does not consider the valuation of such enterprises for the time being). Equity investors usually pursue the maximization of investment return (that is, as financial investors rather than strategic investors), and do not pursue the control of the enterprise. Therefore, retailers still have the management right of the enterprise after equity financing (even if the investor's shareholding ratio exceeds 50%). Retailers can also maintain absolute control of the enterprise through concerted action agreement, voting right entrustment and dual ownership structure. Therefore, when retailers take equity financing for market development, investors maximize their return on investment by setting the optimal valuation level, and then retailers maximize their equity value by setting the optimal effort level.

For enterprises with high P/B ratio (when investors exit from IPO, P/B ratio of the industry can represent P/B ratio of exit to a large extent), high growth enterprises and asset light enterprises, investors will give them higher P/b valuation. This is because when the P/B ratio of the industry is high, investors' equity value will increase and get higher return on investment; Enterprises with higher growth potential can achieve rapid growth through equity financing, and the value-added of enterprises improves the return of investors. The price to book ratio of asset light enterprises should be higher than that of other asset heavy enterprises with the same conditions, otherwise it will reduce the enthusiasm of enterprise managers and hinder the development of enterprises. Investors will increase the price to book ratio to encourage enterprises to take equity financing to capture growth opportunities to achieve rapid development, so as to shorten the investment cycle and improve the return on investment.

4   Numerical Analysis

This section intends to analyze through numerical simulation: 1) the advantages of the valuation model compared with the traditional price to book ratio method; 2) the impact of investors' exit price to book ratio and the growth of enterprises on investment and financing decisions and operation decisions; 3) the impact of changes in fixed assets of retailers. Is it effective for enterprise management to increase fixed assets before financing in order to improve enterprise valuation? The traditional price to book ratio method predicts the current enterprise valuation level through the industry valuation level of listed enterprises. For non listed companies, the valuation level is about 40%–70% of the industry valuation level. For start-ups, the valuation level still needs to be based on this.

Compared with the traditional P/B method (dotted line in the figure), the valuation level given by the valuation model in this paper (solid line in the figure) is higher, which can encourage retailers to make more efforts to accelerate the growth of the enterprise, so as to improve the equity value of retailers and the return on investment of investors. However, when the retailer's growth or exit price to book ratio is high, the valuation model shows its advantages. For the high growth enterprises or enterprises in the overvalued industry, reasonable improvement of the valuation level will help to improve the enthusiasm of the enterprise management. Help enterprises to achieve leapfrog development, so as to achieve a win-win situation for both sides of investment and financing [4].

Before listing, in order to obtain listing qualification and higher valuation, enterprises usually take the initiative to divest some low-quality fixed assets. However, at the start-up stage, some entrepreneurs usually increase "useless fixed assets" in equity financing in order to obtain higher total valuation of enterprises and increase their shareholding ratio. This part of the increased fixed assets usually has little or no effect on the development of the enterprise. Because this kind of "useless" fixed assets can not substantially boost the development of the enterprise, equity investors usually ask retailers to divest this part of the assets or lower the valuation level of the enterprise after obtaining the information. As a result, the retailer's effort level, order quantity, financing amount, and the investor's return on investment and return on investment will not change, as shown in Table 2. However, in the process of equity financing, both sides of the investment and financing will benefit each other and fight each other, which will easily lead to the estrangement between the two sides and may have adverse effects on the normal development of the enterprise. On the contrary, it will waste the value of its fixed assets.

5   Conclusion

Enterprise valuation is an important part of equity financing, which is directly related to the investor's return on investment and the financing Party's control rights and other core interests:

1) The valuation model improves the traditional P/B method and gives the P/B method the theoretical basis of game theory. The valuation model considers the exit behavior of the investors who pursue the maximum return on investment, and carries out enterprise valuation based on the perspective of enterprise value creation and the game between investment and financing parties. Compared with the traditional P/B method, it has more theoretical significance, especially suitable for the valuation of enterprises with high growth or high P/B ratio of industries.

2) The "most undervalued value" of enterprises is not the "optimal valuation" of investors. Enterprises with high growth, light assets and high industry P/B ratio should be given a higher valuation level. Investors pay more attention to the industry (high P/B ratio) and the future development of enterprises (high growth). Improving the P/b valuation of such enterprises helps to encourage the management to make more efforts to develop the market. To accelerate the growth of enterprises and achieve win-win cooperation, we should not blindly lower the valuation of enterprises to seek temporary benefits.

3) Managers' increasing fixed assets before financing can improve their equity, but it will not have a positive impact on the growth of the enterprise. This practice may deepen the contradiction between the investment and financing sides, and may have a negative impact on the healthy and rapid growth of the enterprise. In addition, it will not increase the proportion of management's shareholding under other valuation methods, but will waste the value of new fixed assets.

Acknowledgement. Yunnan University of Business management, 2020 key professional curriculum construction project "Securities Investment", project number: 3011701362009.

References

1. Jinhui, Jin, X.: Based on PE/Pb of China's new third board information technology enterprise value evaluation. Bus. Res. **62**(2), 96–101 (2016)
2. Dong, P., Li, H., Pan, H.: Research on valuation method of venture capital enterprise based on P/E ratio model. Syst. Eng. Theory Pract **22**(6), 121–125 (2002)
3. Liu, P.: Comparison of A-share, red chip and American stock market pricing. Syst. Eng. **27** (1), 44–49 (2009)
4. Hu, X., Zhao, D., Kong, Y., et al.: Heterogeneous enterprises and empowerment of comparable companies (2018)

Detection of Abnormal Power Consumption Mode of Power Users Based on BP Neural Network

Ben Wang[✉], Qing Zhao, Bo Li, and Xiaosong Jing

Beijing China Power Feihua Communication Co., Ltd., Beijing 100000, China
wangben@sgitg.sgcc.com.cn

Abstract. In order to restrain the abnormal electricity consumption behavior in power transmission and distribution system, a new abnormal electricity consumption pattern detection model based on hybrid neural network of wavelet and long-term and short-term memory is proposed. Firstly, the abnormal power consumption simulation algorithm is proposed to generate the abnormal power consumption data sequence; secondly, the feature extraction network is constructed by using the long-term and short-term memory network to extract different sequence features from the power consumption data; finally, the pattern mapping network is constructed with the wavelet neural network as the core to realize the mapping from sequence features to power consumption patterns and complete the detection of abnormal power consumption patterns.

Keywords: Short term memory · Wavelet neural network · Anomaly detection

1 Introduction

In recent years, scholars at home and abroad have carried out extensive research on the problem of reducing non-technical losses in the field of abnormal power consumption detection. The fuzzy clustering method based on c-means is used to find the power users with similar elimination characteristics, and the fuzzy classification is carried out by using membership matrix and Euclidean distance to the cluster center. The experimental simulation results of the above detection model meet the detection requirements in the aspect of abnormal power consumption pattern classification, but there is a lack of theoretical basis in the aspect of feature extraction and threshold selection, and the detection accuracy of the model needs to be improved [1]. In this paper, a hybrid neural network model based on Wavelet and long-term and short-term memory is proposed to detect abnormal power consumption. The model is mainly composed of three parts.

J. C. Hung et al. (Eds.): FC 2021, LNEE 827, pp. 1271–1275, 2022.
https://doi.org/10.1007/978-981-16-8052-6_180

2   Related Network Structure

2.1   Recurrent Network

The current neural network (RNN) is a fully connected. Where: X is the system input; h is the hidden layer output, and the current time (t time) output h is obtained by the hidden layer input through the weight matrix and the activation function; 0 is the system output; L is the loss estimation; y is the real value given in the training set; U, V, W are the weight matrix, which is obtained by subsequent training. Through the CER smart metering project data set test, the proposed abnormal power consumption detection and higher Bayesian detection rate compared with the traditional network model.

2.2   LSTM Network Structure

The LSTM network model is an improvement of the deep RNN. By adding new unit states in the hidden layer for information transmission, the computing nodes are redesigned to effectively control the long-distance information, which can effectively avoid the abnormal growth of gradient values in the process of training the deep network. The LSTM network structure is shown in Fig. 1. A complete LSTM network consists of a memory unit for storing historical information, an input gate for controlling the input information at the current time, a forgetting gate for adjusting the input weight of historical information, and an output gate for controlling the output information at the current time. In order to solve the problem of manual classification, this paper proposes a one class SVM model based on the high frequency load measurement data in am system to establish the normal behavior pattern classifier of users [2]. By combining the anti stealing evaluation system with BP neural network, the anti stealing model is constructed.

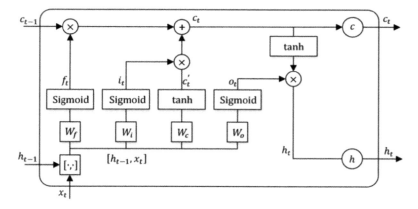

Fig. 1. LSTM structure

Current moment forgetting gate:

$$f_t = \sigma_f(W_f[h_{t-1}, x_t] + b_f) \tag{1}$$

Input gate at current time:

$$i_t = \sigma_i(W_i[h_{t-1}, x_t] + b_i) \tag{2}$$

$$C_t = \tanh(W_c[h_{t-1}, X_t] + b_c) \tag{3}$$

3   Abnormal Power Consumption Detection Model

3.1   Hybrid Neural Network Model

When mining the power consumption data of power users and identifying abnormal power consumption behaviors such as power stealing, the model needs to process the power consumption data of high latitude users into sequence features and map them to specific power consumption patterns. Therefore, analyzing the internal relations of power consumption data and extracting the characteristics of power consumption information are the premise of realizing abnormal power consumption detection, and these relations and characteristics can effectively reflect the characteristics of data. LSTM not only retains the ability of RNN model to extract deep level features, but also solves the problem of gradient disappearance in the process of deep network training. LSTM has obvious advantages in dealing with time series related problems. As shown in Fig. 2.WNN is widely used in pattern recognition and signal classification because of its fault tolerance, anti-interference and strong adaptability [3]. Therefore, a hybrid neural network is proposed to feature extraction pattern mapping in pattern detection,. The weekly data is filtered out by multi-layer feature extraction network, and then the extracted sequence features are input into the pattern mapping network to realize abnormal power consumption detection.

3.2   Abnormal Power Consumption Simulation Algorithm

The common abnormal power consumption mode is to attack the communication interface of smart meter, damage the hardware structure of smart meter or attack the communication network to tamper with the power consumption data. The power consumption data may be directly tampered with to zero or cut according to a certain proportion, or the total amount of power consumption data may be kept constant, and the power consumption curve may be adjusted according to the peak. In this paper, The algorithm gives the exact definition of the possible abnormal power consumption mode, there are six modes.

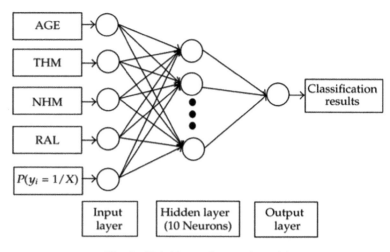

Fig. 2. Hybrid neural network model

In the algorithm, π represents the value of normal power consumption data at t, and π is an abnormal value. Considering the possibility of different abnormal power consumption modes, for the first and fifth abnormal power consumption modes, the probability is 0.1, and for the second, third, fourth and sixth abnormal power consumption modes, the probability is 0.2.

3.3   Abnormal Power Consumption Detection Model

The whole flow chart model proposed in this paper is shown in Fig. 6. First of all, according to the different data sets of power users, it can be divided into three groups: household power consumption, commercial power consumption and other power consumption. Each group uses the abnormal power consumption simulation algorithm to generate abnormal power consumption data. The training set and test set are divided according to the ratio of 7:3. Then, a hybrid neural network model based on WNN and LSTM is established, and the parameters of the network are initialized and optimized by Adam algorithm. Finally, the model after training to get the classification results and evaluate the effect.

4   Experimental Result

First of all, considering the influence factors of the proposed model, the overall performance of the algorithm is very sensitive to the number of LSTM layers in the feature extraction network. If the number of layers of the network is too small, the data abstraction ability is not enough. If the number of layers of the network is too many, it will cause over fitting [4]. At the same time, with the increase of the number of layers of the network, the time complexity of the model increases exponentially, Therefore, it is necessary to select an appropriate number of LSTM network layers. The layer number of LSTM is 1, 2, 3 and 4 respectively. The performance of the algorithm is shown in Fig. 10 and Fig. 11. Increasing the number of LSTM layers will increase the detection rate, but the increase will decrease, and the false detection rate will decrease. However, when the number of LSTM layers reaches, the false detection rate will slightly increase. algorithm is better when the number of LSTM layers in the feature extraction network is three. Support vector machine: according to the long-term power consumption of power users, support vector machine is used as a classifier to complete the classification of different power consumption patterns.

Hierarchical neural network: the hierarchical neural network structure is adopted, and the data processed by the classifier is taken as the input feature to realize the screening of illegal power users.

5   Concluding Remarks

In order to extract data features from high-dimensional data more effectively and solve the problem of feature matching, neural network on Wavelet and for abnormal detection. Through the CER smart metering projec data set experiment, it is verified that the detection model proposed in this paper has better effect than the traditional network model. In the follow-up work, we will study how to speed up the learning speed of deep learning model and reduce the time required for model generation.

References

1. Xiao, H.: Analysis of power situation in 2018 and prospect in 2019. China Prices **358**(2), 15–18 (2019)
2. Ge, Y.: Adhering to "two unswervingly" is a basic experience. Bingtuan Daily (Han), 18 October 2018
3. Jian, F., Cao, M., Wang, L., et al.: Research on power abnormal detection in AMI environment based on SVM. Electr. Measur. Instrum. (6), 64–69 (2014)
4. Hao, Q.: Accelerating the construction of modern industrial system. Learning Times, 4 December 2017

Design of Solar Cell Defect Detection System

Wang Jin[✉]

Urban Construction School of Xi'an Kedagaoxin University,
Xi'an 710109, Shaanxi, China

Abstract. Solar cell power generation has environmental protection and high efficiency, which is favored by many countries. At present, solar cell welding is mainly divided into manual welding, single welding and series welding. In this paper, the solar cell positioning and defect detection system is developed based on the solar cell welding machine and visionpro of Cognex company. The vision system is based on visionpro, using C # programming language, using the image toolkit in visionpro to process the image, to carry out visual positioning and defect detection of solar cells, and through communication with OMRON PLC and Epson robot, so as to realize the full automation of solar cell series welding machine.

Keywords: Solar cell · Machine vision · Visionpro · Location defect detection · Communication

1 Introduction

Solar automatic series welding machine is mainly used for automatic welding of solar cells, and the production capacity is required to be greater than or equal to 1200 pieces/h. The positioning accuracy of the battery directly affects the welding effect of the series welder, and the defect detection of the battery affects the subsequent detection process. In order to solve the problems of low speed, low accuracy and poor real-time performance of manual detection, a battery detection system based on machine vision is proposed to replace manual detection. The feature of machine vision system is to improve the flexibility and automation of production. In the process of large-scale industrial production, using machine vision location detection method can greatly improve the production efficiency and the degree of automation of production, and machine vision is easy to realize information integration, which is the basic technology to realize computer integrated manufacturing. Nowadays, there are some mature vision development software in the field of machine vision, which encapsulates many reliable and efficient algorithms and tools. The second development based on these softwares can not only ensure the stability of the system, but also shorten the development cycle. Visionpro, a machine vision software developed by Cognex company, is a PC based vision system software development package, which is mainly used in the field of complex machine vision. It integrates the tool library for locating, detecting, identifying and communicating. It can be developed by C #, VB and VC. This paper is based on visionpr <, uses C # language to locate and detect the defects of solar cells, and transmits the coordinate value of the center point of the solar cells and the environmental information of the appearance defects to the industrial manipulator,

J. C. Hung et al. (Eds.): FC 2021, LNEE 827, pp. 1276–1282, 2022.
https://doi.org/10.1007/978-981-16-8052-6_181

so as to realize the automation of the welding process. 1 overall structure design of vision positioning and defect detection system [1].

1.1 Visual Positioning and Defect Detection Content

It is 156 mm × 156 mm polycrystalline silicon cells that need to be positioned and tested. The positioning requirements for the battery is to accurately find the center point of the battery, get the coordinate x, y value and angle ankle value of the center point, and the coordinate error is within 0.05 mm, so as to ensure the accurate placement of the battery in series welding. The process of series welding is shown in Fig. 1. The detection requirement of the battery is that the defect area of any side of the four sides of the battery reaches 0.5 mm × 0.5 m. When the side length of any corner of the four corners reaches 0.3 ∼ the defect can be detected accurately. The processing time of the whole vision system is less than 300 ms [2].

Fig. 1. Process diagram of battery slice series welding

1.2 Hardware Composition of Vision System

The hardware structure of solar cell positioning and defect detection system is mainly composed of image acquisition equipment, execution equipment and industrial computer (image processing). In order to get the image of the object, the image acquisition hardware of the visual positioning and detection system is composed of image acquisition card, light source, lens and industrial camera. The camera signal is controlled by OMRON PLC and the actuator is Epson manipulator. The hardware composition of the system is shown in Fig. 2. This system mainly uses Germany Basile: industrial camera aea2500–14 gm, Japan Kowa megapixel industrial lens, Japan CCS strip light source, linear matrix backlight, tp link tg-3269 C adaptive PCI Ethernet card, 5-port Gigabit switch, Lenovo industrial computer, omroncp1h series PLC and Epson LS3 manipulator [3].

1.3 Vision System and Processing Flow

Industrial camera is connected with industrial computer through PCI Gigabit Ethernet network port to ensure the transmission speed and quality of photos. Om-ron PLC and Epson manipulator are connected with industrial computer Ethernet through Gigabit

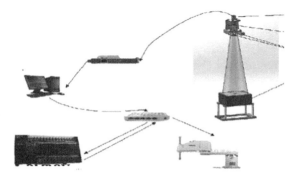

Fig. 2. System hardware composition diagram

switch. OMRON PLC controls the photography signal of the industrial camera. The collected image data is processed by the visual program in the industrial computer, and the calculated data is sent to the Epson manipulator to execute the action.

2   System Control Software Design

The development of visionpro visual program is generally divided into four development modes: (1) quickbuild visual + wizard generated operation interface; (2) quickbuild visual + modified operation interface; (3) quickbuild visual + custom operation interface; (4) custom application program. The difficulty of development from easy to difficult, from simple to complex, the requirement of programming is also higher and higher. This system adopts the third development mode. The steps are to write the visionpro program in quickbuild vision and save it as toolgroup. In C #. Net, we can call the toolgroup program through cogserilizer method. In this way, we can call each image toolkit freely in C #. Net, modify the parameters of each Toolkit, obtain image processing data, result output, and so on [4, 5].

The main function of this system is to locate and detect the defects of solar cells, process the image and transmit the data. It is mainly divided into three parts: image acquisition, image processing and data communication. Therefore, these three parts are mainly developed in C # program.

2.1   Image Acquisition Part

This system uses a CCD camera with Gigabit Ethernet interface. In visionpro, two kinds of CCD cameras are provided in C #Net, that is, tool level and operator level. This system creates a cogacqeifotool image acquisition tool through the tool level interface to collect images, and changes the attributes of the camera, such as exposure time, brightness and contrast, by using its operate attribute. Through this acquisition tool, we can get cogimage8 gray format images, which can provide effective image processing for other image tools.

2.2   Image Processing Part

The image processing of the system is mainly divided into two parts: the positioning of the battery and the defect detection. In most cases, the two visionpro programs are written in a tool group, but the program is cumbersome and difficult to understand; Therefore, it is considered to write location and defect detection in two tool groins, first to locate, and then to detect defects. At the same time, some parameters in defect detection are set based on the results of the location program, which makes the program concise and clear, and the function more clear.

Due to the processing technology and other factors of the battery, the printing effect of the battery may be poor, which makes the standard 156 mm × 156 mm square battery appear irregular shape. Through the experiment, it is found that there is a large deviation when using the intersection of the diagonal lines formed by the four vertices of the battery to locate. Considering that the main purpose of series welding is to cover the strip on the grid line of the cell, the grid line is used for positioning. The positioning schematic diagram is shown in Fig. 3.

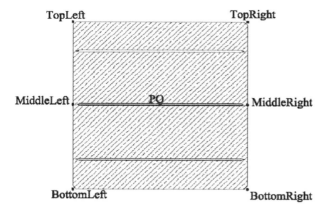

Fig. 3. Schematic diagram of solar cell positioning

First of all, using vinionprc, two calipe: toolkits in the software to get two segments of the left line, and then fit the left line of the battery, and get the other three edge lines of the battery. By using the Internet toolkit, we can make the edge lines intersect each other, and get four top lines: top left, bottom left, tool right and bottom right.

Point, we can get the rough positioning of the battery chip, and also make a good positioning foundation for the next defect detection. Then find the center line of the middle grid line through findline toolkit, and find the intersection point of the center line and the left and right sides by using Internet toolkit. The average value of Y is the center of the two cells. The angle value is obtained by the rotation angle between the center line and the standard X axis, which has a positive and negative relationship, and can be determined according to the needs in the later output. Usually, it is necessary to calibrate the camera coordinate system to the manipulator coordinate system, use the standard checkerboard provided by conice company to calibrate it, but there is such a correction function in the EPSON manipulator program, so the system adopts the correction method in the manipulator program, so that it can be debugged and changed later.

2.3   Defect Detection of Battery

The defect detection of the cell mainly includes the defect values of 4 edges and 4 chamfers. Its principle is to determine the starting point and ending point of the left defect detection toolkit findline according to the two vertices of top left and bottom left. As follows:

Left_Line. RunParams. ExpectedLineSegment.
StartX=TopLeft. X
Left_Line. RunParams. ExpectedLineSegment.
StartY=TopLeft. Y+, ornerSize+4. 0
Left_ Line. RunParams. ExpectedLineSegment. EndX
=BottomLeft. X
Left_Line. RunParams. ExpecteciLineSegment. EndY
=BottomLeft. Y-cornerSize-4. 0

Among them, eornersize is the chamfering size of the cell chip; 4.0 is the pixel value, which is the safety margin.

Furthermore, the processing results of findline toolkit can be analyzed, such as pointtodistance to determine the defect value defect y in Y direction, and the angle value of calmer toolkit can be determined by the angle value of the left line. Set a certain search length and projection length to determine the defect value of defect X in X direction. The settings are as follows:

Caliper Left. Region. CenterX=(TopLeft. X+BottomLeft. X)/2. 0
Caliper_Left. Region. CenterI'=(TopLeft. X+BottomLeft. Y)/2. 0
Caliper_Left. Region. SideXLength=1720
Caliper_Left. Region. SicieI'Length=20
Caliper_Left. Region. Rotation=tR

Among them, 1720 and 20 are pixel values; TR is the coordinate angle value obtained by positioning.

In this way, a defect with an area of defect x * defect y can be obtained. Due to the appearance of crystal spots and solder joints in the process of printing, the size of defect X and defect y can be set to ignore such spots, so as to reduce the error rate. In the same way, the defect values of the other three sides and four corners can be obtained, and the appearance defects of the cell can be determined by analyzing the defect values.

3   Data Communication

Data communication is an important part of the system, which undertakes the role of data exchange between visual software and the external environment. This system mainly has two communication parts: the communication between OMRON PLC and vision software, and the communication between Epson manipulator and vision software.

This system uses Omron CP1H series PLC to control the camera taking signal and the camera preparation and processing signal. Considering the transmission rate, FLC communicates with visual software through finstcp Ethernet mode, and the model of finstcp Ethernet communication protocol is shown in Fig. 4: first handshake is needed for finstcp communication, and then read and write data. After the handshake signal is successful, the camera acts as the client, carries on the network communication through the connect method in tcpclient class, and uses the networkstream class to carry on the network data transmission. As shown in Fig. 4.

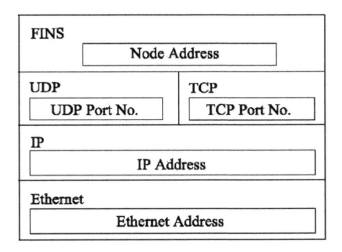

Fig. 4. Fins Ethernet communication protocol mode

Another part of this system is the communication between Epson manipulator and vision software. Because Epson supports the standard TCP/IP Ethernet communication, the system uses socket class to make the camera act as the client and the manipulator act as the server, and uses socket class and connect method to realize communication. Because socket class network communication data transmission is byte stream, using encoding. Utf8. Getbytes() method, the string of visual software processing results is converted into utf8 format byte stream to realize data transmission.

4   Conclusion

This system is based on visionpro, visual software, using C #. Net programming to develop a solar cell positioning and defect detection system, through the test proved that it can achieve high positioning and detection accuracy. Using the secondary development function of visionpro software, using the image toolkit, matching with ordinary industrial cameras, and configuring the communication function between PLC and manipulator, machine vision is applied to the production process of battery welding.

Acknowledgement. Natural science special project of Shaanxi Provincial Department of Education, Study on preparation process and photoelectric properties of solar cell ch3nh3pbi3. NO: 21JK0772.

1. References

1. Wang, Y.: Research on solar cell defect detection system based on machine vision. Jiangsu University (2019)
2. Zhujiang: Software design of solar cell defect detection system. Huazhong University of Science and Technology (2011)
3. Wang, Y.: Design of solar cell defect detection system based on DSP. Huazhong University of Science and Technology (2009)
4. Shan, Y., Cai, B., Wang, X.: Effect of cluster ion gun on ch_(3)NH_(3)PbI_(3) research on the influence of perovskite film [J/OL]. Analytical Laboratory 1–8 (2021). https://doi.org/10. 13595/j.cnki.issn1000-0720.2020
5. Li, W.: Case study on r real estate group's overall acquisition of W company's hotel assets. South China University of Technology (2020)

Information Management Analysis of Prefabricated Building Based on BIM in Construction Stage

Li Wang[✉]

TaiShan University, Tai'an 271000, China

Abstract. Under the theory of smart site, the combination of modern information technology and traditional assembly construction stage management process can effectively improve the assembly efficiency and reduce the difficulty of schedule management, cost management, safety management and quality management. Based on the description of radio frequency technology, virtual technology and BIM Technology Used in zhicao construction site, this paper explores the application of zhicao construction site information management platform, and puts forward a series of suggestions for optimizing information management, so as to coordinate various resources of construction site, and provide scientific reference for the design, production and Construction Research of prefabricated architecture.

Keywords: Intelligent construction site · Construction stage · Prefabricated building · Information management

1 Introduce

Aiming at the problems of low assembly efficiency of prefabricated structure and inadequate management in the construction period of prefabricated building, we can combine prefabricated building with engineering geography, and establish a management model of prefabricated building in the construction stage based on Zhiyi site theory, so as to promote the development and application of intelligent construction site information management in the construction stage of prefabricated building. Through the comprehensive application of advanced technologies such as BM, cloud computing, big data, Internet of things, RfD, mobile application and intelligent application, the construction site will get a more thorough perception, the interconnection will be more comprehensive, and the intellectualization will be more in-depth, so as to improve the efficiency of prefabricated construction, which is conducive to the real-time monitoring of the key elements of "human, machine, material, method and environment" in the construction site, Comprehensive and intelligent monitoring and management can effectively support the on-site personnel, project manager, coordination and management, accurately grasp the enterprise managers, formulate the construction schedule and work line, improve the accuracy and efficiency of prefabrication construction, reduce losses, and ensure the improvement of the information and digital management level of prefabricated buildings. In the process of project construction, the project

J. C. Hung et al. (Eds.): FC 2021, LNEE 827, pp. 1283–1289, 2022.
https://doi.org/10.1007/978-981-16-8052-6_182

progress can be tracked in real time according to the wisdom of the on-site quality information system, Timely discover potential safety hazards, standardize quality inspection and testing, ensure project quality, realize quality traceability and service system management, promote the establishment of big data integrity, effectively support the competent departments of the industry, and supervise the safety and quality of construction site, personnel quality and integrity and service.

2   Application of Smart Site Theory in Construction Projects

2.1   Radio Frequency Technology

Under the theory of intelligent construction site, bran free RF technology is the main technology to analyze and count the potential safety problems in the construction site of prefabricated buildings, and it is also the key to ensure the safety of the construction site staff. The technology control goal is mainly realized by three parts: handset, intelligent chip and information management platform. Before using the handset to sweep the energy chip of the assembly construction personnel, the management personnel can easily copy the anger of the construction site personnel [1]. In case of potential safety hazard, they can access the records in the handset, which is conducive to the implementation of the responsibility of answering the problem. In addition, the technical system also includes the operation film and explanation matching with the construction behavior, in which the stored records can also solve the hidden safety problems of the mold machine, so as to improve the safety degree of the construction site.

2.2   Virtual Technology

According to the actual situation of the construction site, 3D technology can be used to simulate the virtual field personnel in line with the real scene. After wearing VR equipment, they can feel the consequences caused by various safety problems. When experiencing, using objects to hit the head of the personnel can make the feeling more real, so as to guide the construction personnel to establish the awareness of safety construction, and lay a good foundation for reducing the incidence of safety accidents in prefabricated buildings.

Assuming that the UWSN network clustering algorithm, n nodes, clustering K cluster heads, the number of ordinary nodes in the cluster is NK − 1, then the energy EH of the cluster head node receiving and transmitting data is:

$$E_{CH} = lKP_0A(d_{toCH}) + l(N - K)P_r \tag{1}$$

The energy consumption E_{CM} of the member nodes in the cluster is as follows:

$$E_{CM} = l(N - K)P_rP_0A(d_{toCH}) + lP_rK \tag{2}$$

The total energy consumed by the network is etot:

$$E_{ToCH} = E_{CH} + E_{CM} \tag{3}$$

According to the formula (1), when the value of M is too large, it can make the energy consumption of inter cluster network uniform and reduce unnecessary waste; when the value of n is too large, it can reduce the energy consumption of intra cluster network and make the distribution of nodes more uniform; when the value of O is too large, it can avoid the phenomenon of energy hole and effectively prolong the network life cycle. Clustering protocol based on Improved Gray Wolf algorithm and clustering algorithm based on GWO, the transmission mechanism is periodic transmission, and each transmission cycle consists of two stages: cluster establishment and data transmission [2]. In the cluster establishment phase, the network clusters all sensor nodes unevenly; in the data transmission phase, the node is responsible for transmitting the data collected and received by the sensor. In each round of transmission, the base station is responsible for calculating the network clustering and node transmission direction. Finally, the base station needs to broadcast the information to all nodes.

2.3 BIM Technology

BIM Technology is a new technology that can break the two-dimensional management model and improve the three-dimensional model. When building the intelligent construction site of prefabricated building, BIM Technology is used to verify the three-dimensional visualization of the existing masonry facilities layout, which can further improve the rationality of the resources layout of the intelligent construction site, such as machinery, equipment and materials, and fundamentally guarantee the realization of the prefabricated building engineering goal. At the same time, the application of bin technology can also simulate and analyze the construction part with high risk. The analysis results show that it is reasonable and can be used as a safe construction scheme, which not only improves the efficiency of the design stage, ensures the construction progress, but also improves the safety and feasibility of the whole project.

3 Application of Intelligent Site Management Model in Prefabricated Construction Stage

3.1 Construction Personnel Management

Construction personnel management includes: (1) subcontract information management, management of personal status of subcontractors and implementation of construction tasks, so as to ensure that the management personnel can fully grasp the basic data of subcontract; (2) labor personnel information management. With the help of the intelligent site management platform, the input personnel information list and other information, the attendance service and personal basic information are provided, and the final project is closed management, so as to realize the real name management of personnel, which is conducive to the improvement of construction efficiency. (3) Daily management of personnel. Intelligent site management platform will continuously supervise and manage the behavior of construction personnel according to the daily behavior of construction personnel, attendance time, construction location information, etc. Accommodation management. As shown in Fig. 1. Comprehensively record the accommodation of construction personnel, and dynamically supervise the accommodation of personnel and the use of infrastructure.

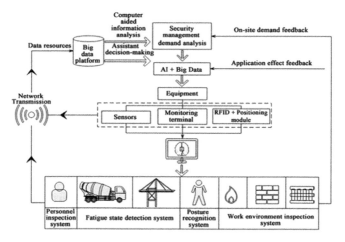

Fig. 1. Construction personnel management model

3.2   Engineering Material Management

Material management includes: (1) material purchasing management. The BIM platform is imported into the engineering materials management system, and after automatic identification by the system, the prefabricated construction materials list is generated, which provides comprehensive and complete materials information for relevant personnel to make procurement plans. (2) material management. The platform will take photos of the weighed materials and the original list, and the relevant personnel can guide the material purchase and transfer in data from the mobile app, so as to improve the efficiency of engineering material inventory and lay a good foundation for improving the quality management level. (3) Material use management [3]. According to the quantity, time and use of materials recorded in the intelligent construction site management platform, the management personnel can comprehensively control the use of materials, so as to pave the way for data management at the end of construction.

3.3   Intelligent Construction Site Information Management

The level and function of the model in the construction site of Zhilan, the perception layer is the foundation. Using Internet information technology in the terminal layer of construction project can realize management informatization, effectively improve the supervision of site construction, and provide comprehensive information guarantee for the whole system. It is a bridge for classified calculation of the data of construction projects, and provides support for flat data. In addition, all managers and operators can communicate and communicate on the platform, and realize the transmission and sharing of information. The application layer is the core, which reverses the management process of Kezi usage and the site management of project management module to make the information management platform meet the standard requirements, so as to integrate all mobile application devices into the terminal layer.

4 Suggestions on Strengthening Smart Construction Site Management of Prefabricated Buildings in Construction Stage

4.1 Information Management Platform and Innovation of Production Technology

The development of science and technology is an important driving force for the development of various leading industries, and also an important guarantee for promoting the construction of prefabricated buildings. In order to further improve the application level of Zhihui site management platform and the innovation of production technology, first of all, universities and enterprises are encouraged to cooperate to give corresponding preferential policies for the lack of data, backward software, stable structure and node in the production of prefabricated components, so as to create an ecology of the wide application of prefabricated building intelligent site theory. Secondly, actively looking for production units with high capacity to solve the problems of construction technology and technology. The design units are encouraged to use innovative thinking to find the construction scheme suitable for the smart building of prefabricated buildings, implement Jinxiang standard, and constantly optimize the production process, so as to produce the component model with high precision and strong applicability, and ensure the smooth development of the construction of prefabricated buildings. Finally, we should strengthen the exploration of BIM information management, conduct BIM academic conferences, expert lectures, exhibitions, etc. in the whole country, strengthen the publicity of the application of smart site theory in prefabricated construction, give full play to the influence of various media activities, so as to create a good atmosphere to support the application and development of smart site management platform for prefabricated construction in the whole society.

4.2 Strengthen the Training of Intelligent Site Management Talents

In addition, we can gain a strong sense of safety from the experience, and strengthen the self-improvement awareness of the management personnel. On the basis of recognizing the application advantages of Liuhui construction site, we can increase the amount of our own professional knowledge, so as to lay a good foundation for the application of related technologies, and provide a cold tolerance for strengthening the construction management, so as to create a prefabricated construction training base in terms of warm and less winter [3]. To provide a talent training platform for professional education in Colleges and universities, give full play to the resource advantages of colleges and universities, society and enterprises, and introduce BIM Technology. As shown in Fig. 2. Internet technology, big data technology into daily teaching activities, so as to lay the foundation for the enrichment of human resources in prefabricated construction, make the construction of intelligent site management platform more targeted, and create learning exhibition hall. By introducing large TV screen, VR technology and touch screen into Xihui exhibition hall, the assembly construction personnel can experience the environment, equipment, construction quality and safety in the construction stage.

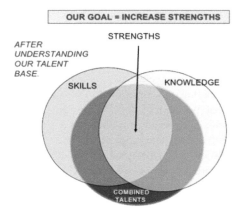

Fig. 2. Cultivation of intelligent field management talents

4.3 Security Management

Operator positioning supervision: combined with the prefabricated building information management mode of digital information platform and mobile Internet technology, a site personnel positioning management system suitable for site management application is developed. The personnel positioning system adopts ultra wideband (UWB) wireless position recognition technology. The helmet can transmit UWB signals, install electronic tags, and install a certain number of base stations in the designated area. By connecting the tag to the base station, the distance between the two is recognized by the system algorithm, and the position of the helmet tag can be obtained accurately. At the same time, the digital information platform can track the position of workers in real time and record the historical track of workers' activities. The digital information platform can immediately send a report to the management for employees who are not on the scene or whose position changes suddenly when they go to work.

5   Conclusion

To sum up, the acceleration of urban planning and construction puts forward higher requirements for many project management. Therefore, in order to further improve the efficiency of prefabricated construction projects, it is necessary to give full play to the advantages of intelligent construction technology, such as virtual, BIM, radio frequency and so on. At the same time, it is necessary to continuously learn the perception, network layer and application of the construction site system by using the Internet, cloud computing, wireless and other technologies, In order to promote the development of prefabricated construction towards intelligent, information and digital direction.

References

1. Liu, H., Yan, L.: Safety management of construction site based on "smart site". Intell. Build. Smart City (10), 90–91 + 99 (2020)
2. Zhang, Z., Sun, H., Liu, Y., et al.: Exploration and practice of prefabricated building information management. J. Eng. Manag. **32**(3), 47–52 (2018)
3. Qiu, T.: Research on Data Information Collaborative Management of Smart Construction Site. South China University of Technology, Guangzhou (2019)

Modeling and Simulation of Time Encroachment Propagation Dynamics Based on SIR Model

Tengfei Wang[✉]

Shandong University of Science and Technology, Qingdao 266590, Shandon, China

Abstract. Based on the basic SR epidemic model, this paper proposes a time encroachment propagation model considering the spontaneous infection rate and external organization sequence, and discusses the propagation of time encroachment in Er random network, NW small world network, WS small world network and BA scale-free network, On this basis, this paper focuses on the analysis of the factors influencing the spread of time encroachment on scale-free networks. It is found that (a) the propagation of time encroachment is affected by pressure and fairness coefficient. Within a certain range, the smaller the pressure is, the greater the probability of time encroachment is, and the faster the propagation process is, on the contrary, the greater the pressure is, The smaller the probability of time encroachment, the slower the propagation process. (B) the propagation of time encroachment is closely related to the probability of insertion and spontaneous infection. The larger the degree of initial node is, the faster the propagation is. On the contrary, the more rapid the propagation is.

Keywords: Behavior propagation · Complex network · Spontaneous infection rate · External communication

1 Introduction

With the rapid development of Internet and communication technology, the forms of time encroachment are becoming more and more diversified, the frequency of time encroachment is also increasing, and the negative impact on the organization is becoming more and more serious. This phenomenon has been widely concerned by all walks of life. In recent years, scholars at home and abroad have studied time encroachment from different perspectives.Most of these studies focus on the factors that induce the occurrence of time encroachment and the nature of time encroachment, but few scholars mention the spread of time encroachment and its path and effect, as well as how to inhibit the spread of this behavior. In fact, in a certain pressure and unfair environment, individuals will spontaneously produce time occupation behavior, and this behavior will spread in a certain space [1]. Once this behavior occurs, the whole group will follow suit, which greatly reduces the efficiency of employees. Therefore, it is of great theoretical and practical significance to study the spread of time encroachment and its prediction, management and control for the improvement of employees' work efficiency.

J. C. Hung et al. (Eds.): FC 2021, LNEE 827, pp. 1290–1294, 2022.
https://doi.org/10.1007/978-981-16-8052-6_183

2 The Mechanism and Model of Time Encroachment

Relevant studies show that the spread of time encroachment is mainly influenced by social impact theory and matching of effort theory. Among them, social influence theory mainly discusses the spread of time occupation behavior from the perspective of pressure, and holds that the work pressure caused by external situational factors is the key factor leading to the occurrence of time occupation behavior. The greater the individual's work pressure is, the higher the individual's effort will be, and the less likely the time occupation behavior will occur. On the contrary, due to the diffusion of responsibility, when the individual's own pressure is small, it is easy to produce social inertia, and the probability of time occupation behavior will increase. According to the effort comparison theory, individual's perception of fairness is an important factor affecting individual behavior. In organizations, when individuals find that the group work efficiency is not high or the free riding behavior is more common in the group, they are more likely to show time occupation behavior. In order to clarify the boundary of the problem, the following basic assumptions are made based on the above theory:

Hypothesis 1 in a group, affected by external environment such as pressure and fairness, individuals will spontaneously change from susceptible to infected individuals. Based on social influence theory and effort comparison theory, hypothesis 1 reasonably reflects the authenticity of the spread of time encroachment, It has a certain practical significance to assume [2]. In the process of time encroachment spreading, the infected individuals will be randomly transformed into immune individuals with a certain probability. After time encroachment occurs, the organization will take a series of measures, such as standardized time management, perfect incentive measures and strict punishment measures, to effectively inhibit and reduce employees' time encroachment, The infected individuals who are experiencing time encroachment behavior will stop the transmission behavior, no longer have an impact on the transmission process, and then become immune individuals.

Based on the above assumptions and analysis, the time encroachment behavior propagation model constructed in this paper is shown in Fig. 1.

Fig. 1. Time encroachment behavior propagation model

3 Time Encroachment Behavior Propagation Model of Complex Networks in Uniform Networks

The propagation rules of time encroachment in complex networks are described as follows: (a) the susceptible individuals become infected individuals with probability PS after contacting with the infected individuals; (b) the susceptible individuals spontaneously become infected individuals with probability a under the influence of external environment such as pressure and fairness; (c) the infected individuals become immune individuals with probability P.

$$a_1 = (K \times P_i + 1)^{ci-2} \qquad (1)$$

In a uniform network, the density of infected nodes, susceptible nodes and infected nodes are expressed by $i(t)$ and $s(t)$, respectively. Obviously.

$$i(t) + s(t) + r(t) = 1 \qquad (2)$$

4 Analogue Simulation

4.1 Related Network Topology and Characteristics

Related network topology and characteristics based on the above analysis, this paper selects 100 nodes to generate Er random network, NW small world network, WS small world network and BA scale-free network by pajek software combined with related algorithms. In the uniform network, the average degree of nodes k = 6, the probability of random edge connection P = 0.01. In the scale-free network, the initial node is 5, the number of random edges is 3, and the average degree of nodes k = 6.

The results show that the compactness of nodes in Er random graph is relatively small, and the arrangement of nodes is relatively regular, and the edges of nodes are relatively random. The compactness of nodes in NW small world network is relatively large, and the overall structure is relatively scattered, and the edges of nodes in WS small world network are relatively uniform, and the overall structure is relatively compact. However, in BA scale-free network, the edges of nodes are not uniform, Based on the above analysis, this paper selects Er random network, NW small world network, WS small world network and BA scale-free network as its propagation network, and discusses the propagation law of time occupation behavior in these four networks [3]. It is assumed that the edges of each network are undirected and have no right.

4.2 Simulation Analysis of Time Encroachment Behavior Propagation

Based on the above analysis, the largest node in each network is selected as the initial propagation node, and the other nodes are susceptible nodes. The susceptible node is assumed to become infected node with the probability of PS = 0.6, at the same time, it spontaneously becomes infected node with the probability of A1 = 0.5, and finally becomes immune node with the probability of PR = 1. The parameter A2 is taken as 0.8, the number of experiments is carried out 50 times, and the average value of the results is taken.

The density curves $s(t)$, (t), $R(t)$ of susceptible nodes, infected nodes and infected nodes in the above four networks with time. As can be seen from Fig. 4, after a given initial propagation node, as the total number of nodes in the network remains unchanged, the susceptible nodes continue to become infected nodes, and the susceptibility density $s(t)$ rapidly decreases to 0 in a short time, indicating that time encroachment behavior spreads very fast in social networks. All people can catch this behavior and spread it in a short time. The node density L (T) first increases, then decreases, and finally tends to 0. After reaching the vertex, the individual gradually loses interest in this kind of propagation, which causes the attenuation of the curve and finally decreases to 0. The density of infected nodes R (t) increased from 0 to 0, and all susceptible nodes became infected nodes.

4.3 An Analysis of the Factors Influencing the Spread of Time Encroachment

In order to further explore the propagation law of time encroachment and the related factors affecting its propagation, this paper selects BA scale-free network as an example, and analyzes the related factors influencing the propagation of time encroachment on BA scale-free network. Through the simulation, we can find that in scale-free BA network, the size of different initial node degree can affect the propagation of time encroachment behavior. The nodes with large initial node degree can make the time encroachment behavior spread quickly, and make the infection density of infected nodes reach the peak rapidly [4]. The above research shows that the topological structure of scale-free network is conducive to the rapid spread of time encroachment behavior, which is also consistent with the conclusion. In order to verify the propagation threshold of time encroachment in complex networks, this paper selects scale-free network as the propagation network to simulate the propagation of time encroachment under the conditions of propagation rate,It can be found that when the propagation rate is $\lambda = 0$, time encroachment will not propagate in the complex network; when the input is > 0 (in this paper $= 001$), time encroachment will propagate in the population, at this time, time encroachment will gradually spread in the whole population. The above simulation verifies the analysis of propagation threshold of time encroachment in complex network, and basically confirms the accuracy and feasibility of the model.

5 Concluding Remarks

Inter occupation is a kind of anti production behavior that takes up working time and engages in other activities. This kind of behavior will infect others in a certain space under the pressure theory and fairness theory, resulting in the waste of working time. In order to explore the transmission law of the above behaviors, this paper constructs a time encroachment behavior propagation model based on the basic SR propagation model, and simulates the above infection models in Er random network, NW small world network, WS small world network and BA scale-free network under different situations, Finally, the relevant factors affecting the spread of time encroachment are analyzed on scale-free network. Through the construction and Simulation of time encroachment propagation model, it is found that work pressure and fair working environment have a certain impact on the propagation of time encroachment. Therefore, creating a good working environment and establishing a fair management system play an important role in effectively reducing the occurrence of time encroachment, actively improving production efficiency and giving full play to the personal value of employees. In addition, due to the fact that there are many factors affecting the spread of time encroachment, this model does not fully consider them, so it has certain limitations. But this limitation also points out the research direction for the further improvement of the spread model of time encroachment and the research of other influencing factors in the future, Future research can also focus on changing the layout of nodes to achieve the goal of restraining the occurrence of time encroachment.

References

1. Wang, Y., Lin, X., Zhang, J.: Review on the research progress of time encroachment in organizations. Foreign Econ. Manag. **37**(9), 45–66 (2015)
2. Lu, J., Fu, X., Meng, Z.: Identification of influential nodes in microblog network. Comput. Appl. Res. **32**(8), 2305–2308 (2015)
3. Zhu, Y., Zhang, F., Qin, Z.: Spread of immunosuppressive virus. Comput. Appl. Res. **32**(5), 1496–1499 (2015)
4. Wu, Y., Du, Y., Chen, X., et al.: Research on reversal of internet public opinion based on newly exposed conflicting news. Acta Physica Sinica **65**(3), 630501 (2016)

A Study on Value Evaluation System of Engineering Consulting Based on VETS Process

Zhiwen Wei[1(✉)], Jiangshen Yu[2], Shunxiong Deng[3], Jingfeng Ou[3], and Kanghua Li[3]

[1] Guangdong Power Grid Co., Ltd., Dongguan Power Supply Bureau, Dongguan 523000, China
[2] Guangdong Power Grid Co., Ltd., Dongguan Power Supply Bureau Capital Construction Department, Dongguan 523000, China
[3] Guangdong Chengyu Engineering Consulting Supervision Co., Ltd., Dongguan 528200, China

Abstract. In the traditional construction mode, the design, construction, supervision and other units are responsible for different links and professions, which separates the internal links of the construction project. It is easy to break the information flow and "isolated island" of information, and it is difficult for the owners to get complete construction services. Based on the comparative analysis of different consulting service modes, combined with the characteristics of the whole process engineering consulting, this paper puts forward the whole process engineering consulting value evaluation system based on V ETS. According to the sequence of stage task event value, the value of the whole process engineering consulting event is generated, and the whole quantitative value of the construction project is formed by summarizing the value events task stage level by level, so as to provide reference for the owners to understand the whole process engineering consulting value.

Keywords: Construction project · Whole process engineering consultation · Value evaluation · Vets · Quantification

1 Introduction

The highly integrated service content can help the project achieve the goals of faster construction period, less risk, less investment and higher quality. It is also the embodiment of policy guidance and industry progress [1].

In February 2017, "cultivate a number of international level whole process engineering, and formulate technical standards models engineering consulting services"; in order to implement the above opinions, On April 26, 2017, the Ministry of housing and urban rural development pointed out in the main tasks of the 13th five year plan for the development of the construction industry that "the development quality of the engineering consulting service industry should be improved"; then, in May 2017, the Ministry of housing and urban rural development issued the notice on carrying out the

© The Author(s), under exclusive license to Springer Nature Singapore Pte Ltd. 2022
J. C. Hung et al. (Eds.): FC 2021, LNEE 827, pp. 1295–1299, 2022.
https://doi.org/10.1007/978-981-16-8052-6_184

whole process engineering consulting pilot work, by selecting qualified regions and enterprises to carry out the whole process engineering consulting pilot work, Improve the whole process of engineering consulting management system, cultivate internationally competitive enterprises.

The implementation of the whole process engineering consulting service is an important measure to deepen the organization and implementation reform of construction enterprises, improve the construction management level, increase the industry concentration and ensure the engineering quality. Investment benefit and standardization of construction market order are the best measures for China's survey enterprises in design, construction, supervision, business structure adjustment, transformation and upgrading planning, and strength improvement. How to evaluate the whole process of engineering consulting service quantitatively. The value index system has not been established, and the improvement trend of engineering consulting ability in the whole process of construction project is not obvious.

2   Analysis and Comparison of Engineering Construction Mode

2.1   Traditional Construction Mode

The traditional construction engineering consulting service adopts the "one to many" mode, and the design, supervision, consulting, cost and bidding are completed by different consulting service subjects entrusted by the owners. Objectives, plans and controls are mainly targeted at different consulting service subjects, which separates the internal relationship of the project by stages and parts, resulting in the poor information flow of the project and the isolated information island phenomenon, resulting in the lack of a unified planning and control system for the whole construction project, the difficulty of owner management, and the inability to get complete and effective consulting services [2].

2.2   Project Management + Professional Services

In order to reduce the difficulty of owner management and project risk, the "one to one to many" mode of project management + professional service consultation is favored by the owner of government invested projects. The owner entrusts a project management company to carry out the whole process project management on behalf of Party A, and is responsible to the owner for design, supervision, consultation, cost, bidding, etc. The disadvantage of this mode is that although the scope of management of the owner unit is narrowed, the service mode which is mainly on behalf of Party A with other single or multiple items still does not fundamentally solve the inherent defects of dispersion and separation between traditional construction modes, and party a still faces different "specialized but incomplete" consulting service subjects, The problem of information isolated island caused by stage and locality has not been solved effectively.

2.3   Whole Process Engineering Consultation

Compared with the former two, the "one-to-one" service mode of the whole process engineering consulting service (as shown in Fig. 1) enables the owner to complete the consulting service tasks originally completed by different consulting bodies such as design, supervision consulting, cost, bidding, etc. by entrusting a consulting unit with stronger strength. It should be noted that the whole process of Engineering Consulting is not a simple superposition of the original consulting, but a reengineering of consulting business [3].

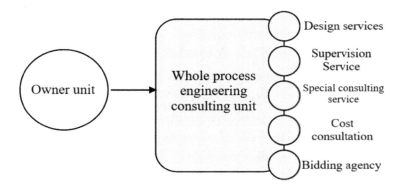

Fig. 1. "One to one" mode of whole process engineering consulting service

In this mode, the main task of project management is to carry out the target control of the project in the project implementation stage, that is, investment control, progress control and quality control. Compared with project management, the target control of whole process engineering consultation has three characteristics.

(1) It changes the angle of target control, from simple project target control to project value-added management with wider vision;
(2) The span of target control is extended from the project implementation stage to the project decision-making stage, and then to the project operation stage;
(3) The dimension of target control is added, from the investment, progress and quality control in the implementation stage to the function of meeting the end users in the operation stage, reducing the operation cost of the project, being conducive to environmental protection, energy saving and project maintenance.

3   Evaluation Method of Consulting Value Based on VETS

3.1   Value Evaluation Model of Consulting Based on VETS

Essence is to evaluate the value, which is the core of the model; value is reflected through events, which is the carrier of value; event is the specific embodiment of a task, and the task is defined and stipulated in the contract terms, which is a kind of requirement and agreement; for the task is often phased, and its value is often reflected in phases [4]. The task, event and its value can be better defined by stage division.

3.2 Value Estimation Path

Consultation of project includes early decision-making, design, construction, operation and other stages. The BIM contract of each stage includes different tasks. In the specific application process of BIM, each BIM task is completed by different combination of application items. From t value of the application item is calculated quantitatively. On this basis, the quantitative value of a construction project is formed through the value event task stage upward summary layer by layer.

3.3 Value Composition of Whole Process Consultation

For the whole process engineering consultation is also so, how to stand in the owner's perspective, the value of the whole process of construction project consulting evaluation is the problem this paper hopes to solve. According to the literature survey and expert investigation, compared with the traditional consultation mode of "one to many" and "one to one to many", the value of the whole process engineering consultation is mainly reflected in the following four aspects:

1) Cost saving (including design cost, construction cost, bidding cost, contract management cost and operation cost, etc.);
2) Shorten the construction period (including shortening the owner's decision-making time, shortening the design cycle, reducing the bidding time, reducing the construction site coordination time, reducing the project rework, reducing the number of information requests, reducing the conflict between design and construction, optimizing the process arrangement, etc.);
3) Improve quality (including reducing design errors, optimizing design scheme, optimizing specialty connection, improving construction quality, improving project delivery quality, etc.);
4) Improve management efficiency (including reducing management omissions and defects, promoting communication and coordination among all participants, refining management process, simplifying contract relationship, improving cross stage data transmission efficiency, helping to form data assets, improving operation and maintenance efficiency, etc.).

3.4 Value Estimation Matrix

Starting from the four value dimensions of owner project management, i.e. saving investment, shortening construction period, improving quality and improving management, according to the value estimation path, each value dimension is decomposed in the order of stage task event value, and the value of each whole process engineering consulting event is estimated, and then it is reversely summarized layer by layer through value event task stage, Form the quantitative value of a construction project.

4 Empirical Analysis of the Factors Influencing the Core Competitiveness

Whole process engineering consulting service enterprises, identify the development trend of enterprises in the future, and put forward corresponding suggestions on this basis. However, it is rather vague to study its development trend directly, so this paper selects the method of case analysis, selects engineering consulting enterprises, and explores the development of enterprises in the whole industry with the influencing factors of core competitiveness of engineering consulting enterprises. When selecting enterprises, we should pay attention to two problems: the types of consulting enterprises should be comprehensive, such as design enterprises and supervision enterprises; the second is that the selected consulting enterprises should be representative.

5 Conclusion

Construction project includes several stages, such as early decision-making, design, construction and operation. The specific tasks of each stage are usually specified in the contract. In the implementation process of the whole process engineering consultation, each whole process engineering consultation task is completed by different event combinations. Starting from the whole process engineering consulting event, the value of the event is calculated quantitatively. On this basis, the quantitative value of the whole process engineering consulting of a construction project is formed by summarizing the value event task stage level by level. The evaluation method has the following characteristics.comprehensive. The whole process of engineering consultation of all construction projects can be evaluated by this method.dynamic. No matter what stage the construction project is in, this method can be used to evaluate. According to the evaluation results, the application short board can be dynamically improved in the follow-up implementation process to maximize the whole process engineering consulting value.

References

1. Zhu, J.: Research on performance evaluation index system and decision analysis of government investment projects. Hefei University of Technology, Hefei (2016)
2. Wei, Y.: Research on comprehensive benefit evaluation method of UHV power grid based on artificial intelligence optimization vague set. Shaanxi Electr. Power (7), 65–69 (2016)
3. Li, L.: BIM Technology to Promote Industrial Project Management Performance Evaluation and Improvement Research. Tianjin University of Technology, Tianjin (2018)
4. Zhang, X.: Research on Performance Management of Manufacturing Enterprise Project Group Based on Dynamic Balanced Scorecard. University of Electronic Science and Technology, Chengdu (2010)

Tourism Route Optimization Based on Clustering and Ant Colony Algorithm

Mingyu Wu$^{(\boxtimes)}$ and Jing Sun

Weifang Engineering Vocational College, Weifang 262500, China

Abstract. The mobile tourism path planning in this paper includes two aspects: obstacle avoidance path planning and TSP path planning. Obstacle avoidance path planning is to find an optimal path from the starting point to the target point to avoid all obstacles in its workspace according to one or some optimization criteria. Tsp path planning problem is to know the distance between several cities, an existing salesman must visit these cities, and each city can only visit once, and finally must return to the departure city, how to arrange the order of his visit to these cities, so that the total length of his travel route is the shortest.

Keywords: Path planning · Genetic algorithm · Traveling salesman problem

1 Introduction

The planning problem originates from the traditional research topic of computer geometry. In the mid-1970s, the demand of Intelligent Tourism Research promoted the research of planning. Because assembly tourism and mobile tourism all involve path planning technology. In the early 1980s, Schwartz moved piano, t.lozano Perez position and posture space method, and r.a.brooks generalized cone method, which made the research of planning problem further developed. He thinks that the expression of environment will affect the search of the path. Khatib and irough proposed artificial potential field method take into account the dynamic performance and direct control of tourism. With the development of the research, the pose space method and artificial potential field method have been widely used and developed into two mature planning methods, and on this basis, many similar methods have been produced. Due to the complexity of path planning, different researchers study some aspects of the problem from different perspectives, and the specific issues are not exactly the same. The typical path planning problem refers to how to find the movement path from the starting point to the end point for tourism in the working environment with obstacles, so that tourism can pass through all obstacles safely and without collision during the process of sports. The research of path planning involves environmental expression, planning methods and path execution. There are two meanings in environmental expression, one is to obtain environmental information effectively, which is related to vision/sensor; the other is how to express environmental information effectively. The planning method is concerned with the effective method to plan the path and optimize the path based on the environmental expression. The execution of the path is closely related to the bottom

J. C. Hung et al. (Eds.): FC 2021, LNEE 827, pp. 1300–1306, 2022.
https://doi.org/10.1007/978-981-16-8052-6_185

control, and considering the dynamic characteristics of tourism, the tourism is controlled to travel according to the set path.

As for the difficult problems of TSP path planning, what we know now is that: any np. complete problem can not be solved by known polynomial algorithm; the study of TSP problem with polynomial algorithm started in the 1950s, such as linear programming algorithm, dynamic programming algorithm, branch and bound method. Although these algorithms can be solved accurately for some small-scale TSP path planning problems, the complexity of calculation increases exponentially with the number of cities n. On the large scale problems, these precise algorithms are powerless and easy to produce combination explosion. Therefore, people retreat to seek the so-called heuristic algorithm (approximate algorithm) which is not perfect to deal with various practical problems. The 1970s and 1980s are the heuristics of the heuristics, but because most of the heuristic algorithms are running according to certain certain certain search rules, the possibility of improving the solution is very small. Therefore, from the late 1980s to the 1990s, some new generation of solutions from other disciplines have emerged, and have achieved considerable success and a series of achievements in the solution of combinatorial optimization problems. For example, Hopfield and tank proposed to solve TSP by neural network association memory technology. In 1983 Kirkpatrick and so on, firstly, the simulated annealing algorithm derived from solid physics was proposed to solve complex combinatorial optimization problems, grenfensentette and others proposed genetic algorithm from biology to solve TSP path planning problem in 1987, and ant algorithm which originated from insect kingdom in recent years. Many scholars in China study the TSP path planning of Chinese cities, which are basically solved by Hopfield neural network [1].

Although TSP still has not found the optimal solution, the algorithm for solving TSP is improved gradually. In 2000, MALIANG summarized and summarized the algorithms of TSP path planning from 1954 to 1996, which can be divided into two types: one is accurate algorithm such as linear programming, dynamic programming, branch and bound method, and the other is approximate algorithm (heuristic algorithm), such as insertion calculation The algorithm includes r-opt algorithm, neural network algorithm, simulated annealing algorithm, genetic algorithm, ant algorithm, etc. At present, there are considerable progress in solving TSP path planning problems at home and abroad. In 2000, Zhou Peide solved the problem of Chinese cargo load with 31 vertices by polynomial time algorithm of convex shell of point set. In 1998, Dan gusfile of University of California, USA, proposed a new TSP with degree and degree of 2 by using green algorithm according to the characteristic that there is a Hamilton path in digraph g of both the output and the entry degree. In 2000, Vladimir gldeineko proved that maxtsp (the Hamilton circle of maximum weight traversing all cities) is np. when distance meets demidenko matrix, it has been proved that TSP is polynomial time solvable under this condition. Xu Zhihong and others use ant algorithm and simulated annealing algorithm to solve large-scale TSP, compare the performance of the two algorithms, analyze the reasons for the performance difference, and put forward suggestions for improvement. Jiang Tai and others used genetic algorithm with constraint optimization to solve TSP, and discussed the application of penalty method in genetic algorithm according to the constraints in actual TSP.

2 Related Work

The traditional optimization methods of obstacle avoidance path planning (such as gradient method, feasible direction method, etc.) use gradient information to find the optimal value of the objective function, which is suitable for solving the optimization problems with continuous functional relationship, but can not solve the discrete variable optimization problems that can not obtain gradient information. Genetic algorithm and simulated annealing algorithm are effective algorithms to solve global optimization problems. Both of them are stochastic methods to solve large-scale optimization problems by simulating some phenomena in nature. They do not require the continuity and differentiability of the objective function. Moreover, they are simple and easy to implement. Although genetic algorithm can find the global optimal solution in a random way in the sense of probability, it may also cause some problems in the process of practical application. Among these problems, the main ones are poor local optimization ability, slow convergence speed, easy to fall into local extremum and so on. On the other hand, simulated annealing algorithm has strong ability of local search and getting rid of local optimum. Therefore, the combination of genetic algorithm and simulated annealing algorithm is an effective way to solve the above problems.

2.1   Genetic Algorithm

Genetic algorithms (GA) is a relatively complete theoretical method formed by John Holland, University of Michigan, USA, from the late 1960s to the early 1970s. The paper attempts to explain the complex adaptation process of organisms in natural systems and simulate the mechanism of biological evolution to construct the artificial system model. After more than 20 years of development, great achievements have been made in application and theoretical research. Especially the evolutionary computing upsurge in recent years, computational intelligence has been an important research direction of artificial intelligence research and later artificial intelligence With the rise of life research, genetic algorithm has been widely concerned [2]. As a modeling method with high performance computing with system optimization, adaptation and learning, genetic algorithm has become more and more mature.

The flow chart of genetic algorithm is shown in Fig. 1, which can understand how genetic algorithm expresses the idea of biological evolution.

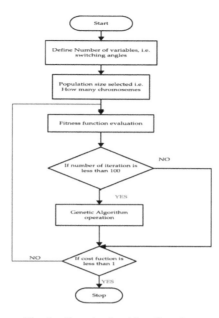

Fig. 1. Genetic algorithm flowchart

3   Application of Genetic Algorithm in TSP Path Planning

Traveling salesman problem a typical combinatorial optimization path planning problem. The number of possible paths increases exponentially with the number of cities n, so it is difficult to find the optimal solution accurately. Therefore, it is of great theoretical significance to find an effective approximate algorithm. On the other hand, many practical problems, after simplified processing, can be transformed into traveling salesman problem, so the research on solving methods of traveling salesman problem has important application value. When the number of cities n is large, the so-called "combinatorial explosion" problem will occur. The TSP path planning problem has been widely used to evaluate the performance of different genetic operations and selection mechanisms. The reasons are as follows:

Because of its typicality, TSP path planning problem has become a standard for the introduction and comparison of various heuristic search and optimization algorithms.

Traveling salesman problem (TSP), also known as the freight man problem, has two kinds of formulation. A simple description of TSP path planning is: if a businessman wants to sell goods in n cities, the distance between I and j of each two cities is DIJ. How to choose a path to make the businessman walk in each city and return to the starting point, the shortest path will be taken. The mathematical model is to find a Hamilton cycle GH on the weighted graph, such that:

$$W(C_h) = \sum_{e \in C_h} w(e) = \min\{eachHamilton\} \tag{1}$$

The classic definition of TSP path planning is narrower than the actual problem, because in fact, the traveling agent can travel on a certain distance, which often saves the total cost of the whole tour (mostly expressed by distance). In addition, if there is a TSP as shown in Fig. 2, the solution can not be found with the above classical definition, but in practice, the traveling salesman can still choose an optimal route, only vertex 3 has to be visited twice.

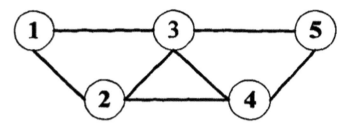

Fig. 2. Schematize diagram of TSP

So another formulation of the traveling salesman is that the businessman starts from a certain city, traverses each city at least once, returns to the original starting point, and designs a route to minimize the total distance. In this case, the mathematical model is to find a generating circuit C on the weighted graph G (V, e). Theoretically speaking, these two expressions are quite difficult [3].

From the problem to the type of graph, TSP path planning usually has two basic classifications: (1) the paths between any two cities are equal or may not be equal; it can be attributed to undirected graph or directed graph problem. (2) There is a path between any two cities, or at least a single path between two cities; this type can be attributed to the problem of complete graph or non complete graph. By combining the above two cases, the following four TSP path relationships can be obtained: (1) fully digraph; (2) completely undirected graph; (3) incompletely digraph; (4) incompletely undirected graph. There are three types of TSP path planning from the strength of the constraints of the problem itself. The first type only gives the distance matrix to find the minimum loop without any restrictions; the second type requires that the distances satisfy the triangular inequality (for a traveling salesman problem, care about three cities, P, Q, R, if the travel cost from P to R is not lower than that from P to R The traveling salesman problem (TSP) satisfying triangle inequality has many excellent properties and can be easily calculated. On the other hand, it has been proved that symmetric traveling salesman problem satisfying triangle inequality is NP complete. For symmetric traveling salesman problem satisfying trigonometric inequality, the approximate solution can be obtained in polynomial time. The third kind is defined in the Euclidean plane TSP, namely Euclidean TSP, which uses the coordinates and Euclidean distance on the Euclidean plane to give the coordinates and distance between cities.

4 Simulation Analysis

In the practical application, the starting point of mobile tourism can be set as (87,7), and the other 29 points can be regarded as the points that mobile tourism path planning must pass through. Select data trial (called tsp30 problem), that is, dig 30. Take the population size m as 200, stop the evolutionary generation D, increase from 500 generations to test, the crossover probability is 0.7, the mutation probability is p [4]. The following simulation show the simulation results of generation 500 and generation 6000 respectively. Typical simulation results are shown in Figs. 3 and 4.

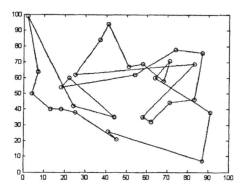

Fig. 3. 500 generations simulation diagram of basic genetic algorithm for TSP path planning

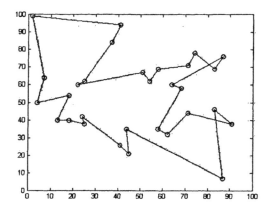

Fig. 4. 6000 generations Simulation diagram of basic genetic algorithm. m for TSP path planning

Only 11 of the 20 simulations performed by the basic genetic algorithm can reach the best solution, which takes 6000 generations and the total length of the final path is 451.5253. It can be seen that when the basic genetic algorithm is used to solve the TSP path planning problem, the success rate of outputting the best solution is low, the convergence speed is very slow, and it is difficult to achieve satisfactory planning results.

5 Conclusion

This paper introduces the basic principle and implementation method of genetic algorithm in solving TSP path planning problem. This paper analyzes and discusses some parameters that have an important impact on the performance of genetic algorithm, and uses the basic genetic algorithm to simulate the TSP path planning problem with population size of 200, analyzes and discusses the shortcomings of the basic genetic algorithm in solving TSP path planning, so as to lay a foundation for the improvement of the algorithm in the future.

References

1. Min, K., Ge, H., Zhang, Y., Liang, Y.: Hybrid algorithm based on ant colony and particle swarm optimization for TSP. J. Jilin Univ. (Inf. Sci. Ed.) (04), 402–405 (2006)
2. Fu, Z.: Research on open vehicle routing problem and its application. Central South University (2003)
3. Tang, L., Cheng, W., Zhang, Z., Zhong, B.: Research on vehicle routing simulation based on improved ant colony algorithm. Comput. Simul. (04), 262–264 (2007)
4. Du, R., Yao, G., Wu, Q.: A travel agent problem solving method based on ant colony optimization algorithm. Comput. Sci. (06), 158–160 (2006)

Application Research of Green Building Design System Based on BIM Technology

Xiong Na[⊠], Fangyan Yu, Li Fei, and Chen Liang

Nanchang Institute of Technology, Nanchang 330044, China

Abstract. At present, all countries in the world are faced with environmental problems such as shortage of resources, serious environmental pollution, global warming and so on. Now, the development of all countries in the world has begun to take environmental protection as the primary construction goal, and maintain ecological balance as the goal of the construction mode, so as to jointly improve the global environmental problems. The development of all enterprises should take maintaining ecological balance as the primary development goal, and the construction industry is no exception For the sake of harmonious coexistence between human and natural environment, the construction industry began to develop green buildings. BIM Technology was used to analyze all kinds of information in the whole life cycle of buildings, and a convenient platform was established for architectural design to collect all kinds of building information in time, make statistics in time, and analyze building data accurately. It is also suitable for the current development of green buildings in the construction industry New ideas of architectural design.

Keywords: BIM technology · Green building design · Application

1 Introduction

Architecture is a complex system, including building function (space) system, maintenance structure system, building equipment system, etc. Architectural design is also a complex process, architects must consider a series of complex and contradictory factors in the design process. The traditional architectural design process considers more about function, form and space, but less about the factors such as climate, maintenance structure, equipment system and so on. With the continuous development of green building, the traditional design method has been difficult to meet the requirements of green building. Therefore, it is necessary to put forward new design methods for green buildings.

BIM (building information modeling) originated from the construction industry. If you ask an architect, a building systems engineer, a building professional, or a civil engineer what BIM is, you'll hear the answer: it's a comprehensive process of digitally exploring its key physical and functional characteristics before a project is implemented [1]. They will point out that BIM helps deliver projects faster and more economically, while minimizing the impact on the environment and providing better working and living conditions for residents. In the whole process, we use coordinated and unified information to design innovative projects, better visualize and simulate the appearance, performance and cost of the real world, and create more accurate drawings. Although the application of Bim in China is still in the embryonic stage, it is an irresistible trend to recognize and develop Bim and realize the information transformation of the industry.

2   BIM Technology and its Significance in the Field of Architecture

2.1   BIM Technology

BIM is a concept, a technology. There are many concepts about BIM. Gu and London believe that BIM is a technical means based on Information Technology (IT), which can establish and maintain the complete electronic information model of each stage of the whole life cycle of engineering project. All building information in the whole life cycle of engineering project exists in the form of database, which can be geometric or non geometric. The national BIM standard of the United States defines it as: BIM Technology is the information expression of the dimensional characteristics and building functions of building products.

As a new type of design, it is different from the traditional two-dimensional design. It can build an information system in line with the actual needs by means of three-dimensional modeling. The system has the advantages of reducing the error rate, improving the production efficiency, saving working time and so on, which can significantly improve the economic benefits of the project. The main features of this technology are as follows.

(1) Visibility. BIM can use 3D Studio vit, 3dmax and other technologies to form architectural renderings, collect architectural parameters, and automatically generate construction drawings.
(2) Simulation. In the design stage, BIM Technology can be used to simulate the parameters and optimize the scheme, such as ventilation and 8-light noise.
(3) Coordination. BIM Technology can combine many specialties such as building structure, mechanical and electrical, etc. into one model, so as to avoid conflicts between design and construction due to poor communication.
(4) It can be optimized. BIM Technology can be optimized by rule model and parameters. BIM is not only a 3D building model, but also can be connected with cost and time. It can give full play to 5D dynamic management technology, generate engineering quantity statistics table for virtual model, and conduct model approval for construction process.

BIM is one of the most important research fields in virtual reality technology. Although the concept of building information model has been put forward for a short time, BIM Technology has become one of the most important means to improve the efficiency of design and construction in the construction industry (especially in developed countries).

2.2 Significance of BIM Application in Construction Field

The 3D dimensions of various building components can be derived, and various reports, project progress and budget can be automatically generated. The accuracy of BIM is directly proportional to the accuracy of modeling. BIM's powerful database function also provides great convenience for future project repair, maintenance, reconstruction and even demolition. Because the model can accurately calculate the quantity of each material, and even directly generate the quantity report, it makes the construction cycle very easy. Relevant personnel only need to adjust the report automatically generated by the model according to their own professional experience to get a more accurate estimation. The three-dimensional digital simulation model based on BIM Technology can realize the virtual design, visual decision-making, collaborative construction and transparent management of construction engineering. It will greatly improve the level of project decision-making, planning, survey, design, construction and operation management, reduce errors and shorten the construction period. It will significantly improve the informatization level of construction industry, promote the development of green buildings and realize the construction industry Transformation and upgrading.

3 Application Advantages of BIM Technology in Green Building Design

3.1 Optimize the Building Environment

Wind, light and sound are important parts of green building. In order to analyze them, we can take advantage of BIM Technology to build 3D building model [2]. The air pollution distribution can be calculated by BIM. The lighting color and light environment are analyzed to meet the lighting demand. BIM Technology can accurately simulate the outdoor noise and evaluate it in combination with relevant regulations and standards; in the analysis of indoor environment, BIM Technology is used to simulate and combine with various indoor parameters to obtain scientific and effective analysis results and realize the optimization of the overall building environment. The simulation model of indoor noise is as follows:

$$P_t = p_0(1 + \delta_1) + A_i \sin(\frac{2\pi t}{T_i} + \varphi) \tag{1}$$

Where, δ_1 is the average growth rate of noise, p_0 is the fuzzy weight factor, Ai is the noise center, Ti is the average period, φ is the phase correction angle; P_t is the indoor noise; t is the noise index.

The calculation formula of noise elimination is as follows:

$$\delta_i = \sum_{i=1}^{m} w_i \delta_i \tag{2}$$

Where, w_i is the noise weight; m is the total noise; δ is the noise period.

3.2 Improving Design Quality and Economic Benefit

Institute. In this way, the relevant problems can be found and solved in advance before the construction, which effectively improves the design quality and is conducive to the control of project cost and construction period, as shown in Fig. 1.

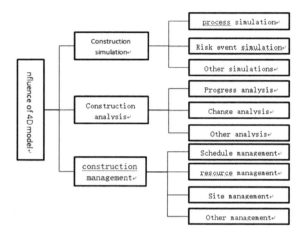

Fig. 1. Influence of BIM on construction process

4 Application of BIM Technology in Green Building Design

Before the application of BIM Technology, it is necessary to build a building model to analyze the influence of environmental factors around the building on the building, and on this basis to explore technical measures to optimize the comfort. In the design process, BIM Technology is used to analyze the simulation model, and the best effect is achieved through repeated comparison. In the process of three-dimensional design, firstly build the building information model, including HVAC, water supply and drainage, electrical and other aspects, and then the relevant units carry out reasonable division of labor, test the accuracy of the model, analyze and correct the collision points, and form the final building data model. The construction of the model can provide the basis for green building analysis, and the analysis results can provide guidance for the model until the best construction drawings are formed.

4.1 Conceptual Design Stage

This stage mainly carries out the initial design of the building structure, which has strong innovation, and the proposed design scheme is also very rich and diverse, which can have a direct impact on the subsequent construction of the building. Therefore, the design of this stage is very important. Combined with BIM Technology, green building related subjects are involved in the conceptual design, including engineers, architects, etc., to analyze all aspects of the construction project, such as lighting, daylighting, etc., in order to achieve the goal of building green buildings. For the lighting, daylighting and other aspects of the building, they can be optimized in the follow-up of the building. However, because the basic tone of such construction is provided at this stage, such as site selection, these can not be solved in the follow-up [3]. Therefore, in the conceptual design stage, we should actively improve the coordination mechanism to ensure that the conceptual design is consistent with the green building standards to the maximum extent, so that the architectural design and construction are more scientific and reasonable, and the requirements and responsibilities of all parties are more clear, so as to provide great convenience for all parties to exchange and communicate.

4.2 Drawing Stage

Complete the drawing work. Based on this technology, 3D model can be built to ensure that the elevation, plane, section and traditional report in the 3D model are included. The operator can expand the visual interface to make it consistent with the floor of the required drawing, and then the plan can be exported, As shown in Fig. 1, the construction of 3D model can accurately observe the floor plan. In building construction, construction drawings play a guiding role. Under the background of the continuous deepening of BIM Technology, construction drawings can fully describe the construction information, and the construction of models will make the description innovative, and the application of 3D models will make the drawing design more accurate and reasonable.

4.3 Detailed Design Stage

Compared with the previous design patterns, the application of BIM Technology in the deepening design stage can make the design details more perfect and promote the green building function to give full play. The technology has the characteristics of cross specialty, and it does not need to repeat modeling in the application, which can make the program more concise [4]. At the same time, the BIM Technology has a variety of interfaces, The model can change a number of parameters in the building, so that a variety of software information input needs can be met. In the design process, the requirements for the degree of architecture are more strict, and the specific software is transferred to BIM software, which can analyze the details of the building, including tracking records, detail processing, alarm, etc., so as to make the architectural design more profound and the design method more scientific and reasonable.

5   Conclusion

To sum up, under the premise of the continuous development of information technology, BIM The birth of technology provides a new design management concept for construction engineering. Through the application of BIM Technology, it can not only improve the construction progress and quality of the project, but also save the construction cost and the consumption of building materials, which has a great role in promoting the development of construction engineering. Nowadays, many large-scale buildings in our country are trying to apply BIM Technology, which has great potential in the future.

References

1. Baichang, L.V., Zhang, Y.: Research on the application of virtual reality technology in green building design. J. Hebei Inst. Archit. Eng. **37**(1), 76–79 (2019)
2. Zhang, S.: Research on the Application of BIM Technology in Green Residential Building Design. Anhui Jianzhu University, Hefei (2018)
3. Qian, J., et al.: Analysis of the significance of building a comprehensive observation network of natural resources elements [J/OL]. China's land and resources economy, 23 May 2021, pp. 1–12 (2021). https://doi.org/10.19676/j.cnki.1672-6995.000621
4. Zeng, J., Huang, G.: Science of science and technology research, 23 May 2021, pp. 1–18 (2021). https://doi.org/10.16192/j.cnki.1003-2053.20210521.002

Design of Quality and Safety Supervision and Management System Based on GPO Optimization Algorithm

Yanyan Yan[✉]

Zaozhuang Vocational College of Science and Technology,
Zaozhuang, Shandong, China

Abstract. At this stage, the development speed of the Internet has been greatly improved in our society, which promotes the reform of various industries. However, the application level of the Internet in the field of construction engineering is not very high, especially in the field of construction engineering quality and safety supervision and management, the level of information construction is very low, Based on the actual work experience and relevant literature records, the author analyzes the actual situation of informatization construction of construction project quality and safety supervision and management in China at this stage, and puts forward some informatization construction methods, hoping to play a certain reference role in the future when the relevant staff analyze this problem.

Keywords: Construction engineering · Quality and safety · Supervision and management · Information construction · Status · Optimization algorithm

1 Introduction

At present, the speed of China's social and economic development is gradually becoming stable, and the scale of engineering construction is showing a trend of continuous expansion. There is a high probability of various construction quality problems in the Internet. Therefore, the quality of construction projects has gradually received the attention of relevant people in various fields of our society. It is difficult for the traditional construction project quality detection method to meet the requirements put forward in the process of China's social development at the present stage. It is a problem that should be solved at present to effectively improve the strength and efficiency of quality and safety supervision and management, and promote the construction project quality management to be standardized, The problem of information construction should be given full attention [1].

J. C. Hung et al. (Eds.): FC 2021, LNEE 827, pp. 1313–1317, 2022.
https://doi.org/10.1007/978-981-16-8052-6_187

2 The Actual Situation of Information Construction in the Field of Construction Project Quality and Safety Supervision and Management at the Present Stage

2.1 General Situation of Information Construction

In modern society, the speed of Internet development and application has been greatly improved, which makes the overall style of our society change to a certain extent and guides the engineering supervision work to develop in the direction of informatization. In 2003, China's Ministry of Construction issued the "2003–2008 national construction informatization development outline", which represents the support of the government for the construction of China's engineering supervision informatization. The outline puts forward higher requirements for the informatization of construction industry and the informatization of construction engineering quality and safety supervision and management, including the guiding ideology, development focus and safeguard measures [2].

In the past, the information construction in the field of construction project quality and safety supervision and management has been paid enough attention. The practical application of information mode, timely follow-up of each project, fair and fair supervision and management of each project, and effectively improve the supervision efficiency and quality. However, compared with developed countries such as Japan and Germany, there is a big gap at this stage in China. For example, the economic development level of various regions in China is different. Some regions have a higher level of economic development, but others have a slow speed of economic development. Therefore, the speed of information construction in backward regions is very slow, The effect of information management is also different. The management mode of high-level development areas has gradually achieved the goals of data docking and dynamic supervision, which can ensure the overall quality and safety of China's construction projects.

2.2 Future Development Direction

At this stage, "cloud technology" in the field of computer gradually gets people's attention. Cloud technology can provide certain technical support in the process of supervision and management informatization construction of construction project quality and safety zone. With the support of cloud technology, advanced information technology is widely used, and the difficulty of achieving dynamic supervision and management objectives is relatively low.

Unifying the informatization construction of construction project quality and safety supervision and management can not only effectively control the informatization construction cost of supervision and management departments at all levels, but also integrate the informatization supervision software development, promote the data and information sharing of supervision departments at all levels in various regions, and promote the development of supervision mode towards the direction of unification, In the process of actual operation, each department should give full play to its own functions, and strive to improve the efficiency of government supervision and management and service quality, so as to lay a solid foundation for the smooth development of the follow-up linkage mechanism construction, and promote the unification of data content, format standards and other functional modules, The deployment and implementation of quality and safety supervision and management of construction projects are completed in a unified way. In the actual operation process, the supervision department directly communicates with the responsible subject, so that the problems existing in the project can be found out in time, and the adaptive measures can be used to solve the problems, so as to reduce the unnecessary communication links between the supervision department and the responsible subject as far as possible, Effective control of information and data transmission time can greatly improve the efficiency of supervision, at the same time, it can effectively control the cost of supervision, and gradually achieve the goal of information supervision and management. In the future, in the process of informatization construction of construction project quality and safety supervision and management, it will develop in the direction of dynamic joint supervision [3].

We can get the cost of waiting for students in unit running time:

$$C_1 = \frac{1}{2}b_2 \times \sum_{i=1}^{N}\sum_{j=1}^{M} q_{ij} \times \Delta t \tag{1}$$

The average satisfaction was as follows:

$$w = \frac{\sum_{i=1}^{N}\sum_{j=1}^{M} p_{ij}}{(M-1) \times K_0 T} \tag{2}$$

The cost is:

$$C_2 = b \times w = \frac{b \times \sum_{i=1}^{N}\sum_{j=1}^{M} p_{ij}}{(M-1) \times K_0 T} \tag{3}$$

2.3 Advantages of Information Construction

With the support of information technology, the digital supervision and management system is gradually constructed to optimize and adjust the government supervision mode, and the supervision process is changing to the direction of refinement, so that the government supervision and service level can be greatly improved. In the process of practical application of information supervision system, information and data can be shared between the government and the responsible subject, so that the difficulty of supervisors can be effectively controlled, and the safety and quality of engineering projects can be guaranteed. In the process of information construction, the standards in the field of engineering supervision should be made clear, the whole process supervision and management should be completed according to the established standards, and the project quality should be placed in the first place, so as to ensure the normal creation of economic benefits for construction projects. In the process of information construction in the future, we must have a certain sense of urgency, so as to effectively protect the lives and property of the people in our country.

3 Information Construction Method of Construction Project Quality and Safety Supervision and Management

Construction project safety supervision and management system is still in an imperfect state, and various problems are prone to occur in the process of practical application. On this basis, we need to take the actual development of the industry as the basis, gradually make up for the defects in the quality and safety supervision and management system, and attach various detailed rules. In addition, the safety supervision and management system should develop towards the direction of standardization and specialization to ensure the efficiency and quality of supervision and follow-up investigation. If problems are found in the process of actual work, they should be reported in time. Key monitoring areas, key monitoring units and key projects should be placed in obvious positions, so as to ensure the efficiency and quality of quality and safety management on the construction site.

To investigate the construction engineering market from various angles and apply the market mechanism practically, so as to give full play to the role of market resource allocation and promote the realization of the goal of informatization construction of construction engineering quality supervision and management. In addition, the organizational role of industry associations should also be fully valued. We should hold regular discussion meetings, invite various well-known construction enterprises and scholars to discuss the informatization system, and greatly increase the capital investment in the field of informatization knowledge popularization. On the basis of widely using various channels, we should complete the knowledge promotion work, and carry out informatization technology training for technical personnel and staff, In order to ensure the professional quality level of quality and safety supervision team [4].

Carry out attendance management for the key positions of the main responsibility. In the construction site, the face or fingerprint attendance system should be applied to the staff of key positions such as project manager and director, and the attendance structure should be input into the engineering quality supervision information system. In the information system, only the image and video transmission function can be added. For the supervision organization, in the process of supervision and law enforcement, the attendance structure should be input into the engineering quality supervision information system, The problems in the construction site can be uploaded to the system by image or video format, so that they can be used as evidence in the future investigation. For each participating subject, in the process of rectification and feedback work, pictures or videos can be used to prove that they have completed the rectification work and can carry out follow-up work. Gradually develop mobile information system. Through the mobile phone client, during the process of inspection work in the construction site, the supervisors can take photos of the places with potential quality and safety hazards in the construction site, complete the picture and video transmission in real time, and record the problems in the site in time, so as to play a certain reference role in the process of future rectification work.

4   Conclusion

In the process of information construction in the field of quality and safety supervision and management, it can promote the smooth development of our people's daily production and life. However, information construction needs a fixed time. In the process of China's social development, many innovative information construction methods will emerge, To promote the efficiency and quality of information construction in the field of construction quality and safety supervision and management has been greatly improved, which can ensure the safety and stability of the operation of buildings in China, and finally make a certain contribution in the process of China's social and economic development.

References

1. Zheng, L.: Supervision and management of construction project safety production. Mod. Vocat. Educ. **24**, 183 (2017)
2. Chen, J.: Potential problems and solutions in quality and safety supervision of electric power construction engineering. Commun. World **16**, 127–128 (2017)
3. Hailong: Analysis of strengthening construction quality and safety supervision on site management. Rural Staff **15**, 214 (2017)
4. Song, X.: Analysis of construction quality and safety linkage supervision work. Farm Staff **15**, 172 (2017)

Application of Multi Sensor Image Fuzzy Fusion Algorithm in Image Recognition

Ying Yang$^{(\boxtimes)}$

Chongqing Vocational Institute of Engineering, Chongqing 402260, China

Abstract. In this paper, the basic principle of multi-sensor image fusion based on fuzzy set theory for image recognition is discussed. A fuzzy algorithm of multi-source and multi-level adaptive variable weight image fusion is proposed. An effective method for expressing and processing uncertain information is provided, which makes full use of the redundancy and complementarity among information sources and the reliability information of each information source.

Keywords: Multi sensor and multi-level image fusion · Fuzzy set theory · Target recognition

1 Introduction

The traditional algorithm of target recognition is detection segmentation classification. The detection of the region of interest depends on the statistical difference between the target echo and the background. Generally, the effectiveness of these methods is low in the presence of clutter or natural and man-made shelter [1]. An effective way to improve the performance of target recognition is to fuse the information of multiple data sets in the same scene, In order to get a better understanding of the scene, so as to enhance the trust and accuracy of target recognition estimation and decision-making, reduce the fuzziness of individual estimation, and solve the conflict between single decisions. Multi sensor images can be fused on the signal layer, pixel layer, feature layer and decision-making layer, Many solutions have been proposed, including statistical method, D-S evidence theory, fuzzy set theory and neural network method, In order to solve the problem of image recognition, fuzzy membership function can be used to describe the uncertainty between different object types and corresponding pixels and within each image system quantitatively. The abundant fusion operators and decision rules in fuzzy set theory provide a necessary means for effective image fusion processing.

2 The Basic Principle of Image Fuzzy Fusion

The basic principle of image fusion fuzzy algorithm applied to image recognition is to use fuzzy membership function to quantify the relationship between different object types and corresponding pixel values.

J. C. Hung et al. (Eds.): FC 2021, LNEE 827, pp. 1318–1323, 2022.
https://doi.org/10.1007/978-981-16-8052-6_188

The possibility measure is defined to represent the overlapping degree between the above two membership functions, so as to determine each type of boundary in the image. The function of introducing this parameter is: ① due to the influence of noise, the real pixel value may deviate from its type interval. Therefore, the possibility measure of the pixel is used to explain the possibility that the pixel belongs to a certain type, In order to reduce the negative effect of noise, image fusion has its own characteristics, so it is impossible to eliminate the influence of noise by simple arithmetic average fusion operator. Otherwise, the fused image of two input images will lose type I [2]. However, the possibility measure of all input images associated with type I is significantly greater than that of other types, These possibility measures can be dealt with in many ways, including simple averaging operators. By fusing the possibility measures of all images related to type ᴵ , a global possibility measure for type I can be obtained. According to one or more of the following rules, a decision about type I is made (let C denote the number of types in the image):

$$\mu_i(x) = \max_{k=1}^{c} \mu_k(x) \qquad (1)$$

$$\mu_i(x) \geq d \qquad (2)$$

Once the boundary of each image type is determined, the internal contrast of each type can be restored or even emphasized by using blur technology.

3   Method of Acquiring Obstacles in Front of Unmanned Ship Based on Vision Sensor

Millimeter wave radar can accurately detect the distance, azimuth and relative radial velocity of obstacles in front. However, in the process of autonomous navigation, USV needs to know the specific size and coverage of obstacles in addition to the above information. Visual sensor can intuitively perceive the environment, so as to obtain the contour information of the object in the environment. For this reason, a method of USV front obstacle acquisition based on vision sensor is proposed.

3.1   Mean Shift Algorithm Smoothing Filter

FCM algorithm needs to initialize the number of clusters and cluster centers, and the clustering results are easily affected by the location of the initial cluster centers. When USV performs tracking, search and rescue, patrol and other maritime tasks, it is difficult to directly set initialization parameters suitable for all images because the real-time extracted video image content is completely unknown and the complexity is high. Therefore, the MS algorithm is selected to filter the acquired image, and the clustering results are used as initialization parameters to initialize the FCM algorithm to meet the practical application requirements.

In order to meet the needs of target extraction in the subsequent image processing, the FCM clustering of the image is carried out. Due to the unknown and complexity of the application scene, it is impossible to use a unified threshold to binarize the gray image. Therefore, Ostu is selected for the global automatic selection of threshold. OSTU method is an automatic threshold selection method in image segmentation using clustering idea. This method realizes the separation of foreground and background by maximizing the total difference of bilateral gray values (Guo Jia et al. 2014). Because the image segmentation algorithm proposed in this paper is a pre order operation for the subsequent image processing algorithm, it only needs to pay attention to the foreground objects, and there are few interferences in the water environment. Therefore, using Ostu algorithm for image binarization can maximize the object segmentation from the background.

3.2   Summary of This Chapter

This chapter focuses on USV. The simulation results show that the proposed algorithm can complete image segmentation and target extraction in a variety of scenes, and the segmentation effect is stable; compared with other methods, the segmentation accuracy is greatly improved, and the operation time is shorter; It can effectively suppress the influence of image noise caused by shadow, water surface reflection, uneven illumination, wavy facula and light reflection and other water environment factors on image segmentation; for multi-target water surface background image, it can also extract the target independently. The simulation results show that the proposed algorithm can complete image segmentation and target extraction in a variety of scenes, and the segmentation effect is stable; compared with other methods, the segmentation accuracy is greatly improved, and the operation time is shorter; It can effectively suppress the influence of image noise caused by shadow, water surface reflection, uneven illumination, wavy facula and light reflection on image segmentation; for the water surface background image of multi-target, it can also extract the target independently to study the environment perception based on visual sensor during navigation, Firstly, the selection of video sensor is introduced, and then msf-fcm algorithm is proposed. The algorithm uses MS algorithm to filter the color image, and transfers the result as initialization parameter to FCM algorithm for clustering. Finally, Ostu method is used to binarize the image to complete the image segmentation. On this basis, the contour of the image is extracted. The simulation results show that the proposed algorithm can complete image segmentation and target extraction in a variety of scenes, and the segmentation effect is stable; compared with other methods, the segmentation accuracy is greatly improved, and the operation time is shorter; It can effectively suppress the influence of image noise caused by shadow, water surface reflection, uneven water surface illumination, wavy facula, light reflection and other water environment factors on image segmentation; for multi-target water surface background image, it can also extract the target independently [3].

4 Method of Obstacle Perception in Front of Unmanned Ship Based on Millimeter Wave Radar and Vision Sensor Fusion

Millimeter wave radar can sense the distance information of obstacles in front of us V, and the range information of obstacles can be obtained by image segmentation and target extraction method based on visual sensor. In order to make USV fully acquire and analyze the environmental information in the process of navigation, the above two aspects of information are fused. The effective target azimuth information obtained by millimeter wave radar is used to construct the region of interest in the image, and then the range and area of the target are extracted from the image by the vision system. In order to achieve the fusion of the two kinds of sensor information, it is necessary to fuse in space and time.

4.1 Spatial Integration

First of all, spatial fusion is needed, that is to realize the unification of millimeter wave radar, video sensor, video image and real world 3D coordinate system. In order to complete the integration of coordinate system, it is necessary to understand the ranging principle of the two kinds of sensor equipment, and extract the corresponding coordinate transformation model, so as to find the transformation equation between the coordinate system of the two kinds of sensor equipment and the world coordinate system. According to the transformation equation, the coordinate system of the two kinds of sensing equipment can be transformed into the world coordinate system.

4.2 Test of Simulated Water Surface Environment

Due to the limitation of test conditions, there is no suitable USV platform for transplantation at present, so the sensor platform is fixed on the shore to detect the target on the water surface. At the same time, because it is difficult to measure the distance and azimuth of the target at a fixed point on the water surface, the existing USV is used as the target to obtain the relative position information through real-time GPS information conversion, The experimental environment and assumptions are as follows: (1) set the radius of the corresponding region of interest (r = 0.8 m) in the fusion process. (2) considering the open space, set the maximum radius of the millimeter wave radar detection range (r = 30 m). (3) the equipment height is: the distance between the millimeter wave radar signal receiving and transmitting port and the horizontal plane is 0.3 m, and the distance between the camera lens center and the horizontal plane is 0.4 m; (4) The signal transceiver of millimeter wave radar is on the same straight line with the center of the outer edge of the camera lens, and this straight line is perpendicular to the ground, and the optical axis of the camera is parallel to the central axis of the millimeter wave detection sector; (5) the site is open without other obstacles [4].

4.3 Summary of This Chapter

Multi sensor fusion is used to realize USV's overall perception of water environment. Firstly, the spatial fusion is completed by coordinate transformation, and the camera calibration adopts Zhang Zhengyou monocular camera calibration method. Secondly, the temporal fusion is completed by multithreading. On this basis, a multi-sensor fusion method based on region of interest is proposed, and the region of interest is constructed according to the orientation information of obstacles detected by millimeter wave radar, The location and coverage of the effective target are determined by matching with the minimum circumscribed rectangle of the obstacle contour obtained by the vision algorithm. Field tests were carried out in land and simulated water environment, and the results show that the fusion method can detect large static objects in land and water environment which are obviously different from the background, and the detection results are close to the real value.

This paper designs and implements a method of obstacle perception in front of unmanned ship based on millimeter wave radar and vision sensor fusion. Through the transformation of coordinate system to complete the spatial fusion, the camera calibration uses Zhang Zhengyou monocular camera calibration method; uses multi thread processing to complete the temporal fusion; proposes a multi-sensor fusion method based on the region of interest, and constructs the region of interest according to the orientation information of obstacles detected by millimeter wave radar, The location and coverage of the effective target are determined by matching with the minimum circumscribed rectangle of the obstacle contour obtained by the vision algorithm.

5   Conclusion

In this paper, the basic principle of multi-sensor image fuzzy fusion based on fuzzy set theory for image recognition is studied, and a multi-source and multi-level adaptive variable weight image fuzzy fusion algorithm is proposed. Its basic ideas are as follows: one is to make full use of the redundancy among the input images to make more definite decisions; the other is to make full use of the complementarity among the input images, The factors that fuzzy technology is suitable for this kind of task are as follows: first, they can solve the fuzziness of image data better; second, all kinds of fuzzy fusion operators allow to consider several types of information and many different scenarios in the same model, especially the reliability information of information source can be introduced into fuzzy fusion operators. An effective obstacle acquisition method based on millimeter wave radar is designed and implemented. The next cycle information is predicted by Kalman filter to judge the continuity of the target; the life cycle theory in biology is introduced to judge the effectiveness of obstacles combined with the continuity of the target. In order to overcome the false alarm caused by environmental factors and obtain the obstacle information accurately. For the detection of static large obstacles, there is no corresponding experimental verification for dynamic obstacles, so the detection effect can not be determined; for smaller obstacles, due to the limit of detection accuracy, this paper chooses to "discard" them by setting the threshold. In the future, we need to further test the dynamic target, improve the accuracy through the algorithm improvement to complete the detection requirements of more obstacles.

References

1. Ma, D.: Overview of environment sensing technology for driverless vehicles. Automob. Driv. Maint. (5), 122–123 (2017)
2. Mu, K.: Development status and future prospects of driverless vehicles. Electron. Technol. Softw. Eng. (21) (2017)
3. Peng, D.: Basic principle and application of Kalman filter. Softw. Guide (11), 32–34 (2009)
4. Song, J., Wen, J.: Application of unmanned ship technology in maritime. China Marit. (10), 47–50 (2015)

Application of Museum Information Dissemination Based on VR Technology

Bo Zhang[1] and Weiqing Sun[2(✉)]

[1] Shanghai College of Publishing and Printing, Shanghai 200093, China
[2] Shanghai Publishing and Printing College, Shanghai 200093, China

Abstract. With the rapid development of all kinds of film and television art means, the integration of virtual reality art into cultural relic information display has become a more common means in museum operation. 5G the opening of the era, for the realization of these art forms to provide a broader space. However, some virtual reality means under the blessing of museum display too much rely on technology, technology as a display gimmick, lack of artistic content on the deep ploughing and other issues have greatly affected the audience's experience. This paper will start from the narrative language of virtual reality art in the context of all media, the interactive narrative strategy under virtual reality art, and the spatial narrative strategy under virtual reality art, and try to study and summarize the new ideas and new paths of narrative virtual reality art for museum cultural relics information dissemination and display, and provide some research strategies for the future cultural heritage protection and communication development.

Keywords: Virtual reality art · Interactive narration · Spatial narration

1 Introduction

All along, the museum is a non-profit permanent (fixed) institution that integrates knowledge learning, thought exchange, academic research, cultural heritage and aesthetic education, serves the society and its development and is open to the public. To some extent, museums carry the historical accumulation and cultural temperament of a region. The development of museums in different times, its social role and function also have different changes. At present, the educational and cultural heritage function of the museum has attracted more and more attention from the society, and the attraction to the public and the number of visitors have become an important standard for the museum to consider [1].

With the rapid development of all kinds of film and television art means, the integration of virtual reality art into cultural relic information display has become a more common means in museum operation. 5G the opening of the era, for the realization of these art forms to provide a broader space. However, some virtual reality means under the blessing of museum display too much rely on technology, technology as a display gimmick, lack of artistic content on the deep ploughing and other issues have greatly affected the audience's experience. This paper will start from the narrative language of virtual reality art in the context of all media, the interactive narrative strategy under virtual reality art, and the

J. C. Hung et al. (Eds.): FC 2021, LNEE 827, pp. 1324–1328, 2022.
https://doi.org/10.1007/978-981-16-8052-6_189

spatial narrative strategy under virtual reality art, and try to study and summarize the new ideas and new paths of narrative virtual reality art for museum cultural relics information dissemination and display, and provide some research strategies for the future cultural heritage protection and communication development.

2   Narrative Language of Virtual Reality Art in Full Media Context

VR is the abbreviation of Virtual Reality, Known as virtual reality or psychic technology, Is a combination of digital image processing, computer graphics, multimedia technology, pattern recognition, network technology, artificial intelligence, sensor technology and high-resolution display technology, With vision, hearing, touch, Information integration technology system for generating realistic 3D virtual environment. As a new form of media, virtual reality art (VR) has changed the way people experience information through text reading in physical venues, With an immersive audio-visual feeling to complete the traditional era of text communication to the background of the media era of image communication change. March 2014, Fackbook $2 billion acquisition Oculus, virtual reality (VR) technology equipment maker VR the tide of development is open. And in the next five years, Virtual reality (VR) technology has experienced from prosperity to low hovering until steady progress. 6 June 2019, China Mobile, China Unicom, China Telecom, China Radio and Television, Marks the full media under the high speed, the low power consumption, the low delay 5 G first year officially opens, This also constructs a good network environment for the normal use of virtual reality art.

It is not difficult to find out from the history of tracing back the development of media narration that the mode of constructing narration is from the original oral narration, the printed paper media narration to the later drama narration, the film and television narration, until now the virtual reality narration in the context of the whole media. Different from the early mode of simply relying on technical means to complete narration, the narrative language of virtual reality art in the context of all media has new connotation characteristics: first, the breakthrough of narrative time and space. Virtual reality art in the context of all media can use multiple carriers to break the inherent limitation of time and space, fully display the charm of "narration" under the frequency resonance of technology and media platform, and complete the presentation of time through ancient and modern times. The breakthrough of time and space makes narration no longer limited to limited frame space and specific time. The space-time dimension under audiovisual narration has been extended to the greatest extent. As a narrative object the identity of the audience has also changed significantly in the new narrative grammar. In the past narrative mode, the viewing audience participated in the narrative as a passive viewer, lacking an autonomous experience. More is the virtual reality in the context of the whole media, so that the ornamental object completely gets rid of the subordinate position, from the object of the informed state to the object that can directly participate in the narrative subject; again, the breakthrough of the narrative mode, as shown in Fig. 1.

Scholars Cai Dongna and Jia Yunpeng believe that "an interactive narrative plot can be regarded as a potential plot event in a set of databases. Audience selection instantiates these potential plot events and gives them the order of play. This process, in turn, builds

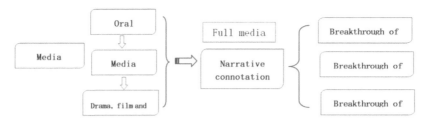

Fig. 1. Development of media narrative

stories in the minds of the audience". From the single linear one-way narration in the original tradition to the interactive narration which can achieve two-way communication —the author, the work and the receiver's role orientation in the past narrative process is completely subverted, which makes them form a common narrator in a sense.

3   Interactive Narrative Strategies in Virtual Reality

With hypertext novels, interactive drama, interactive film, video games and other literary types have become new members of the literary field, the concept of "interactive narration" is on the scene. Interactive narration is a method of interactive entertainment, which enables users to make decisions and directly affect the direction and results of narrative experience generated by computer systems [2].

Museums under the support of traditional virtual reality art show more to provide a one-way information dissemination, simply using VR technology to form an indoctrination presentation mode, the audience in the process of receiving the lack of interaction and dialogue, so that the use of technology for rigid, lack of vitality, to some extent, this way of communication only completed the information "receive" function, not the real sense of "accept". The idea of interactive narration has changed the subordinate position of the audience in the past, so that it can actively participate in the process of narration, give full play to the technical advantages of digital media, and increase the pleasure and fun of viewing.

In the process of interactive narration, the presentation and expression of information content is produced when users interact with the system in real time. Therefore, the node design of interactive narration is particularly important. As an important carrier of history and culture, museum resources shoulder the responsibility of cultural inheritance in addition to their own artistic value. The interactive narrative node of virtual reality art should also be closely consistent with cultural content. Referring to the structure of classical drama in narratology—"build-development-conflict-high tide-end", the narrative interaction node of virtual reality art in museum can also follow this structure to carry out situational design and emotional paving. For example, in the process of displaying the historical origin of bronze ware, the traditional display forms rely too much on the multi-angle display of the collection itself. It is difficult to interact and resonate between the audience and the collection. The virtual reality art under the interactive narration can make full use of the historical story behind the collection and divide it into nodes such as "build", "develop", "conflict", "climax" and "ending", explain the time, place, character and future shield conflict of historical events at the "build" and "development" node, and act as the opening button of the next narrative node through interactive means such as light effect, sound and even behavior action, so that the audience can completely break the boundary and distance between "exhibition" and "viewer" in the traditional mode, and place the audience in the space and time described by the interactive narrative node set up under the virtual technology, The combination of "collection" and "event" allows the exhibition model to be rebuilt. The marriage of virtual reality art and narrative space-time art enables the audience to complete the multi-dimensional understanding of the collection in the narrative scene. Maximize the effect of audience experience [3].

4   Space Narrative Strategies Under Virtual Reality Art

Under the background of the whole media era, with the rapid development of digital technology, network technology and transmission technology, the information dissemination of the whole society shows a fragmented development trend. As Daniel Bell said," post-modern society" is a new era of modern information and technology. The fragmentation of information will lead to the fragmentation of presentation space, and the fragmentation of space will directly lead to the fragmentation of narration. The fragmentation of each part makes virtual reality art unable to better create a real and intuitive sense of immersion and substitution. This is also one of the real pain points of virtual reality art. Therefore, how to present a complete and unified narrative space in the process of audience experience, so as to focus on the attention of the audience and create a real sense of "immersion" is the development direction of virtual reality art in the process of museum construction [4].

The early virtual reality art created a pure immersive effect of transforming the plane text picture into a more real three-dimensional space by scientific and technological means, although it achieved the enhancement of the presentation mode. But it ignores the complete unity of space content. To a certain extent, the enhanced space under the separation causes the disconnection of narration and forms a fragmented organizational structure, which distracts the attention of the audience, and the immersion brought by virtual reality art is greatly weakened. British scholar Bryce once put forward: "the purpose of VR technology is completely different from multimedia technology. The most important principle of virtual reality is to establish an immersive experience. The success of virtual reality demonstrations or works is to make the audience produce an experience that is divorced from their real world and immersed in another environment. This immersive feeling must be produced by the sensory organs alone, not by the psychological imagination. As a result, the essence of VR technology is to build a realistic experience that makes the senses fully convinced."

Compared with the narrative space of literature, architecture and film, the spatial form under virtual reality art has the characteristics of multi-dimensional, multi-angle and boundless, which also provides a certain possibility for the narrative of virtual reality space. Free from the limitation of time and regional boundary, virtual reality presentation in museums can use technical means to form a unified and complete narrative space, guide the audience into the situation, and effectively interact with exhibits and environment, so as to achieve an emotional resonance. For example, in the exhibition strategy of "etiquette Jiangnan—Jiangnan ritual jade ware", the creators can make full use of the artistic skills of space narration to transform the display of objects such as bird lines and jade pieces from the introduction dimension of collection under simple three-dimensional enhancement effect to restore the spatial dimension under historical scene. Through the creation of spatial artistic conception combined with wind and tree movement, life and labor scene, the visitors can return to the Liangzhu cultural era more than 4300 years ago, and feel the real life scene of Liangzhu people as one of the important birthplaces of jade ritual ware, enriching the space presentation effect with techniques such as light, shadow, sound and color, In addition, creators can use virtual reality art to complete the setting of scenes in space, such as reshaping a story background that conforms to the history and culture of exhibits in the space scene. The identity of the audience is transformed from passive visit to active exploration, breaking through the narrative habits of time dimension, making visitors feel the value of the exhibits with the unique characteristics of the environment and age under the infection of space atmosphere.

5   Conclusion

The singing and singing of all kinds of digital art means have brought all kinds of changes to the social life, information exchange and even the mode of thinking of the masses. We can not ignore the double-edged sword attribute while waving flags for the birth of new technology. The virtual reality under the whole media is not the vassal technology of the museum media, but the art that complements and coexists harmoniously with it, at the same time, it also gives the cultural relic information more true interpretation and expression. The shaping of the characteristics of "scene feeling", "immersion feeling" and "reality" of museum cultural relic information dissemination needs the narrative language of virtual reality art to carry. Only by giving it the artistic tension of multi-dimensional narration, breaking through the inherent habit of using technology, upgrading the pure "technology first" virtual reality art 1.0 mode to the 2.0 mode of combining technology with narrative language art, Can better help cultural relics information to convey more profound, more unforgettable narrative connotation, to meet the audience's spiritual and cultural aesthetic demands.

References

1. Jia, Y., Cai, D.: Exploration of interactive narrative form based on plot interaction. Film Art (3) (2013)
2. O.Riedl, M., Stem, A., Dini, D.: Mixing story and simulation in interactive narrative. http://www.aaai.org/Papers/AIIDE/2006/AIIDE06-037.2009-04-05
3. Shi, P., et al.: Bryce, multimedia and virtual reality engineering. Trans, pp. 4–5. China Film Publishing House (2000)
4. Wang, Z.: A Study on the Narrative of Digital Media, p. 39. Communication University of China Press, Beijing (2012)

The Key Technology of 5G Communication Network D2D Network Communication Architecture

Xiaoling Zhang[1], Bin Zang[2], and Qianqian Yue[3(✉)]

[1] Anhui Sanlian College of Robotics Engineering, Hefei 230601, Anhui, China
[2] Hefei Thermal Power, Hefei 230601, Anhui, China
[3] Anhui Sanlian College of Electrical and Electronic Engineering,
Hefei, Anhui, China

Abstract. How to carry out D2D communication in the heterogeneous network with limited spectrum resources is an important problem to be solved in the future 5g network to realize D2D communication, and an efficient network architecture is conducive to better integration of D2D technology in 5g communication system. SDN/nfv, the key technology of 5g communication network, can effectively solve some challenging problems faced by D2D communication. The combination of D2D technology and SDN/nfv technology will improve the network performance, more effectively improve the utilization of spectrum resources, and also bring certain benefits to operators, which makes the realization of D2D communication in business possible.

Keywords: D2D · 5G · Internet of things · Edge computing · MMO

1 Introduction

With the popularization and diversification of intelligent terminals, explosive growth of mobile users and multimedia services leads to the overload of services, and the shortage of spectrum resources. Therefore, how to effectively improve the efficiency of spectrum resources has become one of the key issues in the fifth generation mobile communication research. And the D2D (device to devicecommunication) which can greatly improve the frequency efficiency has attracted the attention of many scholars at home and abroad.

D2D communication technology refers to the direct communication technology between two terminal devices within the allowable close range, such as WiFi direct, Bluetooth ZigBee and so on. In the application scenario proposed in the literature, once the communication link between terminal devices is established successfully, the data transmission does not need the assistance of external infrastructure. Therefore, D2D communication technology has the potential to improve the spectrum utilization, reduce the pressure of base station, reduce the end-to-end transmission delay, and improve the system network performance. At the same time, it can also provide the following gains, namely, hop gain, multiplexing gain, proximity gain, etc. [1]. At

J. C. Hung et al. (Eds.): FC 2021, LNEE 827, pp. 1329–1334, 2022.
https://doi.org/10.1007/978-981-16-8052-6_190

present, D2D communication technology has been listed as one of the important potential wireless key technologies of the fifth generation mobile communication.

2 Challenges and Related Research

2.1 Challenge

Due to the complexity and hybrid characteristics of 5g heterogeneous network, the standardization of D2D communication based on proximity service is also gradually formed, which makes the integration of D2D communication into the future heterogeneous network face great challenges.

First, the traditional cellular network is too closed and rigid. In order to support D2D communication in heterogeneous networks, many components of the network need to be upgraded to meet the requirements. At the same time, the control plane and data plane of the wireless access network and its core network of the cellular network also need to be modified;

Second, spectrum resources are scarce. The sharing of spectrum resources in heterogeneous networks will cause great interference between D2D users and between D2D users and cellular users;

Third, low latency, energy efficiency and scalability are crucial to 5g network. The ultra dense heterogeneous network increases the coverage density and re coverage area. The simultaneous access of large-scale D2D users will lead to high signal overhead and extended end-to-end waiting time.

2.2 Related Research

The core idea of SDN lies in the centralized control of data plane logically and the management chart of network state by distributed controller. It makes the control plane and data plane independent by decoupling control and data plane, which can adapt to the increasing network traffic, Of course, its biggest advantage is that it can simplify the management complexity of large-scale network, centralize the control of wireless resources, improve resource utilization, and improve network reliability.

The domestic 5g white paper network pointed out that in terms of network architecture, SDN/NF ∨ will give full play to its advantages, so as to realize flexible networking in the new infrastructure and enhance its security. At the same time, a large number of literatures have described the integration of sdnnfv in the network. The literatures put forward an intelligent model based on SDN, which can effectively manage heterogeneous infrastructure and resources, and provide a variety of solutions to improve user management, flow control and resource allocation. In the literature, SDN and nfv are integrated into the network architecture, and an architecture that can meet the requirements of 5g network is proposed, which can make more effective and optimal use of resources in the backhaul, and solve the problem of current network closure and rigidity [2]. Aiming at the scalability and performance problems of 5g heterogeneous network, this paper applies it to the network architecture design based on the concept of SDN/nfv and mobile cloud computing.

3   Key Technologies in 2D2D Communication

3.1   Basic Principles

The basic principle model 1 of D2D communication is shown in Fig. 1. In the DD communication mode, the UE (user equipment) no longer relay communication through the BS (base station), but directly connect and communicate with each other. In heterogeneous networks, there will be traditional ue-bs connection and D2D connection at the same time, covering local broadcast communication, Internet of vehicles and other fields. In addition, D2D communication also includes D2D local network (d2dlan). It can be seen that D2D covers a wide range and involves a variety of scientific and technological problems, The key technologies include D2D device discovery, resource allocation, D2D cache network, edge computing and d2d-mimo. According to the demand and research status of D2D discovery, this paper summarizes the main key technologies into four categories: D2D device discovery and session establishment, resource allocation and interference management, D2D cache network and d2dmmo, and summarizes the technical principles and development status. On this basis, it conceives and plans the future development of D2D, focusing on the technical combination and key technologies of D2D and 5g, Internet of things and Sdn.

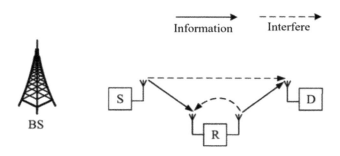

Fig. 1. D2D communication model in heterogeneous networks

3.2   D2D Device Discovery and Link Performance

In D2D communication, how to discover each other and initiate D2D connection between devices is the basis of all kinds of homogeneous or heterogeneous D2D communication in heterogeneous networks. According to the existing research results, this paper divides the research points into device discovery and link performance.

Device discovery and session are based on the research scheme of D2D device discovery, which can be divided into two categories: restricted discovery and public discovery. For restricted discovery, UE is not allowed to be detected without explicit permission. Users are forbidden to communicate with unfamiliar devices to ensure the privacy and security of UE. For public discovery, as long as the current UE is a neighbor of another device, it may be detected, and then establish a connection. Compared with the limited discovery mode, this mode has poor user privacy, but lower connection complexity [3]. The former is suitable for better network environment and more choices, so privacy is a more important issue; the latter is suitable for rescue and emergency communication, such as poor coverage of the main network, when the connection is more important. From the perspective of network, device discovery can adopt tight control and loose control mode of base station.

4 Solutions

According to the above challenges, some solutions for the network architecture of the integration between D2D communication and sdnnf are summarized as follows.

4.1 SDN Solution

For D2D communication, it is an open question whether to use cellular spectrum or unlicensed spectrum. If the cellular spectrum is used, the dedicated spectrum can be allocated to D2D communication, and the spectrum resources can also be shared with D2D communication. The latter helps to improve the spectrum utilization and solve the problem of spectrum resource shortage. However, the interference between D2D users and cellular users has become a challenging problem. However, if unauthorized spectrum is used, the interference between D2D users will become a challenging problem.

Under DDoS attack, the attack host generates a large number of false source address packets with hidden real addresses in a short time, which leads to a significant increase in the generation speed of attack flows and the average number of packets contained in each flow:

$$F_T = (\sum_{j=1}^{n} P_j)/n \tag{1}$$

(1) Average number of bytes under DDoS attack, sending a large number of packets means that the number of bytes of packets in a short time also increases sharply, so the average number of bytes of packets in each stream table increases:

$$B_T = (\sum_{j=1}^{n} M_j)/n \tag{2}$$

The port number generated by the computer conceals the real port number, resulting in obvious changes in the port growth rate:

$$P_T = K_T/T \tag{3}$$

On the network side, the SDN global controller mainly coordinates the mobility management detector (MME) to allocate the radio resources for each D2D pair, while the D2D server is mainly responsible for the connection with the cellular network, and assists the ENB to carry out network related operations for the mobile terminal carrying out D2D communication, For example: management of D2D equipment identification, D2D service identification, old p address allocation, determination of UE distance, etc. [4]. At the same time, in the mobile terminal, as a background service, the SDN local controller is responsible for selecting an appropriate interface for each application, and is responsible for monitoring the status and controlling openvswitch and wireless resource mapper. Open vswitch is a forwarding entity with forwarding function. In the process of D2D communication, it mainly selects the appropriate port for related applications.

4.2   Combination of SDN and nfv

At present, the spectrum interference problem of the same operator's D2D communication can be reasonably solved by adding SDN controllers at the network end and the user end. However, when multiple operators share the network in the future, network sharing includes spectrum sharing and infrastructure sharing, D2D communication will face a new challenge. Nfv is integrated into software defined D2D communication. Abstract, slicing, isolation and sharing are carried out by integrating various wireless access technologies of heterogeneous networks (such as 16, 26, 3G, 4G, WiFi). The wireless network resources can be divided into multiple pieces and assigned to different virtual networks, which can support a variety of communication schemes, thus reasonably solving the problem of network sharing. However, there is still a performance problem in software defined D2D communication in virtual wireless network. The SDN controller needs to have a global view of the network state to optimize the network performance. In addition to packet delay and loss, the network state information (NSI) is not ideal. The network state information is composed of channel state information and queuing state information, which is not an ideal network.

5   SDN/NFV Combined with Cran Scheme

5.1   Principle Explanation

In ultra dense heterogeneous networks, the increase of coverage density and re coverage area will lead to high information overhead, and also increase the cost of operators. The c-ran proposed in 5g white paper 0 of China's future mobile communication forum can provide high frequency spectrum efficiency and high energy efficiency in wireless communication system, and also reduce the capital and operating expenses of operators and equipment suppliers. The application of h14cran technology in D2D communication can effectively reduce signaling overhead and communication waiting time, The integration of D2D technology and SDN/nfv technology combined with cran technology can promote network performance more effectively.

The proposed scheme is based on the concept of layered architecture, which is different from the SDN scheme mentioned above. Its layered architecture concept is to deploy the global SDN controller in the core network and the local SDN controller in the base station, access point and other infrastructure. Because the local SDN controller deployed in the base station can appropriately reduce the burden of centralized control of the global SDN controller deployed in the core network. Therefore, the hierarchical architecture can greatly reduce the pressure on the macro base station in the ultra dense mobile sensing environment, and also can overcome the single node.

5.2   Open Questions

Compared with SDN/nfv, the integration of D2D technology and SDN/nfv technology can bring more benefits to 5g heterogeneous networks, and further improve the spectrum resource efficiency. However, the relevant integration schemes are usually for network assisted deployment scenarios, and usually focus on single hop D2D communication. However, in reality, there are non network assisted public security deployment scenarios, which may need to go through multi hop. At the same time, the power control of D2D communication still needs to be further studied under the network assisted fusion scheme. Therefore, these problems are also the key areas of the integration of D2D technology and SDN/NF V technology in 5g communication system.

6   Conclusion

The network architecture of the integration of D2D communication technology and SDN/nfv technology has the advantages of high frequency utilization and reducing the cost of operators. It will become one of the important combinations of future communication networks. However, only a few solutions for its architecture design are proposed at present. Therefore, this aspect has the value of in-depth research. This paper focuses on the problems faced by the integration of D2D communication into 5g communication, systematically combs the SDN/nfv technology, summarizes the related research solutions and research results of the integration of D2D technology and SDN/nfv technology, and points out some problems with research value in this field.

References

1. Li, J., Zhang, C., Bao, Z., Li, Q.: Research on scalable video multicast algorithm based on D2D network communication. J. Electron. Measur. Instrum. **30**(06), 923–930 (2016)
2. Qi, M.: Design and implementation of online dialogue game based on Cocos2d-x. Guizhou Normal University (2015)
3. Guo, R.: hx_D2 and HX_ comparison and analysis of D3B locomotive network control system. Technol. Innov. Prod. (11), 43–44 (2014)
4. Wang, X.: Design and implementation of 2D online game engine based on resource management. Southwest Jiaotong University (2007)

Decoupling Optimization of Vehicle Powertrain Mounting System Based on Genetic Algorithm

Hongbo Zhu[(✉)]

Henan College of Industry and Information Technology,
Jiaozuo 454000, Henan, China

Abstract. In order to improve the vibration isolation performance of the powertrain five point mounting system of parallel hybrid electric vehicle, the dynamic model of the powertrain five point mounting system is established. The optimization objective is to decouple the six degree of freedom energy of the powertrain and distribute the natural frequency reasonably. The stiffness of the five mounting points is the design variable. The genetic algorithm is used to optimize the mounting system. The suspension system of a parallel diesel electric hybrid vehicle is optimized by the above method. The results of dynamic simulation and real vehicle test show that the steering wheel jitter is eliminated after the suspension optimization, which verifies the rationality of the proposed method. At the same time, the genetic algorithm overcomes the shortcoming that SQP is easy to converge to the local optimal solution. The decoupling performance of the suspension system is excellent, and the optimization result is stable and reliable.

Keywords: Powertrain mount energy decoupling · Optimization · Genetic algorithm

1 Introduction

The NVH performance (noise, vibration and harshness) is an important performance index to measure the quality of a car. The noise and vibration of a car mainly come from the road excitation and the powertrain (engine, clutch, transmission), so the vibration isolation performance of the powertrain mounting system directly affects the NVH performance of the car. At present, the optimization methods for powertrain mounting system of non hybrid electric vehicle mainly include frequency shift method, energy decoupling method, total transfer force minimization method, etc. at the same time, some researches combine ADAMS software to analyze and optimize the vibration isolation performance of mounting system, among which energy decoupling method has been widely used. Some scholars use the energy decoupling method to identify the static stiffness parameters of the mounting system which contribute a lot to the change of the natural frequency and decoupling rate of the mounting system, which provides a certain reference for the design of the mounting system [1].

In parallel hybrid electric vehicle, because an ISG motor is paralleled at the output shaft of the powertrain, the vibration of the powertrain will be more complex. At this time, the vibration isolation performance of the mounting system will be particularly

J. C. Hung et al. (Eds.): FC 2021, LNEE 827, pp. 1335–1341, 2022.
https://doi.org/10.1007/978-981-16-8052-6_191

important. The five point mounting system of parallel hybrid electric vehicle is shown in Fig. 1. In this paper, the powertrain mounting system of hybrid electric vehicle is designed and optimized according to the energy decoupling method.

Traditional optimization algorithms rely on gradient information, and the results are easy to be local optimal solutions. However, as a global optimal search method, genetic algorithm is widely used in many engineering fields, and also in the design of mounting system. At the same time, the mathematical model of the energy decoupling method of the mounting system is extremely complex, so this paper takes the minimum six degree of freedom coupling degree of the mounting system as the optimization objective, takes the five point mounting stiffness of the powertrain as the optimization variable.

2   Control Structure of 4-DOF Manipulator

2.1   The Composition of Control System of 4-DOF Manipulator

The control system of Gugao four degree of freedom manipulator consists of SCARA four degree of freedom industrial manipulator, driving electrical box computer (including DSP motion control card and upper computer control software). For the multi axis control system, the chip with powerful data processing ability is needed, so the main control chip of the core motion control card is dsp2181 high-speed processor. The general controller needs to deal with complex matrix operation, and the selection of the main control chip must be high-speed and high-precision. In the aspect of communication, standard ISA bus and PCI bus are provided. The real object can be seen in the figure below, in which joints 1, 2 and 3 are rotary joints, and joint 4 is vertical lifting joint. This manipulator provides a design mode of combining computer and development type motion control card. The developer does not spend too much energy on the drive design, but focuses more on the design and implementation of control strategy. Figure 1 is the outline drawing of the 4-DOF manipulator body and the electrical control box [2].

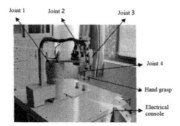

Fig. 1. Outline drawing of four DOF manipulator body and electrical control box

Due to the lighter load of moving objects, there are not too many requirements for the selection of motor control system. In order to study the selection of servo motor of mechanical arm in general, we do some basic research on the drive motor. Now the servo motor brands which are widely used in the market are Panasonic and Sanyo. In general, the appropriate motor and its control card should be selected for the occasion. Firstly, the torque on the reducer is calculated, and the corresponding servo driver and accessories are selected according to the driving capacity of the control card, and the analog voltage or pulse signal is output.

2.2   Mechanical and Electrical Part of Four Degree of Freedom Manipulator

The manipulator model is simplified as shown in Fig. 2.

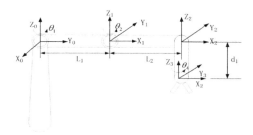

Fig. 2. Schematic diagram of simplified model of manipulator

Estimated motor demand parameters:

Safety factor: n = 1.5
Torque calculation formula: T = Jan
Weight of machine arm parts in handling operation: 500 g

The moment of inertia of the four joints to the base of the manipulator is:

$$J_1 = \tfrac{1}{3}m_1 L_1^2 + m_{10}L_1^2 + \tfrac{1}{3}m_2\left((L_{20}+L_{21}+L_1)^2 - L_1^2\right)$$
$$+ m_{20}(L_{20}+L_{21}+L_1)^2 + m_{21}(L_{20}+L_1)^2 = 2.8\,\text{kg} \cdot \text{m}^2 \tag{1}$$

Similarly, the maximum moment of inertia of the fourth joint to the motor axis of the third joint is:

$$J_2 = \frac{1}{3}m_2(L_{20}+L_{21})^2 + m_{20}(L_{20}+L_{21})^2 + m_{21}L_{21}^2 = 0.56\,\text{kg} \cdot \text{m}^2 \tag{2}$$

According to the demand, the maximum speed of the first joint is x/2rwd/s, and it needs 0.5 s to accelerate to the maximum angular speed. Combined with the motor type of googol company, the motor model can be determined as Panasonic AC msma5aza1g servo motor.

2.3   Software Implementation of Control System for 4-DOF Manipulator

The software of local manipulator control system is a program in the form of dialog box, and the human-machine interface is universal. In the program design, the robot base class is the core, and there are controller class and planner class. The data structures of manipulator attributes such as degree of freedom, joint state, joint motion speed, joint motion acceleration and joint position are defined in the base class. In addition to the data structure definition, various motion functions of the manipulator are also defined, such as single step motion, S-mode motion, etc. the four degree of freedom manipulator class derived from the robot base class is used to create specific objects [3]. The controller class is equivalent to the set of interface functions, which is mainly for the needs of future maintenance. The virtual function declares the corresponding public properties and interfaces of the motion control card, and derives the SG controller class that matches the actual control card. In the SG controller class, the control command response function 14 is specifically defined, and the selection of the shutdown motion mode, such as the servo power on, the control command response function 14 is defined, On/off control card, S/T joint movement mode, origin/limit point capture, etc. The planner class mainly designs the trajectory planning of the manipulator. The class design diagram of the 4-DOF manipulator control system is shown in Fig. 3.

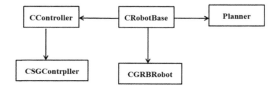

Fig. 3. Class design diagram of manipulator control system

Crobotbase is derived from the base class CObject. It defines some common variables, such as joint angular velocity and joint velocity, current joint position, and the array of temporarily saved variables. Two very important pointers * m controller and * m planner are defined. The first pointer points to the controller class, that is, the motion control class, calls the driving function, and the second pointer points to the planner class, that is, the planner class. In addition to these two class pointers, it also defines the member functions that DG needs to use, such as playback function, save teaching file function, and some function functions, such as the function connected to the controller class. This class mainly describes some common characteristics and functions of the manipulator. The cgrobot class is derived from Cobot base class. The member functions of this class in the control system of the manipulator are all virtual functions. Generally, it is only when solving the kinematics problems that it generates subclasses. The idea of subclass is to add the characteristics of the object to make the object more targeted. The kinematic solutions of the manipulator with different degrees of freedom are also different. The differences in the solutions can be added by subclass.

3   Kinematics Analysis and Trajectory Planning Method of Manipulator

The kinematics of industrial manipulator is an important field in the study of manipulator, which is the basis of manipulator design and control. Manipulator kinematics and robot kinematics are consistent, that is, the movement of the end effector of the manipulator relative to the base coordinate system is regarded as a function of time, regardless of the force and torque between the joints, and the relationship between the joint amount and the pose of the end effector is mainly studied. The kinematics of the manipulator is generally divided into forward solution and inverse solution. The forward solution of kinematics is to know the coordinate value and attitude from the beginning to the end of the industrial manipulator, and to calculate the position and attitude of the end effector relative to the base coordinate system from the coordinate point value in the whole trajectory process. The inverse kinematics solution is the inverse operation of the forward solution, which is to solve the joint angle variables that can drive the desired position of the end effector [4].

A robot arm can be regarded as a kinematic chain composed of a series of rigid bodies connected by joints. These rigid bodies are usually called connecting rods. The connecting rod has four basic attribute parameters, namely the connecting rod attribute parameters and the connecting position parameters of the two connecting rods. Specifically, the common normal distance a and the angle g perpendicular to the two axes in the plane where a is located describe the connecting rod itself; the relative position D of the two connecting rods and the angle between the two connecting rod normals can describe the position relationship as shown in Fig. 4.

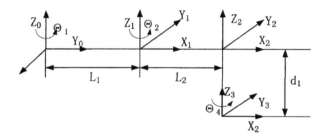

Fig. 4. Connecting rod diagram of SCARA type handling manipulator

4   Concept of Control Trajectory Planning for 4-DOF Manipulator

The trajectory planning of manipulator does not involve artificial intelligence, but studies the method of trajectory generation of terminal actuator in space based on the kinematics and dynamics of manipulator, and describes the trajectory generation with a series of interpolation points. Trajectory not only refers to the generalized route of the end effector, but also includes the displacement, velocity and acceleration of the manipulator in the process of motion. Trajectory planning calculates the desired trajectory according to the needs of customers. Firstly, the motion path of the manipulator, such as the space position of the starting point, the state of the manipulator, such as the space pose required by the starting end, is modeled. Secondly, the path interpolation point to the terminal and the motion speed of the interpolation point end are determined. The coordinate parameters of the starting point are input into the trajectory planner to generate the trajectory. The velocity and acceleration in the process are only expressed in the planning process [5].

There are many common methods to deal with trajectory planning problems. Complex systems can be established to unify all planning problems. The current method has default to black box model as the standard to deal with this kind of problems in manipulator industry. The black box can be used as a platform to apply to all kinds of manipulators. The input variables of path setting, path constraints and dynamic constraints are used as the input of the black box, and the planner calculates the path coordinates under all kinds of constraints. The starting position and attitude lead to the middle position and attitude on the path of the manipulator, and the terminal attitude determines the operation mode of the manipulator. The traditional trajectory planning is divided into two kinds, one is joint space planning, the other is Cartesian space planning. Trajectory planning in joint space refers to the angle change trend of joint amount in each trajectory segment, and the operation time of each trajectory segment is the same. These joint functions and their first and second derivatives are used to describe the expected motion of the robot. Joint space planning method is often used in some high torque occasions, mining heavy industry commonly used joint planning to calculate the loadable capacity, at the same time, it can obtain the desired pose of the middle point of the trajectory, and it can directly use the controlled variables to plan the trajectory; Cartesian coordinate space trajectory planning is mostly used in the actual path needs, because its trajectory is more intuitive and can be simulated in the workplace, Due to the strict constraints on the spatial position, the operation accuracy is high.

5   Conclusion

Firstly, this paper investigates the development trend and status of industrial manipulator at home and abroad. According to the needs of the subject, on the local system of Gugao four degree of freedom manipulator, the trajectory planning of straight line and arc of four degree of freedom manipulator is designed under the development environment of VC + + 6.0, and the trajectory of the manipulator end effector is simulated on MATLAB.

Acknowledgements. Application and research of FP-growth algorithm in data mining", Natural Science Research Projects in Anhui Universities in 2020 (Project No. KJ2020A0806).

References

1. An analysis of Wang Lu's ways to cultivate college students' entrepreneurial ability. J. Ankang Univ. **23** (2011)
2. Saeed, B.: NIKU, Introduction to Robotics. Electronic Industry Press, Beijing (2004)
3. Cai, Z.: Robot Guidance. Tsinghua University Press, Beijing (2009)
4. Bi, S.: Development status of industrial robot at home and abroad, mechatronics of door **2**, 6–9 (2006)
5. Yuan, K.: Development status and trend of industrial robots. Mech. Eng. **7**, 5–7 (2008)

Design of Regional Tourism Culture Integration System Platform Based on Big Data and Cloud Computing

Ming Zhu and Shan Cao[✉]

Applied Technology College of Soochow University, Suzhou, Jiangsu, China

Abstract. With the development of economy, the progress of society, the improvement of national policies and the improvement of people's living standards, China's tourism industry has gradually entered the era of popular tourism, experience economy, smart tourism and national leisure. Fierce market competition, diversified tourist demand and goal setting of interest sharing have become important factors to accelerate the development and reasonable reorganization of tourism resources, diversified presentation of tourism product forms, continuous renewal of tourism organization forms and personalized emergence of tourism operation modes. The innovation and optimization of these economic, social and political environments provide a strong driving force, support and traction for the development of the tourism industry, and promote the deepening, extension and renewal of the cultural connotation, category and form of the tourism industry in the new era.

Keywords: Integration · Cultural integration · Internal mechanism · Integration path

1 Introduction

Economic globalization is a global organic economic whole formed by international trade, capital flow, technology transfer, service provision, interdependence and interconnection. It is mainly manifested in the rapid development of international trade, the free flow of huge funds among countries, the rapid growth of international direct investment, and the spread of transnational corporations all over the world, International organizations have established and formed international economic management system, including the globalization of commodity, service, technology, information, labor, monetary capital and market competition. Economic globalization has accelerated the free flow and optimal allocation of production factors in the global scope, promoted the global industrial restructuring, and profoundly affected and reshaped the global urban regional development pattern [1]. Regional integration is a cross regional economic development mode that local governments use policy means to achieve a common goal, and it is also a high-level way of regional economic cooperation.

© The Author(s), under exclusive license to Springer Nature Singapore Pte Ltd. 2022
J. C. Hung et al. (Eds.): FC 2021, LNEE 827, pp. 1342–1348, 2022.
https://doi.org/10.1007/978-981-16-8052-6_192

2 The Theoretical Basis of Tourism Culture Change in the Context of Regional Integration

2.1 Conceptual Interpretation

Integration refers to the process of integration in order to seek the redistribution of factor resources, market and industrial system in the overall space and obtain more comparative, complementary and selective interests. The emergence of integration has its special background: under the background of economic globalization, the integration of market and industrial system has its own characteristics, With the increasingly detailed division of labor, the intensified market competition and the unbalanced distribution of resources in the world, countries, regions, enterprise groups and organizational alliances gradually break through the limitation of regional space and transcend the nature of the country, Seek the industry, agriculture, finance, service industry and other industries in the capital resources, talent technology, market information and other aspects of exchange and cooperation, in order to promote the effective use of resources, efficient operation of funds, effective development of the market, so as to achieve the ultimate goal of interest sharing. Therefore, the rapid exchange, effective allocation and high integration of comprehensive elements in the whole global industry brought about by integration have given birth to the emergence of a number of international organizations, industry alliances and regional institutions, which is the product and performance of conforming to the development of the times.

Tourism culture, as the sum of cultural phenomena and cultural relations with tourism activities as the core, has the characteristics of comprehensiveness, regionality, contradiction and diversity. According to the perspective of experts and scholars, this paper defines tourism culture as a dynamic evolution result of the deepening of internal elements and external objective performance formed by the interaction of tourism subject, tourism object and tourism media in the process of tourism activities or tourism development. It is the general name of tourism material culture, tourism spiritual culture and tourism institutional culture. In the process of cultural transmission and diffusion, there will be cultural fusion, innovation or assimilation disappear, more or less change. The Encyclopedia of China defines "cultural change" as the change of cultural content or structure, including the growth of new culture and the change of old culture caused by the accumulation, dissemination, integration and conflict of culture, including voluntary change and forced change, limited change and unlimited change, natural change and planned change, cultural mutation and gradual change.

2.2 Correlation Theory

As an important part of cultural anthropology, the theory of cultural change pays more attention to the relationship between human beings and the natural environment, but more emphasis on the function of culture. Through the internal distribution and reorganization of culture, it produces a new cultural form. The occurrence of cultural changes will change the original cultural structure, or produce a new cultural structure to replace the original cultural structure. Its essence is the loss and extinction of the old culture, or the formation of a new cultural model on the basis of the original cultural structure, or the change of the original cultural structure due to the contact and immigration of different cultures [2]. The factors that lead to cultural changes can be divided into geographical factors, cultural communication factors, biological evolution factors and psychological changes by combining the knowledge of human geography, cultural communication, ecology and psychology.

3 An Analysis of the Influence of Regional Integration on the Change of Tourism Culture

3.1 Development Conditions of Regional Tourism Integration

With the deepening of economic globalization and the gradual formation of free market, the scale of tourism industry in the world is expanding and the content is gradually enriched. As a result, a single tourism enterprise or destination is facing the huge challenge of how to maintain its effective competition while facing the global market opportunities and sharing huge profits, This makes the regional integration strategy of "resource sharing, complementary advantages and win-win cooperation" an inevitable choice for countries and regions in the world to develop tourism. Since the mid-1980s, China has gradually started to establish regional tourism economic circles such as Beijing Tianjin Hebei, Yangtze River Delta and Pearl River Delta. In addition, on the basis of economic integration, Changsha Zhuzhou Xiangtan, xiazhangquan, Zhengbianluo and other regions have gradually expanded and deepened the scope and field of cooperation, and gradually formed more influential regional tourism cooperation. The gradual formation and development of these regional tourism economic circles make them reach a consensus in the aspects of resource complementarity, product mix, market co construction, etc., and realize the development goal of benefit sharing while continuously strengthening the regional competitive strength.

3.2 Influencing Factors of Cross Regional Tourism Cultural Change

Under the background of economic globalization and regional integration, regional tourism integration has gradually become a political consensus, industry recognition and academic research hotspot, and has gradually carried out extensive contacts and cooperation between countries, countries and regions, regions and regions, It is under this background and environment that the cross regional tourism cultural change takes place, which is the result of the comprehensive influence of such factors as the economic exchanges between the administrative regions, the optimal combination of resources, the optimal allocation of products, the mutual docking of industries, the guidance of policies and regulations, and the pull of market demand. Therefore, the tourism cultural change of a single administrative region or tourism destination is essentially different from that of the tourism cultural change under the background of regional integration in terms of geopolitical space, cooperation consensus, interest basis and change degree.

4   RNN and LSTM

4.1   Input Participation

Suitable for processing sequence data, that is, when the current input is related to the previous input data, RNN often has a good performance. As shown in Fig. 1, this is a simple RNN, which consists of an input layer, a hidden layer and an output layer. RNN has a reusable neuron, and the weight value s of this neuron depends not only on the current input value x, but also on the value ST1 of the last hidden layer. The weight matrix w records the value of the last hidden layer and participates in the calculation as the weight of this input. It can be seen that the output value of RNN is affected by previous input values, which is why RNN can look forward to any number of input values when processing current data.

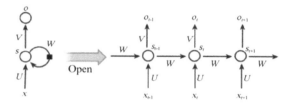

Fig. 1. RNN diagram

The above characteristics make RNN occupy a huge advantage in processing sequence data, but RNN also has defects [3]. Because the gradient descent algorithm of RNN is an exponential function, if the exponent is too large, that is, the reference input value is too far away, the value of the corresponding error term will grow or shrink very fast, so there will be the corresponding 6 gradient disappear or gradient explosion problem.

4.2   Gradient Explosion

Gradient explosion is easier to deal with. When the gradient explodes, a null value will be received, and a gradient threshold can be set here. When the gradient exceeds this threshold, it can be intercepted directly; and the disappearance of gradient is more difficult to detect and process. The disappearance of gradient means that the gradient has almost decreased to 0 from a certain moment, and then the gradient (almost 0) obtained from that moment will not make any contribution to the final gradient value, which means that no matter what the network state is before that moment, it will not have an impact on the training, That's why the original RNN can't handle long-term dependencies. Therefore, LSTM came into being. Considering that the original RNN has only one hidden layer, it is very sensitive to short-term input. LSTM adds a state in the hidden layer to save the long-term state, which solves the problem that RNN cannot handle the long-term dependence.

The formula of "input gate" is:

$$i_r = \sigma(W_i[h_{t-1}, x_t] + b_i) \tag{1}$$

$$c_t = \tanh(W_c[h_{t-1}, x_t] + b_c) \tag{2}$$

$$c_t = f_t C_{t-1} + i_t C_t \tag{3}$$

5 Analysis of the Possibility of Cultural Integration in the Yangtze River Delta Regional Integration

For the analysis of the possibility of cultural integration in the Yangtze River Delta regional integration, we can learn from the "push pull theory" in the field of population migration.

5.1 On the Impetus of Cultural Integration in the Yangtze River Delta from the Supply Side

Culture and the development of human society are accompanied by each other, and different regional conditions give birth to different characteristics of regional culture. The integration of culture is based on all kinds of subcultures in the region [4]. The Yangtze River Delta has a superior geographical location, dense flow of people, and developed economy since ancient times, which provides the popularity and geographical advantages for the birth and cultural integration of all kinds of subcultures.

The four provincial-level cities included in the Yangtze River Delta integration are closely related. They are connected from north to south, from east to west, and from the outside to the inside. The dense inland transportation network and developed water system in central and Western China further promote economic activities and people to people exchanges in the region. They are one of the regions with the earliest start, the best foundation and the highest degree of regional integration in China, The blending of culture has been steadily carried out under the congenital geographical conditions for a long time.

The formation of the Yangtze River Delta can be traced back to the Eastern Jin Dynasty in the 4th–6th century, the southern and Northern Dynasties and the Southern Song Dynasty in the 12th–13th century, as well as the construction of the inner road transportation network in the later period. It gradually became China's famous "land of fish and rice" and "land of silk". On this basis, it formed four subcultures: haowenhua in Jiangsu, Yue Culture in Zhejiang, Shanghai culture and Hui culture in Anhui. Wu and Yue culture, with its characteristic of being meaningful and introverted, especially the quality of wisdom and sincerity in Wu culture and the spirit of utilitarianism and high quality and effectiveness in Yue culture, had an important impact on the development of Jiangsu and Zhejiang in modern times. It also gave birth to practical achievements such as "Southern Jiangsu model" and "Eastern Zhejiang School", which made Jiangsu and Zhejiang model become the object of study in the central and western regions of China since modern times.

5.2 From the Demand Side

The pulling force of cultural integration in the Yangtze River Delta on the growth of demand for cultural products is not only the performance of consumption upgrading, but also the proof of the continuous improvement of people's living standards. In turn, the huge cultural demand market further promotes the development of cultural productivity. According to the statistical bulletin of national economic and social development of the people's Republic of China, in 2017, the per capita consumption expenditure of national residents was 18322 yuan, and the per capita consumption for culture, education and entertainment was 2086 yuan, accounting for 114% of the per capita consumption expenditure. As an endogenous driving force for the development of cultural industry, from 2013 to 2015, the growth rate of cultural consumption has accelerated, and its contribution to culture and related industries has become increasingly prominent. The proportion of cultural consumption in the added value of culture and related industries has increased year by year, from 37.51% in 2013 to 3973% in 2015, an increase of 2.22% points [5]. At the same time, the proportion of cultural consumption in GDP is also increasing year by year, from 1.32% in 2013 to 1.52% in 2015, an increase of 0.2% point. It can be seen that cultural consumption drives the development of culture and related industries, and then plays an increasing role in promoting economic growth.

6 Concluding Remarks

To promote the high-quality development of the Yangtze River Delta, cooperation in the cultural field is also an important part. At the key node of the Yangtze River Delta's core cities about to enter the era of urban integration, cultural integration will become an important thrust to deepen the high-quality development of the Yangtze River Delta region and create high-quality life, as well as an important carrier to enhance the urban energy level and core competitiveness, As a strong soft power serving the people and the overall development of the Yangtze River Delta.. The theory of "push pull" in population migration holds that the factors that are conducive to improving living conditions become the pull of population flow, and the unfavorable living conditions that flow out are the push.

Acknowledgements. The Project of Philosophy and Social Science Research in Colleges and Universities in Jiangsu Province;Study on temporal and spatial characteristic transmutation of tourism route network structure in Yangtze River Delta; 2019SJA2127.

References

1. Han, L.: Thoughts on improving the soft power of regional tourism culture. Tour. Overv. (Second Half) (22), 50 (2016)
2. Zhan, L.: Research on regional tourism culture development strategy based on cultural brand image – taking Shen Congwen and Zhao Shuli as examples. Hunan Soc. Sci. (05), 195–198 (2016)

3. Ma, L.: Analysis on the role of tourism handicrafts with regional cultural characteristics in promoting the development of tourism cultural creative industry. J. Guilin Inst. Aerosp. Technol. **21**(03), 430–433 (2016)
4. Chen, Y.: A study on the regional tourism cultural change of ancient towns in Jiangnan based on tourists' perception. Shanghai Normal University (2016)
5. Yang, Z.: Analysis of the impact of tourism cultural phenomenon in events and festivals on regional economic development – taking Henan Jiaozuo International Taijiquan annual meeting as an example. Urban Geogr. (02), 247 (2016)

Application of Improved Ant Colony Algorithm Based on Slope Factor in Wireless Rechargeable Sensor Networks

Yan Huang[(✉)]

Chongqing Chemical Industry Vocational College, Chongqing 401228, China

Abstract. In recent years, with the deep application of wireless sensor network technology, the problem of energy consumption has become an important factor to limit its development. In the process of reducing the energy consumption of wireless sensor networks, the ant colony algorithm can maximize the charging utility of wireless rechargeable sensor networks (WRSNS). Aimed at the shortcoming of ant colony algorithm, the intelligent autonomous path planning for optimum speed is slow and the initial pheromone shortage, this paper puts forward an improved ant colony algorithm based on slope factor (SFACA). Increase the random initialization pheromone, pheromone concentration in the process of dynamic adjustment algorithm iteration, to speed up the convergence speed of the algorithm, reduce energy consumption on the road and during charging. The simulation results show that the improved ant colony algorithm can significantly improve the convergence speed, and it is easy to find the global optimal path between charging anchors in WRSNS.

Keywords: Ant colony algorithm · Slope factor · Wireless rechargeable sensors · WRSNS · SFACA

1 The Introduction

Power transmission, the problem of energy bottleneck has made a new breakthrough. WPT can be used to provide power supplement for WSN in the network. model. WRSNS mainly wireless charging sensor nodes (SN) and wireless charging vehicle (WCV). The collecting data from the sensors and quickly replacing the batteries for the WCV. The WCV is equipped with a wireless power transmission device that can be used to power the sensor nodes wirelessly. Theoretically, the wireless rechargeable sensor network can keep working state permanently, but due to the constraints of space, time and energy factors, the traditional scheduling scheme can only meet a small number of charging request nodes, which leads to a relatively short life cycle of WRSN in a busy network environment. Based on the Maximum Average Gain Coverage (MAGC) algorithm in Literature [1], this paper screened out the set with the largest Average Gain of directed Coverage nodes in the network, forming one charging anchor point to cover each sensor node in the network. The improved ant colony algorithm based on slope factor was used to reasonably schedule the charging line, and all sensor nodes along the way were charged wirelessly through charging anchor points.

© The Author(s), under exclusive license to Springer Nature Singapore Pte Ltd. 2022
J. C. Hung et al. (Eds.): FC 2021, LNEE 827, pp. 1349–1355, 2022.
https://doi.org/10.1007/978-981-16-8052-6_193

2   The Model of Wireless Rechargeable Sensor Networks

2.1   Network Model

Can be expressed as G = (V0, Vk, E). Where, V0 stands for the data center, which provides energy for the sensor nodes and receives the data returned by the sensor. Vk is the sensor nodes randomly distributed in the plane, and each sensor is equipped with a rechargeable battery. This paper mainly describes cruises to charge the sensors in the model. The network system l is shown in Fig. 1.

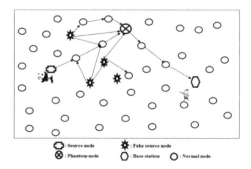

Fig. 1. The model of WRSNS network

2.2   Charging Model

The classification models, existing research results are mainly divided into one-to-one charging models and one-to-many charging models. In the one-to-one model, WVC starts from the data center, provides one-to-one charging service for sensor nodes along the way, and finally returns to the data center. The disadvantages of this method are mainly the low charging efficiency, and the implementation is limited by the specific environment. In order to further improve the charging efficiency of mobile charger, one-to-many charging mode has been discovered and developed well. This scheme mainly uses wireless charging vehicles to visit each docking station in WRSN regularly and charge the nodes within the coverage range. Charging efficiency is improved by reducing points optimizing driving the mobile charger. In this paper, the one-to-many charging mode is mainly adopted, and the wireless charging line vehicle (WVC) charges the charging anchor points along the way according to the scheduling requirements of WRSN, as shown in Fig. 2. In this paper, the wireless charging trolley traverses the charging anchor points along the way through reasonable scheduling algorithm to form the shortest line for wireless charging of all sensors. The 3D dimensions of various building components can be derived, and various reports, project progress and budget can be automatically generated. The accuracy of BIM is directly proportional to the accuracy of modeling. BIM's powerful database function also provides great convenience for future project repair, maintenance, reconstruction and even demolition. Because the model can accurately calculate the quantity of each material, and even directly generate the quantity report, it makes the construction cycle very easy. Relevant personnel only need to adjust the report automatically generated by the model according to their own professional experience to get a more accurate estimation. The three-dimensional digital simulation model based on BIM Technology can realize the virtual design, visual decision-making, collaborative construction and transparent management of construction engineering. It will greatly improve the level of project decision-making, planning, survey, design, construction and operation management, reduce errors and shorten the construction period [2]. It will significantly improve the informatization level of construction industry, promote the development of green buildings and realize the construction industry Transformation and upgrading.

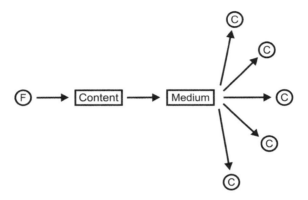

Fig. 2. One-to-many charging model

3   Slope Factor in Wireless Rechargeable Sensor Network

3.1   Main Application of Ant Colony Algorithm

Mechanism to solve complex problems, relying on pheromone cooperation rather than the information exchange mechanism between individuals, so that the algorithm has a high scalability. The ant colony algorithm has great advantages in combinatorial optimization and path planning because of its good robustness, strong positive feedback mechanism and easy combination with other algorithms. Of course, in the process of using ant colony algorithm, researchers also found its shortcomings: (1) When the algorithm is running, the system will spend a long time searching for the optimal solution. The searching movement of ants is random and arbitrary, and when the population size of ant colony is huge, it will consume much computation time to find a good search path. (2) The algorithm is easy to fall into local optimal: Ant Colony Algorithm is easy to fall into stagnation in the process of searching for the optimal solution. Sometimes, after searching for a certain period of time, its solutions tend to be consistent, leading to the failure of searching for the global optimal solution.

3.2   Ant Colony Algorithm Based on Slope Factor (SFACA)

We find that the key to improve the performance establish "Exploration" means that the exploration space of the ant colony must be large enough to easily find the possible optimal solution space. "Utilization" means that the effective information currently possessed by the ant colony should be fully utilized so as to converge to the global optimal solution with a greater probability. In view of the above analysis, slope factor is introduced to dynamically adjust pheromone concentration in this paper, so as to speed up the optimization process of ant colony algorithm. The following formula is used to calculate the slope factor.

(1) The slope between node j and node i

$$S_{ij} = \frac{h_{ij}}{PL_i} \qquad (1)$$

Where, represents the slope length at the node i. The calculation formula , represents the distance between node i and adjacent nodes, represents the slope height corresponding to each path.

(2) The base of slope factor

$$M_{ij} = \begin{cases} 4, & \left(S_{ij} < 0.5\right) \\ 3, & \left(0.2 < S_{ij} \leq 0.5\right) \\ 2, & \left(S_{ij} \leq 0.2\right) \end{cases} \qquad (2)$$

Where, is the weighting coefficient. It can be concluded from Equation that the slope factor on the path with shorter distance is larger, while the slope factor on the path with longer distance is smaller. In the process of ant optimization, it is easier to find the shorter path and abandon the longer path when using the roulette algorithm to make a choice.

(1) Adjust the concentration of the initial pheromone [3]. Ant colony algorithm choose the pheromone a nodes position, planning parameter is set the length of the distance between the nodes usually relies on experience, so the early stage of the larger workload, The concept of slope factor is introduced here, it is used to change the randomness of ants choosing the next node in the network and increase the probability of ants choosing the shortest path.

$$\Delta \tau_{ij}^k(t) = \begin{cases} \frac{Q}{L_k} \cdot SLP_{ij}(k), \ d_{ij} \in L_{short} \\ 0, \qquad otherwise \end{cases} \qquad (3)$$

(2) Improve the update rules of pheromones. In ant colony algorithm, all the ants will update pheromone when population after completing a cycle, this case is not fully reflect the optimal path of guide, lead to some poor populations of pheromone. After introducing the slope factor, the change of the slope factor is used to update the pheromone concentration in ant colony algorithm. In each cycle, add more heuristic information, update more useful pheromones and discard invalid pheromones.

Before the application of BIM Technology, it is necessary to build a building model to analyze the influence of environmental factors around the building on the building, and on this basis to explore technical measures to optimize the comfort. In the design process, BIM Technology is used to analyze the simulation model, and the best effect is achieved through repeated comparison. In the process of three-dimensional design, firstly build the building information model, including HVAC, water supply and drainage, electrical and other aspects, and then the relevant units carry out reasonable division of labor, test the accuracy of the model, analyze and correct the collision points, and form the final building data model. The construction of the model can provide the basis for green building analysis, and the analysis results can provide guidance for the model until the best construction drawings are formed.

3.3   Optimal Path Planning of Improved Ant Colony Algorithm

Consists of 120 sensors, 1 data center and 1 wireless charging vehicle. Firstly, 35 sensor charging anchor points were screened out through the Maximum Average Gain Coverage (MAGC) algorithm (and the longitude and latitude of the 35 charging anchor points were given). Then, a wireless charging vehicle (WCV) started from the data center, trawled the charging anchor points in the system, charged all the sensors, and returned to the data center.

Because of the randomness of ant colony algorithm, it usually needs to run many times to get the best running result when solving WRSNS shortest route planning problem with ant colony algorithm. Here we make a comparative analysis between the slope factor At a maximum of 100 iterations, the basic ACA and the improved ACA each run 20 times. Record the shortest route and the length of the shortest route in each run, and select the route with the best result in 20 runs for comparison. The following are the shortest route found by basic and improved ant length e shortest route. See Figs. 3 and 4.

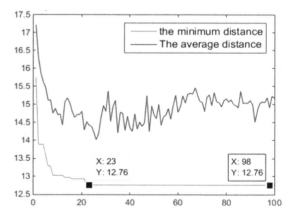

Fig. 3. The minimum distance of ACA

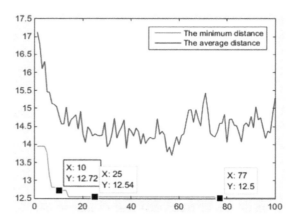

Fig. 4. The minimum distance of SFACA

From the above fig, we can draw the following conclusion:

(1) From the average distance and the shortest distance (Figs. 3 and 4), ant colony algorithm iterative 23 times find local shortest path after 12.76 km, and it can't jump out of local optimum [4]. At the beginning of the iteration, the improved ant colony algorithm based on slope factor can find a shorter path (iterative 13 basic found shorter path length of 12.54 km), the iteration after 70 times still can global, global shortest path length of 12.5 km.

(2) From the shortest route found by the two algorithms there is no intersection between nodes on the shortest route found by the improved sic ant colony algorithm.

4   Conclusion

In this paper, an improved ant colony algorithm based on slope factor is proposed to solve the charging scheduling scheme of wireless rechargeable sensor networks. In this scheme, n charging anchor points that can cover all wireless sensors are selected by using MAGC algorithm according to the location information of sensor nodes in the network, and then the shortest path between charging anchor points is found by using improved ant colony algorithm. This algorithm is suitable for small wireless rechargeable sensor networks with many sensors. In large-scale wireless rechargeable sensor network systems, we can increase the number of charging cars or improve the battery life of sensors.

References

1. Xu, R.: Insect Population Ecology, pp. 61–84. Beijing Normal University Press, Beijing (1987)
2. Zhang, L.X., Li, S.W.: Grey correlation analysis of height growth and climatic factors of populus canadensis and Robinia pseudoacacia seedlings. J. Hebei For. Univ. **6**(4), 259–264 (1991)
3. Gu, D., Zhou, C., Tang, J., et al.: Study on the life table of natural population of Cnaphalocrocis medinalis. Acta Ecol. Sin. **3**(3), 229–238 (1983)
4. Yuan, J.: Grey relational analysis method. For. Econ. (3), 58–62 (1991)

Analysis on the Application of Computer Intelligence System in College Physical Education

Xin Feng[1(✉)] and Xinggang Liu[2]

[1] Chongqing Vocational Institute of Engineering, Chongqing 402260, China
[2] Chongqing Medical Univercity, Chongqing 402260, China

Abstract. In the process of physical education teaching in Colleges and universities, teachers should actively introduce new teaching methods, and constantly improve the quality of curriculum education on the basis of enhancing students' interest in learning. At present, distance teaching and multimedia teaching have been widely used in the teaching of various disciplines, intelligent teaching method is generally recognized by teachers and students, and can improve students' interest in learning, but it is still in the development stage. In view of the shortcomings of current physical education and computer intelligent system, according to the dynamic characteristics of physical education, this paper puts forward a design of intelligent physical education teaching system. In the application, we can find that the design of knowledge-based intelligent physical education teaching system is flexible, teaching is not limited by time and place, and can meet the needs of students in different situations.

Keywords: Computer · Intelligent system · Physical education teaching

1 Introduction

The purpose of offering physical education courses in Colleges and universities is to enhance the physical fitness of college students and cultivate their awareness of physical exercise. In view of the traditional college physical education teaching mode, affected by the objective reasons such as less class hours, students' low interest in learning, teaching difficulty and so on, the teaching efficiency has been difficult to improve.

With the promotion of computer virtual reality technology, the 21st century is recognized as the information age. Computer science and technology has been widely used in many fields, such as social politics, economy, culture and so on. However, in college physical education, due to the limitations of ideological understanding, computer knowledge level and hardware conditions, the application of computer technology lags far behind its application in other fields. In terms of the current development trend, changing the traditional teaching concept and introducing computer technology into physical education is a problem that can not be ignored in college physical education. In college physical education, using computer virtual reality technology to assist teaching has become an effective innovation to improve teaching efficiency.

J. C. Hung et al. (Eds.): FC 2021, LNEE 827, pp. 1356–1361, 2022.
https://doi.org/10.1007/978-981-16-8052-6_194

2   Computer Intelligence System

Intelligence system is a computer system that can produce human intelligent behavior. Intelligent system can not only run on the traditional Neumann computer with self-organization and self adaptability, but also run on the new generation of non Neumann computer with self-organization and self adaptability. It has the following characteristics.

(1) Processing objects
     Intelligent system deals with not only data but also knowledge. The ability to represent, acquire, access and process knowledge is one of the main differences between intelligent systems and traditional systems. Therefore, an intelligent system is also a system based on knowledge processing. It needs the following facilities: knowledge representation language; knowledge organization tools; methods and environment for establishing, maintaining and querying knowledge base; and supporting the reuse of existing knowledge.
(2) Treatment results
     Intelligent systems often use the problem solving mode of artificial intelligence to obtain the results. Compared with the solution mode of traditional system, it has three obvious characteristics, that is, its problem solving algorithm is often uncertain or heuristic; its problem solving depends on knowledge to a large extent; the problems of intelligent system often have exponential computational complexity. The problem-solving methods commonly used in intelligent systems can be roughly divided into three categories: search, reasoning and planning.
(3) The difference between intelligence and tradition
     Another important difference between intelligent system and traditional system is that intelligent system has the ability of on-site sensing (environment adaptation). The so-called scene induction refers to the fact that it may communicate with and adapt to the abstract scene of the real world. This kind of communication includes perception, learning, reasoning, judgment and making corresponding actions. This is what people usually call automatic organization and automatic adaptability.

3   The Role of Computer Technology in Physical Education Teaching Practice

In the past physical education teaching, the teaching of skills and technology mainly depends on the demonstration and explanation of physical education teachers. Therefore, the strength of teachers' demonstration ability directly affects the teaching effect. Teachers' standard and beautiful demonstration can not only make students feel the essentials of learning directly, but also stimulate their interest in learning. Therefore, physical education teachers in the process of teaching skills are trying to do a good job in every demonstration action. However, as we all know, the teaching content of Public Physical Education in Colleges and universities includes track and field, gymnastics, ball games, martial arts and many other items. It is difficult for every physical education teacher to make the demonstration of every action so accurate and perfect. If the computer can be used to synthesize every action, it will be more accurate. This is one of them.

Second, it will be more accurate, Because physical education teaching is different from other teaching, teachers should give students standard demonstration while explaining a technology. Many technical movements are complex in structure and can be completed in an instant. It is difficult for teachers to decompose them and students to understand the main points. Therefore, in physical education teaching, we should make full use of the multimedia technology of computer, projector and video playback, sort out the more difficult movement technology, make it into computer animation, edit it into a repeatable, slow, dynamic and static demonstration, and then match it with concise and clear text instructions that can grasp the key points, so as to replace the teacher's demonstration or technical analysis, It will enable students to clearly understand the technical essentials and action structure of the actions they have learned. Thirdly, students' thinking mode is very active in the process of learning sports, which needs a lot of communication between teachers and students. The design of intelligent sports teaching system mainly realizes the communication between teachers and students. Students can access and query relevant information through the teaching system, and can communicate with teachers. The teaching system can realize students' online learning in the design, so as to improve the teaching effect. The designed intelligent sports teaching system can stimulate students' active learning, and manage students' learning status.

Therefore, the teaching mode of computer intelligent system can be more targeted teaching, teaching can be carried out randomly according to different learning ability, without fixed teaching time or teaching place. Due to the strong practicality of physical education textbooks, the intelligent physical education teaching system is more applicable.

4   The Role of Computer Technology in Physical Education Teaching Practice

The design of any project needs to have clear design criteria to limit and standardize the operation. The traditional physical education teaching system design can reflect less elements of human personality, mainly the application of educational technology. Therefore, the key part of this design is the integration of human personality and educational technology, to achieve a high level of system research.

4.1   Design Criteria of Intelligent Physical Education Teaching System

The design of intelligent physical education teaching system is not only to meet the needs of teaching, but also for future design consideration, that is, the design of teaching system needs to be scalable. In the design, the design of software and hardware should be continuous, and the program design should be able to be modified to facilitate the later maintenance. The design of intelligent physical education teaching system should be practical, which requires that the design function is comprehensive and convenient for the reform of school education. In addition, the design of the system should be easy to understand and control to avoid the waste of resources. In the design of intelligent sports teaching system, we should also pay attention to the safety and stability, which is the most basic principle. Only by ensuring the safety and reliability can we ensure the normal operation of each functional area. When selecting the corresponding products and technologies, we should fully consider the various problems that may be encountered, so as to maximize the stability of the intelligent sports teaching system.

4.2   Design Goal of Intelligent Physical Education Teaching System

The design of the intelligent sports teaching system is based on the campus network, using multimedia design, can realize the user's independent learning, can choose the appropriate content, convenient for students' learning needs. In addition, it is also required to ensure the flexibility of the teaching process, which is not limited by the time and place. The system does not need to get the user's use of CAI courseware through authentication, and can automatically deal with the possible ambiguity in teaching.

5   Design and Implementation of System Structure

5.1   The Structure Design of the System

B/S architecture is adopted in the design of the intelligent physical education teaching system (as shown in Fig. 1). Teachers and students can operate through the browser, and the database server stores the database and related management software needed for teaching. Students can realize the learning process in the computer room. The web server can adjust the teaching strategies according to the different needs of students, and provide suitable teaching information for different students. The teacher can update the teaching content, maintain the organization data and modify the information of the examination database through the browser, and can also check the students' learning level and examination results through the designed intelligent sports teaching system.

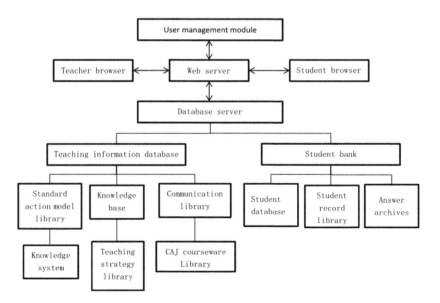

Fig. 1. Mechanism design of intelligent physical education teaching system

5.2   Design and Implementation of Database Server

The data of the server is divided into teaching information database and student database. The design of teaching information database includes knowledge database, standard action model database and student database. The design function of knowledge base is to ensure the design of the system: having the knowledge of sports discipline, which is the most basic design content, can effectively reflect the knowledge system of the discipline, and can realize the selection of different students' teaching materials in the design. The knowledge base designed in this paper is mainly composed of knowledge system, teaching strategy and CAI courseware. The form is premise one, premise two, conclusion one, conclusion two and so on. The knowledge system includes the practice between knowledge points, which is the smallest unit of the course. In the establishment, according to the characteristics of sports discipline, it is endowed with attributes, numbers, importance and other characteristics (as shown in Table 1). The content of CAI courseware library corresponds to the design of knowledge points. The selection of courseware is based on the deep study of teaching materials and syllabus, and the teaching content difficult to understand and demonstrate is selected from the teaching practice as the key point. Therefore, the courseware library is the key point of design, and the design of courseware content should be carried out with the teaching practice as the center. In the teaching strategy database, there are various teaching methods, which are divided into different levels according to the learning objectives and effects.

Table 1. Knowledge system base

Number	Difficulty	Importance	Required mastery	Antecedent knowledge	Postorder knowledge
111	2	2	Imitate	100	120
112	2	2	Imitate	101	120
......	......	......	......	......	......

The main design contents of standard action model library include: Sports normative action, similarity, name and image file, etc. The main content of the design of student database includes students' personal information, learning objectives and mastery. In the design, the student database can be divided into database, record database and answer file database. The contents recorded in the database include students' names, classes, student numbers and other basic information; the learning record database mainly records students' learning progress and mastery degree, and the answer file database mainly records students' grades.

Teachers and students can query the category and name of sports through the system, and can print through the specified action of the browser. The design of review test feedback module can find out the errors and reasons in the learning process of students in time, and then adjust the teaching strategy. In the test, the corresponding sports technology can be adjusted according to the students' ability and cognitive psychology, so as to greatly improve the students' interest in learning. In addition, the online discussion library is added in this design, which is not limited by time and place, and reflects the freedom of learning to the maximum extent.

6 Conclusion

The application of computer intelligence system in college physical education teaching is a teaching innovation in line with the law of college physical education development. The application of computer intelligent system can effectively improve the teaching efficiency on the basis of stimulating students' interest in learning, so that students can master sports skills and love sports more, so as to develop various sports into their own hobbies. It is hoped that in the process of College Physical Education in the future, the majority of college physical education workers can further study the application of computer virtual reality technology, so that students can get a better learning experience, and constantly improve the efficiency of physical education.

References

1. Liu, F., Wang, L., Yu, S., et al.: Design and implementation of intelligent remote multimedia physical education system. Electron. Test. **22**(20), 222–223 (2013)
2. Wu, Q.: Design and implementation of web design excellent course network teaching system based on ASP. Net. Ocean University of China, Qingdao (2011)
3. Hu, Z.: Design and implementation of network distance education system based on video conference. East China Normal University, Shanghai (2012)
4. Chen, M.: Design and implementation of wireless intelligent teaching system in Yuanqing middle school. Dalian University of Technology, Dalian (2013)

Market Economy Forecast Analysis Based on Deep Learning

Yan Wu[✉]

Yunnan College of Economics and Management,
Kunming 650106, Yunnan, China

Abstract. With the continuous development of Internet and computer technology, the focus of enterprise organization in marketing activities has changed, and the horizontal marketing mode has changed from traditional marketing to network marketing mode. This paper first analyzes the connotation of the era of network economy, then points out the change points of marketing environment in the era of network economy, and finally analyzes how to carry out network economy from three aspects The impact of marketing strategy change on enterprises in the new era ψ. The future development is of constructive significance.

Keywords: The era of network economy · Marketing strategy · Change

1 Introduction

In the current network era, network marketing has many advantages, plays a huge role in integrating social resources and connecting economic entities, and gradually changes the development imbalance of traditional marketing, improves the economic benefits of enterprises, and lays a solid foundation for the sound development of enterprises' economy. After entering the 21st century, China's national economy has achieved sustained growth, which makes all kinds of technical means in the field of scientific research also get rapid development in this process. With the help of Internet technology, the public has made great changes in their life style and consumption concept. To change the previous offline marketing mode into online marketing mode, so that consumers can get convenience at the same time, it can also strengthen the stickiness of consumers.

2 The Connotation of Marketing in the Era of Network Economy

2.1 The Connotation of Network Economy

Network economy is an economic system composed of digital economy, information economy and knowledge economy. Because of this special structure, network economy is different from traditional economy in economic system and economic development mode. In terms of economic system, network economy builds an economic system with customers as the main service object based on Internet technology and knowledge reserve. In terms of economic development mode, network economy builds a highly systematic and intelligent economic development mode based on the technical characteristics of Internet open all day, efficient data computing and multiple mobile network terminals [1]. At the same time, the network economy is also facing many challenges, such as in the new network economy environment, how enterprises should carry out market activities. Therefore, we must deeply study the marketing mode in the era of network economy.

2.2 The Connotation of Network Marketing

On the basis of traditional marketing mode, e-marketing integrates customers' needs into the model of pursuing enterprise profit maximization. On the basis of Internet non habitability, it formulates personalized demand scheme for customers through continuous interaction and communication with them, so as to produce products that best meet customers' needs and achieve enterprise profit maximization, The network marketing mode shifts the focus from the pursuit of profit to the satisfaction of customers' needs, including caring about customers' needs and expectations, the cost of customers' purchasing products, the convenience of customers' purchasing products, and the communication between customers and enterprises. On the basis of this kind of work, gradually promote the products and improve the profits of the enterprise. In order to improve the accuracy of marketing activities and achieve accurate positioning, each enterprise must improve its marketing level, because in the environment of network marketing, the products produced by enterprises are basically transparent, so that consumers have greater choice space. Therefore, enterprises must actively cater to consumers' demands for personalized products and services, In order to achieve the stability of the existing customer groups, and on this basis continue to expand their own user groups.

3   The Change of Marketing Environment in the Era of Network Economy

With the advent of the era of network economy, the original advantages, marketing environment, marketing means and marketing strategies of enterprises have changed. According to the commonness of these changes, we can add them into two aspects: the expansion of market scope, the increasingly rich and reliable payment channels.

3.1   Expansion of Market Scope

Generally speaking, the market scope includes two dimensions: time dimension and space dimension. In terms of time dimension, due to the openness of cyberspace, businesses in the e-market can open all day, so the virtual e-market has a larger market scope advantage than the physical stores. In terms of space dimension, people shop in the virtual e-market, such as buying goods on the Jingdong website, Online ordering and offline mailing can save people's cross regional shopping time, and people's daily shopping becomes more convenientp [2]. Due to the wide coverage of Internet technology, the development of network marketing can also promote the development of consumer industry to a great extent. When enterprises develop to a certain scale, they can use network information technology to further expand their market scope. As shown in Fig. 1. In the era of network economy, due to the improvement of social and economic level and the upgrading of people's consumption, the requirements for product quality and service quality are also higher. Therefore, the requirements for the development and marketing of enterprises have also changed to a certain extent, which has a certain impact. For example, consumers show personalized demands for the use of products and services, and enterprises are facing greater market pressure, The competition between modern enterprises is upgrading.

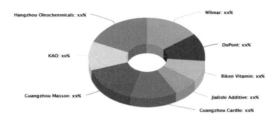

Fig. 1. Expanding market scope

3.2   The Means of Payment are Increasingly Rich and Reliable

Electronic payment has the characteristics of agility and security. In the era of network economy, convenience is reflected in the diversification of electronic payment methods. For example, with the widespread use of mobile phones, electronic credit cards and electronic cash can be used to pay for online shopping. In the era of network economy, security is reflected in the government's policy protection of electronic payment methods, which makes electronic payment methods more reliable. With the improvement of social and economic level, the income level of the public has been greatly increased. In this process, the public has higher requirements for the quality of goods and services, which makes the consumption level of the public continue to improve. From this level, we can see that the public's consumption view has changed dramatically compared with the past. This change is embodied in that the price of goods and services is no longer the only object for consumers to consider. Consumers will also consider other factors in the process of purchasing goods and receiving services. These factors include the added value of products, the cultural connotation of products, the value-added value of products, the value-added value of products, and the value-added value of products The quality of service, etc. In view of this, in the process of conducting and engaging in marketing activities, enterprises must accurately understand and grasp the current consumer values, and in this process, actively revise and adjust the enterprise's marketing plan, such as the adjustment of enterprise development strategy and the accurate division of business areas [3]. Through this series of changes made by enterprises, we can truly ensure that consumers' heterogeneous consumption demands are met and responded.

4 Marketing Change Strategy in the Era of Network Economy

With the change of market environment in the era of network economy, enterprises have new requirements for the way of network marketing. Then, we should first set about establishing the brand image, secondly focus on meeting the needs of consumers, and finally innovate the marketing strategy by using technical means, so as to change the marketing strategy in the era of network economy from these three levels.

4.1 Establish Brand Image

In the era of network economy, there are two trends in the competition of the same industry: the homogenization tendency of super potential products and the shift of competition focus. In the fierce market competition, people in the same industry make full use of the openness of the Internet to search for information. Constantly learning from each other and striving to optimize products, the differences between products become smaller and smaller, and then there is a tendency of product homogeneity, which is the main reason for the shift of competition focus in the same industry. Due to the appearance of product homogeneity, enterprises have to shift the focus of competition to high-quality service. Therefore, in order to actively change the focus of work, enterprises need to start from the service attitude and improve the production capacity of high-quality service, so as to establish a good brand.

$$d = e - 1\{\mathrm{mod}(p-1)(q-1)\} \tag{1}$$

$$c = me(\mathrm{mod} - n) \tag{2}$$

$$m = cd(\mathrm{mod} - n) \tag{3}$$

4.2 Meet the Needs of Consumers

To meet the needs of consumers, we need to shift the focus to user information collection and product pricing. In terms of user information collection, enterprises should make full use of Internet technology to widely collect user information, and apply user information to product design, so that products can fully meet the needs of consumers. In product pricing test, product pricing is the main factor affecting enterprise marketing work [4]. Therefore, both the supplier and the demander can jointly negotiate the product price, In this process, we should not only consider the cost, but also consider the consumer's affordability and consumption level. Through user information collection, product pricing to improve the ability of enterprises to meet the needs of consumers, in order to do a good job in marketing.

4.3   Innovation of Marketing Strategy by Means of Technology

It is necessary to shift the focus of work to the construction of network platform for strategic innovation of sodium marketing by means of technology. Through the construction of network platform, the diversified development of product marketing can be promoted. Consumers from all over the world can communicate with enterprises through the network platform, feed back product problems to enterprises, safeguard their rights and improve the service quality of enterprises. It has the advantages of low cost and high efficiency to spread product information to consumers through the network platform, which promotes the circulation efficiency and marketing strategy of products.

5   How to Change the Marketing Strategy in the Era of Network Economy

In the era of network economy, if we want to change the marketing strategy, we need to use the big data platform scientifically, actively change the marketing concept, actively develop the network platform, change the traditional promotion strategy, strengthen the Internet precision marketing training of marketing personnel, and pay attention to product quality and service, so as to make marketing more suitable for the needs of users.

5.1   Scientific Use of Big Data Platform

Information technology has been widely used in enterprise marketing activities. Among them, cloud computing big data technology provides great convenience for the upgrading of enterprise marketing activities. With the help of this technology, enterprises will be able to accurately understand the demands of consumers, and use the network platform to push two products to consumers. It can not only make the corporate culture understood by consumers, but also create a good corporate image in consumers' mind. For the marketing activities of enterprises, with the help of cloud computing big data extension technology, it will ensure that the book marketing activities of enterprises are more efficient, make the book marketing activities of enterprises avoid uncertainty risks to the greatest extent, and finally make enterprises obtain the expected income. From the perspective of corporate image, corporate image and corporate credit are equivalent to sunk costs. Therefore, when an enterprise adjusts its marketing strategy, it must make macro analysis with the help of big data technology, so as to avoid the influence of the enterprise's status and image in the eyes of consumers due to the omnipresent marketing strategy adjustment. At the same time, with the help of the effective application of big data technology, enterprises understand the demands of consumers of different ages, different incomes, different genders and different degrees of education, which provides a basis for enterprises to formulate scientific and accurate marketing strategies.

5.2   Actively Research and Develop Network Platform, Change the Traditional Promotion Strategy

As the managers of enterprises, they should be deeply aware of the impact and changes of the Internet era on the marketing activities of today's enterprises, allocate special funds and assign special personnel to be responsible for the research and development of network marketing platform, so as to achieve the purpose of promoting enterprise online marketing [5]. If the enterprise's own qualification is small and the scale is limited, it will also seek to cooperate with mature online platforms, such as tmall mall, Jingdong Mall and Amazon mall of Alibaba. This way will enable the enterprise to synchronously realize the parallel of online marketing activities and offline marketing activities, and let the enterprise's development take the east wind of Internet economy. Only in this way can enterprises not derail their development at the same time and keep a place in the market competition.

5.3   Focus on Product Quality and Service

In the era of network economy, great changes have taken place in the way of marketing. However, the core of competition among enterprises is products and services. No matter what opportunities and challenges the era of network economy brings to marketing, as an enterprise, it should pay close attention to the quality of products and services, and constantly carry out products and services. For example, in the actual marketing process, it is necessary to fully consider the needs of consumers, constantly innovate marketing means, analyze the market environment, and realize the scientific positioning of product prices, Meet the needs of consumers, improve the market share of products. At the same time, strengthen the collection of all kinds of information in the market environment, and analyze the collected information resources. In the market environment, consumers are the main body and play an important role. Enterprises should expand the market, fully understand the customer groups in the market, understand the needs of customers, create new products according to the needs of customers, and fully meet the needs of customers.

6   Conclusion

In the face of the changing environment of the information age, enterprises must innovate their marketing strategies in time, and make full use of the Internet for technological innovation, so as to maintain a leading position in the fierce market competition. The advent of the era of network economy has brought great opportunities for the development of enterprises, but also brought many challenges. However, it should be noted that the market has never been flat, and the company must continue to adapt to the economic development trend, carry out snow marketing reform, and use Internet technology to promote its own development. Although the emergence of the Internet makes people's life easier, it also has a huge impact on many traditional industries. If the enterprise wants to meet the current market demand, it must introduce network as the core technology of the overall economic development. The company should seek help from relevant researchers and philosophers, correctly deal with the adverse phenomenon of new environment adaptation, and do a good job in the reform and innovation of relevant content.

References

1. Zhang, B.: On the transformation of marketing strategy in the era of network economy. Mod. Mark. (11), 13–15 (2018)
2. Dong, J.: Exploring the transformation of marketing strategy in the era of network economy. Mod. Mark. (11), 1–7 (2018)
3. Li, N.: On the transformation of enterprise marketing strategy in the era of network economy. Commer. Econ. (8), 3–6 (2018)
4. Zhang, Z.: On the transformation of marketing strategy in the era of network economy. Mod. Mark. (8), 10–11 (2018)
5. Wang, Y.: The transformation of marketing strategy in the era of network economy. Mod. Econ. Inf. (8), 3–6 (2018)

Application of ARMA Prediction Algorithm in Gear Fault Diagnosis of Automobile Drive Axle

Hongbo Zhu[⊠]

Henan College of Industry & Information Technology,
Jiaozuo 454000, Henan, China

Abstract. Based on the gear meshing theory and test data, an analysis method for early fault diagnosis of automotive drive axle is proposed. ARMA model is used to predict and analyze the gear vibration and fault trend. In this paper, the monitoring software of vehicle power transmission system based on ARMA prediction algorithm is designed, which is feasible and effective for the early failure diagnosis of vehicle driving bridge bench test.

Keywords: ARMA · Drive axle gear · Fault diagnosis · Residual life

1 Introduction

In modern production, more and more attention has been paid to the fault diagnosis technology of mechanical equipment. If a certain equipment fails to find and eliminate the fault in time, the result will not only lead to the damage of the equipment itself, but also cause the serious consequences of machine damage and human death. Mechanical fault diagnosis technology is a new subject which develops rapidly in China in recent years. It contains a wide range of contents. With the rapid development of science and technology, mechanical fault diagnosis technology is gradually moving towards a mature development road, which also promotes the progress of automobile fault diagnosis technology [1]. With the development of automobile industry and the progress of electronic technology, the function of automobile structure and electronic system is becoming more and more perfect. Especially in recent years, the development of large-scale integrated circuit and the development of high-performance and multi-functional sensors greatly promote the rapid development of automobile electronic industry. This makes the car in its power, fuel economy, ride comfort and other aspects have been greatly improved. The improvement of automobile structure and electronic system also brings some negative effects. For example, if the electronic system fails, whether it can detect the fault accurately and quickly, and get the fault type, and finally propose the maintenance plan, which is a great test for the special engineering technicians.

The quality of automobile transmission system is directly related to the performance and safety of automobile. It is an important means to install the safety assurance device of automobile transmission system to improve the performance and safety of automobile. However, there are still many problems in the field of automotive powertrain technology, especially in the field of machine condition monitoring and fault

J. C. Hung et al. (Eds.): FC 2021, LNEE 827, pp. 1369–1374, 2022.
https://doi.org/10.1007/978-981-16-8052-6_196

diagnosis using vibration signals. One of them is the early fault diagnosis and residual life prediction of automotive powertrain. Although the vibration parameter analysis (effective value, kurtosis, etc.), FFT spectrum analysis, cepstrum analysis, edge frequency analysis, synchronous average, correlation analysis, wavelet analysis, neural network, adaptive filtering, Hilbert Huang transform, artificial intelligence, expert system and so on have been successfully applied to the fault diagnosis of gearbox and a series of achievements have been achieved, But the effective methods of early failure prediction and residual life prediction of gears and bearings in automobile transmission system have not been determined.

2   Research Status of Early Fault Diagnosis and Life Prediction of Automobile Transmission

In 1996, Ding Kang, Li Yongjian and Xie Ming adopted microcomputer monitoring technology to build the monitoring system of automobile transmission life test bench. In 1999, Zhang Chengbao, Ding Yulan and Lei Yucheng established a set of fault diagnosis method for automobile transmission gear by using neural network. In 2000, Yang Zhengmin and Zhang Chengbao used time series analysis technology to study the fault diagnosis of automobile transmission gear. In 2002, Xiao Shengfa, Zuo Weiwei and Shen Deping studied the fatigue life prediction of light vehicle rear axle. In 2000, Shao Yimin first proposed the logical rule recognition method of "inflection point" of monitoring curve based on test database, and proposed the concept of pprl to improve the prediction accuracy of residual life [2]. At the same time, the theoretical life model and test data life module were introduced into the residual life prediction model, and the multi segment intelligent mode online rolling bearing life accurate prediction method based on neural network was studied.

Due to the complexity of the research on the transmission characteristics of weak impact vibration through multiple interfaces and the calculation method of transmission loss, there are few papers on the early fault diagnosis and life prediction of automobile transmission at home and abroad, and the research results are also few, which are in line with the actual research results. At present, it is difficult to eliminate the environmental noise of automotive power transmission system by using adaptive filtering and noise reduction technology, which can not meet the requirements of extracting its early fault signal. Therefore, it is urgent to develop new noise reduction technology. The research on early fault diagnosis and residual life prediction of automotive powertrain is still blank.

3   ARMA Prediction Algorithm

ARMA (auto regressive and moving average model) is an important method to study time series. It is composed of autoregressive model (AR model) and moving average model (MA model). It is the most perfect prediction method so far. The mathematical description of ARMA model is as follows

$$Y_{(t)} = -\sum\nolimits_{i=1}^{p} \phi_i Y(t-i) + \sum\nolimits_{i=1}^{q} \theta_i a(t-i) + a(t) \tag{1}$$

Where: p and q are the order of autoregressive part and moving average part respectively; ϕ_i (i = 1, 2, ..., p), θ_i (i = 1,2,..., q) are autoregressive coefficient and moving average coefficient respectively; a (t) is independent identically distributed white noise sequence with zero mean value and variance δ^2; Y (t) is stationary sequence.

ARMA model requires the time series to be stationary. In order to reduce the fluctuation of the characteristic parameter series, the recursive method and difference method are used to preprocess to get the stationary series.

The recursive method can remove the interference of random factors in the characteristic parameter sequence, and its mathematical expression is

$$\mu_{x_n} = \frac{n-1}{n} \mu_{x_{(n-1)}} + \frac{1}{n} x_n \tag{2}$$

Where: μ_{x_n} is the recursive value; x_n is the current value of the sequence; n is the number of time series.

After recursive processing, the sequence has obvious linear trend, so the difference method is used to make the sequence stable

$$\nabla x_n = x_n - x_{(n-1)} \tag{3}$$

Among them: ∇x_n is the difference sequence value; x_n is the sequence value after recursive processing.

Figure 1 shows the process of residual life prediction using time series analysis method.

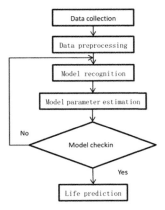

Fig. 1. Forecast process

4   Automobile Drive Axle Gear

Drive axle gear is the key transmission part of automobile. The surface hardness of the gear is high and it has high wear resistance. The center of the gear has high strength and good toughness, which determines the characteristic performance of the spiral bevel gear for automobile. Therefore, low carbon alloy steel must be selected for surface carburizing treatment to meet the requirements of comprehensive mechanical properties of automobile gears and ensure the normal use of the workpiece [3]. The main purpose of carburizing is to obtain high carbon content on the surface of low carbon steel workpiece, and then to obtain high hardness and wear resistance on the surface of workpiece after quenching and low temperature tempering, while maintaining certain strength and toughness at the core.

The common faults of gears are as follows:

(1) Broken teeth, broken teeth is the most common gear failure, tooth fracture generally occurs in the root, because the bending stress at the root is the largest, and it is the source of stress concentration.
(2) Pitting is a common damage form of closed gear transmission, which usually occurs on the tooth root surface near the pitch line. The reason is that the fluctuating cyclic contact stress on the tooth surface exceeds the limit stress of the material. When the fluctuating cyclic contact stress on the tooth surface exceeds the limit stress of the material, fatigue cracks will appear on the tooth surface. When the crack is closed during meshing, the oil pressure in the crack gap increases, which accelerates the crack propagation. In this way, the metal on the surface of the tooth surface will eventually peel off and form pits, that is, pitting.
(3) Wear, tooth surface wear is caused by metal particles, dust and sand particles into the working surface of the teeth. Uneven tooth surface and poor lubrication are also the causes of tooth surface wear. In addition, misalignment, coupling wear and torsional resonance will cause greater torque change at the gear meshing point, or increase the impact, which will accelerate the wear.
(4) Scuffing, tooth surface scuffing (scratch) is due to the meshing tooth surface in the relative sliding oil film rupture, tooth surface direct contact, under the action of friction and pressure, the contact area produces instantaneous high temperature, metal surface local fusion welding adhesion and peeling damage.

5   Design of Vehicle Powertrain Monitoring Software Based on ARMA Prediction Algorithm

The monitoring software can drive the acquisition card to collect data, and display, analyze, save and play back the digital signal after A/D conversion. The real-time monitoring of time-domain statistical parameters can objectively reflect the relationship between the statistics and the set threshold, so as to identify the working state of gears.

5.1 Data Acquisition Module

Before data acquisition, the relevant settings of acquisition parameters are carried out, so that the system can continuously read the signal under the limitation of these parameters. The parameters are divided into two parts: sampling frequency, head end channel and local parameters such as sensitivity of vibration sensor.

5.2 Data Display Module

The data display module consists of five parts: menu bar, digital display, waveform display, data saving progress display and data storage mode selection. The 4-channel original data of vibration acceleration can be displayed synchronously through list control and waveform control. Data storage progress control can display the size of the stored file in the form of progress bar in real time [4]. Signal analysis includes a variety of digital signal processing and analysis methods.

The condition monitoring of automobile transmission system is realized by the analysis method of time-domain characteristics, that is, the real-time calculation and display of multiple time-domain statistical parameters, and the thresholds of different parameters of each channel are set before the start of monitoring, and the fault is judged by comparing the calculated values with the thresholds. Three parameters, root mean square value of vibration acceleration, kurtosis coefficient Ku and peak index C, are used in this monitoring software to identify the faults of automobile transmission system.

5.3 Data Analysis Module

Life prediction is the core function of the monitoring software, mainly through the analysis of the characteristic parameters collected for a period of time, using polynomial fitting or ARMA model trend analysis method to predict the characteristic parameters, so as to know the future working condition of the transmission system [5]. If the characteristic parameter values predicted by polynomial fitting or ARMA model exceed the corresponding threshold, it can be inferred that the automobile transmission system will fail at the next moment and start the software alarm device.

5.4 Data Saving Module

The data saving module provides real-time data saving function for subsequent processing and analysis of data. In the actual monitoring process, it is sometimes necessary to continuously save the real-time collected data, but in most cases, because of the large amount of data, interval saving method is used.

The data playback module is a description of the file to be played back, and also provides parameters for the operation of the playback program.

6 Conclusion

The recursive transformation and differential preprocessing of the time series of characteristic parameters of automotive drive axle ensure the reliability of ARMA modeling and provide conditions for predicting its residual life. The data of drive axle gear is collected through experiments, and the time series model of characteristic parameters is established and predicted. The results show that ARMA model is feasible and effective in predicting the residual life of drive axle.

Acknowledgements. Application and research of FP-growth algorithm in data mining", Natural Science Research Projects in Anhui Universities in 2020 (Project No. KJ2020A0806).

References

1. Zhang, C., Ding, Y., Lei, Y., et al.: Application of artificial neural network in fault diagnosis of automotive transmission gears. Automot. Eng. **21**(6), 374–378 (1999)
2. Yang, D.: Application of stochastic resonance technology in gearbox fault detection. J. Vib. Eng. **17**(2), 201–204 (2004)
3. Zhao, J.: Application research of equipment condition monitoring based on chaos theory. J. Vib. Eng. **17**(1), 78–81 (2004)
4. Zhang, R.: Fault diagnosis device of gear reducer based on DSP. Coal Min. Mach. **28**(8), 190–192 (2007)
5. Zhang, D.: Application of EMD in vibration fault analysis of automobile gearbox. J. Anhui Univ.: Nat. Sci. Ed. **33**(2), 35–38 (2009)

Combination Prediction of Municipal Engineering Cost Based on PCA and NARX

Hongyong Kan[✉]

Shandong Institute of Management, Jinan 250357, Shandong, China
kanhongyong@sdmu.edu.cn

Abstract. Aiming at the problems of many influencing factors, difficult prediction and long period of engineering cost, a new municipal engineering cost prediction method based on hybrid algorithm is proposed by combining principal component analysis with NARX (non linear auto regressive with exogenous inputs) neural network. Principal component analysis is used to process the original data of the main influencing factors of municipal engineering cost to eliminate the correlation, which can effectively reduce the data redundancy and reduce the probability of local minimum in neural network operation. Taking the principal component analysis data as the input and the project cost per unit area as the output, the NARX neural network model constructed by Bayesian regularization algorithm is used to predict the cost of municipal engineering. The example results show that the municipal engineering cost prediction based on PCA and NARX is fast and accurate, which proves that the prediction is effective and feasible.

Keywords: Project cost prediction · Principal component analysis · NARX neural network

1 Introduction

The project cost runs through the whole process of project construction. The project cost prediction in the investment decision-making stage of construction project is an important reference for the preparation of investment plan, application for project loan, project financing, project approval and project investment control. Therefore, how to quickly and accurately carry out the project cost prediction is the key research content of project decision-making and management.

The informatization of construction industry requires the informatization of project cost, and the informatization of project cost is also the development demand of project cost management. Project cost informatization refers to the use of information technology and network technology to form information resources and apply them to the decision-making and prediction of construction projects, so as to improve the accuracy and efficiency and make the project cost management adapt to the development needs [1].

The construction of project cost informatization is to build an information platform to meet the needs of users on the basis of determining the objectives and relevant standards of project cost informatization, The core work is the construction cost

J. C. Hung et al. (Eds.): FC 2021, LNEE 827, pp. 1375–1381, 2022.
https://doi.org/10.1007/978-981-16-8052-6_197

information database and the calculation method database applied to prediction. It is particularly prominent to enrich and enrich the calculation method database and provide the engineering cost prediction method suitable for various majors.

The traditional engineering estimation methods in China include: unit production capacity investment estimation method, production capacity index estimation method, proportion estimation method, Lange coefficient method and engineering construction budget estimation index estimation method.

These methods can meet the requirement of 30% error in engineering estimation, but they can not meet the requirements of the increasing refinement. In view of the wide application of artificial intelligence, it is necessary to establish a scientific method application database suitable for intelligent computing to improve the prediction accuracy and speed.

In order to improve the accuracy and scientificity of project cost prediction, a large number of domestic scholars have made rich achievements. In the aspect of linear regression prediction, the support vector machine is selected to build the housing project cost prediction model, and the structural total least squares method is used to estimate the model parameters to solve the problem that some elements are affected by errors. In the aspect of grey prediction, the GM (1, 1) and GM (1, n) prediction models of project cost are constructed, and the SVR model optimized by PSO is used for prediction. In the application of fuzzy mathematics, the fuzzy mathematics is applied to the cost estimation of hydropower project, and the cost of the proposed project is calculated dynamically and accurately according to the principle of the principle of selecting the nearest. In the aspect of neural network prediction, it optimizes the neural network and tests it with engineering examples, and studies the application of neural network to estimate the main quantities.

The combination of the two methods can solve the shortcomings of a single method: combining random forest with support vector machine to estimate the cost of power transmission and transformation project, using intuitionistic fuzzy analysis method to find out the main engineering characteristics, as the input vector to build a prediction model; Aiming at the slow convergence speed of BP network, Wang Jianping added two parameters of learning rate and inertia factor into the weight and threshold correction algorithm. The grey correlation and neural network are combined to predict the cost of power engineering. Combined with genetic algorithm and artificial neural network, artificial intelligence algorithm GA-BP model is introduced into highway engineering cost estimation. Combining data mining with neural network, applying data mining technology for data preprocessing, combining with the improved neural network fuzzy system for power engineering cost prediction.

2   Combination Forecasting Based on PCA and NARX

2.1   Project Cost Forecast

Simply speaking, the cost of project cost is obtained by multiplying the quantity of divisional and sub divisional works by the corresponding unit price. The difference of project construction scale, project characteristics and function orientation are the essential differences of projects, which determine the type and quantity of project quantity [2].

At the same time, the price will change due to social development and market fluctuation. According to the construction scale and characteristics of the proposed project, the project cost prediction is to select the similar projects in the completed project database, use the appropriate scientific methods in the method database, and combine the subjective experience and judgment ability of professionals to predict the proposed project cost.

The uniqueness of construction projects determines that there are no identical construction projects. How to select similar projects from completed projects becomes the first problem to be solved. In model recognition, feature extraction can effectively alleviate the dimension disaster that often occurs in the field of model recognition.

Project characteristics are the essential characteristics of a project different from other projects, and also the main factors affecting the project cost. There is a complex coupling relationship and information redundancy between these project characteristics that affect the engineering quantity.

Data processing must be carried out to find the main aspects of the problem. The fluctuation of price with time determines that data can not be used directly from database, and the historical dimension characteristics of data should be considered, and the influence of time series should be considered.

This study explores the use of principal component analysis (PCA) to find the main influencing factors of project cost, namely public factors, to eliminate the correlation of the original data. Combined with NARX dynamic neural network, it can reflect the historical state information of the system and has the characteristics of memory function. The main influence factors are taken as the input of neural network, and the project cost per unit area is taken as the output to predict the cost of municipal engineering.

2.2   Principal Component Analysis of Construction Project Characteristics

Feature extraction is the basis of prediction. Feature extraction should be based on the characteristics of prediction project, find a suitable method, extract the optimal project features as the input of prediction model [3]. PCA is a commonly used feature extraction method for pattern classification in model recognition. Its main idea is to obtain principal component variables through linear space transformation, project high-dimensional data space into low dimensional principal component space, and retain most of the variance information of the original data, so that the problem can be best integrated. PCA model can be expressed as:

$$X = t_1 p_1^T + t_2 p_2^T + \cdots + t_n p_n^T \tag{1}$$

2.3   NARX Neural Network

Neural network has good adaptability, and the prediction performance is enhanced with the increase of samples. Its essence is to adjust the connection weights through a large number of samples learning and training, which can simulate the meaning of nonlinear input and output. The disadvantage of neural network prediction method is that the training effect is sensitive to the initial weights of the network. If the network initialization is not ideal, the training will often fall into local optimum [4].

NARX network is called nonlinear autoregressive model with external input. It is a kind of neural network with memory function.

It can be regarded as neural network with time-delay input and time-delay feedback connection output to input.

It has better adaptability to nonlinear data and is very suitable for the characteristics of complex engineering cost and nonlinear time series prediction requirements. Its model expression is shown in Fig. 1.

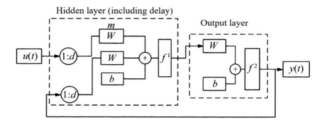

Fig. 1. Structure diagram of NARX neural network model

In Fig. 1, the parameter D is the delay order, M is the number of neurons in the hidden layer, W is the weight vector, B is the bias value, and F is the activation function of the hidden layer of the neural network. The model expression is:

$$y(t) = f[y(t-1), \cdots y(t-y_n), u(t-1), \cdots u(t-u_n)] \tag{2}$$

It can be seen that the next y (t) depends on the previous y (t) and the previous f (·) representing the process function.

Based on PCA and NARX combination forecasting project cost is to extract the influence of completed and proposed project characteristics, principal component analysis, principal component score, according to the time series input neural network, output unit area project cost and actual cost comparison, test whether it can meet the prediction accuracy requirements.

3   NARX Neural Network Prediction

The basic idea of NARX neural network prediction construction of municipal engineering cost: take the score of principal factor obtained by principal component analysis as the input value of NARX neural network, unit project cost as the output value, and select the delay order of NARX neural network. The larger the order is, the more accurate the modeling method of complex system is, and the number of neurons is related to whether the function relationship can be accurately learned by the network,

Therefore, it needs to be debugged repeatedly. Through continuous debugging, the order of delay attachment is 4 and the number of neurons is 12.

Regularization coefficient, its size has a significant impact on the network training effect. This data training uses MATLAB r2016a software for training, and the output result is shown in Fig. 2.

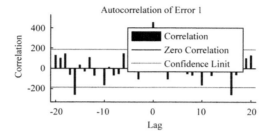

Fig. 2. Error autocorrelation graph of NARX neural network

① NARX neural network training effect judgment: in Fig. 2, except for zero delay, most of the correlation function values fall in the 95% confidence interval, and the goodness of fit in Fig. 3 is close to 1. It is concluded that the model has good generalization ability and prediction performance, and the network characteristics of the judgment model are good, and the technology is feasible.

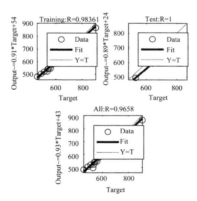

Fig. 3. NARX neural network regression effect chart

As can be seen from Fig. 4, the input and output error of training samples is relatively small [5].

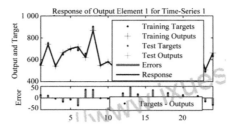

Fig. 4. NARX neural network prediction effect error chart

At the same time, when there is no PCA, 13 main influencing factors are used as neural network inputs. Even if the number of neurons is constantly adjusted, although the data error can be partially reduced, the prediction error is more than 20% when it converges to the local minimum in the second quarter of 2014, which indicates that PCA can effectively solve the problem of data redundancy, After principal component analysis, NARX neural network can be used for better operation training and more effective, as shown in Fig. 5.

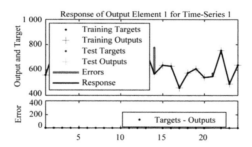

Fig. 5. NARX neural network prediction effect error chart

4   Conclusion

In this study, the 13 main factors affecting the municipal engineering cost are analyzed by principal component analysis, and the main factors are obtained. Taking the score of the main factors as the input of neural network, the NARX neural network is used to establish the municipal engineering cost prediction model of unit area, and the engineering cost prediction based on time series is carried out. The relative error is effectively controlled within 10%, which meets the requirements of engineering investment estimation accuracy, It can also be used for project investment audit, which can effectively improve work efficiency.

References

1. Wu, Y.: Exploration of information construction of project cost management. Ind. Technol. Forum **12**(21), 252–253 (2013)
2. Qin, Z., Lei, X., Zhai, D., et al.: Research on housing project cost prediction based on s V m and ls-s VI m. J. Zhejiang Univ. (Sci. Ed.) **43**(3), 357–363 (2016)
3. Hu, X.: STLS estimation method of multiple linear regression model parameters and its application in project cost prediction. Sichuan Archit. Sci. Res. **42**(8), 142–214 (2016)
4. Hu, Z.: Research on engineering cost estimation model based on fuzzy prediction. Syst. Eng. Theory Pract. (2), 50–55 (1997)
5. Yu, J., Liu, J., Zhang, H.: Fast dynamic estimation method of project cost based on fuzzy mathematics. J. Guangdong Univ. Technol. **24**(2), 107–110 (2007)

BIM Based Prefabricated Building Information Sharing Approach and Method

Na Li[✉]

Anning Kunming Industrial Vocational and Technical College,
Kunming 650302, Yunnan, China

Abstract. The prefabricated building information as the research object, to create p-bim database as the method, information exchange matrix as the way, realize the information sharing in the whole life cycle of prefabricated building. First of all, through the analysis of prefabricated building information, summed up the reasons of the current prefabricated building information sharing difficulties. Then, according to the theory of Bim and P-B I m, this paper studies the method of building pbim database, analyzes the way of information transfer between pbim software, that is, information exchange matrix, and realizes information sharing based on database, software and code of information exchange.

Keywords: Prefabricated building · BIM · p-bim database · Information sharing

1 Introduction

In the development of prefabricated buildings, there are many difficulties. Most of them are due to management problems, mainly information management problems. For example, in the design stage, there is no timely project communication with the component manufacturing unit and construction unit, so that design changes and construction collisions are easy to occur during construction, Finally, it leads to project schedule delay, quality decline and cost increase, which hinder the development of prefabricated building. In order to solve the problem of information management, it is necessary to organize and coordinate the relationship between the participants in each stage, so that the information can be better transmitted. Information sharing is the most important part of information management. A construction project is completed by multiple participants, and a large amount of information will be generated in the whole life cycle. The information is mainly transmitted in the form of paper, so it is easy to cause the lack of information in the process of transmission, and the value of information is difficult to realize, which leads to the decline of production efficiency.

2 Information Analysis of Prefabricated Building

2.1 Definition and Characteristics of Prefabricated Building

Definition of prefabricated building. Prefabricated building is the abbreviation of prefabricated house, which is also called building industrialization or housing

J. C. Hung et al. (Eds.): FC 2021, LNEE 827, pp. 1382–1388, 2022.
https://doi.org/10.1007/978-981-16-8052-6_198

industrialization in other countries. Prefabricated architecture originated from the western industrial revolution [1]. During the Second World War, western countries began to develop prefabricated architecture in order to solve the problem of housing shortage and labor shortage. As shown in Fig. 1. In 1974, the United Nations published the book "government's policies and measures to guide the gradual realization of construction industrialization", which defined the industrialization of Construction: using modern scientific and technological means to transform the backward traditional handicraft production mode of construction industry into an advanced socialized mass production mode.

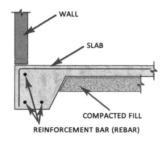

Fig. 1. Definition and characteristics of prefabricated building

2.2   Classification and Characteristics of Prefabricated Building Information

Prefabricated building information classification the whole life cycle of prefabricated building mainly includes five stages: planning, design, contract, construction and operation and maintenance. According to the five stages of the whole life cycle, prefabricated building information can be divided into five categories, namely planning stage information, design stage information, contract stage information, construction stage information and operation stage information. This paper makes a detailed analysis of the main tasks and information contents in the three stages of design, contract and construction of prefabricated buildings. The design stage is to make a detailed plan for the human, material and financial resources of the proposed project from the technical and economic aspects. It is the further detailed deepening of the capital construction plan of the construction project and the main basis for organizing the construction. The prefabricated architectural design is generally divided into two stages, namely the technical design and the construction drawing design. Its main tasks and information contents. The main work of technical design is to segment and dimension the prefabricated components, arrange the reinforcement distribution and design the structure of component connection points according to the structural requirements. As shown in Fig. 2. In the design stage of construction drawings, if there is a standard atlas, it can be marked on the design drawings. If there is no standard atlas, it needs to be completed by the component manufacturer.

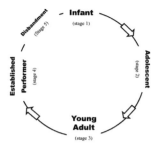

Fig. 2. Five stages

3   BIM Based Prefabricated Building Informations

3.1   Prefabricated Building pbim Database

This project is a prefabricated construction project, which is different from the traditional building construction, so the establishment of p-bim database is also different from the traditional building. According to the establishment method of p-bim database, firstly, the task of the whole life cycle of gangxinyuan project is determined. According to the nature of the construction project, the main tasks are the owner's planning and planning, project design, component production, component installation, property sales and owner's assets management.

Secondly, the general BIM database of the project is determined [2]. According to the product data management, the p-bim database of the project can be divided into six general BIM databases: planning, design, contract, implementation and operation and maintenance. The total BIM database in the three stages of design, contract and implementation of the project is divided into building, structure, equipment, interior, exterior decoration and foundation. Each sub BM database also contains sub BIM databases of various specialties with relatively high degree of information correlation. The sub BIM databases are analyzed below.

3.2   Project Design Phase

Therefore, the architectural design sub BIM database in the general BIM database of the project design stage mainly involves seven sub BIM databases, including wall columns, doors and windows, components, roofs, stairs, floors and others. The structure design sub BIM database includes three sub BIM databases: underground structure design, concrete structure design and prefabricated concrete structure design. Equipment design is divided into five sub BIM databases: building water supply and drainage, heating ventilation and air conditioning design, building electrical design, elevator equipment design and other indoor pipeline design. Interior design sub BIM database includes three sub BIM databases: interior space image design, interior physical environment design and interior decoration design [3]. BIM database of exterior decoration includes curtain wall design, exterior decoration energy saving design and exterior decoration design of other materials. Foundation design is divided into six sub BIM databases, including geotechnical engineering survey, foundation pit engineering design, foundation treatment design, foundation engineering design, slope engineering design and underground waterproof engineering design.

Relative error rate:

$$RER = \sum_{i=1}^{n} (M_i(H)/LB_i - 1) - N \times 100 \tag{1}$$

From the above definition and analysis, it can be concluded that the optimal solution of CMA problem is to find the best value of C (m) in the solution space:

$$C(\pi^b) = \min_{x \in n}\{C(\pi)\} \tag{2}$$

Let P (Jm) Indicates that a part in a known part family is in machine M (m = one, two, M) Processing time on, Sn represents the preparation time of a known part family after the part family and before machining on machine M. Hypothesis C (m) The solution can be expressed as:

$$C(f,j,m) = \max\{C(f,j,m-1), C(r,i,m)\} \tag{3}$$

4   Function Design of Information Resource Sharing System

The function design is mainly to summarize and classify the management characteristics of books and materials, to form an organic whole with differences and connections, so that the management of various departments can not only have a clear division of labor, but also coordinate with each other.

4.1   Purchasing System Module of Books and Periodicals

The main function of this module is to investigate and count the feedback of readers, and according to the book funds and feedback, use the mathematical model to put forward the purchase plan for the leaders to make decisions. According to the purchase plan to achieve book purchasing management, such as the arrival of books, out of stock and other basic information of computer processing, and establish the corresponding database to share with other departments, at the same time, the purchase of books to log in and purchase funds management.

All the family libraries come with Revit, but when the general family library can't meet the requirements, you need to create it yourself. The general creation steps are as follows: select template → define subcategory of family → geometry → dimension → parameter → verification → save and store. After setting the section size and parameters, enter the production interface of the family. In the project browser, adjust the angle of view to the right elevation, draw the designed precast floor in the right elevation, select the wheel friction of the beam side in turn in the stretching command, and stretch the drawing in the 3D angle of view, so that the plane element can be stretched into a solid element. After assembly, the reinforcement details and material list can be output for calculation. This is only a small part in the process of prefabricated architectural design [4]. The whole structure needs the organic combination of numerous components, just like the collaborative work of a team.

4.2   Cataloguing System Module of Books and Periodicals

After the purchase of books and materials, the cataloging department will classify and catalogue them. Therefore, it is necessary to establish a library database, which can be used together with other departments to reduce the cost of development and mainte-nance. For the classification of books, we can establish the classification database of Chinese Library Classification, and establish the query and retrieval function, which is convenient for the staff to input the key words of books directly when classifying books, and then the computer will automatically determine the ownership of books. For some uncertain books such as interdisciplinary classification, it can also provide man-machine dialogue. As for login and inspection, it can provide friendly user interface and realize it by using the collection database.

4.3   Book and Periodical Borrowing System Module

The module is mainly realized by collection database and reader database. Collection database is realized by cataloging department, but reader database needs to be realized by circulation department. According to the borrowing rules, the corresponding knowledge base can be established to judge whether the amount and time of borrowing meet the borrowing requirements. And the historical records of books borrowed by readers are preserved.Provide the maintenance of the library database and reader database. The analysis module of the utilization of library resources carries out sta-tistical analysis according to the borrowing situation of books and finds out the uti-lization rules of various books and materials, which is very necessary to master the basic situation such as whether the book purchase is reasonable or not, and also provides reference for the work of book purchase and borrowing rules making. For the convenience of readers, the system provides retrieval and query functions of library materials, which enables readers to query the materials they need at home or anywhere else where they can access the Internet.

5   Characteristics of Prefabricated Building Modeling

The operation advantages of BMM parts are three-dimensional and visualization, which can simulate the real construction state, evaluate the rationality of the design as soon as possible, and optimize the component layout and construction scheme. The output information is also very easy to be read, especially for the technical disclosure of 3D visualization of complex connection parts and nodes. The construction unit and construction supervision unit especially like this way, which can effectively avoid communication barriers and greatly reduce the drawing meeting and disclosure time. In the design of construction scheme, in addition to conventional projects, dynamic simulation can be used to observe the coordination between different types of work and the use of mechanical equipment, and monitor various construction links.

5.1 Construction and Production

At the same time, consideration should be given to the fabrication, transportation and installation of the components. The monitoring of the ridge information can clearly see the unreasonable phenomena in the process of fabrication, transportation and installation. For example, the prefabricated components need to be transported to the site, so the traffic conditions, means of transportation and stacking methods should be clearly stated and reasonably checked, It is necessary to check the bearing capacity and deformation resistance of members. Because these trusses are welded by steel bars with smaller diameter, it is necessary to consider the influence of stacking on the steel truss. Another difference between prefabricated concrete structure and cast-in-place concrete structure is the connection node. In order to increase the integrity of the structure and fully ensure the bearing capacity of the structure, prefabricated components are often used together with cast-in-place components.

5.2 Ensure Safety

If the precast slab is placed at the edge of the precast beam, it is connected as a whole through the post pouring belt of the surface course. Its advantages are that the precast beam and precast slab can be used as the upper cast-in-place concrete formwork, reducing the site wet operation formwork support, fast construction speed, and good overall performance. The precast floor bears the positive bending moment, the cast-in-place floor bears the negative bending moment, and the embedded steel frame plays the connection role of the composite layer, and helps the floor to resist shear. In addition, the lifting position is designed, and the mechanical performance and deformation performance of the components under the lifting working state are calculated with the help of the structure specialty. The lifting point of the precast floor is set at the quarter point of the floor, and the hook produces two upward and concentrated forces, which is different from the design load of the floor. Both the dead weight and the floor load are vertically downward and evenly distributed in the floor design. There is a large eccentric load in the hoisting of asymmetric components such as balcony canopy, especially when the wind is strong, the structural checking calculation is needed to ensure the safety.

6 Conclusion

How to realize the information sharing among different stages of prefabricated building is a problem faced by the construction industry. BIM is the mainstream idea of the construction industry. It can integrate all the information in the whole life cycle of prefabricated building. Therefore, aiming at the problem that the information of prefabricated building can not be shared timely and effectively, this paper studies the ways and methods of information sharing based on BIM, Taking the public rental housing construction project of Jinan gangxinyuan as an example, the process of informtion sharing is analyazed.

Acknowledgements. Scientific Research Foundation Project of Yunnan Provincial Education Department in 2020 "Research on the talent training model based on the integration of BIM courses and certificates under the 1 + X certificate system" (2020J1337).

References

1. Zhang, Z.: Methods and approaches of archives information sharing among archives under network environment. Arch. Manag. (05), 39–40 (2009)
2. Zou, R.: On the methods and ways of strengthening government banking supervision. Capital University of Economics and Trade (2006)
3. Zhao, L., Li, S., Yang, L., Zhang, H., Secretary, L.: Ways and methods of using information technology to realize the sharing of University Library Resources. J. Hebei Agric. Univ. (Agric. For. Educ. Ed.) (01), 75–77 (2003)
4. Liao, L.: On the strategies, methods and ways to realize the co construction and sharing of literature resources. In: Proceedings of the 2000 Academic Annual Meeting of Fujian Library Society. Fujian Library Society, p. 4 (2000)

Cooperative Scheduling of Renewable Energy Based on Multi-search Optimization Algorithm

Pei Li[1], Jianrong Wang[2], and Changhai Liu[3(✉)]

[1] State Grid Zhejiang Electric Power Research Institute,
Hangzhou 310014, Zhejiang, China
[2] State Grid Zhejiang Changxing Power Supply Co., Ltd., Changxing, China
[3] School of Civil Engineering, Zhengzhou University,
Zhengzhou 450000, Henan, China
liuchanghai@zzu.edu.cn

Abstract. This paper takes the relationship between the uncertainty of scenery and the operation of the system as the starting point, based on wind farm, power plant and other main bodies, establishes the mathematical model, on the basis of which, an algorithm can be put forward to effectively manage the energy. Because chaotic search has outstanding performance in ergodicity and regularity, its application can reduce the corresponding search space of optimization variables, and make the search effect close to the expectation to the maximum extent by speeding up the search speed.

Keywords: Chaos theory · Renewable energy · Searcher · Cooperative scheduling · Optimization algorithm

1 Introduction

Under the background of energy consumption, the application of renewable energy has become the trend of the times. In order to make scientific use of each unit, the key is to adjust the economic dispatching of thermoelectricity, to meet the corresponding constraints of the system as the premise, and to ensure that the heating cost and power supply cost are at the lowest level by fully optimizing the way of heating dispatching and power generation dispatching. In the aspect of solving optimization, the existing intelligent algorithm is easy to appear the local optimal solution. Based on the actual demand, the optimization algorithm based on the original algorithm is put forward. The algorithm can effectively solve the local optimal solution problem and has popularization value.

2 Establishing Mathematical Models

The characteristic of cogeneration is to reduce the amount of polluted. Pjc In the above function, the actual generation power of the unit is represented. Hc Represents the heat generated by the unit. aj-fj Represents unit cost coefficient. Nc Represents the specific number of units.

J. C. Hung et al. (Eds.): FC 2021, LNEE 827, pp. 1389–1393, 2022.
https://doi.org/10.1007/978-981-16-8052-6_199

2.1   Costing Function

The characteristic of cogeneration is to reduce the amount of polluted gas discharged on the supply side on the premise of hot load supply, and to provide guarantee for economic and social benefits. The relevant cost functions are:

$$C_j[P_j^c, H_j^c] = a_j(P_j^c)^2 + b_j P_j^c + c_j + d_j(H_j^c)^2 + e_j H_j^c + f_j H_j^c P_j^c \tag{1}$$

The formula for calculating the operating cost of heat generating units is as follows:

$$C_k(H_k^h) = a_k(H_k^h)^2 + b_k H_k^h + c_k \tag{2}$$

Among them, ak-bk Represents the cost coefficient. Represents heat production. HH_k^{hk} Represents the actual number of heating units.

The supply side of the system includes wind farm, power plant and heating plant. The electric energy and heat energy provided by the supply side are transmitted to the demand side through the corresponding pipeline and line. Except for special cases, the fuel cost can be expressed by quadratic function. If the number of steam turbine valves used in the unit is large, after starting the intake valve, the problem of fuel loss will usually occur, and the fuel cost will increase, and the valve point effect will form [1]. It can be seen that if the cost of generating electricity is to be expressed accurately, the sine function should be superimposed on the basis of quadratic function to ensure that the curve equation has non-convex and non-smooth characteristics.

In this curve equation, the generation power is represented. aP_i^{Pi-ci} Represents the cost coefficient. di and eip_i$^{(P\_min)}$ Represents the corresponding cost coefficient of power generation effect. Represents the minimum output power of the unit i. Np Represents the specific quantity of power supply.

Generally speaking, domestic wind power plants are privately held, operators should buy electricity on the basis of purchase agreement. and the cost paid by the operator, mainly including direct cost, penalty cost and auxiliary cost. The penalty cost means that the operator fails to correctly estimate the cost of wind power output. As we all know, the characteristics of wind power generation are uncontrollable, and it is more common to predict the error. If the actual power generation exceeds the planned power generation, a large amount of wind energy will be wasted [2]. In addition, if the power generated under the previous plan is slightly higher than the available power, the operator will need to purchase backup power through other means to ensure that the load demand is met, which is the auxiliary cost.

2.2   Constraints

The equation constraint objects are wind turbine, fuel unit, generator set and cogeneration unit. The expressions are as follows:

$$C_k(H_k^h) = a_k(H_k^h)^2 + b_k H_k^h + c_k \tag{3}$$

$$\sum_{j=1}^{N_c} H_j^c + \sum_{k=1}^{N_k} H_k^h = H_d \qquad (4)$$

P in the above equation d Represents the corresponding demand power of the electric load. Hd Represents the heat load corresponding to the required power.

The corresponding thermoelectric coupling relationship of the cogeneration unit is better. The region formed by the fixed point line represents the coupling relationship, the calorific value of the unit in the BC section is increasing, the output force is gradually reduced, and the change trend of the calorific value of the unit corresponding to the CD section is decreasing.

2.3 Energy Models

This part of the introduction, mainly on the light, wind speed simulation, the two and power output corresponding relationship.

According to the study, the intensity of illumination is closely related to the distribution of Beta in a specific time range. The functions of the probability density are as follows:

$$f_R(r) = \frac{\gamma(\partial + D)}{\gamma(\partial)\gamma(D)} \left(\frac{r}{r_{max}}\right)^{\partial-1} \left(1 - \frac{r}{r_{max}}\right)^{D-1} \qquad (5)$$

$$\partial = \mu\left[\frac{\mu(1-\mu)}{\partial^2 - 1}\right] \qquad (6)$$

R in this function max Represents the maximum intensity of light. r represents light intensity. μ represents mathematical expectations. σ represents standard deviation.

Based on the probability distribution parameters proposed by wind speed, its function is to count wind resources and provide reference for power generation research. At this stage, there are many models that can be used to represent the probability distribution of wind speed, such as:

$$f_v(v) = \frac{k}{c}\left(\frac{v}{c}\right)^{(k-1)} exp\left[-\left(\frac{v}{c}\right)^k\right] \qquad (7)$$

In the above function, V represents the wind speed variable. v represents the actual value of wind speed. k represents shape parameters and determines curve shape. c represent scale parameters. Here, note that although c can react to the average wind speed, it does not affect the shape of the curve.

3 Optimization Algorithm Description

The algorithm discussed in this paper is based on chaos theory. By integrating the existing algorithms, the related theories and algorithms are introduced in detail below.

3.1 Chaos Theory

The essence of chaos is nonlinear phenomenon, and chaos motion emphasizes on ergodicity and regularity to ensure that it can reach a specific scale corresponding to all states, and its practicability is higher than that of random search. The results show that the characteristics of chaotic search are not only ergodicity, regularity, but also randomness, which is the reason why many optimization algorithms choose to introduce the initial solution with the help of chaotic force. In this way, the ergodicity of the algorithm is enhanced and the optimal convergence speed is accelerated [3]. In addition, non-repeatability determines the chaotic mapping algorithm, which is superior to the traditional algorithm in comprehensive search.

During the initialization of the algorithm discussed in this paper, the Logistic formula is applied to ensure that the generated variables have outstanding chaotic state characteristics. The formula is as follows:

$$Z_{t+1} = kZ_t(1 - Z_t) \tag{8}$$

Z in the formulatIs a variable. k represents control variables. It is proved that the application of chaos theory can keep the initial population unrepeatable and solve the local optimal problem which is easy to appear in the traditional algorithm with the help of multivariate searcher.

3.2 Constraint Processing

The method of introducing constraints to the objective function by the optimization algorithm is the penalty function method. Based on this method, the inequality constraint can be treated, so that the original feasible solution of the equality constraint can be preserved and the overlimit situation can be eliminated. On this basis, the difficulty of subsequent calculation is reduced by transforming the constrained optimization subject from equality to non-equal [4].

3.3 Double Search

The searcher is randomly arranged in each part of the search area, and the related searcher is the global searcher. Based on GS, random arrangement of other searchers in a specific range, the arranged searchers are also called local searchers. The original intention of the LS is to ensure that the GS can perceive the changes of the environment in time and dynamically, and obtain the required data by local search. It is proved that this design can improve the convergence rate significantly and solve the local optimal problem.

r in the above formula min Represents the preset radius minimum. rmax Represents the maximum preset radius. kGS Represents GS level. Optimized GS, its target value is minimum, kGS The value is 0. nGS represents GS quantity. Gbest represents the optimal solution. dist (GS, Gbest) represents the best line distance from the GS. It is proved that the use of double-layer searcher is helpful to achieve the goal of cooperative scheduling of wind thermoelectricity, and the optimization effect is improved significantly on the premise of ensuring rapid convergence to the best advantage.

3.4   Random Walk

After the local search is over, the best advantages are determined by the GS through random walk. At present, many algorithms have applied the rule, and its effect is obvious to all. In addition to special cases, starting with one point and doing random walk calculation to another point, only two variables are usually considered, namely, direction angle and distance. After the GS is based on the relevant rules and obtains a new position by random walk, the relevant personnel can compare the optimal solution with the new position. If the effect of the new position is closer to the expectation, the optimal solution value needs to be replaced.

4   Conclusion

This paper takes the control of operation cost as the final goal, and puts forward an optimization algorithm which can solve the related problems of energy cooperative scheduling on the basis of establishing the optimization model. It has been proved that the value of the algorithm is mainly to obtain the accurate structure of economic scheduling quickly. Because the optimization algorithm can complete the calculation without relying on the mathematical model, the application of the algorithm can solve the problem of non-convex optimization. The focus of future research should be on transfer learning and algorithm integration, combined with wind uncertainty, network loss, continuous optimization of the model to ensure that it has a more ideal universality.

Acknowledgements. State Grid Science and Technology Project: Integrated optimization methodology research and demonstration project construction of rural.
Electrification (5400-202025208A-0-0-00).

References

1. Li, S., Jingxuan, Zhu, L., Sun, Y., et al.: On quantification of operational flexibility and optimal dispatching of renewable energy power systems power grid technology. 1–11
2. Liu, J., Zhao, D., Wang, Y., et al.: Optimization model of multi-energy micro-network scheduling based on J. Uncertain. Renew. Energy Sci. Technol. Eng. 20(35), 14523–14529
3. Zhu, J., Yuan, Y., Wu, H.: Active distribution network optimal scheduling. Consid. Mob. Hydrog. Storage High-Density Renew. Energy Power Autom. Equip. 40(12), 42–50 (2020)
4. Zhao, H.: Research on collaborative scheduling strategy of renewable energy power generation and electric vehicles. Shandong University (2020)

Energy Saving Design of Passive House with Ultra Low Energy Consumption

Zhen Tian[(⊠)] and Yawen Li

School of Environmental and Municipal Engineering,
Qingdao University of Technology, Qingdao 266525, China

Abstract. In recent years, due to the improvement of people's living environment requirements and the aggravation of building energy consumption, passive house, as a new exploration of energy-saving building development, has been widely concerned by the construction industry at home and abroad. Through the research of prefabricated passive experiment, this paper analyzes the problems and key points that should be paid attention to in the design and construction process of prefabricated ultra-low energy consumption passive house from the aspects of building space layout, air tightness design, energy-saving design and renewable energy utilization, so as to provide reference and basis for the development of passive house in cold areas.

Keywords: Passive house · Building energy saving · Prefabricated building · Ultra low energy consumption building · Air tightness: renewable energy utilization

1 Introduction

At present, building energy consumption accounts for more than 30% of the total energy consumption of the society, and the total energy consumption is still rising year by year. In order to reduce China's energy consumption and improve people's living environment, building energy conservation has become an inevitable trend in the development of the construction industry. China's 13th five year plan clearly proposes to accelerate the development of building energy conservation and green buildings, Countries have begun to study the ultra-low energy consumption of passive housing. Germany was the first to promote the development of passive housing, and the building energy saving technology of passive housing has been widely promoted and applied in Europe. At present, 30000 sets of "passive housing" have been built in the world. Passive house embodies the most advanced energy-saving technology and concept. As a new type of energy-saving building, Youli has become the new trend of global residential development in the future.

The design principle of passive house is to minimize the heat loss of the building, which is the core idea of the design concept of passive house, At the same time, the maximum use of renewable energy such as solar energy and fresh air heat recovery system can reduce the consumption of primary energy under the premise of ensuring people's living comfort.

J. C. Hung et al. (Eds.): FC 2021, LNEE 827, pp. 1394–1399, 2022.
https://doi.org/10.1007/978-981-16-8052-6_200

2   Structural Energy Saving Optimization Measures

2.1   Wall Energy Saving Design

The passive room requires high heat transfer coefficient (u0.15wm2k) of the wall. The external wall of the experimental room adopts 300 mm thick sandwich prefabricated sandwich insulation wallboard. The inner and outer panels are reinforced concrete slabs. The thickness of the outer panel is 50 mm, the thickness of the inner panel is 100 mm, and the middle two layers are 75 mm thick high-density polyurethane boards. The insulation coefficient of the whole culture can reach 0.146 w (m2k), which fully meets the wall insulation standard of the passive room in Germany. In order to ensure the air tightness of the wall and avoid the formation of through joints, two layers of high-density polyhydrogen ester board are overlapped by staggered joints [1]. The concrete plates on both sides not only form a good protection effect on the light insulation material which is easy to damage, but also can be used as a good cold and hot material to reduce the fluctuation of indoor temperature.

All walls are prefabricated in factory and assembled on site. Such production and construction mode has the following advantages and energy saving aspects. It can be produced in batch, greatly reducing energy consumption and material saving in the process of building materials production. Factory prefabrication can make full use of raw materials, reprocess waste materials and environmental protection, reduce wet operation on construction site and generate less construction waste, It is conducive to the protection and improvement of the environment. In addition, it can shorten the construction period, improve the construction quality and reduce the project cost.

2.2   Roof Energy Saving Design

As a weak link in the loss of cold and heat in the building, the roof also needs strict thermal insulation and energy-saving design. In addition, the accessible roof also requires the thermal insulation material to have a certain stiffness, so the selected thermal insulation material is different from the wall. The rock wool board with dense pressure resistance, non combustible and non corrosive, and stable chemical performance is adopted. The specific construction method is that the concrete is poured on the ribbed composite board, and the concrete is poured on it, Lay two layers of rock wool insulation board in staggered layers, and then make waterproof layer and finishing layer.

The passive room requires that the cold and hot bridge in the building must be eliminated. The exterior wall of the experimental building adopts sandwich insulation, and the parapet belongs to the overhanging part of the structure. If no special insulation treatment is carried out, the thermal reclamation will penetrate outward along the inner leaf plate to form a cold and hot bridge, which seriously affects the overall insulation effect of the building. Therefore, the parapet part is designed as full insulation [2]. The concrete construction method is to lay two layers of high density polyurethane insulation board for the side board side seam of the inner panel, and lay foam foam glass bricks on top of the parapet, and add metal waterproof cover plates.

2.3   Ground Energy Saving Design

There is no insulation measures in the air floor, through the heat exchange with the ground, will make a lot of cold and heat loss in the building air, so in the passive room design, the ground insulation is also the key point of energy-saving design. The floor insulation system of the experimental building adopts double-layer extruded polystyrene board staggered joint laying. Because the insulation material outdoor is lower than the floor layer foundation, that is, the corner part of the heat insulation material is double layer foam glass, and the gap between the insulation board and the prefabricated sandwich panel is filled with foamed polyhydrogen ester. Because this part of the thermal insulation material needs to be partially buried underground and close to the outdoor apron, the foam glass brick with good thermal insulation performance and waterproof is selected, and the high-density polyurethane insulation board of the wall, the rock wool board of the roof and the extruded polystyrene board of the floor are formed into a continuous thermal insulation layer to provide high-performance thermal insulation structure for the building.

3   Application of Renewable Energy System

3.1   Structure Design of Fresh Air System

Scientific and reasonable, rigorous and precise construction, with good air tightness, its air exchange rate N50 $\leq$ 06h1, can effectively prevent the cold and heat loss caused by air leakage, which is the basis of high-efficiency thermal insulation performance of passive room, but also brings the problem of indoor air quality. The experimental building adopts the replacement heat recovery fresh air system. After preheating and filtering, the fresh air (black arrow) enters the room from above the exhibition hall on the first floor and the office on the second floor. The dirty air in the room is transported to the fresh air unit through the sanitary room on the first floor and the staircase on the second floor for heat exchange. The air volume range of the room air system is 2000 m^3h. It can ensure the fresh air volume and discharge the dirty air from the room, It can also reduce the cold and heat loss caused by air circulation and reduce the heating and cooling energy consumption of buildings [3]. The fresh air system can also eliminate the indoor wet air, prevent the building from mildew and damage, and prolong the life of the building. With the aggravation of haze weather in northern China, the air quality has attracted people's attention. Adding air filter in the fresh air system can ensure the indoor air quality and create a suitable indoor environment even in the case of serious outdoor air pollution.

3.2   Solar Photovoltaic Power Generation System

The 12 m^2 multi grade silicon solar panels installed in the roof garden on the second floor can be used as sunshades to provide suitable space for the garden and make full use of solar energy to supplement the building's power demand. As shown in the figure, the average annual power generation of the solar panel is 2100 kwh, and the conversion efficiency is 15.89%. The small micro grid grid connected power generation system is adopted, and the excess power is incorporated into the campus power grid.

The application of solar photovoltaic power generation system in the experimental building provides data reference for the design of solar energy system in the later assembled ultra-low energy consumption office building.

The mutual radiation between the inner surfaces of the heat storage wall has little effect on the heat load, and the calculation is complex, so the mutual radiation is ignored. Based on this assumption, the heat consumption or gain of each part can be calculated separately. As mentioned above, the instantaneous heat load of the room is:

$$HL(n) = Q_{win}(n) + Q_{wall}(n) + Q_{wind}(n) - Q_{tro}(n) \tag{1}$$

The radiation heat exchange between the wall and the outdoor environment is ignored [4]. The influence of the wall absorbing solar radiation and temperature on the heat load is considered in the form of comprehensive temperature. The heat balance equation of the external surface of the wall is expressed as the solar radiation heat absorbed by the external surface - convective heat transfer to the outdoor air + heat conduction from the internal surface to the external surface of the wall = 0.

$$q_{s,o}(n) - q_{c,o}(n) + q_{ro}(n) = 0 \tag{2}$$

The reaction coefficient method can be used to calculate the thermal conductivity of the inner surface and the outer surface of the wall, as shown in the formula:

$$q_{c,o}(n) = \alpha_{ca}(t_0(n) - t_a(n)) \tag{3}$$

4 Comparison and Analysis of Simplified Calculation Methods

This chapter compares the traditional dynamic calculation method with the simplified calculation method proposed in this study, Thus, the rationality of the simplified calculation method of the heat load of the passive house with heat collecting and storing wall is revealed. There is no error between the simplified calculation method and the traditional dynamic calculation method. Therefore, this chapter first verifies the accuracy of the simplified calculation method of the heat gain of the heat storage wall, The thermal load of the passive room is further compared.

4.1 Comparative Analysis on Simplified Calculation of Heat Gain of Heat Collecting and Storage Wall

Using the simplified calculation method of heat gain of heat collecting and storing wall given the heat gain of typical area is calculated by using the simplified calculation method, and compared with the simulation value obtained by traditional dynamic calculation method. Because the simplified calculation method of the heat gain of the heat collecting and storage wall is to predict the maximum value of the heat transfer and heat conduction through the regression model, then the maximum value of the heat gain coefficient of the ventilation hole and the heat gain coefficient of the heat collecting wall is converted into the convective heat transfer and heat conduction heat transfer, and then the heat gain of the heat collecting and storage wall is obtained by adding the two. Therefore, the following will be divided into three steps to verify the accuracy of the simplified calculation method. In the first step, the maximum values of heat conduction and convective heat transfer obtained by numerical simulation and simplified calculation are compared to verify the accuracy of regression model of maximum values of heat conduction and convective heat transfer; in the second step, the accuracy of heat gain coefficient of ventilation hole and heat collecting wall is verified by comparing the maximum values of heat conduction and convective heat transfer obtained by numerical simulation and simplified calculation; in the third step, the maximum values of heat conduction and convective heat transfer obtained by numerical simulation and simplified calculation are compared, Compared with the numerical simulation and simplified calculation, the heat gain of the heat storage wall (i.e. heat conduction and convection) is obtained, which verifies the rationality and accuracy of the simplified method.

4.2   Comparative Analysis on Simplified Calculation of Heat Load of Passive House with Heat Collecting and Storing Wall

The simplified calculation method of heat load proposed in this paper does not consider the mutual radiation between surfaces. In order to verify the accuracy of the simplified calculation method of heat load accurately. Considering the mutual radiation of each surface, the heat load obtained by coupling solving the room heat balance equations is taken as the comparison object. At present, some commercial software of numerical simulation has this function, and TRNSYS is commonly used to simulate the thermal load of passive room [5]. The software is coupled to solve the room heat balance equations to get the indoor heat load, with high accuracy. TRNSYS software is a dynamic simulation software that modularizes components. It was first developed by solar energy of the University of Wisconsin in the United States, and then developed and improved with some relevant European research institutions. As shown in Fig. 1. As a modular software, its main steps are as follows: the first step is to select and study the relevant modules, and couple the relevant parameters of these modules; the second step is to set the parameter values, input conditions and output results of each module component according to different research needs; the third step is to carry out the operation simulation research.

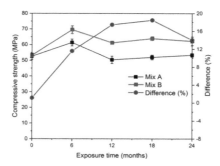

Fig. 1. Heat collection passive house

The simulated value and regression value of the heat gain of the wall are concentrated near the line of "y = x". The relative error between the regression value and the simulation value is relatively large when the heat gain value of the heat storage wall is relatively small, but when the heat gain value of the heat storage wall is relatively small, its impact on the load is also very small, so it also affects the rationality of the simplified calculation method. It can be seen that the regression model rrsme of heat gain of heat storage wall in typical cities is in the range of 0.16–0.22, R2 is in the range of 0.95–0.97, which indicates that the difference between the heat gain of heat storage wall obtained by the simplified calculation method and the numerical simulation method is small, which proves that the dynamic simplified calculation method of heat gain of heat storage wall proposed in this study has high accuracy.

5 Conclusion

The prefabricated passive house starts from the characteristics of the project itself, Climatic conditions of Taiwan, According to local conditions, this paper makes a positive exploration on the prefabricated ultra-low consumption passive house with economic and efficient energy-saving technical measures. The project is based on the parameters in the German certification standard for passive housing, Simulation and optimization of ventilation, lighting and annual energy consumption, Energy saving planning of plane and space layout, Energy saving design of document node, Design and application of renewable energy cooling system, Make the comprehensive energy saving of the building reach the German passive housing standard. To pass this proof, Although the construction cost of passive housing increases a lot in housing, However, a lot of energy consumption is saved in the process of calyx transportation, Extending the life of buildings, To provide people with higher quality of living, From the perspective of building life cycle, Passive housing has good economic, social and environmental benefits, It is a kind of green, environmental protection and sustainable building.

References

1. Wang, D.F., Wu, S.P., Xi, W.H.: Mathematical model and simulation program of passive solar heating house - pshdc of direct benefit type and collector wall type (Part 1). J. Gansu Sci. (1), 1–8 (1981)
2. Wang, X., Nie, J., He, T., et al.: Simulation analysis of annual room temperature of passive solar house in Tibetan areas of Qinghai Province. Build. Sci. (8), 102–106 (2010)
3. Gao, W., Di, H.: Simplified calculation of thermal performance of flower pattern heat collecting and storing wall. Acta Agriculturae Sinica 5(4), 40–45 (1989)
4. Wang, D.: Thermal Process and Design Optimization of Intermittent Heating Solar Buildings. Xi'an University of Architecture and Technology, Xi'an (2012)
5. Wu, Y.: Heat Load Calculation and Enclosure Structure Optimization of Time Sharing and District Heating Buildings. Xi'an University of Architecture and Technology, Xi'an (2015)

Computer Basic Course Classroom Based on Edge Computing

Ying Liang[✉]

Xi'an Haitang Vocational College, Xi'an 710000, China

Abstract. With the development of science and technology and the progress of computer technology, computer technology is widely used in the field of fine arts, which promotes the progress of art and produces new art forms. The application of computer technology in oil painting not only changes the traditional way of oil painting creation, but also has a profound impact on the creative thinking and creative concept of the young generation.

Keywords: Computer aided technology · Oil painting · Creation method

1 Introduction

The progress of science and technology, politics, economy and culture in an era will inevitably lead to the establishment of new aesthetic values, and new art forms will appear in art. There is a close relationship between them. In different historical periods, there are great differences in economic basis. Changes in political system, lifestyle, living environment and content will make people have different views of cognition and establish different aesthetic feelings. Artistic style will be constantly updated and developed, so as to produce new artistic forms that can reflect the characteristics of this era. Today's people in the impact of scientific and technological progress and the continuous influence of mass media information, values have undergone great changes, more and more attention to self and self-awareness. With the rapid development of science and technology, people's aesthetic consciousness and taste have been constantly seeking novelty, change and uniqueness. Compared with any previous era, the artistic creation in the new era under the new technical conditions pursues the individuality, originality and diversity of works [1]. As a kind of painting art, oil painting, of course, is also facing the challenge of new era and new technology. With the progress of science and technology, artists pay more and more attention to the role of computer-aided technology in the process of oil painting creation. This is not only because of the need of rebelling against and innovating the traditional concept, but also because its technology can bring art innovation and promote people's ideological change.

2 Changed the Way of Oil Painting Creation

In traditional oil painting creation, artists usually have a relatively mature creative thinking and need to collect Dacheng's creative materials. Generally, they will go deep into life and nature for sketching. After Dacheng's sketching, they will make oil

J. C. Hung et al. (Eds.): FC 2021, LNEE 827, pp. 1400–1406, 2022.
https://doi.org/10.1007/978-981-16-8052-6_201

painting drafts according to the first-hand materials obtained from sketching, and finally make oil paintings. The characters, animals, landscapes and other visual symbols are all drawn from life. Oil painting is a process from nature to art. The wide application of modern digital photography technology, computer-aided technology, including network technology, has changed the early way of artists to create oil paintings. When contemporary artists collect and organize creative materials, they usually use digital cameras to shoot the materials they need, or even set them purposefully, and then input the obtained image information into the computer. They can also directly access professional websites through the network to download the required resources. Using the computer directly to call these image materials for oil painting draft production, artists can more freely control the picture, image position, size, color can be freely adjusted, and even different styles can be completed in the computer. The finished draft can be printed by printer, or enlarged by professional plotter and printed on the canvas. The painter can draw oil on the canvas directly. Computer aided oil painting, in the early draft drawing process, does not need paper and paint, just a little mouse can be used by the painter. This way is easy and free, and saves time and materials. The painter can be liberated from the tedious and complex draft making, and take more time to carry out artistic conception.

3 Main Application Algorithms

There are many variants of TLF. Smooth truncated Levy flight (STLF) 50 is based on the idea of Mantegna et al., and achieves a good analytical form through smooth exponential regression (gtlf5 is proposed by Gupta et al., which further combines truncation with statistical distribution factor. The step size probability for a given size of the actual system depends on the statistical factor and the physical constraints of the system or its components. The GARCH process and conditional TLF are combined to establish a hybrid model, and the model significantly describes the price changes and the related probability density distribution of volatility in different time periods. Coelho et al. used TLF and population diversity measurement strategy to improve the crossover and mutation operation of DE algorithm, and improved the control parameters Cr and MF, so that the algorithm effectively avoided premature convergence. They apply the algorithm to economic load distribution problem and find that they can get better quality solutions with higher efficiency. The mobility model is very important for the simulation and evaluation of multi hop wireless network protocol. Cao et al. 0 proposed a mobility model based on STLF, and proved that the model is more accurate from the perspective of theory and experience.

Mathematically, Levy distribution can be defined as:

$$L(s, \lambda, \mu) = \sqrt{\frac{\lambda}{2} \exp[-\frac{\lambda}{2(s-\mu)}]} \tag{1}$$

Levy distribution can be defined as Fourier transform:

$$F(k) = \exp[-a|k|^{\beta}], 0 < \beta \le 2 \tag{2}$$

In the algorithm proposed by Mantegna, the calculation formula of length s is:

$$s = \frac{\mu}{|v|^{1/\beta}} \tag{3}$$

4 The Influence of New Era Background on Oil Painting Creation

4.1 The Characteristics of Oil Painting in the New Era

The oil paintings of the 1950s and 1960s have obvious characteristics of the times. Based on the content of the times, the oil painters at that time created such national works as the founding ceremony, the tunnel war and the ramming song. Different painters create social works with unique images through their personal understanding of the times. For example, the ramming song created by painter Wang Wenbin depicts the scene of several working women ramming, and the picture is full of the joy of labor. The ramming peasant girls are placed above the horizon. The characters in the picture are in a radial dynamic layout, giving people a rising visual effect. However, zhixiwen's "the land of the Millennium turned over" vividly shows people's joy after the end of serfdom in Tibet. In the picture, the ox's head is held high, the hostess behind the ox is smiling, and the snow mountain behind it is far away and the sky is blue. This series of vivid images reflect the mood of the laborer as the master. Therefore, oil paintings with different themes and styles contain rich background and realistic content.

4.2 The Influence of New Era Background on the Use of Oil Painting Color

The use of color in oil painting creation will show different styles in different times. Oil painting originated in Europe, the creation of European oil painting has rich and strong characteristics of the times. European painters searched for inspiration from people's clothing, accessories and food, and created a series of dignified and luxurious oils, such as the tour of the island of xitel, the island of love, the wayward woman and the kiss. These oil paintings are very delicate, complex and thick in color, reflecting the gorgeous beauty of the characters and scenery. After oil painting was introduced into China, a series of changes have taken place in its color. The creation of Chinese oil painting in the 1970s and 1980s also reflects the distinctive characteristics of the times. For example, "father" and "Tibetan paintings" are the representative works of oil painting creation in this period [2]. Among them, Chen Danqing's "Tibet group paintings" abandoned the theme creation mode, and instead used realistic oil painting color to depict the daily life of the Tibetan people. In terms of color application, Chen

Danqing used a large number of dark color clumps to express a serious and deep cultural atmosphere. In the picture, the Tibetan people's clothes and manner are very simple, which makes the whole work more realistic.

5   Several Major Aspects of the Main Impact

5.1   Bring Convenience to Composition

In order to express certain thoughts, artistic conception and emotion, the composition of a picture is to arrange and deal with the positional relationship of images and symbols by using aesthetic principles within a certain space, so as to form a convincing artistic whole. "In the traditional oil painting composition, after careful consideration, the painter will arrange the composition on the oil painting sketch, and design the position, size and perspective relationship of each visual element on the space picture. In this process, the painter needs to draw and modify the draft repeatedly, adjust the various relationships between the images, and even draw the multi hat oil painting draft for comparison, and select the final satisfied one for oil painting production. In this way, the painter's time is occupied in the early stage, thus prolonging the whole process of oil painting creation. In computer aided oil painting creation, the painter can use the drawing software provided by the computer, such as Photoshop, to design and make the preliminary draft. In Photoshop, a variety of convenient and easy-to-use tools and commands, such as deformation tools, can adjust the size and angle of the image, and can also tilt, twist and deform the image. The copy and paste of image is also an important part of computer art creation. Figure 1 particle displacement diagram.In the software, the image can be copied and pasted to the required position for countless times [3]. The software also provides tools for cutting, vertical and horizontal flipping of the whole composition, which makes the composition very convenient and fast.

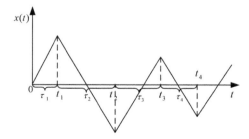

Fig. 1. Particle displacement diagram

5.2   Bring Convenience to Color Scheme Making

The color of the painting is not only the natural presentation of the object, it is the artist's unique feeling and artistic expression of the color image, is the image expression of self emotion. Oil painting is to shape the body with color, color is one of the important artistic language of oil painting. Color is the main means of oil painting performance, which has extraordinary artistic expression and independent aesthetic value. The success of color application will directly affect the success or failure of the work. Giving full play to the expressiveness of color in oil painting is the main factor of the work's personality and the spirit of the times. Painters attach great importance to the grasp of color and tone in painting creation. With the application of computer-aided technology in oil painting creation, painters can not only easily and accurately make excellent drafts, but also easily and freely adjust the color scheme. For example, there are more than 20 methods in Photoshop image adjustment command to adjust the hue, lightness, chroma, color scale and contrast, as well as warm and cold tones of the image, In particular, the induction and conversion of the screen tone is very convenient.

5.3   With Storage and Backup Function

The powerful storage and backup function of the computer provides a strong memory guarantee for oil painting creation. Many of the artist's creative ideas can be quickly realized and saved to the computer, which can be called at any time and displayed on the computer screen for screening and comparison. In the process of painting creation, painters will face many choices. In order to realize a certain idea, they must give up other schemes, because painters do not have the time and energy to paint all the ideas on the canvas. Therefore, the traditional painting process and the realization of imagination are also the process in which imagination is stifled. Using computer to create oil painting can avoid the above regret, the computer can realize all the painter's ideas and save them, and finally choose a satisfactory scheme to implement. The artist's creative ability can be displayed to the greatest extent. We often encounter this situation in the past painting creation. When the painting achieves a relatively satisfactory effect, it is often not as good as the previous effect when it goes further down. At this time, it is impossible to completely recover to the past. It's a pity for the artist. But using computer for oil painting creation is different. In the computer, you can save the currently satisfactory effect and make a backup. You can continue to try the painter's different ideas on the copy [4]. Even if it fails, it will not affect the original scheme.

The application of computer technology to oil painting creation promotes the development of oil painting art, creates new oil painting art style, reduces the repeated work burden of artists, saves the overall time of oil painting creation, and enables artists to concentrate on artistic innovation.

6   The Relationship Between Modern Computer Art and Oil Painting

6.1   The Influence of Modern Computer Art on Oil Painting Creation Concept and Visual Experience

The 21st century is a digital era in which the information explosion and opportunities coexist. Computer art technology has changed the mode of thinking of art. Computer art graphics are easy to use, easy to modify, and fast to copy. They are warmly sought after by the young generation of artists. Due to the use of a large number of image materials by artists in oil painting creation, resulting in the same form of works, subject matter crash, similar style and other phenomena, brain art can simulate various techniques and mechanism effects in traditional art creation, and even the three-dimensional space simulation reproduction of the picture, showing completely different appeal and picture expressiveness of traditional painting. The new ways and means of artistic creation not only affect the traditional aesthetic tendency of artists, but also provide artists with a broader space for artistic expression.

6.2 The Presentation of Modern Computer Art in the New Art Form of Contemporary Oil Painting

Under the great background of the times, the computer generated graphics constitute today's extremely rich popular culture and visual creation generation. The emergence of "Cartoon Generation" is different from the previous generation of artists' thinking. It is bound to inject vitality into the society at that time, and give them cultural significance. It provides "new ways of thinking and basis for new cartoon artists". Cartoon images began in the 1960s, and practice developed in the late 1970s. Under the background of the rapid development of network visual culture, the emerging social culture has become a "common generation". They have the keen insight to see the essence through the phenomenon, and use the language of the picture to feel and express their views on the complex social problems [5]. The main features of cartoonists' works are the flat and exaggerated cartoon style of "game emotion". It is a kind of human instinct that "cartoon" enters into the situation. It has a certain subjective expression that is not tired and helpless of reality and realizes its own psychological needs. The cartoon generation of young artists represented by Li Jikai, Feng Mengbo, Feng Zhongqi, Shen Na, Xiong Yu, Chen Ke, Yang Jing and Xiong Wenyun have created a large number of artistic works in recent years.

7 Influence on the Creation of Young Artists

Curator Yang Weimin said in the foreword of the twelve artists exhibition that art is always developing. In the course of the development of Chinese contemporary art, there have been changes in theme, content, and expression methods. We can feel that the theme of painting has changed from heroes and idols to ordinary people, to themselves and to the present. The new generation of Political Pop "gaudy art" cartoon and other art words emerge in an endless stream, which represents the background and thoughts of several generations. Among these many styles and terms, the new generation and the later generation can be said to show that the most interesting contemporary art vitality of the two generations in China is in progress. In contemporary art, "the younger generation" refers to the new generation of artists born in the Post-70s and post-80s, who have risen rapidly in the art field after the new generation. As painters of new and new humanity, their world outlook is basically established with China's reform and opening up. They live in an era of rapid economic development, stable political situation and rich life. They are deeply aware of western culture, contemporary urban culture, modern electronic and media culture. They are driving the

Internet Express on the information highway. In today's era of economic integration and cultural diversity, there are obvious differences between the young generation of artists and their predecessors in their way of thinking, social concerns and artistic expression. In terms of artistic expression, young artists are no longer limited to the misappropriation and borrowing of images, but to express a special picture effect through "virtual" technique, so that the majority of art audiences can feel the idea that the painter is different from the public. This feature is fully reflected in the works of "Cartoon Generation" and later generations of artists. The application of computer-aided technology in oil painting creation is an indispensable step for them to make use of new technical means for artistic creation and an important technical condition for them to explore new forms of painting expression.

References

1. Qiu, J.: Art in the digital age. Art World (2), 35 (2000)
2. Le, Q.: Art in the age of science and technology. J. Aba Teach. Coll. (1), 88–90 (2003)
3. Zhu, Y.: Analysis of traditional painting and computer art. J. West Anhui Univ. 2L(3), 146–148 (2005)
4. Tan, Y.: On the role of computer painting in oil painting creation. Art Explor. (2), 36 (1998)
5. Zhang, L.: Analysis of the influence of science and technology on contemporary oil painting creation. J. Shanxi Normal Univ. (Nat. Sci. Ed.) 22(3), 173 (2008)

Application of Dassault System 3D Experience Platform in Enterprise Digitalization

Linhao Liu[✉]

Benedictine University, Lisle 60532, USA

Abstract. Dassault's 3D experience solution takes the data management platform as the core, and builds an enterprise level platform integrating modeling, simulation, collaboration and management to boost the digital transformation of enterprises. The management platform provides powerful data standard management functions, supports enterprises or industries to deploy customized data standards according to their own needs, creates models with rich information in the 3D experience platform, and imports models created by a variety of industry software into the 3D experience platform for management.

Keywords: Dassault system · 3D experience platform · Digital transformation · BIM

1 Introduction

Dassault syste MES is the main provider of PLM solutions, and Dassault aviation belongs to Dassault group. Dassault systems has been focusing on 3D technology and PLM solutions for more than 30 years, and has been cooperating with leading enterprises in various industries around the world, ranging from aircraft, automobiles, ships to consumer goods, industrial equipment and construction engineering [1].

Dassault's life-cycle solution is composed of a series of 3D design, analysis, simulation and business intelligence software, serving all processes of the enterprise, including engineering design, optimization, manufacturing, installation, project management, business operation and supply chain management. It contains hundreds of various application modules, and users can choose different modules for combination configuration according to specific business needs. In the construction industry, the most commonly used Dassault software includes the following categories:

(1) Design modeling: mainly based on cata brand, it is a parametric 3D modeling design tool;
(2) Construction simulation: mainly based on delm|a brand, it is a tool for construction simulation and optimization;
(3) Calculation and analysis: mainly based on s|mula brand, it is a general finite element calculation tool;
(4) Collaborative management: mainly based on ENOVIA brand, it is a project management and collaborative tool.

J. C. Hung et al. (Eds.): FC 2021, LNEE 827, pp. 1407–1413, 2022.
https://doi.org/10.1007/978-981-16-8052-6_202

2 Features of 3D Experience Platform

Dassault's success in many industries depends on the advanced 3D design and simulation tools on the one hand, and on the flexible, robust and reliable data management platform on the other. It can manage the data in the whole process from design to manufacturing, installation and maintenance in the same platform. In 2014, Dassault released the latest "3D experience" collaborative platform (referred to as 3D experience platform), as well as a series of industry solutions based on the platform. 3D experience platform has the following characteristics:

(1) Cloud based system architecture. 3D experience platform provides both enterprise cloud version and public cloud version. In the United States, Europe and other markets, Dassault system has provided software services through the public cloud, while in China, Dassault system mainly provides enterprise cloud version, that is, the central system is deployed on the enterprise's own server. However, no matter which way, users' 3D models and project information are uniformly stored in the system server, rather than scattered in each workstation. Therefore, as long as users can access the system through the network, they can get the latest and accurate information to carry out their work no matter when and where.

(2) Manage B | m information in database. Traditional BM software often stores all the information in the file, which is not easy to manage and share, because it is difficult to obtain the accurate information of each component from the outside of the file. The 3D experience platform stores the B | m information as a component in the database, and all the relevant information can be associated with the component object. Therefore, we can browse, query, count and edit the B | m information according to the component list, even without opening the whole 3D model. Query component list and attribute information in web browser [2].

(3) Built in collaborative work mode. Different project members, as long as they log on to the network platform, can carry out concurrent collaborative design, analysis and management according to the corresponding permissions. To this end, the 3D experience platform built-in collaborative work mechanism, including personnel role and authority management, data version management, object lock protection and a series of functions, to ensure that multi professional, multi type of work team can work together online in real time.

(4) Integrate a consistent user experience. On the 3D experience platform, Dassault system realizes the dual integration of the foreground and background: all the data in the background are stored in the same database, different personnel and different software modules share the same data, and there is no need to exchange data or convert format; and each application module in the foreground is based on the same 3D graphics platform, so it can achieve the same operation mode and graphics effect, There is no need to switch between different graphics platforms. Client user interface based on "if we compass".

3D parameter estimation, respectively recorded as:

$$\phi_{s,i} = (n_{s,i} + \beta_i)/(\sum_{i=1}^{V} n_{s,i} + \beta_i) \tag{1}$$

$$\theta_{j,s} = (n_{j,s} + \alpha_s)/(\sum_{s=1}^{K} n_{j,s} + \alpha_s) \tag{2}$$

3   Dassault's Support for IFC Standard

Dassault system 3D experience platform not only provides the support mechanism for BIM data standard, but also preset AEC data standard based on international standard FC in its civil construction industry module, which defines various BIM object types (such as doors and windows, stairs, curtain walls, etc.) and related attributes, and provides FC data import and export interface. The built-in data standard of 3D experience platform is compatible with FC4, while the import / import interface follows IFC2 × 3 standard (considering that most software import and export in the industry are IFC2 × 3 standard). Through the C standard, you can not only create BM models with rich information in the 3D experience platform, but also import BM models created by a variety of industry software into the 3D experience platform for management. The Revita model is imported into Dassault platform through IFC format (As shown in Fig. 1) [3].

For the civil construction industry, the standard FC preset in the 3D experience platform has been able to basically meet the demand of BIM data exchange. For the railway engineering industry, it can be expanded on the basis of the standard IFC, and this work is also in progress. In the r2015x version of "3D experience" released in the first half of 2015, it has been expanded for the bridge field, which is based on the existing draft of the international IFC bridge working group. The r2016x version, which is expected to be released at the end of 2015, will further expand the standards for bridges and tunnels, and more work is still in progress. As shown in Fig. 1.

Fig. 1. Dassault system model

4   3D Experience Platform Leads to Industry Change

Dassault's 3D experience platform is powerful and can integrate BIM environment to provide a highly collaborative method. Through BIM integration technology to promote the design model update, real-time access and understand the real project collaboration. The modeling of field construction, machinery, tooling and equipment box girder of beam making yard is completed by using CATIA application on 3D experience platform. Mr. Wu Jun, deputy general manager and chief engineer of China Railway fourth bureau group, said: "we attach great importance to the promotion and application of BIM Technology. The 3D experience platform of Dassault system will become a revolutionary change in our construction field. Before the construction of the project, you can see the 3D integration effect and experience it, effectively solving some mistakes in the design process." The 3D design module of 3D experience platform is to experience the whole process of product development in a visual way, and integrate the design process of a full set of products. From initial demand to structural design, it not only promotes collaborative innovation among departments, but also stimulates the design of creative products [4].

Mr. Wu Jun, deputy general manager and chief engineer of China Railway No.4 engineering group, said: "3D construction organization management is the core of our project management. Through technical means, our construction organization means have been improved and we have been deeply impressed. Based on the 3D experience platform of Dassault system, we have developed 4D visualization and construction management of Beam Yard, which is the first attempt in China. " The management module of 3D experience platform is a collaborative platform, which ensures the authenticity and unity of information through the sharing and collaboration of construction information, so as to make project management targeted. Through this module, we can compare the construction plan with the actual progress, check the actual progress, process nodes and planned progress, and analyze the construction progress to quickly find out the reasons that affect the construction progress. The construction quality can be managed through the functions of construction record, automatic early warning and diversified inquiry. In addition, the module can also be used for resource management of construction machinery, tooling and equipment, construction materials management, construction labor resource management, etc. Mr. Wu Jun, deputy general manager and chief engineer of China Railway fourth bureau group, affirmed: "at present, our bureau has successively applied information system platforms such as technology management platform, project cost management system, etc., basically realizing full coverage of all business systems, Effectively improve the management level of enterprises. In the future, we will continue to promote the application of information technology, and comprehensively help the internationalization strategy of enterprises.

5   Case Study

The municipal engineering industry is similar to the railway industry, involving a large number of lines, tunnels, bridges and other projects. In this field, Dassault system has carried out close strategic cooperation with Shanghai Municipal Engineering Design and Research Institute. Shanghai Yanggao South Road Underpass project (see Fig. 2) is a demonstration project designed by Shanghai Municipal Engineering Design and Research Institute using Dassault 3D experience platform. The reconstruction project of Yanggao South Road (Century Avenue Pujian Road) extends from the present century avenue to the overpass of Pujian Road, with a total length of 1.95 km. Road, tunnel structure bridge (Zhangjiabang bridge), rainwater and sewage drainage pipe, traffic signs and markings, signal lights, ventilation, monitoring system, power supply and distribution, building, greening and other related facilities and facilities, The construction and installation cost of greening and pipeline relocation is 1.455 billion yuan, with a total investment of 2.47 billion yuan.

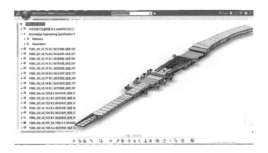

Fig. 2. Shanghai Yanggao South road underpass project

The project adopts Dassault's 3 dexperience r2015 as BM implementation platform. Compared with other software platforms, the platform has the following advantages:

(1) Dassault platform adopts the architecture mode of cloud platform and central server to unify the database of the project and provide reliable protection for the data security of the enterprise project.

(2) It supports the type extension of BIM model. In Yanggao South Road project, the unified deployment method of user-defined type is adopted to uniformly deploy different types of components in tunnel structure, such as side stones, crash barriers, asphalt, etc. (see Fig. 3). It provides great convenience for the later engineering calculation and Simulation of construction.

(3) Multi professional real-time collaborative BM platform, in the process of the project, designers of structure, bridge, pipeline and other different professions can carry out real-time design tasks on the same platform [5]. This synchronous modeling function can immediately find the defects in the design process, and quickly check whether there is interference in the model.

(4) The BM modeling process on Dassault platform is initially formed. The current software format supports large-scale 100 km data model, and the whole BIM scope covers the early scheme design, the middle detailed structural design, the later construction simulation, project report and online browsing based on Internet Explorer.

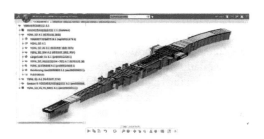

Fig. 3. Unified deployment of user defined types

6 Conclusion

With the increasingly wide application of BIM Technology in the engineering construction industry, the focus of the industry has expanded from the modeling technology itself to how to manage the BIM data structurally, so that it can extend from the design stage to the downstream, and maximize the value in the construction, operation and maintenance stage.

Dassault's 3D experience solution takes the data management platform as the core, builds an enterprise level 3D visualization BIM project management platform integrating modeling, simulation, collaboration and management, and supports industrial and enterprise BIM data standards through flexible data management technology, so as to lay a good foundation for the information management of the whole life cycle of the construction industry.

Dassault system will join hands with Chinese enterprises to promote the realization of construction life cycle management and bring reform and innovation to the construction industry.

References

1. Bai, Y., Liu, Y.: Research progress and future prospect of BIM software secondary development. Sci. Technol. Innov. 81–82 (2019)
2. Yang, K.: Marketing strategy research of Zhongnan survey, design and Research Institute Co., Ltd., Hunan University (2018)
3. Yang, Q.: Application research of visual collaborative design based on BM. Southwest Jiaotong Univ. (2016)

4. Liu, Z.S., Ma, J.S., Wei, Q.X., Xu, R.L.: Application of BIM technology in power station digital design. In: Proceedings of the 14th National Symposium on Modern Structural Engineering of Tianjin University and Tianjin Steel Structure Association. Tianjin University and Tianjin Steel Structure Association: Academic Committee of National Symposium on Modern Structural Engineering (2014)
5. Xiong, B., Zhang, S.: Construction of application platform for design, construction and operation of digital hydropower station. Renmin Changjiang River (2019)

The Form and Evolution of Chinese Ancient Literature Territory Based on Computer Virtual Reality Technology

Qiangzu Liu[(✉)]

Ningxia Institute of Science and Technology, Ningxia 753000, China

Abstract. With the continuous development of computer technology, computer multimedia display system is a kind of software system which takes information output and play as the purpose and information release and transmission as the leading. Through the organic combination of text, picture, animation, video and audio, it forms a continuous picture in real time, and plays it to people through various existing display devices, so as to convey various propaganda information to people. In this paper, the design of the computer-based display system of ancient Chinese literature layout and evolution is conducive to the teaching and dissemination of ancient Chinese culture.

Keywords: Computer · Multi-Media · Ancient Chinese literature · Literature territory · System design

1 Introduction

The development of ancient Chinese literature has experienced the process from germination to development and growth. It is the link between the five thousand years of splendid Chinese civilization and the development of literature and culture. In the process of the establishment, consolidation, development and prosperity of the feudal system, it acts as the vanguard of political reform and the defender of class rule. In ancient China, this relationship is particularly obvious. Since modern times, people often think that literature is subject to or serves politics according to the proposition that ancient literature carries Tao. Chinese literature and politics have always existed together. Their development complements each other and has two sides. The development and characteristics of literature and culture in a certain period are inseparable from the political situation at that time. Therefore, before explaining and discussing the ancient literature of a certain period, we should first understand and explain the political environment of this period. In the teaching of ancient Chinese in Colleges and universities, the teaching of this link is inseparable from the explanation of the political territory at that time. Generally speaking, when explaining the political background of literature, we need to show the literature territory at that time, and then in the subsequent extension and teaching, reveal and explain the literature territory and characteristics at that time, and analyze and explain the typical cases in detail. In order to meet the above needs, teachers often need to prepare a large number of political maps in different periods, which is very inconvenient to use.

First of all, the display of the political map of the authorities is to provide a space for the development of literature and reveal the reasons and laws of its evolution. However, the existing maps and their display methods can only display the territory, but can not reflect the evolution process, and it is not convenient to mark the relevance between the political pattern and the literary pattern, so it is not easy to bring in. Can not better cooperate with the interpretation of ancient literature.

Secondly, in the teaching of literature, because the development of literature is a dynamic process, the political changes before and after will become the driving force of the development and characteristics of literature at that time, so the teaching class often involves the territory of multiple dynasties. The carrying, hanging and replacement of multiple maps bring trouble to teachers, affect the coherence of the classroom, and reduce the teaching efficiency [1].

Moreover, due to the different political situations of different dynasties and periods, it is necessary to have a large number of maps. This brings a lot of work to sorting, storage, maintenance and so on.

2   Computer Multimedia Technology

2.1   Computer Overview

The first digital computer in the real sense is the abbreviation of electronic numerical integrator and calculator. In February 1946, the University of Pennsylvania was put into operation in the United States. ENIAC used about 18000 vacuum electronic tubes, weighing 30 tons, 174 kW, covering an area of 140 square meters, and was added in decimal system and 5000 times per second. It has no keyboard, mouse and other equipment today, people can only switch the huge panel on the countless switches to input information to the computer. The birth of ENIAC laid the foundation of the development of electronic computer, opened up the information age, and pushed the human society to a new era of the third industrial revolution.

Characteristics of computer:

(1) The operation speed is fast. Peak operation of 10 of the top 500 international supercomputers announced in June 2009
   The speed of calculation is more than 300 trillion times, and the Roadrunner made by IBM is more.
   It has reached a very high speed of 1456 trillion times/second.

(2) Large storage capacity. The storage of computer is an important feature of computer which is different from other computing tools.

(3) It is universal. Generality is the basis that computer can be applied in various fields. Any complex task can be decomposed into a large number of basic arithmetic operations and logic operations.

(4) Work automation. The operation operation inside the computer is automatically controlled and executed according to the program that people have prepared in advance.

(5) High accuracy. The reliability of computer is very high, and the error rate is very low. Generally speaking, only in those places where manual intervention can errors occur.

(6) Logical judgment ability. With the help of logic operation, the computer can make logical judgment, analyze whether the proposition is true or not, and take corresponding countermeasures according to whether the proposition is established or not.

2.2 Computer Multimedia Technology

With the continuous development of computer technology, computer multimedia technology gradually steps into human society. Computer multimedia technology is a computer based exchange type comprehensive technology, and integrates digital communication network technology to realize the processing of various media, such as data, text, video, image graphics, sound and other information technologies [2]. Through the use of a variety of software and hardware, The main aspects of the system include: multimedia communication technology, including voice, data, image, video transmission, etc.; Multimedia data processing technology, including video technology, image technology, audio technology, virtual reality, interface design and so on; Artificial intelligence, including hardware and software technology and intelligent technology, etc.

3 The Design of the Layout form and Evolution Display System of Chinese Ancient Literature Based on Computer Multimedia

The computer-based display system of the layout and evolution of ancient Chinese literature is based on computer and touch platform as the controller. With the help of its rich and wonderful content of PPT, the layout and evolution of ancient Chinese literature are fully displayed. And set up small games, teaching in music, is very conducive to the spread of Chinese culture, the system structure is shown in Fig. 1.

Fig. 1. The layout form and evolution of ancient Chinese literature by computer

The computer-based display system of the layout and evolution of ancient Chinese literature is set up as follows:

(1) Touch board system module; The touch board system module includes ppt content module and puzzle game module;

(2) Ppt content module includes layout profile module, evolution process module, representative character poetry collection appreciation module, each layout representative character module;

(3) The puzzle game module includes the literature blank filling module, the literature knowledge connection module;

(4) The layout form introduction module is connected with the intelligent voice broadcast module, which is used to introduce the layout form;

(5) The evolution process module is connected with the intelligent voice broadcast module, which is used to analyze the layout form and figure represent the overall evolution of characters;

(6) The representative poetry collection appreciation module is connected with the intelligent voice broadcast module, which is used to display and analyze the appreciation of the representative poetry collection;

(7) Each layout represents the character module, connects the intelligent voice broadcast module, which is used for the analysis of the representative characters of each layout; The paper introduces the function of literature blank playing module, which connects the intelligent voice broadcast module to fill in the blank games;

(8) The link module of literature knowledge is connected with the intelligent voice broadcasting module, which is used for the connection of literature knowledge;

(9) The touch board system module is equipped with LAN module, which connects ppt content module and puzzle game module.

The link stability and energy mixing model of the LAN module includes:

The structure of local area network is an undirected graph network model G = (V, e), where V represents a group of nodes, E represents a group of edge sets connecting nodes, P (u, u) = {P_0, P_1, P_2, ...P_n} is the set of all possible paths between node u and node v, P_i is the path of node u and v, and the optimal path from node u to node V is selected,

The formula of routing stability and node residual energy is as follows:

$$F_1 = \prod_{e \in P_i} LS(e) \tag{1}$$

$$F_2 = \prod_{e \in P_i} Ce(t) \tag{2}$$

$$Ce(t) = \frac{E_{is}}{E_{iO}} \quad E_{is} > E_{th1} \tag{3}$$

Among them, E_{is} and E_{io} is the residual energy and total energy of node i, and E_{th1} is the energy threshold of node, q_i Denotes the queue length of the node, q_{th} is expressed as the threshold value of the queue length of the node.

The data processing module of the system includes:

(1) FPGA module, through the preparation of precoding method and companding method program, the signal is processed to reduce the PAPR of 0 fdm signal;
(2) A digital to analog converter connected with the FPGA module is used to realize the digital to analog conversion and process the signal with high data rate at the front end;
(3) A power supply module is connected with the FPGA module and the digital to analog converter to provide a DC working voltage;
(4) A series parallel conversion unit is used to transform the input signal and process the inverse Fourier signal;
(5) A precoding unit connected with the serial to parallel conversion unit is used for preprocessing the serial to parallel signal;
(6) An inverse fast Fourier transform unit connected with the precoding unit and the serial parallel conversion unit is used for processing the output signal of the precoding unit;
(7) A signal amplitude estimation unit connected with the precoding unit for estimating the signal amplitude;
(8) A companding unit connected with the serial to parallel conversion unit and the signal amplitude estimation unit is used for reprocessing the parallel to serial conversion output signal;

The fast Fourier inverse transform unit further includes butterfly operation unit, filter, rotation factor unit and Ping Bing buffer structure to realize multi carrier mapping [3]. The butterfly unit uses four pairs of RAM * 2 to store the operands of butterfly operation, which greatly improves the operation speed; The rotation factor unit uses the look-up table method to speed up the implementation of the algorithm; The ping-pong buffer structure is configured as ping-pong structure to further improve the operation speed; The precoding unit adopts excellent zadoff chusesequences: algorithm, which has good autocorrelation and cross-correlation, and is conducive to reducing inter symbol interference. At the same time, the zadoff Chu sequence has symmetry, which can reduce the complexity of sequence generation.

4 Build Excellent Hardware Environment

Based on the whole set of server system of Chinese ancient literature territory form and evolution, the server configuration makes full use of the open source advantage of Linux system and better hardware foundation to build a virtual host, and improves the overall performance on the basis of saving money. In terms of network transmission, relying on the 10 Gigabit Campus Network with hardware firewall, the smooth dissemination of multimedia courseware on campus is preliminarily realized [4]. The external network of the school adopts the three line access of Unicom and telecom education network, which is very conducive to the teaching and dissemination of Chinese culture.

5   Conclusion

In the process of rapid development, computer technology has gradually changed people's production and life. In recent years, new ideas and new technological innovations are constantly emerging, which promotes the rapid development of multimedia technology and presents a rich and colorful situation. Computer multimedia technology has achieved breakthroughs in many fields, especially in holographic image, pattern recognition, sensor technology, etc., Realize the research and development of new technology, promote the multimedia more comprehensive application prospects.

References

1. Rao, J., Wang, J.: Application research of computer multimedia technology. Comput. Telecommun. (12), 3–5 (2010)
2. Dong, J.: Application research of computer multimedia technology. J. Chifeng Univ. (Nat. Sci. Ed.) (08), 11–12 (2013)
3. Introduction to Chinese journal of maternal and child health research. Chin. J. Matern. Child Health Res. 32(05), 646 (2021)
4. The practice of agricultural science and technology innovation to help Henan out of poverty and thinking of preventing poverty. New Agric. (10), 21–24 (2021)

Computer Network Security Analysis Based on Deep Learning Algorithm

Yin Liu[⊠]

Huainan Normal University, Huainan 232038, Anhui Province, China

Abstract. With the development of information technology, computer network information system has been widely used in many fields. However, the security of computer network information system is worrying. In the past decade, the losses caused by the destruction of computer systems have increased sharply year by year, and more than 60% of the users in the world have been invaded by various illegal means. There are great risks in computer network information systems. Based on the deep learning algorithm, this paper will determine the influencing factors in the current computer network security, Then the security analysis is carried out and the security analysis model is established.

Keywords: Deep learning · Network security analysis · Computer

1 Introduction

In recent years, with the rapid development of the Internet, various industries have benefited from it. However, the Internet is also a double-edged sword. In the process of running in with various social industries and systems, this new thing has also given birth to a variety of social problems. More and more disharmonious behaviors, such as personal privacy leakage, fraud, network violence and so on, take the Internet as the carrier, seriously endanger personal living conditions, social order and national stability. Network security has become a social problem that needs to be concerned.

Modern network communication system can be said all over the world. Although many different users use different types of computers and run completely different operating systems, they can still communicate through TCPP protocol series. The establishment of computer model, because the method of attacking computer network is always very complex, and the personal information of enterprises and individuals is easy to leak. Both foreign reports and domestic computer crime reports show that the security system of our network system is very fragile, and the security of the system has become an urgent problem that we need to solve [1].

Network security loopholes are the intentional or unintentional defects in the design, implementation, configuration and operation of information products and information systems. These defects exist in various levels and links of the system in different forms. Once they are used by malicious agents, they will damage the security of the system and affect the operation of normal services built on the information system. Relevant research shows that most of the vulnerabilities are similar in some attributes, such as the attack mode taken against the security vulnerability, the possible

J. C. Hung et al. (Eds.): FC 2021, LNEE 827, pp. 1420–1425, 2022.
https://doi.org/10.1007/978-981-16-8052-6_204

threat to the attacked system, and whether the availability, integrity and confidentiality of the attacked system are harmed. The integration of network security information can ensure the sharing of knowledge and experience related to security vulnerability information. However, some vulnerability publishing platforms established by software manufacturers are mainly aimed at the problems of their own products. Security companies and government departments establish security vulnerability libraries. Although each vulnerability library can contain the vulnerability information of multiple software manufacturers, it integrates the vulnerability distribution to a certain extent. However, these vulnerability libraries focus on different types of vulnerabilities, different data sources, and different information fields. The establishment of network security information management system can ensure the accurate, timely and comprehensive release of vulnerability information. It is of great significance for the improvement of network security.

2   Deep Learning Algorithm

Deep learning is a multi-level analysis of data representation. On this basis, miko1ov1 proposed the source theory of word2vector framework to generate word vectors. Distributed representation solves the problem of exclusive representation. By using fixed length dense vectors to represent words, these generated vectors will form a word vector space, and each vector will be regarded as a point in this space. At the same time, the length of the vector can be set by itself, which is not necessarily related to the size of the dictionary. In the deep learning method, we focus on the problem, determine the feature vector, and then use the training and extraction method to achieve data processing. For example, in Botnet, if we want to identify it comprehensively and effectively, we need to use its related features to establish a model [2].

First of all, the model provides a basic framework to solve the problem of probability calculation, which can be used to calculate the probability of the occurrence of text sequence. For a text sequence $S = w_1, w_2, \ldots, w_t$. Its probability can be expressed as:

$$P(S) = P(w_1, w_2, \ldots, w_t) = \prod_{t=1}^{t} p(w_t | w_1, w_2, \ldots, w_{t-1}) \tag{1}$$

But this kind of calculation needs huge parameter space, and there are not many applications in daily practical applications. People use more simple NGram model:

$$p(w_t | w_1, w_2, \ldots, w_{t-1}) \approx p(w_t | w_{t-3+1}, \ldots, w_{t-1}) \tag{2}$$

In this method, the feature quality needs to be determined first, It directly determines the efficiency of the final learning results, Usually, we need to use the cross vector to carry out the research, This method can improve the dimension of the whole calculation process, This process can be realized by principal component analysis and deep learning algorithm. At present, deep learning algorithm has been widely used in China, including image recognition, speech recognition and language processing, Due to its application in different fields, The research focuses are also different, Part of the focus is on the characteristics of institutional discovery, and part of the focus is on the establishment of nonlinear models, so there are great differences in different fields. Figure 1 shows the structure of deep learning algorithm.

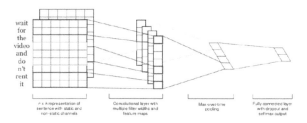

Fig. 1. Structure chart of deep learning algorithm

3   Computer Network Security Analysis

In the information society, the application of computer and network in government, enterprise and people's daily necessities is more and more extensive. At the same time, in the wide application of technology, the network security problems such as data leakage, high-risk vulnerability, intelligent crime are gradually emerging.

Network security related information generally appears in professional forums or websites, but the focus of each forum or website is different, and the direction of network security is not the same. In the face of complex information sources, it is often difficult for technical personnel to obtain the information they want at the first time [3].

Scurity work content, and increase the ability of information sharing and collaborative work. The goal of the system is to effectively collect, sort, identify and classify the security information, knowledge and vulnerabilities scattered on the network, screen out important and urgent information, remind and distribute the team members, and form information sharing.

At present, the security of the computer network in the actual operation process is still unable to achieve comprehensive protection, and the influencing factors include the following: first, computer virus is the main influencing factor in the security of the computer network, once the virus enters the computer network, The internal system program and software of the whole computer will be seriously affected, resulting in the damage or even loss of data in the computer, and the whole computer system can not run normally. If the invading virus is specially treated, its invasiveness will be greatly improved, and the inherent data information in the computer will be changed while destroying the internal program of the computer. Computer virus has a certain incubation period, and also has a strong infectious, through the way of network data continues to spread, until the whole computer system has adverse effects. Second, computer software, at present, all kinds of software are constantly updated. In the process of software design, designers need to study the software design methods and the existing influencing factors from all angles, so as to improve the practical application performance of the software. Due to the differences between the operating environment and infrastructure, there are some influencing factors in the process of software design, which will limit the improvement of software performance. Usually, if there is a vulnerability in the actual operation of the system, the technicians need to repair it at the first time, otherwise the virus will enter into the system through the vulnerability, resulting in the virus infection, resulting in the loss of data in the computer [4].

With the rapid development of computer technology, there are more and more vulnerabilities. If we use manual to classify, it will consume a lot of human resources, So in the classification technology, the use of automatic classification sorting technology can save a lot of time and manpower, and greatly improve the speed and efficiency of classification, help enterprises in a large number of vulnerability information, can quickly pick out the more harmful vulnerability, in the limited time and energy to give priority to deal with the more harmful vulnerability.

4   Analysis of Deep Learning Algorithm in Computer Network Security

Computer network security is the goal that many enterprises and countries want to achieve. We use mathematical methods to choose the appropriate mathematical model according to the actual situation.

$$e_j = -k \sum_{i=1}^{n} f_{ij} lnf_{ij} \tag{3}$$

Replace data with calculation:

$$W_j = d_i / \sum_{j=1}^{m} d_j \tag{4}$$

High quality computer network management can optimize the computer network system and improve the security of the system in actual operation [5]. For example, in the process of establishing the network system, we need to optimize the LAN design scale and network system switch deployment, According to the actual situation, try to increase the number of management switches, this way can continuously narrow the scope of network fault, accurately locate the location of the fault, and then effectively deal with the network system fault in a short time. In addition, the network monitoring software is installed in the core switch to collect and analyze the information in the network, and transmit the final feedback results to the corresponding platform, which can realize the dynamic monitoring of the whole computer network and ensure the comprehensive and effective control of the switch, It avoids the command management mode and reduces the management difficulty of computer network system failure. Computer network security has strong comprehensiveness, so it needs to be analyzed from multiple perspectives to ensure the effectiveness of the final analysis results.

5 Content Security Difficulties

5.1 Safety Problem

The security field is a relatively special scene. We often face problems such as unclear target definition, complex data types, large image quality gap, and often face counter attacks [5]. This scene requires high algorithm capability. In contrast, in 2C marketing scenario, for example, mobile phone scanning product/icon, recall ability is relatively higher for users, and misjudgment is not easy to be found in the use process; In the scene of access control and attendance, due to the high quality of the image collected by the camera, the algorithm needs to solve the effect problem within the limited image quality range. In the field of content security, the normal proportion of online data is high, and there are many types of images, so the problem of misjudgment is very easy to be concentrated; On the other hand, UGC image quality is uneven, image sensitive features are often not obvious, small target, blur, deformation and other problems often appear, accompanied by the development of content security business of e-shield. We deeply explore the application of deep learning image algorithm in this field, and achieve the expected results in the actual scene.

5.2 Simulation Training

After model training and data processing, we will select a model for training. We use induction-v3 as the model selection, mainly considering the balance of performance and effect. We need to start with experiments. We download Imagenet pre training model to local tensorflow slim according to the link of open source, We can call the dtrain image classifier.py script in practice. First, we train the parameters of the lowest level full connection with a larger learning rate, and then adjust all the parameters with a smaller learning rate. The main purpose of network attack is to steal enterprise user data, sales data, intellectual property documents, source code and software secret key. Most attackers use anonymous network, which makes it difficult for security personnel to track the traffic. In addition, the stolen data is usually encrypted, which makes the rule-based network intrusion tools and firewalls ineffective. Recently, anonymity has been used in C & C as a variant of blackmail/malware. For example, onion blackmail uses the tor network to communicate with its C & C server.

6 Conclusion

The attack success rate of the attacker is related to the attack mode, attack and the severity of the vulnerability itself. We should take the initiative to attack the attacker's target by analyzing the attack methods and the characteristics of attack tools based on the existing information. The work of computer network security evaluation model comprehensively checks various security institutions and security elements in the system. It can be used not only for the security analysis of stand-alone system, but also for the security evaluation of network information system, all kinds of security protection strategies should be mixed. At the same time, more and more security solutions will be developed.

References

1. Tian, K.: Security and preventive measures of computer network system in financial industry. Comput. Prod. Circ. (8), 36 (2019)
2. Wang, X.: Computer network security analysis and modeling based on deep learning algorithm. Electron. Technol. Softw. Eng. (16), 195–196 (2019)
3. Wu, Y., Sun, Z., Liu, R., et al.: Research on the construction of network security and protection system: guide (2), 154–154 (2016)
4. Liu, R.: Computer network security analysis (9), 14–15 (2015)
5. Lu, P.: Research on security evaluation standard and design of information security system. Ruoxi Engineering University of Chinese PLA (2001)

Application of Spectroscopy and Spectral Imaging Technology in Food Detection

Yonghui Liu and Xin Zhou[✉]

Shandong Commercial Vocational and Technical College, Jinan 250103, China

Abstract. Food safety is a hot issue for consumers. With the continuous development of computer imaging technology and spectral technology, spectral imaging technology has been applied in the field of food detection. The spectrometer is the core component in the spectral imaging system. It can divide the mixed light with wide wavelength into single wavelength light with different frequencies by means of optical elements, and complete the data acquisition with the help of computer software and hardware. Based on this, this paper mainly discusses the application of spectrum and spectral imaging technology in food detection.

Keywords: Spectrum · Spectral imaging technology · Food testing · Application research

1 Introduction

Food is the most basic material closely related to people's survival and life. From ancient times to now, people have paid great attention to the topic. With the unprecedented development of food industry and market, food detection problems, including food classification, quality identification and food safety, are becoming more and more important, and have been widely concerned by people, industry and the state.

In recent years, with the development of economy and the improvement of industrialization, the use of new technology and chemical drugs in food production and processing industry has exposed various food safety problems in the world. Such as avian influenza, foot and mouth disease, mad cow disease events worldwide; Sudan incident, melamine milk powder, gutter oil event, etc. This also deepen the requirements of improving laws and regulations and improving the optimization of detection technology in food detection and quality monitoring [1].

On the other hand, with the rapid development of optics, spectroscopy and chemometrics, the cross discipline with food science has formed a new interdisciplinary subject with great application value. Whether it is the research and development of new fast and high-precision detection system or the application of analytical algorithm in the qualitative and quantitative detection of specific food, it has attracted the attention of the scientific research and industry. Therefore, in the field of food detection, the research and development of key technologies and application in specific scenarios

J. C. Hung et al. (Eds.): FC 2021, LNEE 827, pp. 1426–1430, 2022.
https://doi.org/10.1007/978-981-16-8052-6_205

have not only practical research value to improve the level of detection technology, but also has important significance for ensuring food quality, people's health and economic and social stability.

2   Spectral Imaging Technology

2.1   Principle of Spectral Imaging

Spectral technology refers to the study of molecular structure and dynamic characteristics by using the interaction of light and matter. That is to say, chemical information related to the sample can be obtained by obtaining the emission, absorption and scattering information of light. Imaging technology is to obtain the image information of the target and study the spatial characteristics information of the target. These two independent disciplines have developed for hundreds of years in their respective fields, but we know that in the 1960s, with the rise of remote sensing technology, space exploration and surface exploration have become a hot spot in scientific research. What people want to get is not only the impact information or the spectral information of the target, but also the image information and spectral information It greatly led to the combination of imaging technology and spectral technology, and gave birth to imaging spectral technology [2].

The essence of spectral imaging technology is to make full use of the absorption or radiation characteristics of different electromagnetic spectra, and add one-dimensional spectral information on the basis of ordinary two-dimensional space imaging. As shown in Fig. 1. Due to the different composition of ground objects, there are differences between their corresponding spectra (fingerprint effect), so the spectra of ground objects can be used for recognition and classification. Spectral imaging technology can obtain many narrow and continuous image data in the electromagnetic band of ultraviolet, visible, near infrared and mid infrared regions, and provide a complete and continuous spectral curve for each pixel.

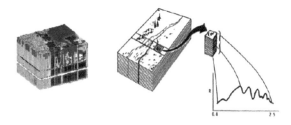

Fig. 1. Schematic diagram of spectral imaging technology

Is the schematic diagram of the imaging spectrum technology. The imaging spectrometer obtains a three-dimensional data cube, and can extract a continuous spectral curve from each spatial pixel. Through the characteristic analysis of the spectral line, it can be used for subsequent exploration and other purposes.

2.2   The System Composition of Imaging Spectrum

A typical push sweep near infrared imaging spectrum system consists of light source, infrared camera, imaging spectrometer, lens, mobile sample station, computer image acquisition system and motion control system. See Fig. 2 for the system composition. The imaging spectrometer is mainly divided into light by grating. The moving sample table is driven by a stepping motor to move at a uniform speed. In order to adjust conveniently, some systems have developed a three-dimensional motion platform, which can move at a uniform speed in the direction of X and Y axes, and adjust the height in the z-axis direction according to the sample size. Some imaging spectral systems have fixed samples, while infrared cameras and imaging spectrometers move at a uniform speed to complete the scanning of spectral images.

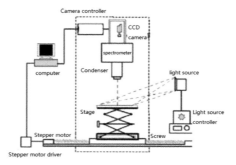

Fig. 2. Composition of imaging spectrum system

3   Data Analysis Method of Imaging Spectrum

At present, principal component analysis (PCA) is widely used to compress high-dimensional hypercube data in imaging spectrum analysis. The principal component (score matrix) t is a linear combination of the original variables. When it is used to represent the original variables, the sum of squares error is the smallest. The first principal component can explain the largest amount of variance of the original variable, the second is the second, and so on. There are many methods of principal component calculation, and the commonly used algorithms are non-linear iterative partial least squares (NIPALS) and covariance matrix decomposition. Through the principal component compression of the hypercube data, the feature information can be obtained. In the principal component space, the differences of imaging spectra of different types of samples can be seen, so as to make qualitative discrimination.

If quantitative prediction is needed, the mathematical model can be obtained by multiple linear regression between the dimension reduced score matrix T and the chemical analysis value matrix Y.

$$Y = TB + E \tag{1}$$

Where: $B = (T^T T)^{-1} TY$ is regression coefficient; E is introduced error.

In the decomposed principal component, the principal component of the front contains most useful information of the X matrix, while the latter is related to noise and interference factors, which is the principal component regression. At present, partial least square (PLS) is also used to establish quantitative models. In the actual data analysis of hypercube, because the data of multiple samples forms a multi-dimensional data set, it is not possible to directly use PCA or pls to calculate. The original hypercube should be expanded in a certain dimension (such as wavelength dimension) in turn, and then the subsequent calculation and analysis should be carried out to extract the sample composition and spatial distribution information contained in the imaging spectrum.

4 Application of Spectral Imaging Technology in Food Detection

4.1 Application of Spectral Imaging Technology in Food Freshness Detection

Food freshness detection is mainly to detect the freshness of agricultural products. Taking the freshness detection of meat products as an example, the method of sensory evaluation of meat freshness is destructive to meat products. After the spectral imaging technology is applied to the freshness detection of meat products, people can use the wavelength region of near infrared spectral imaging system to analyze the quality of meat products with different components and processing parameters. For example, in the process of using spectral imaging technology to analyze cooked meat products, some scholars used the principal component analysis method. After selecting eight characteristic wavelengths of 980, 1061.1141.11174.1215.1321436 and 1241 nm to identify the quality of cooked meat products, the accuracy of the identification results reached 97.20% [3].

For fruits and vegetables, acidity, hardness, moisture, starch content, maturity and internal defects can be regarded as the reflection of their quality characteristics. Taking pork products as an example, color factor can be regarded as an important measure of pork products. According to the research results of some scholars, after the near-infrared spectroscopy imaging technology is applied to pork quality testing, people can use the least square regression method for modeling and analysis, so as to analyze the color reflectance of pork products and the model determination technology. Some researchers have also applied hyperspectral imaging technology to the freshness detection of shrimp. In the detection of shrimp freshness, the wavelength of NIR image is controlled at 400–000 m, and the detection method is combined with continuous projection algorithm and least square regression method. According to the actual detection results, the accuracy of the detection method combined with continuous projection algorithm and least square regression method can reach 98.33%, The prediction accuracy was 95.00% [4].

4.2 Application of Spectral Imaging Technology in Food Biological Contamination Detection

Food biological pollution detection is an important measure to control food pollution. The emergence of cadmium rice incident and Sanlu milk powder incident has sounded an alarm for people in the field of food safety. In the environment of frequent food safety problems, food quality has become a problem of great concern to the public. Take meat products as an example, the temperature fluctuation of cold storage will increase the number of bacteria attached to meat products. Biological pollution detection has become an important factor of consumer safety. From the development status of biological pollution detection system of meat products, people can use near infrared spectroscopy technology to analyze the microbial growth of rotten meat products. The application model based on the partial least squares method is also the research model used by the leading city of food biological pollution detection.

4.3   Application of Spectral Imaging Technology in Food Moisture Detection

At present, the spectral imaging system is mainly composed of hardware system and software system. Sensor is the core component in the hardware system of this system. The sensor includes light source, scanner, control device and other devices; interference imaging spectrometer and grating imaging spectrometer are commonly used in non-destructive testing of food quality. Linear array detector and area array detector are two kinds of CCD array detection equipment commonly used in the field of food detection. The software part of the system includes spectral preprocessing software and data acquisition and processing software. According to the spectral range of hyperspectral image, the spectral range of hyperspectral image in the field of video quality nondestructive testing is near infrared 400–1000 nm and near infrared 1000–1700 nm.

5   Conclusion

With the deepening of research and the continuous development of technology, spectral imaging technology has a broad application space, its application range will be more and more wide, and the requirements of spectral imaging technology are also higher and higher. How to use spectral imaging system combined with data processing software to obtain more accurate data is of great significance to the future development of spectral imaging technology.

References

1. Yan, Y., Zhao, L., Han, D., et al.: Basis and Application of Near Infrared Spectroscopy, pp. 11s–120. China Light Industry Press, Beijing (2005)
2. Li, P., Zhu, J., Liu, Y., et al.: Application and prospect of machine vision in detection and classification of agricultural products. J. Jiangxi Agric. Univ. **27**(5), 796–800 (2005)
3. Wu, L., He, J., Liu, G., et al.: Nondestructive detection of moisture content of long jujube based on near infrared hyperspectral imaging technology. Optoelectron. Laser (01), 135–140 (2014)
4. Sun, J., Wu, X., Zhang, X., et al.: Prediction of lettuce leaf moisture based on hyperspectral images. Spectro. Spectral Anal. (02), 522–526 (2013)

Industrial Internet System and Working Method Based on Deep Learning and Edge Computing

Dahai Wang[✉]

Department of Business Administration, Business School,
College of Humanities, Northeast Normal University, 1488 Boshuo Road,
Jingyue District, Changchun City 130117, Jilin Province, China

Abstract. With the advent of 5g era, its characteristics of low power consumption, high speed, low cost and low delay have brought positive and huge changes to the Internet of things industry. The increasing scale of the Internet of things brings challenges to the energy consumption, transmission bandwidth and processing delay of centralized cloud computing data center. The former is moving from the center node to the edge node with short delay, namely edge computing. Edge computing can share the work of the cloud, reduce the pressure, and meet the needs of users for real-time services. Migration decision and resource scheduling is one of the hot issues in the field of edge computing. This paper introduces the migration decision and resource scheduling of mobile edge computing and the theoretical basis of deep reinforcement learning, and summarizes and analyzes the existing migration decision and resource scheduling schemes and algorithms. On this basis, this paper focuses on how to use deep reinforcement learning algorithm to optimize the migration decision and resource scheduling in two actual scenarios: multi cell multi-user full migration and single cell multi-user partial migration.

Keywords: Edge computing · Migration decision · Resource scheduling · Deep reinforcement learning

1 Introduction

The Internet of things will connect tens billion resource limited intelligent terminal devices However, the scale of the Internet of things is increasing day by day. The amount of data generated by a large number of intelligent terminal devices is also growing rapidly at the exponential level. At the same time, the battery life and computing capacity of intelligent terminal devices are also very limited, which brings consumption, computing center. In mobile network system which needs to ensure real-time, it seems that it is impossible to meet the needs by cloud computing alone. A promising solution is to leverage edge computing (EC) [1]. Edge computing is a decentralized computing architecture, which transfers the intensive computing of applications, data and services from the network center node to the network logical edge node, and the processing delay is shorter. The edge server with low cost and certain information processing and storage capacity shares part of the work of cloud

© The Author(s), under exclusive license to Springer Nature Singapore Pte Ltd. 2022
J. C. Hung et al. (Eds.): FC 2021, LNEE 827, pp. 1431–1436, 2022.
https://doi.org/10.1007/978-981-16-8052-6_206

center server, greatly reducing the pressure of cloud center; meanwhile, as a distributed system closer to the bottom, edge server can handle some of the users' services in time, which meets the needs of users in mobile network for real-time services.

2   Theoretical Basis of Mobile Edge Computing and Deep Reinforcement Learning

Promising solution emerging in recent years. It can liberate the mobile devices with limited resources from the computing intensive tasks, enable the devices to migrate the workload to the nearby MEC servers, and improve the quality of the computing experience. Deep reinforcement learning (DRL), which is suitable for decision making learning. However, compared with cloud computing, edge computing is highly dynamic, so resource management needs to be adaptive [2]. The traditional model-based resource management method is limited in practical application because it involves some assumptions or preconditions. At this time, combining the two, the computing nodes of mobile edge computing can provide a lot of computing resources for deep reinforcement learning reasoning and training. Deep reinforcement learning can also describe the network well for mobile edge computing without any prior knowledge, and then effectively manage the resources of network edge.

2.1   Reinforcement Learning

Reinforcement learning is a learning method to learn how to maximize the reward value in the current scene. It is a learning process of mapping situations to actions. For structure of often a decision-making problem in order to achieve the goal state. It has an agent to observe the state of the surrounding environment, and can execute an action to reach another state in the current environment according to the policy rules, and get a reward (positive and negative feedback). The basic structure of reinforcement learning as shown in Fig. 1 below:

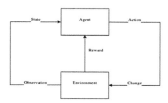

Fig. 1. The basic structure of reinforcement learning

2.2   Deep Learning

A superficial definition of deep learning is "machine learning mainly using deep neural network as a tool". Therefore, deep neural network is the basis of deep learning. The inspiration of neural network comes from the structure of cerebral cortex, in which the most basic unit is perceptron, that is, neurons. A neuron mainly consists of three parts: 1) input weight parameter W; 2) accumulator $\sum$; 3) activation function F. The input weight W acts on the sample input x (which may also be the output of the upper layer network), and then adds it up through the addition gate $\sum$ to get an accumulated result; finally, a response value is obtained through the activation function [3]. The activation function can be regarded as a threshold processing mechanism. Only when the input exceeds a certain value can the output be activated. The schematic diagram of neurons is shown in Fig. 2 below:

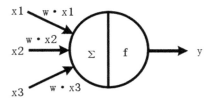

Fig. 2. Schematic diagram of neurons

Neural network can have several layers of interconnected neurons, and the set of neurons in each layer is called a neural network layer. According to the different functions, the neural network layer can be divided into three types: 1) input layer: the layer used to input the sample feature vector data; 2) output layer: the layer used to output the data in the network design; 3) hidden layer between the two. Deep neural network is a kind of neural network with at least one hidden layer. The so-called depth means that the hidden layer in the network is very deep. Similar to the shallow neural network, the deep neural network can also provide modeling for complex nonlinear systems, but the additional level provides a higher level of abstraction for the model, which improves the ability of the model.

3   Multi Cell and Multi User Complete Migration Optimization Algorithm Based on Deep Q Network

However, due to the difference of computing power of different edge nodes in multiple cells, and the different channel conditions between user equipment and different edge nodes in multiple cells, the traditional resource scheduling algorithm can not fully integrate all the complex environmental information to obtain the optimal resource scheduling scheme, It makes the user device unable to reduce the energy consumption to the greatest extent when performing mobile edge computing.

At the same time, deep reinforcement learning technology has fully demonstrated its advantages and ability in decision-making since it was proposed and developed. Therefore, deep reinforcement learning can be used to further optimize the migration decision of mobile edge computing user devices. Compared with traditional methods, deep reinforcement learning does not need to set a fixed parameter model in advance, and can dynamically find the optimal resource scheduling scheme according to the current environmental state, which is more flexible; And because of the introduction of deep learning neural network, the fitting of the optimal scheme can achieve higher accuracy than the traditional algorithm. Therefore, it is of great practical and research significance to introduce deep reinforcement learning into the Migration Decision-making of mobile edge computing.

3.1 Q-Learning Algorithm

The time series differential algorithm calculates the difference of returns between two states in each round (CIC is trained by comparing the difference between the expected difference and the actual difference R). Because the obtained returns are a random variable, and MC method is the sum of the returns under each state, the variance of the actual cumulative returns G obtained by MC method is relatively large, TD only considers the income difference between States, so the variance is small; CNIC seems to only evaluate the quality of a certain action, but in fact it can be directly used to do Q-learning algorithm:

$$\pi'(s) = \arg\max_a Q^\pi(s, a) \tag{1}$$

Therefore, the first step strategy does not adopt the cumulative income of, which must be satisfied

$$V^\pi(s) = Q^\pi(s, \pi(s)) \leq \max_a Q^\pi(s, a) = Q^\pi(s, \pi'(s)) \tag{2}$$

3.2 Existing Problems

Although Q-learning algorithm based on time series difference can solve discrete decision problems, it has some problems. Because the values of States and different actions are stored in the Q table and constantly updated, the state can only be set to a limited state. In the actual edge computing migration scenario, the limited state limits the depth of exploring the best migration scheme. In addition, because the Q table is used for storage, and the training needs to update the Q value of each state, the training efficiency is poor, It takes a long time to search.

4 Industrial Internet Platform Architecture

The architecture of industrial Internet platform includes industrial information physical system, on-site infrastructure and services, various software platforms supporting development and mature general software. The purpose of connecting OT and it through the industrial Internet platform is to improve the information flow of digital workshop, guide the management and decision-making through data, and improve the intelligence and value of products. The industrial Internet platform of this paper integrates the information collection technology of industrial Internet of things, the connection technology of intelligent equipment, and the real-time information transmission technology, General software development framework, mature database system, data analysis and optimization technology, personalized human-computer interaction technology and so on. As a result, the new generation of industrial Internet management and control system realizes the interconnection with production equipment, production facilities, product raw materials and other physical elements, realizes the digital, networked and intelligent production and service mode, and conforms to the future trend of industrial intelligent development [4].

The basis of industrial Internet is perception technology. Through object perception technology, the system can collect industrial data such as identity coding, running status, production information of intelligent objects, and transmit them to the industrial Internet platform through the Internet for data analysis and optimization. Perception technology can be subdivided into four directions: object identification technology, state acquisition technology, scene recording technology and location acquisition technology. Object identification technology is to establish a unique code for each intelligent object, and obtain the code through the corresponding recognition technology. The advent of the industrial Internet era marks that every intelligent object will have its own unique code. Through the tracking of the code, the real-time state of the intelligent object can be obtained. At present, the commonly used coding technologies include two-dimensional code, bar code and RFID. The digital workshop object studied in this paper will add bar code to the material information for coding setting.

Object state acquisition technology refers to the technology of acquiring the state information of intelligent objects through sensors. Machine physical state data is an indispensable part of industrial big data analysis. Through data analysis and optimization, we can know the running state and trend of equipment. As the technical basis of state acquisition, sensor is a device that converts the output signal by sensing the internal physical quantity of intelligent object. The research object of this paper will set up a large number of sensors to obtain the state information of ordinary objects and the sensors of intelligent devices to obtain the real-time state information.

5   Conclusion

In this paper, the migration scheme of mobile and the algorithm based on value function and policy gradient are introduced and summarized in detail. On this basis, the mfbdqn Q network is proposed to optimize the energy consumption of migration decision and resource scheduling scheme for the scenario of multi cell and multi-user complete migration; Aiming at the scenario of single cell multi-user partial migration, this paper proposes spbddpg algorithm based on depth deterministic strategy gradient to optimize the energy consumption of migration decision and resource scheduling scheme. Finally, the proposed algorithms are simulated.

Acknowledgements. This work was supported by Jilin Provincial Department of education "13th five year plan" social science research project "industrial Internet promotes the transformation and upgrading of small and medium enterprises in Jilin Province" Contract No.:JJKH20201302SK.

Achievements of small and medium-sized enterprise development research base in jilin province.

References

1. Liu, Y.: Introduction to the Internet of Things. Science Press, Beijing (2010). ISBN: 9787030292537
2. Yan, S., Peng, M.: Fog Computing: Technology, Architecture and Application. China Machine Press, Beijing (2017)
3. Li, F., Li, Y., Tang, X., et al.: Key solutions and application of MEC. Post Telecommun. Des. Technol. (11), 81–86 (2016)
4. Wu, Y., Dai, J.: Research on 5G oriented edge computing platform and interface scheme. Post Telecommun. Des. Technol. (3), 10–14 (2017)

Platform System and Method of Cloud Component Framework Configuration in Computer Software System

Quanxing Xiang[✉] and Zhuang Chen

Chongqing University of Posts and Telecommunications,
Chongqing 400065, China

Abstract. With the continuous popularization of cloud computing, more and more users begin to run applications on cloud platforms. These applications on cloud platforms are usually composed of distributed components deployed on many virtual hosts. In this architecture, it is quite complex to dynamically deploy, configure and maintain applications on cloud platforms. However, the mainstream technologies now pay more attention to static deployment, Lack of effective support for application dynamic management on cloud platform. In view of the deficiency of the existing work on the cloud platform application dynamic management support, we combine the model driven engineering technology and configuration management technology, through modeling the application on the cloud platform and manipulating its run-time model, realize the model-based cloud application dynamic management system, and deploy and manage the mainstream distributed applications, and verify the effectiveness of the method.

Keywords: Construction system · Multi-objective optimization · Ant colony algorithm

1 Introduction

Cloud computing is an important field of mainstream information technology. As the basic layer of cloud computing layer architecture, infrastructure as a service (as) provides users with computing, storage, network and other infrastructure services. More and more users choose to deploy applications on the cloud infrastructure. These applications on cloud platform usually contain a variety of software components and the dependencies between them. With the increase of the number of components, the management overhead of the application software on the whole cloud platform will also increase rapidly. The mainstream application management technology lacks the deployment and management support for complex applications, which limits its application scope. In order to create application software on cloud platform efficiently and deploy and manage automatically, there must be standard high-level modeling method for components and relationships in application. Our work proposes a model-based method for automatic deployment and management of application software on cloud platform [1]. We use the model to define the application software on cloud platform, and deploy and manage the application software on cloud platform

© The Author(s), under exclusive license to Springer Nature Singapore Pte Ltd. 2022
J. C. Hung et al. (Eds.): FC 2021, LNEE 827, pp. 1437–1442, 2022.
https://doi.org/10.1007/978-981-16-8052-6_207

synchronously through the operation of the run-time model. We use this method to implement the deployment and management of large distributed systems on a variety of cloud platforms, and verify the effectiveness of this method.

2   Research Background and Significance

2.1   Application Software on Cloud Platform

The increasing demand for T system in industry and academia is accompanied by the increasing cost of operation and maintenance management. Whenever new systems and technologies are introduced, the complexity of management will increase significantly. Automated I 'configuration management technology and infrastructure outsourcing can effectively solve this problem. Cloud computing is the supporting technology of these methods, The rise of cloud computing technology has introduced a new way to use and provide infrastructure services, such as the concept of pay as you go for public facilities, just like we use water, electricity and gas today. Instead of managing I resources such as computing, network and storage, people directly host it systems on the cloud platform. This kind of IT system becomes a combination of application software on the cloud platform, eliminating the energy spent in dealing with complex department management and maintenance affairs. Cloud computing is highly expected. People hope to take this opportunity to change the way they use and think about I system. They no longer need to maintain their own infrastructure and applications, but focus on the business itself. From the user's point of view, automatic management of application software on cloud platform is a crucial part of specific business, because from the perspective of capital and time, management and maintenance of application software on cloud platform has become one of the biggest cost factors.

2.2   Model Driven Engineering

Model driven engineering is a development method to create an abstract model of the target system and systematically transform it into concrete implementation. Model Driven Engineering (MDE) mainly uses systematic transformation technology to support the transformation from requirement level abstraction to software implementation to narrow the gap between requirements and software implementation and reduce the complexity of development. Using models to describe complex systems from various perspectives and abstract levels can solve the difficulty. Software development is viewed from the perspective of model driven engineering. Model has become the primary work piece in development. Developers rely on computer technology to transform models into runtime systems. Model driven engineering technology is currently mainly used to generate specific implementation and deployment functions from detailed design models, However, the need for complex systems to adapt to the external changes in the running environment has inspired researchers to consider the use of runtime model to monitor and manage the runtime system. The change agent uses the runtime model to change the runtime system in a controllable way. At this time, the model is used as the management interface by the change agent to maintain, expand

and improve the runtime system. The model driven engineering improves the abstraction level of the target system, and the developer can transfer the precision from the source code to the high-level model. The development model is used to model requirements, architecture, implementation and deployment; the run-time model is used to map the state of the run-time system, which is the real-time abstraction of the run-time system and can be used to make real-time changes to the system.

3   Key Technologies

Analysis of this chapter introduces the research content, related research work and key technologies. Our work is based on cloud computing as platform, configuration management technology and model driven engineering technology. Finally, it introduces the work and achievements of the academia and industry in cloud application management.

3.1   IAAs Platform

Open stack is a free open source cloud computing software platform. Users mainly use open stack to build their own private cloud services. Open stack includes a series of interrelated projects, which are used to control the resource pool of computing, storage and network resources in the data center [2]. Users can manage these resources through web user interface command-line tools or rest API. Figure 1 The Diagram of Openstack Architecture.

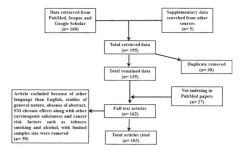

Fig. 1. The diagram of openstack architecture

The full name of AWS is am AZON web service, which is a public cloud service launched by Amazon. AWS has 11 data centers in the world. EC2 (Elastic Compute Cloud) and S3 (simple storage service) are the most well-known services, which provide computing resources and storage resources respectively. AWS also provides many other types of cloud computing resources. Amazon route53 provides highly available DNS services, allowing users to modify DNS records through programming interfaces. Amazon elastic block Oracle (EB) provides the original block storage device service, and users can attach block storage devices to EC2 instances to increase the storage capacity of instances.

3.2   Model Driven Development Model

Driven development is a high-level Abstract development method with model as the main artifact. With the support of tools, the model is transformed into code or runnable configuration as the core asset. Now there are many kinds of model driven development methods in the software industry. In the past years, software development has faced many challenges. New requirements and existing systems are growing, and the systems are becoming more and more complex, so that it is difficult for us to build them in time. In order to solve these problems, there are many new methods, the most prominent one is model driven development. Model driven software development represents a set of theoretical and industrial software development method framework. In the whole life cycle of software development, the system uses model as the main work piece, which is mainly to solve the two fundamental crises of software: complexity and change energy. Model Driven Engineering 13151631339 is a software development method, After the developers create the abstract model of the system, they transform it into concrete implementation through model related technology. In Model Driven Engineering, model is the most important work-piece of development. Developers transfer the focus of related methods from the bottom code to the upper system model, which improves the abstract level of complex system. Model is the bridge between the system and the concrete implementation. Different models of the same system can let us view and realize the system model from different perspectives [3]. In the model driven engineering, we are most concerned about two types of models: development model and run-time model. The development model is the abstraction on the code level, and the runtime model presents the view of the runtime system at a specific level, which is the abstraction of the system at the runtime.

$$\Delta = \frac{1}{2} \sum_{i=1}^{m} (f_\beta(x_i) - y_i) \tag{1}$$

$$E(R, t) = \frac{1}{n} \sum_{i=1}^{n} \| q_i - (Rp_i + t) \|^2 \tag{2}$$

$$\mathrm{cov}(x_i, x_j) = E[(X_i - \mu_i)(X_j - \mu_j)] \tag{3}$$

4   Analysis and Design of Platform Independent Strategy for Application Software on Cloud Platform

4.1   Short Board in Current Field

Although the current public cloud, private cloud and hybrid cloud all allow users to configure and deploy the cloud resource description through text format files such as JSON, they have common shortcomings. They only focus on the resource deployment of IAAs layer, and lack the support for the application deployment of PAAS and SaaS layer. This will lead to the deployment only by embedding scripts into the single virtual machine, However, there is no high-level abstract description for distributed applications. This paper proposes a method to describe the components of application software on cloud platform through extensible markup language (XML) and the mutual references between components, and to deploy and maintain the application software by using the management planning described by business process execution language (BPEL). This method of high-level abstract expression of application software on cloud platform through extensible markup language allows users to manage application software on cloud platform through high-level view description. However, single extensible markup language does not have comprehensive and meticulous description ability of abstract, and lacks effective definition, At the same time, there is no combination of components and dependencies between components to configure deployment operations, and there is a lack of mature tools to support modeling. Users must implement business process to deploy application software on cloud platform indirectly rather than directly, which undoubtedly increases the difficulty of use and causes a lot of inconvenience.

4.2   Platform Independent Support Technology

The last section lists five problems in the process of studying software platform independence on a variety of cloud platforms. This paper aims to use model driven technology to put forward overall solutions to these aspects.

Different levels of models can be generated in the process of research on the platform independence of application software on cloud platform: the target object of research is the actual application software on cloud platform in the real world, and the runtime model is established through abstract expression, so as to sort out the factors that need to be considered in the process of platform independence research, Meta-model is the abstraction of the model at a higher level, which is used to describe the definitions and constraints that the model needs to follow in the run-time. Conversely, the model is an instance of the metamodel. Metamodel is the abstraction of the metamodel at a higher level, which is used to describe the definition of the metamodel. Conversely, the metamodel is an instance of the metamodel. The models of different levels defined here are compared with each other. The meta model is just like the meta grammar of context free grammar, which is used to describe the symbol set of computer language grammar and the normal way of formal language.

The previous section describes the definition of application software on high-level cloud platform described by different hierarchical models, including various components of application software, interdependencies between components, and management operations. The access model has the ability to call the specific implementation of the underlying management operations, so as to assist the system to complete the management work. The current research work is only implemented at the bottom through the use of script, which is not only difficult to maintain, but also does not have the common adaptability on a variety of cloud platforms. Configuration management technology can code the configuration deployment and management operation of application components, and define the running state of components as the abstract expression of configuration attributes. The defined component states are platform independent, which can solve the incompatibility problem of heterogeneous platforms.

In short, it is the configuration management language that solves the heterogeneous differences of platforms.

4.3   Environmental Adaptation

In the research of environment adaptation tools, in order to overcome the problem of adaptability of application software on high-level Abstract cloud platform to heterogeneous underlying platform described in the previous section, this section integrates the management operation toolkit that is called by comparison engine for application software to adapt to cloud platform. The corresponding cloud platform toolkit selects jc1oud, which supports the creation of virtual hardware resources (computing, storage and network) on more than 30 kinds of cloud platforms, such as a, to realize the adaptive operation of application software on a variety of heterogeneous platforms [4]. At the same time, in order to provide high adaptability and user interface for application software that may be installed on any version of operating system, It also provides a software toolkit developed by chef for deploying and adapting software components. In order to meet the above two main requirements, chef is used to develop cross platform software toolkit and specific component environment management methods for distributed environment and single node respectively.

5   Summary and Prospect

This paper introduces and analyzes the existing cloud computing services at different levels. Aiming at the industry weakness of the existing configuration and deployment management technology of application software on cloud platform, a method based on platform independent model is proposed. By defining the meta model and access model of application software on cloud platform, and combining with cloud platform toolkit and software toolkit, a comparison engine is formed, Through the comparison engine, developers can synchronize the operation of the runtime model to the application software on the cloud platform to realize the real-time control and management. In this paper, we have experimentation on the application of Web and Hadoop distributed application, and demonstrate the effectiveness and advanced nature of the method.

References

1. Fan, D., Wang, X., Lei, Y.: Design and implementation of software system for ZY-1 02D satellite. Spacecr. Eng. **29**(06), 43–50 (2020)
2. Ma, Y.: Research and development of two axis driven unmanned patrol system for river environment. Zhejiang University of Science and Technology (2020)
3. Wang, P.: Overview of software system design of Qinhuangdao comprehensive management (emergency) command center. Radio Telev. Netw. **27**(11), 97–98 (2020)
4. Li, Y.W., Liu, N., Zhang, S.P.: Design of integrated software system for shipborne marine dynamic environment elements sensor. Ship Mater. Mark. (11), 18–20 (2020)

Application of Internet of Things Technology in Aviation Warehousing and Logistics

Yang Yang$^{(\boxtimes)}$

Chongqing Aerospace Polytechnic, Chongqing 4000021, China

Abstract. After in-depth study of the Internet of things technology and the construction of China's aviation logistics, this paper proposes to apply the Internet of things technology in the management of aviation logistics, and complete the data collection and transmission through RFID and ZigBee technology, so as to achieve positioning and tracking; At the same time, through the establishment of server and database, the data processing and interaction are completed; Then, the website technology is used to realize the specific method of intelligent management of air cargo. The system test results show that the method can be used for air cargo security inspection, inquiry and location tracking.

Keywords: Internet of things · Zigbee · RFID · Database · Physical distribution management

1 Introduction

Internet of things (lot) has become the fourth information revolution after computer, Internet and mobile communication. It is a high technology to realize the connection of things. Internet of things technology mainly includes sensing technology, RFID technology, network and communication technology, information processing technology. With the rapid development of these technologies, Internet of things technology is gradually becoming mature, and the application field is also increasing rapidly. The Internet of things will be the next "important productivity" to promote the rapid development of the world and another trillions market after the communication network. Now when it comes to the application of Internet of things, many of them focus on smart home or environmental monitoring. It is the first time to apply it to aviation logistics management to locate and track the whole process status of freight goods based on Internet of things [1].

In recent years, with the rapid development of China's aviation logistics, various supporting facilities have been continuously optimized and improved, but there are still many problems in the information construction, mainly as follows:

(1) Lack of industry level, one-stop aviation logistics information platform. At present, almost all large and medium-sized aviation logistics enterprises have introduced or developed logistics information management systems, but they are still in a relatively independent state, lack of interconnection between each other, unable to achieve the whole process of cargo information interaction and resource

sharing, serious information island phenomenon, it is difficult to achieve the positioning, tracking, monitoring and management of the whole process of cargo.

(2) Advanced information technology and business model have not been widely used. Compared with the developed countries in the world, there is still a big gap in the management means, management methods and information basis of domestic air logistics, which is far from being able to adapt to the rapid growth of air logistics.

Most of the existing systems of domestic aviation logistics related enterprises are traditional and single point information systems, which lack the support of new technologies, and restrict the improvement of enterprise operation efficiency and service quality to a considerable extent. Moreover, many business links rely on manual operation and management, with low efficiency, poor quality, many security risks, and lack of artificial intelligence/expert system, communication With the application of advanced information technology such as bar code and scanning, enterprises can not fully and accurately grasp the logistics information of all parties, and can not realize the integration of internal and external logistics in order to seek the optimization and rationalization of logistics system. These aspects are far behind the international level.

2   System Principle

RFID tags are pasted on cargo boxes or passengers' luggage, and RFID readers are installed at the luggage conveyor belt, security check office and cargo warehouse of the airport counter. RFID readers are used to obtain the RFID ID tag information, and ZigBee wireless communication custom transmission protocol is used to transmit data, so as to complete the query and management of air cargo, understand the status, location and distribution place of the cargo, and realize real-time and reliable data transmission Accurate positioning and tracking. After entering the warehouse, we can bind the ZigBee terminal for some particularly important items. We can directly define the ZigBee terminal transmission protocol as the only coding method, and directly transfer the information of the items to the server through the ZigBee terminal equipment, so as to realize multi-mode positioning and tracking of the goods, which is more real-time and accurate The data exchange between the database and the air logistics management system is carried out through the LAN, and the UD network programming is used for data transmission [2].

3   System Design and Implementation

3.1   Hardware Design and Implementation

With the increasing popularity of RFID sensor technology, RFID tags are cheap and reusable. Data transmission is convenient, and user-defined encryption algorithm can be processed based on user's requirements. The master-slave module of RFID ZigBee includes RFID tag, RFID reader and ZigBee wireless network. The RF module of RFID reader adopts trf7970a produced by TI company, and connects 900m antenna to reader antenna port. The microprocessor adopts ultra-low power msp430f2370 [3]. The microprocessor connects RF module through SPI bus interface and ZigBee terminal module through RS232. The microprocessor of ZigBee wireless network module adopts CC2530 produced by TI company. It is the second generation system on chip that meets the 24 GHz IEEE802.15.4 standard. It provides first-class selectivity, coexistence, excellent link budget, and up to 125° The interface circuit of lcc2530 is shown in Fig. 1.

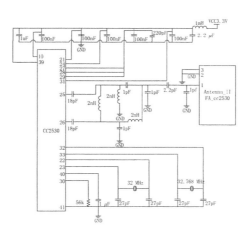

Fig. 1. CC2530 interface circuit diagram

3.2   Software Design and Implementation

RFID protocol design is completed under the environment of IAR embedded platform kickstart, which is an integrated development environment (IDE) used to build and debug embedded applications for isp430 microcontroller. Debugger fully integrated for source and disassembly level debugging, support complex code and data breakpoints. RFID reader function is in the integrated development environment of V isual Studio 2008, using MFC (Microsoft Foundation classes, Microsoft Foundation classes Library) to provide the application framework based on dialog box for program development. According to the standard protocol and instructions of so15693, the RFID reader realizes the functions of card searching, reading and writing, locking and reset.

The design of ZigBee radio frequency data communication software is developed with iarembedded workbench tool. IAR software is a set of embedded development software which is used to compile, debug and download programs in assembler, C language or ￠ +. Coordinator networking, terminal equipment and routing equipment to discover the network and join the network. Coordinator: give feedback information to ZDO layer network, send network start event to Zapp layer, and then go to zappevent loop() function to start network event; Terminal and router: when a network is found, the network layer will give feedback to the ZDO layer to discover the network, and then the network layer will initiate a request to join the network. If joining the network is successful, the network layer will give feedback to the ZDO layer to join the network [4].

4 Application of Internet of Things Technology in Intelligent Logistics System

4.1 Application of Internet of Things in Logistics Transportation

The application of IOT vehicle terminal in the transportation vehicles of logistics companies can effectively improve the visibility and real-time information, so that logistics managers can timely grasp the status of products and vehicles, timely find the road conditions and plan more scientific transportation methods. Improve the route, improve the transportation efficiency, reduce the transportation cost and freight loss. For example, if traffic congestion occurs in the process of transportation, real-time traffic information can be obtained through the Internet of things system. This provides an effective reference for logistics managers to re plan the transportation route, and improves the transportation efficiency to the greatest extent. It can also use the monitoring system to monitor the whole life of the goods, use the environmental monitoring system to track the changes of the transportation environment, provide timely feedback, or reduce the damage caused by poor storage.

4.2 Application of Internet of Things in Warehouse Management

Warehouse storage is an important link that affects the operation efficiency of logistics companies. Centimeter level IOT indoor high-precision positioning technology can accurately control the safety of warehouse management and operation of various equipment and personnel. The efficiency of logistics has been greatly improved. Internet of things technology enables logistics companies to reduce the labor and time cost of finding and identifying products in the management of import and export products, making it easier for managers to find materials. The use of two-dimensional code, RFID technology, wireless sensor and other technologies and equipment can greatly improve the inspection speed, and the goods can be put in and out of the warehouse without the administrator's unpacking inspection. The enhanced Internet of things technology significantly improves the speed of the logistics company's putting in and out of the warehouse, and greatly reduces the labor management cost, which leads to the improvement, Standardized and transparent warehouse management mode. After extending the function of RFID, it can guide the equipment to load and unload materials, such as RF terminals on the stacker and forklift, so that the equipment can select and arrive at the empty location, and update the data information after placing materials. It can also print inventory list, which is very useful for warehouse manager to check and verify materials. The use of Internet of things technology in inventory management can reduce the manpower and material resources required for inventory checking, and achieve faster inventory checking. It also enables managers to track inventory information, replenish inventory according to actual demand and contact sellers, so as to reduce inventory management cost and improve inventory management ability and revenue.

4.3   Application of Internet of Things in Transportation

In the process of transportation and distribution, employees of logistics companies have direct contact with customers, and service quality directly affects customer experience. In order to reduce the ability of traditional freight companies to collect and process freight information in the logistics process, slow down the feedback speed of transportation information, reduce transportation reliability and improve product safety, logistics companies must start to use the delivery management system based on the Internet of things instead of the traditional manual operation, and use RFID, wireless induction and other technologies to automatically classify and distribute goods. After the logistics link information is automatically uploaded to the Internet of things information sharing platform, logistics companies, suppliers and customers can get real-time feedback of logistics information, which greatly improves service satisfaction.

According to the probability calculated by Formula 1, the following bee searches a honey source locally:

$$P(i) = \frac{fit(i)}{\sum_{i=1}^{SN} fit(i)} \tag{1}$$

In this stage, simulated annealing algorithm is introduced to update the honey source. When the fitness of the new honey source is lower than that of the old honey source, the new honey source is accepted with a fixed probability p, where:

$$p = \exp(-\frac{fit'(i) - fit(i)}{T}) \tag{2}$$

4.4   The Internet of Things Supports the Interoperability of Logistics Information and Establishes a Sharing Platform

In order to solve the problem of insufficient logistics information interoperability and asymmetric logistics supply and demand information in China, promote the interaction of logistics information among enterprises, improve the logistics information open standard system, and effectively collect and process the warehousing, transportation and distribution. National Engineering Logistics Industry Association to achieve interoperability and share technology and applications. Internet of things technology has laid a good foundation for logistics interoperability and sharing platform. The application of two-dimensional code, RFID and other IOT sensing technologies in intelligent logistics system can further enhance the ability of organizing and collecting information on the platform, so as to improve the efficiency of logistics companies. For example, the platform is compatible with GPS positioning and tracking system, supports real-time transportation monitoring of vehicles, and provides users with flexible vehicle scheduling and deployment tools to effectively control users in all aspects of transportation. On the other hand, the central dispatcher can find the temporary vehicles nearby by monitoring the GPS transportation on the large screen according to the actual needs. On the other hand, headquarters scheduling can perform operations such as handling, checking and tracking exceptions, which provides an effective tool for users to check transportation routes. The GPS transportation monitoring system is used to display the nearby vehicles. The administrator can search the license plate number or waybill number to complete abnormal monitoring and early warning, so as to search the past transportation track of the target vehicle and a series of related parameters (speed, direction, driving status). Combined with charts and reports, the platform can provide the basis of path planning service, provide evidence of abnormal verification, and provide more accurate mileage for vehicle payment.

5   Conclusion

With the full implementation of IOT structure concept and data collection method, warehouse data collection equipment can provide real-time feedback on product status and information in various warehouse processes, so as to comprehensively monitor the warehouse process. When the technical concept and scale of intelligent storage reach a certain level, automatic storage management and decision-making can be carried out. The information collection and processing process are combined with the storage scheduling decision to further process the collected real-time cargo data information. Then, the large-scale information is classified and processed, which can quickly obtain real-time and comprehensive cargo information in the warehouse. By referring to these data and information, managers can more accurately judge the logistics situation and realize the intellectualization and informatization of warehouse management. The Internet of things technology is still an independent and local intelligent warehousing system, which is only applied in the internal networking of independent warehousing and distribution center. With the help of Internet of things technology, these independent intelligent warehousing systems will be connected to each other, breaking the information island, realizing interconnection and forming a real warehousing Internet of things, which is to produce new changes on the basis of intelligent warehousing, which will bring the revolution of warehousing informatization. If we really realize the Internet of things in warehousing, we can realize the Internet of things and intelligence of warehousing system in a larger warehousing and logistics network based on this, make things have certain intelligence in the large system of warehousing and Internet of things, let the things in logistics know where to go and where to store, etc., and establish the logistics Internet of things system under this concept, It is very different from the original Internet of things system. This will completely break the original structure of logistics information system, and even have a huge impact on modern logistics technology and equipment in the process of logistics operation, and bring revolutionary changes to the structure of modern warehousing and logistics center. This will be the real smart logistics, and truly realize the transformation of smart logistics.

References

1. Jin, J.L.: Research on the application of Internet of Things in logistics warehouse management. Mod. Mark. (Bus. Ed.) (9), 126 (2019)
2. Zhang, H.R., Jiang, Y.L.: Research on logistics warehouse management based on Internet of Things. China Manag. Informatiz. **22**(13), 60 (2019)
3. Wang, L.: On the application of Internet of Things in logistics engineering management. Modern Mark. (Next Issue) (5), 163 (2019)
4. Bian, Z.: Application of Internet of Things in logistics engineering management. Logist. Eng. Manag. **41**(2), 83–84 (2019)

Design of High Efficiency Fruit Picking Robot Driven by Rigid Flexible Compound

Zhuoyuan Yang[(⊠)]

University of Southern California, Los Angeles, USA
zyang205@usc.edu

Abstract. With the rapid development of Chinese economy and the arrival of aging society, a large number of labor force from rural to urban, traditional manual labor has been unable to meet the needs of social development, the demand for artificial intelligence equipment also increased. This paper mainly designs a rigid and flexible combination of efficient fruit picking robot. A 5-degree-of-freedom manipulator is designed for the McNum wheel. The manipulator is matched with a gripper made by artificial muscle technology, and the spatial coordinates are obtained by ZED 2 binocular vision system. The positioning of the manipulator and the acquisition of video images are realized. A strawberry training set was created in this paper to realize the automatic recognition and picking of strawberry, and the yolov 3 algorithm was used to train the data and the strawberry recognition model was obtained. The recall and precision of large strawberry and immature strawberry reached 90%, and the detection recall and precision of small target strawberry reached more than 80%. Meet the requirements of real life picking.

Keyword: Artificial muscle target recognition yolov3 binocular vision raspberry pie machine vision motion control

1 Present Situation and Development Trend of Picker at Home and Abroad

Foreign research status: in the late 1970s, thanks to the rapid development of computer and automatic control technology, the United States began to study all kinds of agricultural robots. After more than ten years of research and experiment, Japan, the Netherlands, Britain and other developed countries have successfully developed a variety of picking robots. For example, a five-degree-of-freedom joint manipulator developed by Kyoto University in Japan in the 1980s [1]. And an apple-picking robot developed by Baeten from Belgium. The whole picking manipulator and its control system are installed in the rear of the tractor. Moreover, France's Pedenc and Motte have developed a model called MAGAL" and the apple harvest robot, which uses the CCD camera to collect the fruit image, slips the picked fruit into the fruit collection box through the hollow arm.

Due to the factors of topography and science and technology development, China started late for picking robot, but developed very quickly. Since the 1970s, the auxiliary machinery for human work has gradually appeared, mainly by hand-held airbag and

J. C. Hung et al. (Eds.): FC 2021, LNEE 827, pp. 1450–1454, 2022.
https://doi.org/10.1007/978-981-16-8052-6_209

electric fruit collector. Overall inefficiency. China's first multi-function orchard operating machine LG-1 was developed by Xinjiang Institute of MachineryAfter that, researchers at various colleges and universities also focused on the visual system of intelligent recognition. For example, Yang Changhui and others used the YOLOV3 algorithm to study the recognition of citrus in natural environment, and the success rate of picking was 80.51%. LiteratureA recognition method based on R-CNN (region-convolutive neural network) is proposed to identify apple branches.

2 Mechanical Arm Design

The distribution of strawberries on their stems and leaves is often full of randomness. In order to successfully pick strawberries, it is necessary for the gripper to reach any position in the designated workspace and to complete the successful picking of strawberries in any attitude. The 6-DOF manipulator in series can reach any position in its designated workspace with six degrees of freedom considering that strawberries droop down due to deadweight and do not require much attitude adjustment, the four-degree-of-freedom manipulator can be used to reduce the system complexity and the complexity of the manipulator control (Fig. 1). at the same time, the ZED2 binocular camera is fixed with the gripper to update the relative position of the gripper and strawberry in real time.

Fig. 1. McNum wheel with DOF mechanical armor

The McNum car is used to carry the manipulator to the position of the target strawberry in its working space, and then the manipulator is controlled to send the gripper to the designated position to pick the strawberry. Finally, the manipulator is controlled to take the strawberry to the designated collection point to loosen the strawberry.

The connecting rod and connection of the manipulator need a certain hardness, as the table Aluminum is abundant and cheap on earth, and its hardness can meet the low strength demand of strawberry picking [2]. As shown in Table 1. A connecting rod is welded by arc welding of multiple plates, the joint shaft seat and the connecting rod are fastened with bolts and nuts, and the joint shaft and its sleeve are matched by sliding bearing. The connecting rod can rotate around the shaft.

Table 1. Experimental record of manipulator connecting rod.

Project	Parameters
M Pa yield limit	≥ 325
Tensile strength/MPa strength	≥ 470
Elongation%	≥ 10
Section shrinkage/%	≥ 29.23

3   Strawberry Detection Algorithm

3.1   Experimental Environment

The system is developed on the window10 operating platform. Anaconda is a variety of versions that can install the same software on a computer, very convenient and fast, so I use it to install the environment, development language for python3.6, in-depth learning environment to build the need for graphics cards, CUDA and CUDNN, specific detailed computer environment version as shown in Table 2.

Table 2. Software and hardware environment table.

Operating system	Windows 10
Development of languages	Python
CPU	Intel (R) Core (TM) i78750HQ CPU @2.2 GHz 2.91 GHz
Video card	GTX1080Ti
CUDA	CUDA 8.0
CUDNN	CUDNN6.5

To accelerate the development of deep learning, The convolution neural network I use is built on a python basis using pytorch. pytorch is also a Python package with two advanced features, With tensor calculations as strong GPU acceleration as Numpy, The deep network can also be built on an automatic tape-based adjustment system, pytorch integration acceleration library will be extensible [3]. After building a deep learning graphics card environment and anaconda, Click the combo button Win + R enter the windows run, Enter cmd into the command line window, Enter the above command for a moment to automatically download the installation.

3.2   Strawberry Dataset

A part of the strawberry dataset in this experiment is from the strawberry dataset in the imagenet, But because some of the strawberry data sets are not labeled, That leads to incomplete data, As a result, this article also uses the Python crawler to obtain some pictures in Baidu pictures, And then use the labelImg tagging software to mark. A dataset of 416 images from Imagenet and 692 crawling images from Baidu, Average of three strawberry targets per picture, Because of the limited strawberry data, In this paper, we do an augmented processing of strawberry data, The way of augmented processing is horizontal flip and up flip, The corresponding annotation file should be modified, Since each image is scaled to 416×416 pixels before entering the yolov3 model, So the target shouldn't be too small, Therefore, this paper will delete the original target pixel less than 300 tagging information. Each image of the data set is annotated by labelme tagging software, Finally, a xml file was generated, The sample original diagram and the annotation effect diagram are shown in Fig. 2.

n07745940_530.
JPEG

n07745940_530
_h.JPEG

n07745940_530
_w.JPEG

Fig. 2. Photo of strawberry

4   Trajectory Planning

With the rapid development of science and technology, the development of robot is becoming more and more deep. The path planning problem of manipulator is directly related to the performance of robot. for the robot system, in order to simplify the description of manipulator motion, the complex function of the trajectory point is allowed to be completed by the system itself. it is only necessary to determine the target position and attitude of the end actuator, and the system can judge the duration, path, speed and so on.

The working area of the end of the robot arm is the range of motion of the manipulator. Ideally, the range of motion of the end of the manipulator can reach a spherical range. During the operation of the manipulator, the trajectory must be as continuous and smooth as possible, so an optimal path should be planned under these constraints at the same time [4]. Therefore, in the process of robot configuration design and trajectory planning, it is necessary to increase the working space of the manipulator and improve the motion flexibility of the mechanism on the premise of ensuring the reasonable movement of the manipulator.

During the robot operation, it is necessary to determine the motion path in more detail, not only to simply give the final pose, but also to give a series of required intermediate points in the path description —— also known as the path point. In this way, in the process of completing the journey, each track point determines its position relative to the target position, and its motion time can also be specified when describing the path. In order to realize the smooth and stable motion of the manipulator, it is necessary to specify a continuous smoothing function, which has a continuous first derivative or a continuous second derivative, and any smoothing function passing through the path point can be used to plan the path. In this paper, joint space and Cartesian space trajectory planning are used respectively.

5   Binocular Vision Positioning Principle

Binocular stereo vision technology is an important branch of machine vision. It is a method based on parallax principle to obtain 3D coordinate information through multi-fault images. The binocular vision system is generally used by two cameras to obtain two images of the space to be measured from different angles. The three-dimensional information of the object is restored by parallax calculation, and the contour and position of the object are reconstructed. It has a wide application prospect in robot and three-dimensional measurement [5]. Therefore, as long as we can find the corresponding points in the space on the image surface of the left and right cameras, we can obtain the internal and external parameters of the camera by camera calibration, and then determine the three-dimensional coordinates of the space points. Based on the principle of binocular vision and the ZED 2 binocular vision system, the spatial coordinates are obtained to realize the positioning of the manipulator and the acquisition of video images.

Z ED depth camera is a small binocular vision system developed by Stereolabs for mobile development platforms such as raspberry pie and Jetsen TK. It outputs high-resolution video side by side on the usb3.0, with two synchronous left and right video streams [6]. Therefore, binocular images can be obtained at the same time through the processor for further depth operations. the depth can be captured in a longer range up to 20 m; the rame rate of depth capture can be as high as 100 FPS; the field of vision up to 110°(H) x 70°(V).Its resolution can reach 4416/1242 pixels, 15 frames per second, compatible with a variety of operating systems and development environment, can obtain three-dimensional information through the OpenCV library. meanwhile ZED SDK can interface with a PyTorch project to add 3 D localization of objects detected using a custom neural network. Based on the above analysis, the sensor meets the requirements of development environment and detection accuracy, so Z ED2 generation camera is selected as the visual system of this design.

References

1. Jiang, L., Chen, S.: A review on the research of fruit picking robot agricultural equipment technology. 32(1), 8–10 (2006)
2. Baeten, J., Donné, K., Boedrij, S., Beckers, W., Claesen, E.: Autonomous fruit picking machine: a robotic apple harvester. In: Laugier, C., Siegwart, R. (eds.) Field and Service robotics. Springer Tracts in Advanced Robotics, vol. 42, pp. 531–539. Springer, Heidelberg (2008). https://doi.org/10.1007/978-3-540-75404-6_51
3. Yang, W.: Design and analysis of manipulator of apple picking robot. Structure Zhenjiang: Jiangsu University (2009)
4. Liu, X., Bahti, Z.H.: Research and development of Mu Shen LG-1 multifunctional orchard operating machine agricultural mechanization in Xinjiang. (1), 42–44 (2009)
5. Yang, C., Liu, Y., Wang, Y., et al.: A study on identification and locating system of citrus picking robot in natural environment. J. Agric. Mach. 50(12), 14–22 (2019)
6. Zhang, Q., Zhang, X., Zhang, J., et al.: Branch detection for apple trees trained in fruiting wall architecture using depth features and regions-convolutional neural network (R-CNN). Comput. Electron. Agric. 155, 386–393 (2018)

Design of Multi Function Robot for Active Video Detection Based on STC MCU

Jinyong Yu[(✉)]

Guangxi City Vocational University, Guangxi 532200, China

Abstract. This paper introduces a design scheme of multi-function robot for active video detection based on stc12c5a6082 MCU. Compared with the traditional wheeled and legged detection robots, the blade wheeled detection robot has the advantages of simple structure design and easy control, strong adaptability to complex terrain, and can complete the detection task in amphibious environment.

Keywords: Deformable impeller · Wireless communication · Pulse width control

1 Introduction

At present, the driving mechanism of the detection robot vehicle developed at home and abroad can be roughly divided into three types, i.e. wheel type, leg type and crawler type. The crawler type driving mechanism is easy to be worn by the soil on the ground surface, so it is seldom used. At present, the most researched detection machines are wheel type and leg type. The wheel type detection vehicle has the advantages of simple mechanical structure, fast speed and easy control, However, it has poor adaptability to complex landforms (such as thick dust, soft soil or dense rock surface), and is good at detecting complex landforms. Some of the detection robots are equipped with legged mechanisms, which have strong adaptability to landforms, and are very suitable for moving in the environment of concave complex rock or soft soil surface, It has good mobility and can be divided into humanoid bipedal, mammalian quadruped, insect like hexapod and crab like Octopod. However, the leg detection robot has complex mechanical structure, slow motion speed, large energy, narrow inspection range and complex control. The deformable bladed wheel machine has the advantages of simple driving structure, strong passing ability and simple motion control, which fully satisfies the shortcomings of the above two kinds of traditional mechanical vehicle driving structure, and can fully meet the needs of complex terrain detection tasks [1].

2 The Practical Application Scope of Robot

2.1 Planetary Exploration Robot

The research based on planetary exploration robot is of great significance in the development of planetary science and the improvement of national defense capability,

J. C. Hung et al. (Eds.): FC 2021, LNEE 827, pp. 1455–1460, 2022.
https://doi.org/10.1007/978-981-16-8052-6_210

because: (1) robot is a powerful tool for landing detection and sample retrieval in planetary science research. At the same time, people who stay in space for several months will seriously lose phosphorus and calcium, This may mean that humans can't fly for 6–9 months or more without gravity, but robots don't have such problems. Therefore, the research of planetary exploration robot is the premise of human on-the-spot investigation of planetsThe cost of detection is saved to a considerable extent. Take Mars exploration as an example. According to a rough estimate, a manned exploration will cost 50–100 times more than an unmanned exploration. Therefore, it is quite reasonable to use robots to carry out unmanned aerial missions based on scientific exploration. It has a considerable effect on improving the level of national defense automation and international status. Therefore, the research of planetary exploration robot is highly valued all over the world [2].

2.2 Ocean Exploration Robot

As for the ocean, most of our images are vast and turbulent, with a lot of available resources, but only based on superficial understanding. Therefore, in order to further explore the ocean, marine robot applications emerge. Marine robots can not only replace human beings to explore the environment in all kinds of dangerous places, but also play a very important role in underwater archaeology, sunken ship salvage, marine biological observation and scientific research. Therefore, it has become an important research topic to make the robot better complete the tasks assigned by human beings.

2.3 Land Environment Detection Robot

As a big coal producing province in China, Shanxi is also a high incidence area of coal mine accidents in China, so we need to attach great importance to the safety of coal mine production. This kind of detection robot can detect and prevent the hidden danger accurately and timely before the occurrence of danger, and carry out rescue and other important dangerous tasks after the disaster. The application scope of detection robot is much wider than our imagination. In the military field, we have developed anti tank mine detection robot, medical detection robot and so on.

3   Selection and Introduction of Hardware

3.1   8051 Single Chip Microcomputer

About 8051 single chip microcomputer, the front has a simple introduction to it, more specifically, it is compatible with 8031 instruction of single chip microcomputer. With the rapid development of science and technology, SCM has become more practical, the cost has been compressed very low, so a large number of products are using 8051 series SCM. At the same time, many other companies have launched a compatible, more superior performance MCU. This design uses a stc89c52rc based on 8051 MCU produced by Taiwan Hongjing company, which is an 8-bit microcontroller with High Performance CMOS, low power consumption and 8K in system programmable flash

[3]. Although it is an 8-bit microcontroller, Hongjing company has done a lot of optimization, which makes it highly flexible in embedded development and can provide more effective solutions. The single chip microcomputer is shown in Fig. 1.

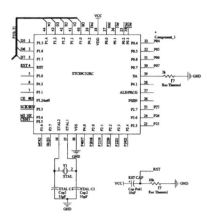

Fig. 1. Minimum system of single chip microcomputer

3.2 Wireless Transmission Chip

About the wireless module, this design uses rf24l01 wireless transmission chip, its working frequency belongs to 24 GHz 2 ghzism international free frequency band. It has the enhanced "schockburst" mode controller which is not available in general chips. It has a high transmission rate of 2 Mbps, and can be multi-channel transmission. It is very convenient to achieve one to many and many to many modes. By using SPI protocol interface, it can almost connect to various MCU first. However, its power consumption is very low under its powerful function. In the transmit mode, the current consumption is 11.2 Ma when the transmit power is odbm, while in the receive mode, the current consumption is 123 Ma, which can run for a long time without worrying about its energy consumption. Especially in power down mode and standby mode, the current consumption is almost negligible. It is widely used in various fields using wireless functions.

3.3 Temperature Detection Chip DS18B20

When choosing temperature sensor, netizen recommended DSI8B20 to me. After a lot of information inquiry, I found that it has the characteristics of convenient wiring and strong adaptability to the environment, which is more than enough for the design. DS18B20 can be used for temperature measurement under various non extreme temperatures. It can be directly connected with MCU without peripheral circuit to achieve the purpose of temperature measurement. It can also be used for multi-point temperature measurement with multiple devices in parallel, and its body can be found in all kinds of confined spaces and control fields [4].

3.4   Pf8591t A/D Conversion Chip and Photosensitive Resistor

In the design of light intensity detection, pcf8591 digital to analog conversion chip and photosensitive resistance are used to form the measurement circuit. Pcf8591 chip is a device with independent power supply and 8-bit CMOS data acquisition. It has one analog output, four analog inputs and one I2C bus communication interface. Three address pins can be used for hardware address programming. Eight pcf8591 chips can be accessed on one bus without additional hardware. The output and input address, data signal and control signal on pcf8591 are transmitted in serial mode through two-way I2C bus. The maximum conversion rate of pcf8591 is determined by the maximum rate of 1C bus, and the photoresistors are universal.

4   Main Circuit Design

The circuit schematic diagram and PCB production of this design use Altium designer winter 09 [3]. Because the design is based on the products in the laboratory, the conditions for selecting devices are relatively loose. After selecting relevant devices for the purpose of design, the relevant schematic diagram is designed (as shown in Fig. 2).

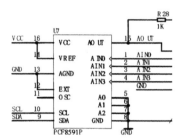

Fig. 2. Pcf8591t A/D conversion chip

4.1   STC89C52RC Single Chip Microcomputer

MCU uses stc89c52rc, the circuit system uses the smallest system, this design uses pqfp44 package, because it needs to test the chip burning program for many times, so in addition to P30, P3.1 add pull-up resistance, improve the stability of data input and make it have enough output capacity.

4.2   Temperature Detection Chip DS18B20

For the convenience of debugging, rf24l01 in this design uses the module that has completed the peripheral circuit design, and the pin has been led out, so it is designed to the socket of header5 * 2, as shown in Fig. 3.

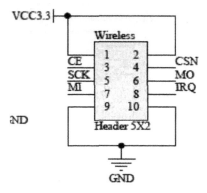

Fig. 3. Socket of header5 * 2

4.3 Display Device LCD1602

This design uses 1602 with character chip, as shown in Fig. 4. For the convenience of debugging, the socket is used.

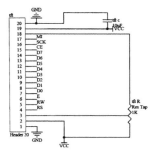

Fig. 4. 1602 of character chip

5 Program Design and Implementation

The C language based on C51 is used in this design. Keil Uvision3 is used to write the program and generate the program burning file. The programming file is divided into the sending end and the receiving end. In order to read the program and maintain and upgrade it later, the modular programming idea is adopted. Each function is edited in blocks. The unified programming idea is adopted. After market research and modification, the related products can be generated directly.

Wireless chip rf24l01 uses SPL bus to transmit data and commands, so the program first designs the driver of SPI bus, SP sending and receiving uses a function to realize the function, which improves the driving ability of the chip. According to the data manual of the chip, the corresponding driver is written. Because it is module programming, both the sending end and the receiving end use the same code. If put into commercial use, we can choose different antenna packages to achieve the purpose of product design.

DS18B20 only uses one data line to send and receive data. It can be driven by parasitic power supply of data line. This design uses external power supply. Through reading the data manual and referring to the driver on the Internet, write the following driver, which can successfully drive ds8b20 and make the work more stable. If it is put into commercial use, this chip is also suitable.

6   Conclusion

This design can not only be used and extended in the mobile environment, but also be further extended in the very popular Internet of things. Nowadays, the requirements of human beings for their own environment are increasing, so it is necessary to monitor the surrounding environment. This design can be expanded into a star network or a linear network layout on the existing basis to achieve a regional wireless environmental data monitoring. By using the actual high-power antenna, it can achieve a larger range of remote environmental monitoring. Star network can be used in small and medium-sized environmental monitoring, which is composed of one receiver and multiple transmitters; In large-scale environmental monitoring, line relay mode can be used to meet the requirements. If you need to access the Internet, you can also write related protocol drivers, such as TCPP and other network protocols, and then you can access the Internet to achieve remote data reception conveniently.

Acknowledgements. Social science research project of Chongzuo in 2019: Study on the docking mechanism between vocational education and key industries under the background of One Belt And One Road Taking Intelligent Engineering College of Guangxi City Vocational University in Chongzuo as an example.

References

1. Geng, L.: Development and application of smart home in China. China Youth (2) (2017)
2. Yu, Z.: Discussion on smart home and its key technologies. Hous. Real Estate (36) (2016)
3. Ruan, R.: Current situation and future of smart home market. China Public Secur. Acad. Ed. **12**, 66–67 (2015)
4. Guo, W., Huang, X.: Exploration on the understanding of modern smart home. Tomorrow's Fash. (2016)

Credit Evaluation Model and Algorithm of C2C E-commerce Website

Jun Zhang[✉]

Weifang Engineering Vocational College,
Qingzhou City 262500, Shandong Province, China

Abstract. With the continuous advancement of vocational education and the rapid development of C2C, e-commerce has experienced the process of germination, exploration and adjustment, and has developed rapidly in the world. In e-commerce, C2C e-commerce website is the most popular, the most widely used and the most concerned. However, due to the large number of C2C e-commerce transactions, the frequency of transactions, and the number of credit evaluation of the transaction subject, the credit environment of C2C e-commerce is poor, which leads to credit problems and hinders the healthy development of e-commerce. This paper studies the existing C2C e-commerce credit evaluation model, proposes a new credit rating, and establishes a credit evaluation index system. The new model provides more effective trading information for new traders, further reduces the trading risk, and helps traders make correct trading decisions.

Keywords: C2C e-commerce · Evaluation model · Trust evaluation algorithm

1 Introduction

With the continuous popularization of personal computer applications, as well as the continuous progress and maturity of Internet technology, e-commerce has a broad space for development, especially C2C e-commerce. In 1999, eBay was established to fill the gap in China's C2C e-commerce market. In 2003, Taobao was established, and the number of C2C shopping websites has been increasing, such as paipai.com, macaulan.com and No.1 store In recent years, China's C2C online shopping market has developed rapidly and become an important support for the Internet economy. E-commerce mode is traditionally divided into B2B, B2C and C2C. Among them, C2C mode refers to the online transactions between individuals, and it is more common to worry about the security of online transactions. Because the participants of C2C transaction are usually individuals or small enterprises, and they are not well-known, and the online transaction adopts anonymous system, the two sides of the transaction do not know each other, and they are unable to know the identity and credit status of the trading partner, so it is easier to produce the problem of mistrust and hinder the transaction. At this time, the establishment of a sound credit evaluation system is particularly important for C2C e-commerce website. It can provide the trading partner's past trading experience and evaluation, and greatly improve the success rate of the transaction. The consumption trend of e-commerce is shown in Fig. 1 below.

© The Author(s), under exclusive license to Springer Nature Singapore Pte Ltd. 2022
J. C. Hung et al. (Eds.): FC 2021, LNEE 827, pp. 1461–1466, 2022.
https://doi.org/10.1007/978-981-16-8052-6_211

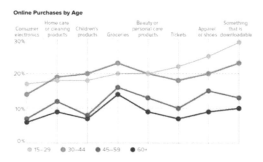

Fig. 1. E-commerce consumption trend chart

In the wave of development of e-commerce, C2C mode of e-commerce development is particularly rapid. C2C e-commerce with its free space and time constraints [1]. competitive prices, low threshold of non payment of rent to allow more consumers and businesses to participate. However, in the process of the rapid development of C2C e-commerce in China, many problems have been exposed, such as the inconsistency of payment and receiving time and space, asymmetric information transmission, which lead to the passive position of consumers participating in online transactions. In fact, most of the victims of online credit fraud are buyers. Personal credit is the cornerstone of social credit system. How to measure people's credit has become an increasingly concerned issue. E-commerce credit problem has become a serious bottleneck affecting its own development, and it is an urgent problem to be solved at this stage. Therefore, it is urgent to establish a credit evaluation mechanism to meet the needs of C2C e-commerce development.

2   E-commerce Credit Evaluation

The popular explanation of electronic commerce is to make use of the Internet, so that the two sides of the transaction can carry out all kinds of business and trade activities without meeting each other.

C2C (consumer to consumer) e-commerce is one of the basic modes of e-commerce. C2C_ The website provides an Internet online platform for buyers and sellers to trade. The platform only provides trading space, does not touch goods, does not participate in currency settlement, is not responsible for inventory, and does not bear freight. Buyers and sellers can independently auction on the trading platform and choose goods to bid [2].

C2C e-commerce is the most active e-commerce mode at present. The rapid development of C2C e-commerce is due to the fact that C2C e-commerce itself is superior to the traditional trading market, which is manifested in the following aspects:

lower transaction cost, reduced transaction links; operation scale is not limited by time and space; convenient information acquisition: highly electronic means of payment. C2C e-commerce has its own characteristics, which virtually increases the economic benefits of both buyers and sellers, and has broad market prospects and development potential.

2.1 Composition of Credit Evaluation System

Credit evaluation system is an organic whole formed by a series of evaluation systems, evaluation index systems, evaluation methods and evaluation standards related to risk evaluation. As a complete system, it should include the following aspects: the elements of credit evaluation are related to the understanding of the concept of credit, reflecting the understanding of the concept of credit, including the ability to perform responsibility and the degree of trust. The index of credit evaluation. Credit evaluation index refers to the specific items that reflect the elements of credit evaluation. Through the evaluation index, we should reasonably and fully reflect the content of credit evaluation and reveal the credit situation. The grade of credit evaluation refers to the symbol and grade reflecting the level of credit status. Some adopt grade 5, some adopt grade 9, and some adopt grade 4. Generally speaking, the credit evaluation system with big fluctuation adopts a wide range of grades.

2.2 The Necessity of Credit Evaluation

Online transactions occur in a virtual environment, the two sides of the transaction will not face-to-face trading activities, the degree of uncertainty and risk is greater than traditional transactions. In C2C e-commerce, buyers and sellers can not identify each other's real identity. The individual identity of transaction is complex and the contradiction is more prominent. The main reason is the credit problem. Therefore, it is necessary to make an objective and reasonable evaluation and Analysis on the credit of the buyer and the seller, so as to take it as the basis of the trust of both parties in the transaction and ensure the credit in the process of transaction [3].

The evaluation and analysis of credit can regulate the order of e-commerce market, prevent transaction risks, and enhance consumers' confidence in online shopping. As an important means of commodity trading, e-commerce credit evaluation system, once established, is conducive to promoting the establishment of a good and orderly market order. There are still many disharmonious problems in the e-commerce market, such as counterfeiting, false advertising, business fraud, which seriously affect the process of e-commerce transactions in China. Evaluating the credit of each individual in the transaction, to a certain extent, constrains the behavior of both parties in the transaction, prevents transaction risks, especially credit risks, enhances the confidence of all parties in participating in e-commerce, expands the scope and quantity of transactions, and improves social participation, which is conducive to the standardization and stability of e-commerce market.

3   C2C E-commerce Credit Evaluation Model

The main functions of establishing credit evaluation model are as follows: first, it can restrict the behavior of transaction subject, ensure the fairness and justice of online shopping, and effectively avoid transaction risks such as online fraud; second, it is convenient to understand the credit status of the other party, so as to make reference for transaction decision and judgment, and improve the success rate of transaction; third, it can reduce the transaction cost of businesses, The credit evaluation score is more effective and economical than the advertisement propaganda of the business [4]. the fourth is that the credit evaluation is an effective form to encourage, educate and warn the dishonest transaction subjects; the fifth is to establish the credit evaluation model, which is a beneficial supplement to the social credit system. In the existing C2C e-commerce credit evaluation model, the general evaluation process is: the buyer and the seller register their personal information on the trading platform to obtain the initial credit score, which is generally zero. By browsing the goods and referring to the previous user's evaluation of the store and goods, the buyer makes a decision whether to buy or not. After the decision to buy, complete the online transaction with the seller and wait for the goods to be received. After the buyer receives the goods, he confirms the receipt on the Internet, and scores the seller's service according to certain scoring rules. At the same time, the seller scores the buyer. After mutual evaluation, both parties receive a credit score, which is included in the credit records of both parties. As shown in Fig. 2 below.

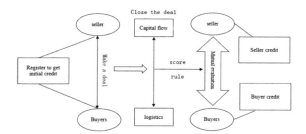

Fig. 2.   C2C e-commerce credit evaluation flow chart

There are two common credit evaluation models for e-commerce websites: one is cumulative credit model, the other is average credit model [5].

At present, the cumulative credit model is often used in C2C e-commerce websites. Its calculation principle is to simply accumulate the credit scores of the evaluated users, and its calculation formula is as follows:

$$R_n = R_{n-1} + r_n, r_n \epsilon \{-1, 0, 1\} \tag{1}$$

Where R_n. Represents the current credit value of the user, R_{n-1} refers to the recent credit value, r_n is the score of the nth credit evaluation.

The average credit model accumulates all the credit feedback scores of the user, and then divides them by the number of scoring times, which is the overall trust degree of the user:

$$R_n = \frac{R_{n-1} * (n-1) + r_n}{n}, r_n \epsilon \{-1, 0, 1\} \tag{2}$$

Where R_n. Represents the current credit value of the user, R_{n-1} refers to the degree of credit when $n-1$, r_n is the nth credit feedback score.

4   Improved Method of Credit Evaluation

The establishment of credit evaluation system from the relationship between managers and online traders in the virtual market, e-commerce credit risk problems also need managers to carry out effective credit supervision on online traders. China's C2C credit supervision system uses foreign methods for reference, and also serves the establishment of credit evaluation system from four aspects: identity authentication, third-party payment platform, community power and legal environment.

The introduction of third-party identity authentication and identity anonymity is one of the root causes of e-commerce integrity problems. Although each major website has its own identity authentication system, it is relatively loose. One of the best solutions is to carry out the third party identity authentication and credit authentication by the national authority [6].

To solve the problems of product quality and after-sales service according to several survey reports of C2C online shopping in China, the problems of product quality and after-sales service are increasingly becoming the main factors that seriously affect C2C e-commerce transactions. The rampant counterfeit goods seriously damaged the credit of C2C e-commerce.

In order to solve the problem of credit Island, there is no authoritative personal credit evaluation and management organization in our country at present. Each website generally independently evaluates and manages the credit of users who trade on its own website.

5   Conclusion

C2C e-commerce transaction is the most closely related e-commerce mode to every consumer. However, it is the consideration and worry of both sides of the transaction about credit that hinder the timely and smooth progress of the transaction. Based on the analysis of the common credit evaluation model, this paper improves the model and advocates that the historical credit of both sides of the transaction and the credit generated by the transaction amount should be included in the evaluation system. The evaluation algorithm calculates the weighted average credit score and credit degree of the evaluated user by considering the credit degree of the opposite party and the number of transactions and the transaction amount, And then determine the credit rating of the users, solve the main problems of the current C2C website used in the cumulative credit evaluation algorithm, make the evaluation more scientific and reasonable. Of course, the credit evaluation of C2C e-commerce is a more complex issue. The suitability of indicators and the choice of evaluation methods are still a topic that needs to be discussed and studied continuously.

Acknowledgements. Research on talent training mode reform of e-commerce major in Higher Vocational Colleges from the perspective of innovation and entrepreneurship studio, NO 201804.

References

1. Li, Q., Liu, C.: Empirical research on C2C e-commerce trust based on system. Econ. Manag. (4) (2008)
2. Taobao online three diamond sellers sell fake. High credibility deceives online buyers. http://tech.tom.com
3. Analysis of the establishment of CtoC e-commerce website credit system. http://BBS. Kingbuy.com.cn
4. Jiang, Y.: Research on the credit management of domestic auction websites based on CtoC model. J. Chengdu Univ. Electron. Sci. Technol. (6) (2005)
5. Lei, P.: Research on credit evaluation system and algorithm. Modern Commer. **36** (2009)
6. Meng, C., Wang, X.: Research on E-commerce credit evaluation model based on C2C mode. J. Jinan Vocat. College (12) (2008)

Tourism Route Planning Method Based on Economic Cost

Yuzhong Zhang[(✉)]

Shanghai City Construction Vocational College, Shanghai 201415, China

Abstract. With the improvement of living standards, tourism has gradually become one of the most popular outdoor activities. In the process of tourism, we can not only feel the beauty of nature, but also appreciate the cultural atmosphere and local customs of different places. Aiming at the problem of tourism route planning, this paper uses the accounting compression dynamic programming method based on economic cost to realize the economic tourism route planning, and gives the actual solution and process. This method can complete the user's request and return the corresponding results in a very short time, and its time is far less than the ordinary search algorithm.

Keywords: Economic cost · Travel · Route planning

1 Introduction

With the improvement of people's living standards, tourism has become a wide range of entertainment mode. In recent years, the tourism industry has been booming, and now it has become a strategic pillar industry of the national economy. In 2017, the number of domestic tourists has reached 5 billion, and the domestic tourism income has reached 4566.1 billion yuan. In 2018, the National Tourism Administration and the Ministry of culture merged to form the Ministry of culture and tourism, which indicates that the overall management of culture and tourism and the allocation of resources are more reasonable. In the context of the continuous development of smart tourism, tourism has entered the era of popularization and globalization.

Whether it is group tour or free travel, the planning of tourism route is an important link. The design of scientific and reasonable tourist routes is not only conducive to improving the satisfaction of tourists, but also can give full play to the characteristics of each tourist destination and promote the long-term development of tourist destinations. Despite the increasing importance of tourism, travel itinerary planning and route design is still a challenging research. Nowadays, the tourism routes launched in the tourism market are almost the same. A single choice of scenic spots leads to the immobilization of tourism routes, which not only seriously reduces the satisfaction of tourists, but also makes travel agencies constantly reduce the price and cost in order to survive, It leads to the decline of service quality and has a great impact on the image of tourism destination. For self-help travel, although there are many online resources, in the face of such huge, chaotic and complex data information, tourists are often submerged in a large number of information retrieval and product comparison and selection, even so, it

J. C. Hung et al. (Eds.): FC 2021, LNEE 827, pp. 1467–1472, 2022.
https://doi.org/10.1007/978-981-16-8052-6_212

may be difficult to obtain useful data information. How to effectively arrange tourist routes, improve the quality of tourism and travel satisfaction of tourists is an important part of route design.

Under the background of the rapid development of Internet technology and online tourism information platform, by referring to the travel experience shared by ordinary users, tourists can get more and more travel suggestions, which provides a representative and reliable reference for the development of tourism planning. Most of the information is uploaded by the user spontaneously, which reflects the user's cognition of the tourism destination and the characteristics of the user's travel behavior [1]. These information are different from the general text, most of which have geographical location and time information. For example, according to the travel route data shared by users, combined with POI data with geographical location, the user's travel trajectory can be restored, and the travel planning that meets the user's time demand can be provided according to the time attribute of the route.

2   Research Status of Tourism Route Planning Methods

With the rapid development of Internet, user generated content including all aspects of life is increasing rapidly. In the field of tourism, a variety of forms (GPS trajectory, Beidou navigation information, check-in records, etc.) of tourism spatio-temporal trajectory data have been formed. These data and a large number of travel experiences and travel photos shared by users form tourism big data. Reasonable use of these data for tourism route planning is a hot spot in recent research. The advantage of this kind of work is that it can quickly get the feasible solution in line with the actual situation, and help users with travel planning, but the difficulty lies in how to reasonably use multi-source data to accurately mine the user's historical behavior trajectory.

GPS track data is mainly collected by mobile phones, vehicles and other self-contained GPS devices, from which a large amount of information can be mined, such as the user's residence time or the sequence of the user's visit to a place. GPS trajectory data is also widely used in the research of route search [2]. However, it is difficult to obtain a large number of user data based on GPS trajectory data. The optimal route searched by using GPS trajectory data is usually based on the popularity of routes between scenic spots, but it also needs to study the heat of scenic spots to meet the needs of users.

3   Analysis of Tourism Route Planning Based on Economic Cost

When traveling, people often think about how to spend the least money to travel to the most places and what kind of tour route can make them most satisfied. This requires us to establish appropriate mathematical models to solve these problems.

First of all, we know that this problem belongs to the optimization of tourism routes. In order to establish the model, we should first transform the routes of scenic spots into a set of points and lines in pure mathematical form for graph theory analysis.

3.1   Proposal

This problem should be solved in two aspects: (1) the cost of tourism should be as low as possible;(2) There are as many tourist attractions as possible. According to these requirements, we can start with the following two solutions:

(1) This paper establishes a multi-objective optimization model and considers the hierarchical sequence method. The first objective is to minimize the cost of tourism and the second objective is to maximize the number of scenic spots. In the feasible solution set of the first objective, the route satisfying the second objective is searched, and the route is regarded as the most suitable travel route.

(2) In the same way, a multi-objective programming model is established to transform the multi-objective into a single objective through appropriate fitting or linear weighting.

In order to increase the practicability of the model, the tickets of each scenic spot and the average daily consumption are also taken into account here. Analysis. The above two schemes show that the tourist route obtained in scheme 1 is not necessarily the best route, because when the travel cost is as low as possible, the route obtained does not necessarily meet the second condition, that is, there are as many scenic spots as possible. Therefore, there are some problems in scheme 1. We choose scheme 2 to transform the multi-objective into a single objective solution model through appropriate fitting.

3.2   Traveling Salesman Problem

Under the condition of abundant representative time, all the scenic spots can be visited, and appropriate models are required to make the transportation cost as low as possible. At this time, the representative is to start from the beginning, arrive at each scenic spot one by one (can not be repeated), and return after watching all the scenic spots.

According to the above analysis, the problem belongs to the traveling salesman problem. In order to establish the model, the scenic spots should be transformed into a set of pure mathematical points and lines, and the graph theory analysis should be carried out. The definition of TSP is given as follows: a seller starts from one of n cities, does not repeatedly walk through the other N-1 cities and returns to the original starting point, and finds the shortest path among all possible paths. TSP is described in mathematical language, that is, given a group of n cities and the direct distance between them, to find a closed journey, So that each city just passed once and the total travel distance is the shortest. Using graph language to describe TSP: give a graph $G = (V, E)$ with non negative weight w (e) on each side $e \in E$, and find Hamilton cycle C of G, so that the total weight of C minimum. TSP is a typical combinatorial optimization problem, and its possible search paths grow exponentially with the increase of the number of cities n, which belongs to NP complete problem.

$$W_c = \sum_{e=E(c)} w(e) \qquad (1)$$

After understanding the above knowledge, we are more sure that the problem is the traveling salesman problem. But in the actual processing, we take the toll of two scenic spots as the weighted value w (E). To a certain extent, the distance between the scenic spots is directly proportional to the one-way toll between the two points, so it is feasible to take the toll of two scenic spots as the weighted value w (E). Set up the West Lake Longjing tea shop in the golden section of the commercial street, promote brand management, stimulate tea consumption, open up the stagnant links in the tea industry, activate the tea market, and promote the virtuous circle of the tea industry. When tea production and sales maintain the original high quality and characteristics, we should also strengthen the creativity and conception of the sales packaging and design of the West Lake Longjing tea, fully reflect the long and charming history and culture of the Chinese nation, attract people to drink, get more contact with the mysterious ancient oriental countries, and in turn attract more people to join the ranks of drinking West Lake Longjing tea.

3.3   Data Sources

The rapid development of online tourism platforms (such as horse beehive, Ctrip, trip advisor.) has brought massive tourism data, which provides a lot of data support for the research of tourism routes. There are various sources of tourism information and different data forms, so it is necessary to determine the source of each data according to the specific steps of the research and different data needs.

Qunar.com, the world's largest Chinese online travel website, is selected as the source of tourist attractions data. As a tourism search engine with large number of domestic users and high popularity, the website contains not only tourism POI with complete attribute information, but also a large number of valuable user generated data [3]. Therefore, the website is also used as the tourism route data source for the construction of scenic spot Association map in this study. Due to the time constraints of short-term travel route planning, it is necessary to select accurate and wide coverage sources for travel time data. In this study, we choose the path planning API of Gaode map web service with open data interface as the source of travel time data acquisition. Gaode Web Service API provides HTTP interface for all developers to use. With the API, we can obtain various types of geographic data services, and the final results are returned in JSON and XML formats. The path planning API provides an interface to query the time of walking, bus and driving in the form of HTTP, which is used to obtain the content of path planning.

Taxi track data is a GPS sampling point data sequence with fixed frequency generated by on-board GPS receiving equipment. It contains location, speed, direction, vehicle operation status and other information, and can accurately express the vehicle position and driving status. In this study, taxi trajectory data is used as the data source to construct the association graph.

4   Basic Idea of Travel Route Planning Algorithm

The purpose of economic cost based travel route planning algorithm is to provide tourists with a number of travel routes that meet their time requirements, meet their travel preferences and minimize the cost [4]. In order to plan the short-term tourism route, this paper uses the solution as shown in Fig. 1. The detailed steps of the solution are as follows:

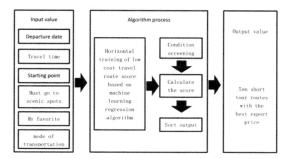

Fig. 1. Struts frame structure

First of all, according to their own schedule and travel preferences, users input six input values, including departure date, travel time, departure place, must visit scenic spots, interested scenic spots and transportation mode.

5   Conclusion

In this paper, the model is established for the design of main tourist attractions, which is suitable for the route planning of tourist attractions in different cities. Using the model, we can not only find out the minimum cost of the route, but also find out the shortest distance, the shortest time and other problems. With the development of economy and the improvement of people's living standard, tourism has become one of the important life styles and social activities. No matter what form people travel, they can't do without tourist routes. Tourism route is an important part of tourism products and an important link between tourists, tourism enterprises, relevant departments and tourism destinations. It is of great significance to the development of regional tourism, the survival and development of tourism enterprises, and the tourism experience effect of tourists. At the same time, it is also to enhance the learning and communication between students, improve the comprehensive quality of students, build a stage for the majority of students to show themselves, and enrich the campus cultural life.

References

1. Tang, J., Wang, T., Wang, J.: Shortest path approximate algorithm for complex network analysis. J. Sofw. **22**(10), 2279–2290 (2011)
2. Song, Q., Wang, X.: Survey of speedup techniques for shortest path algorithms. J. Univ. Electron. Sci. Technol. China **41**(2), 176–184 (2012)
3. Ma, R., Guan, Z.: Summarization for present situation and future development of path planning technology. Modern Mach. **3**, 22–24 (2008)
4. Wu, P.: A review of rural revitalization strategy in China since the 19th CPC national congress. Agric. Econ. **01**, 38–40 (2021)

Quality Evaluation of Machine Translation Based on BP Neural Network

Xueling Zhang[(✉)]

Nanchang Institute of Technology, Nanchang, Jiangxi, China

Abstract. From the perspective of adaptation theory, this paper discusses the translation defects and improvement measures of Jiangxi tourism publicity materials. Relevant means are used to reflect the original culture as faithfully as possible, and to conform to the cognitive ability and aesthetic expectation of tourists. Such as transliteration + interpretation. Also, the improvement measures based on BP neural network is shown in the paper with a semantic mapping model for eliminating errors in English translation of tourism publicity materials, which can supplement the quantity, history, culture and aesthetic information of the landscape, and help the development of tourism in Jiangxi Province.

Keywords: Adaptation theory · Tourism publicity materials · Translation requirements · Improvement measures · Machine translation

1 Introduction

The development goal of Jiangxi tourism includes two basic points: to carry forward the traditional culture of Jiangxi by using scientific and technological innovation; to attract as many tourists as possible to achieve the goal of enriching the people. At present, the typical method of combining science and technology with industry is "Internet+", which is especially suitable for the tourism industry, especially for foreign tourists. When they choose the tourism destination, the most convenient and quick way is to study the tourism information on the Internet, and the information on the tourism official website is particularly authoritative in their view. Therefore, for Jiangxi tourism industry, it is necessary to have high quality English publicity materials for Jiangxi tourism official website. This paper will discuss the translation of these materials from the perspective of Adaptation theory and the improvement measures based on BP neural network.

2 The Concept of Adaptation Theory

Under the theoretical framework of Verschueren's Adaptation Theory, translation involves a continuous making of linguistic choices, consciously or unconsciously, for linguistic and extra-linguistic reasons. These choices can be situated at any levels of linguistic form, such as lexical, syntactical and discourse level. It also includes the choices of different translating strategies and methods. It has four research perspectives.

J. C. Hung et al. (Eds.): FC 2021, LNEE 827, pp. 1473–1477, 2022.
https://doi.org/10.1007/978-981-16-8052-6_213

(1) Contextual adaptation. In the face of Jiangxi traditional culture, designers should be faithful to the original culture, at the same time, they should also pay attention to the selection of the essence that can stimulate the readers' interest, and use the variability, negotiation and adaptability of the language to interpret the original culture in a way that is convenient for readers to understand and resonate with.

(2) Structural adaptation. At the text level, there should be a three-dimensional sense to effectively realize strong and weak communication; the sentence level requires the translator to conform to the English syntax; the vocabulary level requires the translator to choose the words that readers can understand and feel to interpret the original culture. If the transliteration of the names or characteristics of some scenic spots can not be understood by the readers, it is necessary to make the communication smooth.

(3) Dynamic adaptation. It requires the translator to understand the development and change of readers' cultural level and cultural background. For example, with the globalization and the improvement of readers' cultural level, now readers have a certain understanding of Jiangxi traditional culture, so the designers of foreign publicity materials do not need to make too many modifications. This can not only be faithful to the original culture, but also let readers appreciate the true face of Chinese culture.

(4) The translator's consciousness [1]. As an intermediary communicator, we should strive to be objective and have a sense of the overall situation. Therefore, when selecting and interpreting materials, we should always focus on the needs of the readers and the differences between English and Chinsese.

3   Translation Requirements of Tourism Publicity Materials

3.1   Understanding the Cultural Differences Between China and the West

The premise of understanding the cultural differences between China and the west is to have a comprehensive and in-depth understanding of both Chinese and Western cultures [2]. Such a translator can not only accurately grasp the Chinese culture that the original author wants to promote, supplement the tourist cultural interest points missed by the original author, achieve the best relevance to the original text, but also comply with the cultural cognitive ability and expectations of the target readers and guide them.

3.2   Understand the Differences Between English and Chinese

If the purpose of understanding the cultural differences between China and the west is to achieve contextual relevance and adaptation, then to understand the differences between English and Chinese is to achieve language relevance and adaptation.

Vocabulary level: In the promotion of tourist attractions and culture, Chinese highly praises gorgeous rhetoric, and is keen on using idioms and even poems to make Chinese people fascinated by tourist attractions, but this is only limited to the aesthetic expectations and language habits of Chinese people. English pays attention to refinement and practicality. There is no ready-made correspondence of many Chinese idioms and even poems in English. If we insist on keeping them, it will be long and meaningless [3]. Therefore, the translator should grasp the original author's intention, that is, to convey the beauty and culture to achieve the best relevance. The use of refined language can help the tourists understand the uniqueness, aesthetic feeling and cultural value of China's tourist attractions, and effectively stimulate their interest in visiting China.

It is generally recognized that hypotaxis and parataxis is the most distinctive difference between English and Chinese: the meaning is restricted within the form in English and the form is dispeled by the meaning in Chinese. Translators should master the conversion between the two. Otherwise, if the tourists are not comfortable reading the articles or even do not understand them, how can they take the initiative to appreciate the beauty of Chinese culture.

3.3 The View of Approbation and Adaptation in Translation

Only when the translator recognizes the adaptation view of translation, pays attention to convey the original author's intention, and understands to adapt to the cultural background, aesthetic expectation and cognitive ability of the target readers, can the translator make good use of his translation skills, fully and accurately publicize the beautiful landscape and profound cultural heritage of our country, and Tourism Publicity Materials also make the readers understand and accept it, and be deeply attracted by the landscape culture of China.

4 Improvement Measures of Translating Tourism Publicity Materials

From the perspective of Adaptation theory, the improvement measures are mainly composed of two parts: modifying the existing information and semantic mapping relations based on BP neural network.

4.1 Modify Existing Information

Sometimes, there is only one name for a certain original cultural factor, so we can only transliterate it in order to seek the best correlation. However, in order to adapt to the tourists' cognitive level and achieve the purpose of attracting tourists, it is necessary to add necessary explanations. For example, when introducing Haihunhou in Jiangxi, if the translator takes the transliteration only, it does not highlight the historical value of the scenic spot [4]. Many foreign tourists don't understand the historical significance of it. Therefore, the author thinks that the more suitable translation method here is "transliteration + interpretation", which is translated as "Haihunhou (an imperial tomb)". The reason why it is not specified in detail is that the name should be as simple as possible, and many tourists do not understand the status of various dynasties and royal family members in China, but they know the weight of imperial family, so this interpretation shows its historical status and value. Interested readers will naturally read the followed details for further understanding.

4.2   Semantic Mapping Relations

In this paper, the semantic similarity of English is analyzed by using the method of semantic tree and semantic transformation. A semantic mapping model for eliminating errors in English translation is constructed, as shown in the Fig. 1.

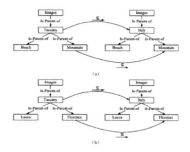

Fig. 1. Semantic mapping model of error elimination in English translation

In the mapping model of English translation error elimination shown in Fig. 1, the association rules are mapped according to each concept information in two different document ontologies, and fuzzy feature analysis method is used. The structure mapping relation of semantic concept analysis is established, and the automatic optimization control marking method in the process of English language conversion is realized by semantics. Let $\{(s_1, a_1), (s_2, a_2), \ldots (s_n, a_n)\}$ is the structural relationship between concepts, assuming that $O = <N, E, F>$ is the distribution of triple words that exclude Chinese translation errors. The fuzzy correlation feature vector $\omega = \omega_1, \omega_2, \cdots \omega_n^T, \omega_j \in [0, 1]$ is obtained by cross compiling control method in the tag set of association rules. Aiming at the strong mapping function of semantic expression ability, the deep learning method is used to modify the process. The function description of the compilation process of natural language processing is as follows (1):

$$\bar{s}, \bar{a} = \varphi_1((s_1, a_1), (s_2, a_2), \ldots, (s_n, a_n)) \tag{1}$$

$$\bar{s}, \bar{a} = \Delta \sum_{j=1}^{n} w_j \Delta^{-1}(s_j, a_j) \tag{2}$$

5   Conclusion

When introducing some scenic spots, some translators failed to the landscape characteristics and significant historical and cultural values, but simply used the transliteration method, which could neither show the physical aesthetic form of the original cultural elements nor make tourists understand it, let alone arouse tourists' interest. Under the guidance of adaptation theory, this paper comprehensively discusses the translation of tourism publicity materials from the perspective of Adaptation theory, and some examples are analyzed to show the importance of adopting proper translation skills to make the translation more informative and readable to the target language readers, such as transliteration + interpretation. Also, the improvement measures based on BP neural network is shown in the paper with a semantic mapping model for eliminating errors in English translation of tourism publicity materials to better play its functions.

Acknowledgements. Study on Chinese Culture "Going Global" from a Perspective of Translational Communication – A Case Study of Marquis Haihun (Humanities and Social Science Project of Universities in Jiangxi Province)

Project Number: YY21220

References

1. Jiemin, B.: A study on the pragmatic model of relevance adaptation. J. Northwest A & F Univ. (Soc. Sci. Ed.) **3**, 136–140 (2004)
2. Shao, H.: On the thick translation of tourism texts from the perspective of cognitive translation. J. Jiangxi Vocat. Univ. **2**, 28–30 (2017)
3. Yanchun, Z.: The explanatory power of relevance theory on translation. Modern Fore. Lang. **3**, 276–295 (1999)
4. Yi, W.: On the translation of scenic spots' names from the perspective of Cultural Translation – a case study of Slender West Lake in Yangzhou. Sci. Educ. **7**, 176–178 (2017)

Design and Research of Experimental Platform for Solar Semiconductor Refrigeration System

Hairong Wang[(⊠)], Jingmei Zhao, and Min Li

College of Optoelectronic Engineering, Yunnan Open University,
Kunming 650223, Yunnan Province, China

Abstract. As a kind of clean energy, solar energy has attracted people's attention because of its huge reserves, renewable and free from geographical restrictions. Solar semiconductor refrigeration has the advantages of no refrigerant and low noise. Therefore, it has a very broad application prospect. This paper describes the basic principle of solar semiconductor refrigeration and estimates the cooling load of the experimental system. On this basis, the component selection of solar semiconductor refrigeration system experimental platform, the design and construction of the system and its performance experiment are completed, and the experimental results are analyzed. The effects of various factors on the performance of semiconductor refrigeration under three cooling modes, i.e. air cooling, water cooling and heat pipe, are preliminarily obtained.

Keywords: Solar energy · Semiconductor refrigeration · Cooling mode · Experimental research · Refrigeration performance

1 Introduction

In recent years, with the rapid increase of energy consumption in the world, the traditional energy export is gradually exhausted. Energy crisis and environmental pollution have emerged in many countries to varying degrees, which seriously affect the healthy development of society. In China, the per capita energy resource ownership in the world is at a low level. The per capita coal resource ownership is equivalent to 50% of the world average level, and the per capita oil and natural gas resource amount is only about 1/15 of the world average level. It can be seen that China is quite deficient in energy resources. In 2014, China's total energy consumption ranked first in the world, accounting for 55.4% of the world's total energy consumption [1]. According to preliminary estimates, by 2020, China's total primary energy consumption will reach 3 billion tons of standard coal, by then, China's total oil consumption will reach more than 450 million tons, and the import dependence will reach 60%, The utilization of new energy (solar energy, wind energy, geothermal energy, etc.) has gradually become a hot topic in social research.

There is a nuclear reaction in the sun from hydrogen to helium, which radiates huge energy into space. According to the rate of nuclear energy produced by the sun, the storage of hydrogen is enough to maintain for tens of billion years. Compared with the limited period of human development history, solar energy is an inexhaustible energy

© The Author(s), under exclusive license to Springer Nature Singapore Pte Ltd. 2022
J. C. Hung et al. (Eds.): FC 2021, LNEE 827, pp. 1478–1482, 2022.
https://doi.org/10.1007/978-981-16-8052-6_214

source. At present, the utilization of solar energy mainly includes three aspects: 1) using the photothermal effect, that is, converting the solar radiation energy into heat energy; 2) using the photovoltaic effect, that is, converting the solar radiation energy into electrical energy; 3) using the photochemical effect, that is, converting the solar energy into chemical energy. In the process of development and utilization of solar energy, like wind energy, tidal energy and other clean energy, almost does not produce any pollution. It is an ideal energy substitute for human beings.

2   Basic Principle of Solar Semiconductor Refrigeration

2.1   Basic Principle of Solar Photovoltaic Power Generation

Figure 1 shows the schematic diagram of solar cell. When the solar panel is illuminated by light, it absorbs photons with constant energy and produces unbalanced carriers (electron hole pairs) [2]. Under the action of the internal electric field of semiconductor, holes flow from N area to P area, and electrons flow from P area to N area. As shown in Fig. 1. Because of the internal electric field, electrons and holes generate drift current from N area to P area, that is photogenerated current. If it is connected to the load, there will be an electric current through the load.

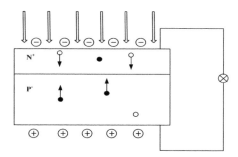

Fig. 1. Schematic diagram of solar cell

2.2   Basic Principle of Thermoelectric Refrigeration

Seebeck found that after two different materials of conductors or semiconductors are connected to form a closed loop, if the temperature of the two interface positions is kept different, then a magnetic field will be generated. Through further research, it is found that there will be electric current flowing through the circuit, that is, there is an electromotive force between the two conductors, which is called Seebeck EMF. This phenomenon is called Seebeck effect or thermoelectric effect.

Experiments show that: as long as the temperature T_1 and T_2 are maintained at the two interface positions of the loop, assuming that $T_1 > T_2$, there will be Seebeck EMF E_{12} between points C and D in the figure at the opening position of conductor 2:

$$E_{12} = \alpha_{12}(T_1 - T_2) \tag{1}$$

Where α_{12} is the Seebeck coefficient (also known as the thermoelectromotive force rate).

Peltier found that when the direct current passes through a circuit connected by two conductors or semiconductors, not only irreversible Joule heat will be generated, but also endothermic and exothermic phenomena will occur at the junction with different current directions.

The discovery of the Peltier effect makes it possible for thermoelectric refrigeration and heating. When the current passes through the circuit, the cold end emits heat and transfers to the hot end, so the cold end temperature decreases and the hot end absorbs heat temperature increases [3]. The experimental results show that the heat release and heat absorption at the loop node are determined by the current direction, while the heat absorbed and released is related to the current:

$$Q_p = \pi_{12}I \tag{2}$$

Where, I is the loop current, A; π_{12} is the Peltier coefficient, W/A.

Like the Seebeck coefficient, the Peltier coefficient depends on a pair of materials rather than one material, and there are positive and negative ones. The positive and negative of the Peltier coefficient are related to the current direction. Further experimental studies show that there is a close relationship between Peltier coefficient and Seebeck coefficient:

$$\pi_{12} = \alpha_{12}T \tag{3}$$

Where T is the absolute temperature at the node.

3   Experimental Results and Analysis of Solar Semiconductor Refrigeration

After the completion of the solar semiconductor refrigeration system experimental platform, four groups of experimental studies were carried out: 1) forced air cooling experimental research. When forced air cooling is adopted at the hot end of the cooling sheet, the influencing factors of the performance of the cooling sheet are analyzed by changing the voltage of the cooling plate and the fan at the hot end, the auxiliary device of the cooling sheet, and the model of the cooling plate. 2) Experimental study on water cooling and heat dissipation. When the cooling mode is used in the hot end of the cooling plate, the influence factors of the performance of the cooling plate are analyzed by changing the voltage of the cooling plate, the total amount of circulating water and the type of the cooling plate. 3) Experimental study on heat dissipation of heat pipe. By changing the voltage of the cooling plate and the heat pipe fan, the influencing factors of the performance of the cooling plate are analyzed when the heat pipe is used at the hot end of the cooling sheet. 4) Research on the best cooling system. By comparing and analyzing the experimental results of different cooling methods, the best cooling system is determined [4].

Several explanations need to be made: first, the circuit connection mode of the experimental device (refrigeration sheet, fan, circulating water pump, etc.) is in parallel; second, the temperature of the cold and hot ends of the semiconductor refrigeration sheet needs to be measured in real time during the experiment. If the thermocouple test end is directly installed on the surface of the refrigeration fin, it will lead to poor contact between the fin and the hot end of the refrigeration fin, and the heat can not be released to the environment in time, The refrigeration effect is seriously affected. Therefore, the test end of the K-type thermocouple is installed on the cooling fins at the cold and hot ends, and the temperature at this place is taken as the cold and hot end temperature of the refrigeration fin, but this will cause some errors in the measurement data, which is also the aspect that needs to be improved in the future work; thirdly, the experimental environment temperature is 30 °C, and the cold end fan voltage is 12 V; fourthly, in order to make the expression more clear, The results described in this chapter are based on the experimental data of a single semiconductor cryostat. Fifth, the voltage in this chapter is the voltage of the refrigerator except the fan voltage indicated.

4   Conclusion

This paper describes the basic principle of solar semiconductor refrigeration and estimates the cooling load of the experimental system. On this basis, the component selection of solar semiconductor refrigeration system experimental platform, the design and construction of the system and its performance experiment are completed, and the experimental results are analyzed. The effects of various factors on the performance of semiconductor refrigeration under three cooling modes, i.e. air cooling, water cooling and heat pipe, are preliminarily obtained. The main conclusions are as follows:

(1) In the forced air cooling experiment, with the increase of the cooling film voltage, the cooling capacity gradually increases, while the actual refrigeration coefficient decreases after increasing; the optimal voltage of the hot end fan is 10V, at this time, the actual refrigeration coefficient of the refrigeration sheet is the largest, and the refrigeration capacity of the refrigeration sheet is also large; when the auxiliary device of the refrigeration sheet adopts the combination of nylon screw and thermal conductive silicone grease, the refrigeration performance is optimal; Compared with tec1–12704, the cooling capacity of tec1–12708 is larger than that of tec1–12704, but the actual refrigeration coefficient is small when the voltage of most of them is 6–14 V.

(2) In the experiment of water cooling and heat dissipation, when the voltage of tec1–12708 is 4–8 V, the refrigeration capacity is larger when the voltage is 12–14 V; the optimal total circulating water volume of the cooling system is 2.5L; compared with tec1–12704, the cooling capacity and coefficient of refrigeration of tec1–12708 are larger.

(3) In the experiment of heat pipe heat dissipation, when the voltage of tec1–12708 is 4–6 V, the refrigeration coefficient is larger, and when the voltage is 12–14 V, the cooling capacity can be obtained; when the optimal voltage of heat pipe fan is 10V, both the cooling capacity and the actual cooling coefficient can be obtained.

References

1. Wang, Z., Ren, Y.: Utilization status and industrial development of solar energy resources in China. Res. Ind. **02**, 8992 (2010)
2. Zhou, Z.: Energy Environ. **06**, 9–10 (2008)
3. Lu, G.: Discussion on the current situation and development trend of new energy in China. Manager **18**, 271 (2014)
4. Dai, Y., Wang, R.: Latest research progress of solar air conditioning refrigeration technology. Acta Chem Sinica **S2** (2008)

Design and the Architecture Cloud Platform of Enterprise Income Tax Assessment System

Chen Yao[✉]

Yunnan Technology and Business University, Yunnan 651701, China

Abstract. This paper introduces the architecture design and technology of enterprise income tax assessment system based on J2EE framework. The system is applied to the system of Zhejiang Provincial State Administration of Taxation, which greatly improves the efficiency and quality of the collection and management software.

Keywords: J2EE algorithm · System architecture · Weblogic

1 Introduction

The "enterprise income tax assessment system" focuses on the five aspects of determining the assessment object, audit and analysis, evidence confirmation, identification processing and result reflection. By reconstructing the tax assessment system platform, we can improve the assessment quality and efficiency, reduce the resource competition with daily business, and effectively improve the operation efficiency and quality of the comprehensive tax collection and management software. Security is an important aspect of web application system.

BEA Weblogic application server can protect network applications with optional encryption, authentication and authorization functions based on RSA secure socket layer (SSL), X.509 certificate and access control table (ACLS) [1]. All beaweblogic devices can be used safely through HTTP tunnel, CORBA IIOP or HTTP (HTTPS) variable SSL and firewall.

1) Through the SSL V3 protocol supported by Weblogic, it provides HTTP secure access channel;
2) Through the access control list function on bea Weblogic, the security granularity of Weblogic users' access to application resources can be defined. And this function can use the level of security integration with operating system and database to provide the security barrier of application layer, so as to protect the access security of enterprise data resources at the back end.
3) Through BEA, Weblogic can also integrate with popular LDAP servers (such as the corresponding products of Netscape, Novell, Microsoft, etc.) to provide authentication, permission access, etc. for application system users.

J. C. Hung et al. (Eds.): FC 2021, LNEE 827, pp. 1483–1489, 2022.
https://doi.org/10.1007/978-981-16-8052-6_215

2   System Architecture Design and Technology

2.1   System Architecture

Enterprise income tax assessment system is developed based on J2EE three-tier architecture, using B/S structure to interact with users. The whole system is divided into user interaction service layer, task control layer and J2EE server layer.

The user interaction service layer constructs a set of fixed form activities according to business logic rules through modeling tools. The system is a user interface constructed by dynamic JSP/Servlet Technology. The user can initiate a request in this layer and transfer it to the task control layer for flow. The task control layer feeds back the results to the user interaction service layer in various ways according to the request category.

Task control layer is the core part of the system, which is based on J2EE application server. It is mainly composed of task processing flow, business logic flow and data processing flow. Task processing flow is a set of task processing rules pre-defined by developers through the process design of modeling tools, which is the basis of task processing flow. Business logic process is to provide a standardized and effective processing method and storage specification for the data of each work link in the task processing process. Data processing flow is a set of methods and processes related to data processing, which is built on the basis of tax assessment integrated database and combined with business logic rules. The three processes in the system are independent and interdependent, which embodies the main design idea of the system and determines the actual operation effect of the system.

J2EE server layer is the basic platform of J2EE application [2]. This system supports mainstream J2EE server such as Weblogic/WebSphere, and can be expanded and clustered according to the selection of web server.

2.2   System Technology System

The enterprise income tax assessment system adopts B/S mode, and the web server is Weblogic. The background database is Oracle 9i.

2.2.1   Oracle 9i

Oracle has always been in the forefront of technology in the field of database server, with the distinctive characteristics of high quality, high stability, advanced and mature technology, which has been widely praised and approved by users all over the world.

Oracle 9i, the latest version of Oracle database server, is a new flagship product in Oracle database server family. Oracle 9i is a database for Internet computing environment. It changes the way of information management and access, and integrates new features into the traditional Oracle database server technology, thus becoming a database for high-end enterprise applications and web information management.

Oracle 9i can process more data, accommodate more users, improve the performance in many aspects, further reduce the maintenance cost, and is the best in security and stability. Oracle 9i supports the parallel processing mode of multi server cluster, and supports the operation of dual computer or multi computer system. In the parallel processing mode, Oracle implements a shared database on each node of the cluster structure, and automatically realizes parallel processing and load sharing. When the server fails, it can realize fault tolerance and non breakpoint recovery, and ensure that the front-end applications are not affected. The real application cluster technology of Oracle 9i database is in the leading position in the industry. Oracle 9i Enterprise Edition provides efficient, reliable and secure data management for a variety of applications, including large load online transaction processing system and query based data warehouse application system. Oracle 9i provides a perfect and easy-to-use system management tool, which can realize the centralized management of network computing environment. In a complex distributed environment, Oracle 9i can easily and effectively distribute data, and improve the ability of transparent and efficient access to distributed data. Oracle 9i's scalable and reliable architecture provides the scalability, availability, and high performance required by unmatched mission critical OLTP systems. Oracle 9i and Oracle 9i real application server can make full use of all hardware system resources, from single processor, parallel multiprocessor, cluster system to massively parallel system.

2.3   MPP System

In order to achieve high performance of transaction processing, Oracle 9i's multi-threaded, multi server architecture can coordinate thousands of concurrent user requests. Individual requests are queued and processed by a minimum number of server processes.

Oracle 9i provides many different access paths for locating transactional data quickly and effectively, including fast full table scan, B-tree single column and concatenated index scan, clustered (pre linked) table, hash clustering (using a single column or a specific SQL hash function), and unique rowidentifier)。 Oracle 9i cost based optimizer dynamically selects the fastest Accessible path and satisfies the query request directly from the index when possible.

Oracle 9i provides high-performance data access capability through a series of advanced technologies. Oracle 9i adopts complete and unlimited row level locking for data and index, and in order to ensure the maximum data access capability, it never upgrades the lock. Oracle 9i's high performance and scalable serial number generator eliminates the competition of traditional transaction processing applications to obtain a unique numeric key value. The reverse key index reverses the bytes at the entry of the index and distributes some inserted continuous keys to different blocks, thus eliminating the need for hot spot insertion.

3 Weblogic

In order to build the system successfully, we divide the application logic structure into three layers according to the J2EE application standard model, namely client presentation layer, server web presentation layer and server application logic layer, and data information layer.

Among them, the client presentation layer can be any browser tool, Java supported client (such as mobile phone) or Java application program; the server web presentation layer is implemented by java servlet/JSP technology, which is similar to CGI in working principle, simple and mature, and its performance is highly adjustable; the server application logic layer is implemented by enterprise Java in WebLogic ServerBeans implementation. EJB has become the de facto standard of a new generation of reusable components. WebLogic Server provides extensible, secure, distributed and easy to manage running environment for EJB developers, which facilitates the development and deployment of application logic and enhances the reliability of application logic. It also makes developers free from the consideration of ensuring the accessibility, reliability, dynamic loading and other service functions of application components, and focus on the business logic of application, so as to simplify the complexity of development and shorten the development time. In the whole logical architecture, the java servlet running on the web server in the second layer is responsible for processing the requests submitted by the client, queuing the requests, and submitting the jobs in the queue to the optimal server according to the load of the application server[3]. The application logic of the second layer is implemented by the corresponding EJB object, and provides access to the data source of.

3.1   The Three-Tier Application Architecture has the Following Advantages

1) The customer platform is independent, which is convenient for the application to expand to any kind of terminal market. 2) Distributed computing mode is adopted. Make full use of the computing power of the client and application layer to reduce the workload of the host. To improve the expansibility of the system in function and performance, especially using object-oriented pure java development. 3) It can easily achieve load balancing in the application layer and improve the response speed of the system. 4) It is convenient to improve the availability of the whole system in the application layer. 5) Simplify deployment/management. 6) Commercial application is separated from data processing.

3.2   BEA Weblogic is Used as the Application Server of the System

First of all, let's take a look at the problems that should be considered when selecting an application server for e-tax and the technical solutions provided by BEA Weblogic.

The application system should consider the problem of distributed implementation, at the same time, with the development of business, we must consider the problem of expansion and upgrade. Therefore, the system should have good scalability, scalability and maintainability as far as possible in the hardware and software level.

In terms of scalability, we can use the software clustering technology provided by BEA Weblogic to ensure that when the scale of the system expands, we can easily add new hardware and software devices to the cluster to cope with the new traffic.

In terms of scalability and maintainability, on the one hand, we can use the leading enterprise Java component technology EJB (Enterprise Java) provided by BEA Weblogic.

On the other hand, through the complete J2EE application implementation based on bea Weblogic, we can achieve the goal that the software can be reused, encapsulated and assembled easily.

The design interface of the software is clear, and the structure and level standard are reasonable, so as to minimize the maintenance cost in the future.

The final implementation of the application system should consider and must meet: regardless of the user's access scale, it needs to provide an acceptable response time, as well as transparent and continuous access.

In this regard, the common technology we use is software clustering provided by the application server. BEA Weblogic provides advanced and comprehensive software cluster functions for key web and e-commerce applications.

BEA Weblogic mainly adopts two technologies to ensure the high performance and reliable processing required by e-commerce applications.

Performance: bea Weblogic uses load balancing technology to optimize the dynamic access requests of directional users, so as to balance the load of background application server and provide acceptable response time for users' requests; reliable processing: bea Weblogic uses key web user runtime data and EJBIn other words, the application service provided in the Weblogic cluster server can reach the level of 24 * 7 to customers, and ensure the reliability and continuity of users' access to the service.

In addition to providing cluster technology, BEA Weblogic application server can also support a large number of clients by carefully managing threads and connections. Using bea we blogic, all two-way communication can share a client/server connection, regardless of the request type and the number of remote objects to be accessed. Database connections can also be shared so that as many clients as possible can be supported at the same time. As shown in Fig. 1. BEA Weblogic caches the database query results, and can automatically update the cached data in real time when the back-end DBMS is modified.

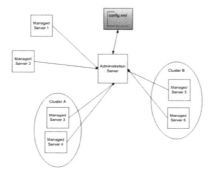

Fig. 1. Weblogic clustering

The following is a detailed description of the key e-tax applications supported by BEA Weblogic cluster technology. WebLogic Server runs on a group of independent servers at the same time to form a webcluster, which ensures the scalability, high availability and fault-tolerant service in web and Java deployment.

"Cluster" refers to the ability to copy application presentation layer and business logic among multiple machines, so that the application load can be evenly distributed among multiple machines. BEA's WebLogic Server is the only web application server that can provide web page and EJB parts cluster.

Web page cluster implements transparent replication, load balancing and error recovery of presentation layer logic generated for web client. Component cluster completes complex replication, load balancing, EJB (business logic) error recovery, and provides object state recovery.

"Fault tolerance" means that WebLogic Server provides transparent and uninter-rupted services for clients under the following circumstances: the server system cra-shes; the server does not work normally due to running errors or malicious interruption; the message loss due to unreliable transmission protocol; the network connection interruption causes part of the cluster to fail to work normally.

Under normal circumstances, each server in the cluster will periodically send a signal to indicate that it is working normally[4]. Once the signal is lost, the work on this server will be automatically transferred to other servers, and the user's appli-cation will not be interrupted.

BEA WebLogic Server provides dynamic JDBC pool, which can dynamically create, modify and delete JDBC connection pool, so that the number of connections in JDBC pool can change dynamically according to needs. Ensure that Weblogic provides 7 * 24 working ability.

4   Conclusion

In terms of management, the latest bea WebLogic Server 6.1 provides an advanced centralized management tool based on browser. In the web management tool provided by BEA Weblogic, system administrators can remotely monitor and update, configure and adjust the status of existing Weblogic applications and WebLogic Server Clusters, and manage the whole enterprise's WebLogic Server Cluster safely and con-veniently from a single remote browser console.

On the other hand, BEA Weblogic application server automatically records diag-nosis information and security check information, and provides interfaces for appli-cations to record their own abnormal conditions. Users can choose to record HTTP communication in normal format. These records can also be viewed remotely from BEA Weblogic's web browser based management console.

In addition, by supporting SNMP protocol, BEA Weblogic can also integrate with popular system management tools (such as HP OpenView, CA Unicenter, TNG, Tivoli, etc.).

References

1. Zhang, R.: Research and Implementation of Enterprise Income Tax Assessment and Analysis system. Xi'an University of Electronic Science and technology, Xi'an (2010)
2. Ding, T.: Design and Implementation of Foreign Income Tax Assessment Information System Based on B/S Structure. Jilin University, Changchun (2009)
3. Taotao, F., Haiyan, G., Ruihua, L.: Research on enterprise income tax management under the framework of the new law. North. Econ. **06**, 75–76 (2009)
4. Guan, X., Bi, X., Gong, P.: On the architecture design and technology of enterprise income tax assessment system. Comput. Knowl. Technol. **5**(07), 1609–1612 (2009)

Design and Implementation of Agricultural Technology Extension and Development Monitoring System Based on J2EE

Xue Bai and Yang Yu[✉]

Jingchu University of Technology, Jingmen 448000, China

Abstract. The rapid development of F Internet provides a new way for people to obtain information and services.The demand of interactivity and transaction promotes the evolution of network from an information publishing platform to a network application platform. The traditional application and calculation based on single machine is gradually extended to the application based on Web.The agricultural technology extension and development monitoring system of the Ministry of agriculture has changed the traditional way of data reporting and adopted a flexible three-tier architecture based on Web and Internet.

Aiming at the requirement of flexibility, stability and expansibility of the system, this paper adopts the design method based on J2EE multi-layer architecture and MVC design pattern. In the process of programming, rational rose, a case tool based on UML, is used for object-oriented software analysis and design, and the system function is successfully realized in a short time.Based on the analysis of the technology, principle and method used in the system design and implementation, this paper introduces the specific process of the system design and implementation. This process has universal applicability for the development of web application based on J2EE architecture.

Keyword: J2EE · MVC · Design pattern · UML activity report

1 Introduction

Based on the deep analysis of the related technologies involved in the development of web application system based on J2EE architecture, this paper introduces the specific process of system analysis, design and Implementation Based on J2EE architecture, following MVC design pattern, and using UML based visual modeling tool rational rose [1].

At the same time, according to the requirements of the system for the verification of activity report and balance relationship, this paper introduces the design idea and design scheme of activity report, as well as the data structure, database structure and balance relationship expression structure of activity report and the corresponding algorithm, so as to realize the flexible reporting of report structure and balance relationship without changing the web page, program and database structure.The first chapter is the design and implementation of agricultural technology extension development monitoring system based on j2eb framework.

© The Author(s), under exclusive license to Springer Nature Singapore Pte Ltd. 2022
J. C. Hung et al. (Eds.): FC 2021, LNEE 827, pp. 1490–1495, 2022.
https://doi.org/10.1007/978-981-16-8052-6_216

2 Research Overview

2.1 Research Background

With the vigorous development of Internet and web technology, the network has developed from an information publishing platform to a network application platform (Yao Shaowen et al. 2001). People can not only get the information they need by browsing the web pages on the Internet, but also provide more and more interactive services for people, such as online shopping, online medical treatment, online booking of tickets and hotels. The development of Internet and web technology has gradually changed people's daily life and work style, and become the driving force of the development of the information age [2].

Web application is increasingly valued by enterprises for its openness, wide access and flexibility. The market competition makes enterprises have to adapt to the changing business process in a short period of time, and develop and distribute application system quickly to gain competitive advantage as soon as possible. Web application is undoubtedly the best solution (Jiang Zuowen et al. 2001). Enterprises not only need to build a new web system to adapt to the new changes, but also should retain the legacy system, original data and business logic. The implementation method of fat client based on client/server two-tier structure has many disadvantages, such as poor scalability and maintainability. Compared with the flexible requirements, the contradiction between them is becoming increasingly prominent. However, the development of multi-tier applications involves many system level service coding, so the workload and complexity are very large. J2EE (Java 2 Enterprise Edition) technology is born in response to this demand of enterprises. It provides a component-based multi-layer distributed method to design, develop, assemble and deploy enterprise level applications, and provides a middle tier integration framework, which enables developers to focus on the development of business logic. The middle tier application server is responsible for system level services. Greatly reduce the development time and cost. (Moniea pawlan 2001) at present, most of the systems based on J2EE architecture are adopted by large and complex applications such as bank, post, telecom and website (Li Yongqiang CEN Yanqiang 2001). However, the application based on J2EE architecture is reliable, extensible, scalable, portable, and not too high workload and complexity. It not only attracts enterprise users, but also attracts more and more attention of medium-sized and complex applications. Under this technical background, this paper adopts the web application solution based on J2EE framework in the design and development of agricultural technology extension and development monitoring system of the Ministry of agriculture. Make full use of advanced technology, while ensuring the flexibility, reliability and scalability of the system, reduce the development cost and system complexity.

At present. The data report based on web is usually implemented by embedding the report structure into the web page, and the back-end database structure corresponds to the front-end report structure. Once the structure of the report is changed, the corresponding web page, back-end database and the program of transferring data in the middle will be changed, so this fixed report implementation method can not adapt to the reportReport content often changes, and the balance relationship control of report data is difficult to achieve. In this context.

In view of the demand of agricultural technology extension and development monitoring system for flexible reporting report, this paper designs activity report and balance relationship expression, realizes flexible reporting report and balance relationship verification, and meets the specific requirements of the system.

2.2 Agricultural Insurance Components

Through the comparison of the characteristics of various clustering algorithms, it is considered that the K-Modes clustering algorithm should be implemented and used first. The reason is that the original data of agricultural insurance products obtained from the network or other ways are based on text. After word segmentation and deletion, the data in the agricultural insurance data table can be divided into numerical data and separable type data. When clustering, it mainly processes the separable type data, such as agricultural product type, planting land and other information. In terms of insurance expense and amount, it can be realized by setting simple sorting when displaying recommendation results. Therefore, when analyzing such data, we should first use the calculation method of non numerical data when calculating the difference degree [3]. Non numerical data here is also called classifiable data. K-Modes algorithm is an improvement of K-means algorithm in order to process classified data. The formula of K-Modes for calculating the dissimilarity between two objects is as follows:

$$d(x, y) = \sum_{j=1}^{p} \delta(x_j, y_j) \tag{1}$$

When using K-Modes, first of all, we need to define the K class. When processing agricultural insurance data, we set the K value to the number of types of insurance objects in the actual data, that is, the types of agricultural products that can be insured. Then we calculate the most frequent value in the cluster according to the accepted insurance area as the class mode. According to this method, we calculate the class mode of other attributes, and finally we can get the class mode Q, As the initial cluster center. Then, according to the class pattern Q, each object is classified into the nearest class, and then the next q is updated and repeated in turn. When there is no change, the whole calculation process is completed.

Input of hidden layer:

$$net_i^{(2)}(k) = \sum_{j=0}^{M} w_{ij}^{(2)} o_j^{(1)}(k) \tag{2}$$

3 Rental Components

The function of the leasing component is to help farmers rent suitable agricultural machinery and equipment to help the sowing and harvesting of agricultural products. Avoid the cost of purchasing the whole equipment for short time use. As the sowing and harvesting of agricultural products have a certain periodicity, it is not necessary to use single function agricultural equipment at every stage. Purchasing alone not only reduces the utilization rate of the equipment, but also increases the planting cost of farmers. The most important thing is that most farmers can not afford and do not need to buy such equipment. Therefore, the provision of agricultural machinery and equipment rental is a very practical functional component for the majority of farmers.

LibLime koha is chosen as the implementation system of the leasing component. At first, LibLime koha is not a professional equipment leasing system, but an open source library management system. As early as 1999, it was used in nearly 1000 libraries around the world. Later, with the continuous expansion of its functions, it began to be applied to industrial enterprises. The lending and return of books are similar to that of mechanical equipment. Therefore, the secondary development and improvement of LibLime koha can meet the needs of agricultural machinery equipment rental function. But on the whole, it can save a lot of time.

In the past, financial leasing was used in enterprises to rent large-scale machinery and equipment. The purpose of providing this leasing function in the agricultural e-commerce platform is that when some more advanced or expensive machinery and equipment are needed by farmers and there is no place to rent, the e-commerce platform can purchase the equipment and then provide it to fixed farmers for renting, In the end, the user can take ownership of the device. Of course, farmers can also contact the manufacturer selling the equipment or contact the financial leasing company to negotiate specific financial leasing services through the user communication component provided by the e-commerce platform. E-commerce platform can provide necessary e-bills and related services for lessees and financial leasing companies through the financial leasing function [4].

4 Virtual Reality Experience Component

The function of virtual reality experience component is to help some users of agricultural e-commerce platform or other interested users to use the component of virtual planting crops to realize planting crops and experience the hardship of farmers planting crops. At the same time, because the mountain users pay the rent of the land, the expenses related to crop planting and the labor income of farmers' friends. Therefore, the employed farmer friends can reduce the planting risk, and the risk is borne by the users and players. The farmer friends only need to plant according to the planting methods of the users or players in the game, and no matter what the planting results are, they can guarantee to obtain income. This not only ensures and improves the income of farmers' friends, but also enables many white-collar workers and bosses who grow up or work in cities to experience the real process of planting agricultural products. Especially now many schools encourage students to experience planting crops, which is another feasible way.

Here, we introduce pulpcore as a part of virtual reality experience component. Pulpcore is an open-source 2d rendering and animation framework, which can be used to make 2D web games based on Web. Through pulpcore, we can reconstruct a game interface similar to happy farm. Chuanhu can grow vegetables and other crops in the virtual world, and also can carry out other operations, but the most distinctive thing is that, The growth status of the virtual crops is determined according to the growth status of the actual agricultural products of the real farmland compared with the virtual farmland. At the beginning of the game, players need to pay a fee to support the management cost of the real farmland corresponding to the virtual farmland they use. This fee is mainly used for the income, labor income and rent of the farmer as the real farmland. Later, the farmer will irrigate the farmland according to the planting instructions of the game user, and transmit the real planting picture to the game user through the camera in the form of facts, which will be automatically determined by the computer, Or farmers can set the agricultural situation in the game according to the agricultural situation and reflect it to users. In this way, users can experience the process of planting crops through the virtual reality experience component of e-commerce platform.

First, the farmers who plan to participate in the virtual reality experience component need to sign an agreement with the e-commerce platform. The general content of the agreement is that farmers should prepare the farmland for the virtual experience agreed by the e-commerce platform, and divide into specified areas according to the requirements and promise to cultivate and irrigate the farmland according to the instructions provided by the virtual reality experience component, And according to the actual situation, the situation of farmland is reflected in the virtual reality system. If there is any condition, web cameras should be installed for each farmland to provide real-time farmland images to users. Then, the farmland situation of contracted farmers is evaluated by e-commerce platform, and the users who participate in the experience are divided into different difficulty levels according to the specific situation of farmland. Users who participate in virtual reality experience can choose different difficulties, and different difficulties correspond to different farmland in e-commerce platform. Then, signing an electronic agreement is about including specific rules of the game. Sichuan households have the right to obtain any crops planted in farmland, and users shall pay certain experience expenses after selection. In the farmland where web cameras are installed, users can not only understand the cultivation of farmland through game pictures, but also can see real-time images of farmland. In the farmland without web camera, users can only understand the situation of farmland through the game picture.

5   Conclusion

The development of e-commerce in the field of agriculture has been highly valued by the country and the computer field from the beginning, and has achieved great development in the process of implementation and promotion. Based on the J2EE framework, the agricultural e-commerce system in this paper combines SOA and component design, so that the overall design and development cost of the platform has been effectively controlled. In addition to the basic functions of the traditional e-commerce platform, it also focuses on the problems faced by the planting, sales and transportation of agricultural products, and solves the problems one by one in the form of extended components. These functions that need to be realized and should be realized in the agricultural e-commerce platform are reflected in each extension component. The whole agricultural e-commerce platform is designed to effectively promote the sales and circulation of agricultural products, provide farmers' income, reduce the risk and cost of agricultural products planting. At the same time, it also has the function of expanding components, which can effectively enrich people's food basket while improving farmers' income. All of these make this paper appear to be of great economic and social value.

References

1. Zhang, F.: Introduction to e-commerce, pp. 13–14. Tsinghua University Press, Tsinghua (2004)
2. Fang, Z., Sun, B., Wang, J.: E-commerce Course, p. 6. Tsinghua University Press, Tsinghua (2004)
3. Ge, Z.: E-commerce Application and Technology, p. 5. Tsinghua University Press, Tsinghua (2005)
4. Shen, F., Liu, J.: Alibaba e-Commerce Primary Certification Course, pp. 5–7. Tsinghua University Press, Tsinghua (2006)

An Empirical Study on Early Warning Mechanism of Financial Business Cycle Based on BP Algorithm

Yao Chen[✉]

Yunnan Technology and Business University, Yunnan 651701, China

Abstract. The development of China's financial economy has obvious cyclical characteristics. The early warning mechanism of financial business cycle based on BP algorithm mainly includes: dividing the financial business cycle into four stages, and introducing economic growth rate index to distinguish different stages; coding different stages of the cycle, and training BP network with historical data to establish a cycle early warning model that meets the requirements; using time series method to obtain the input data of the model. And input the early warning model, so as to identify the cycle stage of China's financial fluctuations in a certain period in the future. Empirical research shows that: in the first quarter of 2010, China's financial and economic cycle is still in the expansion stage.

Keywords: Financial and economic cycle · BP algorithm · Artificial neural network · Economic growth rate

1 Introduction

The financial system has a significant impact on the operation of the business cycle. In the 1980s, the pioneering research of Bernanke and others criticized the "neutral theory" of money and securities, which made a breakthrough in the theory of financial business cycle. A series of subsequent studies laid the foundation of the theory of financial business cycle and established a general theoretical framework. Financial business cycle is a relatively new concept, which mainly refers to the continuous fluctuation and periodic change of financial and economic activities under the internal and external shocks through the financial system transmission. The expert group of the Central Bank of France defines the financial business cycle as follows: the real and sustained fluctuation of the economy is measured by the financial variables closely related to the long-term equilibrium level of the economy. China's financial and economic development has obvious cyclical characteristics. Early warning of financial and economic cycle is helpful for the government and other market participants to make scientific decisions [1].

J. C. Hung et al. (Eds.): FC 2021, LNEE 827, pp. 1496–1501, 2022.
https://doi.org/10.1007/978-981-16-8052-6_217

2   Research Ideas

Financial business cycle early warning is to judge the cycle stage of China's financial fluctuations. The research idea of establishing early warning mechanism of financial business cycle based on BP algorithm is as follows: firstly, according to the actual situation of China's financial fluctuation and combined with the stage characteristics of financial business cycle, the index value reflecting the situation of financial business cycle is corresponding to each development stage of the cycle; secondly, the index value influencing the fluctuation of financial business cycle is taken as the input of BP network. The term used to describe each development stage of the cycle is coded and used as the output of BP network, which is used as the training sample to train BP network and form the basic framework of BP network. Finally, the time series analysis method is used to predict each impact index value of the financial and economic cycle in a certain period in the future, which is used as the input of the trained BP network to get the cycle identification result code. The code corresponds to a certain development stage of the financial and economic cycle [2].

3   Index Selection

Enter the indicator. Financial fluctuation is usually understood as the fluctuation of specific financial variables, such as money supply, stock price index, credit balance, interest rate and exchange rate, and is regarded as a cyclical financial performance. Therefore, according to the availability of data, this paper selects eight financial variables as the impact indicators of China's financial and economic cycle, including cash balance in circulation, currency balance in narrow sense, currency balance in broad sense, monthly yield of Shanghai Composite Index, deposit balance, loan balance, benchmark interest rate of RMB current deposit, and exchange rate of RMB against US dollar, as shown in Table 1.

Table 1.   Impact indicators of China's financial and economic cycle

Category	Variable	Symbol
Money supply	Cash balance in circulation	M
	Narrow monetary balance	M1
	Broad money balance	M2
Stock market index	Quarterly yield of Shanghai Composite Index	I
RMB credit balance	Deposit balance	C1
	Loan balance	C2
Interest rate and exchange rate	Benchmark interest rate of RMB current deposit	R1
	Exchange rate of RMB to us dollar	R2

4 Research on Intelligent Control Algorithm of BP Neural Network

4.1 Neural Network Structure Model

The earliest famous British neuroscientist Sherrington compared the brain to a telegraph machine. The work of the brain is composed of numerous connections between nerve cells for information transmission and processing. It is composed of soma, dendrites and axon. Dendrites can grow in any direction to receive signals. Axons can grow very long, and their terminals contact with the dendrites of other nerve cells to form synapses, and then transmit the generated signals to other nerve cells through axons and synapses, and add the signals received by all dendrites. If the sum of all the signals exceeds a certain threshold, the nerve cells will be excited, and then the electric signals will be generated to transmit the excited state. On the contrary, if the sum of the signals does not exceed a certain threshold, the nerve cells will not be excited. Because of this layer by layer transmission relationship and based on a large number of combinations, the brain has the characteristics of plasticity, good at induction and promotion, high processing efficiency and unsupervised learning. In the process of studying its model, people use a series of 0 and 1 to operate the electro-chemical signals transmitted by each nerve cell [3].

T is the excitation threshold of neurons; y is the output of neurons. The output of neurons is a binary function.

$$y = \begin{cases} 1, & \sum_{i=1}^{n} w_i x_i \geq T \\ 0, & \sum_{i=1}^{n} w_i x_i \leq T \end{cases} \tag{1}$$

For the control of integral time T_i and differential time T_d, the relationship between input and output is as follows:

$$u(t) = k_p \left[e(t) + \frac{1}{T_i} \int_0^t e(t) dt + T_d \frac{d(t)}{dt} \right] \tag{2}$$

4.2 Structure and Algorithm of BP Neural Network PID

The traditional PID control is linear control, which is a widely used feedback closed-loop control. It is composed of proportional unit P, integral unit I and differential unit D. the proportional control is the basis, the integral can eliminate the steady-state error, and the differential can increase the corresponding speed. It compares the collected data with the reference value, and then calculates the error. The new input finally reaches the expected value or remains at the expected level. That is to say, it can adjust the input according to the historical data and error to make the system more accurate and stable. It is the most widely used and simplest control method in industry. In practical application, the basic control law can be used alone or in combination. The common control law is proportional control. Also known as differential control, the system input deviation and output constitute a positive proportional relationship, that is, the output increases with the increase of the deviation. In the actual control, it is difficult to control the scale properly. If it is too large, it can't control the system variables better. Otherwise, if it is too small, it will make the control too strong, which is easy to cause oscillation and lead to the instability of the system. Therefore, for the system with strong amplification capacity, the scale should be smaller to improve the stability; on the contrary, for the system with weak amplification capacity, the scale should be larger, which can increase the system sensitivity and reduce the residual error.

Although the integral control can eliminate the system residual, it can not control efficiently. Because of the existence of bias changes, it will accumulate and then control the system, so it can not solve the problem of system stability. However, PI control can not only control the system efficiently, but also eliminate the system residual.

5 Design of Inverter Control Software

On the basis of hardware, the addition of software can form a complete working system, which is an inseparable part of the two parts. The relationship between them is mainly: interdependence. Hardware is the material guarantee for software to play its role, and the correctness of software is the only way for hardware to play its role. The whole system can work normally only when it cooperates with each other, and at the same time, it can play its own functions. There are no boundaries between them. At present, the field of science and technology is highly developed, some functions can be realized by hardware or software, so there is no real difference between the two in a certain level. Mutual development. In the system, as the requirements of technical parameters become higher and higher, and the functional requirements of the system become higher and higher, it means that the software needs to be improved with the hardware requirements. At the same time, the continuous improvement of software also promotes the hardware update iteration, so the two parts complement each other and promote each other. Therefore, the inverter designed in this paper can make the hardware work as expected on the basis of software drive. The software designed in this paper is written in the most popular C language, including main program, interrupt program and service subroutine [4].

5.1 The Design of Interrupt Program

Interrupt is a signal triggered by hardware or software. When the interrupt signal is generated, the control chip will temporarily stop running the executing program and jump into the interrupt service subroutine. It is mainly divided into the following ways: when the service program issues commands IV NTR, trap and reset, the system will enter the interrupt as required. When the system receives a software interrupt request during normal operation, if it is an intr instruction, the interrupt mode (INTM) of the status register ST1 is set to 1 to prohibit other maskable interrupts. Trap instruction, the interrupt program can be interrupted by other hardware interrupts without affecting the INTM bit. The reset instruction is a non maskable software reset operation. When peripheral devices generate interrupt signals according to system requirements, hardware interrupt is required, including external interrupt generated by external interrupt port and internal interrupt generated by on-chip circuit. The external interrupt is processed in priority mode, mainly through level and edge trigger.

5.2   The Selection of System Control Chip

In the whole control system, the control chip is also a very important part. A good control chip can improve the working efficiency, stability and reliability of the system. In the early inverter, the single-chip microcomputer was used as the control chip. Because of the storage space and processing speed of the single-chip microcomputer, the control efficiency is not ideal. The emergence and application of digital signal processor (DSP) has greatly improved the efficiency and storage problems. The earliest DSP chip s2811 was introduced by AMI company. Then intel2920 was developed by inte company, which is the first DSP chip out of micro processing structure, and also a milestone of DSP chip. After the introduction of DSP and tms32012 Series in 1982, the functions of DSP and tms32012 series of instruments were more and more successful in the field of aviation. In the aspect of control algorithm, BP neural network control algorithm is added in this paper. The advantage of this algorithm is that it can update the weight and threshold value through training, and force the expected value at a faster speed. Through the simulation of the system, it can be seen that the output of the whole system is obviously faster than that of the ordinary PD control system after the system is started, In terms of accuracy and efficiency, it is far better than the ordinary control system, and the traditional PI control can barely satisfy the control of small systems. For more complex systems, the control ability of the system with specific control objects is not ideal, so the research on neural network PD control system can promote the development of control algorithm, However, this paper only makes a simple model simulation in the aspect of control algorithm, and the specific and more complex models need to be studied and studied in depth.

6   Conclusion

Based on BP network, this paper establishes an early warning model of China's financial and economic cycle. Firstly, the financial and economic cycle is divided into four stages, and different stages are distinguished by the economic growth rate. Secondly, the different stages of the cycle are coded, and the BP network is trained by using historical data to establish a cycle early warning BP model that meets the requirements. Finally, the input data of the time series prediction model is used to input the BP model. Therefore, it is necessary to identify the cycle stage of China's financial fluctuation in a certain period in the future, which has a certain early warning effect on scientifically grasping the development trend of China's financial and economic cycle.

References

1. Wang, X.: Principle and development trend of switching power supply. Sci. Technol. Inf. **11**, 18–21 (2007)
2. Liu, C.: Application and development of inverter welding machine. Sci Tech Inf. **30**, 43 (2007)
3. Lei, Y., Wu, S.: On the development of switching power supply technology. Commun. Power Supply Technol. **25**(4), 75–77 (2008)
4. Luo, G., Gao, R., Jiang, S.: Research on digital control technology of switching power supply. Exp. Sci. Technol. **4**(7), 30–34 (2009)

OTN Technology in Power Communication Network Service Routing Optimization

Lipo Gao[✉], Wei Kang, Junkui Hao, Jianhua Zhao,
and Hongmei Zhang

Economic and Technological Research Institute of State Grid Hebei Electric
Power Co., LTD., Hangzhou 050000, China

Abstract. The rapid development of smart grid technology not only spawns many new power communication services, but also promotes the evolution of power communication services towards the direction of high speed and large capacity. Due to the lack of overall consideration and long-term planning in network routing allocation, with the rapid increase of traffic, the main risk of network services in power communication OTN is no longer sudden congestion, but excessive concentration of services on a few transmission links. Therefore, it is of great significance to further study the routing optimization method of power OTN service by comprehensively considering the impact of OSNR on service deployment, which can improve the reliability and risk operation level of the network. This paper proposes a reliability oriented routing optimization algorithm, which aims at the risk balance of the whole network, and considers the constraints of delay and optical signal-to-noise ratio. In the implementation of the algorithm, based on genetic algorithm, the idea of simulated annealing and niche evolution is integrated in the process of population evolution to improve the operation efficiency of routing optimization algorithm and obtain higher solution quality at the same time, and then the global planning and re deployment of service routing of existing services in the network are carried out.

Keywords: Power OTN network · Service routing optimization · Optical signal to noise ratio · Genetic algorithm

1 Introduction

The With the development and maturity of smart grid technology, it integrates synchronous digital hierarchy (SDH) and traditional dense wavelength division (WDM) Optical transmission networks (OTN) technology with the advantage of multiplexing (WDM) technology is more and more widely used in power communication scenarios. OTN has the characteristics of large granular scheduling, large capacity transmission and adapting to a variety of services. It not only gives birth to many new power communication services, but also greatly improves the processing performance of large capacity and high rate services in the network. Power communication OTN service is related to the production scheduling and management control of power system, and has specific requirements for service reliability, service quality and transmission stability.

© The Author(s), under exclusive license to Springer Nature Singapore Pte Ltd. 2022
J. C. Hung et al. (Eds.): FC 2021, LNEE 827, pp. 1502–1507, 2022.
https://doi.org/10.1007/978-981-16-8052-6_218

However, due to the lack of overall consideration and long-term planning in network routing allocation, with the rapid increase of traffic, the reliability of OTN is affected [1].

The main problem is not sudden congestion, but the risk of network operation caused by the uneven distribution of services. Therefore, it is of great significance to limit the occurrence of high-risk events and distribute risk differentiated services evenly in the network to improve the reliability and risk operation level of OTN.

Aiming at the routing optimization problem of power OTN service, the classical routing algorithm or heuristic algorithm is mainly used at this stage. In reference, a KSP multi-path comprehensive cost algorithm is proposed. Dijkstra is used to find the K shortest paths between two points, and the best one is selected for service deployment. Because KSP can only deploy services in turn for a single time, the results are quite different due to different deployment orders, so it can not be applied to the global routing optimization problem. In reference, routing selection is transformed into an integer linear programming problem, and multi-objective routing optimization is solved by genetic algorithm. Although the multi-objective optimization is realized, the convergence and randomness of the algorithm are poor, and the algorithm is limited to the local optimum, so it can not effectively find the optimal solution. Literature uses ant colony algorithm to solve the routing and resource allocation problem, but when considering the routing availability, it does not involve the unique optical signal-to-noise ratio constraint in OTN, and only realizes the load balancing of services from the perspective of resources, without considering the industryIt is suitable to solve the risk of service imbalance in OTN.

Aiming at the problems of the above algorithm, this paper proposes a reliability oriented routing optimization algorithm, which takes the risk balance of the whole network as the optimization objective, and comprehensively considers the network transmission delay and optical signal-to-noise ratio constraints. In the implementation of the algorithm, based on the principle of genetic algorithm, the idea of simulated annealing and niche evolution is integrated in the evolution process, which aims to improve the operation efficiency of the algorithm and obtain higher solution quality, so as to carry out the global planning and re deployment of the existing business routing in the network. The reliability and routing optimization ability of the algorithm are verified by simulation. The algorithm can effectively optimize the risk balance of the whole network in large-scale network topology, and improve the reliability of OTN network operation.

2   Problem Description

The power OTN network architecture is based on the power communication network, and the power communication network architecture is based on the power grid, so the power OTN topology is relatively stable. Moreover, the power OTN belongs to the static network [1]. The data flow in the network basically remains unchanged after the establishment, and the operation risk of network services changes from the traditional sudden congestion to the excessive centralized distribution of services on the transmission link. Therefore, different from the traditional method of increasing resource redundancy, the optimization of power communication OTN reliability is more inclined to explore the routing allocation mechanism from the perspective of service routing, which can balance the network risk and improve the security and reliability of network operation [2].

The reliability oriented routing optimization algorithm designed in this paper mainly aims at the problem of unreasonable service distribution in the network, real-locates the service routing in the current network from the global perspective, and reduces the network risk and improves the reliability of power OTN by evenly distributing the risk differentiated services in the network. As shown in Fig. 1. In this chapter, the algorithm design process, based on genetic algorithm, takes the risk balance degree as the goal, comprehensively considers the network transmission delay and OTN optical signal-to-noise ratio constraints, ensures that the algorithm tends to select the optical channel that meets the delay constraints and has good optical signal-to-noise ratio parameters in the routing process, distributes the high-risk traffic to the transmission link with low risk value, and balances the local traffic of the networkThe network reliability is improved because of the difference of risk between different networks.

Model-1 Redundancy Protected Technology

· ??????? 20 msec/packet
· ? Redundant Voice Packets ????? Packet Stream
· ????? Packet Stream ??? Stream

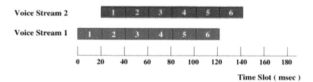

Fig. 1. Problem description

3 Model Establishment

The business risk equilibrium degree of the whole network can reflect the balanced distribution of business risk carried by each side of the network. The lower the index value is, the smaller the difference of comprehensive service risk of each channel in the network is, the more balanced the service distribution in the network is, the better the overall reliability of the network is, and the lower the network operation risk is; the higher the index value is, the worse the balanced distribution of services in the network is, the more the number of services carried on some links or the concentration of key services, and the less the load on some links. The risk equilibrium degree of the whole network can be expressed as Eq. (1).

$$R_B = \frac{1}{2m} \sqrt{\sum (R(e_{ij}) - R(e_{ij}))^2} \tag{1}$$

4 Basic Flow of Algorithm

The main steps of the algorithm are summarized as follows:

Step 1: parameter initialization. Initialize the set of business requests to be optimized, set the maximum number of iterations (termination conditions), individual fitness function f (x) and related calculation factors, initial temperature t0 and cooling factor Δ in simulated annealing mechanism.

Step 2: sample space initialization. The first generation population is randomly generated by a routing strategy and its fitness is calculated. The first generation population was sorted according to the fitness, and divided into n subsets according to the f (x) gradient. For the individual XK (k = 1,2,...,m) with the largest fitness in each subset, n) [3].

Step 3: in the contemporary population, the individual XK with the highest fitness directly enters into the next generation candidate population, and carries out a series of standard genetic operations such as selection, crossover and mutation on the remaining individuals in the niche to generate the next generation candidate population. It should be noted that niche means population isolation, and the treatment processes of populations in different niches are independent of each other.

It represents the optimal action taken in state s, while in a certain strategy, the formula is:

$$V(s) = r + \gamma \sum_{s' \in s} p(s, s') V(s') \tag{2}$$

5 Detailed Design of Algorithm

The proposed algorithm is designed and explained in detail, including the specific implementation of genetic algorithm, the use of simulated annealing mechanism and niche control.

5.1 Coding and Population Initialization

In this paper, the chromosome coding method based on path representation is used to encode the chromosome for each service independently, so as to form the chromosome coding segment. The location of gene in each chromosome segment represents the node through which the service route passes. All chromosome segments are spliced into a chromosome, representing a complete set of service route allocation scheme in the network. Secondly, the initial population is generated by the following path solving method.

Step 1: determine the current network topology state and business requests, and bind the business requests with the same source and sink nodes into a group.

Step 2: according to the source and sink nodes of each power OTN service request, all the end-to-end paths are obtained by using the path search algorithm as candidate paths.

Step 3: calculate the osnrjk and TJK of each path in the candidate path set, and delete the path that does not meet the requirements of osnrjk and TJK from the candidate path according to the constraints of osnrjk and TJK. For each service request, the maximum K paths are reserved as alternative routes, and on this basis, a set of alternative paths J (s) is constructed.

Step 4: for each service, randomly select a path from its corresponding alternative path set J (SK) to form an individual. In this way, different individuals are generated to form the initial population.

5.2   The Coordinate is the Risk Balance of the Routing Scheme

According to the comparison results, it can be seen that in the above simulation scenario, when the shortest path algorithm is applied, the risk balance degree of the whole network service is high, and the problem of unbalanced service distribution is more serious; the risk balance degree obtained by the application of genetic algorithm and the proposed method is relatively low, and the problem of centralized service routing distribution is improved to a certain extent, and the objective function calculation results of the two methods are satisfactoryThe latter is slightly better than the former. The fundamental reason is that although the shortest path algorithm has great advantages in solving the shortest distance problem and solving efficiency, it does not consider the risk of too concentrated traffic distribution, so it tends to choose a few seemingly "good performance" paths for routing [4].

Service routing leads to the congestion of service distribution and high risk of local network operation. The genetic algorithm and the method proposed in this paper both control the risk over concentration from the global perspective, and the algorithm proposed in this paper combines the genetic algorithm and simulated annealing mechanism. Compared with the simple genetic algorithm, it can investigate more abundant sample space, and has good performance.

6   Conclusion

Aiming at the problem of unbalanced network risk, this paper proposes a reliability oriented routing optimization algorithm for power OTN service, which takes the network risk balance as the optimization objective, and comprehensively considers the delay constraint and the unique optical signal-to-noise ratio constraint of OTN network. The algorithm combines the global search ability of genetic algorithm with the local search ability of simulated annealing mechanism, and uses the niche idea to obtain a group of service routing solutions with better risk balance in the whole network with faster convergence speed. Finally, the simulation results show that the algorithm can effectively reduce the cost, optimize the whole network service routing, achieve the goal of optimal risk balance, and also has good algorithm efficiency and convergence.

References

1. Lin, G.: Customer psychology. Taiwan J. Commun. **89** (2001)
2. Mao, J.: The center of Relationship Marketing - customer loyalty. Bus. Res., 165–176 (2002)
3. Quark market research. Retail: getting out of your way. Sales and marketing, pp. 108–111 (2000)
4. Li, C.: Research and Implementation of Customer Segmentation Model Based on Clustering Technology. Harbin Institute of Technology, pp. 11–24 (2006)

Technological Innovation Capability of High Tech Industry Based on Knowledge and Decision System

Peng He[(⊠)]

Liaoning Jian Zhu Vocational College, Liaoning 111000, China

Abstract. With the arrival of the era of artificial intelligence, high-tech industry, which integrates high intelligence and high-tech innovation, is becoming the key industry for countries to improve their international competitiveness. The key to increase the competitive advantage of high-tech industry is to improve its technological innovation ability, so as to achieve higher intelligence. This paper mainly studies the influence of scientific research funds, human investment, industrial agglomeration and fixed asset investment on the technological innovation ability of high-tech industry, and finally comes to the conclusion that scientific research funds and human investment are very important to the improvement of innovation ability, and the industrial intensity and fixed asset investment will not affect the industrial technological innovation ability when they reach a certain degree. Based on this conclusion, the paper puts forward feasible policy suggestions.

Keywords: Artificial intelligence · High tech industry · Technological innovation ability · Influencing factors

1 Research Background

Artificial intelligence has gradually become the core force of a new round of scientific and technological revolution and industrial change, and high-tech industry, which integrates knowledge and technology, has become the focus of artificial energy reform, the focus of economic development of various countries, and the core of international competitiveness of various countries. From 1995 to now, China has issued a large number of policies related to high-tech industry, aiming at providing comprehensive protection and support for the development of high-tech industry [1, 2]. With the increase of China's investment in technological innovation of high-tech industry year by year, the technological innovation ability and have been greatly improved. The high-tech industry is becoming more and more mature. However, there is still promoting the development of hi-tech industry.

2 Journals Reviewed

Sun Jingjuan et al. (2007) analyzed the rapid and steady development of China's high-tech industry in recent years, pointed out the problems existing in the technological innovation of China's high-tech industry, and put forward some

J. C. Hung et al. (Eds.): FC 2021, LNEE 827, pp. 1508–1514, 2022.
https://doi.org/10.1007/978-981-16-8052-6_219

suggestions. Huang Xiaoyi (2008) believed that transforming high-tech into actual productivity, has become a modern industryIt is a new source of economic growth and an important position of international competitionThe productivity index method analyzes and evaluates the problems existing in China's high-tech industry and its various industries, and puts forward suggestions. Deng Lu (2010) uses the stochastic frontier production function to measure the technological innovation efficiency of China's high-tech industry, and thinks that the starting point of technological innovation efficiency of domestic enterprises is low, and only when the technological innovation level of domestic enterprises is improved, can it bring about a fundamental improvement.

Technological innovation ability is the power source to determine whether the high-tech industry can obtain competitive advantage, and then affect the intelligence. This paper will focus on the influencing factors of the technological innovation ability of high-tech industry, improve China's technological innovation ability, enhance the competitive advantage and intelligence of high-tech industry, and finally achieve the goal of faster and better development of China's economy.

3 Overview of High Tech Industry

High tech industry refers to the industrial group that produces high-tech products with modern cutting-edge technology. Among them, cutting-edge technology refers to artificial intelligence, knowledge engineering, intelligent machine guidance and other technologies. It is the representative of the new generation of productivity and will lead to the intelligent revolution. Internationally, there are two main indicators to measure high-tech industry: R & D investment intensity should be more than 5%; R & D personnel investment intensity should be more than 30%. Therefore, the high-tech industry is different from the general manufacturing industry, it is a high-tech industry, regardless of its production process, or the final product has enough high-tech content.

3.1 In the Scope of High-Tech Industry, Technological Innovation Plays an Increasingly Prominent Role in Human Production and Life

The development of high-tech industry has increasingly become the driving force of economic development. With the continuous development of technology and economy, the content of high-tech industry in China has been constantly deleted, added, separated and merged, which can be divided into the following six categories: information chemicals manufacturing industry [3]. These six types of industries possess China's cutting-edge technology and are the core forces that determine China's long-term economic development and international competitiveness. In recent years, with the continuous development of China's high-tech industry, the income is increasing year by year, and the proportion of GDP is also rising. As shown in Fig. 1. Even if the financial crisis broke out in 2008, the income of high-tech industry also reached 15% of GDP. It can be seen that high-tech industry plays an important role in the national economy. Vigorously developing China's high-tech industry is an important measure under the new normal economy.

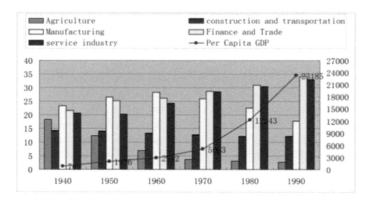

Fig. 1. Main business income of high tech industry and its proportion in GDP

3.2   Characteristics of High Tech Industry

As an industry group with advanced technology, high-tech industry needs to constantly research and develop more advanced technology to make itself competitive in the world. Its essence is the input of high intelligence and the output of new knowledge. Therefore, it is inevitable to invest a lot of scientific research funds. In 2000, the research and experimental development funds of high-tech industry were 11.1 billion yuan, and in 2017, the research and experimental development funds increased to 264.47 billion yuan, an increase of 23.8 times. There is no doubt that high investment must be accompanied by high risk. In the early stage of R & D and production of new technology and new products, it is full of uncertainty. Once it is not accepted by the market and consumers, it will face high investment in vain. But if it is recognized by the market and consumers, it means that it will bring huge benefits. According to the statistical yearbook of high-tech industry, the main business income of high-tech industry was 1003.37 billion yuan in 2000, and the main business income increased to 1593.58 billion yuan in 2017. The rate of return on investment far exceeded that of traditional industries and became an important industry for national economic development.

4   High Tech Industry has High Agglomeration

High tech industry is highly concentrated. We usually use the number of enterprises to measure the degree of industrial agglomeration. In 2000, there were 9758 enterprises in China's high-tech industry [4]. By 2017, the number of enterprises increased to 32027, an increase of 22269. High clustering can not only promote the concentration of regional innovation resources, form a good regional innovation environment and innovation network, create conditions for enterprises to obtain innovation resources quickly and at low cost, so as to promote regional technology upgrading; it can also form knowledge spillover, provide knowledge resources for enterprises in the agglomeration area, reduce the cost of enterprise technology learning, and promote the development of enterprisesTechnology and product innovation, so as to promote the overall level of regional innovation.

5   Model Establishment and Analysis

5.1   Variable Selection

It is very necessary to study the influencing factors of technological innovation in high-tech industry. The influencing factors considered in this paper include scientific research funds, human investment, industrial agglomeration and fixed asset investment. These factors will have a certain impact on the ability of industrial. The technological is measured by the amount of patent authorization (y). The amount of patent authorization is a kind of knowledge-based output of R & D investment in the industry, which can best reflect the value of technological innovation in the industry. The more the amount of patent authorization, the stronger the technological innovation ability of high-tech industry. The scientific research fund ($\times 1$) is measured by the research and experimental development fund, which can more intuitively reflect the value of enterprises. Economic support for technological innovation activities; human input ($\times 2$) is measured by the full-time equivalent of research and experimental development personnel, which can reflect the intensity and actual level of personnel input in the process of research and experimental development activities; industrial agglomeration degree ($\times 3$) is measured by the number of high-tech industrial enterprises, and the more the number of enterprises, the higher the industrial agglomeration degreeAmong the influencing factors, one of the important external economic effects of industrial agglomeration is the knowledge spillover effect, which can effectively reduce the R & D cost of enterprises and improve the output of technological innovation; the fixed asset investment ($\times 4$) is ignored by many scholars, in fact, its impact on industrial technological innovation is obvious, and the requires a lot of investment in instruments and equipmentIt is a necessary condition for technological innovation to be supported by investment.

5.2   Modeling

This model studies the technological innovation ability (y) of high-tech industry from the four factors of scientific research funds ($\times 1$), human investment ($\times 2$), industrial agglomeration ($\times 3$) and fixed asset investment ($\times 4$).

Let the model be:

Take the relevant data from 1999 to 2016, do OLS analysis, and get the results as shown in Table 1.

Table 1. Model regression results

Variable	Coefficient	Std. error	t-Statistic	Prob
C	−17.79843	2.768250	−6.429490	0.0000
LNX1	0.636145	0.291569	2.181800	0.0481
LNX2	1.232867	0.326307	3.778242	0.0023
LNX3	0.029824	0.332057	0.089817	0.9298
LNX4	0.314613	0.226467	1.389222	0.1881

According to the regression results, R2 is close to 1, so the goodness of fit of the equation is high. The logarithm parameter of scientific research funding (×1) is 0.636145, which is significant at the level of 5%, that is, every 1% increase in scientific research funding (×1) will increase the number of patent grants by 0.636145. Therefore, the improvement of technological innovation ability of high-tech industry must rely on a large amount of investment in scientific research funds, and sufficient scientific research funds provide a firm economic foundation for technological innovation; the logarithm parameter of human input (×2) is 1.232867, which is significant at the level of 1%, and the impact of human input on technological innovation is more obvious. Every 1% increase in human input results in the number of patents authorizedIt will increase by 1.232867. On the other hand, it reflects the lack of scientific and technological talents in our country. Technological innovation, we must put the cultivation of first place; the influence of industrial agglomeration (×3) measured by the number of enterprises on the ability of technological innovation is not significant, and industrial agglomeration has different effects on different industries, so we should consider the high-tech industry as a wholeThe impact of agglomeration degree is not very ideal; the impact of fixed asset investment (×4) on technological innovation is not very significant, because China's high-tech industry has invested enough fixed assets after more than 20 years of development, and some enterprises even have repeated construction and waste of resources. Further increase of investment will have less and less impact on technological innovation, so we can't blindlyObjective to increase the investment in fixed assets, the investment amount should be determined according to the specific situation of a specific industry.

The technological innovation ability of high-tech industry can also be measured by the sales revenue of new products. The main purpose of technological innovation is to develop new products. The sales revenue of long-term development of high-tech industry. According to Fig. 2, from 1999 to 2016, the sales revenue of new products of China's high-tech industry has been on the rise [5].

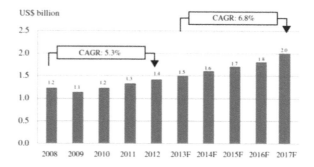

Fig. 2. Sales revenue of new products in high tech industry

In 1999, the sales revenue of new products was 15.257 million yuan. In 2017, the sales revenue of new products increased to 440 million yuan, 28.6 times of 1999. In less than 20 years, China's high-tech industry has made brilliant achievements, even in 2008When the financial crisis appeared in 2009, it did not hinder its development, only maintained a low growth rate in 2009, and then restored its technological innovation ability. Based on the continuous enhancement of technological innovation ability, China's high-tech industry is developing rapidly, which makes China's international competitiveness continuously enhance, and truly realize the transformation from made in China to created in China.

6   Conclusion

After more than 20 years of rapid development, under the background of continuous promotion of artificial intelligence, the technological innovation ability of China' s high-tech industry has gradually improved, and the high-tech industry has grown into a strategic leading industry of the national economy.

The government should also try its best to solve the capital bottleneck in the process of technological innovation in high-tech industry. Most of the government's scientific research funds are invested in Colleges and universitiesScientific research institutions lack the understanding of the importance of high-tech industry innovation, and the transformation rate of high-tech industry achievements is far higher than that of universities and scientific research institutions. The government should increase its scientific research investment, such as establishing a stable financial allocation support mechanism to ensure the financial support of enterprise technological innovation, optimizing the structure of technological innovation through financial allocation, and giving priority to support has the core significance and strategy. When enterprises borrow loans from banks, the government should also do a good job in the middle, build bridges, and provide a fair and just financing environment for enterprises' technological innovation.

References

1. Research on technological innovation ability of high tech industry under the background of artificial intelligence
2. Zhou, S.: Teaching management of Higher Vocational Education under the background of artificial intelligence. J. Zhengzhou Rail. Vocat. Techn. Coll. **33**(01), 48–51 (2021)
3. Liu, L., Ren, X.: Change of skill demand and cultivation of skilled talents from the perspective of task. Adult Educ. **41**(03), 71–76 (2021)
4. Hu, X.: On the legal regulation of algorithm risk in the age of artificial intelligence. J. Hubei Univ. (Phil. Social Sci.) **48**(02), 120–131 (2021)
5. Wang, F.: Application of artificial intelligence in computer network technology. Office Autom. **26**(06), 20–21 (2021)

The Application of Knowledge Base in Enterprise Resource Management Strategy

Fan Peng[(✉)]

Hainan College of Vocation and Technique, Hainan 570100, China

Abstract. According to the characteristics of the private cloud computing platform of power grid system, when the internal user program applies to use the public resource pool, in order to improve the computing efficiency of the resource pool and reduce the user's time cost, a resource load balancing scheduling algorithm based on average probability is proposed. Through calculation, the algorithm allocates appropriate virtual machines to the user's program blocks and reduces the queue of virtual machines. The simulation results show that the algorithm has good effect.

Keywords: Cloud computing · Open beble platform

1 Introduction

The advantages of cloud computing are: low cost of hardware, strong computing power and large storage capacity. Users can get high computing performance in a short time with low cost investment. They don't need to invest in expensive hardware equipment, carry out frequent maintenance and hardware and software upgrades, and save human resources and hardware and software costs. Among them, resource allocation is the key technology of cloud computing technology, which means that computing resources and storage resources are allocated to external users on demand through the Internet.

Users, the efficiency of resource allocation is very important, which has an impact on the performance of cloud computing platform.

Since the birth of computer, the research of resource allocation algorithm is an important field of research. However, the technology of resource allocation for the internal operating system of computer and the early network environment can not meet the needs of the current cloud computing environment [1]. Therefore, many researchers propose to solve the problem of cloud computing environment.

The algorithm of resource allocation is proposed.

2 Purpose of the Experiment

According to the resource scheduling methods in cloud computing environment, different scholars put forward their own strategies. For example, Soto Mayo of the University of Chicago builds a virtual resource lease manager on the platform of open

© The Author(s), under exclusive license to Springer Nature Singapore Pte Ltd. 2022
J. C. Hung et al. (Eds.): FC 2021, LNEE 827, pp. 1515–1521, 2022.
https://doi.org/10.1007/978-981-16-8052-6_220

bebu, and virtual resources are provided to users in the form of lease in different ways. Scholars such as Kong proposed an effective task scheduling method based on fuzzy prediction in the virtualized data center. This method takes time and reliability as the goal, establishes a fuzzy rule prediction model, and proposes a scheduling algorithm based on fuzzy rule prediction. Iyer studies the competitive relationship between a large number of virtual resources, analyzes the performance degradation of virtual machines due to the competitive relationship, and proposes a model to predict the competitive state of virtual resources. Song et al. Proposed a scheduling model with multi-layer scheduler as the core in the virtualized data center. From these studies.

It can be seen that resource scheduling is the core research topic in cloud computing.

In terms of management, cloud computing management can be divided into three layers: user layer, mechanism layer and detection layer. The mechanism layer is mainly responsible for various cloud management mechanisms, including operation and maintenance management, resource management, security management, etc. Among them, resource management is the core of mechanism layer, which is mainly responsible for the deployment and scheduling of resources, accepting resource requests from cloud computing users, and allocating specific resources to resource requesters, so as to reasonably schedule the corresponding resources and make the cloud computing system more efficient.

The task has to be carried out. Resource management includes resource model, resource discovery, scheduling organization, scheduling strategy and so on, among which scheduling strategy is the key to resource management.

Resources in cloud computing environment are dynamic heterogeneous [2]. When scheduling resources for large-scale tasks, we should try to improve the throughput and optimal span of the system, and consider the security of resources, load balancing and other issues. Resource scheduling is a NP problem, and all kinds of heuristic algorithms can be used in resource scheduling.

3   Scheduling Model

Where app stands for application (here, platform software business unit, represented by PSU), VM stands for virtual machine, and host stands for actual host. Users don't need to know which physical host the application is executed on, they just need to submit it to the virtual machine resource for execution. The physical host can have one or more virtual machine names. In other words, each physical host can be divided into one or more virtual machines; at the same time, the virtual machine can also point to one or more physical hosts.

Consider a situation in which each virtual machine has multiple tasks waiting to run, and each task can only proceed to the next task after running, but the performance of each virtual machine is different. This is because the architecture, hardware and software execution speed, capacity and throughput of each physical host may be different in the whole cloud environment. At the same time, the network transmission speed is inconsistent with the ideal situation, so it will be very difficult to calculate the end time of the physical host to complete the execution. Therefore, in cloud computing, assigning tasks to the virtual machine with the highest efficiency and the least queue will greatly improve the overall performance.

Define the execution condition of each PSU, which is composed of CPU performance, memory and bandwidth.

Define full load factor FL: set on VM; waiting to run. As shown in Table 1.

Table 1. Properties of physical host

Virtual host	CPU main frame	Memory capacity	Network bandwidth
1	0.5	0.2	1
2	0.5	0.2	1.5
3	0.5	0.4	2
4	0.8	0.8	1
5	0.6	0.4	1.5
6	0.4	1.0	2
7	0.2	1.5	2
8	0.3	0.4	2

On the basis of full load rate, we get the average probability resource load balancing scheduling algorithm as follows:

The time complexity of the algorithm is.

The idea of the algorithm is that when the number of tasks (PSUs) is less than that of virtual machines, a virtual machine is randomly assigned to run each PSU;When the number of tasks (PSUs) is more than that of virtual machines, the waiting queue of each virtual machine will become longer. The ideal state is that the waiting queue has the same length. However, some virtual machines are associated with better physical hosts, so the waiting queue shrinks faster. New tasks (PSUs) are automatically assigned to the virtual machine according to the algorithm.

Queue. There are the following situations:

Case 1: there are enough virtual machines, and the performance of the associated physical machines is excellent. After each task comes, it will be digested quickly. This is because the queue length is 0. In this ideal case, no algorithm can see the advantages.

Case 2: the performance of virtual machine is insufficient [3]. Many tasks are queued up for the virtual machine to run and process. At this time, when the task arrives, which virtual machine will be allocated is very particular. If the allocation is optimal, the queue length of the queue is not different, and the final total time is less.

4 Experimental Simulation

In the infrastructure pilot project of China Southern Power Grid Corporation Based on cloud computing technology, this paper uses claudim to realize the scheduling algorithm of resource load balancing algorithm based on average probability, and compares it with the scheduling algorithm recommended by cloudsim in the same simulation environment.

Test process: randomly generate PSU units in a certain period of time, and the execution cycle of each PSU running unit is 0.5 s. The basic conditions of PSU running unit are that the CPU running frequency is between 0.20–0.6 ghz, the memory is between 0.2-i.ogb, and the network bandwidth is between 1.0–2.ombis.After generating PSU randomly for 6 s, the waiting queue status of each virtual machine is shown in Fig. 1. Because the virtual machine at this time has not completed the task.

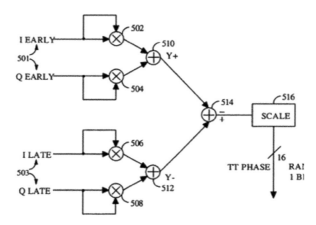

Fig. 1. Random generation processing unit

There are accumulated PSUs in each queue, but the performance of each virtual machine can be seen from the length of the queue. If the performance of each virtual machine is equal, after a random time, the queues of each queue are almost average. Therefore, due to the performance difference, the queue distribution is uneven.The experiment focuses on the queuing situation of the queue and the efficiency of the whole cloud computing task.

It can be seen that, due to the guiding effect of Algorithm 1 on PSU, the performance of virtual machine No. 3} 4 is better, and the queue at this time is the shortest. Slow down the generation speed of random tasks between 6S and Los, and then observe the queue. The number of queues is shown in Fig. 2. Due to the average arrival, the number of tasks available for each queue should be equal without considering the allocation of the algorithm. However, due to the intervention of the algorithm, the machines with good performance get more tasks. If the supply of the corresponding PSU is stopped, the queue consumption of the machines with good performance is fast, which is also in line with common sense. As can be seen from Fig. 2.

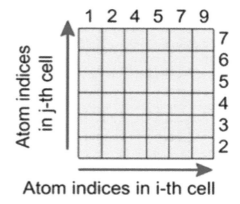

Fig. 2. Queuing status of virtual machine at 10 s

The running performance of the third and fourth virtual machines is still good, indicating that the number of tasks processed per unit time is the largest.

On this basis, the completed PSU unit is compared with the algorithm of claudim, as shown in Fig. 3. When the number of tasks is small, the two algorithms can not distinguish the advantages and disadvantages, which is consistent with the conclusion of case 1 we analyzed before. However, with the increase of the number of tasks, the average probability resource load balancing scheduling algorithm further reflects the advantages, specifically in that the total time to complete all tasks has a significant downward trend.

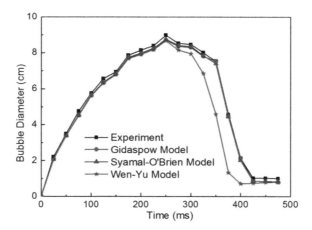

Fig. 3. Comparison of completion time under large task simulation

The simulation results show that the average probability resource load balancing scheduling algorithm is significantly better than the cost optimal scheduling algorithm in the total time of task completion (as shown in Fig. 3).

The analysis results of the trustworthy data of high reputation enterprises formed by the third party have a positive correlation with the return of enterprise operation, which better promotes the enterprises to further perform their duties and keep their promises. The virtual network transaction can better create an honest and trustworthy e-commerce atmosphere on the basis of the construction of the integrity supervision system, so as to ensure the independence and fairness of the regulatory authorities.

Therefore, strengthening the audit of market access, qualification and supervision methods of credit rating agencies will inevitably become an important guarantee for the third-party supervision. To speed up the construction of information service platform for credit evaluation, make full use of big data analysis means and modern network technology, and capture dynamic data in real time, we should not only have constraints, but also pay attention to process supervision [4]. We should accelerate the construction of credit evaluation system, guarantee the independence of third-party institutions with institutional constraints, and improve the credibility and recognition of credit evaluation institutions.

In a word, mobile commerce, a new business model with rapid growth, provides the possibility for small and medium-sized enterprises to participate in market competition and gain competitive advantages. In order to better promote the benign operation and development of new formats, it is necessary to establish and improve the enterprise credit evaluation mechanism in the face of the incentive mobile commerce competitive environment.

The establishment and formation of a good credit evaluation environment under mobile commerce is an important factor for the stable development of mobile commerce, and an important guarantee for the rapid and stable development of market economy. The key of credit evaluation is to provide a safe and reliable mobile commerce environment, and security has always been the most important factor for mobile commerce enterprises.

One of the important evaluation indicators, especially in the B2C e-commerce development today, people's life is more and more inseparable from the influence of mobile commerce, fast-paced life also constantly urges people to adapt to and accept mobile commerce, the convenience and personalized advantages of mobile commerce also bring great satisfaction to people's life.

The proportion of mobile commerce in the national economy is more and more heavy, the simple real economy has been difficult to meet the needs of people's efficient and convenient life. Take the clothing industry as an example, clothing as a highly personalized product, under the traditional business model, people are keen to make continuous selection in different brands and different physical sales places. However, the emergence of mobile commerce makes people can spend moneySmall time cost to complete consumption activities, and the choice is far from comparable to the regional physical stores.

5   Conclusion

In order to improve the utilization rate of cloud computing system resources and emphasize the viewpoint of "those who can do more work", this paper proposes a scheduling algorithm, which is implemented in the South China Grid cloud resource management system (CRMs) developed in the pilot project, so that the virtual machine with good performance can play a better role, and improve the efficiency of the whole cloud computing platform, which can not only meet the performance requirements of the application running,At the same time, it also reduces the total running time of the system, improves the resource scheduling rate, and finally achieves the requirements of energy saving and environmental protection.

References

1. Liu, H.: Research on internal control and risk management of enterprise financial system. Account. Learning (06), 187–188 (2021)
2. Wang, T., Li, J.: Enterprise performance appraisal system planning and application. Yangtze River Technol. Econ. **5**(S1), 159–160+163 (2021)
3. Ye, G., Wang, J.: Construction and application of digital monitoring system for high risk operation in power generation enterprises. Electromech. Inf. (06), 61–62 (2021)
4. Jiang, Z., Jiang, Z.: Discussion on the application of management information system in furniture enterprises. Forest. Ind. **58**(02), 83–85 (2021)

Resource Management and Scheduling Algorithm in Hybrid Cloud Environment

Hang Wang[(✉)]

Sichuan University Business School, Chendu 610064, Sichuan, China

Abstract. In the development of cloud computing, public cloud and private cloud are two mature ways to provide cloud computing services, but the public cloud is trans-parent to users, and the cost of private cloud is high, so many users choose to use hybrid cloud.It can also solve the problem of peak workload.

Keywords: Hybrid cloud · Resource management · Ant colony algorithm · Resource scheduling · Cloud sim

1 Introduction

The most critical part of the use of hybrid cloud is how to manage and allocate the resources in the hybrid cloud. At present, there are few mature products in the hybrid cloud, and the resource management solutions are different according to the needs. Therefore, this paper proposes a hybrid cloud resource management scheme and resource scheduling algorithm suitable for small and medium-sized enterprises.

Based on the research of hybrid cloud resource management, virtualization technology and resource scheduling algorithm, this paper proposes a resource management model in hybrid cloud environment, and analyzes the specific solutions of resource organization, resource monitoring and resource scheduling. Aiming at the special environment of hybrid cloud, a resource scheduling algorithm based on ant colony algorithm is implemented. The algorithm is implemented on cloud computing simulation software cloudsim, and the experimental results are analyzed [1].

Enterprises use hybrid cloud resource management scheme to deal with business needs, which can make enterprises maximize the potential of IT infrastructure. By abstracting physical resources into virtual resources, enterprises can manage heterogeneous resources uniformly, improve resource utilization, reduce costs, and solve the problem of resource shortage due to workload peak in a short period of time.

2 The First Chapter is Introduction

2.1 Research Background

At present, cloud computing and its related services have become one of the hot spots in computer science research and enterprise application development. With the growth of enterprise and individual consumers' demand for big data storage, efficient computing and personalized application services, cloud computing is constantly improving

J. C. Hung et al. (Eds.): FC 2021, LNEE 827, pp. 1522–1527, 2022.
https://doi.org/10.1007/978-981-16-8052-6_221

its own characteristics and building different levels of application services from multiple perspectives. The bottom layer of cloud computing platform uses virtualization technology to build a large virtual cluster, which can dynamically organize heterogeneous computing resources, isolate specific hardware architecture, and provide users with a variety of application service platforms. In the environment of cloud service platform, users can get powerful computing power, massive data storage resources, and high-performance distributed computing power. In the aspect of application services, cloud computing provides users with rich personalized services, good user experience and various timely, efficient and smooth application services [2].

Cloud computing products are also mature with the development of cloud computing technology, and gradually penetrate into people's daily life. Network disk can be said to be a typical representative of cloud storage. It stores personal user's documents, pictures, address book information and other files in the cloud, so as to synchronize the data on different terminals of users, including PC, web and Android. At present, some well-known enterprises have provided permanent free network disks with different capacities, such as Microsoft Windows Live SkyDrive, Huawei network disk dbank, Jinshan fast disk, 115 network disk, etc., and professional users have made a comparative analysis of the advantages of different network disks. In addition, cloud security is also a service close to individual users. It makes use of parallel processing and unknown virus behavior judgment technology in cloud computing. It can obtain the latest information of Trojans and viruses in the Internet through anomaly detection of different software on a large number of clients in the network, and then share the virus database among different users. It can check and kill Trojans and viruses more quickly and comprehensivelyTo provide a secure network environment for users. Although cloud security was controversial when it was first proposed, McAfee, Jinshan and 360 security guards have proposed their own cloud security solutions.

2.2 Research Significance

The reason why the public cloud and private cloud are combined to form a hybrid cloud for users is that the hybrid cloud has better solutions in information control, scalability, sudden demand and failure transfer. Hybrid cloud can enable enterprises to maximize the potential of IT infrastructure, so that it can be fully utilized, and the surplus IT infrastructure can be saved, which can reduce the budget. Hybrid cloud can combine the existing infrastructure to provide a variety of resources through different needs. Enterprise users can choose the best platform for storing their data and programs according to their own situation, and then distribute processing resources more reasonably. Moreover, we don't need to consider the security constraints, performance requirements, and reasonable planning of public cloud computing in everything. Therefore, at present, hybrid cloud has become the best choice for enterprise and individual users.

In addition, hybrid clouds help provide on-demand expansion. Using the resources of public cloud to expand the capacity of private cloud can be used to maintain normal usage in case of rapid workload fluctuation. When using cloud storage to support Web2.0 applications, hybrid cloud should be the most appropriate choice. Hybrid clouds can also be used to handle expected workload peaks. For example, during the

2012 Olympic Games, tens of millions of people from all over the world will come to London to watch various competitions and book tickets from the Olympic ticket system. In a short period of time, tens of millions or even more users visit the ticketing system and complete ticket booking, payment, cancellation and other operations, which is a huge challenge for the ticketing system. Suppose that the company in charge of the booking system has its own private cloud platform, which is usually used to handle some ordinary booking business. If it suddenly undertakes the processing of massive data such as the Olympic Games, it may need to increase the number of servers in the cloud platform, which will lead to the failure.

Increase the company's operating costs. After the Olympic Games, the newly added servers will be idle again. However, if the company adopts the hybrid cloud operation mode, when the data processing capacity is small, it will use the internal private cloud for processing. When there is a large amount of data, it will put a large number of data storage and computing tasks on the public cloud platform, and only need to pay the corresponding fees for the rental of the public cloud when it is used. This can not only cope with the sudden peak workload, but also reduce operating costs [3].

2.3   VMware

VMware's virtualization technology has always been in the leading position in the industry, except that we are familiar with it.

Besides the virtual machine products, there are also mature products in the virtualization of cloud computing technology. In July 2011, VMware released vSphere 5 and cloud computing infrastructure suite, aiming to help customers accelerate the development of more efficient automated cloud infrastructure and management, improve the way to manage and protect resources, promote public cloud, private cloud and hybrid cloud applications, and ultimately improve the relationship between it and its service business. The new VMware cloud infrastructure suite includes the latest VMware V cloud director 1.5, which enables enterprises and service providers to create and operate secure hybrid clouds.

At the 2012 JBoss World Congress, red hat announced a number of hybrid cloud solutions that it will open in the next few months. This indicates that red hat will win a place in the cloud computing competition through hybrid cloud solutions. At this conference, red hat proposed four solutions: openshifttm enterprise PAAS solution: cloud application platform for enterprise users. It will combine red hat cloudforms, Red Hat Linux Enterprise Edition, Red Hat Enterprise Virtualization Technology, JBoss Enterprise Middleware and other tools to build itops PAASPlatform, for enterprise developers to bring efficient and agile at the same time, solve the enterprise governance and operation and maintenance needs. In the future, this solution will also develop to be able to createOpenShift.com There is a total PAAS cloud fully compatible Devops PAAS environment.

3 Mapping of Ant Colony Algorithm to Hybrid Cloud Resource Scheduling

In the virtual machine resource scheduling algorithm, the hardware resource of virtual machine is used to describe the pheromone of a node. The description of virtual machine resources is as follows:

$$VM_i = \{m_i, p_i, r_i, h_i, b_i\}, i > 0 \tag{1}$$

The number of CPUs, P is the processing capacity of CPU (unit: MPIS), R is the memory capacity, h is the hard disk capacity, and B is the network bandwidth. We set a threshold value for each parameter (the maximum value of the parameter in all resources can be used as the threshold value). If the threshold value is exceeded, the threshold value will be used for calculation. As shown in formula 2:

$$m_{max} = m_0, p_{max} = p_0, r_{max} = r_0, h_{max} = h_0, b_{max} = b_0 \tag{2}$$

3.1 Cbudsim Simulation

On April 8, 2009, the grid lab and gridbus project of the University of Melbourne in Australia announced the launch of cloud computing simulation software, called cloudsim. Cloudsim inherits GridSim's programming model, supports the research and development of cloud computing, and provides the following new features: (1) supporting the modeling and Simulation of large-scale cloud computing infrastructure; (2) A self-contained platform supporting data center, service agent, scheduling and distribution strategy. Among them, cloudsim has the following unique functions: one is to provide virtualization engine, which aims to help establish and manage multiple, independent and collaborative virtualization services on data center nodes; the other is to be able to flexibly switch between time sharing and space sharing when allocating processing cores to virtualization services. Cloud platform helps to accelerate the development of cloud computing algorithms, methods and specifications. Cloudsim's component tools are all open source. The software architecture framework and components of clouds IM are shown in Fig. 1:

Cloud IM is developed on the basis of grid IM model, which provides the characteristics of cloud computing and supports resource management and scheduling simulation of cloud computing. A significant difference between cloud computing and grid computing is that cloud computing adopts mature virtualization technology, which virtualizes the resources in the data center into resource pool and packages them to provide services to users. Cloud im embodies this feature, and the extended part implements a series of interfaces to provide virtualization technology based on the data center and modeling and simulation functions of virtual cloud [4].

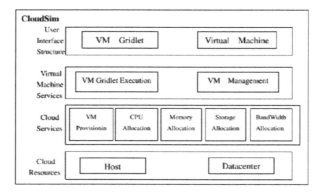

Fig. 1. Cloudsim architecture

3.2   Description of Scheduling Algorithm

In the hybrid cloud environment, for applications in small and medium-sized enterprises, users' jobs are first assigned to the private cloud built by the enterprise for processing. When the resources of the private cloud can not meet the needs of users, they apply to the public cloud operators for resources. The reason is that the private cloud resources built by the enterprise come from the internal IT infrastructure and belong to the existing hardware infrastructure resources of the enterprise, The use cost is low, and the public cloud requires enterprise users to pay a fixed rent. From the perspective of reducing the overall cost, we follow the principle of giving priority to the use of private cloud resources. Then, when to apply for public cloud resources and use hybrid cloud to process user jobs is the key of resource scheduling strategy in hybrid cloud.

4   Conclusion

This paper focuses on the research of hybrid cloud services in enterprise applications, and comprehensively analyzes the development and research status of cloud computing at home and abroad. In depth understanding of virtualization technology and resource scheduling strategy in cloud computing. On this basis, a resource management model in hybrid cloud environment is proposed, which is introduced in detail from the overall design to the functional design of main modules. This paper proposes a resource scheduling strategy for the collaborative module in hybrid cloud, which is a virtual machine resource scheduling scheme based on ant colony algorithm. The algorithm is implemented on cloud computing simulation software clouds, and the efficiency and feasibility of the algorithm are obtained according to the experimental results.

References

1. Wu, Z.: Analysis of Cloud Computing Core Technology. People's Posts and Telecommunications Press, Beijing (2011)
2. Zeng, X.: Implementation and application of cloud computing management platform based on Gridsphere. Master's thesis. Beijing University of Posts and Telecommunications (2011)
3. Hu, J.: VMware vSphere Operation and Maintenance Record. Tsinghua University Press, Beijing (2011)
4. Zhu, J.: Research and design of resource scheduling mechanism for hybrid cloud platform. Master's dissertation. Jiangsu University (2012)

Analysis on Key Technologies of New Energy Internet Information Communication

Huifeng Yang, Qiang Fu, Li Shang, Yong Wei, and Ming Kong[✉]

Information and Communication Branch of State Grid Hebei Electric Power Co., Ltd., Shijiazhuang 050000, Hebei, China

Abstract. PickAt present, with the development of science and technology, how to realize the sustainable development and full utilization of energy has become one of the important tasks in the new era. In order to ensure the full utilization or reuse of energy, the concept of "energy Internet" is gradually put forward under the support of information and communication, which takes the power system as the core and is accompanied by automatic control technology, information technology and renewable energy technology. With the development of advanced technologies such as natural gas, oil, coal, transportation, etc., renewable energy units are mainly distributed and centralized. Under information technology, data can interact in two directions, and the system network presents multiple forms, such as covering natural gas, oil, coal, transportation, etc., gradually forming a new energy utilization system, which is also the basic idea of energy Internet. This paper will analyze the key technologies supporting energy Internet information communication.

Keywords: Information and communication · Energy internet · Key technology · Demand

1 Introduction

Nowadays, the society has entered the information age. With the development of information technology, information has become an important social resource, which is used in many industries and plays an important role in the development of various undertakings. For example, the informatization of power grid, relying on information and communication technology, can promote the interaction and intelligence of the industry. The so-called information and communication technology includes two important technologies of communication and information,There are not only a series of information data collection and processing processes, but also information data transmission. Through information access, alternation, special communication and other aspects of processing, information analysis and sharing can better meet the needs of social development [1]. In the field of energy Internet application information communication, the following first analyzes its communication architecture, combs its characteristics and needs, and explores energy Internet information on this premise.

J. C. Hung et al. (Eds.): FC 2021, LNEE 827, pp. 1528–1534, 2022.
https://doi.org/10.1007/978-981-16-8052-6_222

2   Energy Internet Information Communication Framework

The energy Internet is not a simple Internet + energy, nor is it the superposition of electric power or other energy by using the Internet as a tool. The most important thing is to reconstruct the energy thinking mode. In the new era of social development, energy Internet is a good vision. In 2008, energy technology emerged in Silicon Valley venture capital circle + The concept of information technology, and quickly fermented into a "smart grid", energy Internet first in 2011. In, Jeremy Rifkin, a famous American scholar, put forward that with the pollution and depletion of fossil fuels to the environment, after the second industrial revolution, the scale of utilization of fossil fuels is large. It is predicted that this industrial mode will come to an end. In the future, information technology and energy technology will be deeply integrated, and a new energy utilization system will be formed with this as the feature. This is the reason.

It is the primary concept of "energy Internet". The analysis of its main characteristics includes four points: ① the main primary energy is renewable resources; ② supporting access to super large scale distributed generation system and distributed energy storage system; ③ relying on Internet technology to achieve energy sharing in a wide range of fields; ④ supporting electrification energy of transportation system. It can be seen that the proposed energy Internet mainly uses Internet technology to achieve the coordination among a wide range of energy, load and energy storage equipment. The preliminary definition of energy Internet is: energy Internet takes power system as the core, Internet and other cutting-edge information technology as the basis, distributed renewable energy as the main primary energy, and is closely related to natural gas network, transportation network and other systemsComplex multi network flow system formed by close coupling. Its basic structure and components. As shown in Fig. 1.

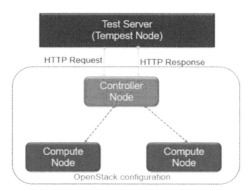

Fig. 1. Basic architecture and components of energy Internet

3   Characteristics and Demands of Energy Internet Information Communication

3.1   Analysis of Energy Internet Information and Communication Characteristics

In the past, the traditional energy network was basically in a closed state, while the energy Internet was on the contrary. Through the information network, information openness can be realized, and real-time access to energy exploitation, utilization and other information can be transmitted to various regions to achieve a balanced state of energy use.

In the past, the structure mode of energy network was top-down, while the energy Internet was bottom-up. Each decentralized small unit was in a completely equal position. In the Internet structure, no matter the users, or energy storage stations, power stations, etc. were in a situation of equal existence. Any unit of energy was adopted, it could ensure its high utilization, and the structure was more flexible.

Compared with the traditional energy network, the energy Internet in the new era is more diversified and intelligent. The platform can interact with users and lay the foundation for the realization of multi energy utilization. Its efficiency is mainly reflected in system flexibility and rapid decision-making.

3.2   Combing the Information and Communication Needs of Energy Internet

Based on the definition and characteristics of energy Internet, the realization of energy Internet information communication has the following requirements: energy Internet information communication technology must have good adaptability, on the one hand, it must have a variety of information collection functions, on the other hand, it must have the ability of flexible network access, that is, networking function. High capacity storage and efficient network transmission power lay the foundation for information sharing. Good information and data processing ability, including data screening, data analysis, etc. High efficiency decision-making ability, to make energy integration, redistribution and utilization, high efficiency decision-making is the most important. Network and information security, energy security is related to personal interests, to achieve the relationship between national interests, so we must have good network security.

4   Analysis of Key Technologies of Energy Internet Information Communication

4.1   Information Data Management Technology

Data management is mainly a series of processing of information data, including the management of the process of collection, collation and analysis [2]. Data information collection refers to the collection and collation of information from various data sources, including the collection and collation of network interface information, and the synchronous data collection and collation of a variety of online application systems. In order to ensure the good quality of the collected information data, the key is to rely on the cataloguing of information resources and the maze of data processing, establish an automatic data quality detection and control system architecture, carry out quality detection on each source data, and the whole process of detection includes the formation, use and abandonment of data, so as to ensure the quality of the collected information. In order to realize the synchronous management of data, it is based on the data model management mapping relationship and information data mapping relationship, and uses metadata to realize the mutual synchronization and migration of various data sources.

4.2   Perceptual Control Technology

From the energy Internet architecture, it can be seen that there are many types of distributed device access in the system structure, such as power grid system, complex environment and large number of system equipment, and the corresponding energy Internet equipment should have good stability and accuracy in terms of operation status or perceptual monitoring, and so on. Based on this feature, the energy Internet information system has high requirementsThe sensor network and sensor for communication needs should have independent control function chip technology. The composition of distributed equipment should also involve voltage transformer, optical current transformer and other sensor equipment, so as to improve the stability of the system. At the same time, it integrates power grid specific reliability chip technology, high-precision and low-power power communication integrated power design technology, etc. to implement accurate monitoring of power grid environment, lines, equipment, etc., so as to realize the intelligent management of the system and improve the intelligent level.

4.3   Remote Monitoring Technology

Based on the information and communication management technology of oil field, the key technology of remote monitoring is formed. Digital oil field can realize remote monitoring of energy exploitation. Through the monitoring system and supported by network technology, many information can be monitored, transmitted and analyzed, such as oil well drawing, temperature alternation, pressure change, power change, current change, etc. It can realize real-time monitoring and judgment for both oil production and exploitation progress. Based on remote monitoring technology, it can calculate the output and analyze the electric energy consumption. Through scientific calculation, it can realize the balanced operation of electric energy and pumping equipment. In general, it can provide reliable information for the comparison of overall operation parameters and optimization of production, and formulate appropriate solutions in time.

4.4   Internet of Things Technology

Internet of things (IOT) technology is gradually applied in the energy oil Internet. It can simplify the oilfield construction work, such as remote monitoring, data information collection, material management and so on. The application of IOT technology solves this complex problem to a great extent, reduces the workload, improves the efficiency, and realizes cross regional collaborative work, The Internet of things has closely linked up the trivial links in the project, so that a number of businesses can be scientifically and orderly integrated, the business process can be more optimized, and it is conducive to broaden the exploration business. It can be said that with the development of society, the Internet of things technology is a development direction of digital oilfield, The combination of information collection and intelligent technology is conducive to the deeper and more rapid development of exploration [3]. In the future, the Internet of things technology will be more mature, which can better serve the oilfield development and improvement.

The rapid development and growth of mobile commerce has brought opportunities and challenges to the development of Chinese enterprises, and also provided opportunities for enterprises to adapt to the new market competition mode and explore business opportunities. Facing the new mode of enterprise transformation, there must be many problems. As the most important challenge to mobile commerce, enterprise credit has become a bottleneck restricting the transformation of enterprises, especially small and medium-sized enterprises. This requires us to be able to find a solution to the problem. Based on the virtual nature of e-commerce and the characteristics of transactions that do not meet, it is particularly important to establish a fair and reasonable enterprise credit. Therefore, an objective and fair evaluation of enterprise credit has become the key to solve the problem. Based on the above reasons, it is of great practical significance to study the problem of enterprise credit evaluation in the mobile commerce environment, to prevent the lack of credit in the transaction, and to help Chinese enterprises face the market competition under the new business state and avoid credit risk.

The establishment and formation of a good credit evaluation environment under mobile commerce is an important factor for the stable development of mobile commerce, and an important guarantee for the rapid and stable development of market economy. The key of credit evaluation is whether it can provide a safe and reliable mobile commerce environment. Security has always been one of the most important evaluation indicators for mobile commerce enterprises, especially in China. Today, with the vigorous development of B2C e-commerce, people's life is increasingly inseparable from the influence of mobile commerce. The fast-paced life also constantly urges people to adapt to and accept mobile commerce. The convenience and personalized advantages of mobile commerce also bring great satisfaction to people's life. The proportion of mobile commerce in the national economy is more and more heavy, the simple real economy has been difficult to meet the needs of people's efficient and convenient life. Take the clothing industry as an example, clothing as a highly personalized product, under the traditional business model, people are keen to make continuous selection in different brands and different physical sales places. However, the emergence of mobile commerce makes people can spend moneySmall time cost to complete consumption activities, and the choice is far from comparable to the regional physical stores.

Therefore, for small and medium-sized enterprises, how to win their own development space in the increasingly fierce online sales market is an urgent problem for every enterprise. One of the important factors to solve this problem is to establish a good enterprise credit.For the whole mobile commerce market environment, only a good credit system in the market can guarantee the overall rapid development of the future mobile commerce market.With the rapid development of e-commerce market, the credit evaluation of enterprises, as a new and fast-growing market, has also been widely concerned in recent years. As an important part of modern finance, the development of credit evaluation industry has already been focused on in western developed countries and has achieved rapid development. Although China's credit evaluation industry started late, with the rapid development of e-commerce, especially mobile e-commerce in China, as the world's leading e-commerce development country, it objectively promotes the development and progress of credit evaluation industry. As a third-party evaluation service enterprise independent of enterprises and consumers, it is striving to provide accurate, authentic and customer-oriented services for both suppliers and sellers. The evaluation of e-commerce concept is more conducive to the healthy operation and development of e-commerce industry, to the e-commerce platform with high credit, to the safer consumption transaction network environment for enterprises and consumers, and to the steady growth and cultivation of consumption market scale.

5   Strengthen the Specialization of Power System Maintenance

With the development of the times, the large-scale consumption of non renewable resources, seeking resource reuse and efficient utilization has become an important problem that the energy industry urgently needs to solve, and it is also the inevitable demand of social development. All energy Internet is a complex multi network flow system based on power system and information communication. It aims to achieve the new goal of renewable energy as the main primary energy and wide area energy sharing. To achieve this development situation, it must rely on advanced technologies such as information data management technology, perception control technology, remote monitoring technology and Internet of things technology. Energy Internet to achieve efficient application of energy.

At present, communication colleges and universities have basically not set up communication power supply specialty, which leads to the fault of power supply maintenance team and the lack of professional maintenance personnel in communication enterprises [4]. In order to solve the maintenance difficulties of communication power supply, including a variety of systems (high voltage, low voltage, rectifier, inverter, battery, oil engine, grounding) and professional equipment (room air conditioning), we can entrust professional management institutions and technical personnel specialized in communication power supply to provide guidance and technical services for power supply maintenance, so as to solve the current dilemma of the shortage of professional maintenance personnel for communication power supply.

6 Strengthen the Specialization of Power System Maintenance

It is necessary to formulate a detailed and feasible emergency plan to deal with emergencies, be busy when problems arise, assign the responsibility to every maintenance technician, and concentrate everyone's strength and wisdom to do a good job in the management and maintenance of communication power supply. To sum up, the stability of communication power supply work is the basic requirement for communication power supply work. Professional technical management and maintenance personnel should improve their theoretical knowledge level and accumulate practical work experience,To develop a good habit of diligent thinking, the power management and maintenance work, from different angles, study the problems existing in equipment maintenance work, continuous improvement and development, in order to ensure the safe and stable operation of communication power supply, that is to ensure the safe and stable operation of communication network.

References

1. Sun, J.: Research on signaling system of video communication system based on H.248 protocol. Shandong University of Science and Technology (2007)
2. Ren, J.: Key technology analysis and application of wireless communication transmission system in container terminal. Shanghai Maritime University (2006)
3. Fang, X.: Analysis, design and implementation of key technologies in physical layer and data link layer of IP based SDH management network protocol stack. University of Electronic Science and Technology (2005)
4. Luo, C.: Analysis and research on some key technologies for improving the reliability of CDMA mobile communication system. Hunan University (2003)

Research and Application of CRM Model in Enterprise Management System

Zhe Yang and Xuebin Huang$^{(\boxtimes)}$

Hainan Tropical Ocean University Sanya, Sanya 572022, HaiNan, China

Abstract. Customer Relationship Management (CRM) is a new management mechanism which aims to improve the relationship between enterprises and customers, improve customer loyalty and satisfaction. It is also a kind of management software and technology. Data mining technology can mine a large number of data and information, find potential high value relationships, and use patterns or knowledge to predict customer needs, help enterprises make decisions. Successful data mining is the key to the success of customer relationship management system.

Keyword: Customer relationship management · Data mining

1 Introduction

With the gradual aggravation of market economy, the products have no further differentiation, the customers' purchasing behavior has become more rational, and the customer relationship management has been paid more and more attention by enterprises. For enterprises, whether they can make rational and effective use of customer data has become the basis and key for them to gain advantages in the competition.

2 Data Mining Process in CRM

Before the implementation of data mining, in order to make the CRM system to establish a good model, it is necessary to formulate what steps to take and achieve the goal of each step. Only with a good plan can we ensure the orderly implementation of data and achieve success. Although the steps are arranged in order, it should be noted that the process of data mining is not linear. To achieve good results, we need to repeat these steps [1].

2.1 Building Data Mining Database

Collect all the data to be mined into a database. Note that this does not mean that you have to use a database management system. According to the size of the amount of data to be mined, the complexity of the data, and the different ways of use, sometimes a simple flat file or spreadsheet is enough [2]. Another reason why you need to build an independent database is that the data warehouse may not support the data structure that

J. C. Hung et al. (Eds.): FC 2021, LNEE 827, pp. 1535–1539, 2022.
https://doi.org/10.1007/978-981-16-8052-6_223

you need for complex data analysis. This includes statistical query, multidimensional analysis and various complex charts and visualizations.

2.2   Modeling

Modeling is an iterative process. We need to carefully examine different models to determine which one is most useful for business problems. What we learn in the process of finding a good model will inspire us to modify the data and even change the original definition of the problem. In order to ensure the accuracy and robustness of the model, we need a well-defined training verification protocol, sometimes called guided learning. The main idea is to use part of the data to build the model, and then use the remaining data to test and verify the model.

2.3   Implementation

The knowledge represented by data mining model is applied to practical work to provide support for decision-making. For example, some triggers can be set according to the obtained knowledge, and special processing can be carried out when the conditions are met. In the establishment of CRM application, data mining is often a small but significant part of the whole product.

In order to verify the optimization strategy of customer relationship management based on data mining technology, a prototype verification system sm-crm (supermarketcrm) is designed based on a supermarket management system. The idea of customer relationship management runs through the whole implementation process of the system. The system solves the problems of data inconsistency, easy to lose, only discrete simple data query in traditional customer relationship management, verifies the feasibility of the application of the proposed model in enterprise management system, makes a useful attempt for the application of data mining technology in the commercial field, and provides a feasible technical way to improve the timeliness of customer relationship management. Supermarket management system sm-crm (supermarketcrm) is developed under the environment of Windows 2000, SQL Server 2000 and C#, which adopts the current popular C/S structure. The front-end client system is developed by C#, and the back-end database is developed by SQL Server 2000. The system structure of sm-crm is shown in Fig. 1. The following is a brief description of the basic functions of sm-crm system: the basic functions of supermarket customer relationship management system sm-crm include four functional modules: sales management, customer management, market management, analysis and decision-making, as shown in Fig. 2.

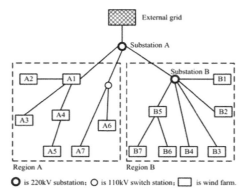

Fig. 1. System structure of sm-crm

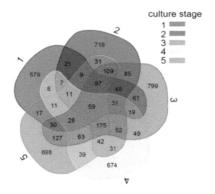

Fig. 2. Basic function diagram of sm-crm

3 Sales Management

Work management involved in order signing and sales completion with customers or partners. Order is the main data carrier of enterprise sales business, and also the main data analysis source of customer relationship management, so order management is an important part of customer relationship management system. Return management is mainly used to define, maintain and query the customer's return order. In the sales process, sm-crm can provide customers with relevant information in time to increase the understanding of the salesman to customers.

3.1 Customer Management

Customer information contains great value. Through the analysis and mining of customer information, we can deeply understand the needs of customers; find the rules of customers' transactions; find the rules of the composition of value customers and so on. These information will be of great significance to the correct decision-making and the improvement of business. Customer information management module can complete the function of inputting, modifying and querying customer information. Through the feedback processing, it mainly completes the input, maintenance, closing and query functions of customer service feedback. At the same time, a customer feedback can be decomposed into multiple tasks. Through the function of customer object mining and the analysis of the historical transaction information of customers and partners, customer care can find out the customers and partners value customers, which are crucial to the enterprise's turnover and profit generation; it can find out the customers and partners value changing customers, which have increased or decreased transactions with the enterprise. We can find out the customers and partners who are dissatisfied with the products or services of the enterprise problem customers [3]. At the same time, the system makes timely suggestions on customer care through analysis, which makes it possible for enterprises to consolidate old customers and improve their satisfaction and loyalty.

3.2   Market Management

Market management is the behavior management of opening up channels for sales and creating pre-sale, sale and after-sale environment. According to the analysis of the market situation, build the market activity project, and edit the feedback information of the project.

Objective to evaluate the implementation. Market information mainly completes the information input, maintenance and query functions of enterprise market activities through new market activities.Competition management is the unified management of competitors' information, which is mainly used to define, maintain and query the enterprise information that forms a business competitive relationship with the enterprise.Including the basic information of competitors, competitors' products and product comparison.Price positioning management is mainly used for product pricing on sales orders. This system mainly provides five methods, namely partner price, member price, cash price, promotion price and employee price.

3.3   Analysis of Resolution

Data mining, statistics and analysis of various information of customers, partners, competitors, market, sales, services and products are carried out to provide decision-making basis for enterprise development.Decision making is the work that enterprise managers must do frequently.Decision making is of great significance to enterprises. It can be said that the level of decision-making determines the competitiveness of enterprises.Sm-crm uses OLAP and data mining methods to provide users with a variety of analysis and prediction tools [4]. It can easily classify and count customers, products, processes, tasks, budgets, plans, expenses and other information, so as to analyze the operation of sales, markets and services, and make scientific and correct decisions.

4   Epilogue

Integrity is not only an important part of China's socialist core values, but also a necessary condition for enterprises to maintain long-term vitality in response to market competition.Even under the restriction of legal norms, the cultivation of moral behavior of enterprise integrity is still an important link in the construction of the guarantee mechanism.The construction of good faith credit mechanism is an important part of the closed-loop system which forms a good credit system outside the legal constraints. In the history of our country, the famous micro business, Shanxi business, Zhejiang business and so on in the commercial culture, although there are many differences in the region, business mode and so on, but no other inherits the excellent culture of the Chinese nation about good faith, and it is only after hundreds of years of inheritance and follow that they are constructedLater generations have paid close attention to these problems.

References

1. Yang, M.: Design and implementation of customer relationship management system of postal savings bank based on .Net. University of Electronic Science and Technology (2020)
2. Yang, J.: Research on as company's CRM system satisfaction. Hunan Normal University (2020)
3. Liu, L.: Research on marketing information management of LH company. Yanshan University (2020)
4. Zhu, Y.: Research on sales logistics business process optimization of M company. Beijing Jiaotong University (2020)

Synchronous Leader Election Algorithm in Hierarchical Ad Hoc Networks

Wenmei Yu[✉]

Department of Social and Cultural Studies, Weifang Municipal Party School,
Weifang 261000, China

Abstract. In recent years, ad hoc network has been widely studied and applied because of its convenience, quickness and not restricted by network infrastructure. However, in the process of video information transmission in hierarchical ad hoc network, due to the mobile nodes and network conditions, the cluster head nodes may be missing, which affects the normal communication of the network. Aiming at the poor mobility of high-level nodes in hierarchical ad hoc network, a synchronous leader election algorithm based on hierarchical ad hoc network is designed and implemented to solve the above problems. The system also introduces the vice chairman mechanism, and the experiment shows that when there are vice chairman nodes in the network, the election time will be significantly shortened. The algorithm can also be applied to small temporary video conference system to solve the problem of host missing and replacement.

Keywords: Network · Distributed algorithm · Synchronous leader election algorithm · Video conference

1 Introduction

Leader election algorithm is a classic algorithm in distributed computing. When several nodes in a group need to choose a leader or when an existing leader needs to recommend a new leader unexpectedly, it can run some algorithm to elect a new leader properly for coordination and management.

Leader election algorithm is widely used in many fields, such as process scheduling in wired network, multiprocessor computing, etc. The running environment of the algorithm is mainly divided into synchronous network and asynchronous network, and the complexity of synchronous network is lower than that of asynchronous network. In view of the poor mobility of high-level nodes in hierarchical ad hoc network, this paper designs and implements a synchronous leader election algorithm based on hierarchical ad hoc network, which is used to solve the leader election between relatively fixed and static nodes in ad hoc network,It shows the advantages of high-level synchronous election algorithm in hierarchical ad hoc network [1].

In recent years, with the popularity and development of wireless network, more and more applications begin to be based on wireless network. The emergence and development of ad hoc network also put forward a new topic for leader election algorithm. How to apply leader election algorithm in ad hoc network environment has become a

J. C. Hung et al. (Eds.): FC 2021, LNEE 827, pp. 1540–1545, 2022.
https://doi.org/10.1007/978-981-16-8052-6_224

hot topic. At present, some journals and academic conferences have published articles on leader election in ad hoc networks.

2   Theoretical Research on Leader Election Algorithm

2.1   Leader Election Algorithm

Leader election comes from the research of token ring. In this kind of network, a "token" circulates in the network, and the current holder of the token has the unique privilege to initiate communication. If two nodes in the network try to communicate at the same time, they will interfere with each other. However, sometimes the token is lost, so it is necessary for the process to execute an algorithm to regenerate the lost token.

Generally, an election process needs two stages: the stage of selecting a leader with the highest priority and the stage of informing other processes who is the leader (winner).

Most of the election algorithms are based on global priority, and each process assigns a priority in advance. These algorithms are also called extreme finding algorithms. The disadvantage is that once a leader is selected, it must be a correct choice. Recently, more work has been done on priority based algorithms, in which the election is more generally based on individual preference, such as locality, reliability estimation and so on.

2.2   Synchronous Network

The simplest model in distributed network is synchronous network model, in which all nodes (processors) communicate and compute in the synchronous wheel. In a network environment, it is assumed that all nodes are on graph G and communicate with its neighbors by sending messages along the edge of G.

The synchronous network system consists of a group of computing elements located at the nodes of the directed network graph. These elements can be hardware (processor) or logical software (process). The description of synchronization network can be expressed by state set and message function. Each process has a set of states [2].

The whole synchronization system starts when all processes are in any start state and all channels are empty. After that, all processes repeat the two steps of sending messages to neighboring nodes and receiving messages at each time step, which are collectively called round. In addition, the model discussed in this paper is deterministic, so if a specific set of initial states is given, the calculation will be carried out in a unique way.

For synchronous distributed algorithms, two complexity measures are usually considered: time complexity and communication complexity. Generally speaking, time measurement is more important in practice. This paper will also measure the overall performance of the leader algorithm by time complexity.

There are many leader algorithms based on synchronous network architecture, such as LCR algorithm, time slice algorithm and so on.

2.3   Application of Hierarchical Leader Election Algorithm

1) Application in small ad hoc video conference;
2) Application in the field of rescue and disaster relief.
3) Establishment of hierarchical ad hoc network.

The establishment process of the network is described as follows:

1) Each node detects the surrounding nodes and records the connectable nodes as its own neighbor nodes.
2) The root node sends a setup message containing the number of network layers to be constructed to its neighbor nodes. For each node, there is a setup visited variable to identify whether the node has received a setup message.
3) After receiving the setup message, the node first compares it with its own clusterid. If it is not equal, it indicates that it is not in the network layer to be built, then discards the message; if it is equal, it sends the setup message to the root node for confirmation, and registers the relevant information of the root node in the leader structure. After receiving the setup message, the root node also stores the information of the child node in the children table.

Finally, the system network structure is established as shown in Fig. 1.

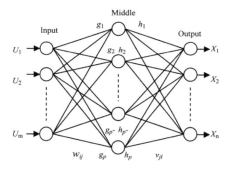

Fig. 1. System network structure

3   Leader Election Algorithm in High Level Ad Hoc Networks

3.1   System Requirements

1) Node identifier

For the hierarchical structure of ad hoc network, because different levels will lead to different algorithms and priorities, cluster ID and node ID are used to identify a node at the same time.

2) Node weight nwv (nodeweightedvalue)

The weight of each node is used to indicate its own state, and its function is to provide the basis for nodes to compare with each other and then elect the local and global leaders [3].

3) Reliable network transmission

In the high-level leader election algorithm, the network topology should be stable, that is, no nodes leave (there is no limit before and after the election), and each node remains relatively static. The communication between nodes is bidirectional, and the performance of each node's network and host is considered to be good, without packet loss due to buffer size or network reasons.

4) Clock synchronization

Since the structure of the layered ad hoc network discussed in this paper has been specified during the network initialization, the first leader, the root node, can carry out the initial clock synchronization. After that, every time the leader command is sent, the network clock synchronization should be carried out first to ensure the smooth operation of the synchronous network.

The evaluation value of the achievement degree of the limited course group is obtained by weighted summation according to its weight proportion:

$$D_{i,j} = \sum_k \left(W_{i,j,k} \cdot D_{i,j,k} \right) \tag{1}$$

The evaluation value of achievement degree of each decomposition index point is obtained by taking its minimum value.

$$D_i = \min_j \left(D_{i,j} \right) \tag{2}$$

3.2 High Level Leader Election Algorithm

1) Node identifier

For the hierarchical structure of ad hoc network, because different levels will lead to different algorithms and priorities, cluster ID and node ID are used to identify a node at the same time.

2) Node weight nwv (nodeweightedvalue)

The weight of each node is used to indicate its own state, and its function is to provide the basis for nodes to compare with each other and then elect the local and global leaders.

3) Reliable network transmission

In the high-level leader election algorithm, the network topology should be stable, that is, no nodes leave (there is no limit before and after the election), and each node remains relatively static. The communication between nodes is bidirectional, and the performance of each node's network and host is considered to be good, without packet loss due to buffer size or network reasons.

4) Clock synchronization

Since the structure of the layered ad hoc network discussed in this paper has been specified during the network initialization, the first leader, the root node, can carry out the initial clock synchronization. After that, every time the leader command is sent, the network clock synchronization should be carried out first to ensure the smooth operation of the synchronous network [4].

3.3   High Level Leader Election Algorithm

The algorithm mainly includes the following six kinds of messages:

1) Hello message, which is used by the child nodes in the high-level network to detect whether the leader node is available.
2) ACK Hello message is the response of leader node to Hello message.
3) After detecting the leader loss, it sends an election message to the neighbor node.
4) Confirm message, compares the newv values in all the element messages, and sends a confirm message to the final selected node with different ID numbers to confirm that it is its parent node.
5) The ad opt message is used to respond to the node that has sent the confirm message.
6) The leader message is used for the leader node to broadcast that it is a leader in the network.

4   Conclusion

In order to solve these problems, this paper proposes an online management scheme based on the concept of engineering education professional certification, which uses information technology to manage the information related to the certification process, The evaluation system of graduation requirement achievement and continuous improvement is constructed to realize the information management of certification work, improve the efficiency of data maintenance, and provide effective data support for decision-making.

Establish a management mechanism to achieve graduation requirements. The weight model of curriculum and graduation requirements and the relationship model of achievement items and curriculum graduation requirements are established to provide model support for the calculation of achievement degree. To achieve the teaching class curriculum achievement degree, curriculum achievement degree, professional achievement degree, administrative class achievement degree and personal achievement degree value calculation, select the appropriate data visualization technology to display the results, so that users can get more information through chart analysis. Establish continuous improvement evaluation management mechanism. According to the requirements of continuous improvement in the process of engineering education professional certification: establish the evaluation mechanism of graduation requirements, regularly evaluate the graduation requirements; establish the quality monitoring system of teaching process, regularly evaluate the continuous curriculum reform; establish the tracking feedback system of graduates, regularly evaluate the achievement of training objectives. According to the characteristics of high-level network in hierarchical ad hoc network, this paper designs a high-level synchronous leader election algorithm with vice chairman mechanism by comparing and improving the classical leader election algorithm. Through experiments, the performance of this high-level synchronous leader election algorithm is analyzed, which shows that its advantage is that for the relatively static network topology, the leader election time is predictable and controllable, and the introduction of vice chairman mechanism will improve the performance of the election algorithm several times or even dozens of times.

References

1. Ni, K., Jin, S., Sun, C., Zhang, Y., Sun, Y.: Certification system of higher engineering education in developed countries and its enlightenment. High. Sci. Educ. (05), 51–55 (2011)
2. Li, Z.: Analysis of the concept of continuous improvement of engineering education accreditation. China High. Educ. (Z3), 33–35 (2015)
3. Ning, B.: Promote the construction of "double first class" with professional certification as the starting point. High. Educ. China (z1), 24–25 (2017)
4. Wang, L., Chen, H., Xie, Y., Wu, S., Fu, W.: An overview of the evaluation method of graduation requirements for engineering education certification. Educ. Teach. Forum (10), 70–71 (2019)

Research on Recognition of Accounting Information Distortion of Listed Companies Based on Classification Regression Tree Algorithm

Xiuhua Zhang[✉]

Shaanxi Fashion Engineering University, Shaanxi 712000, China

Abstract. Wusing 26 financial variables to establish the classification regression tree model to identify the accounting information distortion, the results show that the correct recognition rate of the established model for the accounting information distortion enterprises is more than 80%, and the second type error rate can be controlled below 20%. It is also found that companies with retained earnings less than 2% of total assets are prone to accounting information distortion.

Keywords: Accounting information distortion · Classification regression tree · Data mining

1 Introduction

Accounting information is the data and information that people use accounting theory and method in the process of economic activities and obtain through accounting practice to reflect the value movement of accounting subject. These data and information reflect the value movement of enterprise.

The profitability, solvency and capital flow of the industry. Accounting information distortion refers to the loss of relevance, reliability, timeliness and importance of accounting information due to the subjective or objective reasons of the company [1]. Most of the methods used are logit, probit model and artificial neural network (ANN): Beasley (1996) used logit regression model to calculate the possibility of accounting information fraud. Hansen (1996) established a generalized qualitative response model integrating logit and probit technology. Summers and Sweeney (1998) established the model with cascading logit method, and the prediction accuracy reached 67%. Bell (2000) also made use of Logit to identify the distortion of accounting information.

2 Empirical Research

The neural network constructed by green (1997) can identify up to 85% of accounting information fraud enterprises and 79% of accounting information real enterprises. Fanning (1998) used four methods in empirical research, namely, step log it, linear

J. C. Hung et al. (Eds.): FC 2021, LNEE 827, pp. 1546–1552, 2022.
https://doi.org/10.1007/978-981-16-8052-6_225

discriminant, quadratic discriminant and neural network. The comparison results show that the neural network model has higher accuracy than the other three methods. Domestic scholars have also made quantitative research on accounting information distortion by using various models.

2.1 Research Hypothesis

Based on the hypothesis of Zhang Ling (2006). This paper classifies the companies with non-standard audit opinion (2) into the companies with accounting information distortion (sample 1), and the listed companies with standard unqualified audit opinion (sample 0).

2.2 Classification Regression Tree

Cart is a supervised learning algorithm. Users first provide a learning sample set to construct and evaluate the cart, and then use it. The learning sample set is as follows:

$$L = \{X_1, X_2, ..., X_m, Y\} \tag{1}$$

$$X_1 = (x_{11}, x_{12}, ..., x_{1t}), ..., X_m = (x_{m1}, x_{m2}, \tag{2}$$

$$..., X_m = (x_{m1}, x_{m2}, ..., x_{mt_n}) \tag{3}$$

$$Y = (y_1, y_2, ..., y_k) \tag{4}$$

The establishment of classification regression tree model consists of the following three parts: step 1 constructs classification regression tree. In this paper, Gini index is used to find the best branching rule. If the set t contains n categories of records, then the coefficient is equation

$$Gini(T) = 1 - \sum_{i=1}^{N} p_j^2 \tag{5}$$

If the set t is divided into two parts N1 and N2 under the condition of X, then the Gini coefficient of this partition is equation.

3 Data Classification

Starting from the root node, the segmentation is repeated recursively for each node. Firstly, the optimal segmentation point of each attribute is selected for each node. If Gini split (XI) = min, then Xi is the optimal segmentation point of the current attribute. Then, among these optimal segmentation points, the one with the smallest formula is selected as the optimal segmentation rule of this node;Finally, we continue to segment the two nodes [2]. The segmentation process continues until any of the following conditions are met: each node is very small; pure nodes (y of samples inside nodes belong to the same category); only unique attribute vector is used as branch selection. The pruning of Step 2 classification regression tree. Over fitting occurs in the process of complete growth of classification trees [3]. You need to cut off the leaf knots.

The model can classify the new data more accurately and effectively. In this paper, the minimum cost complexity pruning principle of cart system is used to prune the classification regression tree. Finally, a finite ordered subtree sequence with decreasing number of nodes is constructed. The measurement is as follows:

$$Ra(T) = R(T) + a * |T| \tag{6}$$

4 Sample Selection

4.1 Classification Regression and Prediction

The sample selection of accounting information distortion companies in this paper is based on the following criteria: one of the three opinions in 2006: qualified opinion with explanatory note, negative opinion and refusal/inability to express opinion; if unqualified opinion with explanatory note is issued in 2006, there are at least two non-standard audit opinions from 2002 to 2006. Finally, 98 enterprises with false accounting information in 2006 are selected. The matching samples were selected according to the following criteria: listed companies that were issued with standard unqualified opinions from 2002 to 2006; listed companies that were not listed by St/Pt from 2002 to 2006; A-share or B-share had the same attributes. According to the above criteria, 98 paired samples were selected. In order to test and compare the recognition ability and prediction accuracy of classified regression tree model and multiple discriminant analysis model on accounting information distortion of listed companies, the author took the annual audit reports of A-share and B-share of Shanghai Stock Exchange and Shenzhen Stock Exchange in 2007 as the standard, and made classified regression and prediction on the samples in 2007. A total of 1624 valid samples were selected, as shown in Table 1.

Table 1. Statistics of training samples and test samples

Training sample training sample test sample test sample Quantity percentage of category (%) quantity percentage
Accounting information authenticity (0) 98 50.00 1523 93.78
Accounting information distortion (1) 98 50.00 101 6.22
Total 196 100.00 1624 100.00

4.2   Modeling Financial Index Selection

The index variables reflecting the distortion of accounting information mainly include two categories: one is the quality index reflecting the internal control of the enterprise; the other is the quantitative index indicating the operation status of the enterprise and generally obtained through the accounting statements of the enterprise. Based on the research results of domestic scholars, this paper focuses on the use of quantitative indicators of business operation to establish a classification regression tree model. As shown in Fig. 1. Based on this, this paper summarizes 26 financial indicators with high recognition ability for accounting information distortion. Among them, x1–x5 is the debt paying ability index, x6–x9 is the asset management ability index, x10–x15 is the profitability index, x16–x18 is the development potential index, x19–x22 is the equity structure index, and x23–x26 is other indexes with accounting information content.

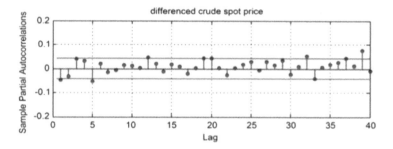

Fig. 1.   Modeling financial index selection

5   An Empirical Study on the Identification of Accounting Information Distortion

5.1   Classification Regression Tree Model

The classified regression tree model does not need to select variables in advance, and can identify outliers. There are no strict requirements for variable attributes, and there are no requirements for the distribution of training samples. Therefore, this paper takes all 26 financial indicators as the attribute variables of the classified regression tree for modeling and analysis directly [3]. The research data are mainly from tianruan financial statistical database (tinysoft financial analysis.Net), sina.com, etc.

Step 1: growth of classified regression tree as shown in Table 2. Taking Gini coefficient as the splitting criterion, the maximum classification regression tree with 9 leaf nodes is obtained, which takes 7 indexes such as X3, X6, x9, X10, X12, x18 and X25 as the splitting variables. The empirical results show that x18 (retained earnings/total assets) has the most significant ability to identify the distortion of accounting information, and is selected as the root node branch variable by the model.

Step 3: classification regression tree evaluation. Through the improvement of the model, we can see that the over fitting phenomenon of the model has been effectively

Table 2. Variable index name

Code indicator name code indicator name
X1 interest cover factor X8 inventory turnover rate
X2 asset liability ratio x9 inventory/current assets
X3 working capital/total assets X10 net profit on sales
X4 quick ratio X11 return on total assets
X5 current ratio X12 main income/total assets
X6 accounts receivable turnover x13 main profit/total profit
X7 growth rate of accounts receivable x14 gross profit rate of sales
Proportion of three related party assets occupied by X22
X16 growth rate of main income x23 effective income tax rate
X17 long term investment growth rate x24 Z value

controlled; the recognition ability of the model has been significantly improved, and the recognition accuracy of the company with distorted accounting information has reached 80.2%, and the recognition accuracy of the company with real accounting information has reached 81.16%. The accuracy of model classification is shown in Table 3.

Table 3. Comparison between this study and Foreign Studies

Number of research method variables type II error rate (%)
2007 2006 2005 2004 2003
This study classified regression tree 1
19.80 21.26 16.67 24.06 20.69

5.2 Validation of Results

Reuslt 1: in the process of empirical research, the author finds that the proportion of retained earnings in total assets is 2%, which can be used as the dividing point of accounting information distortion or not, that is, the ratio of (retained earnings/total assets) ≤ 0.02.

Enterprises are judged as accounting information distortion, and enterprises with (retained earnings/total assets) ≥ 0.02 are judged as accounting information authenticity. In order to verify whether the result is universal and generalizable, the author extracts all the valid data of listed companies from 2000 to 2007, and makes a robustness analysis of the conclusion [4]. It can be seen that the overall accuracy recognition rate of the results for the eight years data is relatively stable, which can basically reach 80%; for the eight years (2000–2007) of accounting information distortion, the accuracy recognition rate of enterprises can be stabilized at more than 64%, and for the past five years (2003–2007) it is basically stable at more than 75%; the second type error rate of the model in the past five years (2003–2007) is basically controlled at about 20%,This is difficult to achieve in previous studies. In a word, the classification regression tree can find indicators with such high recognition ability in the process of data learning, which fully shows the strong learning ability of the classification regression tree model.

This paper makes an empirical study on the distortion of accounting information of Chinese listed companies by using the classified regression tree model (CART), and tests the classification ability of the model by using the financial data from 2000 to 2007. The author finds that: (1) the model has a high ability to identify the distortion of accounting information of Listed Companies in China; (2) the second type error rate of the model is relatively low, which can be controlled below 20%; (3) the classified regression tree model is a nonparametric nonlinear method, which has no requirement on the probability distribution of the target variables, so it is easy to use and the results are easy to understand; (4) it is found that the proportion of retained earnings in total assets is quite different between enterprises with accounting information distortion and real enterprises. Through the empirical test of data from 2000 to 2008, it is found that this conclusion is very stable. When the index value does not exceed 0.02, the regulatory authorities should focus on whether there will be accounting information distortion. The conclusion of this study is worth promoting. It can provide a very convenient and practical method for audit regulators and investors, improve audit efficiency and reduce audit cost.

6   Summary

Based on multi label classification and multi-objective regression, this paper attempts to study and propose the application of multi label classification method to multi-objective regression prediction. More importantly, whether the multi label classification learning method can be extended to multi-objective regression prediction depends on whether the basic target classification algorithm can deal with the exchange of classification and regression data. In this paper, we propose two methods MTS and ERC to solve the multi-objective prediction, and propose their improved algorithms MTSC and ERC.

References

1. Wang, S.: Research on face detection algorithm based on image samples. Nanchang University (2016)
2. Wang, Z.: Customer segmentation and loss prediction based on data mining. Beijing University of Technology (2016)
3. Wang, X.: Research on parallelism of classification regression tree algorithm based on spark. Chongqing University (2016)
4. Gai, W.: Intelligent driving of high speed train based on integrated classification regression tree algorithm. Beijing Jiaotong University (2015)

Statistical Analysis of Ethnic Minority Villages in Border Areas Based on Genetic Algorithm

Yingjie Zhang[✉]

Party School of the Yuxi Municipal Committee of the Communist Party of China
Co-management Teaching and Research Office, Yuxi 653100, Yunnan, China

Abstract. Since 2009, the National Committee for democracy and people's livelihood and the Ministry of finance have implemented the protection and development project of villages with ethnic characteristics, many villages have become the main industry of tourism, which has greatly promoted the development of local economy, culture and other industries. The natural and cultural environment of the villages has been improved. However, there are still some problems in the process of village development, such as unreasonable and inadequate development, excessive reliance on social financing for the development of cultural industry, and damage to the natural environment of the village. To solve these problems is the key to realize the sustainable development of the village.

Keywords: The status quo of ethnic minority characteristic villages · Problems

1 Introduction

The village with ethnic minority characteristics refers to the natural village or administrative village with relatively high proportion of ethnic minority population, relatively complete production and life functions, and obvious cultural and settlement characteristics of ethnic minorities.

Firstly, the villages with ethnic characteristics are an important part of ethnic culture, and the protection and development of the villages with ethnic characteristics is an important part of the development of ethnic culture. Secondly, the villages with ethnic characteristics are an important industry of ethnic economy. Coordinating and planning the development of villages can greatly promote the economic development of ethnic minorities, and promote the realization of poverty alleviation and prosperity in ethnic areas [1]. The gap between urban and rural economic development is small.

2 The Development Status of Villages with Ethnic Characteristics

2009 In 2005, the relevant departments of the state began to implement the protection and development project of villages with ethnic characteristics, and then set up a special fund for the development of ethnic minorities. At the same time, it attracted various funds, and carried out pilot projects in 370 villages in 28 provinces,

© The Author(s), under exclusive license to Springer Nature Singapore Pte Ltd. 2022
J. C. Hung et al. (Eds.): FC 2021, LNEE 827, pp. 1553–1559, 2022.
https://doi.org/10.1007/978-981-16-8052-6_226

autonomous regions and cities across the country, and achieved remarkable results. According to the statistics from the State Commission for democracy and people's livelihood, as of 2018, more than 2000 characteristic villages construction projects have been implemented nationwide, and less than 1057 "Chinese ethnic minority characteristic villages" have been named and listed by the State Commission for democracy and people's livelihood. This paper will briefly describe the development status of ethnic minority characteristic villages from the perspectives of economy, culture and environment.

2.1   Tourism Becomes the Main Industry

In recent years, with the opening of high-speed railway, the construction and improvement of highway network, and the strong support of national policies, tourism has become an emerging industry in many ethnic minority villages. For example, "Nongjiale" in Zhongliao village, Sanya, Hainan, "mujiale" in Sanhe Village, Ulan hada Town, Inner Mongolia, "folk tourism" in Jindalai village, Yanbian. The success of village tourism has also successfully promoted the development and prosperity of other industries in the village. The vast majority of villages have integrated the unique minority culture, developed the cultural industry with regional and national characteristics, ecological tourism, national handicraft processing industry, unique traditional ethnic skills, etc., promoted the economic development of the village and improved the living conditions of the villagers. The market demand of traditional Zhuang brocade is shrinking. The traditional national textile technology has strong practicability and is one of the life skills that people rely on. It can not only meet the needs of people's life, but also create rich and colorful cultural wealth. However, with the development of social economy and the improvement of people's living standards, the aesthetic level and spiritual consumption level of the public are constantly improving. At the same time, a large number of high-quality and low-cost industrial textiles appear on noodles, which has become a new mainstream of consumption fashion. Modern industrial products greatly surpass the traditional handicrafts in quantity and variety. Due to the high price and old style, the traditional Zhuang brocade handicrafts have a smaller and smaller market share [2]. Zhuang brocade handicrafts no longer have an advantage and gradually fade out of the market.

2.2   The Ethnic Culture of the Village is Well Preserved

The culture of villages with ethnic characteristics is mainly reflected in the national culture. The minority culture preserved in the village is rich and characteristic. The material culture includes food and clothing, village buildings and production tools; the spiritual culture includes minority language, writing, art, philosophy, religion, customs and festivals. Ethnic culture is well preserved and displayed in the villages, such as GouLan Yao village, Lanxi Yao Township, Yongzhou, Meicun village, Jinxiu County, Guangxi, chatiao village, Antu Korean Village, Yanbian, Jilin, etc. To introduce ornamental sports events with Korean characteristics is an innovative move to carry out sports targeted poverty alleviation in Yanbian border minority areas. Ball is a business card of Yanbian people. "In Yanbian, football has a broad mass base, and the concentration of football talents is called 32". In addition, Shanghai Lanjiang football characteristic town "is the only key base of characteristic football in Jilin Province. Therefore, it can link the targeted poverty alleviation of sports in Yanbian border minority areas with watching football games, and promote the consumption of football games by watching football games, Drive economic development and help get rid of poverty. In addition, such as wrestling, autumn tau and other Korean folk sports, which are difficult to operate, can not only introduce and develop the Korean sports culture, but also promote the development of primary and tertiary industries while watching the events. This is an effective way to make full use of the characteristics of Korean national sports to improve poverty alleviation, At the same time, it is also an effective channel to broaden the economic income of the poor.

3   Connotation and Practical Significance of Sports Targeted Poverty Alleviation in Yanbian Border Minority Areas

3.1   Connotation of Sports Targeted Poverty Alleviation in Yanbian Border Minority Areas

The precise poverty alleviation from sports helps to realize China's Rural Revitalization Strategy, innovate poverty alleviation model, improve the efficiency of poverty alleviation and consolidate the achievement of poverty alleviation. The connotation of sports precision poverty alleviation is guided by Xi Jinping's socialist ideology with Chinese characteristics in the new era, especially with the strategy of promoting rural revitalization and poverty alleviation and development. We should firmly establish the development concept of innovation, coordination, green, open and sharing, give full play to the comprehensive driving effect of sports, help to get rid of poverty by introducing sports events, developing sports industry, assisting in the construction of infrastructure, and carrying out mass fitness. We should build a "sports ←" or " ← sports" model in the poverty-stricken areas of Yanbian border ethnic minorities, and create a good situation in which education helps Rural Revitalization and targeted poverty alleviation, So that the poor people and poor areas can get rid of poverty and become rich, realize rural revitalization, and join the whole country in entering a well-off society in an all-round way.

3.2   Yanbian in Economic Difficulties

There are five poverty-stricken counties in border minority areas. Among them, there are 4 key counties for national poverty alleviation and development work, 1 poverty-stricken county designated by the province, and 50829 poverty-stricken people have been registered. Since 2015, Yanbian minority areas have made great efforts to build a four beam and eight column policy support system, A series of policy documents, such as the implementation plan for comprehensively promoting poverty alleviation in Yanbian Prefecture, the management measures for income distribution of industrial poverty alleviation projects in Yanbian Prefecture, the 13th five year plan for poverty alleviation in Yanbian Prefecture, and the guiding opinions on poverty alleviation in non poor villages in Yanbian Prefecture, have been published, We will gradually implement various industrial poverty alleviation projects at all levels, with a total investment of 9.006 billion yuan [3]. At the same time, we should innovate the "distribution according to work" mode, implement the "five party linkage mode" of "leading enterprises + specialized cooperative organizations + poor people + financial institutions + insurance companies", establish the "982" poverty alleviation product identification mechanism, and carry out the "four not picking" drip irrigation "and" menu "assistance mode, so as to reduce the incidence of poverty from 37% in 2016 to 0% in 2020 (Fig. 1).

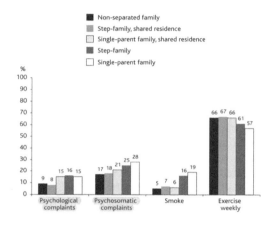

Fig. 1. Main models of Yanbian economic difficulties.

For a long time, the development of Yanbian border minority areas is relatively backward due to the constraints of economy, society, culture, history, nature, geography and so on. When carrying out sports targeted poverty alleviation, there are still several difficulties in the system level: 1) sports targeted poverty alleviation is still lack of accuracy, which is reflected in the lack of complete sports targeted Poverty Alleviation Policies, security system and rules and regulations. 2) Due to the lack of publicity, people usually only associate sports with health, and their awareness of sports targeted poverty alleviation is low. 3) The sports targeted poverty alleviation functional departments "fight alone", and there is a phenomenon of "acting independently and talking to themselves" with other departments, and there is a lack of cooperation and collaborative governance between departments. 4) In the process of sports targeted poverty alleviation, the lack of Internet big data application, it is difficult to understand the information of people's sports needs at the first time. 5) Due to the relatively slow development of sports targeted poverty alleviation, there is still a lack of long-term system of sports targeted poverty alleviation in Yanbian border minority areas.

4 Improvement of Natural Environment and Living Environment in Villages

Environmental improvement is an important part of sustainable development of villages. Since the implementation of the protection and development project of villages with ethnic characteristics, in addition to the special funds set up by the central government, the local government has also actively called on private capital and villagers to invest in the environmental improvement of villages. All local governments have implemented targeted environmental improvement measures in line with local conditions, carried out five changes including water, road, kitchen, toilet and circle improvement, and three construction works including home, garden and pool construction, and vigorously carried out ecological home improvement work.

The ancient village, which stands with a thousand year old building, is glowing with new vitality. The natural environment and living environment of the village have been improved and perfected to a certain extent.

4.1 On the Development of Villages with Ethnic Characteristics

The development of tourism has promoted the construction of economy, culture and other aspects of ethnic villages, but there are still some problems in the process of development.

1. The development of villages is unreasonable and inadequate

First, regional development is unbalanced. Sufficient funds is an important condition for the protection and development of villages with ethnic characteristics. Due to the remote location, inconvenient transportation and limited financial resources of local government, the collective economy of some villages in economically backward provinces is weak, so it is difficult to attract investment and enterprises to develop tourism resources. Secondly, some ethnic cultural handicrafts with low added value, low popularity and low market demand are difficult to attract businesses to invest and develop. Finally, in terms of income distribution, the village income distribution is unreasonable. The contracted development companies take most of the income, and the village residents only get a small part, which seriously frustrates the villagers' enthusiasm to participate in the village construction.

4.2 The Development of Village Cultural Industry Still Relies on Social Financing

Although the traditional culture of the village is well preserved, it is hard to reflect the value of national culture due to the great impact of modernization and foreign culture. Most people's understanding of the value of traditional culture is in a vague state, and the number of teenagers willing to inherit their national culture is decreasing, which leads to the decline of many valuable national cultures, let alone the transformation of industrialization.

$$K_1 = \sqrt{1 + \frac{W_Y}{W_\Delta}(3\cos\theta + \sqrt{3}\sin\theta)} \tag{1}$$

$$w_{m+1,i} = \frac{w_{mi}}{Z_m}\exp(-\alpha_m y_i y_m(x)), i = 1, 2, \cdots, N \tag{2}$$

At present, the development of the vast majority of villages in China still relies heavily on social financing. Although the local government actively provides technical guidance, market development and other services for the cultural industrialization of villagers, but in the process of transforming cultural assets into commercial assets, most villagers still face the problems of lack of experience, ability and technology due to their low cultural level and weak acceptance ability.

4.3   The Destruction of Village Environment has not Been Eliminated Yet

Most of the village buildings are mainly made of wood and bamboo, which are not easy to prevent insects and fire, and are subject to natural erosion all the year round. Although the central and local governments attach great importance to environmental improvement, the environment of some villages is still damaged [4]. On the one hand, in order to reduce the cost and frequency of repairing buildings, some villagers renovate and repair the buildings with brick concrete structure, which changes the architectural style of the original villages; on the other hand, in the peak period of tourism, some village scenic spots carry too many tourists, because the village ecology is fragile, the environmental remediation and protection work is not in place, which leads to serious human destruction of the village environment. However, according to the current level of economic development in Yanbian border minority areas, in the future, both financial and social capital concerns will still focus on the improvement of the basic living level and economic development, and the emphasis on sports will be very low and lack of support. To a certain extent, the restriction of economic conditions has restrained the people's demand for sports in Yanbian border minority areas, and made them pay more attention to their basic life. The demand for sports has not been fully stimulated, which leads to the lack of endogenous power for the development of sports in Yanbian border minority areas and affects the development of sports targeted poverty alleviation.

5   Conclusion

It can be seen that the revitalization of rural areas and the development of ethnic minority economy is no longer simply the pursuit of economic growth rate and various indicators, "ecological environment protection" has been mentioned as a strategic position. Therefore, to develop characteristic villages, we should put "ecological environment protection" in the first place, continue to maintain and give play to its advantages in natural resources and minority cultural resources, coordinate and plan the development and ecological environment protection of villages, so that the construction of characteristic villages can better promote the development and prosperity of minority rural areas.

References

1. Wang, S., Xie, P.: Research on rural tourism development from the perspective of rural revitalization: a case study of jinping ethnic town. Agric. Technol. **41**(04), 150–152 (2021)
2. Jiang, L.: On the development of Yuxi ethnic music. Ethnic Music (01), 53–54 (2021)
3. Wang, R.: Legal protection of tourism cultural resources in ethnic areas. China Press (04), 54–55 (2021)
4. Wei, Z., et al.: Analysis of spatial and temporal distribution characteristics and influencing factors of ethnic minority villages in Guizhou. Guizhou Ethnic Stud. **42**(01), 113–121 (2021)

Explanation of Fui Value by Virtual Simulation Technology

Houhan Wang[1,2(✉)]

[1] Perception Media Corp, New York, USA
[2] School of Art and Design, Hainan University, Haikou 570228, China

Abstract. FUI is a type of conceptual UI design that is capable for cross-field academic researches. Since FUI can both exists in fiction and reality, its values are different from other types of UI. It contains specific academic, publicity, and economic values that are beneficial for both theoretical and practical fields. With the FUI in Crew Dragon Demon-2 as a key example, it'll help to reveal the value of FUI.

Keyword: FUI · UI · Value · Reality · Cinema

1 The Value of FUI

As FUI can both exists in the fictions and reality, it become a bridge that connects the fiction and reality. Its flexibility allows it to be a vehicle for cross field researches. From the cinema and game production perspective, FUI can reflects how technology and design influence human society which shows its academic value. FUI can also project certain cultural impression to others that reflects its publicity value. Additionally, FUI is capable to show its economic value which stimulates commercial behaviors.

1.1 Academic Value

The academic value can often be explored from cinema or game productions. In the case of episode Nosedive from Black Mirror, it offers a good reference to analyze the social rating system FUI design. In the episode Nosedive, FUI assists to illustrates an alternative reality which generalized social rating system. This FUI is constructed with the recognizing section and rating section. The recognizing section is operated by eye-implanted device to identify people's social rating. The rating section allows people to rate each other through rating applications Fig. 1. rating interaction in-between characters, FUI can assists the audiences to understand the mechanics in the show [1]. It helps to illustrates how the rating system influences people's social status and privileges, such as house-purchase benefits. In the story, the social rating FUI has infiltrates into human daily life that manipulates their world view indirectly. Based on the information, analysis from following perspectives are possible:

J. C. Hung et al. (Eds.): FC 2021, LNEE 827, pp. 1560–1564, 2022.
https://doi.org/10.1007/978-981-16-8052-6_227

Fig. 1. Black mirror season

1) From media studies perspective, to evaluate how does this fictional FUI operates in the fictional human society.
2) From cinema studies perspective, it can reflect the social system behind the FUI design as well as the danger of unbalanced democracy.
3) From HCI perspective, it can be analyzed for its practical potential.
4) From futurology perspective, it can represent the cyborg technology and its influence on potential moral threats.
5) From transhumanism perspective, it symbolized the transhumanistic expectation of how technologies can assist human.

These are potential directions for FUI academic researches. As for cyborg or mind digitizing, such as Ghost in The Shell or Altered Carbon, may leads to complex philosophical discussion. Especially for media studies, FUI can be treated as a medium that connects and binds the fiction and reality. These proves FUI is a valuable vehicle for academic researches.

1.2   Publicity Value

Regardless of fiction or reality, FUI can introduces the scientific development for its place of production to others. With Sci-Fi production in specific, the scientific development is the foundation of FUI design. For instance, Eurocentric Sci-Fi productions emphasizes the aerospace, aeronautics, and computer science and technologies which reflects in famous Sci-Fi productions, such as Star Wars or Star Trek; Japanese Sci-Fi productions tend to address robotic or cyborg technologies which reflects in GUN-DAM, Ghost in The Shell, Mental Gear Solid, and other famous Sci-Fi productions. FUI become the medium to introduces the science and technology culture of the place of production to others. It helps to illustrates and solidifies the scientific impression of the place of production which reflects the publicity value of FUI.Western culture often hints its technological superiority through FUI. The FUI of Crew Dragon Demo-2 is a successful example [2]. Through live-stream and official simulator by SpaceX, its FUI allows worldwide audiences to get familiar with the aesthetic and interaction method of Dragon 2 FUI. As it is similar to some FUI design in cinema or game productions, it creates an illusion that merges the fiction with reality which strengthens the western technological superiority impression.

2   The Orientation of Publicity Value

Dragon 2 FUI's publicity value relates to its content. Since Dragon 2 is the first spacecraft utilize touchscreen control panel, its design is simpler than earlier spacecraft, such as Orion and Soyuz [Figure 4, 5]. Dragon 2 FUI is more user friendly and its aesthetic is similar to Sci-Fi productions. This type of FUI has been generalized in western Sci-Fi production. Earliest in 1987 Star Trek: The Next Generation, touchscreen FUI design appears [Figure 6]. The 2017 film Valerian and the City of a Thousand Planets also contains massive touchscreen based FUI design. It's hard to confirm whether some FUI designs in Video games are touchscreen based or not, but they share similar aesthetic, such as Everspace or Elite: Dangerous. The aesthetic similarity constructs a technological superiority impression of western culture.

In addition, Crew Dragon Demo-2 is a historical event that marks the first crewed spacecraft launch by U.S. private company SpaceX. Combine with SpaceX founder Elon Musk's public figure as the "real life Ironman," Dragon 2 FUI helps to shapes the prospect of western technological superiority as well as the leading role of U.S. in aerospace and aeronautics technologies. It reflects Dragon 2 FUI's publicity value from cultural, political, and commercial aspects.

With the example of Dragon 2, it's clear that the publicity value relates tightly with its content and historical background. FUI is neutral that functions as the publicize vehicle which carries different content. For example, the publicity value in 2019 film The Wandering Earth is different from Crew Dragon Demo-2. The Wandering Earth is a milestone Sci-Fi production that marks a breakthrough for the establishment of Chinese scientific impression [3]. It also represents an improvement of Chinese Sci-Fi movie production industry. As shown in Fig. 2. However, due to the lack of experiences, the in-cinema FUI design is not as precise and cohesive Although there are areas of improvement, it successfully conveys a new impression for Chinese scientific development as well as the Sci-Fi production industry. Therefore, FUI is neutral and the orientation of its publicity value depends on its content and historical background.

Fig. 2. Elite: Dangerous (2014), screenshot

3   Economic Value

Due to lack of researches, there are not enough of statistics to prove the economic value of FUI, but there are some indirect references relate to FUI's performance in design field. First, design firms get more business because of their FUI design in cinema or game productions, such as Territory Studio and Perception Media Corp; Second, the FUI production represents the integrity of Sci-Fi production industry which points to economic benefits. In fact, FUI production is an essential segment of Sci-Fi production industry [4]. As the birthplace of Sci-Fi production, western cinema industry has matured FUI production industry that focuses on visual presentation and design strategy. From technical perspectives, FUI design requires graphic design, animation, As shown in Fig. 3. VFX and other professional level skills to execute. These relates tightly with Sci-Fi cinema or game industry.

Fig. 3. The Wondering Earth (2014)

In the case of the wrist device for Mysterio from Spider-Man: Far From Home, its scriptwriting emphasizes the device's military background that is a touchscreen-based device for control weaponized drones.. Therefore, its design strategy focuses on its usability with legible geometrical visual presentation in pre-production phase. During production phase, the client chose to only uses green, yellow, and red colors to indicates three types of drone status that represent good, damaged, and destroyed. Then the designers animate and composite the FUI accordingly to actors' raw footages. In post-production, designers continue to add VFX with compositing and editing process to create cohesive design for final delivery. In fact, this is only one of many FUI design in Spider-Man: Far From Home. The entire FUI designs involve top FUI design studio, such as Perception Media Corp and Territory Studio, and top VFX company, such as The Mill, Framestore, and Luma Pictures. It is never achievable by single studio or company. The economic benefit is self-evidence from production perspective. This type of cross-field collaboration is common in western cinema or game productions which reflects FUI's economic value in cinema or game industry.

4 Conclusion

FUI is capable to support cross field researches as a supplementary segment. Through case studies, it helps to explore FUI's values from multiple perspectives. Especially for media studies and cinema studies, FUI offers a new angle to analyze the impact of technology and design. Further systematic researches are highly recommended to expands FUI's potential as a vehicle for cross field researches in both theoretical and practical fields.

References

1. SPACEX: ISS Docking Simulator. SPACEX. https://iss-sim.spacex.com/. Accessed 5 June 2020
2. Yuen, J.: FUI: How to Design User Interfaces for Film and Games: Featuring Tips and Advice from Artists That Worked on: Minority Report, the Avengers, Star Trek, Interstellar, Iron Man, Star Wars, the Dark Tower, Black Mirror and More. South Carolina: CreateSpace Independent Publishing Platform (2017)
3. Feng, X.: Planning and design of fuliwan Wetland Park in Kunming. Hous. Real Estate (12), 70–71 (2021)
4. Tian, S.: Fuli Huating branch of Xinjian Road Primary School in Xinghualing District, Taiyuan City: cultivating campus culture brand, nurturing growth spirit. Shanxi Educ. (Manag.) (03), 43–45 (2021)

Design of Property Payment Mode Recommendation System Based on Credit Evaluation Mechanism

Jie Xu[✉]

Kunming Metallurgy College, Kunming 650033, China

Abstract. In order to improve the automatic management level of property payment and optimize the design of property payment mode recommendation system, this paper proposes a property payment mode recommendation system based on credit evaluation mechanism, constructs the information evaluation information collection model of property payment mode recommendation, and filters and analyzes the information evaluation information of property payment mode recommendation combined with big data mining and information fusion technology, This paper extracts the statistical information features of the credit evaluation recommended by the property payment mode, analyzes the big data features of the credit evaluation of the property payment mode combined with the big data analysis and clustering feature detection method, adopts the distributed reconstruction method of association rules, realizes the design of the property payment mode recommendation algorithm based on the credit evaluation mechanism, and develops the system software in the embedded platform. The simulation results show that the intelligent payment mode can improve the reliability of property information recommendation.

Keywords: Credit evaluation mechanism · Property payment mode · Recommendation system · Big data

1 Introduction

With the development of artificial intelligence technology, it is of great significance to use the information processing technology and automatic control technology of artificial intelligence to design the property payment mode recommendation system, improve the information level of property payment management, and study the optimization design method of property payment mode recommendation system, so as to promote the information optimization and network design of property payment mode [1]. In the Internet application platform (such as Facebook, Google, app, etc.), optimize the design of the property payment mode recommendation system, establish the information integration model of the property payment mode recommendation, adopt the information management mode, process the data information of the property payment mode recommendation, establish the personalized information fusion model of the property payment mode recommendation, and adopt the database intelligent

© The Author(s), under exclusive license to Springer Nature Singapore Pte Ltd. 2022
J. C. Hung et al. (Eds.): FC 2021, LNEE 827, pp. 1565–1570, 2022.
https://doi.org/10.1007/978-981-16-8052-6_228

information management method, To realize the property payment mode recommendation and information construction, the research on the design method of the relevant property payment mode recommendation system is of great significance in promoting the information management of property payment.

2 Evaluation Index System and Information Collection of Property Payment Mode Recommendation

2.1 Evaluation Index System of Personalized Recommendation of Property Payment Mode

In the cloud resource scheduling environment, by constructing the resource tree management system of property payment mode, the information of property payment mode is provided to each server in the computing cluster, The description formula of statistical characteristic quantity of personalized recommendation of property payment mode is as follows:

$$M_v = w_1 \sum_{i=1}^{m \times n} (H_i - S_i) + M_h w_2 \sum_{i=1}^{m \times n} (S_i - V_i) + \\ w_3 \sum_{i=1}^{m \times n} (V_i - H_i) \tag{1}$$

The sparsity scheduling of information resource data of property payment mode is carried out, and the multi-step optimization of cloud property payment mode information resource scheduling method is used to optimize the property payment mode. According to the number of users in the MAC layer of the current property payment mode resource layer, the statistical information characteristic quantity of the credit evaluation recommended by the property payment mode is extracted [2].

According to the above analysis, the personalized recommendation model of property payment mode is constructed, and the recommendation tree structure model is obtained, as shown in Fig. 1.

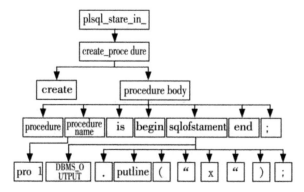

Fig. 1. Personalized recommendation index evaluation system of property payment mode

The personalized features of users in the resource information database of property payment mode are extracted, and the interference features are co filtered. The feature mining model of personalized recommendation of property payment mode is described as follows:

$$\text{minimize} \left(\sum_{i \in V_{w,+}} c_0^w \left| b_{i,0}^w - b_i^w \right| + \sum_{i \in V_{w,-}} c_0^w \left| b_{i,0}^w - b_i^w \right| \right) \tag{2}$$

This paper reconstructs the phase space of the collected bit stream of the recommended credit evaluation information of the property payment mode, extracts the associated characteristic quantity of the recommended credit evaluation information of the property payment mode, and uses the quantitative recursive analysis method to analyze the regular scheduling of the property payment mode recommendation under the cloud computing mode, so as to improve the personalized recommendation ability of the property payment mode [3].

2.2 Information Collection of Property Payment Mode Recommendation

On the basis of the above evaluation index system of personalized recommendation of property payment mode, the credit evaluation information of property payment mode recommendation is collected. The collected credit evaluation information sequence of property payment mode recommendation is mapped into high-dimensional phase space, and the feature set of reconstructed credit evaluation information of property payment mode recommendation is read in the data center network, Information collection of property payment mode recommendation.

The statistical pool of property payment mode recommendation credit evaluation information is composed of N statistical servers, which transmit the characteristic quantity of association rules in the link layer. Therefore, according to the times field in the link table item, the statistical characteristics of property payment mode recommendation are analyzed. By using the sampling feature analysis method, the statistical analysis model of property payment mode recommendation is constructed, and the probability distribution of massive property payment mode recommendation statistics is analyzed. For n property payment mode recommendation sampling sequence points, the distributed detection method of association rules is adopted.

3   Chapter Four Empirical Study of the Effect Brought by the Existing Credit Evaluationmechanism on Consumers' Behavior

3.1   Design of Survey Method

From the actual situation, credit evaluation has an impact on customers' purchase behavior, but the accurate measurement of this impact remains to be studied. Therefore, this paper puts forward the following assumptions: consumer behavior on the Internet is affected by the real life environment. 2. Different credit rating sellers have different service levels. The sampling process includes different stages: (1) determine the population; (2) determine the sampling framework; (3) determine the number of samples; (4) determine the sampling method; (5) obtain samples. The overall situation of this survey is very clear: Amoy For all the customers of the three C2C e-commerce platforms in China, the research object is the credit evaluation system of the three platforms, and the three e-commerce platforms account for 95% of the market share, so their credit evaluation system is very representative, because the population is all the users of the three platforms, so the sampling framework is the whole user list. But the user database is dynamic, changing every day. It is not possible to get the entire list of users [4].

3.2   Time Limit of Credit Evaluation

At present, the research on Yak slaughtering technology, meat loss reduction technology, meat quality and other technologies in Qinghai Province is very mature, and the meat loss and quality control system has been established, and the yak slaughtering performance and meat standards have been formulated. And in Yak breeding, breeding, disease, processing and other aspects of the formation of a mature model project, yak reproduction and production performance has made many major breakthroughs. However, the development of these technologies and research is inseparable from the growth and grade evaluation of yak. In the traditional sense, yak with excellent growth and development characteristics can more effectively improve the growth and breeding performance of offspring. However, the Yak with small size and slow growth and development has a greater influence on the feed proportion, breeding and slaughtering process. Therefore, it is particularly important to study the yak grading according to the attribute data in the growth process of yak. At present, the yak body shape score and equivalence classification are mainly manually marked. Generally, professional researchers or breeding personnel will analyze the yak from the aspects of yak weight, body height, body length, chest circumference, tube circumference, coat gloss, limb and hoof growth, genital development status according to the previous experience, The yaks with lower scores or grades were eliminated and not used as breeding cattle candidates, but were raised as beef cattle. The rest yaks with higher scores or grades were raised as breeding yaks, In order to improve the probability of passing good genes to offspring. Due to the manual labeling needs professional personnel, the workload is large, the labor input is large, and the lack of more accurate standards to measure, the error is difficult to avoid. Generally, the genetic and other factors of the parents are not considered in the scoring or grading.

4   Simulation Experiment and Result Analysis

In order to test the application performance of this method in the realization of property payment mode recommendation, an experimental test is carried out. Tms320vc5409a is used as the digital information processing terminal. The data collection scale of property payment mode recommendation is 2000, the training set scale is 120, and the sample number is 120, 200, 220, 240, 400, 500189, 432, 134 respectively. The sample set of credit evaluation features recommended by all kinds of property payment modes is randomly divided into three samples. Combined with similarity feature analysis and association rule mining method, the sample information collection and credit evaluation of property payment mode recommendation are carried out. The property payment management recommendation performance test is carried out in the equal subset S; (j1,2,3), and the test recall rate is obtained. The test results are shown in Fig. 2, The precision test results are shown in Fig. 3.

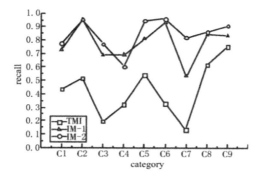

Fig. 2. Recall test

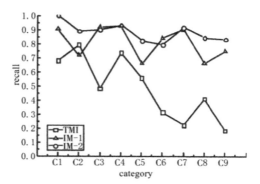

Fig. 3. Precision test results

The analysis results show that the recall rate and accuracy rate are high, which indicates that the accuracy of property payment management and credit evaluation is high, and the reliability of credit evaluation is high, which improves the automation level of property payment mode recommendation.

5   Conclusion

This paper proposes a property payment mode recommendation system based on the credit evaluation mechanism, constructs the credit evaluation information collection model of the property payment mode recommendation, combines the big data mining and information fusion technology to filter and analyze the credit evaluation information of the property payment mode recommendation, and extracts the statistical information characteristics of the credit evaluation of the property payment mode recommendation, Combined with big data analysis and clustering feature detection method, this paper analyzes the big data feature of property payment mode credit evaluation, adopts the distributed reconstruction method of association rules, realizes the design of property payment mode recommendation algorithm based on credit evaluation mechanism, and develops the system software in the embedded platform. The results show that this method has better intelligence, higher reliability of credit evaluation, and improves the automation level of property payment mode recommendation.

References

1. Jiang, Y., Hu, B.: Recommendation of optimal e-commerce based on decentralized credit evaluation index. Control. Eng. **25**(4), 682–688 (2018)
2. Chen, Y., Lu, S., Liu, B., et al.: Mobile sensing path selection algorithm for mobile wireless sensor networks. J. Sens. Technol. **32**(1), 117–126 (2019)
3. Jiang, Y., He, G.: Data control synchronization transmission algorithm for mobile wireless sensor networks based on likelihood estimation compensation mechanism. J. Xinjiang Univ. (Nat. Sci. Ed.) **35**(4), 465–472 (2018)
4. Chen, W., Cheng, X.: Resource scheduling and optimization in cloud computing. Laser J. **37**(6), 115–118 (2016)

Parameter Setting of LM-BP Algorithm in Financial Stock Index Prediction

Zhengsong Ma[✉]

Yunnan Technology and Business University, Kunming 651700, Yunnan, China

Abstract. In view of the complexity of the internal structure of the stock price system and the variability of the external factors, this paper analyzes the principle of stock market prediction based on BP network, establishes a prediction model for the stock market by using three-layer feedforward neural network, and discusses the topological structure of the network, the principle of determining the number of hidden nodes, the selection and pretreatment of sample data, and the determination of initial parameters. In order to avoid the network falling into the local minimum and improve the convergence speed of the network, the improved LM-BP algorithm is adopted and compared with other BP algorithms. Taking the most representative shanghai stock index as an example, the simulation experiment shows that the LM-BP algorithm can effectively and stably predict the short-term trend of Shanghai stock index after learning the selected samples and training the prediction model.

Keywords: Stock index forecast · BP neural network · The number of hidden layer nodes is determined · LM algorithm

1 Introduction

In the research of financial system forecasting, especially stock forecasting, it is difficult to establish a sufficiently accurate model to describe such a system, and it is also difficult to use traditional statistical methods such as time series. In recent is an information processing system, it has many kinds of models. One of them is decision support and so on. The outstanding advantage of BP algorithm is the accuracy of optimization, so BP network is often used in stock forecasting [1].

But in practical application, BP algorithm has the following limitations: ① the error decreases slowly, the adjustment time is long, and the number of iterations is many, which affects the convergence speed and causes slow convergence speed. If the convergence speed is accelerated, it is easy to produce oscillation. ② There is a local minimum problem, but the global optimum can not be obtained. It is difficult for the training to converge to a given error because it is trapped in a local minimum. ③ The selection of hidden node number and initial value is lack of theoretical guidance. ④ The influence of sample selection on system learning is not considered. These limitations greatly reduce the advantages of using BP algorithm to predict financial stock index, so it is necessary to improve BP algorithm on the basis of new theory.

LM (Levenberg Marquardt) algorithm, also known as damped least square method, uses Gauss Newton method to produce an ideal search direction near the optimal value, so as to maintain the characteristics of fast descent speed. It adaptively adjusts the network weights [2]. This paper combines the two methods to form a hybrid training algorithm LM-BP algorithm, which is applied to the prediction of financial stock index to improve the accuracy of prediction.

2 The Prediction Model and Parameter Setting of Stock Price Index

2.1 Model Structure Setting of BP Algorithm

Feedforward neural network has the ability to deal with complex nonlinear signals. Theoretically, it has been proved that feedforward neural network has function approximation function: for any function $f : [0, 1]^n \subset R^n \to R^m$, there is a three-layer feedforward network, which can approach f with any precision; Moreover, feedforward neural network has strong generalization ability. Therefore, neural network can predict time series nonlinearly, and do not need to assume such as stability of time series. Only one neural network is needed to fit time series [3].

The prediction of stock market price by neural network can be divided into two steps: training (fitting) and forecasting. In the process of learning and training of network, the network first randomly generates a set of connection weights between neurons to get an output result through forward propagation. Compare it with the value, it will enter the reverse propagation process, modify the connection weights of the network to reduce the error, and the output calculation of f reaches the requirements, the satisfactory connection weights and thresholds are obtained; The prediction process of network is to input the sample to the network, and predict the sample by using the obtained stable network structure (including training parameters), connection weight and threshold. From the above discussion, it can be seen that a three-layer feedforward can fit any nonlinear continuous function. Therefore, the n-r-1 neural network structure model The topology is shown in Fig. 1.

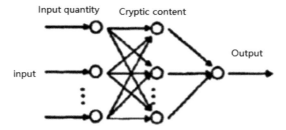

Fig. 1. n -r-1 BP neural network structure diagram

2.2   Sample Selection Range Setting

Because the price of a single stock is easy to fluctuate and sensitive due to individual factors, and the comprehensive index of stock market reflects the development of national economy, which is relatively stable. The training learning input data. In view of the close relationship between the fluctuation of stock price and the following macroeconomic indicators, the monthly total retail sales of social goods, price index (all converted into 100 points of base value in the same period of last year), fixed assets investment are selected into input data of neural network (the weekly data corresponds to corresponding monthly data), formed.The macroeconomic indicators selected have the following effects on the stock price:

On the one hand, the total retail sales of social goods reflect purchasing power, on the other hand, the total retail sales of each company will increase, profits will increase, investors' income will increase and stock prices will rise; Price index, on the one hand, reflects the price level and has a great influence on the psychological state of investors. Because the price rise not only leads to the increase of the cost for purchasing durable consumer goods, but also the decrease of the real return rate of investors, which leads to the decrease of the share price. On the other hand, the price rise increases the profits of the enterprise, and the income of investors increases, which makes the stock price rise; The investment amount [4].

Considering that BP algorithm requires normalization of data, unlike traditional problems, the upper and lower bound of prediction data can be determined according to sample data in advance. Shanghai stock index can not determine the maximum and minimum value of prediction data in advance, In this way, the model after sample training may encounter the failure to restore the predicted data to real data through the anti normalization function (when the actual data x_{min} and x_{max} are larger than the sample data). For the above reasons, this paper proposes the principle of selecting sample data time range when using BP algorithm to predict Shanghai Stock Index: the sample data of time range should contain at least one complete stock market cycle. According to the above principles.

2.3   Initial Network Weight and Threshold Setting

The weights from input layer to hidden layer and from hidden layer to output layer are generally set in three-layer BP neural network, and the thresholds of hidden layer and output layer are random numbers of [0, 1]. However, in the research of financial system forecasting, because of the need to solve the problem of the stability of the actual forecasting, the setting of random initial weights and thresholds obviously can not meet the requirements of the stability of the forecasting results, that is, the forecasting results change greatly every time. In addition, if the initial weights and thresholds are generated randomly, the efficiency of the algorithm will be reduced if the difference between the optimal weights and thresholds corresponding to the samples increases.

According to the characteristics of BP network: if the initial weight is too small, the gradient is very small and the learning speed is very slow; If the initial weight is too large, the sigmoid function is saturated and the gradient is very small. Therefore, this paper adopts the following strategy to set the initial weights and thresholds: divide the training data into three groups according to the length of time, respectively conduct BP algorithm simulation prediction, and preliminarily determine the value range of weights [0.5, 0.75].

2.4   Optimal Hidden Layer Node Number Algorithm

Firstly, according to $n_1 = \sqrt{n+m} + a(n_1$ is the best number of hidden layer nodes; n is the number of neurons in the input layer of the network; m is the number of neurons in the output layer of the network; a is a number from 0 to 10) to determine the selection range of hidden layer nodes; According to the number of each hidden layer node, the same training sample is selected to train the topology of BP network, and the same test sample is used to test. Based on the minimum error criterion, the optimal number of hidden layer nodes corresponding to the problem is found.

3   LM Algorithm

In this paper, we use batch processing method to learn the samples and improved adaptive (Levenberg Marquardt, LM) algorithm to train the network. Different from the gradient descent method, the algorithm based on numerical optimization uses not only the first derivative information of the objective function, but also the second derivative information of the objective function. The specific algorithm of LM is as follows:

$$\begin{cases} f(X^{(k+1)} = \min_{\eta} f(X^{(k)} + \eta^k S(X^{(k)})) \\ X^{(k+1)} = X^{(k)} + \eta^k S(X^{(k)}) \end{cases} \tag{1}$$

Among them: X^k is the vector composed of ownership value and threshold value of the network; $S(X^k)$ is the search direction of vector space composed of X components; η^k is the step size that makes $f(X^k)$ reach a minimum in the $S(X^k)$ direction. In this way, the optimization of network weight can be divided into: 1) first, the best search direction of the current iteration is determined; ② In this direction, the optimal iteration step length is found.

4   Simulation Experiment

The above LM-BP algorithm is programmed on PC pentium4/CPU3.0 GHz/RAM 1.0 GB by MATLAB. In order to test the effectiveness of the algorithm, the prediction of Shanghai stock index is compared with the Shanghai stock index in reality.

(1) LM-BP network simulation prediction curve trend is consistent with the actual trend of Shanghai stock index, as shown in Fig. 2: LM-BP algorithm prediction curve trend of Shanghai stock index is consistent with the actual trend of Shanghai stock index within 20 weeks (2007-11-02–2008-03-14), and has achieved good results in inflection point prediction. By setting the initial weights, LM BP network can quickly find the optimal weights and thresholds through 37 steps of iteration, and can obtain stable prediction results, achieving the effect of practical application.

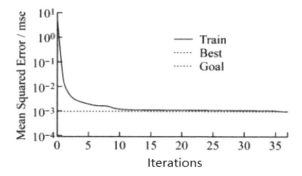

Fig. 2. Convergence trend curve of BP model simulation based on LM algorithm

(2) From November 2, 2007 to February 1, 2008, the error was kept within 10%; From February 5, 2008 to March 14, 2008, the error between BP prediction value and the actual value of Shanghai stock index basically remained within the fluctuation range of 20%. In the following time, the prediction error of Shanghai stock index began to enlarge. Therefore, it is suggested that the prediction time interval should be within 20 weeks.

(3) The regression effect of LM-BP algorithm prediction results is shown in Fig. 3, $R = 0.99797$, Output $\sim$ = 1 * target + -0.0005, which shows that the model established under the principle of parameter setting proposed in this paper has good fitting prediction effect.

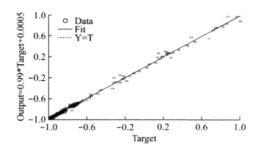

Fig. 3. Regression chart of LM-BP algorithm prediction results

5 Conclusion

In this paper, BP algorithm is used to predict Shanghai stock index, and the parameter setting strategy of BP algorithm is mainly discussed, including sample data selection strategy, initial network weight threshold selection strategy and hidden layer node number determination strategy. Based on this, a three-layer BP model is established, and the actual prediction of the Shanghai stock market closing index is carried out, which achieves the stability of the prediction effect.

References

1. Liu, W., Liu, B.: Prediction model of chaotic dynamics in Chinese stock market. Syst. Eng. Theory Pract. **20**(1), 3438 (2002)
2. Yu, J., Sun, Z., Kroumov, V., et al.: Stock market modeling and decision based on BP neural network. Syst. Eng. Theory Pract. (2003)
3. Zhao, Z., Xu, Y.: Foundation and Application of Fuzzy Theory and Neural Network. Tsinghua University Press, Beijing (1996)
4. Yao, H., Sheng, Z.: Research on wavelet neural network method in stock market forecasting. J. Manag. Eng. **23**(2), 21–26 (2002)

Research and Design of Capital Agricultural Products Micro E-commerce Platform Based on Intelligent Recommendation and Logistics Planning

Juan Tan[1]([⊠]), A. Yumeng[2], and Zhengsheng Zhang[3]

[1] Beijing Technology and Business University, Beijing 100089, China
[2] Beijing Information Science and Technology University,
Beijing 100081, China
[3] Beijing Jiaotong University, Beijing 100089, China

Abstract. In recent years, with the rapid development of agricultural industry and the rapid popularization of mobile Internet, the transition of agricultural product sales e-commerce to mobile Internet is imminent. The construction of wechat e-commerce platform for agricultural products is of great significance to improve the timeliness of agricultural products sales and promote the development of agricultural industry. In view of the sociality of e-commerce operation mode based on wechat platform and the timeliness of agricultural products logistics distribution, this paper studies and analyzes the application of Intelligent Recommendation Algorithm and intelligent logistics planning algorithm in agricultural products wechat e-commerce platform, and completes the design of agricultural products micro e-commerce platform based on intelligent recommendation and logistics planning.

Keywords: Intelligent recommendation · Logistics planning · Agriculture products · Micro E-commerce

1 Introduction

With the development of agricultural industrialization, more and more high-quality agricultural products need to find a broader sales market. Due to the high operation cost and flow acquisition cost of traditional agricultural products business platform, ordinary farmers can't operate alone, and then go to the old way of relying on middlemen for agricultural products. As a result, many characteristic agricultural products are limited in the production areas, and it is difficult to establish safety reputation among consumers, and it is also difficult for consumers to intuitively confirm the quality of agricultural products, and it is even more difficult to realize consumption to guide production, which makes it difficult to adjust the agricultural industrial structure and increase farmers' income. Based on this situation, agricultural products business can only continue to look for other ways of operation. With the rapid development of mobile Internet and the continuous maturity of technology, it provides a new direction for the development of e-commerce of agricultural products. The development

J. C. Hung et al. (Eds.): FC 2021, LNEE 827, pp. 1577–1583, 2022.
https://doi.org/10.1007/978-981-16-8052-6_230

of e-commerce has an important impact on the development of the whole society. E-commerce effectively cuts down the intermediate links of commodity trading and improves the circulation efficiency of commodity trading. As one of the trading commodities, agricultural products are also essential consumer goods in people's daily life. E-commerce of fresh agricultural products has become one of the main directions of the current e-commerce development. For all kinds of agricultural operators and agricultural practitioners, how to use the current trend, with the help of e-commerce to achieve transformation and upgrading, promote agricultural efficiency, increase farmers' income and rural development is a topic that must be considered at present [1].

2   Research on Micro E-commerce Platform for Agricultural Products

2.1   Design Purpose

The booming market prospect of agriculture has brought unlimited business opportunities, and the restrictions of traditional sales mode on the sales of agricultural products are constantly emerging. Agricultural products have the characteristics of seasonality, regionality and policy, while the demand is relatively scattered in the spatial region. It not only leads to the information asymmetry between supply and demand, but also leads to the phenomenon that consumers "buy expensive but not good", and farmers "increase production but not increase income". Therefore, for the sale of agricultural products, there is an urgent need for an e-commerce platform of agricultural products that can always communicate and transfer accurate information. With the continuous maturity and development of information technology and mobile Internet technology, wechat Internet has been gradually introduced into the field of e-commerce, and wechat e-commerce model has gradually emerged. Combined with the social characteristics of wechat e-commerce platform's operation mode and the timeliness of agricultural products sales, the application of Intelligent Recommendation Algorithm and intelligent logistics planning algorithm in agricultural products wechat e-commerce platform needs to realize: first, commodity system, which can realize commodity management and shelf management. Second, the trading system, can realize the sale of goods, transaction management. Third, the payment system can realize the commodity sales and payment management based on wechat payment. Fourth, order system, which can realize order summary and order management. Through the wechat e-commerce platform for agricultural products, consumers can directly place orders, or contact farmers for field investigation. With the deepening of trust, they can also cooperate in the member mode for a long time. Consumers put forward demands, and farmers can plant agricultural products on demand, so as to realize the virtuous circle mode of consumption guiding production [2].

2.2   Feasibility Analysis

At present, the capital has nearly 1000 online stores of various agricultural products, with online sales of agricultural products reaching 4.263 billion yuan. There are more than 100 Taobao villages in the city, and the number of clusters ranks first among prefecture level cities in China. A municipal e-commerce Industrial Park for agricultural products has been built, and an agricultural e-commerce association has been established. At the same time, postal service, supply and marketing, media and other aspects have actively invested in the field of agricultural e-commerce, At present, various counties in the capital have established county-level e-commerce platforms for agricultural products, which has a certain foundation for the development of agricultural products e-commerce in the capital.

In terms of human resources and technology, capital daily group, as the construction unit of the platform, has specially set up an agricultural products e-commerce operation team for the project. The team has 25 members, 15 direct operators and 10 cooperation personnel. The platform core developers have more than three years of programming experience, and are proficient in J2EE, WebService, Oracle and other related technologies, which can better complete the operation task. At the same time, with the help of the storage and logistics transportation team of the publishing team of the newspaper group, as well as the third-party express such as three links and one access, Shunfeng, etc., the real-time distribution service can be realized in the capital region.

3 Research on Intelligent Recommendation and Logistics Planning Algorithm

On the basis of collecting the type and quantity of agricultural products purchased by consumers and the consumption cycle of agricultural products, combined with the comprehensive analysis of social data, consumption data of agricultural products and consumption cycle of agricultural products, this paper studies the application of user based collaborative filtering recommendation algorithm in agricultural products micro store platform. On the basis of collecting the fresh-keeping period, shelf life and other data of different varieties of agricultural products and analyzing the distribution range of agricultural products, according to the characteristics of short fresh-keeping period of agricultural products, this paper studies the application of a * algorithm based on greedy strategy in the distribution route planning of agricultural products according to the fresh-keeping period of agricultural products, categories of agricultural products and regional distribution of users [3].

3.1 Recommendation Algorithm

Because the recommendation algorithm based on collaborative filtering algorithm is widely recognized and widely used in e-commerce recommendation system, this paper uses the recommendation algorithm based on collaborative filtering algorithm as the recommendation algorithm of agricultural products micro e-commerce platform. Considering that the types and cycles of agricultural products purchased by consumers will change seasonally over time, the similarity of users should also change with this cyclical trend. Therefore, the recommendation algorithm adopted in this paper is a user based collaborative filtering algorithm, and takes the user's preference for browsing items, the period of browsing items, the preference for purchasing items, and the period of purchased items as the similarity vector, as shown in Fig. 1.

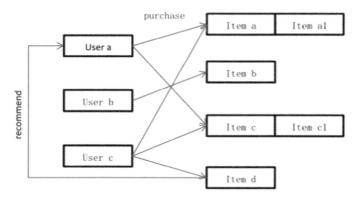

Fig. 1. Struts frame structure

3.2   Research and Analysis on Intelligent Planning Algorithm of Agricultural Product Distribution Route

Due to the diversity of application scenarios, there are many kinds of route planning algorithms, and the shortest path planning scenario is the most widely used. The shortest path planning is the optimal path planning problem from the known starting node to the target node under the condition of known node number, path parameter information, topology and other path information [2].

(1)   A * algorithm

A * algorithm needs to hold two sets at the same time, open set and closed set. The open set is composed of all the nodes that have not been investigated, while the closed set is composed of all the nodes that have been investigated. When the exploration algorithm has checked all the nodes connected with a node, it calculates their F, G, H values and puts them into the open set, For further investigation, this node is called the node under investigation.

The formula is as follows: F (n) is the evaluation function of node n from the starting point to the target point, G (n) is the actual path cost from the starting node to node n in the state space, and H (n) is the evaluation cost of the best path from node n to the target node.

$$f(n) = g(n) + h(n) \tag{1}$$

A * algorithm is one of the best shortest path solving algorithms. It is suitable for small-scale, large-scale and super large-scale path solving problems, and has become a popular shortest path algorithm. Therefore, this paper uses a * algorithm as the shortest path planning algorithm of agricultural micro e-commerce platform.

(2) Optimal path principle of greedy strategy

An example is given to illustrate the principle of the optimal path of greedy strategy. From the starting point a, agricultural products should be delivered to B, C and D, and the fresh-keeping period of agricultural products delivered to B and D is short. If the shortest path from a to B and D is 1 and 5 respectively, priority should be given to delivery to B, then from B to D, and finally to C, as shown in Fig. 2.

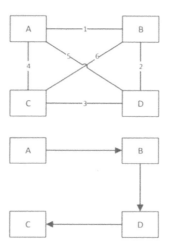

Fig. 2. Time sequence diagram of query process

Based on the above analysis, considering that in the distribution of agricultural products, the process of considering the route from the current location to the next location in a certain delivery address will not affect the previous delivery location and status, but only related to the current location and status, What we want to get is a local optimal route of one of the next several delivery addresses according to the same grade of preservation degree of agricultural products. Therefore, the route planning algorithm used in this paper is a greedy strategy based on a * algorithm. The shortest path from the current location to other delivery address is obtained by a * algorithm, and a greedy strategy is used to select a path as the next level distribution path.

4 Design of Capital Agricultural Products Micro E-commerce Platform Based on Intelligent Recommendation and Logistics Planning

4.1 Overall Structure of the System

Through analysis, the platform of agricultural micro e-commerce includes two platforms: mobile platform of agricultural micro e-commerce and background management platform of agricultural micro e-commerce. After integrating the needs of users, combing the data structure and dividing the functional modules, we have a clear understanding of the overall structure of the system. Based on the systematic and standardized classification of the two platforms needed by users, the platform is subdivided into five systems and seven modules. The agricultural micro e-commerce mobile terminal platform is divided into five systems: commodity system, transaction system, payment system, payment system, payment system and so on Order system, distribution system. Agricultural products micro e-commerce background management platform is divided into seven modules: user management module, sales management module, commodity management module, order management module, inventory management module, recommendation management module, distribution management module [4].

4.2   Function Module Design

The agricultural products micro e-commerce platform is divided into two major platforms from the function, which are agricultural products micro e-commerce mobile terminal platform and agricultural products micro e-commerce background management platform. The two platforms include the following parts. Agricultural products micro e-commerce mobile platform: divided into five systems, commodity system, transaction system, payment system, order system, distribution system. Agricultural products micro e-commerce background management platform: divided into seven modules, user management, sales management, commodity management, order management, inventory management, recommendation management, distribution management. There are two use modes of agricultural products micro e-commerce mobile platform, one is browsing mode, the other is login user mode. Browsing mode can use the query function to view the agricultural products sold on the platform. But we can't buy and sell agricultural products, and we can't use the related functions of transaction, payment, order, distribution and other subsystems. Login user mode can use all the functions of commodity view, collection, purchase, commodity transaction, payment, order, delivery, etc.

4.3   Platform Data Management

For data management, it is the general term for the management of all data in the system. It mainly includes the data of user ordering goods, storage and transportation, and user terminal management. In the process of using, these data will be stored at any time. After a long time of accumulation, a large amount of data information will be formed, and then all the data will be filtered to select some useful information purposefully, so as to improve the utilization rate of platform information.

5   Conclusion

Based on the research and analysis of the advantages of the existing e-commerce platform operation mode of agricultural products at home and abroad, combined with the actual problems of high cost and asymmetric supply and marketing information of the current e-commerce operation mode of agricultural products in China, this paper studies the social advantages and service ability of using mobile Internet, The operation mode of e-commerce platform for agricultural products based on wechat platform, which adds characteristic services such as intelligent recommendation of agricultural products and intelligent logistics planning and distribution of agricultural products. The system requirements of the platform were analyzed in detail. Combined with the system design scheme and feasible technical support, the capital agricultural products micro e-commerce platform based on intelligent recommendation and logistics planning was designed and implemented.

Acknowledgements. The National Social Science Fund of China: Research on influencing factors and mechanism of knowledge remixing in network innovation community, Beijing Social Science Fund: Research on the innovation and development of Beijing e-business in the era of digital economy Number: 18CSH019, 20JCC096.

References

1. You, T.: Development experience of foreign e-commerce and Its Enlightenment to China. J. Heihe Univ. **8**(05), 27–28 (2017)
2. Wu, L.: Research on problems and strategies of developing e-commerce for organic agricultural products enterprises. Capital University of Economics and Trade (2014)
3. Li, X.: Analysis on influencing factors of e-commerce mode selection of agricultural products. Huazhong Agricultural University (2014)
4. Chen, S.: Development of foreign agricultural e-commerce and its enlightenment to China. J. Agric. Libr. Inf. Sci. (09), 112–114+154 (2008)

Overall Performance Optimization Design of Aeroengine Based on Multi Objective Genetic Algorithm

Xu Zhang[(✉)]

Sanya Aviation & Tourism College, Sanya 572000, Hainan, China

Abstract. Multi objective genetic algorithm is a new optimization algorithm developed in recent years for multi-objective optimization problems. Because of its high efficiency and practical characteristics, it has been paid more and more attention by the academic circles. Aiming at the modeling error in the design of flight control system controller and the influence of external interference in the flight process, PID control method is used to complete the design of a combat fuel engine control system controller. In order to improve the control accuracy of fighter and solve the limitation of the difficulty in selecting the weight matrix Q and R in LQG/LTR robust control method, genetic algorithm is added to optimize the weight matrix on line. At the same time, the system under different flight conditions and the parameter uncertainty caused by the system modeling process are simulated respectively, and compared with the traditional PID and the PID control based on genetic algorithm.

Keywords: Multi objective genetic algorithm · PID control · Aircraft engine · Optimal design

1 Introduction

Aeroengine has strong nonlinearity in the whole envelope. It is a complex aerothermodynamic system composed of multiple functional components. At different operating points, if the aerodynamic parameters related to stability, sensitivity and robustness change, the engine will show very different characteristics. In such a complex situation, aero-engine controller is also required to be able to timely and accurately respond to the commands issued by the pilot under various flight conditions, while ensuring the stability of the whole system. Usually, to design a controller for this kind of system [1], a series of working points should be selected first, and then the controller should be designed according to the specific requirements of each working point and a series of performance indexes. Due to the requirement of actuator action continuity, the engine also requires the controller to ensure a smooth transition between each working point in the whole working range.

Aeroengine performance optimization is a complex nonlinear optimization problem, which needs to meet multiple performance indexes at the same time. The objective functions used to calculate these performance indexes have different dimensions and compete with each other. The optimization effect of one of the objective functions is

J. C. Hung et al. (Eds.): FC 2021, LNEE 827, pp. 1584–1589, 2022.
https://doi.org/10.1007/978-981-16-8052-6_231

often improved, It is usually at the cost of the degradation of the optimization effect of other objective functions. Therefore, the optimization problem of aeroengine control system often can not get the unique optimal solution, but a set of non dominated solutions with the same optimization level. Experts and scholars at home and abroad mainly use gradient method, linear programming and other methods to solve this problem. However, the traditional optimization methods all take the gradient information of the optimization target to the control parameters as the search information. It is difficult to get good optimization effect for the optimization problem with multiple peaks, especially for the optimization problem with multiple control parameters or unable to solve the gradient information. Because genetic algorithm does not rely on gradient information in the whole search process, and uses the way of population to organize the search and exchange information with each other, it is a robust intelligent search algorithm with global search ability, which is suitable for dealing with complex nonlinear problems that traditional search methods are difficult to solve. In view of the many advantages of genetic algorithm and the ability of multi-objective genetic algorithm to continuously optimize the current optimal solution set, some experts and scholars in the world have studied and tried to use genetic algorithm to solve this problem.

2 Multi Objective Genetic Algorithm

Genetic algorithm (GA) is an optimization method which imitates the process of biological evolution proposed by Holland in 1962. Based on the biological evolution principle of "survival of the fittest, superiority and inferiority" in nature, genetic algorithm first generates an initial population in a random way. Each individual in the population corresponds to a possible solution of the problem to be optimized. Each individual is actually an entity with chromosome characteristics.

An advanced multi-objective genetic algorithm optimizes two multi-objective functions to test the performance of the algorithm.

First, we investigate a typical optimization problem of finding the minimum of double objective function

$$\begin{cases} f_1(x_1, x_2) = x_1^2 + (x_2 - 1)^2 \\ f_2(x_1, x_2) = x_2^2 + (x_1 - 1)^2 \end{cases} \quad 0 \le x_1, x_2 \le 1000 \quad (1)$$

In this optimization problem, it is expected that the two objective functions will reach the minimum value at the same time, but there is obvious competition between the two objective functions. When one objective function gets a smaller value, the other objective function will get a larger value, that is, the optimization effect of one objective function will be improved, It is a typical multi-objective optimization problem to take the degradation of the optimization effect of another objective function as the cost. Therefore, this optimization problem can not get a unique optimal solution, but a set of non dominated solutions composed of multiple non dominated solutions [2].

In this optimization process, floating-point coding method is used to optimize the two parameters X_1, x_ The evolutionary algebra of the algorithm is 150, the population size is 100, the crossover probability is 0.9, and the mutation probability is 0.1. The NSGA-II algorithm is used for optimization, and the optimization results are shown in Fig. 1.

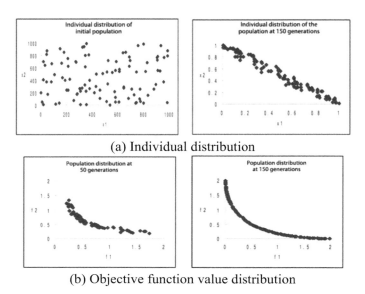

(a) Individual distribution

(b) Objective function value distribution

Fig. 1. NSGA-II optimization multi objective function optimization results

In the figure above, as shown in figure (a), the individuals in the initial population are randomly generated in a large range of values. After 150 generations of evolution, the individuals in the whole population have been concentrated in a small range, that is, they have evolved to the optimal solution set. It can be seen from figure (b) that the population has converged to a small area when it evolves to 50 generations, but many individuals still have mutual dominance. When it evolves to 150 generations, the individuals in the whole population have no mutual dominance at all, and they are evenly distributed in the optimal solution set, and the boundary individuals are well preserved, and good optimization results are obtained.

$$
\begin{cases}
f_1(x_1, x_2, x_3) = \sum_{i-1}^{n-1} \left(-10exp\left(-0.2\sqrt{x_i^2 + x_{i+1}^2} \right) \right) \\
f_2(x_1, x_2, x_3) = \sum_{i-1}^{n} \left(|x_i|^{0.8} + 5\sin(x_i)^3 \right)
\end{cases}
\quad -5 \le x_1, x_2, x_3 \le 5 \quad (2)
$$

This optimization problem has three parameters to be optimized, and it is still an optimization problem of finding the minimum of double objective function.

3 Parameter Optimization Design of PID Controller for Aeroengine

3.1 PID Control Principle

In analog control system, PID control is the most commonly used control law. The principle of analog PID control is shown in Fig. 2. The system is composed of analog PID controller and controlled object.

PID control is a kind of linear controller, which forms the control deviation according to the given value Rin (T) and the actual output value yout (T)

$$error(t) = rin(t) - yout(t) \tag{3}$$

The control law of PID is as follows

$$u(t) = K_p \left(error(t) + \frac{1}{T_1} \int_0^1 error(t)dt + T_D \frac{derrort(t)}{dt} \right) \tag{4}$$

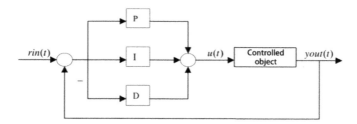

Fig. 2. Principle block diagram of analog PID control system

In order to improve the quality of the system and meet the needs of different control systems, a series of improved algorithms are produced, such as integral separation PID control algorithm, anti integral saturation PID control algorithm, trapezoid integral PID control algorithm, PID control algorithm with dead time, etc. [3].

3.2 Optimization of Aeroengine PID Controller Parameters Based on NSGA-II

The optimization work takes the nonlinear component level model of turbofan engine as the controlled object, the control quantity is mainly fuel flow WFM and the area of tail nozzle A8, the controlled amount is low-pressure rotor speed n1 or high-pressure rotor speed N2 and turbine drop pressure ratio pit. The incremental digital PID controller is used to control the generator model. As described in the previous article, NSGA-II is a more advanced multi-objective genetic algorithm in the world. Compared with previous algorithms, NSGA-II has made many improvements and has better ability of optimization. Therefore, the optimization work is to use the multi-objective genetic algorithm NSGA-II to solve the PID controller parameters of the engine in the specified working state, so that the combination of the performance indexes of the speed step response process can reach the optimal under the effect of the optimized controller parameters. Considering that the differential link has a poor ability to restrain high frequency noise and will produce bad control effect on the engine, we choose the differential coefficient of 0 and the PI control is used in practice.

In the optimization process, the evolutionary algebra of the algorithm is set to be 50, the population size is 100, the cross probability is 0.9, and the mutation probability is 0.1. The parameters of PI controller are optimized for the flight states of the engine at ground and high altitude.

At the current operating point, the engine model is in the intermediate state single variable control. The main fuel flow $wfm = 1.05822$, the controlled amount is the low-pressure rotor speed NL. The parameters of the controller are optimized by the set NSGA-II algorithm. The algorithm ends and the three target function values corresponding to each individual in the population are well converged. Because the final optimization result is an optimal solution set, which contains an optimal individual of population size, we should choose a set of optimal controller parameters according to the needs of specific problems and the preferences of the optimizer. In order to weigh the importance of three objective functions, a group of optimal controller parameters K with short rising time and zero overshoot are selected in the termination population $K_p = 1.002271, K_I = 1.95407$ The optimization results are shown in Fig. 3.

The fuel flow and rotor speed in the following optimization results are dimensionless.

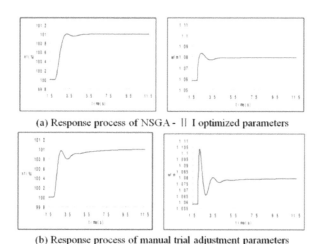

(a) Response process of NSGA - II I optimized parameters

(b) Response process of manual trial adjustment parameters

Fig. 3. Optimization results of controller parameters based on NSGA-II

In this paper, the multi-objective genetic algorithm NSGA-II is applied to the parameter optimization of PID controller for Aeroengine model. NSGA-II optimization algorithm is used to optimize the PID controller parameters of the engine model in multiple flight states on the ground and at high altitude [4], and the simulation results are compared with the traditional manual trial results. The actual simulation results show that NSGA-II optimization algorithm obtains satisfactory results in all flight states, and has excellent performance optimization ability. It is a powerful multi-objective optimization algorithm.

4   Conclusion

In this paper, multi-objective genetic algorithm is used to solve the parameter optimization problem of aeroengine controller. Taking the nonlinear component level model of turbofan engine as the controlled object, the multi-objective genetic algorithm NSGA-II is used to optimize the PID controller parameters of the model. The optimization work is carried out for the single and dual variable control systems of the engine in multiple flight states on the ground and at high altitude. The simulation results show that the parameters of the controller optimized by NSGA-II algorithm can optimize all the performance indexes of the speed step response process, and achieve better comprehensive control effect. Moreover, compared with the traditional parameter tuning method, the multi-objective genetic algorithm has obvious advantages in method and optimization results. It can be seen that the multi-objective genetic algorithm can search the appropriate parameters in the specified range, and successfully solve the decoupling problem of the bivariate control system, even for the engine, which is a large nonlinear complex system with single and bivariate control at the same time.

References

1. Chen, G., Wang, X., et al.: Genetic Algorithm and Its Application. People's Posts and Telecommunications Press, Beijing (1996)
2. Editorial Board of Aeroengine Design Manual: Aeroengine Design Manual, vol. 5. Aviation Industry Press, Beijing (2001)
3. Liu, J.: Advanced PID Control and MATLAB Simulation. Electronic Industry Press, Beijing (2003)
4. Sun, J.: Prospects for aero power control in the 21st century. J. Aero Power (2001)

Construction of Enterprise Management Framework Based on Edge Computing

Mei-ling Yang$^{(\boxtimes)}$

Guangzhou College of Technology and Business, Guangzhou 510850, China

Abstract. With the rapid development of China's e-commerce and Internet business, many domestic enterprises have entered the era of computer network management, and thus improve the management efficiency and market competitiveness. But at present, some enterprises still stay in the original accounting management stage. With the process of global economic informatization and the successful realization of WTO, enterprises are facing unprecedented opportunities and challenges. Under such drastic social situation and fierce market competition, more and more enterprise managers realize the importance of efficient management and scientific management. Based on computer technology, this paper studies the construction of enterprise management thinking framework, hoping to better help enterprises carry out modern management.

Keywords: Computer technology · Business management · Thinking frame

1 Introduction

With the further popularization and development of the Internet, computer technology is widely used in enterprise management. If an enterprise does not use computer, and it is a "High Pavilion", it cannot be called a modern enterprise, and it cannot be invincible in the fierce market competition. Because of this, almost every enterprise is equipped with computers, and has purchased relevant design and management software for production and management of enterprises. However, at present, many enterprises in the management, the role of computers have not been fully played, even some computers become only a modern office equipment. Therefore, in order to adapt to the development trend of management modernization and make computer play a leading role in enterprise production management, it is necessary to emphasize that computer is very useful in enterprise production management [1].

From the production process of the enterprise to the whole process of enterprise management and organization, there is a place for computer technology. By using computer technology, the production cost of the enterprise can be greatly reduced and the production efficiency and quality of the whole enterprise can be improved. Improve the internal management ability of the whole enterprise. At present, Chinese enterprises are in the transition stage. By applying computer technology, the modernization and efficiency of enterprise management can be realized quickly. How to strengthen the management of enterprises through computer technology, strengthen the core competitiveness and realize continuous innovation is of decisive significance to the growth of enterprises.

J. C. Hung et al. (Eds.): FC 2021, LNEE 827, pp. 1590–1595, 2022.
https://doi.org/10.1007/978-981-16-8052-6_232

2   Computer Technology

2.1   Introduction of Computer Technology

Computer technology is used in the field of computer technology methods and technical means. Computer technology has obvious comprehensive characteristics. It is closely combined with electronic engineering, applied physics, mechanical engineering, modern communication technology and mathematics, and develops rapidly. The first general-purpose electronic computer ENIAC was based on radar pulse technology, nuclear physics electronic counting technology and communication technology. The development of electronic technology, especially microelectronic technology, has a great influence on computer technology. The two are closely integrated. The achievements in Applied Physics provide the conditions for the development of computer technology: vacuum electronic technology, magnetic recording technology, optical and laser technology, superconducting technology, optical fiber technology, thermal and photosensitive technology, etc., which are widely used in computers. Mechanical engineering technology, especially precision machinery and its process and measurement technology, is the technical pillar of computer external equipment. With the progress of computer technology and communication technology, as well as the increasing demand of the society for the computer to form a network to realize resource sharing, computer technology and communication technology have been closely combined, which will become a strong material and technological foundation of the society. Discrete mathematics, algorithm theory, language theory, cybernetics, information theory, automata theory and so on provide an important theoretical basis for the development of computer technology. Computer technology has emerged and developed on the basis of many disciplines and industrial technology, and has been widely used in almost all fields of science and technology and national economy.

2.2   The Necessity of Computer Technology in Enterprise Management

(1) It is helpful for leaders to make decisions quickly and improve office efficiency
The office is the place for leaders of various industries to make decisions. Leaders make decisions and issue instructions, which is not only the exchange of documents, but also the collection, storage, retrieval, processing and analysis of information. Correct and timely decisions will have an inestimable effect on the development of enterprises. The efficient office work of employees is the core of leaders' accurate and timely decision-making. Therefore, how to liberate employees from the tedious and tense office environment, improve the creativity of personnel, and enhance the office efficiency is a difficult problem to be solved in each enterprise management.

(2) Improve the level of enterprise management
The development of computer technology is possible to improve the management level of enterprises. The most important reason is the advanced nature of computer technology. Compared with manual operation, enterprise management by computer technology will more accurately feedback data [2]. The implementation

of enterprise management through computer technology will better avoid the adverse consequences of decision-making due to cognitive errors at the leadership level.

3   Enterprise Management Thinking

3.1   Concept of Management Thinking

Management thinking is the most effective thinking mode derived from long-term management behavior. The thought method, which helps to make management decisions in turn, is a result of twice the effort for the management behavior.

3.2   Connotation of Management Thinking

(1) 2.1 people oriented. In management, we should realize that the essential object of management behavior is human. The essential purpose of face-to-face management is to improve the efficiency of the organization. So, in a comprehensive way, it is to guide the planner's behavior in the organization. Improve the efficiency of the whole organization. Since the core is human. We should recognize some of the characteristics of human beings. For example, people have self-consciousness and freedom consciousness. Each person is an independent individual with its own needs. So in the traditional management mode, people are regarded as cold machines, and hope that everyone can work efficiently as a closely connected gear. This management mode ignores the independence and self-identity of human body. It ignores people-oriented and contrary to human nature. Finally, this management mode suppresses human creativity and suppresses people's work efficiency. So the important characteristics of enterprise management are people-oriented, pay attention to the humanistic care in the enterprise, respect the personal dignity and legitimate rights and interests of employees, and help the employees to improve their self-worth, reality and present self-worth. The second level of human-oriented significance is to attach importance to the management and development of human resources. The product and resources of an enterprise may be limited, but there is no limit to the growth of an enterprise and the development of human resources. The essence of the enterprise growth is the composition of the enterprise - the growth of employees. Therefore, enterprises should pay attention to help the employees to carry out learning planning and career planning, pay attention to the training of employees. The improvement of employee ability is the promotion of enterprise value. The third meaning of people-oriented is that we should give full play to the subjective initiative of human beings. Forcing a person to do tasks naturally causes people to work negatively. If one wants to do it, the light and heat that one can't match when he or she is on the shelf [3].

(2) Results oriented. Avoid inertia of thinking. In the process of enterprise management, there will be various problems, but enterprises should firmly grasp the key to the problems, because market economy is the result is the king, only looking at the

results without asking the process. Successful managers generally have their own mature thinking mode. Through this existing experience and model, the decision of managers can be more mature and greatly improve the efficiency of management decision. But at the same time, because of the continuous progress of science and technology, if managers have been enjoying the convenience brought by the inherent mode, then it will eventually fall into the trap of human inertia thinking.

(3) Pay attention to cultural induction. Enterprise culture is an important content in modern enterprise management. The resources of enterprises are the hard strength of enterprises. The culture of enterprises is the soft power of enterprises. It is the embodiment of the comprehensive values of enterprises. Through this value concept, the production and management system of enterprises can be guided to the daily performance of employees. Corporate culture often reflects the values pursued by enterprises through the environment, special rituals, role models and so on.

(4) Grasp the overall situation and think systematically. As the manager of the enterprise, the leader of the ship, what we need to do in the process of management is to see the essence through phenomena, to see the whole situation through the events, to grasp the dynamic statically and to grasp the long term in the near future.

(5) Ensure the cohesion and harmony of the organization. The enterprise should be a complex composed of a whole, not a multi sector. So the friendly and harmonious relationship between employees can not only promote positive work, a good atmosphere of cooperation, but also enhance the sense of belonging of employees.

(6) Keep pace with the times, pay attention to innovation and reform. In different market situation, in different development stages of enterprises, enterprises need to adopt different enterprise management thinking to achieve different management objectives. For example, when the enterprise has only dozens of people in the early stage of entrepreneurship, it is different from the organizational structure adopted when the enterprise develops to a certain extent with thousands of people. Therefore, as an enterprise manager, it is necessary to summarize and reflect, update management strategy, update management methods and management organization.

4 The Construction of Enterprise Management Thinking Framework Based on Computer Technology

Strengthening enterprise management and improving scientific management level are the internal requirements of establishing modern enterprise system, and also an important way for state-owned enterprises to turn losses and increase profits and improve their competitiveness. We must attach great importance to and strengthen the management of enterprises, strictly manage enterprises, realize management innovation, and change the situation of a large number of enterprises' decision-making, lax system, loose discipline and low management level as soon as possible [4].

Based on the above, we can integrate the development direction, management methods and management structure of an enterprise to construct a basic framework of organizational management thinking, as shown in Fig. 1.

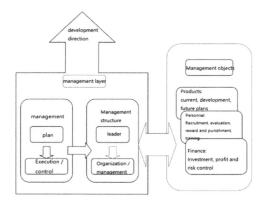

Fig. 1. Thinking frame of enterprise management

On the premise of environmental analysis, the enterprise organization determines the development direction and operation plan, and makes the decision-making plan into the operation budget by authorization; The management control system decomposes the operating budget into the specific tasks of each subordinate department (team) through the redistribution of power, and carries out and implements it; The basic framework of management control system and the structure of enterprise organization are an organic whole. The main role of incentive mechanism and reward and punishment system in the whole process of budget implementation is to make each working team and individual have the power to undertake specific tasks, and guide and constrain the working team and individual to complete the tasks according to the expected direction and objectives; Performance measurement and evaluation is the component of management control system and the core of management control system. It is the basic premise to evaluate the effectiveness of strategic choice and management control system of enterprises. Therefore, there is an inherent logical relationship between the competitive strategy, management control system and performance evaluation system (evaluation index system).

5   Conclusion

The products and resources of an enterprise are only the carrier of its development, while the management of an enterprise is the core driving force of its development. With the development of modern science and technology, the advent of the information age, the rise of the era of knowledge economy, enterprise management based on computer technology is also constantly improving and changing, but some management thinking has the same color, pointing to the essence of management. With the convenience brought by the information age, the future market becomes more complex and uncertain. The long-term development of enterprises should be based on themselves and strengthen their own management, so as to remain firm in the tide of the market.

Acknowledgements. The university level scientific research project of Guangzhou College of Technology and Business in 2018 "Research on the construction of SMEs Innovation Ecosystem in Guangdong-Hong Kong-Macao Greater Bay Area" (Project No: ka201819).

References

1. Luo, W.: Research on the application of computer technology in enterprise management. China Sci. Technol. Expo. (42), 377 (2015)
2. Zhao, M.: discussion on the application of computer technology in enterprise management. Enterp. Technol. Dev. Mon. **35**(2), 33 (2016)
3. Zhang, C.: On the art of people oriented enterprise management. Mod. Enterp. Cult. (2), 16–17 (2010)
4. Chen, C.: Transformation and innovation of corporate culture. J. Peking Univ. (Philos. Soc. Sci. Ed.) (3), 51–56 (1999)

Development and Application of Computer Teaching and Automatic Management System

Bingyi Qu[⊠]

Nanchang Vocational University, Nanchang 330500, Jiangxi, China

Abstract. With the development of economic and cultural level, computer and network technology has also been developed rapidly, and the popularity is also higher and higher. Computer information management technology has also been widely used in the education system. Among them, the use of computer and network technology in school physical education teaching management is still very low, In order to make the A/intranet management technology widely used and promoted in various disciplines, especially in school physical education teaching and management, this paper focuses on the analysis and research of computer physical education teaching and sports network management.

Keywords: Physical education teaching · Management system · Computer-based teaching

1   Introduction

Important means yearning a better life promote people's all-round development. For teenagers, it is the seed of loving sports, What's more, it's a gift that accompanies them all their lives, and it's the cornerstone of creating the future. Competition is a good mechanism to promote the implementation of these good ideas and good visions [1]. According to the data, since the Ministry of education took the lead in campus football work at the end of 2014, it has selected and established more than 30000 national youth campus football characteristic schools, built 110 national youth campus football star training camps, and selected and identified 6000 National Football characteristic kindergartens. Perfect competition system and build competition system become the hard index. Chinese traditional sports also formed a new situation of "one school, many products".

2   Characteristics of Network Physical Education Teaching System

Convenient courseware or teaching information generation environment. PE CAI software and multimedia PE courseware made by PE teachers can be stored in the network center server for PE teaching call and on demand at different terminals on the network. Provide a variety of sports courseware or physical education information using environment. The system can provide a variety of courseware or educational

J. C. Hung et al. (Eds.): FC 2021, LNEE 827, pp. 1596–1600, 2022.
https://doi.org/10.1007/978-981-16-8052-6_233

information using environment according to the students' different time and place: ① multi-functional classroom or gymnasium. In the multi-functional classroom or gymnasium, teachers can query and demand the sports courseware and relevant teaching information on the network server through the sports courseware on demand system, and hit it on the large screen through the projector to cooperate with the physical education teachers for physical education demonstration, It can fully mobilize students' multiple senses and create a good environment for learning. ② network classroom. In the network classroom, students can query, obtain and demonstrate the multimedia sports courseware on the network server, practice on the Internet and learn to browse the information on the server. Students can also control the learning progress and select different learning contents according to their knowledge level and ability level, The network teaching system provides an open learning environment for PE teaching to explore and discover sports basic and sports technology knowledge [2].

3   Realization of Computer Teaching in by Automatic

Sports literature management system. The computer network integrated management special school sports literature library information management, school education, related literature management and sports research guidance as shown in the Fig. 1.

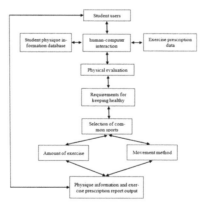

Fig. 1. School sports automation management system

Students' physical health, to evaluate the students' physical function, through statistical analysis, to store, which is more conducive to the students' physical education. Improve students' physical health level, develop good exercise habits.

School sports competition management system.

The system is mainly used in the information release of various school sports competitions, the management and inquiry of students' selected sports events, the arrangement and arrangement of order book, the arrangement and printing of examination sheet, track and field competition forms, etc. students. As an open stadium management system, it regulates and manages the paid use of sports equipment. School personnel file management system.

The system is mainly used for school management of file management, teachers' personal file management, principal file maintenance, principal file maintenance; School assistant; Physical education teaching file management; Evaluation is a comprehensive evaluation of teachers at all levels. Egistration, equipment purchase, file keeping, loss, maintenance of school stadiums. This system is a comprehensive management system of computer network, which is specially used for the selection, management and training of school athletes, the organization of competition, the training of athletes and the organization of sportswear.

4 Network Physical Education Management

4.1 English Teaching in Physical Education

Study, work, lifestyle and way of thinking, and poses a severe challenge to education. Training plan and method of physical education curriculum guidance.

At present, simple, including three balls, three balls, martial arts, fitness, swimming and so on. A two-year compulsory course, 144 h a week, in fact, once a week (2 h), so it is necessary to improve the physical education Simple physical education teaching can not meet the students' full understanding of the organization, guidance and guidance of extracurricular activities. Some characteristics of network education technology can make up for these shortcomings. Because of the asynchronous teaching characteristics of rich teaching materials and unlimited learning time, sports network teaching can undertake the learning outside the sports classroom and become an extension of the sports classroom to expand students' learning content and learning time. Teachers can also guide students' extracurricular exercise and learning asynchronously through network interaction technology, so that students can obtain the same sports knowledge as classroom teaching after class [3].

4.2 The Application of the System in the Management of Physical Education

The application of network technology in college sports management is mainly realized by using sports teaching management system software. The operation of system software includes students, teachers and system administrators, all of which have different operation rights. Among them, students have the right to inquire, select and withdraw courses; Teachers have the login authority to query the results; The system administrator has the operation authority of curriculum objectives, timetable arrangement, class division, grade announcement, establishment of database, establishment of personal information and code of teachers. As shown in Fig. 2 below. Taking the idea of "health first" as the guiding principle.

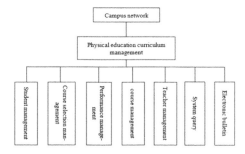

Fig. 2. Sports educational administration management system

5   The Application and Economic of School Sports Automatic Management System

5.1   Application of Integrated Management System for School Physical Education

It is the use of multimedia network teaching, so that teachers from the original stand in the classroom to talk about the content of students cognitive tools, the traditional into students to find problems, research problems to obtain unknown sports knowledge, promote students' sports ability and self-care ability to be gradually improved. When P. E. teachers teach theoretical knowledge of P.E., they only need to use the set Web page in the multi-media classroom to carry out permission teaching. When P.E. teachers set up teaching courseware, the theoretical content should be updated [4].

5.2   Systematic Analysis of School Physical Education Under Computer Comprehensive Management

It is the use of computer network system to help students master the technical action essentials of relevant aspects. Perceive the formation process of technical movement. As shown in Fig. 3. The old teachers do not necessarily have the demonstration in place. At this time, they can consult the relevant information through the computer network system, The use of multimedia technology can fully and clearly show the standardized movement technology to the students, which is convenient for the students to form a deep, clear and continuous impression of technical movement in their brain, so as to master the essentials of technical movement the teaching education.

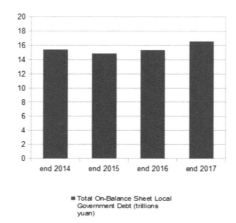

Fig. 3. Analysis of computer integrated system

6   Conclusion

Provides great convenience for all walks of life. To play the functions of information sharing and convenient operation in school physical education, so as to realize office automation in school physical education teaching and better promote the development of students' physical education learning.

To sum up, physical education is an important teaching link it is an important factor to promote the development of students' comprehensive quality. college physical education, we should not only teach students' sports skills and sports methods, of students' lifelong sports concept, combine sports theoretical knowledge with practical activities, and mobilize students' Sports initiative, Let students feel the importance of sports in physical exercise, so that students can adhere to lifelong physical exercise in the future development.

References

1. Huang, H., Ji, K.: Research on the curriculum system reform of physical education undergraduate specialty in Chinese universities. Sports Sci. **24**(3), 51–56 (2004)
2. Peng, W., Luo, Y.: Evolution and enlightenment of American sports thought. Sports Sci. **36**(3), 45–49 (2015)
3. He, Y.: Research on the evolution of American school physical education curriculum. J. Guangzhou Inst. Phys. Educ. **36**(6), 113–116 (2016)
4. Zhang, J., Yang, T.: Development and reform of American Physical Education. Phys. Educ. Sci. **19**(3), 52–55 (1998)

Analysis of the Financing Efficiency Algorithm of the Listed Education Companies Under the Background of the Big Data

Liming Gong[✉]

Xi'an Eurasia University, Xi'an 710065, Shaanxi, China
gongliming@eurasia.edu

Abstract. Building a strong educational country is one of the fundamental projects to achieve the great rejuvenation of the Chinese nation. Nowadays, as the education industry advances rapidly, the financial allocation of government and indirect bank financing can no longer provide it with fully support. It is of particular importance to expand the direct financing channels of listed education companies, so as to increase their financing efficiency. Selecting the educational listed companies located in the United States and Hong Kong as representatives of overseas samples, the paper carries out a comparative study between these listed companies and domestic A-share educational conceptual listed companies. As revealed in the empirical analysis results of financing efficiency, the overall financing efficiency of educational listed companies is relatively lower. In particular, the overall financing efficiency of overseas educational listed companies is higher than that of domestic similar companies. It is found that the scale of enterprise financing is generally small, which needs to be properly expanded to reach the most effective state. In addition, the financing structure should be adjusted and the financing efficiency should be improved along with the appropriate adjustment in the financing scale of companies.

Keywords: Educational listed companies · Financing efficiency · DEA (Data Envelopment Analysis)

1 Introduction

Building a strong educational country is one of the fundamental projects to achieve the great rejuvenation of the nation. The Chinese government highly emphasizes the investment in the education industry and support. As a result of the vigorous support given by government policies and the continuous input of capital, the education investment in China has accounted for over 4% of the GDP in 2018, which is far from enough relative to the thirst of education industry for capital. The expansion of direct financing channels constitutes effective supplement to the development of education industry [1]. In 2018, there were 19 companies to be listed/had been listed in the Hong Kong stock market, while there were 6 companies to be listed/had been listed in the US stock market. A total of 13 education companies achieved IPOS throughout the year. Therefore, it is of great practical significance to explore the financing efficiency of listed companies.

© The Author(s), under exclusive license to Springer Nature Singapore Pte Ltd. 2022
J. C. Hung et al. (Eds.): FC 2021, LNEE 827, pp. 1601–1607, 2022.
https://doi.org/10.1007/978-981-16-8052-6_234

2 Concept Definition and Literature Review

From the perspective of corporate financing methods, the paper conducts a study on the financing efficiency by focusing on the utilization of corporate financing channels and corporate operating benefits. In this paper, several major evaluation methods of financing efficiency are adopted, including Analytic Hierarchy Process (AHP), Fuzzy Comprehensive Evaluation Method, Data Envelopment Analysis (Input-Output Method) and so on. Each method has different emphasis in terms of the analysis of enterprise financing efficiency. The respective weight of analytic hierarchy process and the fuzzy comprehensive evaluation method is determined in accordance with the degree of importance of the selected indicator systems, which can hardly ensure its objectivity in the process. Moreover, such methods are generally applicable to the comparative analysis of the efficiency of several models. In other cases, both methods are not suitable when there are a number of decision making units. Financially speaking, the corporate financing process can be regarded as an input-output process in essence. In this paper, the method of data envelopment analysis is chosen to analyze the financing efficiency of listed companies in education.

France Modigliani & Merton Miller (1958) initiated the MM theorem, indicating that the total value of a company is irrelevant to its financing structure, but is related to its profits creation competence. Integrating the financing theory with life cycle, Berger & Undell (1998) observed that a variety of factors such as enterprise size, information constraint and capital demand are the main reasons for the structural change of enterprise financing in different phases of life cycle. Zeng Kanglin (1993) was the first to propose the concept of "Financing Efficiency" in China, and then he analyzed the cost, output and efficiency based on the comparison between direct financing and indirect. After 2000, the empirical studies on financing efficiency rose which involved the Data Envelopment Analysis (DEA). Data envelopment analysis (DEA) is characterized by high operability as it saves the subjective attaching of weight to variables, which is thus favored by researchers [2]. Wang Xinhong (2007) measured the financing efficiency of high-tech enterprises with the help of DEA method, revealing that there is relevance between financing efficiency and the capital structure of enterprises. Through the measurement of NEEQ (National Equities Exchange and Quotations) enterprises on the basis of DEA-Malmquist method, Ding Hua (2019) discovered that the financing efficiency of the new OTC market was not high as a whole, and the principal restricting factor lies in the pure technical efficiency.

3 Empirical Analysis on Financing Efficiency of Educational Listed Companies

As a non-parametric evaluation method, DEA (Data Envelopment Analysis) performance evaluation mainly evaluates the relative effectiveness of decision making units with multiple inputs and outputs by means of mathematical model. An empirical analysis is performed in the paper on the financing efficiency of private education on the basis of DEA model. CCR model is extensive applied as the basic model of DEA analysis. However, it also has some deficiency in the sense that it assumes that the scale benefit of decision making unit is constant in CCR model. Therefore, the paper is based on the BCC model, which is also called variable scale return model.

3.1 Model Establishment and Indicator Selection

The paper mainly chooses three major input indicators, which are total assets, asset-liability ratio and cost-expense ratio in income respectively. Next, two output indicators are also selected, namely business income and return on equity (ROE). The data is derived from the annual financial statements of 2017 in the WIND database. With the aid of DEAP2.1 software, an empirical analysis of the data is conducted in the paper, and the software requires each indicator to be non-negative value. Based on the results of previous research, the paper introduces the dimensionless method into the non-negative processing of individual negative values in the sample data.

In terms of the samples of study, a total of 13 overseas education listed companies and 29 domestic education listed companies are selected respectively. With the rise of "Internet +" mode in 2014, listed companies in A stock market that are engaged in non-educational fields have also started to turn to the education industry.

According to the education related conceptual module of Oriental Fortune Software and the corporate operating data of WIND database, 29 major A-share listed companies in the line of education are selected with reference to the proportion of education related income in the main business, shown as Hongtao Group, Huamei Holding, Irtouchsystems, Shaanxi Jinye, Xin Nanyang, China Hi-Tech, Vton Group, Baolingbao, Avcon, Iflytek, Qtone Education, Phonix Publishing& Media, Wanxi Media, Changjiang Publishing& Media, Hailun Piano, Oriental Pearl, Kingsun, Guomai Technologies, Hedgon Technology, AVIT, Focus Technology, Kaimeite Gases, Southern Publishing& Media, Tianyu Information, Talkweb Information, The Great Wall Of Culture, Star-Net Communication, Xiuqiang Glasswork, COL Digital Publishing.

The paper takes into consideration the data availability and the correlation between the main business of company and education, the paper selects 13 renowned educational listed companies listed in the United States and Hong Kong (A signifies the listed companies in the United States, while H denotes the listed in Hong Kong) [3]. Maple Leaf Education, Virscend Education, Wisdom Education, Yuhua Education, Minsheng Education, China New Higher are in Hong Kong. New Oriental, Bright Scholar, CDEL, TEDU, Hailiang Education, Nord Anglia, TAL Education are in USA.

3.2 Analysis of Financing Efficiency Evaluation Results of Overseas Education Listed Companies

The calculation is made on the basis of DEAP2.1 software, obtaining the optimal solution and relative efficiency value of each decision unit. In this way, the technical efficiency (TE), pure technical efficiency (PTE) and scale efficiency (E) of listed education companies abroad can be gained.

Amongst the 13 overseas educational listed companies, there are five companies including Maple Leaf Education (H), Hailiang Education (A), New Oriental Education

(A), TEDU (A) and China Distance Education Holdings (A) that has achieved the DEA efficiency of financing efficiency, i.e. TEBCC = 1. Furthermore, the relevant slack variable is zero, accounting for 38.46% of the sample. It indicates that these companies have neither input redundancy (minimum input) nor output shortage (maximum output) in the reference set. Yuhua Education (H) has achieved the purely technical scale but instead of scale efficiency (i.e., weak DEA efficiency), with a slack variable of zero, which accounts for 7.69% of the sample. It suggests that the use of its input capital is efficient under the current financing structure. Nevertheless, the financing scale cannot be matched with the input and output, which cannot reach the degree of constant scale return [4]. A total of 7 companies failed to reach neither technical efficiency nor scale efficiency, accounting for 53.85% of the sample, such as Wisdom Education (H), Virscend Education (H), Bright Scholar Education (A), Nord Anglia (A), Minsheng Education (H), China New Higher Education Group (H), and Tomorrow Advancing Life (A). It can be interpreted that the capital input of these companies fails to achieve the ideal output maximization. Considering all the information above, the financing efficiency of overseas education listed companies is not good enough, and there exists idle or insufficient funds in the use process. What is more, the company scale needs to be adjusted so as to enhance the overall financing efficiency. Shown as Table 1.

Table 1. Overall financing efficiency of 13 overseas educational listed companies

Listed companies	TE (BCC)	
	Number of companies	Proportion (%)
Efficient	5	38.46
Weakly efficient	1	7.69
Inefficient	7	53.85

The following conclusions are thus drawn: Firstly, the financing efficiency of overseas educational listed companies is generally low, based on the fact that only 53.85% of companies have their financing condition above the average level, the financing efficiency of which reaches or is close to DEA, which implies complete efficiency. Secondly, in terms of the financing efficiency of overseas educational listed companies, it is obviously polarized. As shown, 53.85% of sample efficiency values in Region A are relatively high, with 23.08% of the samples in the Region D suggesting a lower sample efficiency value accordingly. The two regions account for more than 75% of the total sample size. Furthermore, the samples in region D are far apart from those in region A and C with a big gap. Finally, a majority of the samples are biased towards Regions A and C. With a relatively higher average PTE value of the samples, it can be observed that the pure technical efficiency value of overseas educational listed companies is relatively high, while the scale efficiency value is relatively lower as a whole.

3.3   Analysis of Returns to Scale

As a result, 5 companies with effective financing efficiency in DEA maintain constant returns to scale, which means the efficient utilization of the integrated funds of the enterprise. In this case, the current output level is reached with the most appropriate input volume. Seven companies whose DEA is not fully efficient are in the state of increasing return on scale, except for one similar company in the state of decreasing return on scale. It suggests that there is still room of improvement in their overall financing capacity, so as to boost the implementation of related projects and the general development of enterprises. Based on the fact that a vast majority of education enterprises belong to asset-light industries, their technology-intensive characteristics are evident. While obtaining greater financial support, they can strive to achieve a higher output ratio by intensifying the research and development of new technologies and new products, transforming the project results and improving the capital operation efficiency, so as to realize the economies of scale [5]. Out of the eight companies without DEA complete efficiency, 1 company reaching the weak DEA efficiency has achieved pure technological efficiency, the financing scale of which is smaller though. Therefore, the key to DEA efficiency is expansion of the financial scale. By introducing new investors through equity financing such as additional issuance and rights offering, or by issuing bonds in the bond market and attracting investment from industrial funds, they can strengthen their financing capacity to give better play to economies of scale. For the other 7 non-DEA efficient companies, their financing structure should be adjusted to enhance their financing efficiency while properly adjusting their financing scale.

4   Analysis of Financing Efficiency Evaluation Results of Domestic Educational Listed Companies

4.1   Overall Evaluation of Financing Efficiency

In general, out of the 29 domestic educational listed companies studied in the paper, 8 companies including Huamei Holding, Wanxi Media, Chnagjiang Publishing and Media, Hailun Piano, Oriental Pearl, Kingsun Science, Kaimeite Gases and Star-Net Communication have realized the DEC efficiency in financing efficiency, namely TEBCC = 1. Besides, the relevant slack variable is zero, which accounts for 27.59% of the sample. It proves that these companies have neither input redundancy (minimum input) nor output shortage (maximum output) in the reference set. The two companies, Phoenix Publishing& Media and Tianyu Information, have achieved pure technical efficiency rather than scale efficiency (DEA weak efficiency). The relevant relaxation variables are both zero, accounting for 6.9% of the sample, which illustrates the efficient utilization of their input capital under the current financing structure [6]. However, the financing scale does not match the input and output. As for the remaining 19 companies, they are neither technically effective nor scale effective, which accounts for 65.52% of the sample, meaning that the ideal output maximization is not reached with regard to the capital input of these companies. By contrast with for foreign counterparts, this indicator (53.85%) is nearly 12% higher, thus revealing that listed education companies in China tend to have lower financing efficiency, and they are idle with insufficient funds to use, with the company size having to be adjusted urgently. In this sense, there is greater room of improvement in their overall financing efficiency.

4.2   Scattered Distribution

The paper comes to the following conclusions: First of all, educational listed companies at home tend to show a lower financing efficiency compared to that of overseas listed companies. The financing status of only 48.28% of the companies is above average, with the financing efficiency reaching or approaching the DEA full efficiency, 5.57% lower than that of overseas listed companies. Secondly, the financing efficiency of domestic educational listed companies is also polarized. Region A contributes to 48.28% of the efficient value sample, while region D has 31.04% of the low efficiency value sample. The samples of the two regions account for 79.31% of the total. Finally, the overall trend of the sample is concentrated in Regions A and B, with relatively higher average SE value being found in the samples. It proves that the scale efficiency value of domestic educational listed companies is relatively higher than that of the overseas listed companies. There exists the problem of lower value of pure technical efficiency and higher scale efficiency value.

4.3   Returns to Scale Analysis

As a result, 8 companies having fulfilled the financing efficiency DEA have constant returns of scale. On the other hand, the 21 non-DEA fully effective companies are in the state of increasing return on scale. Therefore, it validates the relatively smaller financing scale of relevant enterprises in China, which requires proper expansion to reach the most efficient state. Moreover, there are also two other companies showing weak DEA efficiency, and they have achieved the pure technical efficiency. Despite of this, they have the problem of lower financing scale, thus requiring appropriate increase in the financing scale to achieve the DEA efficiency and finally give better play to its scale efficiency. As for the other 19 non-DEA effective companies, their financing structure should be adjusted to enhance the financing efficiency while appropriately adjusting their financing scale.

5   Conclusion

Combined with the empirical analysis of financing efficiency mentioned above, the listed educational companies have a low financing efficiency as a whole, and it is found that the overall financing efficiency of listed educational companies is higher than that of domestic counterparts. Meanwhile, there exists obvious "polarization" in the financing efficiency of such companies. The study shows a higher average PTE value of the sample, suggesting that the pure technical efficiency value is high. However, the overall scale efficiency value is low. The vast majority of companies that are not fully efficient with DEA are in the state of increasing return on scale. It illustrates the relatively smaller financing scale of these companies, which requires proper expansion to reach the most efficient state. In addition, the financing structure should be adjusted to enhance financing efficiency along with appropriate adjustment in the scale of corporate financing.

References

1. Modigliani, F., Miller, M.: Cost of capital, corporation finance, and the theory of investment. Am. Econ. Rev. (06) (1958)
2. Ding, H., Gao, D.: Financing efficiency of new three board listed companies - based on DEA - malmquist method. Friends Account. (02), 21–25 (2019)
3. Song, Z.: Empirical analysis on financing efficiency of listed companies. Bus. Res. (5), 97–99 (2003)
4. Song, X., Liu, Z.: A study on financing efficiency in technological innovation of high-tech enterprises. J. Finance Account. (03), 10–12 (2008)
5. Wang, X.: Research on financing efficiency of high-tech enterprises in China. Northwestern University (2007)
6. Zeng, K.: A brief discussion on direct financing and indirect financing. Financ. Res. (10), 7–11 (1993)

Clustering Analysis of Power Grid Data Based on Dynamic Hierarchical K-Modes

Yu Shi[1,2(✉)], Meiqi Li[2], Yuzhi Zhang[2], and Yuanmei Zhang[1]

[1] Power Economic Research Institute of Jilin Electric Power Co., Ltd.,
Changchun 130000, Jilin, China
[2] Changchun University and Technology, Changchun 130000, Jilin, China
{zyz23777,zhangyuzhi}@ccut.edu.cn

Abstract. With the development of power industry and the continuous improvement of power metering system, the demand side power consumption characteristics of power grid are diversified. How to tap the characteristics of users' electricity consumption behavior, so as to promote the marketization of electricity price, has become a concern. Firstly, the K-Modes algorithm in clustering analysis is introduced, and a dynamic hierarchical K-Modes algorithm is proposed to deal with generic data and give a reasonable H value; Secondly, the data processing method of curve data difference and generic transformation is proposed to better reflect the user curve shape; Finally, the dynamic hierarchical K - modes algorithm is used to get good classification results on simulated data.

Keywords: Power grid system · Cluster analysis · Dynamic hierarchical K - modes algorithm · Curve data

1 Introduction

With the rapid development of hardware technology, people's ability to obtain data is getting higher and higher, the leading city to obtain data is becoming wider and wider, and the data form is also changing from static data to massive and continuous data stream data. For example, call record data flow in communication field, packet flow in network monitoring, transaction data flow in financial field and retail enterprise, and a large number of data flow generated by sensor network developed in recent years [1].

As soon as the data stream is put forward, it has attracted extensive attention of researchers. At present, data stream research can be divided into two aspects: data stream management system (DSMS) and data stream mining [2]. Data stream clustering has become an important research direction of data stream mining because of its important application value in network monitoring and bank transaction data analysis. Several secondary scan clustering algorithms have been proposed, such as stream algorithm, clustream algorithm and hpstream algorithm, as well as some derivative algorithms based on them. But in general, the existing data flow algorithms are basically the improvement of the static data set algorithm.

2 Algorithm Introduction

2.1 K-Mods Algorithm

In 1998, based on the expansion and improvement of the data type of K-means clustering algorithm, Huang proposed the K-Modes clustering algorithm for classified data [3].

The data set processed by Huang's K-Modes algorithm can be formally described as a four tuple classified data information system, which is defined as follows.

A classified data information system can be represented by a quadruple $IS = (U, A, V, f)$, where:

(1) $U = (x_1, x_2, ..., x_n)$ is a nonempty set of N data objects;
(2) $A = (a_1, a_2, ..., a_m)$ is a nonempty set of M classification attributes;
(3) V is the set of all attribute fields, $V = U_{i=1}^m V_i$, $V_{a_i}(i = 1, 2, ...m)$ is the range of attribute $a_i(i = 1, 2, ...m)$;
(4) f is a function of $U \times A \rightarrow V$, which can be formally expressed as: $(\forall x)(\forall y)(x \in U)(y \in A) \rightarrow f(x, y) \in V_{a_i}(i = 1, 2..., m)$.

The K-Modes algorithm mainly improves the k-means algorithm in the following three aspects: ① It introduces a new method to measure the dissimilarity of classified data, that is, a simple 0–1 matching method; ② Use modes instead of means; ③ The idea based on frequency is used to update the cluster center.

2.2 Dynamic Hierarchical K-Modes Algorithm

Furthermore, we hope to use the clustering pedigree to select the appropriate k value. In hierarchical K-Modes, due to the fixed value of k, the division of the pedigree graph generated by p times of K-Modes with the same H value and hierarchical clustering is largely affected by the selected k, which is usually the same as the initial k value. In order to overcome this weakness, we propose to change the k value of each k-mode, so as to weaken the influence of artificial selection of k value. Under this change, p-order K-Modes bring more classification information, so that we can find better k value from a wider range [4].

The main algorithm steps of dynamic hierarchical K - modes are as follows:

(1) Set the value of h as an integer from 2 to $p + 1$, cluster the K-Modes for k times, and get the class mode for each time. Save $\sum_{i=1}^{p} k_i$ k modes of K-Modes for p times as set M;
(2) Hierarchical clustering method is used to classify the modes in set M;
(3) The appropriate k value and corresponding modes were selected by pedigree diagram;
(4) Use the result in step (3) as the initial condition to perform K-Modes once, and return the result.

3 Data Feature Extraction

The curve data is clustered according to the curve shape. Considering that the morphological differences between segments in the time series will be ignored when clustering directly using the original data, the clustering result is unreasonable.

After smoothing the original data, the first-order difference value is taken, and the original matrix X is transformed into the difference matrix D. each behavior d_i is composed of an m-dimensional row vector, which represents the i-th difference sample; Furthermore, the threshold $t(d_i)$ is " $\pm 0.1*$ difference sample range".

$$ca_{ij} = \begin{cases} 1, d_{ij} > t(d_i) \\ 0, |d_{ij}| < t(d_i) \\ -1, d_{ij} < -t(d_i) \end{cases} \tag{1}$$

Where $j = 1, 2, \ldots, m$.

After processing all the difference samples with formula (1), we get the generic matrix C, which is $n \times m$ dimension matrix. Each behavior c_i is composed of an m-dimensional row vector to reflect the shape of the curve. Figure 1 is a data feature extraction graph.

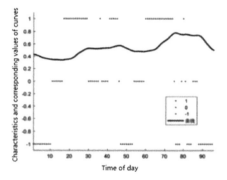

Fig. 1. Feature extraction

The curve in the graph is a dot obtained from a sample X after smoothing and after differential and Eq. (1) processing. It represents three states of the sample curve at each time: rising, flat and falling. The continuous time series data is transformed into discrete state values of generic type, so as to extract the morphological characteristics of the sample, Dynamic hierarchical K - modes algorithm can be used for clustering.

4   Algorithm Experiment

In order to test the advantages of dynamic hierarchical K-Modes compared with traditional K-Modes, a curve data set with 10 dimensions, 3 classes and 500, 200 and 50 samples in each class is simulated by computer, as shown in Fig. 2 to Fig. 3 and Fig. 4, in which the number of bimodal samples is 500, the number of unimodal samples is 200 and the number of single valley samples is 50.

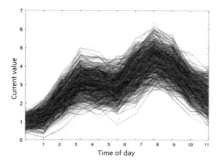

Fig. 2. Bimodal type

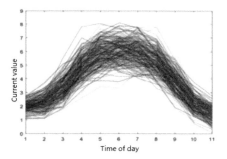

Fig. 3. Unimodal type

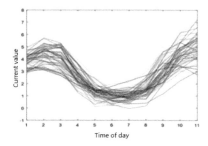

Fig. 4. Single valley type

Before the algorithm comparison, the calculation rule of classification accuracy is given

$$CE = \frac{\sum_{i=1}^{k} n_i}{N} \times 100\% \qquad (2)$$

Where, n_i is the number of the i-th class in the current clustering result that contains the most one class in the correct classification.

For the traditional K - modes algorithm, k must be initialized manually. For a more objective comparison, assume that the number of known classes is $k = 3$, and use K - modes algorithm to cluster the simulation data. The clustering results are shown in Fig. 5.

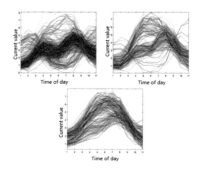

Fig. 5. Ordinary K-modes clustering (k = 3)

As shown in the subgraphs in Fig. 5 and Figs. 2, 3 and 4, the classification accuracy only reaches CE = 47.87%, that is, most of the clustering results fail to accurately correspond to the correct classification. Using dynamic hierarchical K - modes algorithm to cluster the simulated data does not need to determine the $k$value artificially, but only need to determine the desired K value and initial modes through the returned pedigree. Take the number of K-Modes $p = 8$, so the k values calculated for 8 times are 2–9 respectively. In the hierarchical clustering, the "hamming" distance is selected, and the "av-level" class spacing calculation method is applied to obtain the pedigree diagram as shown in Fig. 6.

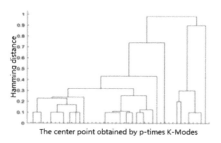

Fig. 6. Clustering pedigree

As can be seen from Fig. 6, the height difference of the cluster tree is the most obvious when k = 3, which can be intercepted from k = 3 and returned to the corresponding modes. On this basis, the hierarchical K-Modes clustering graph as shown in Fig. 7 can be obtained by one K-Modes. As shown in Fig. 7, under the same k = 3 condition, compared with traditional K-Modes, the classification accuracy of dynamic hierarchical K-Modes reaches CE = 76.93%, which is significantly higher than that of traditional K-Modes; And the number of classes K can be determined correctly, so the classification effect of K-Modes is better than that of traditional K-Modes.

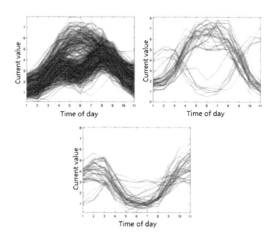

Fig. 7. Hierarchical K-modes clustering

5   Conclusion

The traditional K - modes algorithm combined with hierarchical clustering algorithm successfully transplanted the advantages of hierarchical K - means algorithm to generic data. Moreover, the proposed dynamic hierarchical K - modes algorithm can also determine the K value by clustering pedigree graph, which solves the open problem that the K value must be given initially to a certain extent. The algorithm also has better adaptability and outlier recognition, and can give effective clustering results for real data.

Considering that the dynamic hierarchical K-Modes algorithm is obviously better than the traditional K-Modes algorithm in the results, but there will still be some misclassification phenomenon in clustering. In the follow-up work, the robust idea is introduced. Instead of taking all samples into the clustering iteration process, only a part of important samples are selected for iteration, so as to improve the accuracy of the algorithm. At the same time, in the aspect of feature extraction, we also hope to introduce a more refined adaptive threshold setting to improve the clustering quality from the perspective of data.

Acknowledgements. Research on Key Issues of Market System and Mechanism of Jilin Power lement Allocation in State Grid Jilin Economic Research Institute in 2020.

References

1. Sun, J., Liu, J., Zhao, L.: research on clustering algorithm. Acta Sin. **19**(1), 48–61 (2008)
2. Su, J., Xue, H., Zhan, H.: K-means initial clustering center optimization algorithm based on partition. Microelectron. Comput. **26**(1), 8–11 (2009)
3. Jin, J.: Summary of clustering methods. Comput. Sci. **41**(11a), 288–293 (2014)
4. Zhang, Y., Zhou, Y.: Overview of clustering algorithms. Comput. Appl. **39**(7), 1869–1882 (2019)

Hybrid Optimization Algorithm for Optimal Allocation of PMU in Power System

Yu Shi[1,2(✉)], Hongwei Gao[2], Yuzhi Zhang[2], and Yuanmei Zhang[1]

[1] Power Economic Research Institute of Jilin Electric Power Co., Ltd.,
Changchun 130000, Jilin, China
[2] Changchun University and Technology, Changchun 130000, Jilin, China
{zyz23777,zhangyuzhi}@ccut.edu.cn

Abstract. In this paper, a hybrid optimization algorithm is proposed to solve the Optimal PMU placement problem of phasor measurement units. The hybrid optimization algorithm takes particle swarm optimization as the main body, introduces crossover and mutation operation, and combines with simulated annealing mechanism to control the update of particles. When dealing with the constraint problem of understanding, a heuristic repair strategy based on probability is adopted to avoid the single feature of the repaired solution. The hybrid algorithm is compared with other algorithms on several IEEE standard systems. The results show that the convergence rate of the hybrid algorithm is several times higher than that of the standard particle swarm optimization algorithm on large-scale systems, and the computational complexity is tens of times less than that of the simulated annealing algorithm. It shows that the hybrid algorithm is feasible and efficient.

Keywords: Power system · PMU · Hybrid optimization algorithm · Phasor measurement unit

1 Introduction

With the development of society and the rapid development of power industry, the power system is striding forward to the era of high voltage, large power grid and large units. Compared with the traditional power grid, large power grid has obvious advantages, such as more reasonable development and utilization of resources, greatly reducing investment and operating costs, and improving power supply reliability. Therefore, countries all over the world regard large power grid interconnection as the future development direction of power industry. Phasor measurement unit, short for PMU, is a kind of synchronized phasor measurement equipment developed by using GPS clock synchronization technology. It is an important method to observe the real-time dynamic behavior of power system. PMU can not only measure bus voltage phasor synchronously, but also measure line current phasor, generator power angle and speed. By installing PMU in different places and connecting them with communication network, the synchronized phasor measurement Wan is formed. The operator can use this measurement system to master and control the operation status of the whole power system in real time in the dispatching center.

© The Author(s), under exclusive license to Springer Nature Singapore Pte Ltd. 2022
J. C. Hung et al. (Eds.): FC 2021, LNEE 827, pp. 1615–1620, 2022.
https://doi.org/10.1007/978-981-16-8052-6_236

Hybrid optimization algorithm (HOA) is a hybrid optimization algorithm designed in this paper, which can solve the problem of PMU configuration optimization under the condition that the system state can be observed completely. The main body of the algorithm is particle swarm optimization algorithm, which has the characteristics of simple programming, high search efficiency and easy implementation; The crossover and mutation operations are used to ensure the diversity of the population and improve the global optimization ability; In order to control the updating of particles, the simulated annealing mechanism is specially introduced, which successfully overcomes the shortcomings of falling into some extreme points, and improves the convergence accuracy and speed of evolution. At the same time, when explaining the constraint problem of the solution, the heuristic repair strategy is used on the basis of probability to solve the problem that the solution is easy to have a single local feature after repair. In application, the effect of this strategy is very obvious [1].

2   Observability of Power System

If there are enough measurements in the power system, and the geographical distribution is appropriate, the current state of the system can be solved, then the system is called observable. There are two main methods to investigate whether a system is observable: numerical method and topological method.

In order to judge the observability of the system by numerical method, it is necessary to check whether there is zero principal component in the triangular decomposition of the information matrix, which requires a lot of calculation. Therefore, the idea of topological method is used to judge the observability of the system.

In topology method, if there is at least one full rank measurement spanning tree in the network, and all network nodes can be connected through the branch, then the system is called observable. That is to say, if all nodes in the system can be observed, the power system is completely observable. Among them, several concepts about quantity measurement are as follows.

(1) Direct measurement: refers to the voltage phasor on the bus with PMU and the current phasor on the connected line
(2) Indirect quantity measurement: it refers to the voltage and current phasor calculated by Ohm's law and Kirchhoff's law using known quantity measurement. Four principles of indirect measurement are obtained: ① if the voltage and current at one end of the line are known, the voltage at the other end can be obtained; ② If the voltage at both ends of the line is known, the current of the line can be calculated; ③ If the current of only one of the connected branches of the zero injection bus is unknown, the current of the branch can be calculated; ④ If the voltages of the adjacent buses of the zero injection bus are known, the voltage of the zero injection bus can be calculated.

Using the concepts of direct measurement and indirect measurement, based on the idea of topology observability analysis, the following methods can be adopted to judge the observability of the system: firstly, the direct measurement of the bus with PMU is obtained, that is, the voltage of the bus and the current of the connected lines; Then, according to the network topology and the above four principles of calculating indirect measurement, indirect measurement is obtained from known direct measurement; Then a new indirect measurement is calculated by searching the known direct and indirect measurements. If a new indirect measurement can be obtained, it will continue to search and calculate the new indirect measurement according to the current known quantity until the new indirect measurement can no longer be obtained. At this time, if all buses can be observed, then the whole power system is observable.

This method of system observability judgment based on topology idea is easy to realize in programming, and the amount of calculation is less, which greatly improves the efficiency of the algorithm [2].

3   The Mathematical Model of PMU Optimal Allocation Problem

3.1   Form of Objective Function and Solution

The objective of PMU optimal allocation problem is to minimize the number of PMUs under the constraints of system observability

$$
\begin{aligned}
\min J &= m \\
s, t, U_i &= 1 \quad i = 1, 2, \cdots, N
\end{aligned}
\tag{1}
$$

Where m is the number of PMUs; $U_i = 1$ indicates that bus i is observable; N is the number of system buses.

The fitness function is defined as

$$
f = C_{\max} - \sum_{i=1}^{N} P_i
\tag{2}
$$

Where: C_{max} is a large positive number to ensure that the fitness function value is always positive; If PMU is installed on bus i, then P_i is 1, otherwise P_i is 0. When the fitness function reaches the maximum, the optimal solution of this problem is obtained [3].

The dimension of the solution of the optimization problem is equal to the bus number of the power system, and binary coding is adopted, that is, the value of each dimension can only be 0 or 1. When a bus is installed with PMU, the corresponding position of the solution is taken as 1, otherwise it is taken as 0.

3.2   Treatment of Constraint Conditions

The constraint of the optimization object is that every bus in the system can be observed. In this paper, a heuristic method based on probability is used to repair the infeasible solution when the solution does not meet the constraints in the optimization process, so that the solution of the problem is always strictly limited in the feasible solution space. The specific operation is: in the unobservable bus, randomly select a bus to install PMU by roulette method (the probability of bus I being selected P: = D/2D, where D; Is the outgoing degree of bus J, and S is the number of unobservable buses); Then judge whether the current situation meets the constraints. If not, continue to install PMU in one of the unobservable buses in the above way until the constraints of the optimization problem are met.

In the past, only the bus with the largest degree of outgoing line is selected when the bus with PMU is selected. Although experience shows that setting PMU on the bus with large outgoing degree is helpful to improve the observability of the system, it is always selected to install PMU on the bus with the largest outgoing degree, so that the local characteristics of the repaired solution are single, and some positions of the solution are always 1, which limits the search space. Therefore, the idea of probability is used in the selection of bus with PMU in the repair method, which makes the bus with large degree of outgoing line have a greater chance to be selected, and the bus with small degree of outgoing line also have a certain chance to be selected, so that both the observability effect and the diversity of solutions are taken into account. Practice has proved that this method is effective.

4   Design of Hybrid Optimization Algorithm

Particle swarm optimization (PSO) was first proposed by Eberhart and Kennedy in 1995. In this algorithm, all individuals are assumed to be particles moving at a certain speed in n-dimensional space. The final solution of the problem needs to be optimized is expressed by the specific position of the particle. The particle will continuously change its speed according to its own motion experience and the overall trajectory, and gradually move to a better region, so as to find the optimal solution needed to solve the problem [4].

The so-called simulated annealing algorithm is a combinatorial optimization method, which takes the metal annealing process as a reference. Because the algorithm can make the worse point be identified with a certain probability, it can prevent the solution from falling into the local optimal solution, but the optimization process is long and the efficiency is not ideal. Therefore, this paper combines the advantages of particle swarm optimization algorithm and simulated annealing algorithm, so as to obtain the ideal optimization effect, which is also the purpose of this paper to design a hybrid optimization algorithm.

The main body of the hybrid optimization algorithm designed in this paper uses particle swarm optimization, combined with crossover and mutation operation to increase the diversity of the population, and the update of particles depends on simulated annealing mechanism. The basic calculation process of the algorithm is: using random method to select the initial particle population; A new group of individuals is generated by iterative optimization and particle swarm optimization; Crossover and mutation operations are used to screen particles, and simulated annealing is used to select acceptable particles; The annealing temperature was changed and a new iteration was carried out; Until the exit condition is met, the optimal solution is obtained. In this process, new particles can be generated by the combination of crossover operations, and then the effective search can be carried out in the solution space to reduce the damage to the effective mode; Mutation operation can not only reduce the premature convergence of the algorithm, but also enhance the ability to explore the optimal solution. The specific method of crossover operation is to get some particles from the original particle swarm, and then randomly cross the obtained particles to produce the same number of new particles. In contrast, the strategy of mutation operation is based on probability. The specific method is to randomly select one bus with PMU, and then cancel the PMU to obtain a new particle solution. This strategy can greatly improve the local mining ability of the algorithm and enhance the optimization speed. At the same time, in the crossover and mutation operation, we need to pay attention to the protection of elite individuals to ensure that the better solution is reasonably protected. The hybrid optimization algorithm is shown in Fig. 1.

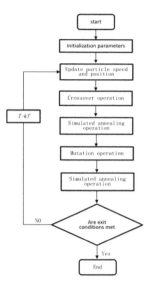

Fig. 1. Flow chart of hybrid optimization algorithm

5   Conclusion

With the rapid development of communication network and phasor measurement technology, PMU is widely used in power system. Under the premise of ensuring that the power system is fully observable, how to minimize the installation and use of PMU is an important issue that we need to pay attention to. To solve this problem, this paper designs a new algorithm model. The theme of the algorithm is particle swarm optimization algorithm, combined with crossover and mutation operation, and simulated annealing mechanism is used to update the master particles. The algorithm is easy to master, easy to implement, and can converge to the optimal solution with less computation.

Acknowledgements. Research on Key Issues of Market System and Mechanism of Jilin Power lement Allocation in State Grid Jilin Economic Research Institute in 2020.

References

1. Cheng, T.: Optimization of PMU configuration for improving state estimation accuracy. Southwest Jiaotong University (2008)
2. Wang, L.: State estimation based on PMU. North China Electric Power University, Beijing (2008)
3. Xing, J.: Contribution of PMU to power flow calculation and estimation. Shandong University (2006)
4. Liu, B.: Power system PMU optimal configuration and feasible scheme evaluation. Hunan University (2009)

Research and Implementation of Student Management System Based on Java EE Technology

Shuolun Song[✉], Illuminada Isican, and Yun Fu

University of the Cordilleras, 2600 Baguio, Philippines
irisican@ic-bcf.edu.ph

Abstract. With the development of information technology, information technology is gradually extended to the campus management, administrative office, the existing student management system can not meet the actual needs because of its adaptability. With the increase of college enrollment, the complexity of student management system is getting higher and higher. It is urgent to use computer to realize student information management and statistics, and provide better services for teachers and students. Based on the Java EE platform, this paper develops a student management system with certain characteristics, which has good reliability, portability and security, and has important reference significance for promoting the development of student management and the realization of informatization in Colleges and universities.

Keywords: Java EE · Student management · Management system

1 Introduction

Student work plays an important role in higher education, and it is an important part to achieve the goal of talent training. With the deepening of national education reform, quality education has been promoted in an all-round way, the scale of each school has been expanding, and teaching resources have become increasingly tense. For the school management department, this not only increases the workload, but also increases the difficulty of work. Manual recording of student information has many problems, such as large amount of information, difficult to find, inconvenient to save, and can not be reused, which will cause a lot of waste of hardware equipment and a lot of unnecessary human, material and financial input, and the low work efficiency can not meet the development needs of the current situation. Education departments and universities have invested a lot of human, material and financial resources in the construction of campus information system and the improvement of network facilities.

Information construction of teaching management in Colleges and universities, the development of a perfect and effective student management information system, work departments for effective management, and save a lot of manpower and material resources for the school, reduce some problems in the work. It has greatly promoted the process of information management of college students, and greatly promoted the development of student management. It is of great significance to solve the problems

J. C. Hung et al. (Eds.): FC 2021, LNEE 827, pp. 1621–1625, 2022.
https://doi.org/10.1007/978-981-16-8052-6_237

encountered in the process of university reform, and the smooth solution of these problems also points out the direction for the promotion of university reform, which is paperless, intelligent and comprehensive, and lays a good foundation for the further realization of the perfect computer student work management system and the whole University Information System [1]. The system can timely provide data and service awareness of inter departmental cooperation, improve work efficiency, improve management level, and contribute to the smooth realization of the grand goal of modernization of higher education management.

2   Java EE Technology Architecture

SSH2 is the abbreviation of spring, Struts2 and hibernate. It is an excellent open source architecture combination based on Java EE and a lightweight development platform. One of the main advantages of spring framework is layered architecture, which allows users to choose the components they need, and provides an integrated framework for Java EE application development. It is powerful, easy to decouple, simplify development, support AOP programming, support declarative transactions, and easy to integrate various excellent frameworks. As one of the open source projects sponsored by Apache Software Foundation, struts implements the application framework of three-tier design pattern (MVC) based on Java EE application. The implementation of MVC framework based on Sun Java EE platform mainly relies on servlet and SP technology. Hibernate is a persistent object relational mapping framework, and provides the source code. It encapsulates JDBC with lightweight objects and provides many useful templates and interfaces for program calls, which enables Java programmers to use object programming thinking to operate the database as much as possible [2].

3   Overall Objective of the System

Student management is a systematic and complex work process, which needs the coordination of various functional departments in Colleges and universities. Any department needs to complete the management work within the scope. After practical investigation, demand communication and research, the organizational functions of various functional departments related to student management are shown in Fig. 1.

The informatization construction of colleges and universities in our country has been out of scale, and the infrastructure construction is more and more perfect. But on the whole, there are still many shortcomings. It is pointed out that many colleges and universities' information construction generally "pay attention to the construction, despise the use". The education department has failed to develop perfect information software for colleges and universities, and it is difficult to design for different schools. The implementation of credit system makes the contradiction between the growing demand for student management and the ineffective management of current software and management methods increasingly prominent. At present, the management system of many colleges and universities is not standardized, and the management system is

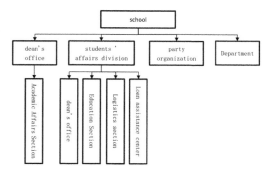

Fig. 1. Student management organization

not perfect. It is imperative to develop a student management information system suitable for colleges and universities [3].

This paper studies and implements a student management system based on the network environment, which can solve the specific problems existing in the process of students' comprehensive information management under the network environment. The main functions include student information management, student status management, enrollment and employment management, curriculum management, performance management, examination management, Party Organization management, loan management, poor students management, dormitory management Class management, reward and punishment management, etc. It is expected to achieve the following goals:

(1) Ensure that students complete personal information management, score query, course selection and other functions; Administrators can complete the basic information management, educational affairs management, dormitory information management and other resource management functions, teachers can complete the performance management, curriculum declaration and other functions, to achieve a comprehensive information management system;

(2) Good versatility, high reliability and openness. The system can accommodate or connect with other systems, provide good data interface with other educational administration management systems, realize information exchange, and lay the foundation for the establishment of educational administration management information system with integrated functions, networking and automation;

(3) Simple and convenient operation, beautiful interface, easy to use: with a variety of query, statistics and reporting functions, to provide decision-making basis for teachers and leaders.

4   System Design

4.1   Hardware Architecture Design

The system architecture design is shown in Fig. 2. Users access the remote web service content through the network, and the web server is connected with the distributed database through the application server. In order to prevent the problems caused by the concurrent access of a large number of students, the access sources are shunted, and the requests are shunted to different web servers. In order to prevent the intrusion of malicious users, a firewall is added to shield the IP address [4].

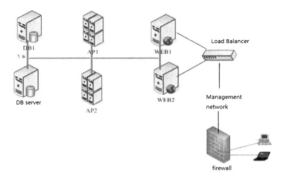

Fig. 2. System architecture diagram

4.2 Software Architecture Design

The system software architecture is shown in Fig. 3. The system software architecture is divided into three layers, the top layer is the view layer, the middle layer is the business logic layer, and the bottom layer is the database layer.

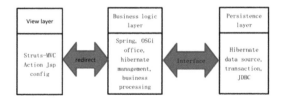

Fig. 3. Software system architecture

The top layer of the student management system is the view layer. The user can access the system through the view of the student management system. The middle layer is the business logic layer, which provides an interface for other integration, and can integrate resource information through the data processing adaptation layer. The business layer is based on Java EE technology architecture. The data layer completes data persistence and provides data operation services for the upper layer. In order to meet the network requirements of the system, the system adopts a distributed architecture and needs to complete the communication through the network communication layer.

4.3   System Function Module Design

Student management system is a comprehensive information management system, including many functional departments of the school. Combined with the reality of the educational administration management system and examination management; Student management module includes enrollment and employment, reward and punishment management, poor students management (including national loans, difficult subsidies), dormitory management, Party committee organization management includes party organization relationship transfer and organization of educational activities, Party member development, etc. system maintenance includes authority management, system initial work, data backup management, log management and online help module.

5   Conclusion

In the situation of popularization of higher education and large-scale enrollment expansion, student management has become extremely cumbersome. The scope of university student management involves all aspects of colleges and universities, including teaching, scientific research and management departments. The management content is cumbersome and extensive. If we only rely on manual management or simple office automation tools to manage the information and data related to students, it can not meet the needs of student management. Information technology plays a great role in promoting the development of the school. The rapid development of network technology, software and computer technology makes it possible to establish an efficient, coordinated and integrated digital office system on the existing campus computer network. This paper studies and implements a student management system based on Java EE. Combined with the school student management business practice, the student management system is promoted and put into trial operation in the whole school, and has achieved good results. The online operation of the management system greatly improves the efficiency of student management, saves a lot of manpower and material resources for the school, enhances the information exchange and interaction among students, classes, teachers and majors, and promotes the development of student management, which has important reference significance for the realization of informatization in Colleges and Universities.

References

1. Wang, L.: Design and implementation of the information management system for college students, pp. 15–17. Shandong University (2007)
2. Wu, Y., Wu, H.: Compiled by Java2 Programming. Science Press
3. Li, W.: Student information management system based on Java, pp. 103–104. University of Electronic Technology (2009)
4. Yangshiqing: Research on student information management system based on network environment, pp. 203–205. Wuhan University of Technology (2004)

Design and Implementation of Student Integrity Information Management System Based on Blockchain Technology

Fei Han$^{(\boxtimes)}$, Lety Epistola, and Shuolun Song

University of the Cordilleras, 2600 Baguio, Philippines
lcepistola@uc-bcf.edu.ph

Abstract. With the continuous expansion of the scale of the school, the number of students has increased dramatically, and the amount of information about students has doubled. Facing the huge amount of information, we need to have a student information management system to improve the efficiency of student management. Based on the blockchain technology, this paper designs the student's information management system. Through this system, we can achieve the standardized management of information, scientific statistics and rapid query, so as to reduce the workload of management.

Keywords: Blockchain technology · Student information · Information management · Management system

1 Introduction

Student management is an important part of university management. The stability of students is a powerful guarantee for the rapid development of the school, which is not only related to the performance of students in school, but also measures the tracking management of a student to a great extent; At the same time, student management is also related to the evaluation of counselors. The importance of student management is self-evident, but student management in the school is very cumbersome, in each department, each student needs to have the corresponding file records. In the school management, student management involves other management content is the most complex, it involves the student personal file management, daily assessment management, performance management and so on. So in the actual management work, often due to the number of records, complex management, poor continuity, resulting in student management confusion. The best solution to this confusion is to record the whole student management and implement electronic management with the help of computer technology and database management system. The purpose of this project is to develop "student management information system" based on blockchain technology. Through this system, we can master the information management of students, realize the electronization of student information management, and provide an electronic student management platform [1].

The realization of the student information management system and the corresponding supporting facilities can make the student management work play out the

© The Author(s), under exclusive license to Springer Nature Singapore Pte Ltd. 2022
J. C. Hung et al. (Eds.): FC 2021, LNEE 827, pp. 1626–1630, 2022.
https://doi.org/10.1007/978-981-16-8052-6_238

maximum efficiency, so as to obtain a huge harvest. Its significance is mainly manifested in: the development and application of student management system can improve the management level of the school. School office efficiency can be greatly improved, providing a good tool for school information management, simplifying the tedious work mode, so as to make the school management more reasonable and scientific. A good management information system saves a lot of manpower and material resources, and also avoids a lot of repetitive work [2].

2 Blockchain Technology

Blockchain is a kind of distributed shared database, which enables all nodes to maintain the same public data account book through decentralization and distrust. Any number of nodes in the blockchain system are associated to generate data blocks by using cryptographic methods. All transaction data in the blockchain system are included in the block for a period of time, and a data "password" is generated to verify the validity of the block information. This "password" is also used to link with the next block [3].

The block header of each block will refer to the hash value of the previous block, which is used to connect the block with the previous block in the blockchain. Such connected blocks form a chain, as shown in Fig. 1.

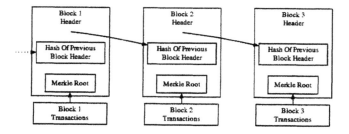

Fig. 1. Blockchain and blockchain in blockchain

3 The Design Principle of the System

From the perspective of system mode, the construction of student information management system in Colleges and universities has just started. Each college's student management has its own characteristics, and there is no fixed mode to refer to. In addition, high system availability and high system throughput are required. Therefore, the following principles should be embodied in the selection and development of system scheme:

(1) Information resource sharing: it can meet the needs of query.
(2) Practicability: Based on the user's needs, realize all the requirements of student management, and ensure the normal operation of student management business.

(3) Advanced Nature: we must ensure the advanced nature of technology and conform to the trend and requirements of future development.
(4) Scalability: the construction of the system should not only meet the current needs, but also have good scalability. With the increase and change of business functions and the number of users, it provides a convenient and fast implementation and upgrade scheme.
(5) Stability and recoverability: to ensure the stability of the system, to ensure that in case of problems can be timely and accurately restore the system.
(6) Manageability: the use and management of the system should be based on the principle of simple, easy to operate, convenient and practical, so as to ensure the high manageability of the system and reduce the cost of system management and maintenance.

The reform of student management, whether in policy or in mechanism, is a work that needs continuous improvement. Therefore, the construction of student information management system also has a process of continuous improvement. From the construction goal of student information management system, it is a socialized system engineering with high technical requirements, strong professional background and long system construction cycle. Therefore, in order to ensure its quality, it is necessary to design, develop and manage according to the method of system engineering, and there must be a set of quality assurance system.

4 Design of Student Integrity Information Management System Based on Blockchain Technology

The overall task of the development of student information management system is to realize the systematization, standardization and automation of student information relationship.

The basic requirement and function of the system development is to realize the management and operation of student information data, including data related to students. After research, it is found that the data management based on blockchain technology database system makes the software have better performance [4].

4.1 Functional Module Division

Student information management system has the functions of examination results and student information query, student performance management, student information management, examination arrangement, class management and subject management. According to these functions, the functional module diagram of the system can be drawn, and the functional module diagram of the client is shown in Fig. 2.

The function of the client is relatively simple, mainly to facilitate students to query personal information and examination information.

It is mainly divided into six parts: basic information management, score management, unified examination management, class management, subject management and administrator maintenance.

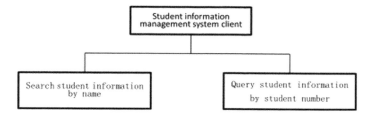

Fig. 2. Division of client function module of student information management system

City ID number: basic information management module: basic information includes student name, student ID, ticket number, ID card number, gender, date of birth, city, area, address, postcode, home phone, bedroom telephone, mobile phone, specialty, class, rewards and punishments, specialty and so on. Through this function module, the student information list can be displayed, the student information table can be updated by entering student information, and the student information can be queried by name or student number.

Performance management module: there are many small function modules in the performance management module, which mainly involves the implementation of performance ranking. The module includes single subject performance ranking and total performance ranking.

Unified examination management, class management and subject management module: these three functional modules are mainly through the operation of the corresponding table to add, modify and delete information.

Administrator maintenance module: when you need to add a new administrator or the old administrator password needs to be corrected, you need to use this module.

4.2 Analysis of Database Conceptual Structure

On the basis of the above data items and data structure, we can design various entities and their relationships to meet the requirements, and then express these contents with entity relationship diagram, namely E-R (entity relationship) diagram.

5 Conclusion

With the continuous development of computer technology, management information system is also a developing concept. On the whole, it develops from electronic data processing data system to management system, and then to more advanced management information system including decision support system. Based on the blockchain technology, this paper designs a student information management system, including student basic information management, student information query management, score information management, unified examination information management, class management, subject management and other practical functions. The modular program design has good scalability. Student information management system is to achieve the management and query of student information, course information and performance information, which has a good auxiliary effect on teaching.

References

1. Shen, H.: Analysis and design of student information management system. J. Read. Writ. **3**(9), 37 (2006)
2. Chang, H.: Design method of student information management system. J. Liaoning Inst. Technol. **24**(3), 28 (2004)
3. Yuan, Y., Wang, F.: Development status and prospect of blockchain technology. Acta Autom. Sin. **42**(4), 481–494 (2016)
4. Zhu, S.: Management Information System Course. Tsinghua University Press, Beijing (2006)

Social Space and Personalized Recommendation System of Education Information Platform

Wei Chen[(✉)]

Yunnan Technology and Business University, Kunming 651701, China

Abstract. With the popularity of the Internet and the improvement of people's living standards, the society has paid more attention to preschool education, and more and more kindergartens have invested in preschool education information management and service. However, many preschool education services only focus on information management, and do not really provide real-time communication and interaction platform. In addition, with the rapid development of the website and the increase of the number of users, the amount of information will expand at a very high rate. How to filter and spread the information in time to improve the user experience is also very necessary. On the basis of discussing the development status of social services in kindergartens, combined with the application of personalized recommendation in social networks, this paper designs and implements the social space system and personalized recommendation system of preschool education information platform. Through the application of personalized recommendation technology, it provides better communication and sharing experience for parents, teachers and principals.

Keywords: Preschool education · Promotion of information technology · Social contact · Personalized recommendation system

1 Introduction

With the development of the times, more and more parents are employed in cities. The proportion of parents taking care of young children in their homes has declined year by year. They have been sent to kindergarten when they are very young, and most of them are only children, which are cared by parents and grandparents. The growth of children in kindergarten is closely related to the hearts of every family. Parents are more expected to know the real situation of their children on campus. Although more and more kindergartens have begun the information management mode, it is urgent to realize the seamless home school interaction expected by people. Providing a social space platform for close communication between parents and teachers in the preschool education information service website is an inevitable trend under the background of kindergarten information management. The mode of real-time interaction in the Internet will also be welcomed by kindergarten and even parents [1].

J. C. Hung et al. (Eds.): FC 2021, LNEE 827, pp. 1631–1636, 2022.
https://doi.org/10.1007/978-981-16-8052-6_239

In addition, because the social space in the preschool education information service website is not the mainstream social network, too little interaction, low viscosity and insufficient activity among users and between users and systems, which can easily lead to a large number of users loss, which is very unfavorable to the development of preschool education information platform, it is necessary to introduce the recommendation system into preschool social space, The introduction of recommendation system can better enhance the attention and communication between parents and teachers, and improve the interaction of social space. Therefore, this paper first designs a social space system suitable for Kindergarten Parents and kindergarten parents to communicate and interact with the kindergarten, then researches the application of collaborative filtering recommendation system in social network, aiming at solving the problem of cold start and data sparsity, and proposes a recommendation algorithm suitable for the social environment and applies it to the spatial system, Provide personalized recommendation for users with corresponding spatial information.

2   System Requirements Analysis

With the improvement of living standard, people pay more and more attention to preschool education, and kindergartens gradually turn their attention to the management mode of website. On the one hand, kindergarten introduces the information website of preschool education service to provide a social space platform for close communication between parents and the kindergarten, which is also an inevitable trend of social informatization: on the other hand, with the increasing of users and functions of the website, Too much useful or useless information makes people unable to meet the needs of others. Users often fail to see the content that is useful or interested in themselves, which will lead to the decrease of user activity, low viscosity between users and users and systems, which will cause a large number of users' loss, which is not conducive to the development of preschool education information social platform, Therefore, it is necessary to add the personalized recommendation function of information to the preschool education information social network [2]. The social space system of this paper is a subsystem of preschool education information platform, which is composed of social space system and personalized recommendation system. Based on social space system, the user experience is improved through personalized recommendation technology. The needs of the social space system for preschool education are as follows:

(1) Provide personalized space for all registered users and manage users
(2) The system role is divided into teachers, parents, employees and administrators, among which the faculty is divided into the director, teacher, etc., which provides the personalized configuration function of personal space for users of different roles;
(3) Provide personal information display, user data display (talk, diary, photo album, video, etc.), class information (class teacher, classmate), friends relationship, dynamic interaction, message reminder and other functions;

(4) To provide one-to-one or one-to-many interaction, notice, announcement and other functions between teachers and students;

(5) Provide the users with the entrance of campus homepage and class homepage, and provide the garden management and class management office platform for the garden director and teachers;

(6) Provide personalized recommendation.

3   System Architecture

The social space system of preschool education information platform is composed of basic social space module, personalized recommendation module and data module. Users store personal information, dynamic information and social relationship data in the database, and interact with the other two upper modules through the data module. The basic module of social space is the main module of social space system and the data source of personalized recommendation module. The overall framework of social space system is shown in Fig. 1.

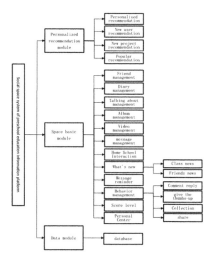

Fig. 1. Framework of social space system of preschool education information platform

When a user logs in, the space system will assign different space to the user according to the user's role, as shown in Fig. 2.

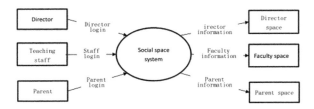

Fig. 2. Overall data flow diagram of social space system

SSH is an integrated framework of struts + Spring + hibernate, which is divided into presentation layer, business logic layer and data persistence layer. The web application has the advantages of clear structure, reusability and easy maintenance. Among them, struts is responsible for the separation of MVC (model view controller) and the control of business jump; Spring uses basic JavaBeans instead of EJB (Enterprise JavaBeans), which provides more functions for enterprise application development; Hibernate provides support for data persistence layer, provides lightweight object encapsulation for JDBC (Java Data Base Connectivity), and can complete data persistence in web application of servlet/JSP [3]. Adopting such a three-tier architecture is conducive to the development and maintenance of the project, as well as the reuse of functional modules.

4   Design of Personalized Recommendation System in Social Space

4.1   Collaborative Filtering Algorithm

Collaborative filtering was formally proposed by Xerox in 1992 at the PARC research center. Since collaborative filtering was proposed, many researchers have been studying it. Collaborative filtering has achieved great success in a short period of time, and has become the most popular technology in all kinds of recommendation technology applications. In addition, collaborative filtering is also one of the most commonly used technologies in social network personalized recommendation system in traditional recommendation algorithm framework [4]. The user model in this paper is composed of the traditional collaborative filtering user interest model and the social network model with friend relationship and subordinate relationship.

$$Model = Model_{interest} + Model_{social} \tag{1}$$

In view of the universal adaptability of collaborative filtering algorithm, the algorithm is widely used in many fields. The core idea of the algorithm is to use the similarity or similarity of user preferences for content recommendation. Collaborative filtering recommendation algorithm mainly includes three types: user based collaborative filtering recommendation algorithm, article based collaborative filtering recommendation algorithm and model-based collaborative filtering recommendation algorithm. Taking the collaborative filtering recommendation algorithm based on users as an example, users a and C have higher similarity in item preferences. As shown in Fig. 3.

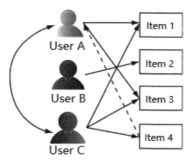

Fig. 3. User based recommendation diagram

4.2 Recommendation System Architecture

The recommendation system includes four modules: personalized recommendation, new user recommendation, new project recommendation and popular recommendation, as shown in Fig. 4. Users can not only generate recommendations according to their interest preferences and historical behavior data, but also view the popular information recommendations of the park. Among them, personalized recommendation module is the core module of the whole recommendation system, new user recommendation and new project recommendation are the modules set up for the cold start problem solving strategy of personalized recommendation, and hot recommendation is the additional module. The recommendation data source of the system is all kinds of personal dynamic data generated by users in the social space. The user interest model comes from the user item rating matrix of Boolean preference generated by users' like behavior. The user social relationship model comes from the real-time friend relationship established in the social space.

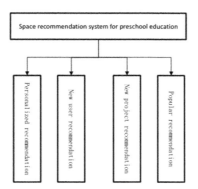

Fig. 4. Recommendation system architecture

5 Conclusion

This paper designs and implements a social space for preschool education information platform, and applies the personalized recommendation system in this space. It also studies and improves some problems existing in collaborative filtering. However, due to the immature application of personalized recommendation, the recommendation effect needs to be improved, In the follow-up work, we need to further develop a more suitable high-quality recommendation algorithm for this platform.

Acknowledgements. The Ministry of education project "Analysis on the construction path of double qualified excellent teachers in preschool education" 2019XQJYKT62; the Ministry of education project "Research on the construction path of preschool education curriculum system under the background of" 1 + X "certificate" 2019XQJYKT63.

References

1. Li, S.: Observation and analysis of "home interaction" column on the website of model kindergarten. Shanghai Educ. Sci. Res. (1), 90–93, 84 (2014)
2. Wang, X., Huang, Z.: Personalized recommendation based on K-means clustering in network resources. J. Beijing Univ. Posts Telecommun. **s1**, 120–124 (2014)
3. Wang, W., Gao, L.: online social network friend recommendation algorithm based on social circle. Acta Computa Sin. **04**, 801–808 (2014)
4. Zhou, C., Li, B.: A recommendation method based on user trust network. J. Beijing Univ. Posts Telecommun. **04**, 98–102 (2014)

Evaluation Method of Posture Balance of Wushu Athletes Based on Higuchi Analysis

Sheng-li Jiao[✉]

Anjiang University of Technology, Nanjing 243031, China

Abstract. In martial arts, the correct recognition and correction of the wrong posture of martial arts athletes can improve the quality of their daily training. In the process of posture recognition and correction, the affine deformation of human body is easy to occur in the process of martial arts, which leads to the appearance of action feature points with low brightness. However, the traditional method is to extract these feature points and compare them with the correct posture to realize the recognition and correction of posture. As a result, the wrong posture of athletes can not be detected and corrected in real time. In order to accurately evaluate the balance of the control of specific movements and postures of Wushu athletes, a balance evaluation method based on Higuchi fractal dimension analysis is proposed. The fractal dimension (FD) of the time series in the front to back and inside to outside directions is calculated by Higuchi algorithm. It can realize the tracking and monitoring of athletes' movements and complete the detection and recognition of Wushu postures, High accuracy and stability.

Keywords: Wushu movement · Higuchi analysis · Balance assessment

1 Introduction

In the World Wushu Championships, coaches and athletes of all countries attach great importance to the competition results, and the progress of each competition result is inseparable from the improvement of athletes' technology. Correct Wushu posture is helpful for athletes to improve their skills. Wrong Wushu posture is not conducive to the progress of athletes' skills and can not achieve the ideal results. Therefore, the accurate recognition and correction of Wushu Athletes' posture is very important for athletes to improve their skills and achieve good competition results.

The coach's oral teaching can not make all athletes improve their skills, and it is difficult for athletes to make accurate imitation. At the same time, it is difficult for athletes to correct their own details of technical movements through hearing, to obtain self "feedback information", and to achieve the purpose of improving their own technology. Therefore, it is very important to apply multimedia technology to the correction of athletes' wrong posture. By means of video recording, comparison of positive and negative action pictures, on-site shooting and replay of video technology, the recognition and correction of Wushu Athletes' posture has become an important research topic of relevant experts and scholars, which has great research value. In the research of recognition methods of athletes' wrong posture, some achievements have

J. C. Hung et al. (Eds.): FC 2021, LNEE 827, pp. 1637–1642, 2022.
https://doi.org/10.1007/978-981-16-8052-6_240

been made. Through the target tracking of Wushu movement, the posture of Wushu athletes is identified and corrected; Higuchi analysis algorithm is used to detect and recognize the wrong posture of Wushu athletes, and complete the recognition and correction of Wushu Athletes' posture.

Aiming at the disadvantages of traditional methods, a new calculation method is proposed [1]. By combining the bone tracking method of depth image acquisition with Higuchi analysis algorithm, the moving target image is tracked and feature extracted. The affine transformation of the image is maintained by adjusting the rotation angle, scale scaling and brightness. The noise is relatively stable. The following Fig. 1 is the image of martial arts. Finally, the real-time tracking of the stability of martial arts posture is completed, so that the athletes can realize the mistakes of their own actions, and correct their actions in time to obtain better results.

Fig. 1. Martial arts image

2   The Relationship Between Posture and Wushu Routine

The changes of body movement and movement direction of Wushu routine are complex, and the body center of gravity of athletes is constantly changing. And in Wushu routine, there are many movements of balance and somersault, so state reflection can play an important role in Wushu routine.

State reflex is a kind of reflex activity that causes muscle tension readjustment when the head position changes. This kind of reflex includes labyrinthine tension reflex and neck tension reflex. The state reflexes were as follows: (1) the tension of the extensors of the upper and lower limbs and the back was increased due to the head backward (2) The tension of the extensors of the upper and lower limbs and the back was weakened, while the tension of the flexors and abdominal muscles was relatively strengthened (3) When the head was twisted or tilted, the tension of the ipsilateral upper and lower limb extensors was increased, while the tension of the contralateral upper and lower limb extensors was decreased [2].

In the swallow style balance movement, the athlete's upper body is bent forward, his back is straight, his left leg is straight, his leg is extended backward, above the level, his feet are flat, and his arms are lifted to the left and right. If the head is not fully tilted back, the leg will be raised on the low side, the back will be bent, and the hips will be protruded. In this way, the stability of the movement will not be maintained, and the beauty of the posture will be affected. Therefore, in the swallow style balance movement, the head fully tilted back can cause the tension of the upper and lower limbs and the back extensors to strengthen, and reflexively strengthen the control of the back muscles, The movement can be completed with high quality.

In the balance movement of lifting the knee, the right leg is upright, the left leg is bent and lifted, the foot is straight, and it is hung in the front of the right leg, and the two eyes are looking straight to the left. This movement requires the head to turn to the left, which can reflexively cause the tension of the left arm extensors to strengthen and the tension of the flexors to weaken, so that the left arm can be fully extended and lifted back. Turning the head to the left will also reflexively increase the tension of the right arm flexors and decrease the tension of the right arm extensors, so that the right arm can be slightly bent and lifted up on the head. When the trunk twists to the left under the action of tension redistribution of the extensors and flexors of the two arms, it twists with the knee flexion and lifting of the left leg, which fully demonstrates the stretch of the movement.

3   Higuchi Fractal Dimension Analysis Method of Athlete Posture

In the traditional assessment method of posture balance, a time domain measurement method of support point movement range is often used, that is, Co P The maximum minus the minimum of the trajectory in a specific direction in the data is taken as a measure.

In the process of posture recognition and correction, $Q = (Q_R, Q_G, Q_H)$ represents the foreground pixel of the athlete's posture image, $W = (W_R, W_G, W_H)$ is the background pixel of the athlete's posture image, and formula (1) is used to locate each joint point of the athlete's Wushu

$$G(x, y, z) = \frac{Q \cdot W}{(\Delta r, \Delta D, \Delta 0, \Delta L0, \Delta L1, \Delta R0, \Delta R1)} \tag{1}$$

Through the analysis of the joint points, we can extract the characteristic points of each joint of the athlete's limbs in Wushu by using formula (2)

$$x^2 + y^2 + z^2 = \sqrt{\frac{shead}{\pi}} \times \frac{(x, y)}{R} \frac{(x - r \cdot \varnothing, y + r)}{\varnothing = (1 + \sqrt{5})/2} \tag{2}$$

Among them, R represents the radius of the athlete's range of motion, $\sqrt{\frac{shead}{\pi}}$ It represents the approximate value of R, $(x - r \cdot \varnothing, y + r)$ represents the position of the left key, and $\emptyset = (1 + \sqrt{5})/2$ represents the X coordinate of the left shoulder. According to the theory of symmetry, the X coordinate position of the right shoulder can be obtained.

The affine deformation of human body is easy to occur in the process of martial arts, which leads to the appearance of action feature points with low brightness. However, the traditional method is to extract these feature points and compare them with the correct posture to realize the recognition and correction of posture. As a result, the wrong posture can not be detected and corrected in real time. This paper puts forward a method of posture balance evaluation of Wushu athletes based on Higuchi analysis.

Based on the Higuchi analysis algorithm, the position of the athlete's target action is estimated, so that the complex prediction problem can be transformed into a relatively easy classification problem. In order to realize the motion target monitoring better, human Wushu posture is taken as the target of tracking and detection.

Fractal is a method used to describe the global and local similarity of irregular data. For the analysis of time series, fractal dimension represents the complexity of data. The main methods to calculate fractal dimension are based on the transformation of time series to phase space series and based on time city. Among them, the time domain methods include Katz algorithm, Castiglioni algorithm and the time domain method, Sevcik algorithm, Higuchi algorithm and box dimension method.

In this paper, Higuchi algorithm is used to determine the fractal dimension of time series cop trajectories. This is because Higuchi algorithm has high accuracy in calculating fractal dimension. Set a time series of length N, $X = X(i), i = 1, 2, \ldots, n$. Firstly, according to the original time series X, a new inclusion is constructed by delay method é The new time series X. X of data is expressed as follows [3]:

$$X_i^m = X(x + jk); m = 1, 2, \ldots, k; j = 0, \ldots, [N - m/k] \tag{3}$$

Where, [] is the integer, K and m are the integers of interval time and initial time respectively, and N is the number of data points.

In this paper, Higuchi algorithm is used to calculate the fractal dimension of cop time series in the two directions of anterior posterior (AP) and medial lateral (ML). In this process, when the fractal dimension reaches stationary after a K, the calculation stops. This k is called the parameter K saxx, which is set to be 20 in all directions.

4 Experiment and Analysis

Eight professional Wushu athletes (age: 18–24 years old; age: 18–24 years old) were selected; Body weight: (64 ± 9.3) kg) and 8 amateur athletes (age: 16–21 years old; Body weight: (68 ± 11) kg) participated in this study. The athletes were all male. The professional group members had at least four years of continuous training, and the amateur group members had more than one year of training, less than two years of training, The force data were collected at a sampling rate of 200 Hz to provide the cop trajectory data in AP and ml directions. If any athlete does not perform the action correctly under the observation of the coach, the data will be deleted, In order to determine whether there is significant difference between the two groups, medcalc software was used to implement independent sample t-test, in which the significance level threshold was 0 α It is set to be equal to 0.05, that is, when the significance $P > 0.05$ α The difference is not significant.

5 Experimental Result

The average fractal dimension (FD) in AP and ML direction of cop trajectory is calculated by the fractal dimension method proposed in this paper. For comparison, ROM in two directions is also calculated by traditional method. For professional and amateur athlete data, the average FD in AP and ML direction is calculated. In addition, the standard deviation (SD) between each athlete data is calculated, as shown in Table 1, The average FD results of professional and amateur athletes in AP and ML direction are significantly different ($P < 0.05$), and the average FD of professional athletes is significantly higher than that of amateur athletes. However, the average ROM obtained by traditional measurement methods is only significantly different in AP direction ($P < 0.05$). An effective analysis method for professional and amateur athletes data, the measurement value should be significantly different.

Table 1. Measurement results of branch dimension and ROM of Wushu Athletes

Athlete level	Branch dimension method		Traditional ROM method	
	Average FD	FD standard deviation	Average ROM	ROM standard deviation
amateur	1.157 8	0.058 3	0.058 3	0.274 5
major	1.347 2	1.106 4	0.058 3	0.058 3
amateur	1.106 4	0.069 6	2.369 6	0.371 1
major	1.358 1	0.0579	1.8927	0.258 3

In traditional measurement methods [4], ROM data of professional and amateur athletes are only different in AP direction. However, the FD obtained by fractal dimension analysis is significantly different in both AP and ml directions. This finding reflects that fractal dimension analysis is more sensitive to movement proficiency, This complexity indicates that more skilled athletes have more complex time series of cop data, which is a sign of the improvement of posture control ability, The FD of the cop time series calculated by Higuchi algorithm is between 1 and 2. When the dimension is close to 1, the degree of freedom of the system is fixed to control itself, which is not a safe control strategy. Athletes may be injured during the performance. When the dimension is close to 2, the degree of freedom of the system is fixed to control itself, The FD of the professional group is significantly higher than that of the amateur group, which indicates that a more dynamic system is used and there are more degrees of freedom in performing the balancing motion.

6   Conclusion

In this study, professional and amateur martial arts athletes carried out the front kick balance exercise, obtained the time series of COP movement through the force detection board, and evaluated the movement proficiency by fractal dimension analysis. In this paper, Higuchi algorithm is used to calculate the FD of cop data directly in time domain, and the ROM of cop trace is compared as a traditional measurement method. The results show that FD analysis can be used to distinguish the training degree of Wushu athletes.

References

1. Ying, L., Zhi, B.M.Y.: Linear and nonlinear evaluation of body posture control. J. Beijing Sport Univ. **38**(5), 68–71 (2015)
2. Zhu, D., Zhao, G., Yao, S., et al.: Exploration of martial arts strength: analysis and effect evaluation of Xingyi boxing. J. Shanghai Inst. Phys. Educ. **38**(1), 89–94 (2014)
3. Li, Q., Li, C., Yang, Y.: Application of self similarity and fractal dimension in wind field analysis. J. Power Eng. **36**(11), 914–919 (2016)
4. Qin, J., Kong, X., Hu, S., et al.: Comparative analysis of fractal dimension algorithms for one-dimensional time series. Comput. Eng. Appl. **52**(22), 33–38 (2016)

Orientation Design and Research of Heavy Bamboo Substrate Considering Genetic Programming and Artificial Intelligence Algorithm

Wei Chen[✉]

Nanchang Institute of Technology, Nanchang 330029, Jiangxi, China

Abstract. As a kind of cheap, easily available and renewable natural material, bamboo is an advantageous resource for sustainable design in today's "circular economy" era. In this paper, a kind of genetic artificial intelligence algorithm is studied for the development of heavy bamboo substrate oriented products. Genetic algorithm is a widely used intelligent optimization algorithm. The selection optimization problem in heavy bamboo product development system is studied. The technology platform for manufacturing bamboo based fiber composites has been built, and a variety of bamboo based fiber composites used in landscape, building structural materials, packaging and transportation materials, interior high-end decoration and other fields have been successfully developed and industrialized.

Keywords: Genetic algorithm · Product development · Heavy bamboo · Intelligent algorithm

1 Introduction

As a natural material, bamboo has been used since ancient times to meet the needs of production and life. Due to geographical advantages, East Asian countries cover most of the world's bamboo resources, and "bamboo civilization" has almost become the synonym of East Asian civilization. China has the most complete set of bamboo products system, including bamboo food, bamboo and rattan fabrics, bamboo crafts, bamboo and rattan furniture, bamboo household utensils, bamboo construction industry, bamboo composite products, etc., and thus produced the most important part of Chinese traditional culture - bamboo culture.

Bamboo based panel is an important part of the bamboo industry with the most vitality. After nearly 40 years of development, new products such as bamboo woven plywood, bamboo plywood, bamboo curtain plywood, bamboo Glulam and Reconstituted Bamboo have been successfully developed and industrialized. Reconstituted bamboo is a new type of bamboo based panel, which is composed of bamboo bundles, assembled along the grain direction and glued by hot pressing (or cold pressing and heat curing). It overcomes the defects of small diameter, anisotropy and uneven material. It has the advantages of high utilization rate of raw materials, designable product performance, natural beautiful texture and excellent physical and mechanical

J. C. Hung et al. (Eds.): FC 2021, LNEE 827, pp. 1643–1648, 2022.
https://doi.org/10.1007/978-981-16-8052-6_241

properties, It is deeply loved by consumers at home and abroad, and is a characteristic industry with great development potential [1].

Recombined bamboo manufacturing technology is a new technology developed by our country with independent intellectual property rights and successfully realized industrialization. After more than 10 years of development, it has made breakthrough progress. The products have been widely used in the fields of wind turbine blade substrate, building beam and column, indoor and outdoor board, landscape material, furniture, container floor and so on.

2   Current Situation and Problems of Heavy Bamboo Substrate

Although bamboo has many advantages, it still stays in the level of using original bamboo and weaving primary functional products such as handicrafts, household utensils and simple furniture, and has not entered into large-scale industrial production like wood. On the whole, the brilliant history of bamboo use has gone far away. The bamboo industry once went into depression, bamboo products gradually faded out of people's lives, and the competitiveness of international export products is not strong and stable. So what is the key to this situation?

First of all, the Aesthetic identity of bamboo encountered a historical bottleneck. Today, Chinese people seem to regard bamboo art products as a thing of the past. Because of their historical "folk" function and lack of changing style, bamboo products are framed in the semantic connection of simple, crude and low-end quality of life. With bamboo becoming a non mainstream material, its audience circle has shrunk to a small number of people who want to experience the "idyllic taste". In the fierce situation of industrial competition, the market decides to survive. Only by striving for wide recognition and re opening the market of products, can we realize the strategic direction of sustainable development of bamboo art industry.

Secondly, "folk art" culture itself lags behind the times. Bamboo weaving technology and all kinds of bamboo products stick to the rules, do not seek change, and stay in the inheritance of traditional technology. The style and pattern are familiar and monotonous [2]. The product features are simple and convergent, and are out of touch with the times. It cannot meet the consumption needs and aesthetic taste of modern people, and it does not compete for the green consumer goods market which is expected to be created with the advantage of natural materials. Finally, the traditional industry mode limits the development of bamboo art. The precious resources of bamboo culture are lack of modern and effective management, and most of the bamboo industry still stays in the simple and extensive workshop mode, with traditional craftsmen or skilled workers as the main force; Even the township enterprises, which have initially formed a sense of modern management, are obviously lack of the grafting of contemporary art and the flow of creative blood, and are always in a low-end market level, repeating the low-cost competition. In this paper, through the study of genetic intelligent algorithm, to optimize the heavy bamboo substrate oriented product development.

3 Development of Heavy Bamboo Products Under Genetic Algorithm

3.1 Overview of Genetic Algorithm

With the increase of the types and scale of problems, compared with the traditional optimization methods, genetic algorithm provides us with a more effective method, and more and more people pay attention to it. This is inseparable from its own characteristics. The basic features of genetic algorithm are as follows:

(1) Intelligence
   Genetic algorithm can search and organize according to information when solving the problem by determining the coding method, fitness function and crossover and mutation operator. Genetic algorithm is based on Darwin's "survival of fittest" natural selection mechanism, and the individuals with larger fitness function will have higher survival probability and find better offspring in the process of cross variation.

(2) Parallelism
   There are two aspects to embody the parallelism of genetic algorithm: one is that genetic algorithm can search in the population to reflect its inherent parallelism. In consideration of improving the speed of genetic algorithm, three schemes of global, independent and decentralized are proposed to implement parallel genetic algorithm; Another is its internal parallel mechanism, which can be calculated simultaneously by using multiple computers, and finally the appropriate solution is selected.

(3) Global optimization
   Genetic algorithm can search the solution of the whole region (groups) simultaneously. For the problem of multi peak and nonlinear complex function optimization, the disadvantages of traditional algorithm are overcome, and the global optimal solution can be obtained, and the disadvantage that the conventional algorithm is prone to fall into local minimum is avoided [3].

(4) Robustness
   The genetic algorithm is simple and efficient, which determines its universal applicability under different environment and different conditions. Because genetic algorithm does not consider the structural characteristics of the problem, it only needs to set different fitness functions according to different problems, and does not need to modify other programs of the algorithm, so it is easy to realize.

(5) Processibility
   Genetic algorithm is to select the optimal individual through a series of processes such as coding, selection, mutation and crossover, and only the termination conditions are set manually.

(6) Polysolubility
   Genetic algorithm can search for multiple solutions in the population when searching for the optimal solution in the population, that is, multiple feasible solutions can be obtained, and finally the optimal solution can be selected.

3.2   Research on the Development of Bamboo Products with Shortest Path

The shortest path problem is to seek the shortest path from s to t in the set of paths from s to t at a given starting point. This is to study the development process of heavy bamboo products. Sometimes people should not only know the shortest path between two specified vertices, but also the shortest path between one vertex and any other vertex. There are not many constraints and restrictions on the solution for solving this kind of problem by genetic algorithm, so the shortest path between any two points and a batch of short paths can be obtained quickly [4].

Genetic algorithm (GA), a bionic theory, which applies biological principles to scientific research, has attracted the attention of science and technology in recent years because of its unique ideas and methods. Genetic algorithm is a stochastic optimization algorithm based on Darwin's "natural selection, survival of fittest" bioevolution. It was first proposed by Professor Holland of Michigan University and his students in 1962. It was aimed at problems rather than coding parameters. In the process of optimization, GA randomly generated multiple starting points in high-dimensional feasible solution space and started searching at the same time. On the one hand, the search direction was guided by fitness function. On the one hand, the search efficiency was high, on the other hand, the search area was wide, Therefore, it is a global, parallel and fast optimization method. Because it does not need to calculate gradient, its objective function is not limited, and it does not need to be continuously differentiable and other auxiliary information. It is more applicable than general optimization algorithm.

4   Application of Heavy Bamboo Products

4.1   Bamboo Based Fiber Composite Materials for Building Structural Materials

Wood structure building has a long history in China. For thousands of years, it has been the tradition of building houses with wood. It is this kind of "civil engineering" that constitutes the essence of Chinese architectural culture. But with the rapid development of urbanization in China, the traditional wooden structure building has been replaced by a large amount of reinforced cement. The bamboo based fiber composite material for building beam and column was created with bamboo based fiber composite material with the length of 12M, static bending strength of 154 mpa, elastic modulus of 22 GPA, formaldehyde release of 0.04 mg/100 g, and the combustion performance level reached GB 8624-2008c-s2, d0 and T1 without adding flame retardant. Bamboo based fiber composite material for building support was created with bamboo and dragon bamboo as main materials, with a length of 5.2 m, static bending strength: 207.9 MPa under vertical loading, 220.6 mpa parallel loading, 27.31 gpa of elastic modulus: horizontal shear strength: 18.7 mpa vertical loading and 29.2 mpa parallel loading; The swelling rate of water absorption thickness was 1.8% in 2 h. Taking Shouzhu, Cizhu and white bamboo as the main materials, bamboo based fiber composite material was created for cement template. The water absorption thickness expansion rate was 2.44% in 24 h, dry: static bending strength: longitudinal 125 MPa, transverse 76 MPa, elastic modulus: longitudinal 14 GPA, transverse 7 gpa: wet: static bending strength; longitudinal 114 MPa, transverse 76 MPa, elastic modulus; longitudinal 13 GPA, transverse 6 GPa. The bamboo materials of the building structure are shown in Fig. 1 below.

Fig. 1. Bamboo materials for building structure

4.2   Bamboo Based Fiber Composites for Packaging and Transportation Materials

Since the reform and opening up, China's economy has grown rapidly, becoming the second largest economy in the world. In 2012, China's total foreign trade volume ranked first in the world. In the process of massive cargo packaging and transportation, it needs to consume a lot of high-quality wood materials, such as for the construction of trains, cars, ships, airplanes and other means of transport, and the construction of containers, packing boxes and other transport containers. These materials usually require strong bearing capacity and good dimensional stability. At present, hardwood or high-quality large-diameter softwood imported from abroad is usually used, which not only consumes a lot of foreign exchange, but also attracts criticism from the world environmental protection organization. The bamboo based fiber composite material for packaging and transportation developed by this technology, with moso bamboo and green bamboo as the main materials, creates all bamboo container bottom plate, with static bending strength of 105.9 MPa along the grain, 59 MPa across the grain [5], elastic modulus of 11 370 MPa along the grain, 6 000 mPAN across the grain, bonding strength of 3.24 MPa, and the performance index meets the performance requirements of container bottom plate; Bamboo based fiber composite for railway car floor was developed with Neosinocalamus affinis and Shouzhu as main materials. The 6 h water absorption rate was 2.5%, static bending strength was 276 MPa, elastic modulus was 28 GPA, compressive strength was 167.1 MPa and impact toughness was 212 kJ/m, which met the requirements of bamboo laminate for railway car. It can replace the imported high-quality wood and be used in the fields of railway car floor, container floor and heavy packaging materials. As shown in Fig. 2 below, the production of bamboo based fiber composite materials.

Fig. 2. Production of bamboo based fiber composites

5 Conclusion

The manufacturing technology of bamboo based fiber composites closely follows the hot spots and major technical difficulties in the field of bamboo research and production at home and abroad, and combines with equipment manufacturers and production application units. This paper studies the development and exploration of heavy bamboo substrate oriented products based on genetic intelligent algorithm. It has built a bamboo based fiber composite material manufacturing technology platform, successfully developed a variety of bamboo based fiber composite materials used in landscape, building structural materials, packaging and transportation materials, interior high-end decoration and other fields, and realized industrialization.

Acknowledgements. Subject Name of Science and Technology Research Project of Jiangxi Education Department: Directed product development research based on heavy bamboo base material (subject No.: GJJ191013).

This thesis is a PhD in scientific research.

References

1. Yu, W., Yu, Y.: current situation and Prospect of the development of wood and Bamboo Reconstituted timber industry in China. Wood Ind. **27**(1), 5–8 (2013)
2. Ma, Y.: An analysis of the problems in the development of the recombinant wood technology. China Wood Based Panel **18**(2), 1–5, 9 (2011)
3. Yao, W.: Genetic algorithm and its research progress. Comput. Digit. Eng. (04) (2004)
4. Wang, W.: Recombinant bamboo new technology and new products. Wood Ind. **3**(4), 52–53 (1989)
5. Yu, W., Yu, Y., Zhou, Y., et al.: Study on the factors influencing the properties of small diameter bamboo reconstituted structural timber. Forest. Ind. **33**(6), 24–28 (2006)

Optimal Operation of Cascade Hydropower Stations Based on Chaos Optimization Algorithm

Xiaolei Zhong[1(✉)], Miao Yu[1], Shuang Lu[1], Ye Wu[2], and Xuesheng Yu[3]

[1] Shenyang Institute of Technology, Fushun 113122, Liaoning, China
[2] Liaoning Share Country Garden Property Co., Ltd., Shenyang 110000, Liaoning, China
[3] LiaoNing Academy of Safety Science, Shenyang 110000, Liaoning, China

Abstract. A medium and long-term reservoir optimal operation model of cascade hydropower stations is established, and chaos optimization algorithm is used to optimize the medium and long-term reservoir operation of cascade hydropower stations. The calculation results show that the algorithm can solve the nonlinear cascade hydropower station reservoir optimal operation problem with complex constraints, with high accuracy and fast convergence speed. It provides an effective algorithm for solving the optimal operation of cascade hydropower station reservoir.

Keywords: Cascade hydropower stations · Optimal operation · Chaos optimization algorithm

1 Introduction

The optimal operation of reservoir is to regulate the power flow reasonably in the process of reservoir water storage under the premise of ensuring dam safety according to the principle of water energy utilization and the primary and secondary tasks of water conservancy and hydropower undertaken by the reservoir, and on the premise of ensuring dam safety, improving the comprehensive utilization efficiency and centralized control of the technology of using reservoir. Under the market economy, public welfare water conservancy projects should be optimized, analyzed and compared according to the market economy law, and the investment benefit should be emphasized. The water conservancy projects with economic benefits as the main part should be established and operated according to the market economy law. Public welfare water conservancy projects with government investment as the main part should be optimized, analyzed and selected according to the law of market manager in the process of project demonstration, construction and management, The economic benefits of investment are emphasized. In order to give full play to the benefits of water resources, the optimal operation of reservoirs must be carried out when constructing hydropower stations [1].

Since 1990s, with the development of hydrological forecasting technology and computer technology, the technology and means of hydropower station and its reservoir have

J. C. Hung et al. (Eds.): FC 2021, LNEE 827, pp. 1649–1653, 2022.
https://doi.org/10.1007/978-981-16-8052-6_242

been greatly improved. The reservoir operation technology and means of hydropower station have been greatly developed. The connotation of hydropower station reservoir operation has gradually expanded from the past one plant to the current cascade operation The compensation and dispatching of the whole power grid and the joint optimization of the water and fire power stations are carried out. The operation of hydropower station and its reservoir has become one of the important contents of the safety and economic operation of the power grid. With China's accession to WTO, the international competition is increasingly fierce. Only when China is in the overall leading position in hydropower quantity, quality, science and technology, management and efficiency, can hydropower become the first power in the real sense. Especially with the deepening of the reform of the power industry system, the cross regional large power grid will be formed gradually aiming at adjusting the power structure, improving the power quality, improving economic benefits and optimizing the allocation of resources. Therefore, the optimal operation of hydropower stations (groups) is more important [2].

Therefore, carry out the research on optimal operation and management of hydropower stations (groups) and improve the level of control and application of hydropower stations.

2   Chaos Optimization Algorithm

Chaos is one of the most important discoveries. As early as the end of the 19th century, Poincare, a French mathematician, had predicted some behaviors of chaos. However, it was only after Lorenz studied the chaos in the atmospheric flow in 1963 that the study of chaos had a great development [3].

Chaos is a kind of cyclic behavior without fixed period, that is, the non periodic phenomenon with gradual self similar order. Chaos has its unique properties: ① randomness, that is, chaos has random variable like chaos performance; ② Ergodicity, that is, chaos can go through all States in the fixed state space without repetition; ③ Regularity, that is, chaos is produced by iterative equations with certainty.

In this kind of algorithm, the optimization variables should be in chaos state in phase space. Generally, logistic chaos model is used to generate chaos variables. The form of logistic mapping is as follows:

$$X_{a+1} = 1, (\mu, X_a) = \mu X_a(1 - X_a) \tag{1}$$

Where $\mu \in [0,4]$The control parameters are: f: $[0,1] \rightarrow [0,1]$;

When $\mu = 4$, $[0,1]$ is the chaos invariant set of f. in the logistic chaos invariant set, when the time is long enough, any orbit of L is dense in it, that is, for $\forall \varepsilon > 0, \forall x \in [0, 1]$ kick off B (x, ε) If we look at it in reverse, any orbit of f can approach all the points in $[0,1]$ with any precision.

For solving the m-dimensional optimal solution problem, it is equivalent to determining a point in the m-dimensional space to minimize the value of the objective function. Therefore, m independent logistic maps are needed to generate m coordinate components of the midpoint in the optimization space, which are called chaotic vectors, So the resulting points will be dense in the m-dimensional unit hypercube. That is to say, the sequence of these points can approach all points in the hypercube with any precision, and of course, it can approach the global optimal solution in the hypercube with any precision. This is the ergodicity of the orbits of chaotic sequences, and it is the fundamental starting point for chaos to be used in continuous optimization problems.

3   Establishment of Mathematical Model for Optimal Operation of Hydropower System

3.1   Objective Function

The objective function of traditional cascade generation optimal dispatch is to minimize the generation cost of thermal power or the total consumption of thermal power units under the condition of meeting the system load and other generation requirements. In this kind of mathematical model, hydropower dispatching, as a means to assist thermal power plants to achieve optimal dispatching, its own optimization is not the goal, so hydropower only appears as constraints.

At present, the optimal operation of hydropower will consider the optimization of hydropower itself more. In short-term optimal operation, given the daily load curve of cascade power stations, the power generation, minimize the total water consumption and maximize the power sales revenue. When the electricity price is the same, the above two goals are the same. When the electricity price of each power station is different, it is more advantageous to take the maximum electricity sales revenue as the goal.

The objective function is expressed as follows

$$E = \max \sum_{j=1}^{N} \sum_{i=1}^{T} N_i^j \Delta t_i \tag{2}$$

Where T is the total period of time calculated in the year (the calculation period is month, $T = 12$); N is the total number of reservoirs; N_i^j is the output of J reservoir in period i, MW; E is the annual power generation, $kW \bullet h$.

3.2   Solution Process

Solution steps:

① The range of water level at the beginning and end of reservoir period is determined and discretized;

② A state variable Zn of the first reservoir is selected to calculate the outflow flow according to the change of reservoir capacity and inflow flow;

③ According to the inflow discharge reservoir water level curve, the outflow discharge downstream water level curve and the inflow discharge head loss, the static head of power generation is calculated, and the downstream water level at the end of the period and the expected output of the power station are calculated;

④ The inflow of the next reservoir is calculated. According to the initial conditions of the reservoir period, the output of the power station is calculated. By analogy, the downstream water level at the end of each reservoir period, the output of the power stations at all levels and the total output of the cascade are calculated.

⑤ If any process exceeds the limit, return to (2) select the discrete point again;

⑥ The objective function value is calculated;

⑦ Turn to (2) select the discrete points again until the optimal objective function is obtained [4]. The water level and output of each station corresponding to the optimal objective function value are the optimal dispatching plan.

The solution process is shown in Fig. 1.

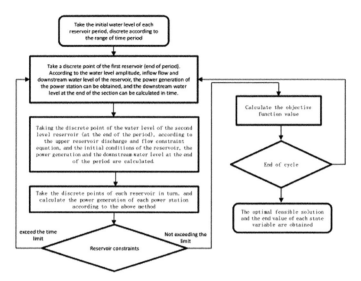

Fig. 1. Solution flow chart

4 Application of Chaos Optimization Algorithm in Optimal Operation of Cascade Hydropower Stations

According to the above-mentioned mathematical model for optimal operation of cascade hydropower stations, the objective function is to find $N \times T$ decision variables $Q_1^1, Q_1^2, \ldots, Q_T^1, Q_1^2, Q_2^2 \ldots, Q_T^2, \ldots Q_2^N, Q_2^N, \ldots, Q_T^N$ has the biggest annual power generation problem.

In order to solve the optimal operation problem of cascade hydropower stations, an initial vector $X_0 = (X_{01}, X_{02}, \ldots X_{0TN})$ of $N \times T$ dimension is randomly selected. Then, by using the randomness of chaotic motion, the logistic equation $X_{k+1,j} = 4X_{k,j}(1 - X_{k,j}), j = 1, N \times T$ randomly generate chaotic sequence $\{X_k\}k = 1, 2, \ldots;$ It is carried to a region containing the feasible city s of the objective function where s is the set of objective functions of cascade hydropower stations satisfying the yodong condition.

5   Conclusion

Chaos directly uses chaos variables to search in the allowed solution space. The search process is carried out according to the law of chaos motion itself, which makes it easier to jump out of the local optimal solution and has high search y.

Applying chaos algorithm to optimal operation of cascade hydropower stations, the following conclusions can be obtained.

(1)   The concept of chaos algorithm is simple, easy to understand, and the calculation process is simple and easy to implement.
(2)   For the same problem, the chaos optimization algorithm is better than other algorithms, which has the advantages of small computational workload, short time and high accuracy.

References

1. Xie, M., Wei, X.: Optimal operation of cascade hydropower stations based on large-scale system gradual optimization algorithm. China Sci. Technol. Inf. (05), 82–84 (2020)
2. Wu, Z.: Optimal operation of cascade hydropower stations based on simulated water cycle algorithm with multiple constraints. Hydropower New Energy **32**(12), 20-25+59 (2018)
3. Yang, X., Huang, Y., Huang, Q.: Optimal operation of cascade hydropower stations based on improved multi-objective cuckoo algorithm. J. Hydropower **36**(03), 12–21 (2017)
4. Zhan, S., Li, K.: Development of rural cultural industry: value pursuit, realistic dilemma and promotion path. Zhongzhou Acad. J. (03), 66–70 (2019)

Mineral Processing Crushing System Based on Fuzzy Genetic Algorithm

Fenlan Peng[✉]

Kunming College of Metallurgy, Kunming 650033, Yunnan, China

Abstract. Ore crushing is a key link in the process of mineral processing industry. The crushing process requires stable material level in the machine cavity to improve the work efficiency of crushing process. In view of the inertia, lag and nonlinear time-varying characteristics of feeder and crusher, a crushing system based on fuzzy genetic optimization algorithm is proposed. The fuzzy adaptive PID control algorithm is adopted for the optimal control of feeding quantity, and the genetic algorithm is used to optimize the parameters of the controller, so as to improve the online optimization ability of parameters.

Keywords: Control of ore feeding quantity · Fuzzy control · Genetic algorithm

1 Introduction

The beneficiation process includes crushing, screening, grinding and separation. Crushing process is the initial part of the whole process, which plays a key role in Beneficiation quality and output. One of the main tasks of the automatic control of crushing process is to realize the automatic adjustment of ore feeding, so as to stabilize the material level in the cavity of the crusher, so as to improve the crushing capacity and the overall operation efficiency of the crusher, and make the crusher work in the best state. Therefore, it is of great economic benefit and practical value to develop the automatic control system for ore feeding.

The amount of ore feeding directly affects the concentration of tailings, overflow concentration and the state of equipment. The time lag is large, and there are many interference factors in the process. The domestic mineral processing industry generally adopts the common logic control and safety interlock protection mode for the operation process of the crusher, while the manual adjustment method is usually used for the feeding quantity control of the crusher system, which is obviously difficult to meet the actual requirements of the project. At present, the conventional PID algorithm is used to control the feed quantity in engineering, but for complex working conditions, it is difficult to coordinate the contradiction between rapidity and stability due to the lack of the ability to quickly suppress disturbance, and the control effect is not ideal. In view of the main problems existing in the feed control of domestic mills, Ren Jinxia and others applied the method of combining adaptive control and fuzzy PID to the feed control system of cone crusher. The simulation results show that the control performance index of this method has been significantly improved, which has a certain engineering

© The Author(s), under exclusive license to Springer Nature Singapore Pte Ltd. 2022
J. C. Hung et al. (Eds.): FC 2021, LNEE 827, pp. 1654–1659, 2022.
https://doi.org/10.1007/978-981-16-8052-6_243

application value. But through the analysis, it is found that these research methods have not carried out in-depth discussion on the online optimization of PID parameter tuning. Therefore, this paper combines genetic algorithm with fuzzy PID control to design a mineral processing crushing system based on fuzzy genetic algorithm, so as to further improve the control quality of ore feeding crushing system, and improve the anti-interference ability and robustness of the control system [1].

2   Brief Introduction of Technology

2.1   Mineral Processing Technology

Beneficiation is to remove the gangue and harmful elements from the original ore, enrich the useful minerals, separate the symbiotic useful minerals from each other, and obtain one or more kinds of useful mineral concentrate products. The ore is mined out from the mine, generally with low grade and often contains several useful components. In addition to the useful components, ores often contain some harmful impurities with low grade. These harmful impurities (such as sulfur and phosphorus in iron ore) must be removed by mineral processing before smelting, so as to avoid affecting the quality of metal due to harmful impurities entering into metal during smelting. Before smelting, it is necessary to separate the useful minerals and gangue in ore by mineral processing, so as to improve the grade of useful ore and obtain the raw materials suitable for smelting or other departments [2].

Therefore, mineral processing is an essential part of metallurgical industry. Mineral processing is a very complex industrial production process. The ore from the mining plant has to go through various operations of the beneficiation process, and finally the concentrate meeting the smelting requirements can be obtained.

2.2   Crushing process

Crushing is a key link in the process of mineral processing industry. There are many factors affecting crushing efficiency, among which feeding control is the most critical. The main goal of ore feeding control is to stabilize ore feeding, improve crushing capacity and overall operation efficiency of fine crushing. It is of great economic and practical value to develop the ore feeding automatic control system of crusher.

The collected raw ore is divided into concentrate and tailings through two major processes of medium crushing and fine crushing. The fine crushed ore is transported back by belt and screened, and part of it is qualified concentrate; The other part, due to its large size, continues to be broken by the crusher in the process of fine crushing, so as to ensure that the particle size of the ore meets the requirements. The design of feeding automatic control system is around how to control the material level of crusher at about 300 mm [3].

3   Basic Scheme of Control System

3.1   Feed Control System

In the process of mineral processing and crushing, the size and hardness of ore are very easy to change. When encountering high hardness ore, the feed setting value of crusher must be reduced in real time. When the ore is small and fragile, correspondingly, the feed setting value of crusher must be increased in real time to improve the efficiency of crusher, Therefore, it is necessary to control the feeding speed properly to compensate for the wide range changes of ore volume and hardness. At the same time, in the actual production process, the different capacity of crusher and the length of belt will also affect the time constant and pure lag time of the object. Therefore, the feed rate control system of crusher is a large pure delay system with uncertain time delay and variable parameters. In the control process, the digital PID control algorithm has a simple structure, which is the mainstream control algorithm of industrial control at present. The fuzzy controller with strong robustness can effectively control the time-delay system with nonlinear, time-varying and multi disturbance, but considering that the initial value of PID, quantitative factor and scale factor have a great influence on the fuzzy controller, which is generally set by human, The genetic algorithm can search the optimal solution by simulating the natural evolution process, which is a more effective search heuristic algorithm for solving optimization [4]. Therefore, this paper optimizes the fuzzy control parameters by genetic algorithm and applies it to the conventional PID control, so as to obtain the basic structure of ore feeding control in the ore crushing system, as shown in Fig. 1.

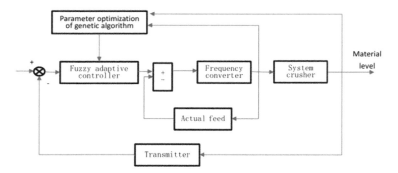

Fig. 1. Basic architecture of control system

3.2   Design of Fuzzy Controller

The block diagram of ore feeding control system based on fuzzy logic control is shown in Fig. 2.

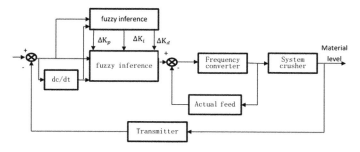

Fig. 2. Structure diagram of fuzzy control system

The initial value of PID parameters and the quantitative factor and the proportion factor of the controller are set manually. Different initial values of parameters and the quantitative and proportional factors will affect the control effect of the controller.

$$G(s) = \frac{5.56}{12S + 1} \frac{4.4}{20S + 1} e^{-16s} \tag{1}$$

For the crusher object shown in formula (1), when $K_p = 5, K_i = 1, K_d = 2$, and the quantization factor of the input variable e, the quantization factor of ec, and the output of the fuzzy controller AP, AK are selected. When the scale factors of AK, AK and AK are all 1, the system tracking curve is shown in Fig. 3

For the crusher object shown in formula (1), when $K_p = 5, K_i = 1, K_d = 2$, and the quantization factor of input variable e, quantization factor of ec, output $\Delta K_p, \Delta K_i$, and scale factor of ΔK_d in fuzzy controller are all taken as 1, the system tracking curve is shown in Fig. 3.

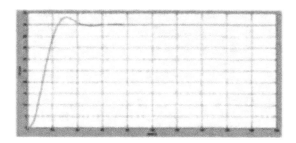

Fig. 3. Tracking curve of fuzzy PID control system after parameter correction

The control system has no overshoot and steady-state error, the control effect is better than the traditional PID controller, but the system response speed is slow. The simulation results show that the control effect of fuzzy adaptive PID is much better than that of ordinary PID, but the control effect is greatly affected by the initial value of PID, quantization factor and scale factor, and the characteristics of the motor will also change in the control link. Obviously, if these parameters are not optimized, the control effect will be difficult to achieve the best, Therefore, genetic algorithm is designed to optimize the parameters.

4 Parameter Optimization of Genetic Algorithm

Genetic algorithm is a probabilistic optimization algorithm, which is a parallel random search optimization method based on the simulation of natural genetic mechanism and biological evolution theory. Based on the principle of biological genetic evolution and survival of the fittest, it is an intelligent algorithm to select, cross and mutate individuals and search for the optimal solution of parameters on the basis of individual fitness. The basic genetic algorithm includes initialization, fitness calculation, selection, crossover, mutation, termination judgment and other operations.

4.1 Coding Method

The existing GA coding methods include: two-level coding, real coding, delta coding, natural number coding, gray coding, symbol coding, linked list coding, tree coding, dynamic variable coding, matrix coding, qubit coding and so on. The code of genetic algorithm is gray code. Because the fuzzy PID controller of feed level has two quantitative factors and three proportional factors, plus the initial value of PID controller parameters, there are eight parameters to be optimized. Each parameter is encoded by 10 bit gray code, and the eight encoded parameters are connected from left to right to form a gray code gene string with a length of 80 bits.

4.2 Decoding Algorithm

The decoding function for converting bit string individuals from bit string space to problem parameter space is described in.

DJC is the decimal representation of the c-th parameter to be optimized in the j-th chromosome, and gji is the i-th position of the gray code of the j-th chromosome gene.

4.3 Convergence

After parameter optimization, the convergence curve of genetic algorithm is drawn, as shown in Fig. 4.

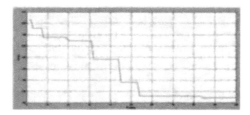

Fig. 4. The convergence curve of single optimization of genetic algorithm

Among them, the horizontal axis represents evolutionary algebra, and the longitudinal axis represents the sum of the mean square of tracking error.

5   Conclusion

In this paper, the crushing system of mineral processing is taken as the specific research object. According to its characteristics of nonlinear large delay and external load disturbance uncertainty, based on the in-depth study of the characteristics of crushing system of mineral processing, the mathematical model of feeding quantity control system is selected, and the fuzzy genetic control method is applied to it. The simulation results show that: Based on the fuzzy adaptive and genetic algorithm, the on-line optimization control strategy of feed quantity can carry out the cycle on-line modeling of the mineral processing crushing system, and optimize the parameters of the fuzzy adaptive controller in real time, so as to ensure that the parameters of the controller are always the optimal value. It has good control quality such as fast dynamic response, small overshoot and strong stability, and can effectively realize fast position tracking. The control effect has been obviously improved, and it has high engineering application value.

References

1. Zhang, X.: Application of advanced control technology in mineral processing control. Northeastern University (1999)
2. Sun, F.: Research and application of automatic control system for feeding speed of crusher. In: Proceedings of the 9th China Iron and Steel Annual Meeting (2013)
3. Tan, L.: Application of fuzzy PID control in ball mill feed control system. Chem. Autom. Instrum. **38**(12), 1434–1436 (2011)
4. Yu, G., Deng, R., Lei, G.: Research on the improvement of the despatch algorithm by gray code. China Sci. Educ. Innov. Guide (12), 26–27 (2007)

Design and Implementation of Cultural Tour Guide System Based on LBS

Wei Niu[✉]

Global Institute of Software Technology,
Yishui County 215163, Shandong, China

Abstract. Internet plus tourism is like a raging fire in the Internet and tourism, especially with the new concept of "Internet plus" in the first two years. With the maturity of mobile network, mobile Internet technology develops rapidly, which leads to the rapid development of LBS Internet services. This will also provide users with more rich mobile experience services. This paper takes the design of cultural tourism guide system based on LBS as the clue, and puts forward the solution to the closure of tourism resources and accelerate the development of tourism culture. The program integrates cultural creativity and national customs perfectly into the current platform.

Keywords: LBS · Cultural tourism · Tour guide system

1 Introduction

Tourism is one of the fastest growing industries in the world in recent years. The tourism industry is comprehensive, the market development potential is great, and the economic radiation and driving effect is obvious. Therefore, the development of tourism has become an important means to promote the local economic development in many cities and regions of our country. However, problems such as poor service quality, incomplete guide information and inadequate personalized service have become the main problems affecting the development of tourism industry [1].

In this context, a social tour guide system based on LBS and web services is designed and implemented by choosing popular technologies such as social network technology, mobile lbs location-based service technology and text to speech conversion technology. The purpose is to make up for the lack of tourism services in modern life and meet the high-level needs of tourism, To provide personalized tour guide service for the majority of self-help travelers, at the same time to provide good technical means for further improving the service quality of tourist attractions, and ultimately promote the development of tourism industry.

2 System Theory and Technology

2.1 Research on LBS System

Location based service is a kind of value-added service that obtains the location information (geographic coordinates or geodetic coordinates) of mobile terminal users

J. C. Hung et al. (Eds.): FC 2021, LNEE 827, pp. 1660–1664, 2022.
https://doi.org/10.1007/978-981-16-8052-6_244

through the radio communication network (such as GSM network, CDMA network) or external positioning mode (such as GPS) of Telecom mobile operators, and provides corresponding services for users with the support of Geographic information system (GIS) platform. The system structure of LBS is shown in Fig. 1:

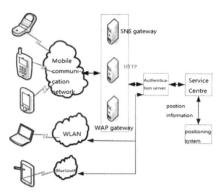

Fig. 1. LBS system architecture diagram

The mobile service center is the connection center of the whole lbs system platform, which is responsible for the information interaction with mobile terminals and the network interconnection of each sub center (location server, content provider), completing the classification, recording and forwarding of all kinds of information, as well as the flow of interest and business information between sub centers, and monitoring the whole network.

Location technology solves the location problem of mobile terminal, and to provide location-related services, we need to rely on GIS technology. GIS is geographic information system (GIS): Based on geospatial database, it obtains, stores, analyzes and manages geospatial data. We usually say that latitude and longitude have no special representation in daily use. It must be put into the geographic information to represent the exact meaning, such as city, location, etc., so as to be accepted by the public. Therefore, in addition to using positioning technology to obtain terminal location information, it is also necessary to convert the longitude and latitude it obtains into geographic information that users care about, such as map service, path search, path navigation and so on [2].

2.2 Android Development Platform

Android is an open source operating system designed by Linux platform, which includes operating system, operation interface and application program. Mainly used in mobile phones and tablet devices, it is now developed and maintained by Google. It has the functions of touch screen, video display and Internet access. Users can take photos on mobile devices, surf the Internet and view pictures. Moreover, it is more widely used than apple. It can be said that it is a single platform integrated into all web applications.

The main advantages of its system and all existing systems are as follows:

(1) Unlimited hardware selection
    This is related to the openness of Android platform. Due to the openness of Android, many manufacturers will launch a variety of products with various features. The differences and features of functions will not affect the data synchronization, or even the compatibility of software. For example, you can switch from a Nokia Symbian style mobile phone to an Apple iPhone. At the same time, you can bring excellent Symbian software to the iPhone. You can easily transfer information such as usage and contacts [3].
(2) Seamless integration with Google and other software
    Now Google has gone through 10 years of history. From search giant to comprehensive Internet penetration, Google services such as maps, e-mail and search have become an important link between users and the Internet, and Android mobile platform will seamlessly integrate these excellent Google services [4].

3   System Functional Requirements

As a social platform of tour guide function, the Android client system function module determined in this paper is shown in Fig. 2.

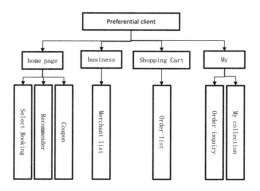

Fig. 2.   System module diagram

According to the figure above, the system is divided into five modules. It includes personal registration and login function, personal management function, business display function, business positioning and route navigation function, online ordering function and dynamic comment and sharing function. The following are the specific functions of Android client:

(1) Basic functions

Retrieval: fine search platform content.

Positioning: recommend shops according to geographical location.

Select Reservation: display in recommended shops.

Coupons: display the coupons of the merchants.

Business classification: classify the small categories under the large categories, such as catering, food, leisure and entertainment, hotel accommodation, rural entertainment, so as to facilitate the search.

Screening conditions: screening according to classification, region and evaluation.

Shopping cart: the user's unpaid group purchase will be listed, and then wait for the user to operate. The operation is divided into settlement or deletion.

My: in this function module, users can view their account balance, order, collection, activity, lottery and comments.

(2) Extended functions

Online booking: online booking business.

Map display: map display of the destination, convenient for search.

Map navigation: map navigation, according to the destination to provide appropriate bus routes, such as the existing bus, driving, walking routes.

Business display: detailed display of business. It is convenient for users to understand.

4 System Overall Design

4.1 System Architecture Design

The tourism guide system based on LBS is mainly to complete the online ordering of users in Android client and navigation according to the ordered goods. At the same time, merchants can upload their own goods and carry out settlement function. And the goods on and off the shelf processing. The overall architecture of the system is shown in Fig. 3: tourists can order goods online through the client. Merchants can upload goods and settle goods through web client.

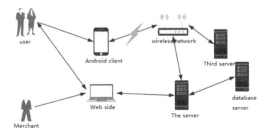

Fig. 3. The overall architecture of the system

4.2   Android Client Architecture of the System

Mobile client architecture is C/S architecture. The design of the interface is based on XML layout file. There is a problem of data interaction between client and server. The server side operates on the database server to complete the data interaction function of adding, deleting, modifying and querying data.

The user's operation on the client is realized through the related activity class to interact with the user. To realize the interaction between the client and the server, the first step is to start the server to call the listening interface. The third-party server here is Baidu server. Because the system integrates Baidu map service, it needs to call the data of Baidu server when starting the map service, and call mkmapviewlistener, the monitoring interface of Baidu service, to start the map monitoring event.

4.3   Web Server Architecture of the System

The web architecture in this paper adopts browser/server mode. It can be divided into three parts: customer layer, function layer and data layer.

In the web design, PHP language is used. The most popular ThinkPHP framework is used to build the background system, and HTML technology is used to process the front page. Using the most typical MVC design pattern, the model is completed by defining the model class, and the controller is divided into the core and action controller. The core controller is responsible for scheduling control, and the application controller is responsible for completing business process. However, there is no relationship between template and framework, and they are 100% separated.

5   Conclusion

In 2010, due to the improvement of people's living standards and consumption quality, the momentum of people's tourism is irresistible. On the basis of ensuring the comfort of tourism and the comfort of climate, tourists also have a new understanding of the convenience of tourism. In recent years, the development of internet tourism has installed a four-wheel drive in the fast lane of tourism, which makes it more efficient to serve tourists. This paper introduces the design and implementation process of cultural tour guide system based on LBS and web services. The system shows good performance in the practice of freshmen campus tour guide, and can basically meet the needs of personalized tour guide.

References

1. Li, Y.: Design and development of android client for LBS based travel information interaction platform. Beijing University of Technology (2013)
2. Chen, X., Shi, X.: Thoughts on LBS based tourism location service. Tour. Dev. Res. 214–219 (2013)
3. Yu, C., Zeng, H., Song, J.: Design and implementation of tourism information service system based on LBS technology. J. Zhejiang Univ. Technol. 28–31 (2013)
4. Wang, T.: Design and implementation of LBS based mobile service framework. Mob. Internet 25–27 (2013)

Research on Evaluation and Management Mechanism of Education Management System Based on Clustering Algorithm

Na Dai[(⊠)]

Shandong Institute of Commerce and Technology, Jinan, Shandong, China

Abstract. In recent years, with the rapid development of information technology and the deepening of education system reform, information management has gradually penetrated into all aspects of education. The construction of teaching network has gradually become the core and development trend of information construction in schools, which is also the practical need of deepening education and teaching reform. In order to share teaching resources and serve students. According to the characteristics of school teaching resources and the development trend of network teaching, this paper develops an education management system based on clustering algorithm. The system establishes an interactive teaching website between teachers and students, and is committed to creating a teaching network widely used and loved by teachers and students in vocational secondary schools.

Keywords: Clustering algorithm · Education management · Pipeline mechanism

1 Introduction

In this highly information age, if China wants to rise in the world, it must pay enough attention to the current situation of education. Although in general, the financial allocation for education has been gradually increased, except for coastal and well-known colleges and universities, the hardware facilities of the school are still quite backward. Naturally, they are far behind the development of the information age. In order to realize the high-end of information technology, Bijing region must be supported by high-tech equipment and talents. First of all, the school is inborn deficient in hardware and equipment. Secondly, it is difficult to attract young people with knowledge, culture, science and technology and ideals with a meager salary. Of course, as of today, the facilities of colleges and universities have been greatly improved, but the corresponding software development is lagging behind.

The application of education management system is still relatively few. Colleges and universities still use manual operation. All kinds of announcements, results sorting and information management have to manually enter documents or forms, and then use transmission tools to live on the network for reporting and distribution. This not only wastes a lot of human and material resources, but also inevitably leads to mistakes, and even serious consequences. And in the process of approval, if there is unqualified

J. C. Hung et al. (Eds.): FC 2021, LNEE 827, pp. 1665–1670, 2022.
https://doi.org/10.1007/978-981-16-8052-6_245

situation, it is necessary to return the data file to each department, and then the Department rearranges, enters and modifies it, and then submits it to the superior again. In this way, there are many times, which seriously reduces the efficiency [1].

In this paper, the education management system based on clustering algorithm is based on the analysis of the disadvantages of the real education management system and the integration of cloud computing platform services, so as to provide basic management functions for university teachers and staff, and expand the relevant functions of the whole education management system. In this way, we can really promote the innovation of traditional education management system with high technology, provide convenience for staff with efficient operation system, and scientifically and reasonably manage key information such as student status information, teaching plan, curriculum arrangement and achievement information.

2   Clustering Algorithm

2.1   Concept of Clustering Algorithm

In natural science and social science, there are a lot of classification problems. The so-called class, popularly speaking, refers to the collection of similar elements. Cluster analysis, also known as group analysis, is a statistical analysis method to study (sample or index) classification. Cluster analysis originated from taxonomy. In ancient taxonomy, people mainly rely on experience and professional knowledge to achieve classification, and rarely use mathematical tools for quantitative classification. With the development of human science and technology, the requirement of classification is higher and higher, so that sometimes it is difficult to classify accurately only by experience and professional knowledge. So people gradually introduce mathematical tools into taxonomy, forming numerical taxonomy, and then introduce multivariate analysis technology into numerical taxonomy, forming cluster analysis. The content of cluster analysis is very rich, including systematic clustering method, ordered sample clustering method, dynamic clustering method, fuzzy clustering method, graph theory clustering method, cluster prediction method and so on [2].

2.2   Classification of Clustering Algorithm

(1) K-means algorithm
     The k-means algorithm accepts the input K; Then, the N data objects are divided into K clusters to satisfy the following requirements: the similarity of objects in the same cluster is high; The similarity of objects in different clusters is small. Cluster similarity is calculated by using a "center object" (gravitational center) obtained from the mean value of objects in each cluster.
(2) Clara algorithm
     K-medoids algorithm is not suitable for the calculation of large amount of data, but Clara algorithm is a method based on adoption, which can process a large amount of data. The idea of Clara algorithm is to replace the whole data with the actual data sampling, and then use the k-medoids algorithm to get the best medoids on these

sampled data. Clara algorithm extracts multiple samples from the actual data. In each sample, k-medoids algorithm is used to get the corresponding (01, 02, 0i, 0k), and then the one with the smallest E is selected as the final result.

(3) Clarans algorithm

The efficiency of Clara algorithm depends on the size of the sample, and it is not likely to get the best result. On the basis of Clara algorithm, clarans algorithm is proposed. Different from Clara algorithm, in the process of Clara algorithm searching for the best medoids, the sampling is invariable. The clarans algorithm uses different sampling in each cycle. Different from the above process of finding the best medoids, the number of cycles must be artificially limited.

3   Education Management System Based on Clustering Algorithm

3.1   Overall Design Scheme

This paper will design the overall scheme as shown in Fig. 1. At the same time, the paper will select the information database of the college, which includes the examination results, student information and course information of all courses.

The first step is data acquisition. In order to ensure the integrity and accuracy of data, we must first select and sort out the original data.

Fig. 1. Implementation diagram of design scheme

The second step is data preprocessing. Data preprocessing is a step-by-step process from the outside to the inside. After four steps of data review, data cleaning, data conversion and data verification, data preprocessing can solve the problems of data conflict and data inconsistency, and finally form a student information table.

The third step is to implement the clustering algorithm. After the mining task is determined, the K-means clustering algorithm is programmed in MATLAB to realize the processing of K-means in student information analysis.

The fourth step is to evaluate the clustering results, to explain and evaluate the information found in the clustering results. After using k-means clustering algorithm, in the student performance evaluation, each class is a performance group. Different classes divide each performance group accordingly, and also give the central scores of different performance groups. These central scores are one of the reference standards for student performance division.

The fifth step is to put forward the corresponding strategies, provide the information to the teaching decision-makers, adjust the teaching strategies, further guide the teaching work, and improve the school management system.

3.2   Algorithm Design Principle Based on K-means

Figure 2 shows the flow of K-means algorithm to study student achievement. In the whole design process, there are two key problems, namely, the expression of achievement and the distance calculation of achievement.

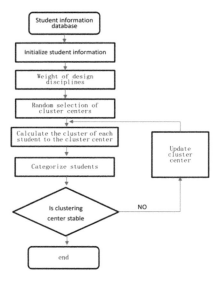

Fig. 2. K-means algorithm research student achievement process

For the first problem, the paper regards each student's examination achievement of each subject as a q-dimensional vector, and records it as X (= (xux.., x), (I = 1,2,.., n), where x Yang represents the achievement of the k-th subject whose student number is I, The score adopts the hundred point system, and gives different weights according to different subjects [3]. For the second problem, the Euclidean weighted distance is used to define the distance between students' grades. If the number of clustering groups is set as P, C (J = 1,2,…, P) is the clustering center, then the distance between grades and clustering center can be expressed as follows:

$$\|x_i - c_i\| = \sqrt{\sum_{k=1}^{q} \omega_k (x_{kp} - c_{kj})}, (1 \leq j \leq p) \tag{1}$$

Where q is the dimension of the particle's attributes, and B is the weight of each attribute.

4 Student Management Strategy Based on Education Management System

4.1 Actively Improve the Concept of Student Management

Based on the big data environment, student management workers should actively adopt the concept of keeping pace with the times, and actively explore the innovative development path of student management suitable for the big data environment. First of all, personnel engaged in student management should have a correct understanding of big data, clearly know the advantages of big data in student management, and then provide scientific and standardized information reference when making management decisions in Colleges and universities. Secondly, to deeply understand the existence value of big data, it is not a simple data information collection technology, but also can deeply analyze the data information, so as to provide valuable information results for the formulation of educational objectives and the development of student management work [4].

4.2 We Should Pay More Attention to the Information Construction of Student Management

The advantage of big data in college student management is more and more prominent, but the application of big data technology in student management is not the only innovation path, but to make college managers have a good sense of management innovation in the context of big data. The survey results show that the application of big data in college student management is still in the research stage, and there is no perfect application system. Therefore, university administrators can try to gradually promote the information construction of student management, and improve the efficiency of student management with the help of information technology, which is not only the performance of innovative management thinking, but also can create a good environment for the efficient use of big data technology.

5 Conclusion

This paper studies the application of K-means clustering algorithm in student information analysis. Through the data preprocessing, using k-means algorithm, using MATLAB tools to process and analyze the data, to make up for the defects of traditional statistical methods. And for different types of students, it gives students' self-development strategies and teaching management strategies, so as to prepare for the later improvement of education management and teaching quality.

References

1. Xi, C., Tan, S., Bai, Y., et al.: Innovative exploration of college student education management in the era of big data. J. Yunnan Agric. Univ. (Soc. Sci. Ed.) **11**(4), 110–114 (2017)
2. Ling, H.: Design and implementation of graduate education information management system. J. Hefei Univ. Technol.: Nat. Sci. Ed. (5), 797–800 (2002)
3. Tan, Q.: Research on test paper score analysis based on K-means clustering algorithm. J. Henan Univ. (Nat. Sci. Ed.) **39**(4), 412–415 (2009)
4. Jiang, S.: Design and implementation of education management information system. Comput. Appl. **22**(9), 17–19 (2011)

Design and Implementation of Real Estate Market Forecasting Model Based on Multiple Regression Analysis

Jie Gao[✉]

Chongqing College of Architecture and Technology, Shapingba District, Chongqing 400030, China

Abstract. In recent years, with the rising of China's GDP index, the problem of real estate market has become increasingly prominent. It is necessary to prevent the real estate market from developing too hard, resulting in the hard landing of the real estate bubble, and to ensure that the real estate market can meet the real needs of the masses. This requires a good real estate market prediction model to accurately predict the sales area of the real estate market and guide the real estate to make reasonable planning. This paper uses multiple linear regression analysis to design a real estate market forecasting model, explores and designs a real estate market forecasting algorithm based on multiple linear regression.

Keywords: Multiple linear regression · Real estate market · Prediction model

1 Introduction

In recent years, the hottest economic topic in China is the real estate issue. Since ancient times, Chinese people have the concept of "settle down and work" and "live and work in peace and contentment". Therefore, the closely related "house" naturally has irreplaceable importance in the eyes of Chinese people. With the continuous development of China's GDP index, the problem of real estate market has become increasingly prominent. The "bubble economy" of the real estate market has also brought a haze to China's GDP, which has brought a huge potential crisis to our economy. The real estate market is not only an economic problem, but also a social problem. It is a big problem that China will face in the process of social development in the future. After years of blind popularity, the bubble in the real estate market has been increasing. The state and government have realized this problem. Therefore, many regulatory measures have been introduced to make the real estate market gradually return to rational growth [1].

This paper mainly studies the design of a real estate market demand forecasting model based on multiple linear regression analysis. On the basis of learning and referring to foreign market forecasting theories, through learning different real estate market forecasting methods, using multiple regression analysis method, selecting appropriate influencing factors, a new multiple regression model is obtained. In the whole multiple regression analysis, this paper combines more market forecasting theory to conduct a comprehensive analysis of the model. The data needed for multiple

© The Author(s), under exclusive license to Springer Nature Singapore Pte Ltd. 2022
J. C. Hung et al. (Eds.): FC 2021, LNEE 827, pp. 1671–1677, 2022.
https://doi.org/10.1007/978-981-16-8052-6_246

regression analysis is obtained on the website through professional data acquisition software panda acquisition software, and a more accurate real estate market prediction algorithm is obtained. According to the prediction model, an experimental platform is designed, which can predict the real estate market demand of different regions, such as the real estate market demand area of different provinces and cities in the next few years.

2 Market Prediction

Market prediction is based on the information obtained from market research, using scientific methods and means to predict and infer the future evolution law and development trend of things. The purpose of market forecast is to minimize the impact of uncertainty on the forecast object and provide the basis for scientific decision-making. The emergence and development of market forecasting can be divided into three stages: market analysis and forecasting technology came into being in 1930s; Since the 1940s, the scientificity of market forecasting has been increasing: from the 1970s to the 1980s, market research and forecasting has developed into an independent new discipline. There are three main functions of market forecast: it is an important means to guide social production and meet market needs; it is the basis for enterprises to formulate business strategies and make scientific decisions; it is the basis for improving management level and economic benefits. Market prediction is an important field of marketing research [2]. There are many forecasting methods. According to the statistics of some western research institutions, there are more than 200 kinds of forecasting methods. There are also 20 or 30 commonly used forecasting methods, which can be roughly divided into two categories: qualitative forecasting and time series analysis.

2.1 Qualitative Prediction Method

Qualitative prediction method is also called judgment prediction method. It is based on the existing historical data and realistic data, relying on personal judgment and comprehensive analysis ability, the forecasters make judgment on the future trend of the market. In the market qualitative forecasting methods, the methods often used are expert meeting method, experience estimation method, customer opinion method and so on. This method is generally used to forecast the sales of new products.

Expert meeting method: it is to invite relevant experts to make evaluation on some forecast events and their development prospects through meeting, and synthesize various opinions on the basis of expert analysis and judgment, so as to make qualitative and quantitative conclusions on the investigation and analysis of events. The biggest advantage of this method is that it can gather wisdom. At the same time, through mutual inspiration and exchange, the participating experts constantly improve their suggestions and "collide" to come up with new ideas and ideas. The biggest drawback is that it is easy to yield to "authority". They are not willing to revise the opinions of others publicly, even if they are obviously wrong.

2.2   Time Series Prediction Method

Time series analysis method is also called historical extension method. It is based on the historical time series data, using certain mathematical methods to extend outward, to predict the future development trend of the market. The time series forecasting method is a group of observation values which equate economic development, purchasing power growth and sales change with one variable. They are arranged according to the time sequence, and then extended outward by certain mathematical methods to predict the future development trend.

The arithmetic mean method can be divided into simple arithmetic mean method and weighted arithmetic mean method. The formula of simple arithmetic mean method is as follows:

$$\hat{y} = \tilde{x} = \frac{\sum x_i}{n} \tag{1}$$

The formula of weighted arithmetic mean method is as follows:

$$\hat{y} = \tilde{X}_F = \frac{X_1 F_1 + X_2 F_2 + \ldots X_K F_K}{F_1 + F_2 + \ldots F_K} = \frac{\sum X_i F_i}{\sum F_i} \tag{2}$$

3   Multiple Regression Analysis

3.1   The Meaning and Types of Regression Analysis and Prediction Method

Regression analysis forecasting method is a kind of forecasting method, which starts from the relationship between various economic phenomena, and calculates the future state and quantity performance of the forecasting object through the analysis of the changing trend of the phenomena related to the forecasting object. The independent variable of regression analysis forecasting method is different from that of time series forecasting method. The independent variable of the latter is time itself, while the independent variable of the former is other variables reflecting market phenomena. Regression analysis is an important market forecasting method. It is a concrete, effective and practical method of market forecasting [3].

3.2   The Conditions of Applying Regression Analysis and Prediction Method

In order to improve the accuracy of prediction, we must pay attention to the preconditions when using regression analysis.

There must be a close relationship between dependent variables and independent variables. Only when we have a correct understanding of the internal inevitable connection and external accidental connection between economic phenomena, not confused by false correlation, and accurately analyze the correlation between them, can we make a correct judgment. The method of judging the close degree of correlation can be through drawing correlation graph and calculating correlation coefficient. The type of correlation can be judged by the correlation chart drawn from historical data. There are several types of correlation graphs.

(1) Zero correlation graph. When there is no correlation between independent variable x and dependent variable y, it is called zero correlation, as shown in Fig. 1.

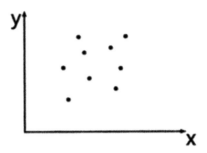

Fig. 1. Zero correlation graph

(2) Strong positive correlation graph. When the independent variable x increases, the dependent variable y also increases, and the distribution of points is concentrated and linear. So there is a strong correlation between them,as shown in Fig. 2.

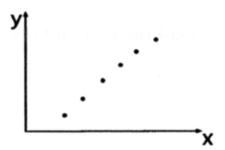

Fig. 2. Strong positive correlation graph

(3) Weak positive correlation graph. When the value of independent variable x increases, the value of dependent variable y also increases, but the distribution of points is not concentrated. There is only a certain correlation between them, which is called weak positive correlation, as shown in Fig. 3.

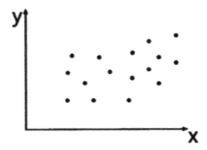

Fig. 3. Weak positive correlation graph

4 Prediction Model of Real Estate Market Based on Multiple Regression Analysis

4.1 Design of Multiple Regression Model

The real estate market forecast should follow four principles, including inertia principle, correlation principle, analogy principle and statistics principle. There are three important indexes in real estate market demand forecast: market demand, market demand potential and sales potential. Market demand refers to the total quantity of real estate purchased by consumers in a certain period and under certain conditions [4].

Among them, market demand is divided into market demand potential and market sales potential. Market demand potential refers to the maximum amount of market demand in a specific marketing environment. Market sales potential refers to the highest quantity that a real estate development enterprise can sell in the market. Market sales potential is an important basis for real estate enterprises to determine and select the target market and mix marketing strategies. The relationship between market demand, demand potential and sales potential is shown in Fig. 4.

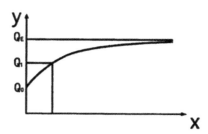

Fig. 4. Relationship between market demand, demand potential and sales potential

Y is the market demand, x is the marketing cost of the enterprise, Q_E is the market demand potential, Q_1 is the market sales potential, and Q_0 is the lowest point of the market sales. It can be seen from the figure that the market demand has a saturation point, and no amount of marketing will increase the market sales when it reaches this critical value.

The basic procedure of the real estate market prediction model includes five steps, as follows:

(1) Determine the forecast target and make the plan. (2) Collect and analyze the forecast data. (3) The prediction method and model are selected. (4) Scientific prediction. (5) Verify the prediction results.

4.2   Realization of Real Estate Market Forecast Platform

In this chapter, through the demand analysis of the real estate market prediction platform, we choose the appropriate development tool for the function of the real estate market prediction platform. Java is selected as the main development language. In the platform, the database is placed on the server, so the database is required to have good network support in operation, And the database has the requirements of metadata reading and data recording, so the requirements of timeliness and concurrency are relatively high, so the database management system on the server selects mysql, uses Tomcat as the server, Tomcat is Apache Company, the foreground display is realized through jsp.html technology, and uses Spring3.0 and mybatis3.0 framework for J2EE background development. The prediction process is shown in Fig. 5.

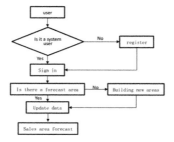

Fig. 5. Forecast flow chart

5   Conclusion

The research of market forecast has always been a hot topic. The market forecast methods at home and abroad are very mature, and there are many real estate market forecasts. However, due to the lack of technical tools, most of these theories and methods only stay in the theoretical discussion, not applied to the practice of market forecast. This paper studies the real estate market forecasting algorithm. The algorithm studies the influencing factors of the sales area of the real estate market through the multiple linear regression analysis method, and establishes a multiple regression real estate market forecasting model, Using the established multiple linear regression model, an experimental prediction platform is designed, which can accurately predict the next year or the next quarter of the sales area of the real estate market in a certain region.

References

1. Liu, L., Skirt, D., Qin, T.: Application of regression analysis in beverage sales forecast. Stat. Prediction (6), 28–29 (2001)
2. Li, Z.: Non economic factors affecting the demand of housing in China. North. Econ. (11), 33–34 (2005)
3. Huang, X.: Application of one-dimensional linear regression analysis in supermarket product sales. Sci. Technol. Inf. (11), 77–78 (2013)
4. Lu, Y.Y., Zhang, G.L.: Hierarchical analysis of the factors influencing the demand of the housing real estate market. Group Econ. Res. **4**, 32–33 (2007)

Research on Optimization Model and Algorithm of Logistics Distribution System Under E-commerce

Bingkai Huang[✉]

Jimei University Chengyi College, Xiamen 361021, Fujian, China

Abstract. E-commerce, a new business model, has become the main trend of the development of human society. At the same time, logistics distribution as one of the key links, will also usher in a good development, but will also face a series of challenges. Under the e-commerce environment, the traditional logistics distribution mode has been unable to meet the actual needs of consumers. It is an important problem for logistics enterprises to consider how to deliver goods at a faster speed, lower cost and within the time acceptable to customers. This paper studies the logistics distribution system model under e-commerce and optimizes the model algorithm, aiming at improving the efficiency of logistics distribution and pointing out the direction for logistics enterprises to solve the practical problems of distribution path optimization, And can enrich the relevant theory of logistics distribution.

Keywords: Electronic Commerce · Logistics distribution · Model · Algorithm

1 Introduction

With the development of Internet and the rapid development of information technology, e-commerce is increasingly affecting the economic life of today. E-commerce is the most favorable business model for the end customer service. It provides the goods directly for the final customers, eliminates many intermediate links, reduces the operating cost, and enables the end customers to enjoy the price benefits; Can communicate with customers face-to-face through the network. E-commerce activities mainly involve information flow, commercial flow, capital flow and logistics, and in these four flows, the research and application of information flow, commercial flow and capital flow have made progress. However, due to the backward logistics distribution in China, it restricts the development of e-commerce, and logistics distribution under e-commerce presents many new characteristics: on the one hand, logistics enterprises serve a large number of personalized end customers, which have the characteristics of small demand, many varieties and scattered location; On the other hand, customers also put forward higher requirements for logistics distribution service, such as fixed point, timing and quantity. Therefore, many logistics distribution systems established in traditional ways can not meet the actual needs of e-commerce development, and become a major bottleneck to hinder the development of e-commerce [1].

J. C. Hung et al. (Eds.): FC 2021, LNEE 827, pp. 1678–1684, 2022.
https://doi.org/10.1007/978-981-16-8052-6_247

Therefore, in the new environment, combined with new characteristics, how to make reasonable decisions on the location, demand distribution, transportation mode and route selection of facilities under e-commerce, and establish an efficient e-commerce distribution system, so as to reduce the cost of distribution, improve the speed of arrival of goods, guarantee the quality of services and improve the service level, It has become an important problem for the development of logistics enterprises under e-commerce.

2   Relevant Theoretical Knowledge

2.1   Basic Concept of Logistics Distribution

The meaning of logistics is: according to the customer's order requirements, organize the distribution and distribution in the logistics center, and then deliver the allocated goods to the customers in the form of modern transportation within the specified time, so as to realize the optimal allocation of resources.

We can understand logistics distribution as a common service mode, which can cover the following aspects: the common use of all kinds of resources; Common adoption of facilities: the common process of management. Specifically, logistics distribution is a unique business form of logistics activities. Its uniqueness lies in that it does not exist alone, but is organically combined with capital flow, logistics and information flow. The flow of these resources exists in the whole process of logistics distribution. It is no exaggeration to say that logistics distribution can contain the necessary factors of logistics activities. On the surface, distribution is a function derived from transportation. In fact, logistics distribution can basically cover all the functions of logistics, which is a complete process. To a large extent, logistics distribution is easily affected by the external environment, with strong randomness. Therefore, it is necessary to establish a complete management and control, and strong theoretical and technical support [2].

2.2   Introduction to VRP

(1) The meaning of VRP

VRP (vehicle routing problem) is the abbreviation of vehicle routing problem (VRP). Its meaning is to meet the basic constraints, such as delivery time, limited truck capacity, customer demand and so on, in order to achieve the established goal (the shortest distribution time: the shortest total travel distance; the shortest distribution time; the shortest distribution time; the shortest distribution time; the shortest distribution time); The total number of vehicles delivered is the least; Through the reasonable optimization of the route, the distribution vehicles can shuttle between the distribution center and pick-up point orderly, and can return to the origin within the specified time.

(2) The source of VRP

The problem of VRP comes from TSP, that is, the traveling salesman problem: a businessman will go to m cities for product promotion. Assuming that the distance between X and Y or the cost is DXY, how should the traveling salesman design his route so that he can travel all the cities (each city only goes once) and return to the origin with the shortest route. In order to show the TSP problem more vividly, the corresponding diagram is given, as shown in Fig. 1. The specific expression of this problem in graph theory is as follows: in graph $F = (C, R)$, C is the aggregate of all cities, R is the aggregate of all driving routes, $F = \{(x, y)|x, y \in C\}$. Assuming the distance between cities x and y, we can know that the traveling salesman starts from any point and finally returns to this point. All points (cities) must walk and can only walk once to find the shortest route. This circuit is actually a closed circuit, which can also be called Hamiltonian circuit. If it's $D_{xy} = D_{yx}$, then the problem is a symmetric TSP problem, if they are not equal, then it is an asymmetric TSP problem. In Fig. 1, ①, ②,... Represent the cities along the way, and ① represent the departure cities.

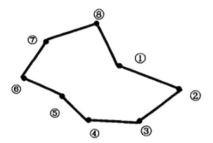

Fig. 1. TSP model

MTSP (multiple traveling salesman problem) is an extension of TSP. It means that there are multiple traveling merchants going to each city from any one city to promote their products. All cities must be passed by and can only be visited by one traveling salesman once. All traveling merchants must finally meet in the starting city, and the goal is to find the shortest route. The main difference between MTSP and TSP is that the number of participants is different. The latter has only one travel agent, while the former has more than one [3]. Here, if there are m traveling agents, the problem can be expressed as a set of M closed loops converging in the initial city in graph theory, as shown in Fig. 2.

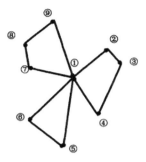

Fig. 2. MTSP model

3 Logistics Distribution Under E-commerce

3.1 Basic Meaning

This mode is a new form of logistics distribution. By using sophisticated Internet technology, advanced hardware equipment, complex software system and flexible management mode, enterprises can make clear the needs of the whole society, strictly abide by the customer's order requirements, sort, assemble the goods, and ensure the safety and efficiency of the goods within the time required by customers Deliver the quality and quantity to the place designated by the customer.

E-commerce has the incomparable advantages of traditional commerce, including low cost, high efficiency and wide range, which brings great convenience to enterprises and is also the main way of future trade. The development of business activities must be supported by logistics distribution, and the relationship between them is inseparable. Each e-commerce transaction includes three basic processes: information release; Commodity trading and settlement; Delivery. The former two can be realized through computer network, while the flow of tangible products can reach customers only through actual transportation. On the surface, e-commerce is a virtual and intangible business activity, but behind it is a large number of physical transportation. Therefore, in a sense, it can be said that logistics distribution determines the success or failure of e-commerce activities to a certain extent. On the contrary, e-commerce also brings new development opportunities and challenges to logistics distribution. E-commerce greatly affects the production environment, consumption environment and business flow environment of distribution activities, and also puts forward higher requirements for the distribution process. E-commerce environment requires that the whole distribution process should be realized with faster speed, lower cost, faster response and higher quality service [4].

3.2   Main Features

Logistics distribution under EC environment is different from traditional logistics distribution, which has four basic characteristics.

(1) The customers are special

In EC environment, the number of customers is large and the transaction volume is small, but the number of transactions is large and the geographical distribution is not concentrated.

(2) The nature of the goods to be delivered is extremely special

There are many kinds of distribution goods in EC environment, including large-scale production equipment and daily consumer goods, which are closely related to people's life. With the development of e-commerce, the number of daily necessities such as clothes, food, books, cosmetics and so on is increasing rapidly.

(3) Distribution cost and quality are very special

The products ordered by online consumers have the characteristics of small quantity, many varieties and scattered distribution, which leads to the high cost of logistics distribution. The whole distribution process is also more complex and difficult. The quality of the whole distribution process is easily affected by time, distance and traffic conditions.

(4) The supply chain system of distribution is very special

The logistics distribution under EC environment must establish a perfect supply chain system, only in this way can we achieve sufficient inventory and rapid response. The distribution process in e-commerce environment can be seen in Fig. 3.

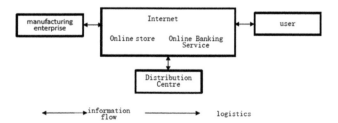

Fig. 3. Flow chart of logistics distribution under E-commerce

4   Multi Objective Logistics Distribution Model and Algorithm in Deterministic Environment

4.1   Construction of Multi-objective Model

In the fierce market competition, enterprise logistics distribution is facing the requirements from all aspects of the market, on the one hand, it should have good sensitivity, on the other hand, it should also have strong flexibility. If logistics distribution enterprises want to win in the competition, they must maximize customer satisfaction on the basis of cost reduction, which requires enterprises to optimize the path and deliver goods to customers within the time allowed by customers, so as to achieve the minimum cost and maximum profit. The logistics distribution route optimization problem is actually a multi-objective problem. The traditional VRP model simplifies the actual situation and establishes a single objective optimization model with the objective of minimizing the total vehicle transportation cost (it is generally assumed that the transportation cost is proportional to the transportation distance) [5].

$$Z = \min \sum_i \sum_j \sum_k C_{ij} x_{ijk} (i \neq j) \tag{1}$$

In model 1, Z represents the total cost, I and j represent any two points, K represents any transport vehicle, and C represents the transport cost from point I to point J. the variable XG is defined as follows, as shown in Eq. 2.

$$x_{ijk} = \begin{cases} 1, \text{The transportation task from i to j is completed by k} \\ 0, \text{otherwise} \end{cases} \tag{2}$$

This paper takes into account the actual situation, such as the original short distance due to traffic congestion and other emergencies, which makes the distribution takes a very long time; The road quality is poor, the truck bumps seriously, and the fuel consumption is also increasing, which will increase the whole transportation cost. Therefore, the total transportation cost can not be simply measured by the transportation distance.

4.2   VRPTW Model Optimized Under Multi-objective

In the actual environment, there is a special situation, which is often encountered by enterprises in the actual distribution. VRPTW, a special case of VRP, is that the customer has strict requirements on the service start time, and cannot be earlier than the earliest acceptable time, but it can not be later than the latest time that can be tolerated. Assuming that the time window of the customer is within the range of $[ET_i, LT_i]$, That is to say, the distribution activities must start in this time period, if the delivery vehicle reaches point i, d_i before et, wait; If later than LT, the service will be delayed. If s is used, it represents the start time of the service, e represents the end time of the service, and δ_i represents the time spent to complete the customer's required delivery task. Here, it mainly includes loading or unloading, then $\delta_i = e_i\text{-}s_i$ (the assumption here is that the customer time window is satisfied).

In this case, the VRP model established above also needs to be revised again to form VRPTW model. In this case, the first three optimization objectives are consistent with the optimization objectives of the previous model. On this basis, another goal is to reach within the allowable time of the customer to reduce customer dissatisfaction.

$$\begin{cases} \min Z_1 = \sum_i \sum_j \sum_k T_{ij} x_{ijk} \text{(the shortest time)} \\ \min Z_2 = \sum_i \sum_j \sum_k D_{ij} x_{ijk} \text{(shortest driving distance)} \\ \min Z_3 = \sum_i \sum_j \sum_k C_{ij} x_{ijk} \text{(minimum cost)} \\ \min Z_3 = \sum_i \{\max[(ET_i - d_i), 0] + \max[(d_i - LT_i), 0]\} \text{(Arrive on time)} \end{cases} \quad (3)$$

5 Conclusion

The continuous change of customer demand and external environment puts forward higher service requirements for logistics enterprises, especially under the background of e-commerce, the number of customer demand increases, while the single demand is smaller. In the face of multiple customers, how to transport the right products to the right place with the fastest speed, the lowest cost and the highest customer satisfaction is the key problem that logistics distribution center should consider. The changing external environment requires the logistics distribution center to re-examine its own distribution network, and constantly optimize the distribution path, in order to ensure the smooth development of logistics distribution work.

References

1. Dai, H., Li, Y.: Mathematical model and genetic algorithm of distribution problem. Comput. Eng. Appl. 41(31), 189–191 (2005)
2. Tan, H.: Development and application of E-commerce oriented logistics distribution system. China Manuf. Informatiz. 36(1), 21–24 (2007)
3. Wang, S.: Brief introduction of enterprise logistics distribution mode. Logist. Manag. (10), 20–22 (2012)
4. Yu, Y., Jin, O.: Modeling of vending machine logistics distribution based on Petri net. Comput. Inf. Technol. 18(4), 14–16 (2010)
5. Huang, H.: Research on logistics distribution mode of retail chain supermarket. Mod. Econ. Inf. (1), 68–70 (2011)

Research on EHB System Algorithm and Controller Implementation of New Energy Vehicle

Haiqiong Li[1(✉)] and Shifei Li[2]

[1] College of Mechanical and Electrical Engineering, Yunnan Engineering
Vocational College, Anning City, Kunming 650304, Yunnan Province, China
[2] Kunming Chuanjinnuo Chemical Co., Ltd.,
Kunming 650500, Yunnan Province, China

Abstract. Fuel cell vehicle (FCV) uses fuel cell engine as the power source and motor as the drive. Therefore, when the traditional braking system is used in FCV, there are two obvious shortcomings: the vacuum booster device loses the reliable vacuum source and the combination of friction braking and regenerative braking cannot be realized. Therefore, it can be said that the vehicle brake by wire system forms the overall design scheme of EHB system; Based on the EHB system and three-way proportional reducing valve, the mathematical model of EHB system is established, and the unknown parameters of the system are identified; The controller module of the electro-hydraulic braking system is designed, which will have a very broad practical application prospect in today's rapid development of hybrid electric vehicles and fuel cell vehicles.

Keywords: New energy vehicles · EHB system algorithm · PID controller

1 Introduction

BBW (brake by wire) is a kind of X-by-wire, which refers to the integration of a series of intelligent brake control systems. It provides the functions of existing brake systems such as ABS, vehicle stability control, power assisted braking, traction control and so on, and organically combines each system into a complete functional system through on-board wired network. It is different from the traditional automobile braking system in that it uses electronic components to replace some mechanical components, so there are few mechanical connections, and the power transmission between the brake pedal and the brake is separated. Instead, it is connected by wires, which transmit energy and data lines transmit signals, and corresponding programs are designed and written in the electronic control system, Through the control of electronic control components, the distribution and size adjustment of braking force can be realized, and the functions of ABS and ASR can be easily realized. The original brake pedal is replaced by an analog generator, which is used to accept the driver's braking intention, generate and transmit the braking signal to the control and actuator, and simulate the feedback to the driver according to a certain algorithm. The EHB system uses the electric signal to transmit the driver's braking command to a microcontroller ECU, which processes other sensor

J. C. Hung et al. (Eds.): FC 2021, LNEE 827, pp. 1685–1690, 2022.
https://doi.org/10.1007/978-981-16-8052-6_248

signals synchronously, and calculates the optimal braking force of each wheel according to the specific driving state [1].

Firstly, the displacement sensor integrated in the electronic pedal senses the travel speed and force exerted by the driver on the pedal to identify the driver's braking intention; Then, the EHB electronic control unit (ECU) calculates the required braking force of each wheel according to the relevant signals transmitted by the system electrical circuit; Then, according to the control command output by EHB ECU, the hydraulic executive unit applies the required braking force to each wheel accurately through the high-pressure accumulator to make the vehicle brake or decelerate more quickly and stably. Therefore, EHB braking is a brake by wire system composed of induction, calculation and electric control. The braking force of EHB system is not controlled by hydraulic or pneumatic device, but by electronic motor. The electronic controller sends a control signal to control the motor of the brake driver. The motor drives the hydraulic pump to pump high-pressure oil into the accumulator to realize the braking force required by the brake. In this system, each wheel can be controlled independently, which ensures the stability of the vehicle in the braking process.

2　EHB System

2.1　EHB System Structure

EHB is flexible controlled by electronic system and powered by hydraulic system. The main parts are divided into three parts according to their functions.

(1) High pressure oil generating part

It is mainly composed of oil accumulator, hydraulic pump, motor, accumulator and check valve. The hydraulic pump driven by DC motor inputs the low-pressure oil in the oil accumulator to the high-pressure accumulator, and the one-way valve is set in the oil circuit to realize one-way delivery. The maximum pressure of the accumulator is 20 Mpa and the minimum pressure is 12 Mpa. The accumulator is a gas spring, and the pressure storage process is an isothermal compression process. Hydraulic pump can choose vane pump or plunger pump, which requires small flow, high pressure and high volumetric efficiency.

(2) Brake pressure regulation part

It is mainly composed of proportional reducing valve and related oil circuit. The current of the coil is proportional to the electromagnetic force. When the proportional reducing valve is opened, the pressure of the brake pipeline is fed back to the rear of the valve. When the feedback hydraulic pressure is equal to the sum of the electromagnetic force and the two spring forces, the control valve is closed and the brake pressure is maintained.

The working process is shown in Fig. 1: the driver presses the brake pedal, and the pedal force acts on the left end of the pedal guide bar through the lever mechanism to move the guide rod to the right. In the empty stroke C, the balance spring is free length, the control valve is closed, and there is no pressure in the brake line. The braking force can be generated by regenerative braking. When the pedal force is high, if the regenerative braking capacity is insufficient, the control valve is opened

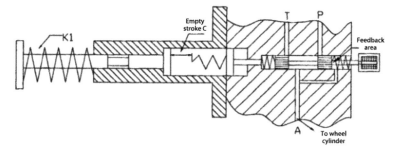

Fig. 1. EHB braking process diagram

by the return spring of the valve valve overcome by electromagnetic force. The high-pressure brake fluid flows into the brake line through port P from the accumulator, and the pressure of the brake line increases. At the same time, the pressure of the brake line is fed back to the rear of the spool valve to generate feedback pressure. When the sum of the feedback pressure and the spring return force is equal to the electromagnetic force, the control valve closes, and the brake line generates corresponding brake pressure to realize the braking process [2].

When the electrical system fails, although the electromagnetic force does not work, only the pedal force is applied to the control valve by the balance spring, but it can still provide a certain pressure of the brake line to realize the backup braking.

(3) Brake lines and wheel cylinder sections

All parts of this part still adopt the original vehicle scheme, and the pipeline, wheel cylinder and so on remain unchanged. The maximum working pressure of the pipeline is 12 mao.

2.2 EHB System Parameter Identification

Proportional pressure reducing valve is the key component of electro-hydraulic loading system, its dynamic and static performance directly affects the control accuracy of the electro-hydraulic loading system. The mathematical model of proportional pressure reducing valve has been established. Because of many factors affecting the parameters of the model, it cannot be completely determined theoretically. Therefore, the experimental method, i.e. parameter identification, must be used to determine the structural parameters of the proportional pressure reducing valve. At present, the theory of system identification has become an important branch of modern system theory. ZAD once defined system identification as: "system identification is to determine a system equivalent to the observed system from a class of systems based on input and output." The system here means generalized, can be a system, can also be a component. The basic process is shown in Fig. 2 below.

There are many methods of parameter estimation, including least square method, maximum likelihood method and prediction error method. This paper mainly uses the least square parameter estimation method.

In the field of identification and parameter estimation, the least square method has been a basic and important estimation method. It can be used in both dynamic and

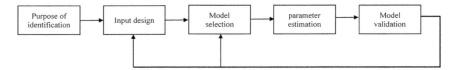

Fig. 2. Basic process of system identification

static systems; It can be used in both linear and nonlinear systems; It can be used for both off-line and on-line estimation. In a random environment, the least square method does not require the observation data to provide information about its probability and statistics, but the estimation results obtained by this method have fairly good statistical characteristics. The least square method is easy to understand and master, and the identification algorithm based on the least square principle is relatively simple in implementation. When other parameter identification methods are difficult to use, the least square method can provide a solution to the problem [3].

Considering the increasing development and popularization of digital computer, in system identification, the input information is often provided by the digital machine and added to the identified system after D/a conversion. The output information of the identified system is returned to the digital machine after measuring device and a/D conversion. The digital machine processes the input and output data according to some identification algorithm, Provide the parameter estimation or mathematical model of the identified system. The principle block diagram is shown in Fig. 3.

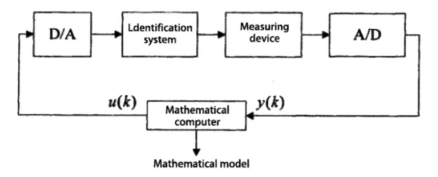

Fig. 3. Principle block diagram of system identification

3   Fuzzy PID Controller Design of Electro-Hydraulic Proportional Valve Control System

For valve digging system, there are usually "dead zone" links, which will worsen the quality of system regulation, affect the output accuracy of the system, whether the pressure deadband of proportional pressure reducing valve and pipeline is accurately compensated, which is directly related to the positioning accuracy and stability of the system.

However, the actual system has some nonlinear errors and has large hysteresis and dead zone. As shown in Fig. 4, the dynamic control can be controlled by both closed-loop and open-loop control. When closed-loop control is used, the pressure or flow of proportional pressure reducing valve can be controlled. According to the different control quantity, pressure or flow sensor is selected as the detection element, and the control function is realized by software algorithm. In order to ensure the fast response and no oscillation and shock, software algorithm can be used to form a variety of curves, and three common control curves, such as slope, integral and sine wave, are used. Make the pressure or flow change according to the control requirements to achieve the purpose of fast and stable control.

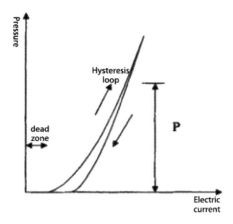

Fig. 4. Current - pressure curve

The purpose of nonlinear compensation is to adjust the output current of constant current source circuit by changing the duty cycle of ECU output to control the opening of proportional reducing valve, so as to reduce the lag effect of the system, increase the response speed of the system and improve the control accuracy for 4 s [4]. The first mock exam is to combine the nonlinear compensation control law and the proportional pressure reducing valve control system, and a digital PID fuzzy control method is adopted.

4   Working Principle of EHB

BBW (brake by wire) is the inevitable choice for the development of fuel cell and hybrid electric vehicle. As the first step to realize BBW, EHB (electro hydraulic braking system) has a very broad practical application prospect in the rapid development of hybrid electric vehicle and fuel cell vehicle.

EHB is a braking system which combines electronic system with hydraulic system. It is mainly composed of high-pressure oil generating part, electronic pedal, electronic control unit (ECU), hydraulic actuator (valve element), etc. The electronic pedal is mainly composed of brake pedal and pedal displacement sensor.

The working principle of EHB is to replace the brake pedal in the traditional brake system with electronic pedal, which is used to reflect the driver's braking intention, generate the braking signal and transmit it to the control unit ECU. The ECU sends the braking command to the actuator, and at the same time, it also feeds back the information of the braking strength to the driver, so as to ensure that the driver has enough pedal feeling. In the braking process, the wheel braking force is controlled by ECU and actuator, and the pedal displacement sensor continuously converts the pedal displacement signal into electrical signal, and inputs it to the electronic control unit. ECU will adjust the opening of the electro-hydraulic proportional valve according to the control signal, and then adjust the brake pressure on the wheel cylinder. In order to achieve the safety of the system, a backup redundant hydraulic system is designed in the system to ensure that the control system still has braking capacity in case of failure and ensure safety.

5   Conclusion

In this paper, according to the new requirements of new energy vehicles on the braking system, the shortcomings of traditional braking system are compared correspondingly. On this basis, the original vehicle braking system is improved to adapt to the most urgent need to solve the constraints of traditional braking system on new energy vehicles.

Acknowledgements. Research on Improving the Training Model of Automobile Maintenance Service Ability of New Energy Auto Repair Major Students in the Scientific Research Fund of Yunnan Provincial Department of Education Project No: 2019J0478.

References

1. Li, Y., Yu, S.: Double closed loop control technology of electro hydraulic proportional valve. Micromotor (6), 35–37 (2005)
2. Kong, X., Tang, D., You, B.: Micro control of electro hydraulic proportional valve. Hydraul. Pneumatic (3), 61–63 (2004)
3. Chen, B.: Electric Drive Automatic Control System. Department of Industrial Automation, Shanghai University of technology, Shanghai (1991)
4. Chen, J.: Automobile Structure. China Machine Press, Beijing (2002)

Dynamic Prediction Method of Casing Damage Based on Rough Set Theory and Support Vector Machine

Canchao Liu[(✉)]

Daqing Oilfield Limited Company,
No. 7 Oil Production Company, Daqing 163517, China
liucanchao@petrochina.com.cn

Abstract. With the continuous development of the oilfield, the casing damage is becoming more and more serious in the world, which has brought huge economic losses to the oilfield and seriously hindered the normal production of the oilfield. Therefore, it is of great significance to extend the service life of casing, reduce the rate of casing damage and realize efficient production by using reasonable prediction methods to make early prediction of casing damage and take corresponding preventive measures. Casing damage problem is a very complex large-scale system, which has the characteristics of uncertainty, fuzziness and time-varying. Therefore, from the perspective of system theory, this paper first introduces the basic theory of forecasting and common forecasting methods. Through the comparative analysis of common forecasting methods, the support vector machine method is optimized, which is especially suitable for solving small sample, nonlinear and high-dimensional problems.

Keywords: Casing damage · Rough set theory · Vector machine · Dynamic forecast

1 Introduction

Oilfield development practice shows that with the continuous extension of production time, the continuous adjustment and implementation of development plan, and different geological, engineering and management conditions, a large number of oil and water well casings are damaged. According to incomplete statistics, as of the end of 2007, more than 22000 casing damaged wells have been found in all oilfields in China. The problem of casing damage is also prominent in foreign countries, such as West Siberia oil field, North Caucasus oil and gas field, Turkmen oil field, Texas oil field, Gulf of Mexico oil field and Gulf of Suez oil field. Casing damage brings many problems to oilfield production, such as direct economic loss caused by production decline; New investment is needed to repair casing damaged wells; It destroys the normal injection and production well pattern and strata, resulting in a huge waste of oil and gas resources; Disrupting the normal production deployment will bring a lot of blindness to macro decision-making and specific management. Casing damage has become one of the key factors restricting production. The mechanism of casing damage is different under different working conditions. The problem of casing damage is uncertain, fuzzy and time-varying. The reliability and practicability of existing prediction models are poor. Therefore, the author establishes a dynamic prediction method of casing damage based on Rough Set (RS) theory and support vector machine (SVM) algorithm. Using the existing geological and production data of the oilfield, the dynamic prediction of casing condition in the process of oilfield development can be realized, which provides the basis for adjusting development parameters and formulating casing protection measures as soon as possible [1].

2   Research Methods of Casing Damage

In recent years, many experts and scholars at home and abroad have devoted themselves to the research of casing damage. According to the theories and models adopted, the main methods are analytical method, numerical method, actual measurement method, etc.

Because both casing and cement sheath are circular and axisymmetric, under the ideal condition (assuming that both casing and cement sheath are circular and uniform in thickness), some theoretical solutions can be obtained by using analytical method. In salt rock, sandstone and mudstone creep formation, casing usually has to bear non-uniform extrusion stress, which is an important reason for casing damage. Therefore, it is an important work to analyze the deformation characteristics of casing under non-uniform external load. Based on the elastic assumption of casing subjected to non-uniform external load, Fang Jun et al. Analyzed the stress conditions of casing and cement sheath under different contact conditions, which provided a preliminary theoretical basis for the study of loading and deformation characteristics of casing and cement sheath.

In this paper, we discuss the plane strain problem of casing external load caused by formation rheology under the action of non-uniform in-situ stress, and verify the accuracy of finite element by comparing the numerical solution and theoretical solution under the action of uniform in-situ stress, It is concluded that the creep formation under the action of non-uniform in-situ stress will produce a non-uniform radial extrusion force greater than the in-situ stress and a non-uniform tangential load on the casing. By using Maxwell solid model and finite element analysis, it is found that the casing stress tends to be stable after a certain period of time. Moreover, under the action of non-uniform load, the stress around the casing presents an oval distribution. If the non-uniform load increases, the tensile stress will appear on the casing wall [2].

Regular inspection of oil pipelines can effectively predict and prevent quality accidents in oil wells. Experts and scholars at home and abroad have developed and improved many engineering logging tools and techniques to detect casing damage from many aspects. The engineering logging technology mostly judges the defects in the pipe indirectly through the detected physical signals. The well diameter series is a conventional detection method for the well bore condition of oil and water wells, which can provide the change of casing inner diameter; In the acoustic logging series, the borehole ultrasonic imaging logging can provide comprehensive and intuitive casing damage condition, while the noise logging is used to detect the formed pipe leakage and channeling; Azimuth series is used to determine the azimuth angle of casing deformation; Magnetic logging well series can check casing deformation, stagger, inner and outer wall corrosion and perforation quality.

3   Research Status and Application of Rough Set Theory and Support Vector Machine

3.1   Rough Set Theory

Rough set theory is a mathematical method proposed by Pawlak in 1982 to deal with fuzzy and uncertain problems. It has powerful data mining and knowledge discovery ability, and can solve complex problems with high nonlinearity and uncertainty.

For an information system $S = (U, A, V, f)$, where $U = \{z_1, z_2, \ldots, z_n\}$ is a set of non empty finite objects (also known as universe); $A = P \cup D, P \cap D = \emptyset$, P is called conditional attribute set, D is called decision attribute set; $f : U \times A \to V$ is a confidence function, which specifies the attribute value of each object Z in U, that is, for $z \in U, a \in A$. Therefore, any simplified B of P can be used to replace P without losing the information to be expressed in the information table, thus a simplified information table can be obtained.

3.2   Support Vector Machine

Support vector machine (SVM) is a new tool in the field of pattern recognition and machine learning in recent years. Its core content was put forward only in 1992. The publication of the natural statistical learning theory by Vapnik in 1995 marks the maturity of statistical learning theory system. Support vector machine (SVM) is based on statistical learning theory, which can effectively avoid the traditional classification problems such as over learning, dimension disaster and local minimum in classical learning methods. It still has good generalization ability under the condition of small samples, so it has been widely concerned [3].

Support vector machine maps the input vector to a high-dimensional feature space, and constructs the optimal classification surface in the feature space. It can avoid some shortcomings that cannot be overcome in the multi-layer forward network, and it has been proved theoretically that when the appropriate mapping function is selected, most of the linear inseparable problems in the input space can be transformed into linear separable problems in the feature space. However, in the process of mapping from low dimensional input space to high dimensional feature space, it is difficult to directly calculate the best classification plane in feature space in most cases because of the rapid growth of space dimension. By defining kernel function, support vector machine skillfully uses kernel function of original space to replace inner product operation in high-dimensional feature space

$$K\left(x_i, y_j\right) = \varphi(x_i) \cdot \varphi\left(y_j\right) \tag{1}$$

In the feature space, the dimension is large enough to make the image of the original spatial data have a linear relationship, and then the linear optimal decision function is constructed in the feature space. The training problem of support vector machine is essentially a quadratic programming (QP) problem. In optimization theory, there are many mature algorithms to solve this problem, such as interior point method, common gradient method. The structure of support vector machine is shown in Fig. 1.

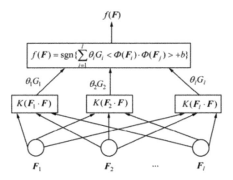

$$f(F) = \mathrm{sgn}\{\sum_{i=1}^{l}\theta_iG_i < \Phi(F_i)\cdot\Phi(F_j) > +b\}$$

Fig. 1. Structure of support vector machine

Because of its excellent generalization performance, SVM has been widely used in pattern recognition (character recognition, text automatic classification, face detection, head pose recognition), function approximation, time series prediction, fault recognition and prediction, information security, power system and power electronics.

(1) Pattern recognition

Pattern recognition is the earliest and most successful field of SVM application, and it is the direct application of support vector machine classification and multi classification. The specific applications include: face recognition and pose recognition, human body recognition, fingerprint recognition, gene data analysis and coding, computer keyboard user identity verification, voice classification retrieval, image retrieval, etc.

(2) Interpolation approximation of type function

The main purpose of this paper is to solve the problem of image fracture and period by using the fractal interpolation of support vector machine fitting function.

(3) Time series prediction

Because of the excellent prediction performance of time series, SVM has been used in casing damage prediction in recent years.

4 Research on Prediction Theory and Method

In the early stage of oil field exploitation, it is only when the normal production encounters problems that the casing damage under the well is understood and studied. In this way, whenever casing is damaged, it is necessary to interrupt production, and spend a lot of manpower, material and time to analyze casing damage, which seriously affects the normal production plan and reduces the efficiency of oilfield production. In order to realize the dynamic judgment of casing condition in oil field development, it is necessary to make forecast analysis. Therefore, the selection of prediction method is especially important, which is directly related to the reliability of prediction results. This chapter introduces the basic theory and common forecasting methods from the perspective of system theory, and analyzes some basic principles and mathematical models of statistical learning theory and support vector machine.

The casing loss is a geological engineering problem related to both static parameters (such as geological parameters) and dynamic parameters (such as development parameters). It is impossible to deal with dynamic parameters by using conventional support vector machine method, and it is easy to face dimension disaster problem under large sample conditions. Therefore, rough set theory and support vector machine integration method are adopted, The dynamic prediction model of casing loss is established by making full use of the advantages of both. Firstly, RS is used to preprocess the data, and SVM prediction system is established according to the information structure after RS processing. The method has three obvious advantages: firstly, RS method is used to reduce the number of features of information expression, reduce the data quantity of SVM input greatly, and improve the speed of system operation; ② After removing redundant information by RS method, the main factors causing casing damage are found out, which can eliminate the interference of noise data and improve the accuracy of the system; ③ SVM is used as the post information processing system, which has better fault tolerance and anti-interference capability. The modeling steps are as follows [4]:

(1) The knowledge base of casing damage is established, which mainly includes the factors affecting casing damage and casing condition.
(2) The sample data preprocessing includes attribute quantification, continuous attribute discretization and normalization.
(3) The original decision table system is constructed, and the decision table containing conditional attribute and decision attribute is established by using the attribute value of quantization and discretization.
(4) The rough set method is used to reduce the attribute of decision table and the minimum conditional attribute set is obtained.
(5) The RS SVM training machine is established. The structure of RS SVM is constructed, and the kernel function is chosen to train SVM with reduced samples.
(6) Verify the forecast effect. The test samples are input into the trained SVM to check the prediction results and verify the effectiveness of the method.

SVM dynamic prediction model of casing loss.

Support vector machine is a machine learning method r proposed by Vapnik based on statistical learning theory. As a post-processing method of the loss prediction model, SVM algorithm has strong fault tolerance and anti-interference ability. The basic idea is to transform the input space into a high-dimensional space by using the nonlinear transformation defined by the inner product function, and find the nonlinear relationship between the input and output variables in this high-dimensional space.

The new sample set is obtained by reducing the sample data by rough set theory

$$(x_1, y_1), (x_2, y_2), \ldots, (x_n, y_n), x \in R \tag{2}$$

To find an optimal function $\{f(x, w)\}$ in a set of functions $f(x, w_0)$ main_ 0) minimize the expected risk of prediction.

$$R(w) = \int \frac{1}{2} |y - f(x, w)| dF(x, y) \tag{3}$$

The rough set theory requires that the data to be processed must be discrete data, while in the loss decision information table, some belong to continuous attributes, some of them are discrete attributes, and they need to be discretized and the discrete attributes are quantified. For the discretization of continuous attributes, there are many kinds of continuous methods, such as equal distance method, equal frequency method, maximum direct method, neural network method and genetic algorithm. On the problem of casing damage, through years of engineering practice and independent research on the influencing factors of casing damage (excluding the influence of other factors), it is found that the degree of casing damage changes with the change of the influencing factors, and has certain regularity. Therefore, various influencing factors can be divided into certain grade standards, and then the characteristic value can be used to replace the attribute value. The discretization attributes include fault condition, cementing quality and casing damage grade, which need to be quantified.

5   Conclusion

(1) The new method of dynamic prediction of casing damage based on the integration of rough set and support vector machine makes full use of their advantages, realizes the dynamic prediction of casing condition in the process of oilfield development, and opens up a new idea for the research of casing damage.
(2) Rough set is used to reduce attributes, delete redundant attributes, and find out the main factors leading to casing damage. The purpose of data dimension reduction in prediction is realized, which is conducive to improving the operation speed and prediction accuracy.
(3) This method completely depends on the field data, and the prediction results are reliable and accurate, which has high theoretical and practical value.

References

1. Zhang, W., Wu, W., Liang, J., et al.: Rough Set Theory and Method. Science Press, Beijing (2001)
2. Deng, J., Wang, Q., Mao, Z., et al.: Hybrid algorithm of support vector regression based on rough set. J. China Univ. Petrol.: Nat. Sci. Ed. **33**(5), 159–163 (2009)
3. Wang, G.: Research on data mining based on support vector machine. Comput. Eng. **34**(8), 47–49 (2008)
4. Zhang, X.: On statistical learning theory and support vector machine. Acta Autom. Sin. **26**(1), 32–42 (2000)

Large Scale Graph Data Processing Technology in Cloud Computing Environment

Liya Liu$^{(\boxtimes)}$, Qingqing Guo, and Yinan Chen

Sanya Aviation and Tourism College, Sanya 572000, Hainan, China

Abstract. Cloud computing is a new Internet related application mode. The emergence of cloud computing has led to a revolution in IT industry. The academia has also conducted effective exploration and Research on various situations and related problems in cloud computing. In recent years, the rapid growth of emerging applications such as bioinformatics network, semantic web analysis and bioinformatics network analysis, It improves the requirement standard of large-scale graph processing function utility, and increases the demand of large-scale graph data processing technology ability. In this paper, a brief overview of the analysis in the cloud computing environment for large-scale graph data processing technology related applications.

Keywords: Cloud computing · Large scale graph data · Treatment technology

1 Introduction

Graph is the most commonly used abstract data structure in computer science, which is more complex in structure and semantics than linear table and tree, and has more general representation ability. Many application scenarios in the real world need to be represented by graph structure, and the processing and application related to graph are almost ubiquitous The citation relationship of scientific and technological literature, etc.; Emerging applications such as social network analysis, semantic web analysis, bioinformatics network analysis, etc.

Although the application and processing technology of graph has been developed for a long time, and the theory is becoming more and more perfect, with the advent of the information age, all kinds of information grow in an explosive mode, which leads to the increasing scale of graph. How to efficiently process large-scale graph has become a new challenge [1].

2 Introduction to Cloud Computing

2.1 The Meaning of Cloud Computing

Cloud computing is a new computing method under the background of the rapid development of Internet technology, and its core is the Internet. Cloud computing was proposed by IBM at the end of 2007, which is the first time that cloud computing

J. C. Hung et al. (Eds.): FC 2021, LNEE 827, pp. 1697–1703, 2022.
https://doi.org/10.1007/978-981-16-8052-6_250

appeared in front of the world. IBM defines cloud computing as follows: cloud computing is used to describe a system platform or a type of application at the same time.

2.2   Characteristics of Cloud Computing

The technical characteristics of cloud computing can be summarized as follows:

(1) Virtualization. Users can use computers, laptops, mobile phones and other tools to connect to the cloud computing server at any location of the connected network to achieve a variety of computing tasks. These cloud computing servers are usually clusters and do not have tangible entities.
(2) Universality and scalability. Under the support of large-scale cluster, cloud computing platform can support and construct a variety of applications, not only for specific applications, and the same cloud computing platform can support different applications running at the same time; The scale of cloud computing cluster has dynamic scalability, which can meet the needs of users and application scale growth, which makes it have good scalability.
(3) High economy. Due to the need to support a lot of computing tasks at the same time, the scale of cloud computing cluster is very large. For example, the number of cloud computing platform clusters of cloud computing service provider Google. Le exceeds 2 million, and other providers such as IBM, Microsoft and Yahoo also have hundreds of thousands of cloud computing clusters. Users can get rich computing resources from cloud computing cluster. Almost all cloud computing cluster servers are cheap devices, because cloud computing itself has super fault tolerance, it can use cheap devices to form a cloud cluster, which not only fully reflects the low-cost advantages of cloud computing, but also lays the foundation for its wide application [2].

2.3   Related Technologies of Cloud Computing

(1) Massive distributed storage technology. Distributed storage is the main way for cloud computing to store data, which ensures the high reliability, availability and economy of data to a certain extent. The high reliability of data storage is mainly achieved by means of redundant storage. Reliable software tools can effectively make up for the shortcomings of hardware. Because cloud computing system needs to meet the needs of a large number of users, it requires that data storage technology should have higher transmission rate and throughput, so as to provide services for all users in parallel, and distributed storage can just meet this requirement [3].
(2) Data management technology. Because cloud computing needs to process and analyze large data sets regularly to complete the computing tasks submitted by users, cloud computing data management technology must have the ability of efficient management of large data sets, and also be able to find specific data from huge data sets, so as to meet the application needs of users. BigTable and HBase are the most well-known and commonly used cloud computing data management technologies.

(3) Virtualization technology. At present, the global IT industry is gradually entering the era of cloud computing. Although a single virtualization technology has brought many benefits to it, it has to be admitted that people pay more attention to its comprehensive virtualization strategy. In the environment of cloud computing, the solution of virtualization is system integration, which integrates server, software, related services, storage system and network equipment. What it brings to people is the real sense of virtualization services.

(4) Parallel programming mode. Because a large number of users provide large-scale tasks at the same time, the programming model of cloud computing must ensure the complex parallel execution and task scheduling in the background, which is the premise and foundation of the effective use of cloud computing resources. At present, cloud computing generally adopts map reduce as the programming mode, which provides a simple, effective and practical solution for data processing of parallel systems. Its advantages lie in the cost of merging, high yield, better performance, better effect and easier deployment.

3   Cloud Computing Large Scale Graph Data Processing

3.1   Large Scale Graph Data Processing

Take the Internet and social networks as an example. In the past decade, with the popularity of the Internet and the promotion of Web2.0 technology, the number of web pages has increased rapidly. According to CNNIC statistics, in 2010, the scale of Chinese web pages reached 60 billion, with an annual growth rate of 78.6%. And the Internet-based social networks also later ranked first, such as Facebook, the world's largest social network, which has about 700 million users. In China, such as QQ space, Renren, etc., The development is also very rapid.

The expansion of entity scale in the real world leads to the rapid growth of graph data scale, which often has billions of vertices and trillions of edges. What this paper refers to is the large-scale of a single graph, which usually contains more than 1 billion vertices. Facing such a large-scale graph, it poses a huge challenge to massive data processing technology.

Taking the PageRank calculation commonly used in search engines as an example, the PageRank score of a web page is calculated according to the hyperlinks between web pages. The web pages are represented by graph vertices, and the links between web pages are represented by directed edges. 10 billion graph vertices and 60 billion edges are stored in the form of adjacency table, assuming that the storage space of each vertex and out degree edge accounts for 100 bytes, Then the storage space of the whole graph will exceed 1TB. For such a large-scale graph, the time and space costs of its storage, update, search and other processing are far beyond the capacity of traditional centralized graph data management. Efficient management of large-scale graph data, such as storage, index, update, search and so on, has become an urgent problem [4].

3.2   Advantages of Using Cloud Computing Environment to Process Large Scale Graph

Cloud computing is the product of grid computing, distributed computing, parallel computing, utility computing, network storage, virtualization and other advanced computer technology and network technology development. It has universal applicability. The development of cloud computing technology has been closely related to large-scale data processing, It is a direction with great development potential, and its main advantages are as follows:

(1) Massive graph data storage and maintenance capabilities. The amount of large-scale graph data can reach hundreds of GB or even Pb level, which is difficult to be stored in the traditional file system or database. Cloud computing environment provides distributed storage mode, which can gather the storage capacity and computing power of hundreds of ordinary computers, and provide high-capacity storage services. It is fully able to store and process large-scale graph data, concurrency control, data management, and so on Consistency maintenance, data backup and reliability control strategies can provide guarantee for the maintenance of large-scale graph data.

(2) Powerful distributed parallel processing capability. Taking advantage of the distributed parallel processing characteristics of cloud computing, a large graph can be divided into several subgraphs, and the processing of a large graph can be divided into several processing tasks for subgraphs. The distributed parallel computing ability of cloud computing can significantly improve the processing ability of large-scale graphs.

(3) Good scalability and flexibility. From the technical and economic point of view, cloud computing environment has good scalability and flexibility, which is very suitable for dealing with large-scale graph problems with elastic changes in the amount of data. Cloud computing environment is usually composed of cheap ordinary computers. With the increasing scale of graph data, nodes can be added to the cloud dynamically to expand storage capacity and computing resources, Without the huge investment of traditional parallel mode.

4   Graph Data Model and Storage Management

The logical expression form and physical storage structure of graph data are the basis of graph processing. This section first introduces the graph data model, and then introduces the storage management of graph and the index structure established for graph data in order to improve the search efficiency. As an important branch of mathematics, graph theory takes graph as the research object. Based on the simple graph, it derives hypergraph theory, polar graph theory, topological graph theory, etc., so that graph can express the real world in many ways. There are many kinds of data models used in current large-scale graph data management. According to the complexity of the nodes in the graph, they are divided into simple node graph model and complex node graph model; According to the number of vertices that an edge can connect, it can be divided into simple graph model and hypergraph model. Whether it is simple graph model, hypergraph model, simple node model or complex node model, their vertices and edges can have attributes. The simple graph model and hypergraph model [5] are introduced below.

4.1   Simple Graph Model

The simple graph mentioned here is not a simple graph in graph theory, but a simple graph relative to a hypergraph. As shown in Fig. 1, an edge can only connect two vertices, and loops are allowed. The storage and processing of simple graphs are relatively easy. For general applications, the expression ability of simple graphs is fully competent, such as pagerant algorithm Pregel HAMA and other systems use simple graph model to organize, store and process large-scale graph data.

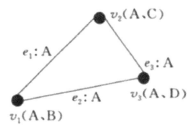

Fig. 1. Simple diagram

4.2   Hypergraph Model

An edge can connect any number of vertices of a graph. In this model, the edges of a graph are called hyperedges. Based on this feature, hypergraphs are more applicable and retain more information than the simple graphs mentioned above, as shown in Fig. 2.

For example, a graph vertex represents an article, and each edge represents two vertices (articles) with the same author.

V 1 (author a, b), V 2 (author a, c), V 3 (author a, d), all the authors of the three articles have a. Figure 1 (a) shows the simple graph storage mode, three independent edges E1, E2, E3 = {V1, V2}, {V1, V3}, {V2, V3} can not directly retain the information that author a is the author of the three articles V1, V2, v3. Figure 1 (b) represents the hypergraph storage mode, super edge E1 = {V1, V2, V3}, V3 keeps the information that a is the author of three articles V1, V2 and v3. For applications with complex relationships, hypergraph model can be used to model, such as social network, bioinformatics network, etc. Trinity and other graph database systems support hypergraph model to manage large-scale graph data.

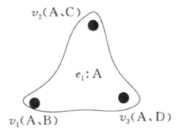

Fig. 2. Storage method of graph data

In the current large-scale graph data management applications, two data models, simple graph and hypergraph, are mainly used. Their organization and storage formats are slightly different. Both models can deal with directed graph and undirected graph. By default, they are directed graph, while the edges in undirected graph can be regarded as two directed edges, that is, a kind of directed graph. In the later discussion, the direction of edges in graph will not be emphasized.

The common storage structures of simple graph model include adjacency matrix, adjacency table, cross linked list and adjacency multiple table. Considering the application requirements of large-scale graph processing and the complexity of maintenance, adjacency matrix and adjacency table are the two most commonly used structures. Adjacency matrix is used to represent the topological structure of graph, which is intuitive and concise, and easy to quickly find the relationship between vertices, But the storage cost of adjacency matrix is high, especially for large graph data.

Large scale graph data storage needs a distributed storage system based on cloud computing environment. There are two kinds of storage systems in cloud computing environment: one is a distributed file system represented by GFS and HDFS, which can store adjacency matrix and adjacency table directly; The other is NoSQL (not only SQL) distributed database represented by BigTable and HbAS.

5 Conclusion

To sum up, based on the cloud computing environment, this paper studies the related technologies of large-scale graph data processing. The conclusion shows that the related technologies of cloud computing can be used in large-scale graph data processing, which fully reflects the wide applicability of cloud computing in data processing.

Acknowledgements. This research was financially supported by Higher Education Reform Research Program in Hainan Province of China (Hnjg2020-161) and Reform Research Program of Sanya Aviation and Tourism College (SATC2020JG-06).

References

1. Du, Q., Yu, C., Ren, F.: Using nested pyramid model to organize tile map data. J. Wuhan Univ. (Inf. Sci. Ed.) (5) (2011)
2. Fang, L.: Research on the theoretical framework and key technologies of cloud computing based land resource service efficient processing. Zhejiang University (2011)
3. He, W., Li, Q., Zheng, Y., Cui, L.: Data distribution and search strategy in community cloud computing environment. Comput. Res. Dev. 47(z1) (2010)
4. Chen, F., Wang, X.: Technology and application of statistical cloud service in Weishu library. J. Univ. Libr. (03) (2010)
5. Li, Z.: Analysis on the potential value of cloud computing in library construction and information service. J. Univ. Libr. (01) (2011)

Design and Implementation of Information Management System Based on Data Mining

Shufeng Li[✉]

Xi'an Siyuan University, Xi'an 710038, Shaanxi, China

Abstract. With the development of science and technology and social progress, computer technology has become a widely used technology. With the rapid development of computer technology, information technology as an important field of computer technology, its development speed is extremely fast, and the role of information technology in many aspects of society is growing. The system is designed based on B/S structure. According to the architecture of BS mode, the system is divided into three layers: presentation layer, data layer and business layer. The presentation layer generally refers to the browser, which is commonly known as the front end. The browser is only responsible for displaying the content of the page, but does not have the authority to modify the content of the page, and cannot participate in the processing of the system; The function of the data layer is to process the data transmission between the database and the server. When the data processing is finished, it will send the results to the server and update the data at the same time; The business layer is mainly responsible for the requests made by users through the browser, and the related operations are performed through the database and server.

Keywords: Mis · B/S structure · Data mining · Design

1 Introduction

Data mining is a process of automatically finding useful information in large data repositories. It is used to explore large databases, discover previously unknown useful patterns, and predict future observations. The purpose of data mining is how to make good use of data, from the operation history of the record reward, mining out the valuable experience. In recent years, with the development of data mining technology, it has played a great role in the field of business intelligence. It becomes a tool to transform the existing data into knowledge and help enterprises make wise business decisions. How to make data mining technology better applied to life production, trade decision-making activities, has high research value.

In order to transform data into knowledge, data warehouse, OLAP tools and data mining are needed. Data mining is a concept refined from KDD. At first, data mining only refers to the process of using algorithms to process data in the process of KDD, which is the stage of learning in KDD and the core technology. The so-called data mining refers to the non trivial process of revealing hidden, previously unknown and potentially valuable information from a large number of data in the database. It is realized by data transformation, core algorithm and result expression. It involves the

© The Author(s), under exclusive license to Springer Nature Singapore Pte Ltd. 2022
J. C. Hung et al. (Eds.): FC 2021, LNEE 827, pp. 1704–1709, 2022.
https://doi.org/10.1007/978-981-16-8052-6_251

extraction, transformation, analysis and modeling of a large number of data in the database, from which the key data for decision-making is extracted. Data mining can help decision-makers to find rules, find neglected elements, predict trends, and make decisions. It is also a high degree of abstraction and generalization of the inherent and essence of data [1].

Data mining is a process of extracting information and knowledge from a large number of, incomplete, noisy, fuzzy and random practical application data that people do not know in advance but are potentially useful. So it requires that the data source must be real, massive and noisy; What we find is that users are only interested in; And the results are acceptable, understandable and applicable. In technology, the discovery of knowledge is not required to be a universal model, but only a result that can be applied to a specific data source.

2 B/S Architecture

This system is designed based on B/S structure. B/S mode refers to browser/server structure, which is a structure used on veb. In BS structure, the browser (client) is consistent for users, and the process of system processing requests is reflected in the server, which is invisible to users. Because this mode is adopted, the development process of the system is simplified. Not only that, for the maintenance and upgrade of the system, it only needs to maintain and upgrade the server, which is an important feature of BS structure superior to s structure.

The architecture of B/S mode is divided into three layers: presentation layer, data layer and business layer. The presentation layer generally refers to the browser, which is commonly known as the front end. The browser is only responsible for displaying the content of the page, but does not have the authority to modify the content of the page, and cannot participate in the processing of the system; The function of the data layer is to process the data transmission between the database and the server. When the data processing is finished, it will send the results to the server and update the data at the same time; The business layer is mainly responsible for the requests made by users through the browser, and the related operations are performed through the database and server.

The system designed with B/S structure doesn't need to install specific software on the computer, just open the browser and input the web address; For the B/S structure system, the browser side is also controlled by the background server side, so the upgrade and maintenance only need to operate on the background server, reducing the cost of maintenance and upgrade; in addition, the B/S structure can match almost any operating system, and there is no need to adopt different development modes for different systems, For programmers, it reduces the cost, and for users, users can use any system and browser, thus increasing the user experience.

Of course, the B/S structure also has its disadvantages. For example, when the business volume of the system is relatively large and the logical relationship between businesses is very complex [2], the pressure of the server will be very large and the load will be very large, so the efficiency of the system will be reduced; In addition, the reliability of HTTP protocol is not very high. For programmers, it is risky to use HTTP protocol to update, maintain and upgrade the system. In addition, the general system now uses embedded modules such as javascri PT to enrich the system configuration, but these modules are distributed in many different pages, This will also cause more workload for maintenance and update. The general picture of BS structure is as follows As shown in Fig. 1.

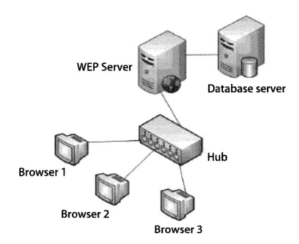

Fig. 1. Overview of B/S structure

3 Design and Implementation of Information Management System

3.1 System Design

The overall structure of the system consists of five parts: Web server, reverse proxy server, database server, front end of CRM system and client. The overall framework of the system is based on the MVC strus + Spring + hibern ATC (SSH) integration framework. Therefore, the system architecture is mainly divided into five layers from the technical level [3], including front-end display layer, request processing layer, service layer spring, Dao layer and entity layer. The system architecture is shown in Fig. 2.

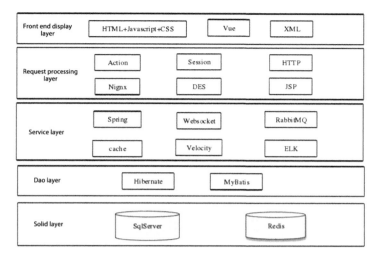

Fig. 2. System architecture diagram

The front-end display layer is responsible for user interaction and web data display, mainly including HTML, JavaScript and Vue technology. The request processing layer is responsible for responding to the requests from the front end, mainly including action object, session management, nignx reverse proxy, DES encryption algorithm and other technologies. The service layer mainly provides the business logic implementation function of the system, including spring, rabbitmq message queue, cache and other technologies. Dao layer is responsible for reading and writing data between logical layer and entity layer, mainly including mybatis, hibernate and other technologies. Entity layer is responsible for data persistence operation, mainly using sqlserver and redis technology.

3.2   Information Management Module

Aiming at the universal function of information management system, this paper develops an information management system based on data mining. The main functions of information management include basic information management, information verification, duplicate checking and data backup. Data is generated by the system. Basic information management includes the functions of data viewing, adding, modifying and deleting. The viewing function includes all data viewing and keyword retrieval data viewing. The verification of information mainly aims at the rule verification of the corresponding field content when creating or modifying data. Duplicate checking is also used to check the name, telephone number and email address of a new building to ensure the consistency of the name and telephone number or the name and email address. Data import and export function: due to business needs, some offline sales need to be manually imported into the system, and the export function needs to export the data to the local in Excel file format.

The class diagram of the management module is shown in Fig. 3.

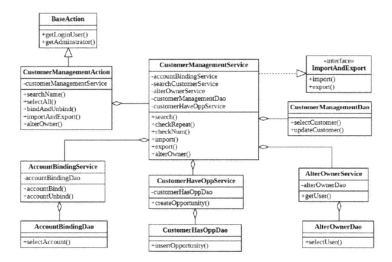

Fig. 3. Management module class diagram

4   Establishment and Training of Data Mining Model

In this design, we need to use the data mining extension language, the significance of which is to define a unified concept and query expression for data mining, just like SQL language in database. The language is composed of data definition language (DDL), data manipulation language (DML), functions and operators [4].

There are many data mining products in the data mining market. Each product describes and builds data mining applications in its own way. Most data mining toolkits have their own algorithms, their own model browsing and storage formats, their own data cleaning tools and even their own report generation tools. In this way, the data mining system is separated from the enterprise operating system, which increases the difficulty and cost of data mining solutions.

In Microsoft SQL server, DMX language is an expression language used to create the structure of data mining models, train these models, and browse, manage and predict models. Using expression language is easy for database developers to understand, so it has better readability and maintainability. Different from the advanced development language, the language does not involve complex algorithms, it only provides a detailed language for data mining problems to be discussed. Only through the mining model trained by the language, can the development of high-level and customized data mining stage be carried out.

Based on the mining stage, it can be divided into three parts. They are model building, model training and model using. We generally believe that the decision tree is to predict the most likely things, but in reality, the operation of accurate prediction is difficult to achieve. Therefore, in a typical application, the proportion of cases that are misclassified occurs. Therefore, the smaller the cost is, the less cases are classified by mixed classification, and the higher the accuracy of prediction is. In the process of building decision tree, the greedy algorithm of cost minimization is followed.

In fact, linear regression model can be considered as a variant of decision tree algorithm. If the linear regression algorithm is selected, the special case of decision tree algorithm with parameters will be called. These parameters will not only limit the algorithm behavior, but also require the type of input data. It has a wide range of uses, especially for the research and analysis of social phenomena that can not be obtained by experimental methods.

For the linear regression algorithm, it is very important to understand the direction and strength of the relationship between the independent variables and the dependent variables, as well as the prediction of the dependent variables by the model established by the independent variables.

5   Conclusion

This paper expounds the technology and method of data mining, studies the development of data warehouse and data mining tools, and finally chooses the development method of combining Microsoft SQL Server 2008R2 and Microsoft Visual Studio 2008 to develop the data mining system. It lays a foundation for the successful establishment of data mining information management system. This paper describes some commonly used data mining algorithms, and focuses on the decision tree and linear regression algorithm. The information model of the system is established by using the method of dimension modeling, and a data mining information management system of B/S architecture is implemented by programming.

References

1. Xu, Y.: Customer relationship management system based on data mining technology, vol. 4, pp. 9–11. Jilin University, Changchun (2001)
2. Chen, Z.: Research and development status of data mining technology at home and abroad. Young Writ. (16), 2 (2009)
3. Dong, N.: Application of data mining technology in CRM. Comput. Eng. Design (6), 1430–1431 (2007)
4. Yan, W.: Application research of digital publishing CRM based on data mining. Xi'an University of Technology, Xi'an (2018)

Research on Multi AGV Control System and Scheduling Algorithm of Warehouse Automation

Liquan Ou[✉], Jinyun Peng, Jia Chen, Xiaoyue Zou, Fengyang Sun, and Bin Yang

Kunming Logistics Service Center of Power Supply Bureau of Yunnan Power Supply Co., Ltd., Kunming 650000, Yunnan, China
{yunlv_yang,Jin_liu,menglin_tian,xinggui_song, xiong_wang}@ddtech.com.cn

Abstract. The domestic logistics industry presents an unprecedented development trend. The traditional logistics industry is undergoing great changes. The position of logistics warehouse in the logistics supply chain is becoming more and more important. It is urgent to upgrade from labor-intensive to intelligent and automatic. In most applications, two or more ACV vehicles are needed to complete the task of transportation. At this time, the control mode based on AGV is difficult to meet the requirements of multi ACV collaborative application scenarios. With the characteristics of saving labor, reducing labor intensity, high efficiency and strong environmental adaptability, it plays an increasingly important role in the field of factory automation and logistics. This paper studies the cooperative scheduling problem of multi AGV system, and analyzes the AGV conflict resolution method in the dynamic scheduling process.

Keywords: AGV · Automation · Scheduling algorithm · Storage

1 Introduction

Since the development of human beings, we have experienced three industrial revolutions, which have had a great impact on the advancement of human civilization. The first industrial revolution (also known as "industry 1.0") began in 1766 when British James Watt improved the steam engine. The appearance of steam engine makes the manufacturing mode of goods changed dramatically, and the industrial power system and freight transportation efficiency have also been improved unprecedentedly "Industry 2.0", that is, the second industrial revolution was born in the late 1860s [1]. The remarkable feature of this industrial revolution is that electricity gradually replaced steam engine and became a new energy system. And on the basis of division of labor, the assembly of products and the production of parts are separated successfully. At the same time, the production line is driven by the power energy system to make the mass production of products. From then on, the mass production mode of products is born. Therefore, the second industrial revolution era is also known as the electrification era.

© The Author(s), under exclusive license to Springer Nature Singapore Pte Ltd. 2022
J. C. Hung et al. (Eds.): FC 2021, LNEE 827, pp. 1710–1716, 2022.
https://doi.org/10.1007/978-981-16-8052-6_252

After "industry 1.0" and "industry 2.0", the development of computer technology, aerospace technology and atomic energy technology led to the birth of the third industrial revolution (also known as "industry 3.0") in the 1940s–1950s. The core and typical products of this industrial revolution were computers and aerospace tools, and some manual work began to be replaced by machines, So that the speed of product manufacturing continues to accelerate, improve the level of automation of industrial production. After the third industrial revolution, mankind has entered the information age. After three industrial revolutions in the world, social productivity has been greatly developed, leading to a new change in the global industrial layout.

In recent years, many scholars have proposed new methods, such as banker based scheduling algorithm and time window based scheduling algorithm. In recent years, some scholars have proposed path planning and scheduling algorithm based on decentralized motion, scheduling algorithm based on harmony search, scheduling algorithm based on mixed regional control mode, hybrid genetic algorithm and particle swarm optimization multi AGV scheduling algorithm, But these methods are still relatively high in time complexity and space complexity. Due to the high efficiency requirement of AGV system in warehouse automation system, it is very important and inevitable to develop a scheduling method with low time complexity and low space complexity. On the basis of previous researchers, especially inspired by the research of banker scheduling algorithm, this paper proposes a dynamic unlock real-time scheduling algorithm based on lock mechanism, which is defined as dynamic unlock algorithm (Dua) in this paper. It effectively solves the path conflict and deadlock problems of multi AGV, and applies the scheduling algorithm to multi AGV system.

2 Multi AGV System of Warehouse Automation

A well-designed modern warehouse automation system can not only save a lot of human and material costs, but also greatly improve the efficiency and throughput of the system. But to establish a relatively complete warehouse automation system, in addition to the need for a lot of manpower, but also need to invest a lot of money. Besides AGV software control system, AGV intelligent hardware, AGV path planning algorithm, task allocation algorithm, scheduling algorithm and path guidance system, some mechanical auxiliary systems are also included, such as loading/picking mechanical equipment (manipulator), conveying equipment (conveyor belt, rolling steel bar, etc.), storage shelves, etc. The realization of warehouse automation system is a huge project, in addition to software development, hardware cooperation is very important and also very strict. Without the cooperation of related peripheral equipment, a complete warehouse automation system cannot be realized, let alone a complete solution.

2.1 AGV Software Control System

The architecture of software control system is client/server architecture. The server is the background, scheduling system and related system service thread, and the client is the AGV, mechanical equipment and other hardware equipment that can receive the server scheduling command. In addition, the system includes basic system functions, such as account verification system, database storage system, AGV status monitoring system, log management system, task plan management system, etc., as shown in Fig. 1.

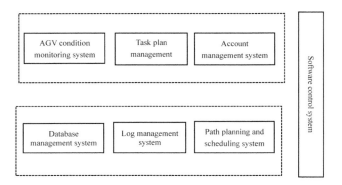

Fig. 1. Architecture of AGV software control system

Account management system is used to manage system users, such as login, logout, registration, modify permissions, etc. Database management system is used to store account data, task order data, etc. AGV status monitoring system is used to monitor the running status of all AGVs in the current environment system, as well as related system variables. Log management system is used to record system errors in the process of daily use, which is convenient for later improvement. The task plan management system is used to manage the order tasks of the warehouse system, and to add, delete, modify and query the task orders.

2.2   Communication Between AGV Software Control System and AGV

For a modern large warehouse, the upper AGV software control system must communicate with the AGV in the actual environment by means of network. If the wired mode is adopted, the AGV must communicate with the upper software control system through the transmission line in the scheduling process, but this will introduce a large number of limited cables to the AGV operation environment, thus hindering the normal AGV scheduling, so the communication mode of this system is wireless communication. ZigBee or wify is the most popular wireless communication mode. However, there are still some problems in wireless communication. In the process of the system implementation, there is a problem that the radiation area of a wireless network device is less than that of a large warehouse due to its limited signal radiation range, as shown in Fig. 2.

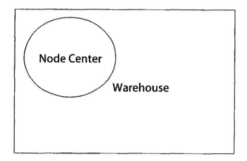

Fig. 2. Wireless signal range

Suppose the length of the warehouse is l, the width is w, and the signal radiation radius of a single network device is r [2],

$$L * W > \pi * R * R$$

Obviously, a single wireless network device cannot cover the entire warehouse area. At this time, if the AGV location is not within the range of wireless network device signal radiation, then the AGV loss will inevitably occur, which means that the software control system will not be able to control the AGV. Therefore, in order to make the wireless signal cover most of the warehouse area, multiple wireless network devices must be added, as shown in Fig. 3.

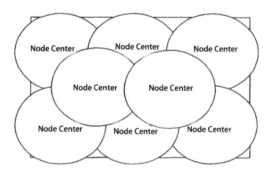

Fig. 3. Full coverage of wireless signal

After adding eight network devices to the warehouse shown in Fig. 3, the signal radiation range of the wireless network is basically close to the area of the warehouse. Of course, if comprehensive signal coverage is needed, the wireless network devices can be added to ensure the full coverage of the signal, so as to prevent the loss of all AGVs and ensure the controllability of all AGVs.

3 Path Planning for Multiple AGV Systems

The modeling of multi AGV system is essentially an abstract representation of path guidance system and AGV. The design and implementation of path planning algorithm, task allocation algorithm and scheduling algorithm are inseparable from the running environment of AGV, namely path guidance system. The path guidance system is composed of running track and station. Through the arc edge and vertex in the analogy graph theory, the path guidance system can be abstracted into graph data structure, that is, the path planning module and scheduling algorithm module are analyzed with the help of graph theory, and then the relevant definitions are given.

For any graph data structure g, it is composed of vertex set node and edge set arc

$$G = (NODE, ARC) \tag{1}$$

$$NODE = \{n_1, n_2, \ldots, n_k, k \in N\} \tag{2}$$

$$ARC = \{a_i, a_2, \ldots, a_k, i \in N\} \tag{3}$$

Where node is the set of all vertices in the graph, n_1 represents vertex i, arc is the set of all arc edges in the graph, a_i is the arc edge numbered I, $|a_i|$ is the weight of the edge, that is, the path length of the edge [3].

Path planning algorithm is used to solve the shortest path search and path sequence generation problem based on graph. The path search algorithms used in graph theory include breadth first search (BFS), depth first search (DFS), Dijkstra search algorithm and a * search algorithm. In this paper, BFS is mainly used to solve the shortest path. BFS algorithm was invented by Dr. knorad Zuse in 1945. The key idea of BFS algorithm is to track an extended ring called boundary and continuously explore the non traversed vertices around the extended ring until all the vertices are traversed. The process of breadth first search can be regarded as "flood filling", as shown in Fig. 4. Take the center as the basic point, and continue to expand outward until all the vertices are traversed.

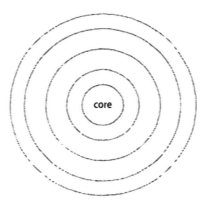

Fig. 4. Breadth first search process

Breadth first search is mainly used for path search operation in graph or tree data structure (constructed by Formula 1, formula 2 and formula 3). It can search the path between any two points (in the case of connectivity). In practical application, it is not enough to only search the target point, but also need to save the search path in order to generate the shortest path sequence.

4 AGV Dispatching

After the task is assigned to AGV by task allocation algorithm, AGV needs to be controlled by scheduling algorithm to execute and complete the task. The essence of scheduling algorithm is to solve the problem of path conflict and deadlock in the execution scheduling process of multiple AGVs. It is assumed that the number of AGVs in a multi AGV system is m. In the case of M = 1, when the AGV path is planned, the path deadlock problem will not occur, and the task can be completed directly without waiting. However, when the warehouse automation system contains multiple (M > 1) AGVs, due to the limited number of track paths in the map, there will be path duplication in AGV path planning, which leads to the problem of path conflict and deadlock collision in the scheduling process of multiple AGVs. Therefore, the research of scheduling algorithm is an unavoidable and even urgent problem [4].

For any storage automation system with AGV, AGV scheduling system determines when, where and how an AGV performs a system task. It also includes the planning path for the AGV to perform the current task, that is, AGV scheduling system has absolute control over all AGVs. The robustness of scheduling system design will determine whether the system can correctly control all AGVs to perform tasks according to the given scheduling trajectory. Scheduling system can be divided into offline planning scheduling system and online real-time scheduling system. The off-line planning and scheduling system plans and schedules all tasks uniformly when all tasks have been distributed, and maximizes the scheduling efficiency of the system by introducing a performance evaluation model. However, the discrete planning and scheduling system has strict system restrictions, such as ensuring that all AGVs will not fail in the process of executing tasks. Therefore, in practical application scenarios, the off-line planning and scheduling system is often difficult to meet the requirements, because in a continuous real-time multi AGV system, the arrival of tasks is discontinuous, which can be called task discrete multi AGV system. Due to the discrete nature of tasks, it is impossible to plan all tasks off-line at one time. In this case, real-time scheduling method arises at the historic moment, real-time scheduling system can overcome the shortcomings of offline scheduling system. When the discrete tasks arrive, the real-time scheduling system can make reasonable planning and arrange the current tasks to the more appropriate AGVs according to the working state of all AGVs in the current multi AGV system. At the same time, the real-time scheduling system can also use the scheduling algorithm to solve the path conflict and deadlock problems in the multi AGV dynamic scheduling process.

5   Conclusion

Firstly, the configuration of AGV and station in the project is briefly introduced. According to this situation, the production line map is established by using the method of topology map, and the AGV scheduling algorithm is selected to realize the AGV path planning, which improves the efficiency of path planning. Through the priority based scheduling method, the dynamic scheduling of multiple AGVs is realized, and the AGV conflict resolution method in the dynamic scheduling process is analyzed.

References

1. Ma, B.: Design of multi AGV scheduling system based on RFID technology. Shandong University (2015)
2. Li, G.: Research on AGV visual guidance system based on RFID. Zhejiang University (2008)
3. Ling, Z.: Research on AGV path planning algorithm. General Research Institute of Mechanical Science (2013)
4. Wang, N.: Research on multi AGV job scheduling based on improved ant colony algorithm. Shaanxi University of Science and Technology, Xi'an (2017)

Preliminary Planning and Design of Power Transmission and Transformation Project Based on Tilt Photography GIS + BIM Technology

Haiyi Pan[✉] and Kuo Li

Guangzhou Power Supply Bureau of Guangdong Power Grid Co., Ltd., Guangzhou 510000, Guangdong, China

Abstract. With the rapid development of national economy, the importance of electric power in social and economic life is becoming increasingly significant. It can not be ignored to realize the safe and economic operation of power system and ensure the reliability and quality of power supply. This makes the city's distribution system is facing enormous pressure. Therefore, it is urgent to change the current situation of weak urban power grid, poor reliability, high loss, unstable power quality and low degree of automation. Therefore, the planning and design in the early stage of the implementation of power grid project has become the top priority. Based on oblique photography GIS + BIM Technology, this paper carries out the planning and design for the early stage of power transmission and transformation project.

Keywords: Power transmission and transformation project · Oblique photography · GIS · BIM

1 Introduction

Due to the rapid development of urban and rural economy, a large number of planned power transmission and transformation projects conflict with local government planning projects, which is mainly reflected in: when government departments at all levels make planning, they often omit the planning projects of power sector, which makes it difficult for power supply departments to select substation site and high-voltage line channel when carrying out the preliminary work of power transmission and transformation projects. Although many substation sites are recognized by the relevant government departments at the initial stage of site selection, due to the involvement of agricultural land or cultivated land, or the acceleration of urban and rural construction in recent years, the government planning has changed, which makes it difficult to determine the substation site and line path, or changes after selection, which leads to the relocation and work, greatly affecting the progress of the preliminary work of the project. Because some substations are very close to industrial parks or expressways, the line corridor is often unable to avoid, and the cross domain industrial parks or

J. C. Hung et al. (Eds.): FC 2021, LNEE 827, pp. 1717–1723, 2022.
https://doi.org/10.1007/978-981-16-8052-6_253

expressways and other issues are complicated, so it is not possible to reach a consensus with the relevant units in a short time, and it also needs repeated communication and modification, which leads to the relatively difficult completion of the preliminary work of the substations near the urban area [1]. Based on the oblique photography GIS + BIM Technology, this paper carries out the planning and design of the early stage of the power transmission and transformation project, aiming at the standardization of the construction management of the power transmission and transformation project, within the reasonable scope and the approved cost limit, in order to maximize the use of human, material and financial resources, and obtain a good return on investment.

2   Oblique Photogrammetry Technology

2.1   Overview of Oblique Photogrammetry Technology

Tilt photography is a high-tech developed in the field of international surveying and mapping in recent years. It overturns the limitation that orthophoto images can only be taken from the vertical angle in the past. By carrying multiple sensors on the same flight platform and collecting images from five different angles, such as one vertical angle and four tilt angles, users are introduced into the real intuitive world in line with human vision. In this paper, the application, technical characteristics, core technology and outstanding advantages of the current tilt photogrammetry technology are described in detail, and the data processing process and existing problems are summarized.

Tilt photogrammetry technology is mainly through carrying multiple sensors on the flight platform, simultaneously collecting image data from different angles such as vertical and tilt. When taking photos, it records some other attitude data, such as speed, heading, altitude, longitudinal overlap and lateral overlap, and then carries out post-processing on the tilt image.

2.2   Characteristics of Oblique Photogrammetry

(1) Reflect the real situation around the features
    Compared with orthophoto, oblique image allows users to observe the ground objects from multiple angles, more truly reflects the actual situation of the ground objects, and greatly makes up for the shortcomings of Orthophoto based applications.
(2) Single image measurement can be realized by tilting image
    Through the application of supporting software, we can directly measure the height, length, area, angle, slope and so on based on the result image, which expands the application of tilt photography technology in the industry.
(3) The texture of building side can be collected
    For various 3D digital city applications, the use of large-scale aerial photography mapping features, coupled with batch extraction from oblique images and paste texture method, can effectively reduce the cost of urban 3D modeling.

(4) Small amount of data, easy to publish on the Internet
Compared with the huge 3D data of 3D GIS technology, the amount of image data obtained by tilt photography technology is much smaller. The data format of the image can be quickly released on the Internet by using mature technology to realize shared application.

3 GIS + BIM Technology

3.1 Basic Concepts of GIS

GIS is the abbreviation of geographic information system or geo information system, translated as "geographic information system", sometimes also known as "geoscience information system". Based on different fields and different application purposes, the definition of GIS is also different. According to Baidu Encyclopedia, GIS is a technical system that collects, stores, manages, calculates, analyzes, displays and describes the geographic distribution data of the whole or part of the earth's surface (including the atmosphere) space with the support of computer hardware and software systems. Like other computing systems, GIS includes four elements: computer hardware, software, data and user, but all data in GIS have geographic reference, that is to say, the data is connected with a specific position on the earth surface through a certain coordinate system [2].

The functions of GIS are as follows:

(1) Data editing and transformation. The existing graphic data, text data and related attribute data are converted into the data form that can be read by GIS system. Data conversion includes data format conversion and data scale conversion.
(2) Data storage and management organization. GIS includes database system, which can update, manage and store geographic data. Whether the database management is proper or not can directly affect the efficiency of data management and use. The function of analysis and update. Proper management and storage of GIS database can provide convenience for users.

3.2 Definition of BIM

As for the concept of BIM, we agree that the building information modeling proposed by McGraw Hill is the whole life process of project design, construction and operation management by creating and using digital model. BIM realizes the transformation from traditional two-dimensional drawing to three-dimensional drawing, which can show the building information more comprehensively and intuitively.

BIM is the digital expression of the functional characteristics and physics of construction projects or facilities, and is the knowledge resource shared by the project participants. Through extracting, inserting, perfecting and updating information parameters in BIM, the participants in different stages of construction projects can achieve joint assistance in common BIM model. In the whole life cycle of construction projects, BIM model provides direct and reliable basis for all decisions in the construction process, so as to achieve the purpose of improving the quality of construction projects, increasing efficiency and reducing risks.

The characteristics of BIM are as follows:

(1) Three dimensional graphics. BIM model is intuitive and easy to accept and understand. Through BIM three-dimensional graphics, we can fully express the spatial position of the building and the geometry of the components. At the same time, the whole life cycle process of BIM building information model is visual, not only the generation of all parameter reports and the display of construction renderings are visual, but also the discussion in the process of project design, construction and operation. Communication and decision-making are also carried out in a visual state.

(2) In BIM model, the information of beam, slab and column can be expressed completely by itself. So as to accurately and perfectly express the three-dimensional geometry of all kinds of building components.

(3) Information parameterization. BIM model building information is expressed and stored in the form of parameterization, so BIM model can be expressed by the following equation: BIM = 3D graphics + n-dimensional parameters + application. The professional BIM application software can use all kinds of parameter information to achieve different functions and uses. Participants do not need to re-enter the same information, and share information parameters at all stages of the building life cycle in a common model [3].

3.3   Examples of Integrated Application of GIS and BIM Technology

The starting and ending mileage of Lixiang zongsi tunnel is dk124 + 770–dk126 + 975, the central mileage is dk125 + 872.5, and the total length of the tunnel is 2205 m. The maximum buried depth of the tunnel is 190 m, and the whole tunnel is single-sided uphill. The 1830 m section of dk124 + 770–dk126 + 600 at the entrance is 24%. The other 375 m section is 10%. The uphill is straight except that the 414.11 m section of the entrance is located on the left curve of r-1600 m, the 893.22 m section of the exit is located on the right curve of r-10000 m, and the other 897.67 m sections are straight lines, as shown in Fig. 1.

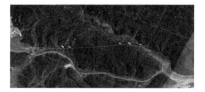

Fig. 1. The location, topography and geomorphology of zongsi tunnel

GIS + BIM Technology is comprehensively applied to carry out special design and Research on the setting scheme of auxiliary tunnel. The special research report on the setting scheme of auxiliary tunnel of zongsi tunnel in Lixiang 6 bid is formulated to comprehensively analyze the setting position, design parameters and construction access of auxiliary tunnel. The comparison and design of construction access scheme using GIS + BIM Technology are shown in Fig. 2 and Fig. 3

Fig. 2. Three dimensional design of construction road scheme for auxiliary tunnel of zongsi tunnel (1)

Fig. 3. Three dimensional design of construction road scheme for auxiliary tunnel of zongsi tunnel (2)

The comprehensive utilization of GIS and BIM Technology can realize the research, comparison and selection of construction technology scheme design under the real terrain environment, change the traditional technology scheme research and design method, and greatly improve the work efficiency and design quality. The real terrain rapid modeling technology (modeling for about 5–10 min at a time) communicated here is based on free DEM data and satellite images obtained from the network. The maximum real terrain model within 200 km^2 can be built in a single time.

4 Preliminary Planning and Design of Power Transmission and Transformation Project Based on Tilt Photography GIS + BIM Technology

4.1 Preliminary Design Stage

The preliminary design stage is the core link of design innovation, which is related to the implementation of innovation scheme. At this stage, the design unit organizes the corresponding designers to carry out the preliminary design of the power transmission and transformation project according to the innovative scheme and standardized design results determined by the design planning, combined with the inclined photography GIS + BIM Technology. After the completion of the preliminary design, the preliminary design review is carried out. As a key link in the power transmission and transformation project, the review plays a key role in the design quality. Focus on whether the design planning documents are implemented in the preliminary design documents, and the design unit modifies and improves the preliminary design documents according to the review opinions [4].

4.2 The Key to the Early Stage Planning and Design of the Project

(1) The design unit needs to cooperate closely with the power supply unit, be familiar with the relevant national laws and regulations, and avoid major design changes.

(2) To strengthen the preliminary work management of the design unit, it is required that the design unit must carry out the preparation of the feasibility study report in strict accordance with the requirements of the depth of the feasibility study design of the power transmission and transformation project of China Southern Power Grid Corporation. If the depth of the feasibility study design is not enough, there will be major changes in the later design technical scheme and large deviations between the project final accounts and the approved budget estimate, and between the budget estimate and the feasibility study estimate, This is the main assessment content of infrastructure projects, which should be avoided by design units.

(3) Strengthen the coordination and communication with government departments, actively track the construction and commissioning progress of the city's industrial parks, timely and reasonably arrange the supporting power supply scheme for the city's industrial parks, and provide strong power support for the city's industrial development.

(4) Strengthen the project feasibility study management, and strive to further improve the management level of the Bureau headquarters and county-level power supply enterprises.

5 Conclusion

Substation construction is in a very important position in the construction of power grid. It has become an important part of modern distribution network integrated system. It combines with power transmission project to form an efficient and reliable power supply system of power grid. The preliminary planning and design of power transmission and transformation project based on oblique photography GIS + BIM Technology, It plays a vital role in ensuring the power supply of power grid. At the same time, we continue to use tilt photography GIS + BIM Technology as the basis of the design scheme, and constantly optimize the typical design concept of the substation, so as to form a series of perfect customized scheme design to meet the needs of urban and rural power grid construction.

Acknowledgements. Guangdong Power Supply Bureau Technology Project of Guangzhou Power Supply Bureau (GZHKJXM20190018).

References

1. State Grid Corporation of China. General design of power transmission and transformation engineering. Modular construction drawing design of 35–110 kv Smart Substation. China Electric Power Press, Beijing (2017)
2. Cheng, J.: Research on BIM of architecture. Anhui Archit. **6**, 19–20 (2013)
3. Chen, S., Lu, X.: Introduction to Geographic Information System. Science Press, Beijing (2000)
4. Qi, D., Song, X.: Project Management. Dalian University of Technology Press, Dalian (2001)

Intelligent Building Management and Control System Based on Intelligent Video Analysis Algorithm for Behavior Recognition

Lu Yuan[(✉)]

Nanchang Institute of Technology, Nanchang 330511, Jiangxi, China

Abstract. Today's world is an era of big data. Intelligent video analysis plays an important role. As one of the key technologies in intelligent video analysis, target tracking has great academic research value and broad application prospects. This paper provides a smart building management and control system based on intelligent video analysis algorithm to realize behavior recognition. Through the video analysis component to identify the emergencies in the monitoring video and generate early warning information, the corresponding emergency plan is called through the emergency service component, and the external access equipment is handled according to the emergency plan through the data acquisition monitoring component, To a certain extent, it can solve the problems that the monitoring and management system can not achieve unattended, missed and misjudged emergencies, and can not quickly and automatically implement response measures after emergencies.

Keywords: Identification and monitoring · Intelligent video analysis · Management and control system · Target tracking

1 Introduction

In recent years, with the development of network information technology and multimedia technology, the world has entered the era of big data. In this era of explosive growth of data, there are text, pictures and videos to present the data of surrounding things. Compared with text and pictures, massive video data is facing new challenges and technologies. Under the condition of continuous growth of video data and limited human and computing power, how to process a large number of video data quickly, accurately and at low cost, and find the characteristics and activity trajectory of related targets, the newborn of intelligent video analysis came into being [1].

Intelligent video analysis (IVA) is a branch of computer vision and artificial intelligence. It uses computer vision technology and digital image processing technology to analyze the video. Firstly, the background and target in the scene are separated, the background interference (such as illumination change and branch jitter) is removed, and the moving target information is extracted, Through feature extraction, we can find the interested targets and events, and then track the targets further, so as to analyze the behavior characteristics. According to the preset template or user set rules, we can analyze and judge the alarm events, so as to further deal with them by light,

J. C. Hung et al. (Eds.): FC 2021, LNEE 827, pp. 1724–1729, 2022.
https://doi.org/10.1007/978-981-16-8052-6_254

sound and other means. The essence of IVA is algorithm, which has nothing to do with specific hardware circuit and system architecture. It can be transplanted to PC platform or embedded devices.

Facing the broad market prospect, there is a great demand for products based on intelligent video analysis technology, which has a strong practical significance for the research of intelligent video analysis; At the same time, the current intelligent video analysis is facing strong challenges, attracting a number of scholars to study it, which has great academic research value and potential.

2 Intelligent Video Analysis Technology

Figure 1 briefly introduces the general process of general intelligent video analysis, from which we can see that intelligent video analysis mainly involves the following key technologies:

Fig. 1. The process of intelligent video analysis

(1) Foreground motion detection technology: it separates the moving area in the video image from the scene background. At present, the main methods are background subtraction, such as frame subtraction, Gaussian mixture background modeling and vibe. Each method has its own advantages and can be applied to different scenes.

(2) Target detection technology: analyze the state or motion characteristics of the violent change area in the video sequence, and then extract the moving target. The implementation methods are: foreground motion detection, feature matching technology, which are used in different occasions.

(3) Target tracking technology: according to the historical motion information of the target, the possible position of the moving target is predicted in the current frame, and the real position of the target is searched near the position. The implementation methods include mean shift, particle filter, template matching, compression tracking and so on.

(4) Target classification technology: for the successfully tracked targets, the target image features (such as contour, size, texture, etc.) are used to realize the

discrimination and classification of target types, such as dividing the tracked targets into people and vehicles [2].

(5) Trajectory analysis technology: data analysis and processing of the target trajectory generated by the tracking module, so that the motion trajectory can more accurately reflect the state of the target. The common processing methods are filtering, smoothing and error correction.

(6) Behavior analysis technology: normal/abnormal behavior analysis and understanding of crowd, traffic flow and other targets. Normal behavior analysis, such as identifying the speed and direction of the target, and counting the number of targets; Abnormal behavior analysis includes illegal parking and behavior of vehicles, fighting in public places, etc.

(7) Event detection technology: the target information and preset alarm rules to determine whether there is a violation of the rules, and then respond: alarm or release processing.

In some embodiments, the camera, i.e. the image acquisition device, is set in the monitoring area of the property, and the monitoring video collected by the video analysis component camera is analyzed in real time. When a preset type of emergency occurs in the monitoring area, the video analysis component can identify it in time, including its specific event type, and then generate the corresponding warning information. The warning information includes but is not limited to: trip line warning information for vehicles or pedestrians, intrusion warning information for vehicles or pedestrians, vehicle retrograde warning information, crowd aggregation warning information for pedestrians.

3 The Application of Intelligent Video Analysis in Intelligent Building Management and Control System

In the construction industry, the development of monitoring technology has gone through three stages and is evolving towards intelligent video monitoring, as shown in Fig. 2. Intelligent video surveillance (IVS) refers to the application of IVA technology in traditional video monitoring. With the help of powerful data processing capacity of the processor carrying IVA algorithm, the useless interference information in video images is filtered, and the key and useful information in video source is automatically extracted and analyzed, The monitoring system not only has "machine eye" (camera), but also "machine brain" which carries human intelligence (computer or special DSP carrying IVA algorithm), which realizes intelligent monitoring, breaks through the original video display or video function, and can automatically think and locate and track the interested targets. In the era of big data, the application of intelligent video analysis technology is making a new round of change and development for video monitoring.

The traditional video monitoring is mainly to watch the real-time video information manually, and retrieve the video records stored in the system one by one when the key information needs to be found. Obviously, human participation can not extract the effective information in video in time, and massive useless video data need to be

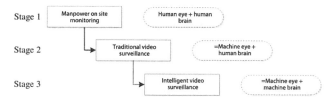

Fig. 2. Development of monitoring technology

transmitted and stored. Intelligent video surveillance has the following advantages in these aspects [3].

(1) Automatically "watch" the video and conduct pixel level analysis to make a quick warning of possible threats (2) Security personnel only need to pay attention to the relevant information to make a rapid response and improve the effectiveness of monitoring (3) Powerful data search and analysis functions can provide fast retrieval and investigation and evidence collection time, and expand the application of video surveillance to non security field.

After years of development, intelligent video surveillance has ushered in the "spring" of application. In foreign countries, the scenes of intelligent video surveillance applications are mainly concentrated in ports, airports, railways, national security and other units and major venues. Domestic demand for intelligent video surveillance is mainly to strengthen the security of building system. In addition, in recent years, there are also demands in production control and business management.

4 Target Tracking Analysis in Intelligent Building Management and Control System

Object tracking is a key technology in intelligent video analysis. The so-called object tracking (OT) refers to the detection, extraction, recognition and tracking of interested people or objects in the video sequence, so as to obtain the position, speed and trajectory of the target. In other words, the task of target tracking is to assign the same ID number to the same target in different video frames. Robust, accurate and fast target tracking will lay a solid foundation for subsequent behavior analysis and other higher visual tasks.

After years of development of intelligent video analysis, there are two kinds of mainstream system connection: PC based video system and embedded video system. The architecture is shown in Fig. 3. PC based video system is the traditional design of intelligent video analysis system. In terms of hardware, only the front-end camera and the PC in the dispatching center are needed. The camera is responsible for image acquisition and local data storage. The PC in the dispatching center reads the video data of each terminal through the network, and then carries out intelligent video analysis on the PC; In terms of software, intelligent video analysis is mainly implemented on PC, and opencv library is generally used to shorten the period of video development. PC based method has the advantages of simple and convenient implementation, fast development speed, low price and cost, so it is generally regarded as the initial research of development; However, due to its own limitations, low flexibility and slow processing speed of the system, it is not suitable for large amount of data and complex on-site monitoring.

The embedded scheme of video monitoring, generally for each monitoring terminal is equipped with an embedded intelligent device lower machine, can carry out video encoding and decoding, video data storage and image processing, after a certain processing, through the network back to the upper computer of the dispatching center, the upper computer generally provides a visual interface, interactive processing. Compared with PC scheme, this design scheme has the advantages of small volume, flexible installation, saving network bandwidth, quick response to on-site crisis, and good real-time performance. It has become the mainstream of current research and actual project development.

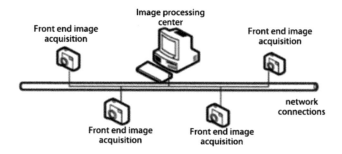

Fig. 3. PC based video system connection

The monitoring and management system can monitor the real-time images of buildings or construction sites through image acquisition equipment. When there are emergencies such as intrusion warning area, vehicle trip line, vehicle retrograde, crowd gathering and so on [4], it can carry out real-time warning and implement counter-measures. In the implementation of some monitoring and management systems, including power supply system, wireless transmission system, front-end detection system and background monitoring system. Firstly, when the front-end detection system detects the abnormal situation on the scene, it starts the voice alarm system on the scene and captures the scene photos at the same time; Then, the alarm information and scene pictures are transmitted to the monitoring center through the wireless transmission system; Finally, the on duty personnel of the monitoring center timely find the alarm information of the monitoring screen and inform the relevant personnel to deal with emergencies.

However, the above-mentioned monitoring and management system requires personnel to be on duty for a long time. In the case of personnel negligence, there will be missed judgment and misjudgment of emergencies. After the emergency warning, the garrison personnel need to inform the relevant emergency handling personnel, and the emergency handling personnel will review and understand the process of the emergency again in the monitoring room, and then formulate corresponding solutions, Then, it is implemented by the property management personnel.

5 Conclusion

The intelligent building management and control system based on intelligent video analysis algorithm to realize behavior recognition has the following beneficial effects on the technical solutions provided: through the video analysis component to identify emergencies and generate early warning information, the unattended monitoring and management system can be realized, and the missed judgment and misjudgment of emergencies can be avoided; Furthermore, the emergency service component can call the corresponding emergency plan according to the early warning information, which can automatically generate emergency solutions; Furthermore, through the data acquisition and monitoring components to deal with the external access equipment according to the emergency plan, the emergency plan for emergencies can be implemented quickly and automatically, and the timeliness of dealing with emergencies can be improved.

References

1. Ma, J.: Video Moving Target Tracking Method. Electronic Industry Press, Beijing (2013)
2. Liang, D., Zhang, Y., Cao, N., et al.: Network Video Monitoring Technology and Intelligent Application, pp. 205–206. Renmin Posts and Telecommunications Press, Beijing (2013)
3. Yan, Q., Li, L., Xu, X., et al.: Review of video tracking algorithms. Comput. Sci. **40**(06A), 204–209 (2013)
4. Cai, R., Wu, Y., Wang, M., et al.: Overview of video target tracking algorithms. Telev. Technol. (012), 135–138 (2010)

Research on the Development of Ship Target Detection Based on Deep Learning Technology

Dongdong Cui[✉], Lijun Guo, and Yongzhen Zhang

Information Science and Engineering College, Ningbo University,
cNingbo, Zhejiang, China
guolijun@nbu.edu.cn

Abstract. The rapid development of inland shipping and marine shipping has promoted the economic growth, but it also makes the frequency of ship collision, ship Landing and other accidents more and more high, illegal fishing, illegal ship parking and other phenomena often occur. Therefore, the realization of automatic detection of ship targets is of great significance to the management of surface ships. Combined with the advantages of deep learning in big data feature learning, this paper proposes a fast ship target detection method based on deep learning.

Keywords: Deep learning · Ship characteristics · Ship inspection

1 Introduction

Since the era of great navigation, it has greatly promoted the exchanges between the continents on the earth, and formed many new trade routes. With the development of land resources becoming increasingly saturated, attention naturally focuses on the ocean, which accounts for 2/3 of the earth's area, and countries have turned to compete for marine resources. As a result, there are many problems, such as illegal infringement of the sea area under the guise of freedom of navigation, confrontation between passing vessels near disputed islands and reefs, and so on. However, at present, when the maritime police ships and fishery administration ships are carrying out cruise surveillance, for the illegal acts of infringing China's marine rights and interests, their rapid detection, timely evidence collection and real-time early warning are the problems faced by the maritime monitoring departments to improve the technical level of maritime rights protection and law enforcement. Therefore, it is imperative to use automatic means to improve the judgment ability and early warning ability of maritime law enforcement ship monitoring system on the target of illegal invasion. The premise of achieving efficient early warning is to accurately locate and quickly detect the suspected targets in the sea area, determine the nature of the target, and carry out subsequent priority ranking and scientific early warning according to the credibility score of target detection. In addition, the sea surface hydrology and weather environment are complex and changeable, and the delay of detection and alarm system in millisecond level can cause irreparable serious consequences. Therefore, the real-time performance of detection is also an extremely important consideration when selecting

J. C. Hung et al. (Eds.): FC 2021, LNEE 827, pp. 1730–1736, 2022.
https://doi.org/10.1007/978-981-16-8052-6_255

the detection system. It is urgent to develop an effective ship target detection system on the sea [1].

In this paper, the ship in the marine background is taken as the research object. The acquisition method of the research object can be through the sea buoy and maritime law enforcement ship. Taking advantage of deep learning in the field of big data target detection, neural network gradually learns and extracts the image features from shallow to deep through training, and studies the fast and accurate ship target detection algorithm Lay the foundation for management.

2 Basic Concepts of Deep Learning

Deep learning, also known as deep neural network (DNN), is a collection of algorithms for modeling high complexity data through multi-layer nonlinear transformation. Multilayer refers to the number of layers of neural network, and the depth of deep learning is more than 8 layers of neural network, the more layers, the deeper the depth. Nonlinearity refers to dealing with complex nonlinear separable problems in practical applications, using complex lie number approximation, and then more detailed characterization of the characteristics of the data. The essence of deep learning is to use multiple hidden layer machine learning models and massive training data to fully represent and learn useful feature information, and then predict or identify the results [2].

2.1 Characteristics of Deep Learning

The advantages of deep learning are: reducing the shortcomings of incomplete information extraction caused by artificial design features; Under the application background of satisfying the characteristic conditions, some machine learning applications based on deep learning architecture show better recognition effect and classification ability than the original algorithm.

The disadvantages of deep learning are: in order to achieve better recognition effect and higher estimation accuracy, deep learning needs a lot of data support and a long learning process. Better software programming skills and better hardware support are the prerequisite to improve the performance of deep learning.

2.2 Model Classification of Deep Learning

Common deep learning models are shown in Table 1. There are the following classification methods:

1) According to the structure level, deep learning can be divided into basic model and overall model. The basic model is composed of basic components, such as full connection model FCM, self encoder AE and so on; The whole model is composed of basic models and contains two or more intermediate layers, such as convolutional neural network (CNN) and deep Boltzmann machine (DBM).

2) According to the training process and mathematical characteristics, deep learning can be divided into deterministic model and probabilistic model. Deterministic depth model adopts deterministic description, and the training objective is to minimize the loss function of a specific form. The common models include convolutional neural network CNN, depth self encoder DAE and so on; The probabilistic model is described by probability, and the training method is to maximize the likelihood function about the training data by maximum likelihood estimation. The common models include deep belief network DBN, deep Boltzmann machine DBM and so on.

3) Deep learning can be divided into unsupervised, supervised and mixed depth models according to whether or not to use category labels. The unsupervised depth model does not use the observation data with class label in the training process, that is, it calculates and analyzes by obtaining the high-order correlation between the observation data. The common models include depth confidence network DBN, restricted Boltzmann machine RBM, etc.; Supervised depth model uses the observation data with category label in the training process, which contains the information of pattern category; The hybrid depth model uses both the observation data with category label and the observation data without category label in the training process. The model includes the production module and the discriminative module. The production module is used to assist the discriminative module. The common model is the depth sparse self coder dsae.

4) According to the way of network connection, deep learning can be divided into adjacent layer connection, cross layer connection and ring connection model. The adjacent layer connection model is the connection mode between adjacent network layers, and it is easy to optimize the model; The cross layer connection model can be connected between adjacent network layers and different network layers, and the structure can model the dependency relationship between cross layers; The ring connection model is that any layer can be connected with each other to form a ring structure.

Table 1. Common deep learning model

Composition hierarchy	Model name	Mathematical characteristics	Training methods	Connection mode
Basic model	Perceptron model	Deterministic supervision	Supervised	Adjacent layer connection
	Autoencoder	Deterministic supervision	Unsupervised	Adjacent layer connection
	Boltzmann machine	Probability	Unsupervised	Annular connection
Overall model	Convolutional neural network	Deterministic supervision	Supervised	Adjacent layer connection
	Depth self encoder	Deterministic supervision	Hybrid	Adjacent layer connection
	Deep decomposition network	Deterministic supervision	Hybrid	Cross layer connection

3 Particularity of Ship Target

Ship is usually a closed or semi closed three-dimensional geometry composed of deck, bottom and port and starboard, which is symmetrical and long and narrow. Its geometric elements include the size and shape of the ship. For the detection of ship target, its shape feature is mainly used, so the discussion of ship feature is mainly focused on its shape. The main parameters of hull shape include principal dimension, shape coefficient and scale ratio, which describe the geometric characteristics of hull size, shape and width. Its configuration is shown in Fig. 1.

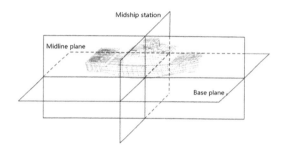

Fig. 1. Conventional hull structure

The midline plane refers to the vertical plane passing through the mid point of the ship width at both ends and along the length direction of the ship. The two sides of the plane are symmetrical port and starboard of the ship, that is, the midline plane is the only symmetrical plane of the ship. The midship station surface (also known as midship section) is a transverse vertical plane passing through the mid point of the ship's length and along the ship's width, which divides the ship into front and rear parts. The base plane is the horizontal plane where the bottom of the ship passes through the upper edge of the hull keel and passes through the intersection of the midship station plane and the centerline plane. The midline plane, midship station plane and base plane together constitute the plane of hull coordinate system.

In target detection, the shape characteristics of ship target should be considered, and the detector should be designed reasonably, so that the detection can better adapt to the shape of target and get more accurate positioning information [3].

4 Ship Target Detection Based on Deep Learning Technology

4.1 Target Detection Based on Regression

Convolutional neural network (CNN) is one of the deep learning models. Aiming at the speed disadvantage of convolutional neural network in image target detection task, the deep learning target detection method based on regression idea has been developed rapidly. At present, the most commonly used network based on regression idea is yo (you only look once). Yo partitions the input image grid. If the coordinates of the center position of an object are contained in a grid, then the grid is responsible for checking the object. Yolo network transforms image target detection into regression problem. After image input, only one detection is needed to get the target position and the corresponding confidence probability, so it is called "you only look once". The efficiency of the whole network is greatly increased, and the speed is far from meeting the real-time requirements [4].

4.2   Yolo Detection Process

The image may contain multiple targets or multiple categories of targets, so it is necessary to judge the multiple categories of each prediction box. The specific detection process is as follows:

(1) The image is divided into grid cells of SXS, and each grid gives B bounding boxes, each of which contains 5-Dimensional information, namely (x, y, W, h, score). The information contained in the candidate borders is shown in Fig. 2. The white dot in the figure represents the center point of the ship, so the grid is responsible for predicting the ship in the image.

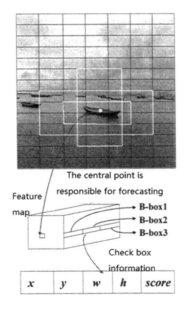

Fig. 2. Yolo network candidate border

Where, (x, y) is the offset of the center of the candidate border from the cell boundary, (w, h) is the ratio of the width and height of the border to the whole image, and score is the confidence *score*, which means the confidence degree of the target contained in the candidate border, as follows:

$$score = \text{Pr}(object) \times IOU_{pred}^{truth} \qquad (1)$$

Among them, *Pr* (*object*) indicates whether the target exists in the cell corresponding to the candidate border. If there is a target in the grid, its value is taken as 1, otherwise it is taken as 0. IOU_{pred}^{truth} indicates the intersection over union of prediction and ground truth, as shown in Fig. 3.

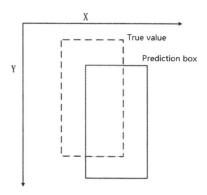

Fig. 3. Diagram of intersection and union ratio

(2) CNN extracts features and forecasts, and predicts targets in the last full connection layer. Each grid gives the conditional probability pr (class | object) of C categories when there are objects, and then obtains the probability of each category in the whole network, The confidence score of a certain category in the SXS grid of the whole image is obtained by multiplying the probability of a certain category in each detection frame with its corresponding confidence.

(3) Set the threshold of confidence score of detection frame, filter the frame whose score is lower than the preset threshold through non maximum suppression, and the remaining frame is the correct detection frame, and output the final judgment result.

5   Conclusion

With the advantages of precision, speed and end-to-end implementation, deep learning has rapidly become the most popular research method in the field of target detection. Ship target detection based on deep learning has important research significance and application value. It can provide an effective technical means to discover, monitor, track and collect evidence of foreign-related infringement targets and behaviors, and significantly improve the technical level of maritime rights protection and law enforcement.

References

1. Zang, F.: Research on ship target detection algorithm in intelligent video surveillance. Ocean University of China (2014)
2. Li, J.: Application analysis of optical remote sensing image in ship target detection. Ship Electron. Eng. **36**(10), 30–34 (2016)
3. Wang, F., Zhang, M., Gong, L., et al.: Fast detection algorithm of ships on the sea in the background of ocean. Laser Infrared **46**(5), 602–606 (2016)
4. Zhu, D., Luo, Y., Dai, L., et al.: Salient object detection via a local and global method based on deep residual network. J. Vis. Commun. Image Represent. **54**, 1–9 (2018)

Design and Implementation of Interactive Network Education Platform Based on Campus Network

Jielan Zhang[⊠]

Jiangxi Teachers College, Yingtan 335000, Jiangxi, China

Abstract. This paper analyzes the problems existing in the process of basic education informatization in China. The application of campus network in education is not in-depth, and high investment fails to produce high efficiency. In view of this problem, this paper designs a general network education platform based on campus network, which creates an ideal environment for teachers' teaching and students' learning, so as to promote the in-depth application of campus network in education and teaching. This paper analyzes the characteristics of network education and summarizes the educational functions of campus network. Based on the constructivism learning theory, this paper discusses the requirements of the "learning based" teaching method to the network teaching environment; Based on the theory of distance learning circle, this paper discusses the relationship between the elements of network teaching, and expounds the essential attribute of network teaching interaction.

Keywords: Network education platform · Interactive · Design and implementation · Campus network

1 Introduction

The university network education resource platform is to integrate all kinds of educational resources that can provide fast and convenient resource storage, sharing, learning and computing ability. The application of campus network in the teaching resource platform not only reduces the cost of the teaching resource platform, but also improves the utilization rate of teaching resources, and realizes and promotes the sharing of teaching resources and the development of teaching education. All colleges and universities use their own network equipment to build their own teaching platform and develop and integrate their own teaching resources, which leads to the repeated construction, investment and development of network resources, hardware facilities and software platform. The teaching platform of campus network aims to realize the sharing of computer hardware resources, software resources, teaching resources and the interaction between teachers and students, so as to build a cloud education platform with complete functions, stability, efficiency and security.

The proposal of campus network provides a lot of reference for the construction of modern network teaching resources. In the future campus network environment, how to build network teaching resource database has become a problem we have to consider

J. C. Hung et al. (Eds.): FC 2021, LNEE 827, pp. 1737–1743, 2022.
https://doi.org/10.1007/978-981-16-8052-6_256

[1]. The construction of resource bank is a long-term project, which must have a long-term plan to make it develop healthily and continuously. The ultimate goal of teaching resource library is to support teaching and scientific research. In the process of construction, we should follow the basic principles of overall planning, division of labor and cooperation, focusing on application and mechanism innovation, make full use of existing resources, reasonably classify, sort out and absorb the latest and foreign outstanding achievements, combine with the IT architecture ideas and services provided by campus network service providers, and constantly innovate to build a scientific and high-quality teaching resource library Efficient and applicable teaching resource library.

2   The Educational Function of Campus Network

Computer network has become an educational network in education and teaching. According to the geographical range covered by the educational network, it mainly includes classroom network, campus network, education metropolitan area network and international Internet.

1. The concept of campus network
   According to the IEEE, LAN technology is "connecting computers, terminals, peripheral devices with large capacity memory, controllers, displays, and network connectors used to connect other networks, etc., scattered in a building or adjacent building, Campus network refers to the general term of computer network system in the campus, which is obviously a kind of LAN. It not only involves the connection of local network of oneortwo buildings, but also the connection between computer LAN in multiple buildings, and also connects with external computer network (such as Internet). Campus network is a computer network information system which provides resource sharing, information exchange and collaborative work for school education and teaching.

2. The educational function of campus network
   With the acceleration of the process of educational informatization, the construction of campus network in the ordinary primary and secondary schools in large and medium-sized cities is becoming more and more popular, and campus network will become an important platform for the reform of primary and secondary schools' curriculum and the overall promotion of quality education. In order to construct and use campus network better, it is necessary to carry out scientific and reasonable function planning for campus network. As shown in Fig 1 below, it is a general functional model that can be used to guide the design and use of campus network education platform. Therefore, the functions of education services provided by campus network are seven parts: teaching, management, resources, extracurricular education, family education, social education and communication. Each part can be regarded as a subsystem. Among them, the teaching subsystem, management subsystem and resource subsystem constitute the basic function structure of the general school; The individualized system of extracurricular education, family

education and social education can be seen as the extension and extension of school function; As the support structure of other subsystems, communication system combines them into a whole.

Fig. 1. Educational function model of campus network

The characteristics of network education.

Using the campus network to implement network education will directly promote the informatization of school education, and form a strong support for the contemporary education reform. It is a consensus to regard the modern information technology based on the network as the breakthrough and fundamental way of the education reform in the 21st century. This requires us to combine advanced educational ideas and methods with modern information technology, and make full use of the advantages of network education to build an innovative educational environment [2].

So what is the general direction of the current world education reform? What is the difference between innovative teaching and traditional teaching? In the report "using educational technology to support educational reform" submitted by American scholar B. means, some characteristics of innovative teaching are put forward, as shown in Table 1.

Table 1. Comparison table of traditional teaching and innovative teaching characteristics

Traditional teaching	Innovative teaching
Teacher guidance	Student exploration
Didactic teaching	Interactive guidance
Single subject, isolated teaching module separated from situation	Multidisciplinary extension module with practical tasks
Individual work	Cooperative operation
Teachers as knowledge givers	Teachers as promoters
Homogeneous grouping (by capacity)	Heterogeneous grouping

It can be seen from the table that the core of modern education reform is to make students change passive acceptance learning into active engagement learning, so that they can learn and accept challenging learning tasks in a pragmatic environment.

3   Design of Interactive Network Education System Platform

The characteristics of online education determine that its teaching structure is student-centered, so constructivist learning theory naturally becomes the theoretical basis for us to build online education platform. Network education is also modern distance education. Therefore, this theory also has important guiding significance for the design of network education platform. The system is designed by using an open source software Eucalyptus on the virtualization layer, which is used to manage the infrastructure cloud service platform, and installing the node control components of Eucalyptus on them, which can not only shut down, check, start and clear the work. To access the underlying virtualization layer and virtual resources, it is necessary to deploy two servers on the management layer and install the corresponding components. The management layer will eventually convert the corresponding requests of the service layer into access to the virtual resources. On the issue of installing and deploying servers, one server installs the cloud controller components, The other is to install cluster controller component and storage control component with nodes connecting service layer and virtualization layer. The overall architecture of the design is shown in Fig. 2.

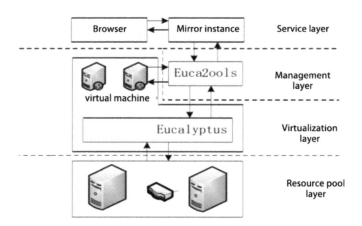

Fig. 2. System architecture design

The whole network education platform consists of two parts: education support platform and education implementation platform. The main function of education support platform is management and service, and the main function of education implementation platform is to realize the whole teaching process.

The education support platform consists of eight subsystems: public information system, enrollment system, student information system, teacher information system, educational examination system, expense management system, learning center management system and platform maintenance system. It provides a complete set of teaching administration management scheme from enrollment, enrollment, school and after graduation. Public information can be released through public information system, and enrollment system can realize online registration, examination, admission and other links. The information system of students and teachers is used to manage and inquire the relevant information of relevant personnel. Educational administration examination system is the key part of the whole teaching management process, which provides accurate information of teaching organization, process management and supervision. In the cost management, we should provide two sets of tuition management programs: charging by credit and not charging by credit [3]. Through the learning center management system, we can supervise and guide the operation of the learning center, and improve the service function. The platform maintenance system is used to ensure the data integrity and reliability of the database, manage various users, and provide the function of managing various subsystems. The education implementation platform consists of eight subsystems: education resource management system, online teaching system, online supervision system, online tutoring system, online homework system, online question answering system, online learning exchange system and online examination and test system. It provides all teaching functions such as teaching, self-study, discussion, guidance, test and examination through network. It also provides a variety of teaching environment and teaching methods to meet the learning needs of different learning groups. It can transmit the contents being explained in the multimedia classroom to other teaching points through the network or satellite. Students can also learn in real time through this live broadcast, creating an immersive learning environment; It can also replay the classroom teaching process or excellent courseware through on-demand, so as to achieve the teaching effect of real-time class. And through a variety of learning environment to collect students' learning behavior, analyze students' learning state, give learning suggestions for different students.

4 The Realization of Network Teaching and Learning Module

The resource management part mainly realizes the upload and download function of learning resources. Users can store learning resources in distributed storage system through the upload service provided by the system, and also download the existing learning resources in the system, and realize the sharing of learning resources among users.

Public information mainly realizes the function of public information publishing and browsing. When the user with permission enters the public information module, it can not only browse information, but also edit information for publishing. Publish information: when the administrator or teacher clicks "publish information" in the public information module, they enter the public information publishing page (noteadd. jsp), and the page editor still adopts the FCKeditor that is obtained as you see. When the user enters the page information and clicks the submit button, the operation of publishing information is mapped to addnoteservlet for processing. The servlet extracts page information and generates entity objects that persist in the data store of gae, with the key code as follows.

```
PublicList<notice>noticeList(start,end){
Query qry=pm.newQuery(Notice.class);
qry.setRange (start,end);
qry.setFilter("categoryType=='news'");
qry.setOrdering("publishTime descending");
return(List<notice>)qry.execute();
}
```

Teachers enter the online teaching system to check whether there is online teaching content. If there is no online teaching content, they will start to create new teaching content and submit them for approval. If there is online teaching content, check whether the audit is passed, if it fails to pass the audit, modify the teaching content and submit it to the audit again. If the online teaching content is approved, then check the online teaching period, If online teaching time has started, online teaching will be conducted. If online teaching time is over, check whether there is any teaching course video, and upload teaching course video if there is no teaching course video [4].

Students can manage their own learning resources on the system platform, view their own learning resources, learn about their learning resources, and also understand the current popular learning resources; At the same time, students can upload their own learning resources, but also download their own learning resources. Students can manage their own learning notes on the system platform, view their own learning notes, and also understand the learning notes of their friends. Students can ask teachers questions about their problems in the learning process on the system platform, and can get the answers and answers from teachers. Students can create their own circle of friends on the system platform so that students with similar interests can help each other, support and encourage each other in the process of learning and life. Students can discuss and communicate on the system platform, and students can publish text, pictures, videos, etc.

5   Conclusion

At present, there are still many problems to be solved and perfected in the modern distance education platform. The education support service system based on the traditional teaching concept can not provide intelligent and personalized services for students, and can not effectively manage the uncertainty in the learning process. This is a common problem in the current educational informatization process, and also the bottleneck restricting the development of educational informatization in China. The purpose of this paper is to promote the application of campus network in teaching, provide a platform for teachers and students to improve teaching and learning methods, improve the interaction between teachers and students, the interaction between home and school, and promote the deepening of education reform through educational information, so as to improve the teaching quality in an all-round way.

References

1. Li, Y., Dong, J.: Realization of information layer accounting in network education system. Electron. Measur. Technol. **33**, 56–60 (2003)
2. Huang, X.P., Wu, J., Zhang, S.Y.: Computer engineering and application. Archit. Des. Netw. Educ. Manag. Syst. **21**, 221–230 (2003)
3. Liu, R.: Design and implementation of computer network teaching platform. Digit. World **56** (07), 118 (2020)
4. Li, Y.: Design and implementation of university network education platform based on cloud computing. Electron. Prod. **71**(20), 59 (2016)

Research on Dance Action Recognition Method Based on Multi Feature Fusion

Jingxian Liu[✉]

Shandong Management University, Jinan 250100, Shandong, China

Abstract. Action recognition is a very challenging subject in the field of computer vision. Because of its high research value, it has become a very popular research direction in recent years. At present, a large number of scientific research institutions and researchers have done in-depth research on this topic and achieved some good results. A video clip retrieval method based on video summarization is proposed. Firstly, the key frame sequence of video is extracted, and the key frame is used to represent the content of each video segment. Then the orb feature is used as the feature of the video segment. When constructing the similarity matching model, the temporal order factor, granularity factor, interference factor and other factors are considered. By comparing the similarity, the video segment similar to the query segment is selected.

Keywords: Feature extraction · Action recognition · Multi feature fusion · Attitude

1 Introduction

With the improvement of computer technology and hardware level, as well as the popularity of smart phones and digital products, the rapid development of Internet and multimedia technology has brought a huge impact on people's daily work and life. At the same time, video has become the main carrier of information dissemination, enriching people's lives, but also brings opportunities for the vigorous development of artificial intelligence and big data industry. Among them, computer vision plays a crucial role. Computer vision is an interdisciplinary subject which integrates computer science, image processing, pattern recognition, machine learning and so on. The purpose of computer vision research is to endow the computer with the visual perception function similar to human beings, so that it can recognize the external objects and analyze the surrounding activities through the visual perception of the surrounding world like human beings.

Action recognition is a very challenging topic in the field of computer vision research. Its purpose is to use image processing and classification recognition technology to analyze video data to recognize human action. Because of its high research value, action recognition is a very popular research direction in recent years. In recent years, it has attracted a large number of scientific research institutions and scholars to engage in scientific research in this direction. Action recognition technology is widely used in various video scenes. At present, it has been widely used in many fields, such as intelligent monitoring, virtual reality, intelligent human-computer interaction and so

© The Author(s), under exclusive license to Springer Nature Singapore Pte Ltd. 2022
J. C. Hung et al. (Eds.): FC 2021, LNEE 827, pp. 1744–1749, 2022.
https://doi.org/10.1007/978-981-16-8052-6_257

on. The application case of action recognition is shown in Fig. 1. According to the latest development in this field, this paper briefly analyzes the application prospect of action recognition technology in related fields.

Fig. 1. Main application fields of action recognition

2 Action Recognition Method

At present, there are two kinds of action recognition methods: single-layer method and hierarchical method. In the single-layer method, actions are usually regarded as the feature categories of video, and classifiers are used to recognize the actions in the video. The image sequence in video is considered to be generated by specific action categories. Therefore, the single-layer action recognition method mainly involves how to represent videos and match them. Hierarchical method is mainly to identify high-level actions by identifying simple actions or low-level atomic actions in video. High level complex actions can be decomposed into a sub action sequence, and actions can be decomposed as high-level actions until they are decomposed into atomic actions. Figure 2 is the classification diagram of action recognition methods.

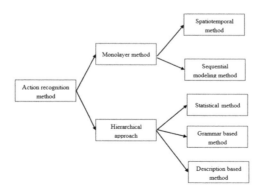

Fig. 2. Classification of action recognition methods

The main idea of trajectory based method is to use the observation sequence of joint point tracking to recognize the action. Trajectories are usually generated by tracking human joints or other points of interest on human parts. At present, there are many kinds of action representation methods and corresponding trajectory matching algorithms for action recognition. Wang proposed using dense trajectories to describe video, sampling dense points from each frame, and tracking dense points through displacement information in dense optical flow field, so as to form dense trajectories. At the same time, local descriptors such as hog, Hof and MBH around interest points are calculated.

The application of local feature in action recognition is extended from target recognition in image. Local feature refers to the description of feature points and the discriminative information around them. Local feature representation methods based on the density of extracted feature points can be divided into two categories: sparse and dense. Harris 3D detector and dollar detector are the representatives of sparse feature points, and optical flow based method is the representative of dense feature points. Bregonzio et al. [1]. called using spatiotemporal interest points to remove the limitations of dollar detector.

3   Dance Movement Recognition Based on Multi Feature Fusion

In order to improve the performance of the model to the complex dancers' movement recognition, this paper designs a dance action recognition algorithm based on multi feature fusion. The new algorithm uses a feature pyramid network (FPN) to extract features, then extract the features of different scales, and finally sample the features to the original image size for feature fusion. As shown in Fig. 3, the residualblock in the figure represents the residual module.

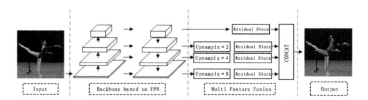

Fig. 3. Schematic diagram of dance action recognition algorithm based on multi feature fusion

In this paper, the directional gradient histogram feature, optical flow directional histogram feature and audio feature are extracted to describe the characteristics of dance action from the aspects of appearance and shape of human dance action in dance video [2], movement of human dance action and assisted by audio feature. Considering the limited ability of each type of feature to distinguish dance action, this paper uses the linear weighted combination method of multi-core learning method to fuse the direction gradient, histogram feature, optical flow direction histogram feature and audio feature, so as to realize the mutual complement of multiple types of features and improve the recognition ability of the classifier. The specific process is as follows: set group kernel function for each feature, and each kernel function has corresponding weight. Finally, combine multiple kernel functions to form a new kernel function in the way of linear weighting, and use support vector machine classifier for multi class classification [3].

Therefore, let's assume that there are p dance movements X in the dance dataset $x_1, x_2, \ldots, x_p$ and category $y_1, y_2, \ldots, y_p$. At the same time, the G kernel functions corresponding to hog features are defined as $k_g(x_i, x_j)$ The f kernel functions corresponding to $g = 1, 2, \ldots, G$, Hof characteristics are defined as $k_f(x_i, x_j)$, $f = 1, 2, \ldots, F$, m kernel functions corresponding to audio signature features are $k_M(x_i, x_j)$, $m = 1, 2, \ldots, M$. In this paper, the linear combination of kernel functions with the above three features can be expressed by (1)

$$k(x_i, x_j) = \sum_{g=1}^{G} \beta_g k_g(x_i, x_j) + \sum_{f=1}^{F} \beta_f k_f(x_i, x_j) + \sum_{M=1}^{M} \beta_M k_M(x_i, x_j) \qquad (1)$$

Since the research on the combination of motion recognition technology and dance has just started, the available dance data set is still relatively small. The open data set is Carnegie Mellon University's motion capture data set, but the data set contains very little dance data, which can not be used for dance motion recognition research; Dancedb541 dance dataset released by Virtual Reality Laboratory of University of Cyprus can meet the requirements of dance action recognition research. Therefore, two kinds of dance data sets are used in the experiment, which are dance DB data set and folk dance data set (Folkdance) produced by my laboratory. In the dancedb dataset, each dance category uses emotion markers; Folkdance dance data set is divided into four groups, each group contains a number of subdivided dance movements, the types of movements are relatively rich, and each group of dance movements are relatively complex and challenging. The sample frame of the dance dataset is shown in Fig. 4.

Fig. 4. Dance dataset sample frame

Image entropy is widely used in extracting key frames. The main work of image entropy is to count the features in the image, calculate the entropy value of each image in the image sequence, and then compare the entropy value of the current frame with that of the previous frame. If it is greater than the threshold value, it is considered as a key frame. The entropy calculation of image frame only considers the characteristics of image, but for dance video, the change of action is the key.

4   Dance Video Segment Retrieval

In the dance video segment retrieval, traditional video segment retrieval method is difficult to be applied in dance video. Although dance video is the same as other video structure, but the dance action is changeable, it is difficult to segment the video effectively by video content.

After extracting key frames for dance video, some image frame sequences which can summarize video content are obtained. Dance video retrieval library is composed of some key frames and their summarized video clips. The formation step is to extract the key frame of dance video first, and then segment each dance video by audio segmentation processing method. When performing dance video segment retrieval, the search segments to be retrieved will match the video clips in the video library. Through the previous video segment retrieval methods, the key frame can be extracted from the video, and the retrieval time can be greatly shortened by comparing the key frames under the lens. This paper improves from two aspects: on the one hand, in the extraction of key frame stage, through the analysis of the two data sets used in this paper, the key frame extraction work is carried out by using the movement amount of each dance action in dance video, which improves the accuracy of key frame extraction, and then improves the accuracy of dance video retrieval. On the other hand, the multi factor measurement is used in retrieval, which makes the similar video obtain have high quality characteristics [4].

Firstly, the key frame is extracted from dance video. Dance video is composed of a series of dance actions. The continuous dance movements reflect more or less movement amount. The motion information in dance video is represented by optical flow, and then the information in each optical flow graph is counted by entropy. The entropy sequence is combined with the music characteristics to obtain a music related entropy sequence. Then, the selection of key frames is made by the threshold value, and when the threshold is set, the optimal threshold is selected by comparing with the key frame set selected by multiple users. Thus, a key frame set of dance video is obtained for subsequent video segment retrieval.

Finally, the similarity model is constructed by considering many factors. Based on the characteristic factors, time order factors and interference factors, the similar model is constructed. Each factor will have different weights according to the video. Through experiments, a weight pair suitable for the video is selected. The most similar video clips are obtained.

5   Conclusion

The main research content of this paper is to retrieve video clips based on the key frame of music dance video extraction. The key frame is extracted from music dance video. By combining music features with motion features of dance actions at frame rate, a video segment retrieval method based on video summary is proposed. Video segment retrieval is generally compared with the key frame under the lens, but the method is not suitable for dance video. Therefore, this paper compares the key frame with the video frame extracted equidistant from the query clip. One or more keyframes can summarize the content of the video, and compare the query video with some key frames, If these keyframes can also summarize the content of the segment, the video clips corresponding to the keyframes are similar to the query clips.

References

1. Fan, T.: Research on Key Technologies of Content-Based Video Retrieval. Beijing Jiaotong University, Beijing
2. Qu, Y.: Research on Shot Segmentation and Key Frame Extraction in Video Retrieval. Shenyang University of Technology, Shenyang (2016)
3. Pang, Y.: Key frame extraction of action video based on priori. J. Henan Univ. Technol. Nat. Sci. Ed. **35**(6), 862–868 (2016)
4. Ying, R., Cai, J., Feng, H., et al.: Human action recognition based on motion block and key frame. Fudan J. Nat. Sci. Ed. (6), 815–822 (2014)

Using Genetic Algorithm to Optimize the Elements of College Students' Physical Training

Chunyue Yang[✉]

Liaoning Engineering Vocational College, Tieling 12000, Liaoning, China

Abstract. In this paper, an improved hybrid genetic algorithm with global optimization ability is designed by combining sorting adaptive genetic algorithm with simplex method. The algorithm is used to optimize the teaching and training elements of College Students' physical strength (based on heart function and lung function), and the experimental research is carried out in college students. The experimental results show that this method has better training effect than the traditional method, and has innovative significance to improve the physical health of college students.

Keywords: Genetic algorithm · College student · Physical fitness · Training elements

1 Introduction

In the process of studying and living, developing physical ability and improving health level is one of the important tasks of growth. Excellent physical quality is the foundation of the comprehensive ability development of college students, and the foundation of stepping into society and taking up the post. Excellent physical quality is the basis of the development of College Students' physical and mental health. The influence of the study environment, living environment and information age of college students in the new period makes the physical and health condition of some college students not guaranteed, which affects the professional learning and talent of college students. At present, in the physical training of college students (based on the heart function and lung function), most of them adopt the traditional teaching training methods. Although the traditional methods can meet the basic teaching training requirements, there is still a lack of an optimized teaching training mode in practice to further improve the physical training effect of college students. Therefore, we use hybrid genetic algorithm to optimize the physical training elements of college students, so as to improve the effect of physical education and training.

J. C. Hung et al. (Eds.): FC 2021, LNEE 827, pp. 1750–1755, 2022.
https://doi.org/10.1007/978-981-16-8052-6_258

2 Genetic Algorithm

2.1 Introduction to Genetic Algorithm

Genetic algorithms (gas) is an adaptive global optimization probability search algorithm which simulates the genetic and evolutionary processes of organisms in natural environment. It was first proposed by Professor J. Holland of Michigan University in the United States. It originated from the research on natural and artificial adaptive systems in the 1960s. Its main characteristics are group search strategy and information exchange among individuals in the group, and search does not depend on gradient information [1]. In the 1970s, do Jong carried out a lot of experiments on the computer based on the idea of genetic algorithm. Based on a series of research work, Goldberg summarized it in 1980s, and formed the basic framework of genetic algorithm.

2.2 Elements of Basic Genetic Algorithm

(1) Chromosome coding method. The basic genetic algorithm uses a fixed length binary symbol string to represent individuals in a population. Its allele is composed of binary symbol set $\{0, 1\}$. The gene values of each individual in the initial population can be generated by the random number of uniform distribution.

(2) Individual fitness evaluation. The basic genetic algorithm determines the chance of each individual in the current population to inherit to the next generation population according to the probability proportional to the fitness of the individual. Genetic algorithm is the only feedback standard based on fitness function, which can judge the quality of chromosome only by the fitness function value, and determine the survival rate of chromosome to be preserved to the next generation

$$\text{Fitness function} = \text{target function} + \text{penalty function}$$

(3) Genetic operator. The basic genetic algorithm uses three genetic operators: selection, crossover, mutation.

Genetic algorithm is random in the whole evolutionary process, but its characteristics are not completely random search. It can effectively use historical information to predict that the next generation expected performance has been improved to the searchable advantage set. This kind of evolution continues to evolve, and finally converges to an individual that is most suitable for the environment, and obtains the final solution of the problem [2]. The basic flow and structure of the standard genetic algorithm are shown in Fig. 1.

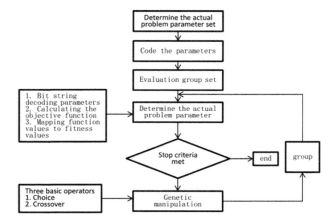

Fig. 1. Basic flow chart of simple genetic algorithm

3   Concept and Elements of Physical Fitness

3.1   What is Physical Fitness

There are many origins of the term physical fitness, and different people have different opinions on its definition. Western scholars have western scholars' way of speaking, while eastern scholars have Oriental Scholars' logic. However, from a macro point of view, the concept of physical fitness is mainly derived from the physical quality of motor skills, which is fully affirmed by both Chinese and Western scholars. Specifically, physical fitness refers to the physical quality of individuals or groups in the process of physical activity. Among them, physical fitness is mainly strength, speed, sensitivity, coordination, reaction, coordination and other aspects. Physical fitness is the comprehensive embodiment of the above aspects of physical fitness, is an important embodiment of the body to adapt to competitive sports.

3.2   Core Elements of Physical Training

Physical fitness is an important branch of physical fitness. In other words, individual health and fitness is an important basis of physical fitness. Therefore, before talking about the core elements of physical training, it is very important to understand that physical fitness is reflected through physical exercise. However, we should also know that physical movement is physical, and it is a manifestation of the body inside and outside. Therefore, the core element of physical fitness is not only physical fitness, but also involves some factors of healthy physical fitness. Based on this, the core elements of physical fitness are composed of cardiovascular function, muscle function, coordination function, agility, balance function and flexibility function [3].

4   Applying Genetic Algorithm to Design Physical Training Mode

Genetic algorithm is used to optimize the training elements of College Students' physical fitness teaching. Fifty college students (25 males and 25 females) aged 18–21 were selected. The average height of male students was 173.1 cm and the weight was 60.6 kg; The average height of female students is 159.9 cm and the weight is 52.1 kg.

The teaching and training effect of College Students' physical strength (mainly based on heart function and lung function) is mainly determined by the amount and intensity of exercise. In this paper, genetic algorithm is used to optimize the design of these two elements. The specific design ideas are as follows.

4.1   Algorithm Design of Genetic Algorithm

Determination of coding method: let the amount of exercise be represented by S_m and the intensity be represented by S_i. In this paper, the decimal floating-point coding method is used to encode the optimized S_m, and S_i parameters. Specifically, let S_m [$S_{m\ min}, S_{m\ max}$], $S_i \in [S_{imin}, S_{imax}]$ arrange the decimal floating-point numbers in the two optimization ranges together to form an individual X(t).

The selection of cross probability P_c and mutation probability P_m in genetic algorithm is the key to affect the performance of genetic algorithm, and directly affects the convergence of the algorithm. In this paper, an improved adaptive genetic algorithm is adopted to make P_c and P_m automatically change with the fitness. The P_c and P_m are adjusted according to formula (1):

$$P_m = \begin{cases} k_3 \frac{(f_{\max}-f)}{f_{\max}-f_{avg}}, f \geq f_{avg} \\ k_4, f < f_{avg} \end{cases}$$

$$P_m = \begin{cases} k_3 \frac{(f_{\max}-f)}{f_{\max}-f_{avg}}, f \geq f_{avg} \\ k_4, f < f_{avg} \end{cases} \tag{1}$$

Where f_{max} is the largest fitness value in the population; f_{avg} is the average fitness value of each generation population, f is the larger fitness value of the two individuals to be crossed, f is the fitness value of the individuals to be mutated: k_1. k_2, $k_3, k_4 \in (0, 1)$. For the sake of intuition, the expressions of P_c and P_m are represented by graphs, as shown in Fig. 2.

The design of simplex local search algorithm: because the genetic algorithm designed in this paper is a hybrid algorithm of genetic algorithm and simplex method, the simplex method with strong local search ability is added into the execution process. In the process of adding the simplex method, we mainly consider the following two aspects: (1) the allocation of time required by the simplex search algorithm: if we simply combine the genetic algorithm with the simplex method, most of the operation time of the whole algorithm will be occupied by the simplex search algorithm, and only a little time will be used for genetic operation, In this way, it is not easy to reflect the

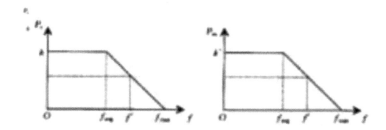

Fig. 2. Schematic diagram of adaptive crossover probability P_c and mutation probability P_m

strong global search ability of genetic algorithm: in addition, the simplex search does not need to be solved accurately [4]. Therefore, this paper limits the search time of the simplex method from two aspects: one is the number of iterations n of the simplex method, which is limited to $N \leq 20$ after many experiments in this paper; The other is the search probability P_i of simplex method (that is, the number of individuals participating in local search is limited). In this paper, after many experiments, we take $P_i \in [04,0.7]$. (2) Parameter setting of simplex method: reflection coefficient y, expansion coefficient a, search coefficient β Set as follows: $\gamma = 1$, $\alpha = 2$, $\beta = 0.75$.

Judge whether the termination condition of the algorithm is satisfied. The initial population generated by the improved method obtains a new generation of population through hybrid genetic operation. The generation of population is substituted into the fitness function again for detection and evaluation operation to observe whether the termination condition of the algorithm is satisfied. If it is satisfied, the optimal solution is output and the optimization ends. If it is not satisfied, the above operation is repeated until it is satisfied.

4.2   Experimental Result

Heart function tester and lung function tester were used to measure the step test and vital capacity index of college students (25 men and 25 women) under two teaching and training modes (mode 1 is the conventional mode, and mode 2 is the genetic algorithm mode). Some comparative test results are shown in Table 1 and Table 2.

Table 1. Comparison table of boys' experimental index

Student	Step test index		Vital capacity index	
	Mode 1	Mode 2	Mode 1	Mode 2
1	57.5	59.3	3750	4025
2	50.1	51.3	4725	4815
3	48.3	50.4	4100	4200

Table 2. Comparison table of experimental index of girls

Student	Step test index		Vital capacity index	
	Mode 1	Mode 2	Mode 1	Mode 2
1	56.8	57.3	3250	3315
2	53.4	55.2	4295	3540
3	40.3	53.6	3.50	3195

The above experimental results show that the teaching and training method designed by genetic algorithm for college students' physical fitness teaching and training elements is better than the traditional teaching and training methods, which shows that the method has obvious effect on Enhancing College Students' physical fitness (cardiopulmonary function), It has practical significance to improve the physical health of college students.

5 Conclusion

The experimental results show that the optimized design method of College Students' physical fitness teaching and training elements based on genetic algorithm has practical value for improving college students' physical fitness. It is an objective, effective and fast teaching and training method, which can be easily extended to other sports teaching and training projects, From a new perspective, this paper puts forward a new teaching and training mode to enhance the physical health of college students.

References

1. Zhou, M., edited by Sun, S.: Genetic Algorithm Principle and Application. National Defense Industry Press, Beijing
2. Chen, G., et al.: The Algorithm of Heritage Transfer and Its Application. People's Post and Telecommunications Press, Beijing (1996)
3. Zhang, L.Z.: The importance and Countermeasures of strengthening physical training in Physical Education in Colleges and universities. J. Beijing City Coll. (03), 62–66–74 (2019)
4. Wang, W.: Genetic Algorithm: Theory, Application and Software Implementation. Xi'an Jiaotong University Press, Xi'an (2002)

Application of Two-Dimensional Filtering in Rain Clutter Filtering

Wei Wang[1(✉)], Guangzhao Yang[1], and Yin Zhang[2]

[1] Sansha Hailanxin Ocean Information Technology Co., Ltd., Sansha, Hainan, China
wangw@highlander.com.cn
[2] School of Computer Science and Cyberspace Security, Hainan University, Haikou, China

Abstract. In navigation radar applications, rain clutter is a common source of interference in the practical application of navigation radars. The existing clutter suppression methods are generally difficult to distinguish between rain clutter and land clutter. The clutter suppression results often lead to the target or land echoes are suppressed together, and the clutter filtering effect is poor, which is difficult to meet the application in sea surface observation, resulting in false detection or missed detection. This paper uses a two-dimensional echo image processing method to filter rain clutter, which has a relatively great filtering effect.

Keywords: Radar · Filter · Rain clutter

1 Introduction

Navigation radar is one of the necessary equipment for seafaring ships. The clutter interference of radar has always been one of the most concerned issues of radar designers and users. For marine navigation radars, radar co-frequency interference, ocean wave interference, receiver internal noise, rain and snow interference, etc. will all have an adverse effect on the detection and recognition of targets. These interference signals are collectively referred to as radar clutter. In the past navigation radars, most of the analog circuit hardware methods are used to achieve various clutter suppression. The disadvantages are that the dynamic range is not high, the effect is not ideal, and it is difficult to adapt to the complex and changeable marine environment, especially the harsh weather conditions [1]. In particular, most current navigation radars still use magnetron transmitters, whose signal features only have amplitude information, and it is difficult to use coherent features to filter clutter in signal processing [2].

The navigation radar echo data display is displayed by two-dimensional scanned images. Through research, it can be found that there are still large differences between the rain clutter and the target echoes such as land, islands, and other targets. Based on the radar echo image, the image processing algorithm can suppress rain clutter to a certain extent. This paper is to use an image filtering processing algorithm to process the navigation radar echo image and get a better processing effect.

© The Author(s), under exclusive license to Springer Nature Singapore Pte Ltd. 2022
J. C. Hung et al. (Eds.): FC 2021, LNEE 827, pp. 1756–1761, 2022.
https://doi.org/10.1007/978-981-16-8052-6_259

2 Principles of Navigation Radar

The navigation radar uses the principle of the time difference between transmitting and receiving electromagnetic wave pulses to realize the ranging of the target in the detection scene, and the radar realizes the measurement of the target direction through rotation. The basic form of navigation radar is consistent with other radars, including transmitter, transmitting antenna, receiver, receiving antenna, processing part, and display. As shown in the Fig. 1, it is the principle of the navigation radar scanning image. With the ship's radar as the center, the surrounding scene is scanned 360°, and the radar echo image is finally formed by scanning and stacking. The crew also uses the echo image to judge the surrounding environment.

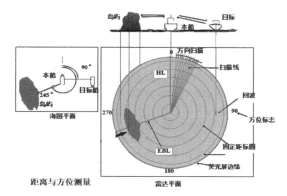

Fig. 1. Navigation radar schematic.

At present, the radar generally uses the constant false alarm rate (CFAR) method to generate the target detection threshold. If the radar echo exceeds the threshold, it is the detected echo. If the radar echo is below the threshold, it is considered as clutter and cannot be detected. It is considered that the clutter is filtered out, but this method is difficult to distinguish rain clutter, land, and islands when detecting sea surface targets, resulting in poor clutter filtering effect, which is difficult to meet the application in sea surface observation, resulting in false detection or leakage detection.

As shown in Fig. 2 is the navigation radar processing flow. After the radar front-end collects the radar echo data, it first carries out preprocessing to remove the co-frequency interference and then generates a threshold detection matrix through the echo matrix, which is used to detect the threshold of the preprocessed radar echo. If it is detected, it is the target, and if it is lower than the threshold, it is the useless information filtered.

This paper provides a radar echo processing method, which is to filter the two-dimensional scanning radar image, through the filtering process, the rain clutter can be well suppressed, and the land, islands, ships and other navigation radar attention targets can be well preserved.

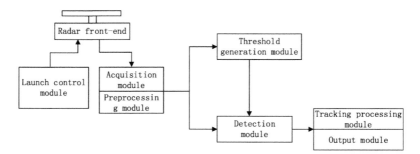

Fig. 2. Navigation radar processing flow.

3   Two-Dimensional Filtering

Through analysis, it is found that there is a relatively large spectral difference between rain clutter echo and radar echoes from land, islands, and ship targets. The two-dimensional image of rain clutter echo has relatively large low-frequency components. The echo has obvious edge jump characteristics, and its high-frequency components are more obvious.

By performing fir high-pass filtering on the two-dimensional spectrum of the radar echo, the rain clutter can be effectively filtered, and at the same time, part of the land or ship target echo in the rain clutter is also filtered. In order to better filter the clutter and retain the target echo, two-dimensional filtering is used to assist the traditional CFAR detection method to process the echo. The process is as follows:

Step 1: Radar collects the original echo video signal;
Step 2: preprocess the radar echo to form a frame of echo matrix data according to the scanning cycle;
Step 3: Perform time-frequency transformation on the echo data, and transform the echo time-domain image into a two-dimensional frequency-domain image;
Step 4: Perform high-pass filtering on the echo image, and this filtering can use FIR filter components;
Step 5: Inversely transform the filtered frequency-domain echo and convert it into a two-dimensional image. At this time, it can be found that the rain clutter has been suppressed, and most of the echoes such as land, islands, and targets are preserved;
Step 6: Use the reserved echoes to mark the echo matrix, that is, the reserved places are marked as 1, and the other filtered places are marked as 0; Step 4: Perform high-pass filtering on the echo image, and this filtering can use FIR filter components;
Step 7: Generate a detection threshold through conventional CFAR detection. For the marked areas in step 6, do not participate in CFAR detection and keep the low threshold. At this time, if the navigation radar adjusts the rain clutter suppression threshold, the marked areas remain unchanged. Threshold adjustments are carried out in the marked places according to artificial adjustments;

Step 8: According to the detection threshold matrix, the echo is detected. The echoes that exceed the threshold are retained, while those that do not exceed the threshold are considered as clutter or noise and are filtered out. So far, through the above 8 steps, we can get a better clutter suppression function in the rain clutter scene (Fig. 3).

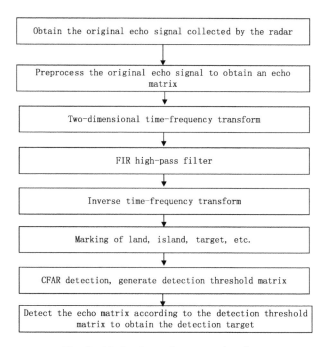

Fig. 3. Navigation radar processing flow.

4 Treatment Effect

In this section, the actual navigation radar data is used for processing. As shown in Fig. 4, the intensity of rain clutter in this image is very high, which belongs to a heavy rain scene; Fig. 5 is the spectrogram of the image, the white and bright in the middle is the low-frequency component with a large amplitude; Fig. 6 is the scene after the filtering of the above processing flow, it can be seen that most of the rain clutter has been filtered; Fig. 7 is the CFAR detection combined with the rain clutter The processing results after wave filtering and marking show that the land, islands, and targets are well preserved, while the rain clutter is effectively suppressed.

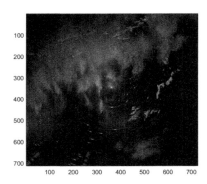

Fig. 4. Navigation radar image with rain clutter

Fig. 5. The frequency spectrum of the original image

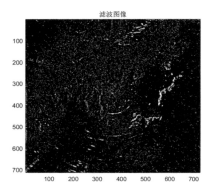

Fig. 6. Filtered image.

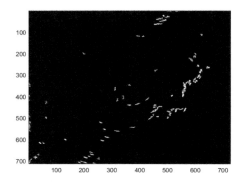

Fig. 7. Final detection result

5   Conclusion

The content of this paper is aimed at the common rain clutter suppression problem in navigation radar applications. Conventional CFAR detection is difficult to suppress this clutter cleanly, especially in the case of such strong rain clutter, target retention and clutter Inhibition is difficult to balance. The two-dimensional image filtering process provided in this paper assists the traditional CFAR detection method, which can filter the rain clutter well, and then obtain a better processing effect, which provides a better research idea for the navigation radar rain clutter suppression.

Acknowledgements. This work is partially supported by Major Science and Technology Program of Hainan Province (ZDKJ201811), Hainan Provincial Natural Science Foundation of China (620RC563), the Science Project of Hainan University (KYQD(ZR)20021).

References

1. Chao, H., Qing, G.: Implementation of Clutter Suppression Algorithm in Shipborne Navigation Radar Signal Processing. University of Electronic Science and Technology (2016)
2. Hong, L.: Research on Radar Rainfall Interference Detection and Suppression Technology. Harbin Engineering University (2019)

Land Detection Method and Application in Radar Video Based on Electronic Chart

Xiaoyi Wang[1(✉)], Xueyang Feng[1], and Fanglin An[2]

[1] Sansha Hailanxin Ocean Information Technology Co., Ltd., Sansha 573199, Hainan, China
wangxy@highlander.com.cn

[2] School of Computer Science and Cyberspace Security, Hainan University, Haikou, China

Abstract. In the case of severe clutter or obscuration, radar cannot accurately detect land information, which will bring safety hazards to ship navigation. In response to this problem, the use of electronic nautical charts and radar images to superimpose can provide a more comprehensive understanding of the sea area surrounding the ship. In this paper, the electronic nautical chart is used to extract the land information around the ship, and combined with the global satellite positioning system to obtain the latitude and longitude data of the ship and the heading data provided by the compass, the land information in the nautical chart is superimposed on the radar image for display, which is the radar video image provide a basis for land recognition. The upper computer software system inserts the extracted terrestrial information into the signal processing algorithm to shield the radar echoes in the terrestrial area. Offline data testing proves that after introducing the land information of the electronic chart, the upper computer software system can effectively reduce the false targets tracked in the background and reduce the false alarm rate. This research has a great reference for the application of land information in radar signal processing significance.

Keywords: Electronic chart · Radar video · Land inspection · Clutter suppression · Target tracking

1 Introduction

The radar obtains the distance and azimuth information of the target by obtaining the echo data of the measured target so that the ship has navigation and collision avoidance functions. However, radar detection data is easily affected by the environment. Under the interference of complex clutter, the quality of its echo becomes worse. It is difficult to accurately identify the target through the radar image, which brings safety hazards to the ship's navigation.

Radar signal processing is mainly divided into two parts: filtering and tracking. The filtering part can effectively remove rain clutter, sea clutter, and false echoes in the radar echo, but in the clutter suppression process, part of the land information echo is also filtered out, and multiple broken land echoes are generated. The radar tracking algorithm will create new targets on these broken land echoes. These targets are confused with the real targets, which affects the tracking performance of the radar.

Electronic nautical charts, as an important auxiliary navigation tool for modern navigation, can provide ships with the latitude and longitude information of surrounding land and islands, as an important basis for whether the ship navigates according to the planned route [1]. The electronic chart based on the S-57 international standard describes the spatial attributes of the objects as points, lines, and planes. The characteristic objects are divided into four types: meta objects, geographic objects, cartographic objects, and relationship objects. These features object mark records and space object mark records occupy most of the chart data capacity [2]. Mingxiao Wang et al. studied the fusion display technology of electronic nautical charts and radar images, and achieved independent control of the two system parameters [3], but only fused and displayed electronic nautical charts and radar images; Liu Jian et al. studied the technology of electronic chart drawing and radar video overlay display [4] and realized the specific functions in VxWorks system. Existing research on the fusion of electronic chart and radar detection technology is mostly in the direction of image overlay display, and there is very little research on integrating electronic chart data into radar signal processing algorithms. The land information in the electronic chart can be used as a priori information for radar signal processing to provide a basis for land recognition in clutter and target shielding in land areas. Therefore, the use of electronic chart information to extract land information from radar images and assist radar signal processing will help improve the radar signal processing system's ability to recognize real targets and play an important role in improving navigation safety and improving radar tracking performance. It has very great research significance and application value.

2 Overall Framework of the System

This paper builds a radar data acquisition and processing system, which mainly includes radar antenna, PCI-E data acquisition card, GPS satellite navigation system, upper computer, display screen, compass, electronic chart, and so on. The overall framework of the data acquisition and processing system is shown in Fig. 1.

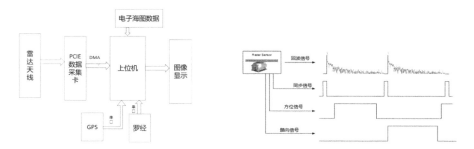

Fig. 1. Overall framework of the system   **Fig. 2.** Data transmitted over the DMA channel

The PCI-E data acquisition card transmits the collected radar echo and synchronization, azimuth, and heading signal data to the designated storage area of the upper computer via the DMA data transmission channel. The data transmitted via the DMA channel is shown in Fig. 2.

The upper computer software optimizes the received radar original echo data, and then converts it into a plane image according to the conversion relationship between range and pixel, and displays it on the display. The upper computer software has developed a target tracking algorithm. After filtering the radar signal, the target is identified, and the target information is superimposed and displayed on the radar echo image. The GPS equipment obtains the ship's latitude and longitude information and transmits the data to the upper computer software through the serial port. The software extracts the land information around the radar from the electronic chart based on the ship's position information, and then transforms the coordinates and superimposes it on the radar image. Use the compass to provide azimuth reference for direction calibration, and further match and display radar image and electronic chart data.

3   Key Technology Research

3.1   Matching of Electronic Chart and Radar Video

Radar detection gives the distance and azimuth of the target relative to the ship, while the description of the land information in the nautical chart is the latitude and longitude information of the land boundary points. To connect the nautical chart data to the radar signal processing system, it is necessary to carry out these latitude and longitude points. Coordinate transformation is converted into distance and azimuth information from the radar. In this paper, the northeast sky geographic coordinate system is selected as the navigation coordinate system (n axis).

Where λ is the longitude and L is the latitude, the coordinate transformation from the earth system to the navigation system can be expressed as

$$\begin{bmatrix} x_n \\ y_n \\ z_n \end{bmatrix} = \begin{bmatrix} -\sin\lambda & \cos\lambda & 0 \\ -\sin L\cos\lambda & -\sin L\sin\lambda & \cos L \\ \cos L\cos\lambda & \cos L\sin\lambda & \sin L \end{bmatrix} \begin{bmatrix} x_e \\ y_e \\ z_e \end{bmatrix}. \tag{1}$$

The conversion formula from latitude and longitude to the earth system is

$$\begin{bmatrix} x_e \\ y_e \\ z_e \end{bmatrix} = \begin{bmatrix} (N+H)\cos\lambda\cos L \\ (N+H)\cos\lambda\sin L \\ (N(1-e^2)+H)\sin\lambda \end{bmatrix}. \tag{2}$$

Among them,

$$N = \frac{a}{\sqrt{1-e^2\sin^2\lambda}}. \tag{3}$$

In the formula, the semi-major axis a of the earth's reference ellipsoid is 6378137m, and the square of the first eccentricity is 0.006694380 [5]. From this, the formula for converting the latitude and longitude points of the land information in the chart into the azimuth θ and the distance r is

$$\begin{cases} \tan\theta = \frac{x_{n1}-x_{n2}}{y_{n1}-y_{n2}} \\ r = \sqrt{(x_{n1}-x_{n2})^2 + (y_{n1}-y_{n2})^2} \end{cases}. \tag{4}$$

3.2 Radar Data Algorithm Flow

The upper computer software processes radar data and chart data in parallel. After the system starts, all parameters of the program are initialized. The data of radar echo is collected by the PCI-E acquisition card at the sampling rate of 62.5 m. The software obtains the data through DMA channel and processes the data accordingly. The chart data processing module automatically loads the corresponding chart according to the longitude and latitude information of the ship's location and then extracts the land information around the ship from the loaded chart. Through coordinate transformation, the longitude and latitude points describing the land boundary information are transformed into the azimuth and distance information from the ship. Finally, the converted land information is sent to the display module and superimposed on the radar map.

The radar data processing flow is as follows. After the program is initialized, the original radar data in the designated memory area is read in a loop, and the data is analyzed to extract the radar echo data of a certain azimuth direction, and then preprocess the data in this azimuth direction to remove the co-frequency interference, filter out the clutter, and transmit the filtered data to the display module, and the display module can realize the display of radar echo through the conversion of range and pixels. The tracking algorithm in the software condenses the filtered data to form echo blocks and then uses these echo blocks to match the target to update the target parameters, and the echo blocks that do not match the target will be used to create a new target.

When establishing a new target for the echo block, the program needs to first determine whether the echo block is in the land area given by the chart. If the area of the echo block belongs to the land area, it will be regarded as an invalid target and screened.

4   Data Processing and Result Analysis

4.1   Radar Data Acquisition and Filtering

In the process of ship navigation, the radar echo data is collected by using the x-band radar equipment of Hailanxin under the condition of moderate rain environment and sea condition less than the first level. the original radar image drawn by the upper computer software is shown in Fig. 3.

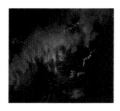

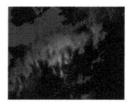

Fig. 3. Radar raw image under rain clutter

Fig. 4. Radar image after filtering

Fig. 5. Superimposed effect of land information in the chart on the original radar image

It can be seen from Fig. 3 that the land clutter information in the radar image under the interference of rain clutter is difficult to distinguish. Adjust the corresponding filtering parameters, filter the original echo, and filter out the clutter on the premise of keeping as many targets as possible. The filtered radar image is shown in Fig. 4.

By comparing the clutter and land radar images in Fig. 3 and Fig. 4, it can be seen that after processing by the filtering algorithm, the clutter in the original radar data can be effectively filtered, but the echo signal of the land will also be attenuated, and the land information becomes incomplete. Therefore, the recognition and detection of terrestrial information cannot be accurately completed by only relying on radar data.

4.2   Display of Land Information on Chart Data

GPS data and compass data are used to locate the ship's position and heading data, and the land information around the ship extracted from the electronic chart is superimposed and displayed on the original radar image. The drawn image is shown in Fig. 5.

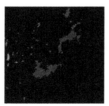

Fig. 6. False targets created by fragmented land

Fig. 7. Tracking effect based on chart data.

Figure 5 plots the land information of the electronic chart on the original radar data map. The red boundary line in the figure depicts the land boundary information in the electronic chart. Even if the clutter in the radar image is large, the user can observe the land boundary line and distinguish the position information of the land relative to the ship on the radar image, and it is not interfered by the radar clutter signal.

4.3   Application and Verification of Land Information in the Nautical Chart in the Tracking Algorithm

When the radar data filter parameters are set unreasonably, the land information is fragmented, resulting in a lot of incomplete land echo data, and many false targets are established in the follow-up software tracking algorithm. These targets are confused with the real targets. In the case of nautical chart data, it is difficult for the user to identify the target to be tracked. Figure 6 shows the false targets generated by the fragmented land after filtering.

The land information in the chart data is transferred to the radar signal processing algorithm, and the upper computer software judges whether the new target of the radar is in the land area. No new target is created on the echo block in the land area, which can effectively avoid the generation of filtering. The false target problem caused by incomplete land echo, the radar target recognition effect after optimization calculation is shown in Fig. 7.

Compared with the target established by the radar tracking algorithm in Fig. 7, the radar tracking effect of introducing electronic chart data for filtering is obviously better than that without chart information. The land after introducing an electronic chart will not be mistakenly identified as a target by radar tracking algorithm. Based on this, the number of targets tracked can be reduced and the quality of target output can be improved.

5   Conclusion

This paper superimposes the land information extracted from the electronic chart to the radar image for display, verifies the consistency between the chart data and the radar video, and connects the extracted data to the tracking algorithm of radar signal processing to realize the tracking of the land. The area shielding function reduces the number of background targets to be tracked and provides a basis for the radar signal processing algorithm for land processing and research.

Acknowledgements. This work is partially supported by Major Science and Technology Program of Hainan Province (ZDKJ201811), Hainan Provincial Natural Science Foundation of China (620RC563), the Science Project of Hainan University (KYQD(ZR)20021).

References

1. Li, D.: Analysis on the application of electronic chart information system in ship navigation. Digit. Technol. Appl. **39**(01), 73–75 (2021)
2. Tang, B.: S-57 Format Electronic Chart Data Reading and Hierarchical Display. Guangdong University of Technology (2011)

3. Wang, M.: Research and Implementation of Fusion Display Technology of Radar Image and Electronic Chart. Chongqing University of Posts and Telecommunications (2017)
4. Liu, J., Ni, X., Meng, F.: An electronic chart drawing and radar video overlay display method. J. Mil. Commun. Technol. **38**(01), 86–89 (2017)
5. Wang, X.: Research and Program Design of Integrated Navigation Algorithm Based on MEMS. Beijing Institute of Technology (2016)

Research on Dynamic Compression Method of Radar Echo Data

Shuanglin Zhou[1(✉)], Guangzhao Yang[1], Jinzhan Gao[1],
Xueyang Feng[2], and Yin Zhang[3]

[1] Sansha Hailanxin Ocean Information Technology Co., Ltd, Sansha 573199,
Hainan, China
zhousl@highlander.com.cn
[2] Sanya Hailan Huanyu Marine Information Technology Co., Ltd,
Sanya 572025, Hainan, China
[3] School of Computer Science and Cyberspace Security,
Hainan University, Haikou, China

Abstract. The amount of radar echo data is usually very large, which often occupies a huge storage space and a large transmission bandwidth during transmission. This will directly affect the data storage method and the performance of the transmission system. Therefore, the radar echo data needs to be compressed deal with. Existing radar data compression methods do not fully consider the characteristics of radar echoes, and it is easy to make it difficult to distinguish strong noise signals from small target signals after data compression, and the data accuracy after compression is not high. Improving the accuracy of data compression usually requires increased computational overhead, resulting in low compression efficiency. The radar data dynamic compression method proposed in this paper, that has a simple calculation formula and fully considers the characteristics of the radar signal to achieve efficient compression of radar data while improving the accuracy of the compressed data.

Keywords: Radar echo data · Radar signal characteristics · Dynamic compression

1 Introduction

With the continuous development of electronic information technology, the radar detection range and detection parameters continue to increase, and the data accuracy continues to improve, resulting in an extremely large amount of data generated during the continuous acquisition of radar echo signals [1]. In the application system, on the one hand, the radar data is required to be transmitted through the network in real-time and quickly, on the other hand, it is necessary to inverse and analyze the massive radar sampled data in real-time. If the huge data file is directly operated, the subsequent transmission and storage will be carried out difficulties [2]. Radar data compression can effectively reduce the overhead of data transmission and storage, reduce the transmission bandwidth, increase the transmission rate, and also improve the efficiency of data back-end processing [3]. Method [4] does not distinguish between the noise signal and the target signal when performing radar data compression, and compresses the received echo signal uniformly, which easily causes it to be difficult to distinguish between a strong noise signal and a small target signal after data compression, thereby improving False alarm rate and missed alarm rate. Method [5] uses a block adaptive vector quantization algorithm to compress radar data, which can effectively improve the compression accuracy, but it will increase the computational overhead and reduce the compression efficiency. In this paper, combining the characteristics of radar signals, different quantization methods are used to compress the target signal and noise signal, so as to achieve high-efficiency compression of radar data and improve the accuracy of the compressed data.

2 Data Dynamic Compression Method

Considering the characteristics of the noise signal and the target signal in the radar echo data, the dynamic compression method of radar data needs to ensure that the compression has a high enough accuracy when compressing the data, and it can effectively show the original nature of the data; at the same time, the dynamic data compression method is also necessary to correctly identify small targets and noise signals, and dynamically compress them into data of different amplitudes, to reduce the false alarm rate and missed alarm rate of the compressed data.

2.1 Data Dynamic Compression Model

In order to meet the requirements of dynamic data compression, this article considers different data compression for the noise signal and the target signal. At the same time, in order to ensure the accuracy of data compression, combined with the characteristics of the noise signal and the target signal in the radar data, the data is preprocessed for interception. The statistical parameters are used to obtain the intensity range of the noise signal, the upper limit of the noise signal threshold, and the maximum value of the target signal intensity to provide the required parameters for signal interception.

The specific implementation steps of the model are as follows:

(1) Input radar echo data X, and use statistical parameters to determine the intensity range of the noise signal, the upper limit of the noise signal threshold, and the maximum value of the target signal; (2) Determine whether the echo signal strength is greater than the maximum value of the noise signal strength, if the echo signal strength is greater than the maximum value of the noise strength, then the current echo X is judged as the target signal, otherwise, the echo signal X is judged as the noise signal; (3) If it is judged that X is a noise signal, it is further judged whether X is greater than the upper limit of the noise signal threshold, and if it is greater than the upper limit of the noise signal threshold, the value of X is set as the upper limit of the noise signal threshold; (4) Calculate the compressed noise signal value according to the linear proportional relationship; (5) If it is judged that X is the target signal, it is further judged whether X is greater than the maximum value of the target signal strength, and if it is greater than the maximum value of the target signal strength, it is set as the maximum value of the target signal strength; (6) Take a logarithmic operation on the

target signal X and calculate the slope coefficient of the linear compression of the target signal after taking the logarithm; (7) The linear equation is used to calculate the compressed target signal value.

2.2   Data Dynamic Compression Scheme

According to the radar data dynamic compression model, the data dynamic compression scheme is designed. This solution needs to obtain the parameters in the radar data first, so before the dynamic compression of the data, the parameter of input echo signal needs to be counted to obtain the amplitude range of the noise signal $[0, Noise_{max}]$, the upper limit of the noise signal threshold $Noise_{threshold}$ and the maximum target intensity $Target_{max}$.

When performing data dynamic compression, it is first necessary to compare the amplitude X_{amp} of the echo signal X with the maximum value $Noise_{max}$ of the noise signal. If the amplitude X_{amp} of the echo signal X is greater than $Noise_{max}$, then X is considered as the target, and less than or equal to $Noise_{max}$ is noise. The corresponding calculation formula is as follows:

$$X = \begin{cases} Noise & X_{amp} \leq Noise_{max} \\ Target & X_{amp} > Noise_{max} \end{cases} \tag{1}$$

Considering that the value after noise compression should not be too large, so as not to cause interference to the detection of small targets, so the maximum threshold value $Value_{noisemax}$ after noise compression is set. If it is judged that X is noise, the amplitude X_{amp} of X is further compared with the signal threshold upper limit $Noise_{threshold}$, and the amplitude of X greater than $Noise_{threshold}$ is set to $Noise_{threshold}$.

When performing data compression on the noise signal at $[0, Noise_{threshold}]$, the noise floor of the conversion should be considered to maintain the change characteristics of the noise, and the noise signal is compressed to range $[0, Value_{noisemax}]$ using the linear proportional relationship. The calculation formula is as follows:

$$data_{noise} = \frac{X_{noise}}{Noise_{threshold}} \cdot Value_{noisemax} \tag{2}$$

The range of the amplitude value of the original radar data is usually very large, and when the data is compressed, the excessively large data can be truncated appropriately to ensure the applicability of the algorithm. Therefore, the target intensity maximum value $Target_{max}$ is set, and if it is determined that X is the target signal, the signal greater than the target signal intensity maximum value $Target_{max}$ is further truncated.

On the other hand, the signal strength of different radar targets is usually very different. Simple linear compression is not suitable for target signal compression, and the square root operation can usually solve the problem of large data differences. Therefore, this paper adopts the n_c power operation of the target signal first, where n_c is obtained by statistics based on the characteristics of the current target signal. Then use linear equations to compress the target signal to obtain the final radar target signal dynamic compression range $(Value_{noisemax}, Value_{targetmax})$, the calculation formula is shown in the following formula 3, where 3 is the set maximum target signal compression, $Value_{targetmax}$ is the linear compression of the target signal slope coefficient.

$$data_{target} = slope_{tar} \cdot \left(\sqrt[n_c]{X_{target}} - \sqrt[n_c]{Noise_{max}} \right) + Value_{noisemax} \tag{3}$$

In summary, the function diagram corresponding to the radar data dynamic compression scheme proposed in this article is shown in Fig. 1:

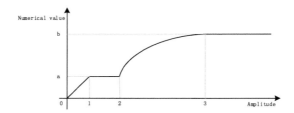

Fig. 1. Schematic diagram of radar echo data dynamic compression function curve.

Where a represents the maximum threshold value after noise compression $Value_{noisemax}$, b represents the maximum value of target signal compression $Value_{targetmax}$, 1 represents the upper limit of the noise signal threshold $Noise_{threshold}$, 2 represents the maximum value of the noise signal $Noise_{max}$, 3 represents The maximum value of target signal $Target_{max}$.

3   Simulation Experiment and Performance Analysis

Table 1 shows the comparison before and after the dynamic compression processing of part of the original radar data, including target data, noise data, and small target data respectively. Figure 2 is a schematic diagram of the original echo data waveforms received by some radars, and Fig. 3 is a schematic diagram of the radar echo data waveforms processed by the dynamic compression method proposed in this article.

From the results of the radar data value and the echo data waveform diagram, it can be seen that the target and noise can well maintain the characteristics of the noise signal

Table 1. Part of the dynamic compression radar raw data.

Number	Data processing	Radar data value					Type of data
1	Before compression	363726	553897	1161959	1784242	2029131	Target
	After compression	130	161	215	247	255	
2	Before compression	40293	26295	14526	15591	23329	Noise
	After compression	9	5	2	3	5	
3	Before compression	55364	64180	65639	64138	71453	Small target
	After compression	10	14	15	14	21	

and the target signal after the data is dynamically compressed, and the small target can be effectively distinguished from the noise after the dynamic compression, improve the detection rate of small targets, reduce the false alarm rate and missed alarm rate of compressed data.

Figure 4 is the echo image drawn after the raw radar data is dynamically compressed. Various objects can be clearly observed from the figure, and the distance and azimuth parameters of the target can be effectively displayed. At the same time, small target objects can be better displayed, ensure the detection accuracy of the radar.

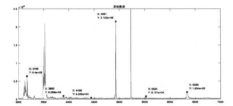

Fig. 2. Waveform diagram of raw echo data received by radar.

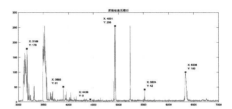

Fig. 3. Dynamic compression of radar echo data waveform diagram

Fig. 4. Dynamic compressed radar echo image.

4   Conclusion

This paper proposes a dynamic compression method for radar echo data. Compared with the traditional radar data compression method, the radar data compression method proposed in this paper has the advantages of simple calculation formula and high compression efficiency. At the same time, the dynamic compression method proposed in this paper fully considers the characteristics of the noise signal and the target signal strength value in the echo data and uses different methods to compress the noise signal and the target signal to reduce the false alarm rate and the missed alarm rate of the compressed data, improve the accuracy of compressed data.

Acknowledgements. This work is partially supported by Major Science and Technology Program of Hainan Province (ZDKJ201811), Hainan Provincial Natural Science Foundation of China(620RC563), the Science Project of Hainan University (KYQD(ZR)20021).

References

1. Tang, Y., Guo, P., Zhang, B., Yangyang, G., Ma, Y.: Current status and prospects of radar target echo simulation technology. Aerosp. Electron. Warfare **36**(06), 28–34 (2020)
2. Zhang, L.: Research on SAR Raw Data Compression Algorithm. National University of Defense Technology, Beijing (2010)
3. Xiao, K.: Research on Radar Real-Time Target Simulation Technology. University of Electronic Science and Technology of China, Beijing (2014)
4. Moureaux, J.M., Gauthier, P., Barlaud, M., et al.: Vector quantization of raw SAR data. In: Proceedings of ICASSP 1994, IEEE International Conference on Acoustics, Speech and Signal Processing, vol. 5, pp. V/189-V/192. IEEE, 1994:
5. Wang, H., Lou, X.: The application of BAVQ compression algorithm to SAR raw data compression——the method of selecting the best vector dimension. Sci. Technol. Eng. **9**(14), 4024–4026+4031 (2009)
6. Wang, Q.: Modulation Recognition Method of Low Signal-to-Noise Ratio and Low Intercept Radar Signal. University of Electronic Science and Technology of China, Beijing (2020)
7. Hu, C.: Research on High-Speed Three-Dimensional Imaging Lidar of Pulsed Semiconductor Laser. National University of Defense Technology, Beijing (2005)

Target Tracking Method of Shipborne Navigation Radar

Wentao Xia[1(✉)], Kongze Xing[1], and Zhen Guo[2]

[1] Sansha Hailanxin Ocean Information Technology Co., Ltd,
Sansha 573199, Hainan, China
xiawt@highlander.com.cn
[2] School of Computer Science and Cyberspace Security, Hainan University,
Haikou, China

Abstract. For ship maneuvering target tracking, this paper proposes a long and short wave gate tracking algorithm based on the mean filter. In this paper, the long and short wave gate tracking algorithm based on the mean filter is introduced firstly. The target tracking gate is created by using the ship prediction state, target historical trajectory information, and target prediction state. In the tracking gate, the target and echo are matched, and the target is updated and tracked. The experimental results show that the target tracking algorithm based on long and short wave gates proposed in this paper can effectively track the maneuvering target.

Keywords: Target tracking · Maneuvering target · Mean filter

1 Introduction

Shipborne radar, as the ship's eyes, plays an important role in the safe navigation of ships. Radar, as the main navigation device, is used to detect and track targets [1]; the clutter information in the echo is filtered according to the filtering algorithm to detect the target. After target detection is completed, the target needs to be measured and tracked. Radar target tracking is an important part of radar data processing [2]. After the Kalman filter theory was proposed in the field of maneuvering target tracking in the 1970s, many scholars devoted themselves to the problem of maneuvering target tracking. Above, many tracking algorithms such as the Kalman filtering algorithm and least square filtering algorithm are proposed [3].

This paper proposes to perform weighted mean filtering based on the target historical trajectory information to complete the estimation update of the predicted speed, matching speed, predicted distance, and azimuth, and achieve target prediction and matching.

2 Long and Short Wave Gate Tracking Algorithm Based on Mean Filtering

As shown in the flowchart of target tracking algorithm processing in Fig. 1, the main process of the shipborne radar long and short wave gate tracking algorithm is the prediction of the ship's state, the prediction of the target state, the creation of the gate, and the target matching.

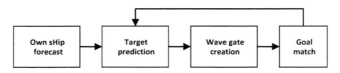

Fig. 1. Target tracking algorithm processing flowchart.

2.1 Target Prediction

Target prediction includes local ship information prediction and target information prediction. According to the prediction information and target historical trajectory information, the relative distance and azimuth of the target relative to the local ship are calculated. Local ship information prediction is to predict the next state of the ship according to the position, speed, and time of the ship output by the sensor. Target state prediction is to predict the state of the target according to the historical trajectory point of the target prediction.

As shown in the target prediction process in Fig. 2, calculate the predicted speed of the target, and calculate the target speed according to the weighted average of the target historical speed and the current measured speed.

Calculate the speed V_x of the target in the X-direction at the current time through the measured speed V_x of the target in the X-direction at the current time and the speed V_{x0} of the target in the X-direction in the previous frame.

$$V_x = \frac{(V_{x0} + 3 * V_{x1})}{4} \tag{1}$$

Calculate the speed V_y of the target in the X-direction at the current time through the measured speed V_{y1} of the target in the Y-direction at the current time and the speed V_{y0} of the target in the X-direction in the previous frame.

$$V_y = \frac{(V_{y0} + 3 * V_{y1})}{4} \tag{2}$$

The target X-direction predicted speed V_x and the Y-direction predicted speed V_y are synthesized into the target speed V.

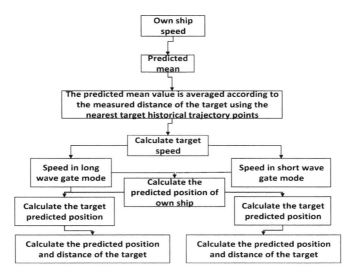

Fig. 2. Target prediction process.

$$V = \sqrt{V_x^2 + V_y^2} \tag{3}$$

2.2   Target Match

As shown in the target matching flow chart in Fig. 3, the target tracking gate is determined according to the predicted azimuth and distance of the target. In the tracking gate, the spot trace and the target are matched according to the track area and position deviation.

① Determine the starting point trace and ending point trace of the point trace search according to the gate position;
② Take the starting point trace as the starting point and the ending point trace as the endpoint to traverse, and calculate the area ratio between the point trace and the target;
③ If the area ratio of the target to the point trace is greater than 0.125 and less than 1, then calculate whether the distance between the target associated point trace and the point trace overlaps;
④ Calculate the distance deviation between the predicted target position and the search point trace, if the distance deviation meets the requirements, calculate the distance deviation between the target associated point trace and the search point trace, and then calculate whether the change of the target speed is within the change range;
⑤ After the above conditions are met, the three best matching points are determined.

Use the long and short wave gate target to predict the relative distance Lr, Sr relative azimuth Lb, Sb to establish the tracking gate, gate width, gate length, gate start distance, gate end distance, gate start azimuth, and gate end azimuth.

Long and short wave gate wave gate width Lw, Sw:

$$Lw = width * 2 + 50 \tag{4}$$

$$Sw = width * 2 + 50 \tag{5}$$

Where width is the measured width of the point trace.
Long and short wave gate width Ll, Sl:

$$Ll = length * 2 + 50 \tag{6}$$

$$Sl = length * 2 + 50 \tag{7}$$

Where length is the measured length of the point trace.
Long and short wave gate starting distance $LStartR$, $SStarR$:

$$LStartR = Lr - length \tag{8}$$

$$SStarR = Sr - length \tag{9}$$

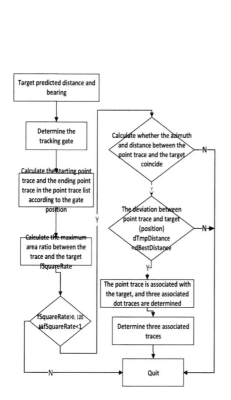

Fig. 3. Target matching flowchart.

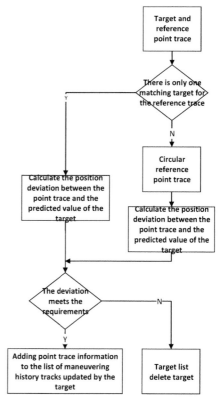

Fig. 4. Target matching flowchart.

After the tracking gate is determined, the point trace is matched with the target in the tracking gate. As shown in the target matching flowchart in Fig. 4, the target is matched based on the distance deviation between the point trace and the target as the main indicator.

① Judge the match between the dot and the target according to the deviation between the dot and trace information and the target predicted value;

② If there is only one matching target, whether the target has been updated is judged. If the target has not been updated, the matching is judged. If the matching meets the requirements, the trace information is added to the target. If not, the target is removed from the matching list;

③ If there are multiple matching targets, the matching list of targets is traversed to judge the matching between the trace information and the predicted value of the target. If the matching meets the requirements, the trace information will be added to the target, and if not, the target will be removed from the matching list;

④ After the trace information is added to the target, check and update the historical track point information of the target maneuver.

3   Experimental Verification

3.1   Experiment Platform

In the offshore area of Ningbo, Zhejiang, the X-band radar developed by the company is equipped on the ship, and the experimental speedboat performs uniform linear motion, turning, and S-curve motion within 5 nautical miles of the ship; the main experimental content is to test the variable gate matching tracking based on the mean filter. The algorithm is aimed at the tracking performance of speedboat targets.

3.2   Experimental Result

As shown in Table 1, the sea test scene in Ningbo, Zhejiang, the speedboat performs uniform linear, radial acceleration and deceleration, tangential acceleration and

Table 1. Zhejiang Ningbo offshore test scene.

Scenes	Description
2020–12–15 12–47–01	Radial linear acceleration + uniform linear motion, the target speed is about 20 kn
2020–12–15 13–01–13	Radial uniform acceleration linear motion + 180° steering, the target speed is about 20 kn
2020–12–15 13–14–44	Radial acceleration and deceleration linear motion, the target speed is about 20 kn
2020–12–15 13–28–14	Radial acceleration + uniform linear motion + steering, the target speed is about 20 kn
2020–12–15 13–37–09	180° steering + tangential uniform speed linear motion, the target speed is about 15 kn
2020–12–15 13–49–06	Tangential acceleration + uniform linear motion, the target speed is about 20 kn
2020–12–15 14–24–50	Radial S-curve movement, the target speed is about 10 kn
2020–12–15 14–54–09	Rotate 2 circles + turn movement, the target speed is about 10 kn

deceleration, uniform circular, curved, and high-speed motions within 5 nautical miles of the radar.

As shown in Table 2 Zhejiang Ningbo offshore test results, the target tracking stability is better based on the average filtering long and short wave gate matching algorithm.

Table 2. Zhejiang Ningbo offshore test results.

Scenes	Is it lost	Lost test times
2020–12–15 12–47–01	No	3
2020–12–15 13–01–13	No	3
2020–12–15 13–14–44	No	3
2020–12–15 13–28–14	No	3
2020–12–15 13–37–09	No	3
2020–12–15 13–49–06	No	3
2020–12–15 14–24–50	No	3
2020–12–15 14–54–09	No	3

4   Conclusion

1. For the uniform moving target, the long and short wave gate tracking method based on the mean filter is stable;
2. For the curvilinear moving target, the long and short wave gate tracking method based on the mean filter is stable;
3. For the moving target with uniform acceleration and deceleration, the target tracking is stable based on the mean filtering long and short wave gate tracking method.

Acknowledgements. This work is partially supported by Major Science and Technology Program of Hainan Province (ZDKJ201811), key research and development program of Hainan Province (ZDYF2020212), the Science Project of Hainan University (KYQD(ZR)20021).

References

1. He, Y., Guan, J., Peng, Y., et al.: Radar Automatic Detection and CFAR Processing. Tsinghua University Press, Beijing (1999)
2. He, Y., Tang, J.: Research on multi-radar integrated tracking algorithm. Res. Rep. Naval Acad. Aeronaut. Eng. (1991)
3. Zhou, H.: Research on correlation region in maneuvering multi-target tracking. Acta Aeronautica et Astronautica Sinica **03**, 296–304 (1984)

Research on Statistical Method of Sea Clutter Detection Curve of Navigation Radar

Jing Yang[1(✉)], Kongze Xing[1], and Zhen Guo[2]

[1] Sansha Hailanxin Ocean Information Technology Co., Ltd., Sansha 573199, Hainan, China
`yangj@highlander.com.cn`
[2] School of Computer Science and Cyberspace Security, Hainan University, Haikou, China

Abstract. The navigation radar sea clutter detection curve is obtained using a combination of statistical models and manual adjustment of the sea value and gain value. Through clutter graph statistics and the method of keeping the detection curve down, it can be obtained in real-time that is consistent with the sea clutter change at the time Basic sea clutter detection curve; by dividing the sectors and counting the maximum distance of sea clutter, the distance of the sea value in each sector can be determined, to match the changes of sea clutter in more detail; in the basic sea clutter based on the curve, by carefully manually adjusting the sea value and gain value, the noise and sea clutter can be effectively suppressed, and the effect of small target detection can be achieved.

Keywords: Sea clutter detection curve · Statistical model · Small target detection

1 Introduction

Navigation radar is a radar installed on a ship for navigation avoidance, ship positioning, and navigation in narrow waterways, also known as marine radar. The navigation radar provides the necessary means of observation for the navigator when the visibility is poor. Its appearance is a major milestone in the development of navigation technology. Navigation radar suppression of sea clutter is one of the key technologies for radar signal processing. Due to different sea conditions and sea clutter changes in sea areas, the processing difficulty of sea clutter is increased. Therefore, how to find a better sea clutter. The wave suppression method makes it extremely important to be able to detect small targets while suppressing most of the sea clutter.

2 Traditional Sea Clutter Detection Curve Model

The signal received by the navigation radar receiver generally includes the target's echo signal, the signal reflected by noise, and sea clutter. The signal reflected by the sea clutter is processed to minimize the false alarm rate. According to the radar equation, the power of the reflected signal from the normal target changes according to 1/R4,

J. C. Hung et al. (Eds.): FC 2021, LNEE 827, pp. 1782–1787, 2022.
https://doi.org/10.1007/978-981-16-8052-6_263

while the sea clutter belongs to the abnormal target, and its effective RCS (the area of the sea illuminated by the radar) changes with the distance. The RCS of the sea area illuminated by the radar is generally given by the following formula:

$$A_s = K \times \theta \times R \times \Delta R \tag{1}$$

In the formula, K is a constant, θ is the beamwidth of the antenna, R is the distance from the target to the radar, and ΔR is an increment. It can be seen that the irradiated area of the sea is proportional to the distance of R, which gives:

$$P = kP_t G^2 A_s / R^4 = kP_t G^2 R \sigma_s / R^4 = kP_t G^2 \sigma_s / R^3 \tag{2}$$

In the formula, P is the power of the sea clutter signal returned by the radar receiver, P_t is the known transmit power, G is the antenna gain, and σ_s is the factor in the radar equation. It can be seen that the change of sea clutter is inversely proportional to the third power of distance, and where there is no sea clutter beyond the radar line of sight, it is inversely proportional to the fourth power of distance. The radar line-of-sight depends on the installation height of the radar antenna, which is usually determined when the radar is installed.

In related technologies, radar usually needs to manually adjust the threshold gain of the signal based on the fixed R3 curve to suppress sea clutter. However, the amplitude distribution of sea clutter can be divided into Rayleigh distribution, Weibull distribution, normal logarithm distribution, and K distribution. The above-mentioned adjusted detection curve does not fit the sea clutter change trend, which will affect the detection of low, small, and slow targets.

3   Statistical Model of Sea Clutter Detection Curve

In related technologies, when processing sea clutter signals received by radars, radar equation calculations are generally used to determine the sea clutter change curve, which makes the curve less adaptable and cannot be applied to different sea areas and different sea conditions. Similarly, the amplitude distribution of sea clutter can be divided into Rayleigh distribution, Weibull distribution, normal logarithmic distribution, and K distribution. The characteristics of sea clutter under each distribution are quite different. This fixed curve corresponds to the change of sea clutter. The low degree of trend fit will affect the detection of low, small, and slow targets. The sea clutter detection curve based on the statistical model can match the changing trend of the sea clutter. By manually adjusting the sea clutter gain value, most of the sea clutter can be effectively suppressed, and a radar image with better echo quality can be obtained.

The establishment process of the statistical model of the sea clutter curve of the shipborne radar is as follows (Fig. 1):

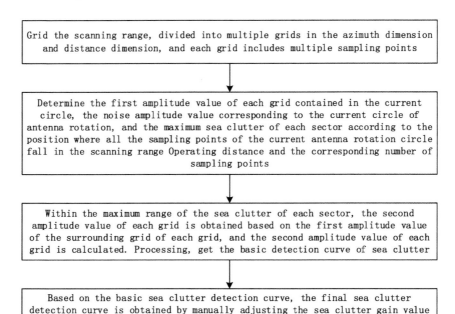

Fig. 1. Process of establishing the statistical model of sea clutter curve for shipborne radar.

3.1   Sea Clutter Distance Statistics

The scanning range is divided into multiple azimuth units in the azimuth dimension, and each azimuth unit is divided into multiple distance units in the distance dimension. According to the position where the current circle of the antenna rotates and falls in the scanning range, it is determined that all the current circle contains The reference amplitude value of the grid, where the reference amplitude value of each grid is obtained by averaging the amplitude values of all sampling points in the grid.

The reference amplitude value of each grid is weighted to obtain the first amplitude value of each grid, $y_{m,n}(i) = a * x_{m,n}(i) + b * y_{m,n}(i-1)$, where a and b represent the weighting coefficient respectively. $x_{m,n}(i)$ represents the reference amplitude value of the grid with the n-th distance unit of the m-th azimuth unit of the i-th turn of the antenna, $y_{m,n}(i)$ represents the first amplitude value of the grid with the n-th distance unit of the m-th azimuth unit of the i-th turn of the antenna, and $y_{m,n}(i-1)$ represents the antenna rotation rotate to the first amplitude value of the grid of the m-th azimuth unit and the n-th distance unit of the $(i-1)$-th turn of the antenna.

The average value of the amplitude of all sampling points in each distance ring is obtained. The minimum value of the amplitude average of all distance rings is taken as the noise reference amplitude value of the current circle of antenna rotation. The noise reference amplitude value is weighted to obtain the noise amplitude value, $Noise(i) = c * x(i) + d * Noise(i-1)$. Among them, c and d are respectively weighted coefficients, $x(i)$ represents the noise reference amplitude value of the i-th turn of the antenna, $Noise(i)$ represents the noise amplitude value of the i-th turn of the antenna, $Noise(i-1)$ represents the noise amplitude value of the $(i-1)$-th turn of the antenna.

For each sector, the first amplitude value of each grid in the range dimension is compared with the noise amplitude value, and when the first amplitude value of a plurality of continuous grids is less than the noise amplitude value, the distance of the first grid in the plurality of continuous grids is taken as the maximum operating distance of sea clutter.

The first amplitude values of four grids around each grid are weighted to obtain the second amplitude value of each grid, $z_{m,n}(i) = \frac{z_{m,n-1}(i) + z_{m,n+1}(i) + z_{m-1,n}(i) + z_{m+1,n}(i)}{4}$, where $z_{m,n}(i)$ represents the second amplitude value of the current grid, $z_{m-1,n}(i)$ represents the first amplitude value of the grid on the current grid, $z_{m+1,n}(i)$ represents the first amplitude value of the grid under the current grid, $z_{m,n-1}(i)$ represents the first amplitude value of the grid before the current grid, and $z_{m,n+1}(i)$ represents the first amplitude value of the grid after the current grid The first amplitude value of the lattice.

The reference sampling points within the maximum operating range of sea clutter in each sector are weighted to determine the sampling points within the maximum operating range of sea clutter, $SeaRSample_p(i) = e * x_p(i) + f * SeaRSample_p(i-1)$, where e and f represent the weighting coefficient respectively, $x_p(i)$ represents the number of reference samples within the maximum operating range of the sea clutter in the sector p of the i-th turn of the antenna, $SeaRSample_p(i)$ represents the number of samples within the maximum operating range of the sea clutter in the sector p of the i-th turn of the antenna, and $SeaRSample_p(i-1)$ represents the number of samples within the maximum operating range of the sea clutter in the sector p of the $(i-1)$-th turn of the antenna.

3.2   Sea Clutter Detection Curve Statistics

The second amplitude value of each grid is broken line processing. In broken line processing, if the second amplitude value of the current grid is greater than the second amplitude value of the previous grid, the slope of the broken line between the two grids is 0, and the basic detection curve of sea clutter with a downward trend is obtained, where $BasicCurve_p(j)$ represents the amplitude value of the basic detection curve of sea clutter at the j sampling point of the p sector.

By manually adjusting the sea clutter gain value, the basic detection curve of each sector is adjusted to determine the detection threshold value of each sampling point within the maximum operating range of sea clutter. By manually adjusting the sea clutter gain value, the extended sampling points within the range of the maximum operating range of sea clutter to the operating range of sea clutter edge are obtained, and by manually adjusting the sea clutter gain value and fixed gain value, determine the detection threshold value of each extended sampling point within the range of maximum operating distance of sea clutter to edge operating distance of sea clutter, synthesize three curves within the range of maximum operating distance of sea clutter, the range of maximum operating distance of sea clutter to edge operating distance of sea clutter, and the range of edge operating distance of sea clutter to the preset range to obtain preset value for the first detection curve within the preset range, the detection threshold value from the start sampling point to the corresponding sampling point of the preset range is set as a fixed gain value to obtain the second detection curve within the preset range.

For each sampling point, the maximum value of the detection threshold value in the first detection curve and the detection threshold value in the second detection curve corresponding to the sampling point are taken as the detection threshold value of the sampling point.

The detection threshold value of each sampling point is smoothed, and the final detection curve of sea clutter is obtained.

4   Statistical Model Application

The statistical model is applied to test the shipborne echo data. The original echo data of the shipborne radar is shown in the Fig. 2, 3 and 4:

Fig. 2. Shipborne radar raw echo data.

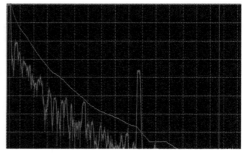

Fig. 3. Curve extraction.

The scan line data and sea clutter detection curve of a certain azimuth is extracted and displayed as shown in the figure below. From the figure, it can be seen that the changing trend of the sea clutter detection curve and sea clutter is consistent. By manually adjusting the gain value of sea clutter, small targets can be better detected and sea clutter can be suppressed.

The processed echo image is shown in the figure below. The green rectangular box describes the detection of small targets and the suppression of sea clutter. The rectangular box near the center of the circle indicates that the suppression of sea clutter in close range is relatively complete. The other two rectangular boxes describe the detection of small targets, especially the buoy small targets in the large rectangular box, which can be completely preserved.

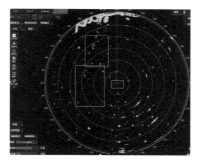

Fig. 4. Echo image.

5   Conclusion

The distribution of sea clutter is different in different sea areas and sea conditions, so it is difficult to describe the actual amplitude distribution characteristics of clutter with an amplitude distribution. The detection curve obtained by the invention is the result obtained by real-time weighted calculation of multiple frames, to reflect the average trend of actual sea clutter amplitude varying with distance.

Compared with the prior art, the method of the invention can make the adjusted curve have strong adaptability, and can be applied to different sea areas and different sea conditions. For different sea clutter amplitude distributions, such as Rayleigh distribution, Weibull distribution, normal logarithm distribution, and K distribution, the curve adjusted by the method of the invention is similar to the sea clutter change law, keeps the curve descending, can be consistent with the sea clutter real-time change trend, and can be applied to the detection of low, small and slow targets.

Acknowledgements. This work is partially supported by Major Science and Technology Program of Hainan Province (ZDKJ201811), key research and development program of Hainan Province (ZDYF2020212), the Science Project of Hainan University (KYQD(ZR)20021).

References

1. Huang, P.: Radar Target Characteristic Signal. China Aerospace Press, Beijing (1998)
2. Ding, L.: Radar Principle. Xidian University, Xi'an (1997)
3. Cai, X.: Introduction to Radar System. Science Press, Beijing (1982)

Ship Target Tracking Method Based on Mean α-β Filtering

Wentao Xia[1(✉)], Zhiming Dai[2], and Zhen Guo[3]

[1] Sansha Hailanxin Ocean Information Technology Co., Ltd.,
Sansha 573199, Hainan, China
xiawt@highlander.com.cn
[2] Sanya Hailan Huanyu Marine Information Technology Co., Ltd.,
Sanya 572025, Hainan, China
[3] School of Computer Science and Cyberspace Security,
Hainan University, Haikou, China

Abstract. Aiming at the traditional α-β filtering algorithm, this paper proposes a ship target tracking algorithm based on the mean α-β filtering. The paper first introduces the traditional α-β filtering and the ship target tracking algorithm based on the average α-β filtering. According to the historical trajectory information of the ship, the average filtering is used to correct the target measurement value, and the corrected target measurement value is input for target filtering and prediction; it is verified that the ship target tracking algorithm based on mean α-β filtering proposed in this paper can effectively reduce the impact of sudden changes in target information.

Keywords: Target tracking · α-β filtering · Mean filtering

1 Introduction

With the development of the shipping industry, marine ships are developing rapidly in the direction of technology, density, and speed, which puts forward higher requirements for target detection and tracking. Radar is one of the important navigation equipment for navigation. It is used to detect ships, islands, icebergs, fishing nets, buoys, and other targets, and provide obstacle information to sailors in time to avoid obstacles and ensure the safe navigation of ships [1].

Radar target tracking is a very important research topic in the process of radar data processing. The difficulty of target tracking lies in the problem of multiple targets and maneuvering targets [1]. In the 1970s, after the Kalman filter theory was proposed in the field of maneuvering target tracking, many scholars devoted themselves to the problem of maneuvering target tracking, and then proposed many tracking algorithms such as the Kalman filter algorithm and least square filter algorithm [2].

J. C. Hung et al. (Eds.): FC 2021, LNEE 827, pp. 1788–1794, 2022.
https://doi.org/10.1007/978-981-16-8052-6_264

2   Tracking Algorithm Based on Mean α-β Filtering

2.1   Traditional α-β Filtering Algorithm

The target filtering algorithm refers to the process of mixing error data in the target data and using the filtering algorithm to remove the target doped clutter data as much as possible to restore the real target data [3]. The traditional α-β filtering algorithm is mainly composed of the creation of the target motion equation, the filtering estimation equation, and the prediction equation.

① Target motion equation

The α-β filtering algorithm is based on a uniform linear motion model, and its motion equation is:

$$x_n = x_{n-1} + v_{n-1} * T \tag{1}$$

In the formula, x_n is the coordinates of the target at time n, x_{n-1} is the coordinates of the target at time $n - 1$, v_{n-1} is the speed of the target at time $n - 1$, and T is the radar sampling period.

② Filter equation

Position filter estimation equation:

$$\widehat{x}_n = \widehat{x}_{n-1} + \alpha_n [x_n - \widehat{x}_{n-1}] \tag{2}$$

$$\alpha_n = \frac{4n - 2}{n^2 + n} \tag{3}$$

Where v_n is the position measurement value at time n, $\widehat{v}_n$ is the position prediction value at time n, v_{n-1} is the position measurement value at time $n - 1$, $\widehat{x}_{n-1}$ is the position prediction value at time $n - 1$, and α_n is the position filter parameter.

Speed filter estimation equation:

$$\widehat{v}_n = \widehat{v}_{n-1} + \frac{\beta_n}{T} [v_n - \widehat{v}_{n-1}] \tag{4}$$

$$\beta_n = \frac{6}{n^2 + n} \tag{5}$$

In the formula, v_n is the speed measurement value at time n, $\widehat{v}_n$ is the speed prediction value at time n, v_{n-1} is the speed measurement value at time $n - 1$, $\widehat{v}_{n-1}$ is the speed prediction value at time $n - 1$, and β_n is the speed filter parameter.

③ Prediction equation

Position prediction equation:

$$\hat{x}_{n+1} = x_n + \hat{v}_n * T \tag{6}$$

Speed prediction equation:

$$\hat{v}_{n+1} = \hat{v}_n \tag{7}$$

④ Filtering algorithm flow

a) Save position prediction value $\hat{x}_{n-1}$ and speed prediction value $\hat{v}_{n-1}$ at time $n-1$;
b) Calculate position filter parameter α_n and speed filter parameter β_n at time n;
c) Calculate position filtering estimation $\hat{x}_n$ and speed filtering estimation $\hat{v}_n$ at time n;
d) Calculate and save position prediction value $\hat{x}_{n+1}$ and speed prediction value $\hat{v}_{n+1}$ at time $n+1$;

2.2   A-β Filtering Algorithm Based on Mean Value

The traditional α-β filtering algorithm is a constant gain filtering method based on uniform linear motion. Because of its small amount of computation, short operation time, and high real-time performance, it has been successfully used in the design of filters in various systems. In the measurement process of radar target information, the instability of echo results in the jump of target measurement information. In this case, the traditional α-β filtering algorithm cannot accurately predict and estimate, resulting in the loss of target tracking.

In view of this problem, a new algorithm based on the mean α-β filtering is proposed. As shown in the flow chart of the process of the algorithm, the historical track information and current measurement values of the target are updated by mean filtering, and the updated measurement value filtering estimation, target update, and target prediction are used (Fig. 1).

Target Filtering
Target filtering is the process of predicting the current position and speed by using the position and speed prediction information of the previous time, the filtering parameters of the current position and speed, and the correction value of the current position and speed.

As shown below, based on the traditional α-β filtering algorithm, the position and speed information of the target correction are used for filtering estimation; the position $\hat{x}_n$ and speed $\hat{v}_n$ at time n are estimated according to the position filtering parameter α, speed filtering parameter β, target correction position $\dot{x}_n$ and target correction speed $\dot{v}_n$ at time n, and the filter estimation position $\hat{x}_{n-1}$ and filter estimation speed $\hat{v}_{n-1}$ at time n−1.

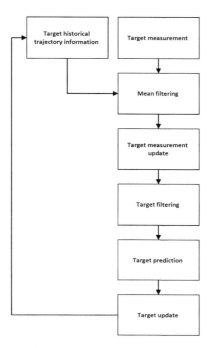

Fig. 1. Processing flow chart of tracking algorithm based on mean α-β filtering.

The estimation equation of the position filter is as follows:

$$\widehat{x}_n = \widehat{x}_{n-1} + \alpha_n[\dot{x}_n - \widehat{x}_{n-1}] \tag{8}$$

$$\alpha_n = \frac{4n - 2}{n^2 + n} \tag{9}$$

Where $\dot{x}_n$ is the position correction value at time n, $\widehat{x}_n$ is the position prediction value at time n, x_{n-1} is the position measurement value at time $n - 1$, $\widehat{x}_{n-1}$ is the position prediction value at time $n - 1$, and α_n is the position filtering parameter.

The estimation equation of the speed filter is as follows:

$$\widehat{v}_n = \widehat{v}_{n-1} + \frac{\beta_n}{T}[\dot{v}_n - \widehat{v}_{n-1}] \tag{10}$$

$$\beta_n = \frac{6}{n^2 + n} \tag{11}$$

Where $\dot{v}_n$ is the speed correction value at time n, $\widehat{v}_n$ is the speed prediction value at time n, v_{n-1} is the speed measurement value at time $n - 1$, and $\widehat{v}_{n-1}$ is the speed prediction value at time $n - 1$, β_n is the speed filter parameter.

Target Prediction

The traditional α-β filter is based on the motion model of the uniform linear motion. The target filter is used to obtain the prediction estimates of the target position and speed at the current time. According to the predicted estimation value, the position and speed of the target at the next time are predicted and used for the next time of the target filtering.

Position prediction equation:

$$\widehat{x}_{n+1} = \widehat{x}_n + \widehat{v}_n * T \tag{12}$$

Speed prediction equation:

$$\widehat{v}_{n+1} = \widehat{v}_n \tag{13}$$

3   Experimental Verification

3.1   Experimental Platform

This experiment is carried out on a real ship to verify the effectiveness of the algorithm. The radar is installed on the real site, real-time detection and processing of real ship data by the radar, and finally real-time printing of traditional data by the background α-β filtering algorithm and mean based algorithm α-β The ship speed information of filtering algorithm is compared with ship AIS data.

3.2   Experimental Result

The experimental site is on zhucha island. The main content of the experiment is the comparative analysis of the prediction value of the target speed based on the traditional α-β filtering algorithm and the average α-β filtering algorithm for the uniform linear moving target, as shown in Fig. 2, 3, and 4 are the predicted values of 0-5kn, 5-10kn and 10-15kn ship targets based on traditional α-β filtering algorithm and average α-β filtering algorithm respectively. Blue represents the predicted value of target speed based on traditional α-β filtering algorithm, and red represents the predicted value of target speed based on average α-β filtering algorithm. According to the comparative analysis, the average α-β filtering algorithm can smooth the predicted value of target speed to improve the performance of target tracking, the mutation value in target information is removed.

Fig. 2. 0–5 kn ship target.

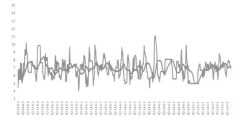

Fig. 3. 0–5 kn ship target.

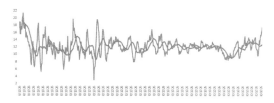

Fig. 4. 0–5 kn ship target.

4   Conclusion

The tradition α-β filtering algorithm cannot effectively remove the mutation of target information. The mean α-β filtering algorithm can remove the mutation value in the target information and improve the target tracking performance. For the time being, the research content of this paper focuses on the uniform linear motion, and the influence on the maneuvering target needs further research and experiment.

Acknowledgements. This work is partially supported by Major Science and Technology Program of Hainan Province (ZDKJ201811), key research and development program of Hainan Province (ZDYF2020212), the Science Project of Hainan University (KYQD(ZR)20021).

References

1. Jiang, X.: Research on Filtering Algorithm of Radar Target Tracking. Dalian Maritime University, Dalian (2008)
2. Lu, W.: Research on Moving Target Tracking Technology of Ship Navigation Radar. Dalian Maritime University, Dalian (2009)
3. Hong, L., Li, F., Zhao, Y.: An Adaptive α-β Filtering Algorithm for Maneuvering Target Tracking, pp. 278–291. Radar Science and Technology (2007)

Research on Radar Target Detection Method Based on the Combination of MSER and Deep Learning

Guanping Fang[1(✉)], Wei Wang[1], Zhiheng Wei[1], and Jun Ye[2]

[1] Sansha Hai Lanxin Marine Information Technology Co., Ltd., Sansha, China
fanggp@highlander.com.cn
[2] School of Computer Science and Cyberspace Security, Hainan University, Haikou, China

Abstract. Aiming at the problems that arine radar cannot automatically output targets and is difficult to deal with false echoes, this paper adopts the method combining MSER and deep learning to detect radar echoes based on the theory of radar echo image processing. Through the off-line radar echo data and the real-time test at the radar station, the method can automatically output radar targets, and the detection rate is high. Therefore, this method can be used as a reference for radar target detection algorithm research and engineering application.

Keywords: Radar echo · MSER · Deep learning · Target detection · Contour extraction

1 Introduction

Marine radar is divided into navigation radar and traffic management radar, its main function is to achieve target detection and target tracking. Navigation radar is mainly to achieve navigation and collision avoidance functions. For ship pilots, radar is equivalent to their "eyes". Shore-based radars are generally installed at ports or beaches to monitor and assist safe entry of ships. Radar has all-weather fairness. According to IEC62388, X-band (9.2–9.5 GHz) and S-band (2.9–3.1 GHz) are mainly used in the field of navigation. At present, marine radar uses traditional signal processing methods to achieve target detection which requires manual adjustment of parameters such as thresholds to obtain object information. There are cases where clutter is detected as a target or small targets are filtered out as a result of the difference of the way of each person's adjustment. It affects the accuracy of radar target detection leading to inaccurate provision of accurate and reliable target information for users [1]. Therefore, the realization of automatic detection of radar targets has important practical significance.

The function of radar automatic target detection can be implemented according to the existing engineering design requirements, without changing the equipment hardware, and ensuring that the data processing time is less than the antenna rotation period. This paper takes a method that combines pre-detection and deep learning which is described in Fig. 1 below:

© The Author(s), under exclusive license to Springer Nature Singapore Pte Ltd. 2022
J. C. Hung et al. (Eds.): FC 2021, LNEE 827, pp. 1795–1800, 2022.
https://doi.org/10.1007/978-981-16-8052-6_265

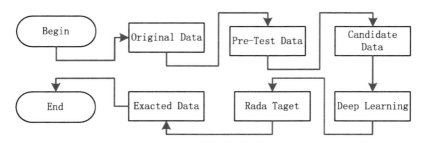

Fig. 1. Data processing flow chart

As shown in Fig. 1, this paper first pre-detects the original radar echo data to obtain the candidate area, then uses deep learning to detect the candidate target to obtain the target, and finally extract the contour of the target.

2   Pre-detection

Maximally Stable Extremal Regions (MSER) was proposed by J. Matas in 2004 based on the watershed concept [2]. The basic idea of MSER is to select a series of thresholds from 0 to 255 to perform binary segmentation of the image and the gray value below the set threshold is 0, but higher than or equal to the set is 255. The image will form a closed region as the threshold increases from 0 to 255. The maximum extremal region is the region with the smallest area change within a certain threshold range [3]. This paper uses method of MSER as pre-detection to obtain candidate target regions.

The mathematical definition of MSER is [4]:

$$q(i) = |Qi + \Delta - Qi - \Delta| / Qi \tag{1}$$

where Qi is the connected region when the threshold is i, Δ is the variation of gray threshold value and q(i) is the rate of change of Qi when the threshold is i. When q(i) is the local limit value, Qi is the maximally stable extremal region.

In addition, this paper separately counts the echo sizes of S-band and X-band long, medium and short radar pulses corresponding to different ships. The radar echo size is limited to: width range [3, 200], height range [3, 60]. The pre-detection targets can be further screened by comparing the stable region with the echo size.

3   Deep Learning

Deep Learning refers to an algorithm that uses a larger number of layers of neural network structures to complete training and prediction. It has been widely used in machine vision such as pedestrian detection, face detection, remote sensing image detection, and medical image processing. This paper adopts the ResNet (Residual Network) network, which is a residual school framework proposed by K. He et al. [5] (Fig. 2).

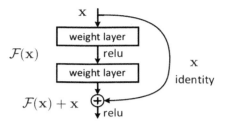

Fig. 2. Residual mapping for ResNet

The residual structure can be simplified as follows [6]:

$$x_{l+1} = x_l + F(x_l, W_l) \tag{2}$$

The expression of the L feature of any layer unit is:

$$x_L = x_l + \sum_{i=l}^{L-1} F(x_i, W_i) \tag{3}$$

If ξ represents the loss function according to the chain rule, we can get:

$$\frac{\partial \xi}{\partial x_l} = \frac{\partial \xi}{\partial x_L} \frac{\partial x_L}{\partial x_l} = \frac{\partial \xi}{\partial x_L} \left(1 + \frac{\partial}{\partial x_l} \sum_{i=l}^{L-1} F(x_i, \omega_i)\right) \tag{4}$$

Based on the above and open source related theories, the dataset used in this paper comes from the original echo data recorded by navigation radar and shore-based radar, then the sample data is labeled and trained. Among them, the sample labeling tool is labelImg, the training environment: CPU is Intel(R) Core(TM) i7 8200X, GPU is GeForce RTX2080Ti, operating system is Ubuntu18.04 LTS, Cuda10.2, and finally the network model is trained with.caffemodel file.

4   Experiments and Results

Based on the above theoretical discussion, the data processed by MSER is used as the input of deep learning ResNet, and then the output radar target is detected. Finally, the offline radar echo data and real radar sites are used to conduct experiments to statistically analyze the number of ship targets within 8nm, target detection rate, false detection rate, and missed detection rate. The method described above is used to process the original radar echo data, and select 8 different radar scene data for testing and analysis. The details are shown in Table 1 below.

Fig. 3. Zhuhai Dangantou radar station experiment

Figure 3 shows the results of the actual measurement at the Zhuhai clubhead radar site, which draws a rectangular frame on the original radar echo image to represent the detected target.

Table 1. Statistical table of real scene test results

No	Data	Target	Detection	Missed	False	Missed detection rate	False detection rate	detection rate
						Missed/Target	False/Target	Detection/Target
1	Scene 1	87	86	1	4	1.15%	4.60%	98.85%
2	Scene 2	36	34	2	7	5.56%	19.44%	94.44%
3	Scene 3	83	77	6	1	7.23%	1.20%	92.77%
4	Scene 4	53	51	2	0	3.77%	0.00%	96.23%
5	Scene 5	62	56	6	9	9.68%	14.52%	90.32%
6	Scene 6	62	61	1	11	1.61%	17.74%	98.39%
7	Zhuhai Dan club head	119	114	5	1	4.20%	0.84%	95.80%
8	Zhuhai Hebao Island	120	111	9	1	7.50%	0.83%	92.50%

As shown in Table 1, scenes 1 to 6 are the echo data scenes recorded by the navigation radar, and scenes 7 to 8 are the measured scenes at the Zhuhai Dangantou and Zhuhai Hebao Island sites. In Table 1, the number of targets is the number of ships and other targets within 8 nautical miles, the number of detected targets is the number of targets detected by the algorithm processing, the number of missed detections is the number of targets not detected by the algorithm, and the number of false detections refers to the number of false detections of clutter as the target by the algorithm. This method has a certain missed detection rate and false detection rate, but its detection rate is above 90%.

5   Target-Based Contour Extraction

For the radar target detected by MSER+ResNet, use OpcnCV to extract the contour of the radar target of which OpcnCV (Open Source Computer Vision Library) is an open source computer vision library [7]. This paper uses the findContours() function in OpenCV to extract the contours in the binary image, and then uses the drawContours() function to draw the radar target echo on the image attributes, and fill it with a custom color, as shown in Fig. 4 below.

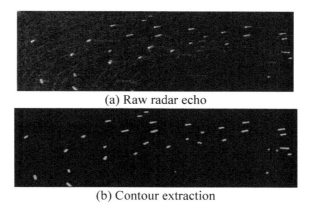

(a) Raw radar echo

(b) Contour extraction

Fig. 4.   Radar raw echo and Contour

As shown in Fig. 4, Fig. 4(a) is the original echo image, and Fig. 4(b) is the target echo image after contour extraction. It can be seen from this that if a high detection rate is ensured, this method has clutter suppression and can improve the visualization effect.

6   Conclusion

Based on image processing related theories, this paper uses a combination of MSER and ResNet to process radar echo data, and tests on offline data and radar sites respectively. In a radar antenna rotation period, proposed method can effectively process a frame of echo data, the detection rate is above 90%, and the radar target is automatically output. And extract the contour based on the radar target echo and customize the fill color to increase the visualization effect of the radar echo. It has certain reference significance to the research and engineering application of radar target detection algorithm.

Acknowledgements. This work is partially supported by Major Science and Technology Program of Hainan Province (ZDKJ201811), Hainan Provincial Natural Science Foundation of China (620RC563), the Science Project of Hainan University (KYQD (ZR)20021).

References

1. Fang, G.P.: A target detection method and device. China Patent: 202011586833.2. Accessed 29 Dec 2020
2. Matas, J., Chum, O., Urban, M., et al.: Robust wide-baseline stereo from maximally stable extremal regions. Image Vis. Comput. **22**(10), 761–767 (2004)
3. Ding, W.R., Kang, C.B., Li, H.G., et al.: Extraction of UAV image building region based on MSER. J. Beijing Univ. Aeronaut. Astronaut. **41**(3), 383–390 (2015)
4. Zhou, P.F.: Research on Text Detection and Recognition Technology in Natural Scene Images. Xi'an University of Technology, Xi'an (2019)
5. He, K., Zhang, X., Ren, S., et al.: Deep residual learning for image recognition. In: 2016 IEEE Conference on Computer Vision and Pattern Recognition (CVPR), pp. 770–778. IEEE (2016-Decem)
6. Zhang, X.X.: Research on recognition of oil spill area on sea surface in SAR image based on ResNet. Dalian Maritime University, Dalian (2020)
7. Mao, X.Y., Leng, X.F., et al.: Introduction to Opencv3 Programming. Electronic Industry Press, Beijing (2014)

IF Signal Processing Based on Matlab

Jubin Chen[1(✉)], Tao Zhai[1], and Zhen Guo[2]

[1] Sansha Hai Lanxin Marine Information Technology Co., Ltd., Sansha, China
chenjb@highlander.com.cn
[2] School of Computer Science and Cyberspace Security, Hainan University,
Haikou, China

Abstract. In the process of radar design, the way of digital signal processing determines the performance of the whole machine. The characteristics of LFM signal are analyzed, and the advantages of digital quadrature phase demodulation are explained. The principle of pulse compression is discussed, and the difference between time-domain pulse compression and frequency-domain pulse compression is simulated.

Keywords: Linear frequency modulation signal · Digital quadrature demodulation · Pulse compression

1 Introduction

In order to coexist the range and range resolution of the radar, the frequency or phase of the signal is modulated at the transmitting end to increase the signal bandwidth, and a large time-width signal is transmitted to ensure the radar range. On the receiving side, the received wide pulse signal is pulse-compressed through a matched filter to obtain a narrow pulse signal, which improves the range resolution, so that the radar system takes into account the range of action and range resolution**Error! Reference source not found.**.

2 Analysis of Radar Echo Signal (LFM)

The main advantage of the linear frequency modulation (LFM) signal is that when the target speed is unknown, the pulse compression of the echo signal can still be achieved with a matched filter.

The mathematical expression of the LFM Signal is:

$$s(t) = rect(\frac{t}{T})e^{j2\pi(f_c t + \frac{k}{2}t^2)}$$

In the above formula, f_c is the carrier frequency, $rect(\frac{t}{T})$ is the rectangular signal, $K = \frac{B}{T}$ is the frequency modulation slope, and the instantaneous frequency of the signal is $f_c + Kt(-T/2 \leq t \leq T/2)$, as shown in Fig. 1.

© The Author(s), under exclusive license to Springer Nature Singapore Pte Ltd. 2022
J. C. Hung et al. (Eds.): FC 2021, LNEE 827, pp. 1801–1804, 2022.
https://doi.org/10.1007/978-981-16-8052-6_266

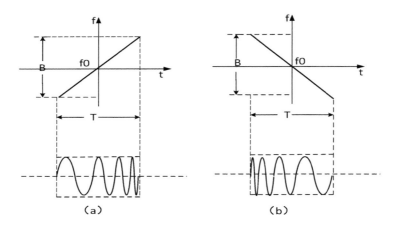

Fig. 1. The signal waveform of LFM (a) K > 0 (b) K < 0

When TB > 1, the characteristic expression of the LFM signal is as follows:

$$S(t) = rect(\frac{t}{T})e^{j\pi Kt^2}$$

s(t) is the complex envelope of signal s(t). According to the properties of Fourier transform, *s(t)* and s(t) have the same amplitude-frequency characteristics, only the center frequency is different [1]. Therefore, when using MATLAB for simulation, only *s(t)* needs to be considered.

3   Pulse Compression Processing

3.1   Time Domain Pulse Compression Processing

According to the matched filter theory, the unit impulse response $h(n)$ of the pulse compression filter is the conjugate of the mirror image of the input signal $x(n)$, i.e.

$$h(n) = x^*(N - 1 - n)\,(0 \leq n \leq N - 1)$$

Then, the output $x_o(n)$ of the pulse compression filter is the convolution of the input signal $x(n)$ and the impulse response h(n) of the filter, as follows

$$s_o(n) = s(n) * h(n) = \sum_{k=0}^{N-1} s(k)h(n - k) = \sum_{k=0}^{N-1} h(k)s(n - k)$$

$$\sum_{k=0}^{N-1} s^*(N - 1 - k)s(n - k)\,(0 \leq n \leq N - 1)$$

The procedure is as follows:
ht = conj(fliplr(st));
s1 = conv(st,ht);

3.2   Frequency-Domain Pulse Compression Processing

As shown in the target matching flow chart in Fig. 2, the target tracking gate is determined according to the predicted azimuth and distance of the target. In the tracking gate, the spot trace and the target are matched according to the track area and position deviation.

In frequency domain pulse compression processing, the Fourier transform of the input signal is used to obtain the signal spectrum X(k), and then X(k) is multiplied with the frequency response H(k) of the matched filter. Finally, the inverse transformation is carried out to obtain the pulse compression result Y(n), as shown below.

$$y(n) = y_I(n) + jy_Q(n) = IFFT[X(k)H(k)]$$

First calculate the frequency response H(k) of the matched filter and store it in ROM. If you need to use a window function to suppress sidelobes, you only need to store the product of the matched filter frequency response H(k) and the spectral characteristics of the window function in the ROM, and no extra memory is needed.

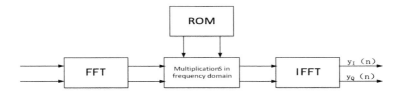

Fig. 2. Schematic diagram of frequency domain digital pulse compression processing structure

The procedure is as follows:

$$YN = fftshift(ifft(fft(In1 + j * Qn1). * HK));$$

Time-domain and frequency-domain pulse compression are compared as shown in Fig. 3.

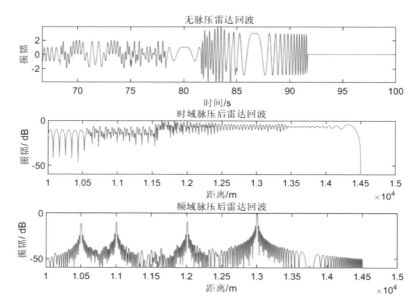

Fig. 3. Comparison of pulse compression in time domain and frequency domain

4 Conclusion

This paper introduces the basic knowledge of IF signal processing of radar, simulates the difference between time-domain compression and frequency-domain compression, and further understands the properties of matched filter by adjusting the output signal-to-noise ratio.

Acknowledgements. This work is partially supported by Major Science and Technology Program of Hainan Province (ZDKJ201811), key research and development program of Hainan Province (ZDYF2020212), the Science Project of Hainan University (KYQD(ZR)20021).

References

1. Chen, B.: Analysis and Design of Modern Radar System. Xidian University Press, Xi'an, 9 (2019)
2. Wu, S., Mei, X.: Radar Signal Processing and Data Processing Technology. Electronic Industry Press, 2 (2008)
3. Yan, W.: Research and Design of Digital Down Converter and Pulse Compression, Xidian University (2014)
4. Liu, L.: Design of Software Radar Simulation Platform Based on MATLAB. Lanzhou University (2008)

An Optimization Scheme for IoT Data Storage Based on Machine Learning

Yin Zhang[✉], Kejie Zhao, Mengying Xiong, and Long Su

School of Computer Science and Cyberspace Security, Hainan University,
Haikou, China
zy0826@hainanu.edu.cn

Abstract. With the development of the Internet of Things, the data of the Internet of Things is bound to grow rapidly in a very short time. However, IoT devices generally do not have the ability to store large amounts of data. Currently, data collected by IoT devices are uploaded to cloud storage servers, so massive data will bring a lot of pressure: 1) Take up a lot of transmission bandwidth; 2) Consume a lot of disk storage space. To this end, this article proposes a machine learning-based IoT data storage optimization scheme. With the help of machine learning's strong adaptability and wide application range, the data compression ratio can be well enhanced. Among them, the combined use of machine learning-based deduplication and data compression technology can weaken the shortcomings of the two technologies to a certain extent and further optimize data storage efficiency.

Keywords: Internet of Things · Deduplication · Data compression · Machine learning

1 Introduction

The rapid development of the Internet of Things has caused a surge in the number of IoT devices. The number of IoT devices will grow to about 30 billion by 2020, and it will exceed this figure twice by 2025 [1]. At the same time, the data generated by IoT devices is also increasing day by day, putting tremendous pressure on cloud storage and network bandwidth. Deduplication and data compression technology can be used to solve the problem of data redundancy, but there are some shortcomings in eliminating data redundancy, discovering redundancy methods, redundancy granularity, and application perspectives.

With the rapid development of machine learning model technology, more and more fields have made good progress through machine learning. The introduction of machine learning in the field of data compression can effectively solve the problems of traditional methods. Machine learning has excellent content adaptability, effective use of a larger receptive field and flexible data processing methods, all of which enable data compression technology to have a better data compression ratio.

© The Author(s), under exclusive license to Springer Nature Singapore Pte Ltd. 2022
J. C. Hung et al. (Eds.): FC 2021, LNEE 827, pp. 1805–1811, 2022.
https://doi.org/10.1007/978-981-16-8052-6_267

The contributions of this article are as follows:

① An effective deduplication model based on machine learning is proposed, which can improve the data block segmentation strategy;

② An effective data compression model based on machine learning is proposed, which improves the data compression ratio and can process data more flexibly.

2   Related Work

Azar et al. [2] applied a fast error-bounded lossy compression procedure to the collected data before transmitting IoT data, without affecting the data quality, the amount of transmitted data has been reduced by 103 times. Wen et al. [3] studied the potential data compression methods of smart meter big data, but with the development of smart meters, it needs to be more optimized to achieve an acceptable balance between efficiency and loss ratio. Barannik et al. [4] improved the video data compression technology on the basis of reducing the structural redundancy under limited loss of visualization quality. Yu et al. [5] proposed a quantum algorithm that compresses a large high-dimensional dataset in quantum parallel by reducing its dimensionality based on PCA. Uthayakumar et al. [6] proposed a new algorithm called Neighborhood Indexing Sequence (NIS) for data compression in WSN, the proposed compression algorithm is not only efficient but also highly robust for different WSN dataset. Ni et al. [7] proposed a new SHM data compression and reconstruction method based on autoencoder structure, which can recover the data with high-accuracy under such a low compression ratio. Zhou et al. [8] proposed a novel battery data storage method which uses the frequency division model of the battery pack, which can effectively save the required storage space. In order to minimize the energy consumption of edge computing, Xu et al. [9] used data compression to compress the offloaded data before transmission to reduce the data size. Huang et al. [10] proposed a stacked auto-encoder (SAE)-based load data mining approach. The SAEs are utilized to compress load data in a distributed way, but the iterative gradient descent calculation is extremely time-consuming. Yang et al. [11] proposed an image data com-pression approach using the hidden Markov model (HMM)/pulse-coupled neural network (PCNN) model in the contourlet domain, which has better compression performance. Wang et al. [12] proposed a deep-learning-based compression method for smart meter data via stacked convolutional sparse autoencoder (SCSAE), the proposed method can attain significant enhancement in model size, computational efficiency, and reconstruction error reduction while maintaining the most abundant details, but the calculation time still needs to be reduced.

3   Deduplication Based on Machine Learning

When carrying out large-scale IoT data storage, we will find that there will be many duplicate data in these data, which brings many disadvantages: 1) Takes up a lot of transmission band-width; 2) Consumes a lot of disk storage space. The proposed deduplication program effectively alleviates the problems caused by a large amount of duplicate data storage. Traditional deduplication schemes are roughly divided into file-level and data block-level. This article is based on data block-level deduplication for research, because more and more IoT data storage schemes tend to use distributed storage technology. Therefore, it is more advantageous to study the data block-level.

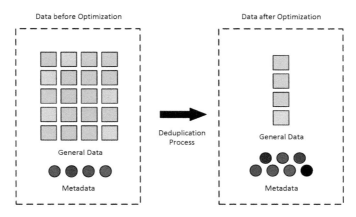

Fig. 1. Principles of deduplication technology.

Data block-level deduplication process: 1) Divide the data file into a set of fixed-length or variable-length data blocks; 2) Calculate the unique fingerprint of each data block, and the data block fingerprint is through Hash such as MD5/SHA1 algorithm for calculation; 3) Data blocks with the same fingerprint can be considered as the same data block, and only in the storage system need to keep a copy; 4) Establish a Hash table with the fingerprint of the data block in order to restore the original data. According to the above process, the purpose of deleting duplicate data and reducing storage space can be achieved, thereby alleviating the storage pressure caused by a large amount of data in the Internet of Things. The principle of traditional deduplication technology is shown in Fig. 1.

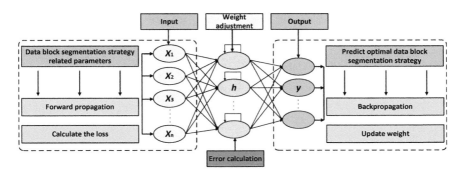

Fig. 2. Deduplication model based on machine learning.

Reading the above process, we can easily find that the efficiency of deduplication at the block-level data is largely affected by the data block partitioning strategy. Machine learning is good at learning from historical data and summarizing relevant laws to predict future trends. How to block the massive data of the Internet of Things to achieve the best deduplication efficiency. For this reason, we propose a high-efficiency data block model. The learning and training process is divided into 5 steps: 1) Forward propagation, calculate the predicted value of the data block of the model; 2) Calculate the error between the predicted value and the actual value; 3) Back-propagation, allocate the error to each data block-level structural unit; 4) Calculate the weight of each weight Gradient; 5) Optimize the algorithm and update the weights. See Fig. 2 for details.

4 Data Compression Based on Machine Learning

Faced with the problems caused by the massive data of the Internet of Things, we also advocate the use of data compression technology to relieve this pressure. The origin of data compression can be traced back to Shannon coding proposed by Shannon, the father of information theory, in 1947. In 1952, Huffman proposed the first practical coding algorithm to achieve data com-pression. The lossless data compression algorithm is implemented by dictionary coding technology, which can be roughly classified into the first type of dictionary method and the second type of dictionary method. The basic process of the first type of dictionary method: 1) Try to find whether the character sequence being compressed has appeared in the previous input data; 2) If it is, replace the repeated character with a "pointer" to the string that appeared earlier string. The basic process of the second type of algorithm: 1) Attempt to create a "dictionary of the phrases" from the input data; 2) When encountering a "phrase" that has appeared in the dictionary during the process of encoding data, encode The processor outputs the "index number" of the phrase in this dictionary, not the phrase itself. The data com-pression coding concept is shown in Fig. 3.

In recent years, the field of machine learning has developed rapidly. Some studies have shown that the use of traditional neural network models has achieved good results in the field of data compression. Data compression model process based on machine learning: 1) Forward propagation, calculate the predicted value of the data compression of the model; 2) Calculate the error between the predicted value and the actual value; 3) Back-propagation; 4) Calculate the gradient of each weight; 5) Optimize the algorithm and update the weights. See Fig. 4 for details.

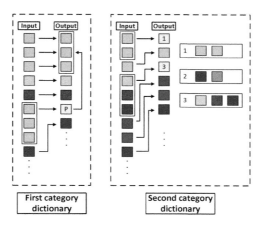

Fig. 3. Data compression coding concept.

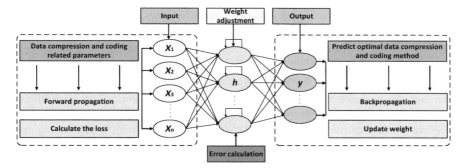

Fig. 4. Data compression model based on machine learning.

5   An Optimization Scheme for IoT Big Data Storage Based on Machine Learning

The data storage optimization in this article is based on the distributed storage scheme of IoT data. The general idea of the scheme is: IoT devices with insufficient computing power will transmit data to a blockchain network composed of edge computing devices. Edge computing devices divide the data according to the principle of secret sharing, and finally store the data blocks separately in the distributed cloud storage device. However, the burden of massive data in the Internet of Things has not been resolved.

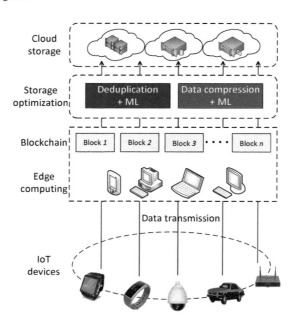

Fig. 5. Big data storage optimization model of Internet of Things based on machine learning.

In fact, the two technologies of deduplication and data compression have different pertinence for data storage optimization, but the combination of the two technologies will achieve a higher data compression ratio. This article will comprehensively apply deduplication and data compression technology, but there is a prerequisite for doing so, that is, deduplication must be per-formed before data compression, so as to reduce the processing requirements of the system and increase the data compression ratio. The machine-based IoT big data storage optimization model is shown in Fig. 5.

6   Conclusion

The combination of machine learning and massive data storage will greatly improve the data compression ratio. In this article, we propose a deduplication model and a data compression model based on machine learning. The machine learning is good at learning from historical data and summing up relevant laws to predict the future trend. For the data block strategy in the traditional deduplication model, machine learning can improve the block efficiency and strengthen the depth of deduplication; For the compression strategy in data compression, machine learning can plan a better compression method from the historical compression methods.

Acknowledgements. This work was supported by the Science Project of Hainan University (KYQD(ZR)20021).

References

1. Internet of Things (IoT) connected devices installed base worldwide from 2015 to 2025 (2016). https://www.statista.com/statistics/471264/iot-number-of-connected-devices-worldwide/

2. Azar, J., et al.: An energy efficient IoT data compression approach for edge machine learning. Future Gener. Comput. Syst. **96**, 168–175 (2019)

3. Wen, L., et al.: Compression of smart meter big data: a survey. Renew. Sustain. Energy Rev. **91**, 59–69 (2018)

4. Barannik, V., Yudin, O., Boiko, Y., Ziubina, R., Vyshnevska, N.: Video data compression methods in the decision support systems. In: Hu, Zhengbing, Petoukhov, Sergey, Dychka, Ivan, He, Matthew (eds.) ICCSEEA 2018. AISC, vol. 754, pp. 301–308. Springer, Cham (2019). https://doi.org/10.1007/978-3-319-91008-6_30

5. Yu, C.-H., Gao, F., Lin, S., Wang, J.: Quantum data compression by principal component analysis. Quantum Inf. Process. **18**(8), 1–20 (2019). https://doi.org/10.1007/s11128-019-2364-9

6. Uthayakumar, J., Vengattaraman, T., Dhavachelvan, P.: A new lossless neighborhood indexing sequence (NIS) algorithm for data compression in wireless sensor networks. Ad Hoc Netw. **83**, 149–157 (2019)

7. Ni, F., Zhang, J., Noori, M.N.: Deep learning for data anomaly detection and data compression of a long-span suspension bridge. Comput. Aided Civil Infrastruct. Eng. **35**(7), 685–700 (2020)

8. Zhou, L., et al.: Massive battery pack data compression and reconstruction using a frequency division model in battery management systems. J. Energy Storage **28**, 101252 (2020)

9. Xu, D., Li, Q., Zhu, H.: Energy-saving computation offloading by joint data compression and resource allocation for mobile-edge computing. IEEE Commun. Lett. **23**(4), 704–707 (2019)

10. Huang, X., Hu, T., Ye, C., et al.: Electric load data compression and classification based on deep stacked auto-encoders. Energies **12**(4), 653 (2019)

11. Yang, G., Yang, J., Lu, Z., et al.: A combined HMM–PCNN model in the contourlet domain for image data compression. PloS One **15**(8), e0236089 (2020)

12. Wang, S., Chen, H., Wu, L., et al.: A novel smart meter data compression method via stacked convolutional sparse auto-encoder. Int. J. Electr. Power Energy Syst. **118**, 105761 (2020)

Design of Pulse Measuring Instrument

Dongfang Jia, Zhicong Liu, and Fanglin An$^{(\boxtimes)}$

School of Computer Science and Cyberspace Security, Hainan University,
Haikou, China

Abstract. New coronary pneumonia has attracted attention to telemedicine technology. It is a scientific research trend to develop an instrument that allows people to test their health. With the advent of the 5G era, the information rate is getting faster and faster, prompting the rise of some telemedicine industries. Because of the non-contact diagnosis, it provides people with a lot of convenience and protects people from the impact of new coronary pneumonia, it is safe and reliable. The technical field of the Telemedicine System relates to a telemedicine system, and particularly to a system with improved operability, suitable for home health monitoring. With the rapid development of science and technology, it is very important to study a set of equipment that can objectively detect pulse. The new coronary pneumonia has made the society aware of the vacancies in telemedicine technology, and the introduction of digital pulse diagnosis technology is non-contact, convenient and fast. The pulse measuring instrument is composed of a main control chip, a collection module, a signal module, an alarm module, and a display module; Compared with other types of designs, it has a simple, efficient, real-time display, and practical pulse test system, and has its own characteristics in terms of practicability and ease of operation.

Keywords: COVID-19 · Telemedicine · 5G · Pulse · Test mode

1 Introduction

The new crown pneumonia has caused a large number of experts and scholars to pay attention to the telemedicine system, involving network connection technology and big data analysis principles, which has greatly changed the modern medical system. In order to ensure that citizen can enjoy a high-quality life, pulse theory has been repeatedly demonstrated by medical practitioners in various generations, and has been continuously developed and improved. It is gradually complete and is used to detect human health. It is necessary to study a device that can obtain pulse information more efficiently and accurately in the future. With the improvement of the quality of life, there is a trend to increase the importance of the health of oneself and family members. In the face of illness, early warning of health conditions is given. People need high-efficiency equipment to learn the physical information of the body, and to scientifically arrange their physical exercises. Clinical data shows that the annual morbidity and mortality of cardiovascular diseases have remained high worldwide [1]. Due to the

J. C. Hung et al. (Eds.): FC 2021, LNEE 827, pp. 1812–1817, 2022.
https://doi.org/10.1007/978-981-16-8052-6_268

characteristics of the pulse, it can better show us the information of the human cardiovascular system, the rhythm of the heartbeat, blood vessel filling, and the elasticity of the arterial wall. There are old people, babies, and young people in the family. The pulse frequency is affected by age, and the pulse information is different for each age group, the advantage lies in the research of a pulse meter that can span all age groups and display heart rate in real time.

2   Research Status and Development Trends at Home and Abroad

Experts and scholars at home and abroad have successively developed pulse measuring instruments in different time dimensions. For pulse measuring instruments, different types have been developed [2]. The Institute of Pathology of Radiology and Ophthalmology Clinic of Sarajevo University Hospital joined the telemedicine experimental project SHARED [3]. Telemedicine is the use of information or communication technology to provide healthcare services. In the current pandemic situation, telemedicine can supplement healthcare services without personal visits [4]. The Indian government recently launched e-sanjeevani OPD, which is a nationwide remote consultation service, and many state governments have made it a mandatory measure for medical service providers [5]. my country has been studying pulse measurement since ancient times. After the new century, piezoelectric chip pulse recorders have been developed, and then three-line pulse loggers have been introduced to measure human pulses. The research on pulse measurement is getting more and more in-depth, and it is divided into two categories, sensor volume method and piezoelectric sensor measurement. The development of portable medical equipment has extremely high development prospects for the early warning of various human diseases. The social development at home and abroad and the analysis of public demand: the pulse meter can shorten the space distance between the doctor and the patient, and it has a great significance for sudden diseases. Alarm function. In the era of coronavirus disease (COVID-19), telemedicine becomes even more important. People are using it because it is easy to obtain, thereby reducing their waiting time in the hospital, easy to obtain and cost-effective. Despite a large number of telemedicine trials, there is little information about its economic costs and benefits. Most telemedicine programs are funded as special projects and are not subject to regular budget procedures.

3   Selection and Demonstration of Sensor Modules

The selection and demonstration of the display module is compared with the dot matrix digital tube display, which can display text and patterns. The store uses a large dot matrix digital tube. If it is used to display the value of the pulse signal, it occupies a large area of the sheet, high relative energy consumption, and low cost performance. Others use LED digital tubes, which are usually seven-segment, and are easy to control. They occupies few pins of the single-chip microcomputer and are used to display numbers. However, due to the dynamic scanning of the digital tube, the time needs to be set, and the shift register is needed for data processing and display. The content of display is very limited, and it is not easy to find its problems in debugging [6], this scheme is not adopted. The display results are clear, the display information is complete, and it saves energy and is environmentally friendly. The display is bright in color and is less harmful to the eyes.

Piezoelectric sensors have a wide range of applications, excellent design, and stable output signals under complex conditions, but they are not suitable for small-scale projects and are suitable for hospitals. Compared to other designs, I use ST188, which is a photoelectric sensor. Lambert Beer's law states that the absorbance of a substance at a certain wavelength is proportional to the concentration [7]. The light source generates luminous flux, and the transmittance and reflectance of light are different, and the pulse can be detected indirectly [8]. The through-beam detection is realized by infrared through-beam diodes, and the reflective type uses ST188 chip. ST188 is convenient and cheap to detect pulse signals and is cost-effective.

4   MCU Control Design

This design uses a single-chip microcomputer as the control module, which has several modules for signal acquisition, signal processing, alarm, main control, and display, so that the pulse can be detected and displayed in real time [9]. After the signal is collected, the DC signal interferes, and the capacitor eliminates the interference signal. The frequency of the pulse signal is in the low frequency range. The signal is processed by a low-pass filter, amplified by a non-inverting operational amplifier, and output after being shaped by a voltage comparator [10]. The output signal has high and low levels, and the diode and the pulse signal are on and off at the same frequency. The single-chip microcomputer collects the pulse signal high and low level cycles, and collects the pulse value in real time [11]. If it is higher than the upper limit of the pulse set by the alarm, or lower than the lower limit, the alarm sounds and the single-chip computer will perform calculations and send the data to the LCD in real time In the display, average the collected effective pulse signal values and display them to the preset position of the LCD screen [12].

The key control part is divided according to the function, which is divided into two parts: keyboard scanning and key setting. The key scan part is put in the timer interrupt, and the timer provides the delay time. The reset button provides circuit error management and re-detection functions, and the menu button is responsible for setting the upper and lower limits of the alarm value and mode switching functions.

5   Classification Adjustment and Optimization

The alarm circuit part is used to give an effective alarm when the pulse value exceeds the range of a certain test. Considering that there will be noise pollution in the test process, the design only alarms the effective pulse value. The effective pulse value is based on data analysis in the medical field. The above research results prove that there are different pulse health levels among young people, the elderly, and babies. Big data analysis shows that the pulse frequency of the elderly is generally small, so the next step can be measured by adjusting the mode conversion appropriately [13]. When detecting the pulse values of the elderly, babies, and young people, we know that the pulse alarm range should be a fixed value. In the free test mode, the upper and lower alarm limits can be modified, which is improved for users with special requirements.

The mode conversion of the alarm circuit designed in this design is more advantageous than other designs. Free working mode and rated working upper and lower limits. The basic working principle of the pulse meter alarm device in this design is to set the upper and lower pulse limits of the pulse value in each working mode, and compare with the real-time effective value detected by the pulse meter, the pulse value exceeds the limit, the buzzer alarms [14]. Hardware circuit resources have limitations. Relying on software logic can break through boundaries and achieve more functions. However, to complete the design, circuit simulation tests are required, and design defects and deficiencies are discovered in time during the test, as shown in Fig. 1.

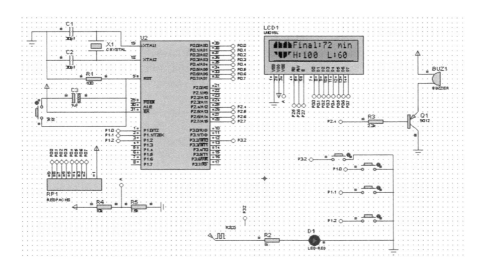

Fig. 1. Circuit simulation diagram.

After related tests and debugging, compared with other designs, the test accuracy rate of this design reaches 98%, which solves the vacancy of the instrument for diagnosing pulse, and designs four relative mode conversions, which are suitable for larger and wider crowds. Finally, consider the future development field is suitable for medical foundation and medical research, and is more suitable for large-scale investment and development than others. The whole design also takes into consideration the setting of the upper and lower bounds of blood and pulse, and finally realizes relative expansion.

6 Conclusion

This pulse meter is based on a single-chip microcomputer, combined with a signal display and alarm function, to achieve pulse measurement. Compared with other designs, it is simple and easy to use, and provides real-time display and average value display. The longer the detection time, the more accurate the value. At the same time, it is suitable for family life. The design has four modes of free test, elderly, young, and baby, which meet the needs of different ages. In the future, it is envisaged to develop and establish a server that can be connected to the Internet for processing, combined with the development trend of 5G, can analyze the data of all people's tests on the Internet, and the general health status of the population at the stage when the data is processed by cloud computing [15]. The development of the field plays a key role. At the national level, we can clearly understand the living conditions of the people across the country and implement corresponding policy changes. People need a healthy and stable social environment and can clearly understand their physical condition. This development prospect is considerable, and it will progress to a deeper level. Since we are doing medical testing, we will assign an ID to each user in the future, upload the measured data to the server, and use the knowledge of big data and data mining to visualize the content and display related to the pulse Analysis, data prediction, analysis of changes in pulse value, timely data reports for each user, and timely warning of physical conditions, there is a lot of room for future research.

Acknowledgements. This work was supported by the Science Project of Hainan University (KYQD(ZR)20021).

References

1. Elliott, T., Yopes, M.C.: Direct-to-consumer telemedicine. J. Allergy Clin. Immunol. Pract. **7**(8), 2546–2552 (2019)
2. Calton, B., Abedini, N., Fratkin, M.: Telemedicine in the time of coronavirus. J. Pain Symptom Manage. (2020)
3. Hatcher-Martin, J.M., Adams, J.L., Anderson, E.R., et al.: Telemedicine in neurology: telemedicine work group of the American academy of neurology update. Neurology **94**(1), 30–38 (2020)
4. Mukherjee, N., Miesse, A., Racenet, D.C.: Surgical instruments including sensors: U.S. Patent, 10, 624, 616 (2020)
5. Hollander, J.E., Carr, B.G.: Virtually perfect? telemedicine for COVID-19. N. Engl. J. Med. **382**(18), 1679–1681 (2020)
6. Nematolahi, M., Abhari, S.: Assessing the information and communication technology infrastructures of Shiraz University of medical sciences in order to implement the telemedicine system in 2013. Interdisc. J. Virtual Learn. Med. Sci. **5**(2), 44–51 (2020)
7. Johansson, A., Esbjörnsson, M., Nordqvist, P., et al.: Technical feasibility and ambulance nurses' view of a digital telemedicine system in pre-hospital stroke care–a pilot study. Int. Emerg. Nurs. **44**, 35–40 (2019)

8. Xu, H., Huang, S., Qiu, C., et al.: Monitoring and management of home-quarantined patients with COVID-19 using a We Chat-based telemedicine system: retrospective cohort study. J. Med. Internet Res. **22**(7), e19514 (2020)
9. Ekeland, A.G., Bowes, A., Flottorp, S.: Effectiveness of telemedicine: a systematic review of reviews. Int. J. Med. Inf. **79**(11), 736–771 (2010)
10. Al-Ali, A.: Low power pulse Oximeter: U.S. Patent Application 16/174,130 (2019)
11. Hurley, D.: Neurologists scramble to respond to COVID-19 with telemedicine: the challenges and opportunities. Neurol. Today **20**(8), 18–19 (2020)
12. Bagot, K.L., Moloczij, N., Barclay-Moss, K., et al.: Sustainable implementation of innovative, technology-based health care practices: a qualitative case study from stroke telemedicine. J. Telemed. Telecare **26**(1–2), 79–91 (2020)
13. Contreras, C.M., Metzger, G.A., Beane, J.D., et al. telemedicine: patient-provider clinical engagement during the COVID-19 pandemic and beyond. J. Gastrointest. Surg. 1 (2020)
14. Zhu, Y., Gu, X., Xu, C.: Effectiveness of telemedicine systems for adults with heart failure: a meta-analysis of randomized controlled trials. Heart Fail. Rev. **25**(2), 231–243 (2020)
15. Atmojo, J.T., Sudaryanto, W.T., Widiyanto, A., et al.: Telemedicine, cost effectiveness, and patients satisfaction: a systematic review. J. Health Policy Manage. **5**(2), 103–107 (2020)

A Survey of Event Relation Extraction

QunLi Xie, JunLan Pan, Tao Liu, BeiBei Qian, XianChuan Wang$^{(\boxtimes)}$, and Xianchao Wang$^{(\boxtimes)}$

Fuyang Normal University, Fuyang 236037, Anhui, China

Abstract. Human beings recognize and understand the real world in units of events. In recent years, events have been used as the basic unit to process unstructured text in the field of natural language processing, but there is often a connection between events and events. Therefore, recognizing the relationship between events and events in unstructured text has become an important task in the field of natural language processing and has attracted more and more researchers' attention. This paper first introduces the evolution of the method of event temporal relation and causal relation in the extraction research, comparing the advantages and disadvantages and method performance; Then, the event relation extraction model based on deep learning can be divided into strong supervision method and weak supervision method, and the extraction methods of event relation are analyzed, compared and summarized respectively, among them, the method of strong supervision based on deep learning can be further divided into pipeline method and joint learning method, and the method of weak supervision based on deep learning can be divided into semi-supervised learning method, remote learning supervised method and unsupervised learning method. Finally, this paper summarizes the methods of event relation extraction and points out the future research direction.

Keywords: Temporal relation · Causal relation · Deep learning

1 Introduction

With the advent of the big data and the development of the Internet, a large number of unstructured texts have been produced on the Internet, among which a large number of unstructured texts contain a lot of useful unstructured information, which are ambiguous and fuzzy, making it difficult for computers identifying and acquiring knowledge. Therefore, how to mine valuable information from these unstructured texts and present it in a way that is "easy to understand" by computers has become a big challenge in the field of NLP.

As a dynamic semantic unit, event has attracted more and more attention. It is generally accepted that an unstructured text is composed of multiple events, which contain various static concepts, such as time, place, participants, etc., and there are generally some semantic relations between events, such as causal relation, temporal relation, etc. Event relation is important and has great research significance in fields of medicine, politics, and aviation safety.

In the early research of event relation, it is usually inclined to use pattern matching and traditional machine learning methods to classify relation. These methods usually

J. C. Hung et al. (Eds.): FC 2021, LNEE 827, pp. 1818–1827, 2022.
https://doi.org/10.1007/978-981-16-8052-6_269

rely on the experience of experts to obtain the corresponding features and external resources, which are time-consuming and labor-intensive, and may ignore some important recessive characteristics, while the accuracy rate is high, the recall rate is not ideal. In recent years, the accumulation of large-scale annotation data and the development of deep learning have promoted the deep neural network and its application in the extraction of event relations. The neural network model not only reduces the cost of domain experts' work, but also uses some hidden features. Although a large amount of annotated data has been accumulated in the corpus at present, the sample number of some relationship types is still too rare, resulting in the failure to identify them by using neural network method and the accuracy cannot be guaranteed. At present, some researchers have put forward deep weakly supervised learning method, which combines rule-based method, feature-based method and neural network model, and has achieved good results. Relation extraction is an important sub-task of information extraction. The main work of this paper is to summarize the methods of event relation extraction for unstructured text.

2   Evolution of Event Relation Extraction Method

2.1   Temporal Relation Extraction Method

The research of event temporal relation extraction (ETE) has been carried out earlier, and with the development of deep learning, various types of neural networks have been successively applied to ETE tasks, such as Convolutional Neural Networks (CNN), Recurrent Neural Networks (RNN), Long-Short Term Memory networks (LSTM) and so on. Dligach et al. [1] proposed the model structure of ETE using CNN with tags as input and Bi-LSTM with richer semantics. Zhou et al. [2] proposed adding the attention mechanism to the Bi-LSTM model to learn long-distance dependencies, while Zhang et al. [3] combined the multi-head attention mechanism with the non-linear network layer to further improve the representation ability of the attention network. Li et al. [4] introduced BERT into the Bi-LSTM structure, and integrated multi-dimensional event information to mine cross-sentence timing information. Compared with the traditional ETE method, the ETE method based on deep learning can learn automatically and has higher performance and stronger generalization ability without relying on manual features.

2.2   Causal Relation Extraction Method

Traditional event causal relation extraction (ECE) research mainly uses lexical features, semantic features, manual construction patterns and other methods to extract causal relation between events, while today ECE methods are similar to ETE, mostly using neural network methods based on deep learning. Silva et al. [5] and Li et al. [6] introduced CNN into ECE, and Dasgupta et al. [7] proposed to use LSTM structure to explore the potential semantic information in the text. Zheng et al. [8] use the idea of divide and conquer to decompose ECE into two sequence labeling tasks, use CRF to complete sequence labeling, introduce BERT and CNN to enhance the expression ability of event causal features, and introduce residual ideas to capture important semantic features of text. It effectively solves the problems of insufficient semantic representation of causal relations and weak boundary recognition capabilities. Li et al. [9] solves the problem of insufficient data through contextual string embedding, and introduces the multi-head self-attention mechanism into the BiLSTM-CRF structure to understand the interdependence between causal words.

3   Event Relation Extraction Model Based on Deep Learning

Although the traditional event relation extraction method has high accuracy, it is difficult to be applied in practice because it is often time-consuming and costly. With the development of deep learning, many methods based on deep learning have performed well in the field of relation extraction. Relation extraction methods based on deep learning can be divided into strong supervision methods and weak supervision methods. At present, strong supervision techniques based on deep learning in the task of event relationship recognition are still the best.

3.1   Strong Supervision Model Based on Deep Learning

Traditional relation extraction methods based on supervised learning mostly rely on feature engineering. In recent years, a variety of strong supervised relation extraction models based on deep learning can solve the artificial dependence problem in traditional methods. The strong supervision method based on deep learning usually regards event relation extraction as a classification problem, and trains the model through the existing labeled corpus to obtain the optimal model, and finally judges the output of the model to achieve the classification purpose. According to different learning methods, the event relation extraction methods based on deep strongly supervision can currently be divided into two types: pipeline method and joint learning method.

Pipeline Method. Most of the early deep strongly supervision methods used in relation extraction are based on pipeline. The idea of the pipeline method based on the event relationship is to use the pipeline form to decompose the relation extraction task into two sub-tasks: event extraction and relation classification. The two sub-tasks are separated from each other and do not interfere with each other. The relation classification task is to classify the relationship between events on the basis of event extraction, and the result of the relation classification depends on the result of the event extraction task.

Since the pipelined model is relatively simple to construct, many scholars have proposed the application of pipelined methods to solve problems in the early NLP. The DMCNN model proposed by Chen et al. [10] is a typical pipe-like event extraction model, which transforms event extraction into two sub-tasks, event recognition and argument role classification. Liu et al. [11] divided the relation extraction problem into two parts: entity extraction and relation classification, and put forward for the first time the use of convolutional neural network to deal with the relation classification between two given entities. Most of the early pipelined methods were based on extended optimization of convolutional neural network and recurrent neural network. Zeng et al. [12] have applied CNN model to relation classification, and used convolution DNN algorithm to replace the traditional feature engineering method, and used convolution depth neural network (DNN) to extract lexical and sentence features, and proposed that the position feature (PF) is used to encode the relative distance between the current word and the target word. Finally, the relation is classified by the hiding layer and the softmax layer, automatic learning can be achieved without any external resources or NLP modules for optimal performance at the time.

In the follow-up research, researchers proposed some models based on CNN structure for relation extraction, such as Multi-Window CNN [13], Multi-Level attention CNNs [14], etc., which have made good progress, but CNN cannot handle global features and time series information, especially for the long-distance dependence of the text sequence, the effect of the CNN model will be worse. Therefore, Zhang et al. [15] proposed to use RNN to learn remote semantic information. Compared with CNN, the RNN model can handle long-distance patterns and is very suitable for time-series feature extraction models. It is difficult to capture long-term time association because of gradient explosion or disappearance in recurrent neural network model with time, and LSTM can solve this problem well. LSTM is an improvement of RNN, RNN can only maintain short-term memory because of gradient disappearance, and LSTM can solve the problem of gradient disappearance to some extent by introducing memory unit and gate control unit to combine short-term memory with long-term memory. Xu et al. [16] proposed the SDP-LSTM model to classify the relationship between two entities in a sentence, and used the shortest dependency path (SDP) between two entities to obtain heterogeneous information, at the same time, the multi-channel LSTM network is used to integrate information from heterogeneous sources in a dependent path.

Relational classification is to classify the relation between entity pairs, and event relation extraction also needs to judge whether there is a relationship between temporal entity pairs or causal entity pairs. Taking into account that temporal relation extraction is similar to relational classification task, Cheng et al. [17] proposed to apply DP and LSTM to ETE in view of the remarkable effect of dependency path (DP) and LSTM in relational extraction task, aiming at the problem of how to represent the dependent paths between the cross-sentence entities, the Bi-LSTM model along the dependent paths is adopted and a "common root" hypothesis is proposed to extend the DP representation of the cross-sentence join, in which both sentences are represented as dependent paths, share a "common root". Ning et al. [18] considering whether ETE can benefit from external resources, a probabilistic knowledge base which is named Temprob is constructed according to the fact that event words themselves contain time information that can be used as prior knowledge, the ETE task is divided into two steps: extracting event words and extracting relationship.

Pipelining method needs to identify event trigger words and related arguments first, and then classify event relations. This method ignores the internal correlation between two sub-tasks, and it is easy to lose information. It is easy for the error of the previous task to affect the subsequent relational classification task, resulting in error propagation. Pairing unrelated pairs of events creates information redundancy. These will interfere with the performance of relation classification.

Joint Learning Method: Event relationship and event information are closely interactive. The joint learning method combines two sub-tasks and optimizes them together in a unified model. This method can use the potential correlation between the two subtasks to solve the problems in the pipeline.

In the study of relation extraction, Miwa et al. [19] proposed an end-to-end neural network model for joint modeling of entities and relations for the first time. Although this model introduced the idea of joint learning, the model learning process was still similar to the pipeline method. Zheng et al. [20] proposed a hybrid neural network model to alleviate the problem of long-distance dependency between entity tags unresolved by Miwa's method. Both of these two models adopt the method of parameter sharing, which can effectively improve the problem of error propagation and neglecting the inherent correlation between subtasks existing in the pipeline method, and improve the robustness of the model, but it is still easy to produce information redundancy.

Aiming at the problem of information redundancy in the shared parameter method, Zheng et al. [21] proposed a new sequence labeling scheme, which converts the joint extraction problem into a sequence labeling problem. Katiyar et al. [22] proposed a new LSTM model based on the attention mechanism to jointly identify entities and relation, and after the current location entity is identified, the attention mechanism is used to compare its similarity with all previous location entities, which is considered to be real joint learning. In response to the problem that the previous joint learning model relied heavily on artificial features and external NLP tools [19–22], Bekoulis et al. [23] proposed to use CRF to model entity recognition and use relationship classification as a multi-head selection problem. The final result also proved that the model is superior to the automatic feature extraction model at that time.

Inspired by the joint learning model of entities and relation, Han et al. [24] proposed a Neural SSVM (Neural Structured Support Vector Machine) model for the first time to extract events and temporal sequence relations between events simultaneously. This model belongs to end-to-end structured joint learning model. In the bottom layer, the word representation obtained by the BERT model is input to the BI-LSTM layer for coding. By sharing embedding in E-E module and relation extraction module, the joint identification of events and event timing relationship is realized by placing it in SSVM to judge whether events and event relations exist or not. In view of the advantages of joint extraction method for sequence labeling in relation extraction, Li et al. [9] designed a causality labeling scheme to directly extract event causality and proposed a SCITE (self-attentional BILSTM-CRF transfer embedding) model, introducing a self-attentional mechanism to capture long-term dependencies between causality. Experiments show that the SCITE model based on the causal labeling scheme is effective.

The joint learning method combines the two sub-tasks of event identification and relationship classification, attaches importance to the interaction and association between the two sub-tasks, and effectively solves the problems existing in the pipeline method. By sharing parameters, the problem of error propagation and information loss is alleviated. Sequence labeling effectively solves the problem of entity redundancy in shared parameters. However, no matter the joint learning method based on shared parameters or sequence labeling is adopted, it still does not have a good effect on the problem of overlapping relationship.

3.2   Week Supervision Model Based on Deep Learning

At present, the deep learning technology based on strongly supervised learning has achieved great success in the field of event relation extraction, with high accuracy and recall rate. However, the model based on strongly supervised learning needs to rely on a large number of hand-marked training data, which is costly, time-consuming and laborious. In contrast, the weakly supervised learning method with low labeling cost has attracted more and more researchers' attention and has been initially applied in the field of relation extraction. Methods based on deep weakly supervision can be divided into three categories: semi-supervision method, remote supervision method and unsupervised method.

Semi-supervision Method: Compared with the strong supervised learning method, the semi-supervised method only needs a small number of labeled samples and a large number of unlabeled samples, better suited to the current era of big data.

The commonly used methods for semi-supervised relation extraction include Bootstrapping, Collaborative training, and annotation propagation, etc. At present, the most commonly used semi-supervised learning method in the field of relation extraction is Bootstrapping, which uses a small number of seed labeled samples to train the model in order to extract more entity pairs of relations, and then iteration training was performed again. Brin et al. [25] first introduced the semi-supervised method based on Bootstrapping in the field of relation extraction and established the DIPRE system to automatically obtain new relationship instances from the World Wide Web. Kipf et al. [26] proposed that graph convolutional networks can be used for semi-supervised classification, which can effectively learn the hidden layer representation of graph structure and node features. As the joint learning method based on deep strong supervision fails to solve the overlapping relation problem well, Phi et al. [27] proposed to creatively put forward the sorting of automatic seed selection and remote supervised data noise reduction tasks in Bootstrapping.

The semi-supervised method only needs to construct the initial seed set manually, which can reduce the dependence of the event relation extraction on the tagged corpus to some extent, but it requires high quality of the initial seed, and the initial iteration can not guarantee absolutely accurate, and there will be an inevitable decrease in the accuracy rate under iteration, that is, semantic drift.

Remote Supervision Method: As early as 2009, Mintz et al. [28] first proposed the application of remote surveillance in relation extraction, which can be aligned with unstructured text using relational instances in the knowledge base, based on the assumption that a pair of entities contains a certain relationship, as long as the sentences containing the pair of entities contain such a relationship, a training corpus and a classifier are automatically constructed to solve the traditional method's dependence on manual annotation. However, the assumption of the remote supervision method is too positive, which can easily lead to the problem of incorrect labeling and introduce a lot of noisy data. At present, the existing literature has proposed a variety of effective solutions to the remote supervision noise problem and the wrong label problem, such as the introduction of multi-instance learning, attention mechanism and other methods.

Aiming at the noise problem, Zeng et al. extended to remote supervision on the basis of the original method [12], and proposed the PCNN model [29]. The piece-wise convolutional neural network (PCNN) was used to improve the original global max pooling and made local Max pooling, and use multi-instance learning to solve the problem of mislabeling. In addition, the previous methods applied the supervision model to the designed features when acquiring labeled data through remote supervision. These features usually come from pre-existing NLP tools. Because NLP tools inevitably have errors, they will cause errors in the features. The extraction continues to propagate or accumulate, and most deep learning methods require sentence-level tags. In this regard, Lin et al. [30] added an attention mechanism to the PCNN model to reduce the negative impact of false labels, and used CNN to express the relationship with the semantic combination of sentence embeddings, so as to make full use of the information of the training knowledge base, and achieved good results.

Unsupervised Method: Unsupervised learning does not need to label training corpus at all. As early as 2004 [31], unsupervised learning has been applied in the field of relationship extraction, and the extraction method is bottom-up. Vo et al. [32] used the Open IE system to automatically extract the relational triplet to construct the event network, and realized unsupervised automatic identification of the potential time and causality between two nodes in the event network by performing a specific form of traversal on the already constructed event network.

Unsupervised event relation extraction is don't need to depend on the manual annotation corpus, also don't need any predefined relationship types, can be automatically extracted in unstructured text event, for strong adaptability in the multidisciplinary event relation extraction, domain migration performance is good, but the current extraction model based on unsupervised event relation of overall accuracy and recall rate are low. For some low frequency instance relation extraction rate, it is difficult to quantify and unify the evaluation standard of event relation extraction.

3.3   Comparative Analysis of Event Relation Extraction Models

At present, both the deep strongly supervised learning method and the deep weakly supervised learning method can achieve good results, and the deep strongly supervised learning method achieves the best effect in the task of event relation extraction. CNN, RNN, LSTM, and later graph neural networks, hybrid neural networks, etc. have become common structures for event relationship extraction models. Introducing attention mechanisms and shortest dependency paths into these structures have also become common practices for event relationship extraction. The pipeline method based on deep strongly supervision has problems of information loss, error propagation and information redundancy due to ignoring the internal correlation between two sub-tasks. A similar joint learning method is proposed in the field of event relation extraction, which combines event extraction and event relation classification to enhance the interaction between two sub-tasks, it can be divided into joint learning methods based on shared parameters and sequence annotation. The shared parameters method can solve the problems of information loss and error propagation in pipeline, sequence annotation can alleviate the problem of information redundancy in shared parameter method.

Compared with the deep strongly supervision method, the weak supervision method only needs a small amount of annotated corpus to achieve the effect close to the strong supervision method, which can save a lot of costs in practical application and is very practical. However, semi-supervised method is prone to semantic drift due to continuous iteration, so it is necessary to improve the quality of seed set. Remote supervision method is easy to lead to mislabeling and noise problems, attention mechanism and multi-instance learning are introduced to reduce noise and error label. The unsupervised method has a low extraction rate for some low-frequency relationship instances. Compared with the strong supervision method, the weak supervision method is still immature in the field of event relation extraction and is still in the early stage of exploration. It is difficult to accurately calculate the accuracy and recall rate, and more effective evaluation methods are still in need.

4 Conclusion and Future Work

In recent years, good results have been achieved for event relation extraction, but it is still in the early stage of exploration and there are still some difficulties to be solved. At present, there is no mature solution for cross-domain, cross-language, cross-data set and other aspects of event relation extraction model. In the future, transfer learning can be introduced into the field of event relation extraction, and the alignment of knowledge base and unstructured text can improve the generalization ability of the model. In addition, compared with the strong supervision method, the method based on deep weakly supervision only requires a small amount of annotated corpus or even no annotated corpus at all, which is of great practical application significance. In the future, the application of weak supervision method in the field of event relation extraction and the accuracy of weak supervision evaluation method should be strengthened.

Acknowledgements. This work was supported in part by the Key Project of Natural Science Research in Anhui under Grant KJ2019A0533, in part by the Innovation Team from Fuyang Normal University under Grant XDHXTD201703, Grant XDHXTD201709 and Grant kyt-d202004, in part by the Fuyang Normal University PhD Research Project under Grant 2018kyqd0027.

References

1. Dligach, D., Miller, T., Lin, C., Bethard, S., Savova, G.: Neural temporal relation extraction. In: Proceedings of the 15th Conference of the European Chapter of the Association for Computational Linguistics, vol. 2, Short Papers (2017)
2. Zhou, X., Wan, X., Xiao, J.: Attention-based LSTM network for cross-lingual sentiment classification. In: Conference on Empirical Methods in Natural Language Processing (2016)
3. Zhang, Y.J., Li, P.F., Zhu, Q.M.: Event temporal relation classification method based on self-attention mechanism. Comput. Sci. **46**(08), 251–255 (2019)
4. Li, J., Li, P.F., Zhu, Q.M.: A Chinese temporal relation identification approach on multi-dimensional information. J. Shanxi Normal Univ. (Nat. Sci. Edit.) **44**(3), 1–9 (2021)

5. Silva, T., Xiao, Z., Rui, Z., Mao, K.: Causal relation identification using convolutional neural networks and knowledge based features. World Acad. Sci. Eng. Technol. Int. J. Mech. Mechatron. Eng. **11**(6), 696–701 (2017)
6. Li, P., Mao, K.: Knowledge-oriented convolutional neural network for causal relation extraction from natural language texts. Expert Syst. Appl. **115**, 512–523 (2019)
7. Dasgupta, T., Saha, R., Dey, L., Naskar, A.: Automatic extraction of causal relations from text using linguistically informed deep neural networks. In: Proceedings of the 19th Annual SIGdial Meeting on Discourse and Dialogue (2018)
8. Zheng, Q.D., Wu, Z.D., Zhou, J.Y.: Event causality extraction based on two-layer CNN-BiGRU-CRF model. Comput. Eng. (2020). https://doi.org/10.19678/j.issn.1000-3428.0057361,1-9
9. Li, Z., Li, Q., Zou, X., Ren, J.: Causality extraction based on self-attentive BiLSTM-CRF with transferred embeddings. Neurocomputing **423**, 207–219 (2021)
10. Chen, Y., Xu, L., Kang, L., Zeng, D., Zhao, J.: Event extraction via dynamic multi-pooling convolutional neural networks. In: The 53rd Annual Meeting of the Association for Computational Linguistics (ACL2015) (2015)
11. Liu, C., Sun, W., Chao, W., Che, W.: Convolution Neural Network for Relation Extraction. In: Motoda, H., Wu, Z., Cao, L., Zaiane, O., Yao, M., Wang, W. (eds.) ADMA 2013. LNCS (LNAI), vol. 8347, pp. 231–242. Springer, Heidelberg (2013). https://doi.org/10.1007/978-3-642-53917-6_21
12. Zeng, D., Liu, K., Lai, S., Zhou, G., Zhao, J.: Relation classification via convolutional deep neural network. In: Proceedings of COLING 2014, the 25th International Conference on Computational Linguistics: Technical Papers, pp. 2335–2344 (2014)
13. Nguyen, T.H., Grishman, R.: Relation extraction: perspective from convolutional neural networks. In: Workshop on Vector Space Modeling for Natural Language Processing (2015)
14. Wang, L., Zhu, C., Melo, G.D., Liu, Z.: Relation classification via multi-level attention CNNs. In: Proceedings of the 54th Annual Meeting of the Association for Computational Linguistics, vol. 1, Long Papers (2016)
15. Zhang, D., Wang, D.: Relation classification via recurrent neural network. Comput. Sci. (2015)
16. Xu, Y., Mou, L., Ge, L., Chen, Y., Zhi, J.: Classifying relations via long short term memory networks along shortest dependency paths. In: The 2015 Conference on Empirical Methods in Natural Language Processing (EMNLP) (2015)
17. Fei, C., Miyao, Y.: Classifying temporal relations by bidirectional LSTM over dependency paths. In: Proceedings of the 55th Annual Meeting of the Association for Computational Linguistics, vol. 2, Short Papers (2017)
18. Qiang, N., Hao, W., Peng, H., Dan, R.: Improving temporal relation extraction with a globally acquired statistical resource. In: Conference of the North American Chapter of the Association for Computational Linguistics: Human Language Technologies (2018)
19. Miwa, M., Bansal, M.: End-to-end relation extraction using LSTMs on sequences and tree structures. In: Proceedings of the 54th Annual Meeting of the Association for Computational Linguistics, vol. 1, Long Papers (2016)
20. Zheng, S., Hao, Y., Lu, D., Bao, H., Bo, X.: Joint entity and relation extraction based on a hybrid neural network. Neurocomputing **257**(000), 1–8 (2017)
21. Zheng, S., Wang, F., Bao, H., Hao, Y., Zhou, P., Xu, B.: Joint extraction of entities and relations based on a novel tagging scheme (2017)
22. Katiyar, A., Cardie, C.: Going out on a limb: joint extraction of entity mentions and relations without dependency trees. In: Proceedings of the 55th Annual Meeting of the Association for Computational Linguistics, vol. 1, Long Papers (2017)

23. Giannis, B., Johannes, D., Thomas, D., Chris, D.: Joint entity recognition and relation extraction as a multi-head selection problem. Expert Syst. Appl. **114**(DEC.), 34–45 (2018)
24. Han, R., Ning, Q., Peng, N.: Joint event and temporal relation extraction with shared representations and structured prediction. EMNLP/IJCNLP (2019)
25. Brin, S.: Extracting Patterns and Relations from the World Wide Web. In: WebDB, pp. 172–183 (1998)
26. Kipf, T., Welling, M.: Semi-supervised classification with graph convolutional networks. arXiv preprint arXiv (2016)
27. Phi, V.T., Santoso, J., Shimbo, M., Matsumoto, Y.: Ranking-based automatic seed selection and noise reduction for weakly supervised relation extraction. In: Proceedings of the 56th Annual Meeting of the Association for Computational Linguistics, vol. 2, Short Papers (2018)
28. Mintz, M., Bills, S., Snow, R., Jurafsky, D.: Distant supervision for relation extraction without labeled data. In: ACL 2009, Proceedings of the 47th Annual Meeting of the Association for Computational Linguistics and the 4th International Joint Conference on Natural Language Processing of the AFNLP, 2–7 August 2009. Association for Computational Linguistics, Singapore (2009)
29. Zeng, D., Kang, L., Chen, Y., Zhao, J.: Distant supervision for relation extraction via piecewise convolutional neural networks. In: Conference on Empirical Methods in Natural Language Processing (2015)
30. Lin, Y., Shen, S., Liu, Z., Luan, H., Sun, M.: Neural relation extraction with selective attention over instances. In: Proceedings of the 54th Annual Meeting of the Association for Computational Linguistics, vol. 1, Long Papers (2016)
31. Hasegawa, T., Sekine, S., Grishman, R.: Discovering relations among named entities from large corpora. In: Proceedings of the 42nd Annual Meeting of the Association for Computational Linguistics, 21–26 July 2004
32. Vo, D.T., Al-Obeidat, F., Bagheri, E.: Extracting temporal and causal relations based on event networks. Inf. Process. Manag. **57**(6), 102319 (2020)

Research on LZW Algorithm Based on AES in Data Backup Under Data Block Compression and Encryption

Tao Zeng[1(✉)] and Shi-bing Wang[2]

[1] Information Center, Fuyang Normal University, Fuyang, Anhui, China
zengtao@fynu.edu.cn
[2] School of Computer and Information Engineering, Fuyang Normal University, Fuyang, Anhui, China

Abstract. With the rapid development of network information technology and digital storage of information, there are more and more data leakage, loss, damage by virus, illegal string change and other security issues, so data security is increasingly valued by people. In order to ensure the security of data, the method of data backup is often used to restore the problem data, but the backup data may also be leaked, damaged and other problems. On the basis of lossless compression LZW data compression algorithm, AES encryption algorithm is used to compress and encrypt data in blocks. This scheme can reduce the storage space of backup data and encrypt the data in blocks. Using the characteristics of LZW block compression, part of the value in the previous compressed block is used as the encryption key of the next data block. Even if the encryption key is leaked for the first time, all encrypted backup data will not be leaked and the security of backup data will be enhanced.

Keywords: Data backup · Lossless compression · Block compression

1 Introduction

With the development of information technology, "big data" has gradually become the symbol of modern society. With the rapid growth of all kinds of data, higher requirements are put forward for the transmission and storage of information, especially in the application of weather prediction, aerospace remote sensing, etc. [1–3]. Therefore, more and more attention has been paid to data compression, especially lossless compression technology, which can not only reduce the storage space occupied by data, but also ensure the integrity of data.

Data compression [2] means that under certain data storage space requirements, the relatively large original data is reorganized into a data set that meets the above-mentioned space requirements, so that the information recovered from the data set can be consistent with the original data, or the same quality of use as the original data can be obtained. Data compression reduces the space required for data storage, which indirectly reduces the time and resource consumption required for data processing. In 1977, Israeli scholars proposed a dictionary-based compression algorithm based on the idea that repeated data could be encoded by short code in data stream, and LZW algorithm [4] was put forward on the basis of this algorithm. Literature [4] improves the LZW algorithm and proposes an improved prefix mapping coding compression algorithm. Based on LZW, Literature [5] further proposed an improved L-H algorithm to improve the compression performance by adding suffix characteristics and combining the advantages of Hoffman coding with run-length coding.

Aiming at the problem that the above compression algorithm cannot guarantee the safe transmission of data, a new data block compression encryption scheme is proposed in this paper. Based on the research of LZW principle, combined with AES algorithm [6, 7], the compressed data of LZW is encrypted. Through the simulation experiments of BAK, EXE, DOC and other different types of data files, it is proved that the scheme has a certain feasibility. On the basis of slightly increasing the file size, the reliability of communication is guaranteed.

2   The LZW Algorithm

2.1   Introduction of LZW Algorithm

LZW algorithm [8] is a lossless data compression algorithm. By analyzing the input data, LZW algorithm obtains a dictionary table without repeating substrings, and each substring corresponds to a code, so as to realize lossless compression of data.

2.2   The Coding Principle of LZW Algorithm

LZW algorithm is a dictionary-based coding method. LZW compression has three important objects, namely data stream, coding stream and coding table. In coding, the data stream is the input object and the coding stream is the output object. In decoding, the encoding stream is the input object and the data stream is the output object. The encoding of LZW algorithm is performed by successively reading and encoding the string S starting from the left [9]. The steps are as follows:

Step1: Initially, the dictionary contains all possible single characters, and the current prefix is null.
Step2: If the current character C is equal to the next character in the character stream, determine whether P + C is in the dictionary.
Step3: If it is in the dictionary, then P = P + C; if not, do the following: ①Outputs the corresponding codeword representing the current prefix P to the codeword stream. ②add P + C to the dictionary.③ Let P = C.
Step4: Determine whether there are any codewords to be translated in the codeword stream. If there are, return to Step 2.If not, the codeword representing the current prefix P is printed to the codeword stream.
The decompression steps are as follows:
Step1: The dictionary table contains all the single characters at the beginning of the decoding.
Step2: Reads the first codeword CW in the encoded data stream.
Step3: Output String.cw to the character data stream Charstream.

Step4: pW = cW.

Step5: Reads the next codeword cW of the encoded data stream.

Step6: Is there a String.cw currently in the dictionary?

Step7: If there is String.cW in the dictionary, do the following:①Outputs String.cw to the character data stream. ②P = String.pW.③Assign the first character of String.cW to C.④Add the string P + C to the Dictionary.

Step8: If there is no String.cW in the current dictionary, do the following: ①P = String. pW. ② Assign the first character of String.pW to C. ③Assign the first character of String.cW to C.④ Outputs the string P + C to the character data stream and adds it to Dictionary, where it is now consistent with CW.

Step9: Is there any Codeword in the encoded data stream? If so, return to step 4 to continue decoding, otherwise, end the decoding.

3   AES Algorithm

3.1   Introduction of AES Algorithm

Due to the small key length (56 bits) of the DES data encryption standard algorithm, it can no longer meet the requirements of data encryption security in today's distributed open networks. Therefore, in 1997, NIST publicly solicited a new data encryption standard, namely AES. After three rounds of screening, the Rijndael algorithm submitted by Joan Daeman and Vincent Rijmen of Belgium was proposed as the final algorithm of AES. This algorithm has become the new data encryption standard in the United States and is widely used in various fields. As a new generation of data encryption standard, AES brings together the advantages of strong security, high performance, high efficiency, ease of use and flexibility. Relatively speaking, the 128 key of AES is 1021 times stronger than the 56 key of DES.

3.2   AES Algorithm Principle

AES is an iterative, symmetric-key grouping encryption algorithm. It can use 128, 192, and 256-bit keys, and use 128-bit (16-byte) blocks to encrypt and decrypt data. Unlike public key encryption, which uses a key pair, symmetric key encryption uses the same key to encrypt and decrypt data. The number of bits of the encrypted data returned by the block cipher is the same as the input data. Iterative encryption uses a loop structure in which permutations and substitutions of input data are repeated.

4   Backup and Recovery of AES-Based LZW Algorithm Under Data Block Compression and Encryption

4.1   The Principle of Backup and Recovery Based on AES-Based LZW Algorithm Under Data Block Compression and Encryption

When the LZW algorithm compresses data, with the continuous addition of new strings, the data in the dictionary table will also continue to increase. If the original data is too large, the label set in the generated dictionary will become larger and larger. At this time, operating on this set will cause efficiency problems. When using the LZW algorithm, if the label set in the dictionary is large enough, a clear flag should be inserted at this position, and the dictionary will be constructed again. At this time, AES is used to encrypt the data compressed. Because the data compressed this time is very small, the time to use AES encryption is very short, and because of the block encryption method, the dynamic generation of the key for each encryption can be used.

4.2   Implementation of Backup Based on AES-Based LZW Algorithm Under Data Block Compression and Encryption

On the basis of the AES algorithm, the LZW algorithm is improved. The implementation of backup under data block compression and encryption is as follows:

Step1: The initialization key is KEY, and the maximum dictionary table length is N. The initial dictionary table contains all the possible number of single characters in the string, denote as NC, and make the current dictionary table use space TN be NC.
Step2: The data string S is compressed using the LZW algorithm, the dictionary table contains all the single characters in the data string S, and the current prefix P is empty.
Step3: The current character C takes the next character in the data string S.
Step4: If P + C is in the dictionary, then P = P + C. If it is not in the dictionary, do the following:① Output the codeword representing the current prefix P to the codeword stream.② Add P + C to the dictionary, and then add 1 to TN.③ P = C.
Step5: Determining whether the dictionary table length TN reaches the last character of data string S. If it does, obtain the compressed string S1 corresponding to the label from TN to N, and then obtain a new key KEY_NEW = NEW_KEY(S1) from S1. Then AES encryption algorithm is performed on the compressed string S1 using Key to obtain the encrypted string S2, whose length is L. And add the length of the encryption string L to the head of S2 to get a new string S3, S3 is appended to the encrypted file after writing.
Step6: Reset TN = NC, KEY = KEY_NEW;
Step7: Repeat steps 2 through 5 until the last character of S, and the final compressed and encrypted file is obtained.

4.3   Recovery of LZW Algorithm Based on AES Under Data Block Compression and Encryption

Based on the AES algorithm, the LZW algorithm is improved, and its recovery under data block compression and encryption is realized as follows:

Step1: Initialize the KEY
Step2: The length of the first crypto string is obtained by the head of crypto string S, and then the first crypto string S1 and the rest crypto string S2 are obtained according to LS.
Step3: Call the AES (KEY) algorithm, decrypt the compressed string S3, and get the new KEY from S3, whose value is NEW_KEY(S3).

Step4: The LZW decompression algorithm is called to obtain the original string S4 from the compressed string S3 and write S4 to the file in an appended way.

Step5: Assign S2 to S. If the S character has not been read, return to step 2 to continue execution. If all the S characters have been read, the algorithm ends.

4.4   Flow Chart of Algorithm Implementation

This algorithm calls AES encryption algorithm when LZW compresses each block, so that the data compressed and encrypted each time is not long, which can reduce the time of compression and encryption, and reduce the memory space occupied during compression and encryption. The flow chart of the algorithm is shown in Fig. 1.

5   Experimental Analysis and Application

In order to verify the effectiveness of the algorithm, the algorithm in this paper is written in C#. Assuming that the length of the dictionary table is 2^16-1, the value of the initial dictionary table is 0 to 255, and 16-bit binary is used for encoding. The subscript of the dictionary table array is the label corresponding to the encoding. Load the source file in a binary stream, and then read a byte from the binary stream in turn, and compare whether the current byte read is stored in the dictionary table. If it is in the dictionary table, read the next byte and merge it into the previous string, and then judge whether the code is in the dictionary. If it is not in the dictionary table, add this code to the dictionary table and write the corresponding label to the compressed string, and so on, until the dictionary table space is full. Then AES algorithm is called to encrypt the compressed string, and the initial key can be set arbitrarily. Since CFB encryption mode can make the encrypted length close to the original length without greatly increasing the size of the compressed string, the AES algorithm in this paper uses CFB encryption mode and Zeros filling mode. Finally, initialize the dictionary table, and continue to compress and encrypt the following characters until all the binary streams are compressed and encrypted (Table 1).

Since AES is block encryption, the length of the block in CFB mode is 16 bytes. If each compressed string is less than 16 bytes, it should be filled to 16 bytes before encryption processing. Therefore, the size of the compressed string after encryption is larger than the size before encryption. From the experimental analysis, it can be seen that the size of the compressed and encrypted file is similar to the size of the compressed file only, so we have achieved the purpose of file compression encryption. It can be seen from the experimental data that the compression ratio of the data file is as high as 1:9. Therefore, this algorithm is suitable for the backup and encryption of the data in modern information system. In particular, when the data is backed up in different places, the data can be compressed small enough to facilitate the rapid transmission to the remote backup end. The backup data is encrypted first and then transmitted, which can improve the reliability of transmission, so this algorithm can well meet the requirements of remote remote database backup.

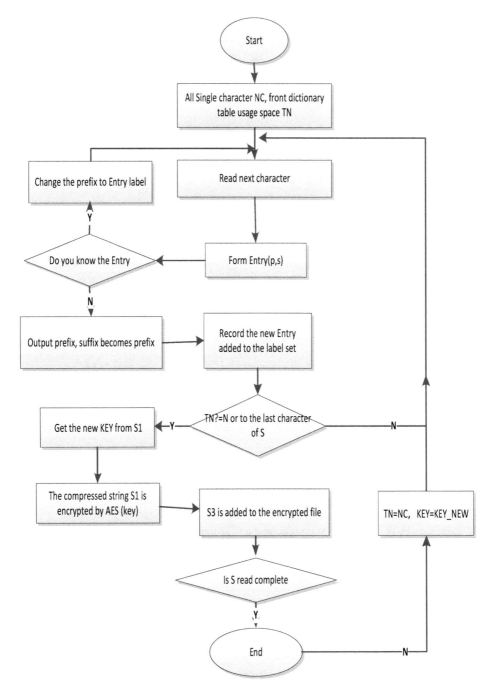

Fig. 1. Flow chart of compression encryption algorithm.

Table 1. Table captions should be placed above the tables.

Source file type	Source file size	The compressed size of LZW	LZW-AES compresses the size after encryption
Bak	10710 KB	1358 KB	1810 KB
Exe	5906	1468	1958
Doc	1189	581	775
Pdf	1159	981	1121
Xls	2306	565	754
Ppt	2401	2051	2396
Txt	530	139	185

6 Conclusion

Based on the research of AES algorithm, a LZW data compression and encryption algorithm based on AES is proposed. The algorithm is slightly lower than LZW in compression ratio and has higher security in data security. It is suitable for compression of text, character, data and other types of files. Moreover, for the original data files with many repeated characters, this algorithm still has a high compression ratio. Because this algorithm adopts block compression encryption, the memory space consumed by this algorithm is less in compression and encryption.

Acknowledgement. This research work was supported in part by the Natural Science Foundation of Anhui Province (No. 1708085MF155).

References

1. Datta, A., Ng, K.F., Balakrishnan, D.: A data reduction and compression description for high throughput time-resolved electron microscopy. Nat. Commun. **12**(1), 1–5 (2021)
2. Knolhoff, A.M., Fisher, C.M.: Strategies for data reduction in non-targeted screening analysis: The impact of sample variability for food safety applications. Food Chem. **350**, 128540 (2021)
3. Shen, L.J., Feng, X.Y., Lin, M.: Application of lossless compression technology in telemetry of launch vehicle noise. Telemetry Telecontrol **39**(01), 6–10 (2018)
4. Yan, H.Z., Xu, B.G.: Optimization and application of lossless compression algorithm LZW prefix coding [J]. Comput. Eng. **43**(03), 299–303 (2017)
5. Liu, C., Li, Y.F., Chen, H.: Improvement and realization of lossless data compression technology based on LZW. Electron. Des. Eng. **27**(24), 51–56 (2019)
6. OuX, G., Liu, X.Y.: Hybrid Encryption Algorithm of Hyperchaos and AES. J. Jiangxi Univ. Sci. Technol. **41**(05), 80–87 (2020)
7. Zhang, M.H., Chen, Z.J., Xu, X.Y.: Design and Implementation of DSP security protection based on AES algorithm. Microelectron. Comput. **36**(10), 32–36 (2019)
8. Ni, X.J., She, X.H.: Improved LZW algorithm for wireless sensor network application. Comput. Sci. **47**(05), 260–264 (2020)
9. Han, B., Zhang, H.H.: Improved LZW prefix coding scheme based on azimuth information. Comput. Sci. **46**(08), 157–162 (2019)

Intelligent Rock Recognition Based on Deep Learning

Hongmei Liu[1], Yihua Chen[2(✉)], and Shiliang Chen[2]

[1] Fuyang Normal University, Fuyang 236037, Anhui, China
[2] Hefei University of Technology, Hefei 230009, Anhui, China

Abstract. At present, the application of deep learning to image recognition has become one of the research hotspots. It can use the least program code and spend the least time to build a deep learning model to train, evaluate accuracy and make predictions. Based on this, after preprocessing the acquired magmatic rock data set such as classification and data enhancement, Keras is used to construct pre-training models such as VGGNet, Xception, and convolutional neural network training models. In the training process, we can get satisfactory accuracy and small loss by adjusting the hyper-parameters of the model and comparing different batch sizes. After the accuracy of the model is stable, the pre-training model and the convolutional neural network model are integrated using the voting method. Compared with a single model, the fusion model has a certain improvement in model accuracy.

Keywords: VGGNet · Xception · Deep learning · Convolutional neural network · Magmatic rock

1 Introduction

For geological work, the identification of rocks and minerals is an important link, which plays a fundamental role in the development of the work. Therefore, improving the method of rock identification to enhance the efficiency and accuracy of identification has certain practical significance for the geological field. Now, the identification of rocks generally relies on human eyes. But the accuracy and efficiency of this method are relatively low [1].

In recent years, deep learning research has become more and more popular, and good results have been achieved in many fields. In terms of image recognition, technologies such as object detection, video analysis and face recognition have been widely used in our real lives [2]. The steps of preliminary identification of rock mentioned above are all based on the image of the rock, we come up with a method to recognize and classify the image of magmatic rock through deep learning technology.

J. C. Hung et al. (Eds.): FC 2021, LNEE 827, pp. 1835–1842, 2022.
https://doi.org/10.1007/978-981-16-8052-6_271

2 Main Models

2.1 VGGNet

VGGNet is a deep convolutional neural network developed by researchers from the Computer Vision Group of Oxford University (VisualGeometry Group) and Google DeepMind [3]. VGGNet explored the relationship between the depth of the convolutional neural network and its performance. By repeatedly stacking 3*3 small convolution kernels and 2*2 maximum pooling layers, VGGNet successfully constructed deep convolutional neural networks of 16–19 layer. Compared with the previous state-of-the-art network structure, VGGNet has a significant drop in error rates. At the same time, VGGNet is very extensible, and the generalization of migration to other image data is very good. The structure of VGGNet is very simple. The entire network uses the same size of the convolution kernel size (3*3) and maximum pooling size (2*2). So far, VGGNet is still often used to extract image features [4].

2.2 Convolutional Neural Network

Convolution Neural Networks (CNN) is designed to process multi-dimensional array data, such as a color picture with three color channels. Many data are in the form of multi-dimensional arrays: 1D is used to represent signals and sequences, such as voice, 2D is used to represent images, and 3D is used to represent videos or images with sound. The basic structure of a convolutional neural network is shown in Fig. 1, which is mainly composed of an input layer, a convolutional layer, a down-sampling layer (pooling layer), a fully connected layer, and an output layer. In fact, the Convolutional Neural Network is a multi-layer neural network, in which the basic operations include: convolution operation, pooling operation, fully connected operation and recognition operation [5].

2.3 Model Ensemble

2.3.1 Ensemble Learning

In machine learning and statistical learning, Ensemble Learning is a method that combines multiple learning method to achieve better performance. The so-called ensemble learning is to predict the data set by using multiple classifiers, to improve the generalization ability of the overall classifier [6].

2.3.2 Multi-model Ensemble

Combining the obtained multiple models can improve the overall accuracy to some extent, and the low-correlation results can be combined to get better results. A single model is easy to overfit, and the ensemble of multiple models can improve the generalization ability; the prediction ability of a single model is not high, and multiple models can often improve the prediction ability.

3   Data Acquisition and Processing

3.1   Data Acquisition

The rock image data set of this paper is obtained from Baidu, Google, National Experimental Teaching Center of Geology, National Rock and Mineral Fossil Specimens Sharing Platform and other websites through the crawler technology, and the quality of the images is high.

3.2   Data Processing

During data preprocessing on the acquired image data set, "dirty" data sets are removed, and data sets with obvious characteristics and higher quality are retained. And then crop the data sets acquired from preliminary processing to a uniform size.

3.3   Data Set Classification

The objects of image recognition in this paper are magmatic rocks, the Magmatic rocks can be divided into many categories, the representative categories of magmatic rocks are selected for classification and recognition.

3.4   Data Enhancement and Coding

In order to improve the classification accuracy, to better extract image features and to generalize the model (to prevent the model from overfitting), A large amount of sample data is required in the training of the neural network. Since the data set used in this paper is relatively small, the Image Data Generator data enhancer is used to enhance the data set. The enhancement method is rotating the image Angle randomly, translating the image horizontally (vertically), changing the image based on shear principle, scaling the image randomly and flipping half of the image horizontally randomly.

4   Model Realization

4.1   Pre-training Model

The pre-training model is using models that have been trained by others for their own project training, fine-tuning the pre-training model can help to obtain a model with higher accuracy. This paper uses VGGNet model to carry out experiments. Initialization parameters of VGG16 and Xception models are obtained from GitHub, such as weight value files, offset, etc.

4.2   VGG16 Model

Just to quote the initialization parameters of VGG16 in the code, use the weight values that have been trained, and set the top 3 fully connected network layers to not retain (Fig. 1 and Fig. 2). It is found that the accuracy of the model is particularly low, and it exists as a scatter graph without rule to follow. So the VGG16 model is needed to adjust to make it is suitable for recognition task.

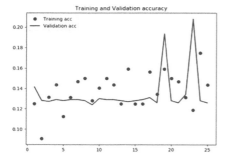

Fig. 1. VGG16 model accuracy diagram

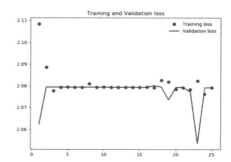

Fig. 2. VGG16 model loss diagram

4.2.1 Freezing All Layers of the Model

The neural network model has large amounts of weight parameters which are variable and constantly change in the process of model training.

By freezing all layers of the model, the accuracy of image recognition does not obviously ascend, but the phenomenon that the verification accuracy is higher than the training precision appears. By freezing part of the layers of the model, the accuracy can reach 63% and gradually meets the training requirements (Fig. 3 and Fig. 4). After the first 20 iterations, the fitting effect of the model is still poor and also has very big promotion space.

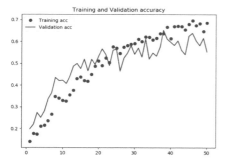

Fig. 3. Part of freezing model accuracy diagram

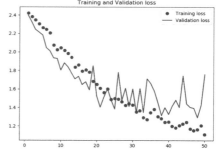

Fig. 4. Part of freezing model loss diagram

4.2.2   Adding Weight Regularization

After adding weight regularization (Fig. 5 and Fig. 6), it can be seen that the precision of the model steadies at around 63%. Fitting effect of the image is better, and the basic requirements of the training are achieved.

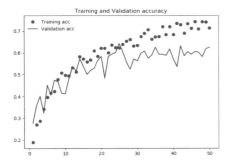
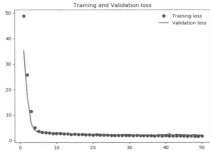

Fig. 5. The regularized model accuracy diagram

Fig. 6. The regularized model loss diagram

4.3   Xception Model

Adding droupout layer and weight regularization (Fig. 7 and Fig. 8), the fitting effect is good, and model accuracy steadies at around 60%, and over fitting phenomenon does not appear obviously.

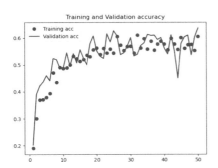
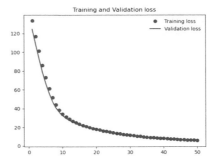

Fig. 7. Xception model accuracy diagram

Fig. 8. Xception model loss diagram

4.4   Convolutional Neural Network Model

Using convolution layer to structure the neural network, six layer convolution is used here (Fig. 9 and Fig. 10). The precision of the model is only about 35%, after 40 times iteration, the verification accuracy is higher than the training precision. Fitting effect is not beautiful.

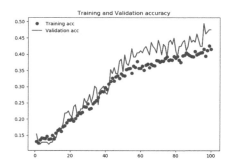

Fig. 9. Convolutional neural network model accuracy diagram

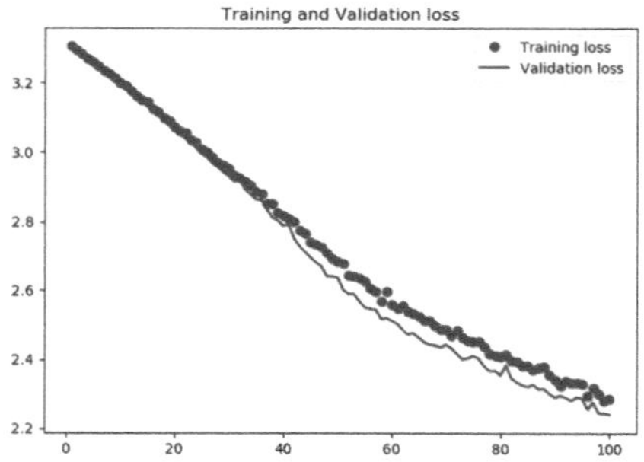

Fig. 10. Convolutional neural network model loss diagram

The principle of the convolutional neural network is to extract the original image features through the convolution kernel to obtain feature information, thereby reducing the number of neurons to obtain a better training model and higher model accuracy. The number of convolution kernels represents the number of types of feature learning, A proper number of convolution kernels can effectively extract image features and reduce the impact of other image "noise". Since some images in the data set of this paper come from websites such as Baidu and Google, some images have large "noise", so we will appropriately adjust the number of convolution kernels in each layer to obtain higher accuracy.

Droupout is added to convolutional neural network model with 6 layers above, and 0.5 is selected as the deactivation rate which can effectively prevent the over fitting problem. After adjusting the number of convolution kernel and joining droupout layer, the precision of image has the obvious rise, and can achieve about 60%. After 80 rounds, slight over fitting phenomenon appears. The overall fitting effect is ideal.

After adding Droupout, there are still over fitting phenomenons. The reason may be that the parameters of the model is too much or the model is too complex. The weight regularization is added in the following, such as adding L1 regularization and L2 regularization (Fig. 11 and Fig. 12). The whole fitting effect of the model has been enhanced. Over fitting appears from the 130 rounds instead of the original 80 arounds.

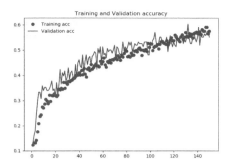

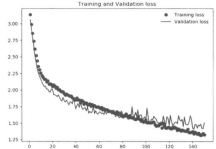

Fig. 11. Convolutional neural network model accuracy diagram

Fig. 12. Convolutional neural network model loss diagram

4.5   Model Ensemble

From the neural network models trained above, the more accurate and stable VGG16 model, Xception model and convolutional neural network model is selected for model ensemble (Fig. 13).

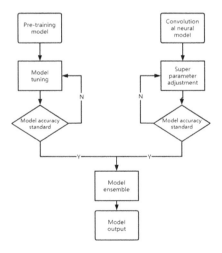

Fig. 13.   Model ensemble flow chart

This paper adopts the voting method for model fusion. From the final training results, we know that the accuracy of the VGG16 model is 63%, the accuracy of the Xception model is 60%, and the accuracy of the convolutional neural model is 60%, so the accuracy of the ensemble model can be obtained as:

$$0.63 * 0.6 * 0.6 + 0.63 * 0.6 * 0.4 * 2 + 0.6 * 0.6 * 0.37 = 0.6624$$

The accuracy of the ensemble model is improved to a certain extent than before, which can improve the identification and prediction of rocks to a certain extent.

5 Summary

According to reference mature pre-processing models such as VGGNet and Xception, and ensemble multiple models, this paper constructs a model to recognize magmatic rocks. In the training process of the model, convolution kernel, the number of convolutional layers and the regularization parameters have been fine-tuned, and the learning rate and loss function have been added, the model's training test fitting effect and accuracy rate reach a high level. Improve generalization and prediction capabilities to achieve better training results.

The network resources of rock images are too poor, resulting in uneven image quality of our data set. Therefore, the quality of image data is the main factor that limits the accuracy of our model. If we can shoot the rocks under uniform conditions in the future, and remove the background of the rocks in batches or replacing it with rock slice data the extraction of feature values and noise removal of the convolutional layer will be more effective, which will improve the accuracy of model training and the credibility of predictions.

References

1. Feng, Y.X., Gong, X., Xu, Y.Y., Xie, Z., Cai, H.H., Lv, X.: Lithology recognition based on fresh rock image and twins convolution neural network. Geogr. Geo-Inf. Sci. **35**(5), 89–94 (2019)
2. Cheng, G.J., Guo, W.H., Fan, P.Z.: Study on rock image classification based on convolution neural network. J. Xi'an Shiyou Univ. (Nat. Sci. Edit.) **32**(4), 116–122 (2017)
3. Wang, T., Li, H., Hu, Z.: Design and implementation of image style migration algorithm based on vggnet. Comput. Appl. Softw. **36**(11), 224–228 (2019)
4. Karen, S, Andrew, Z.: Very Deep Convolutional Networks for Large-Scale Image Recognition. Cornell University, Ithaca (2014)
5. Chen, G.H., Liang, S.S., Wang, J., Sui, S.L.: Application of convolutional neural network in lithology identification. Well Logging Technol. **43**(02), 129–134 (2019)
6. Han, S.: Image feature extraction and classification research based on ensemble learning model. Nanjing University of Posts and Telecommunications (2017)

Author Index

J. C. Hung et al. (Eds.): FC 2021, LNEE 827, pp. 1843–1848, 2022.
https://doi.org/10.1007/978-981-16-8052-6

Printed by Printforce, the Netherlands